Lecture Notes in Computer Science 691
Edited by G. Goos and J. Hartmanis

Marco Ajmone Marsan (Ed.)

Application and Theory of Petri Nets 1993

14th International Conference
Chicago, Illinois, USA, June 21-25, 1993
Proceedings

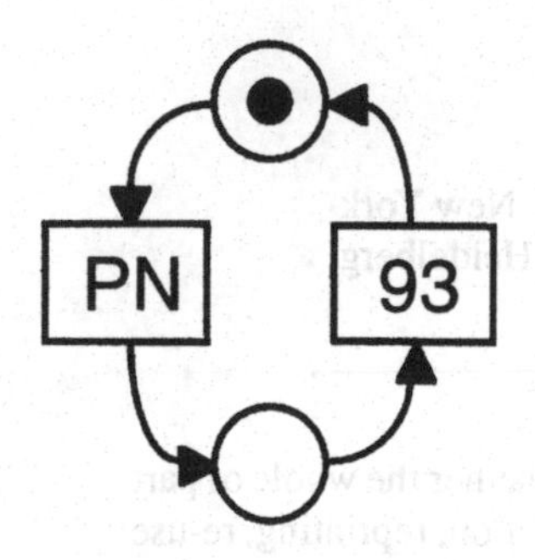

Springer-Verlag
Berlin Heidelberg New York
London Paris Tokyo
Hong Kong Barcelona
Budapest

Series Editors

Gerhard Goos
Universität Karlsruhe
Postfach 69 80
Vincenz-Priessnitz-Straße 1
W-7500 Karlsruhe, FRG

Juris Hartmanis
Cornell University
Department of Computer Science
4130 Upson Hall
Ithaca, NY 14853, USA

Volume Editor

Marco Ajmone Marsan
Dipartimento di Elettronica, Politecnico di Torino
C.so Duca degli Abruzzi, 24, I-10129 Torino, Italy

CR Subject Classification (1991): F.1-3, C.1-2, D.4, J.6

ISBN 3-540-56863-8 Springer-Verlag Berlin Heidelberg New York
ISBN 0-387-56863-8 Springer-Verlag New York Berlin Heidelberg

Printed in Germany

Typesetting: Camera ready by author
Printing and binding: Druckhaus Beltz, Hemsbach/Bergstr.
45/3140-543210 - Printed on acid-free paper

Preface

This volume contains the proceedings of the 14th International Conference on Application and Theory of Petri Nets. The aim of the Petri net conferences is to create a forum for discussing progress in the application and theory of Petri nets. Typically, the conferences have 150-200 participants, one third of whom come from industry, while the rest are from universities and research institutions. The conferences always take place in the last week of June. The previous conferences were held in Strasbourg, France (1980), Bad Honnef, Germany (1981), Varenna, Italy (1982), Toulouse, France (1983), Aarhus, Denmark (1984), Espoo, Finland (1985), Oxford, United Kingdom (1986), Zaragoza, Spain (1987), Venice, Italy (1988), Bonn, Germany (1989), Paris, France (1990), Aarhus, Denmark (1991), Sheffield, United Kingdom (1992).

The conferences and a number of other Petri net activities are coordinated by a steering committee with the following members: M.Ajmone Marsan (Italy), J.Billington (Australia), H.J.Genrich (Germany), C.Girault (France), K.Jensen (Denmark), G.de Michelis (Italy), T.Murata (USA), C.A.Petri (Germany; honorary member), W.Reisig (Germany), G.Roucairol (France), G.Rozenberg (The Netherlands; chairman), M.Silva (Spain).

This 14th conference is organized for the first time in the United States, in Chicago. In addition to the conference, an exhibition of Petri net tools, and several tutorial lectures (both at the introductory and at a more advanced level), are organized.

The number of submissions this year reached the record level of 102; 26 submissions were accepted for presentation as regular papers, and 6 as project papers. Invited lectures are given by C.A. Ellis (USA), M.Silva (Spain), and K.S.Trivedi (USA).

The submitted papers were evaluated by a programme committee with the following members: M.Ajmone Marsan (Italy; chairman), E.Best (Germany), J.Billington (Australia), M.Diaz (France), S.Donatelli (Italy), C.Girault (France), K.Jensen (Denmark), H.C.M.Kleijn (The Netherlands), B.Krogh (USA), A.Mazurkiewicz (Poland), J.F.Meyer (USA), M.Molloy (USA), T.Murata (USA), G.Nutt (USA), K.Onaga (Japan), L.Pomello (Italy), W.Reisig (Germany), M.Silva (Spain), P.S.Thiagarajan (India), W.M.Zuberek (Canada). The programme committee meeting took place at Politecnico di Torino in Italy.

I should like to express my gratitude to all authors of submitted papers, to the members of the programme committee, and to all the referees who assisted them. The names of the referees are listed in the following page. A special thanks goes to Susanna Donatelli, whose help was crucial in the automation of the process for the management of submissions.

For the local organization of the conference in Chicago, all of us should be thankful to T.Murata, S.Shatz, and U.Buy. The support received from the following corporate sponsors: Bull Information Systems, Inc., Fujitsu Network Transmission Systems, Inc., Hitachi Software Engineering Co., Ltd., and SUN Microsystems, is gratefully acknowledged. Finally, I should like to mention the excellent cooperation with Springer-Verlag in the preparation of these proceedings.

Torino, Italy
April 1993

Marco Ajmone Marsan

List of Referees

M. Ajmone Marsan
M. Akatsu
M. Amin
C. Andre
P. Azema
F. Baccelli
W. Bachakene
R. Bachatene
G. Balbo
D. Banard
K. Barkaoui
E. Battiston
B. Baynat
M. Becker
C. Beguelin
L. Bernardinello
G. Berthelot
B. Berthomieu
A. Bertoni
E. Best
A. Bianco
J. Billington
J. Biskup
R. Blumenthal
A. Bobbio
G. Bruno
D. Bruschi
U. Buy
J. Campos
A. Castella
G. Cattaneo
G. Chehaibar
G. Chiola
P. Chretienne
S. Christensen
G. Ciardo
P. Civera
J. Coleman
J.M. Colom
G. Conte
A. Coyle
P. D'Argence
P. Dague
A. David
F. De Cindio
G. De Michelis
F. De Paoli
F. Derrough
G. Desel
M. Diaz
S. Donatelli
C. Dutheillet
C. Ellis
J. Engelfriet
J. Esparza
J. Ezpeleta
G. Findlow
A. Finkel
G. Florin
G. Franceschinis
R. Gaeta
C. Ghezzi
C. Girault
H. Goeman
U. Goltz
D. Gomm
B. Grahlman
W. Grunwald
S. Haddad
N. Hanen
N. Hansen
I. Hatono
G. Hebuterne
A. Heise
W. Henderson
K. Hiraishi
T. Hisamura
S. Honiden
H.J. Hoogeboom
P.W. Hoogers
R.P. Hopkins
S.J. Huang
P. Huber
J.M.Ilie
K. Jensen
R. Johns
J. Jozefowska
G. Juanole
B. Keck
P. Kemper
T. Kerrigan
E. Kindler
G. Klas
H.C.M. Kleijn
S. Kodama
H.J. Kreowski
B. Krogh
D. Leu
H. Linde
C. Lindemann
J. Lobo
K. Lodaya
J. Martinez
M. Molloy
S. Morasca
M. Mukund
T. Muldner
T. Murata
P. Muro Medrano
N. Naga
M. Nielsen
G. Nutt
K. Onaga
A. Pagnoni
N. Pekergin
L. Pomello
L. Portinale
C. Priami
S. Qi
A. Ramirez
P. Redmiles
W. Reisig
M. Ribaudo
G. Ristori
N. Sabadini
D. Saidouni
L. Saitta
P. Samarati
V. Sassone
P. Schwartz
M. Sereno
S. Shatz
K. Shibata
H. Shiizuka
C. Sibertin Blanc
M. Silva
J. Silverthorn
C. Simone
A. Sistla
R. Sisto
U. Solitro
Y. Sugasawa
T. Sukiyama
I. Suzuki
P. Taylor
E. Teruel
P.S.Thiagarajan
S. Tu
N. Uchihira
T. Ushio
R. Valette
F. Vernadat
I. Vernier
J.L. Villarroel
W. Vogler
P. Wagner
R. Walter
J. Wheeler
P. Mc Whirther
D. Zhang
B. Zovari
W. Zuberek

Table of Contents

Invited Papers

Full Papers

Project Papers

Modeling and Enactment of Workflow Systems

Clarence A. Ellis and Gary J. Nutt*
Department of Computer Science
Campus Box #430
University of Colorado, Boulder 80309–0430
skip@cs.colorado.edu
nutt@cs.colorado.edu

March, 1993

Abstract

Petri net models and variants thereof have primarily been used to model structured systems such as computer programs, factory production lines, and engineering hardware. In contrast, this paper discusses the issues and challenges in the modeling of human activity in the workplace. This type of activity frequently has a large component that is unstructured, creative work. It is dynamic and difficult to capture via traditional Petri nets. Our research group at the University of Colorado has been investigating Information Control Nets (ICNs), derived from high level Petri nets, as a tool for modeling office workflow. After carefully explaining the notion of Workflow, this paper presents a formal (and also an informal) definition of ICN. We illustrate the utility of ICNs via an office analysis example.

Besides being a tool for workflow analysis, ICNs have been used as the basis for the implementation and use of office work coordination systems. These systems provide a computer network based environment to help coordinate, monitor, schedule, and assist in the execution of office work items. To succeed in realistic organizational environments, these systems must be reliable, robust, easy to use, and flexible. Organizational flexibility demands a capability for dynamic change and exception handling. We introduce a mechanism for dynamic change and exception handling in the latter part of the paper. We conclude the paper with a summary, and discussion of some research challenges that we feel must be met for the successful implementation of future generations of workflow systems.

***Supported by funding from Bull Worldwide Information Systems Imaging and Office Solutions**

1 Introduction

There are many successful computing tools designed as personal information aids (word processors, spreadsheets, etc.) but fewer tools designed for supporting collaborative work by groups of people. Computer supported cooperative work (CSCW) can be accomplished by a collection of software tools, systems, and environments that are often referred to as *groupware*. Groupware can be informally defined as "systems that support groups engaged in a common task or goal, and that provide an interface to a shared environment." [7].

There are several ways to characterize group interactions (and hence groupware systems), though the *time space taxonomy* is the most widely used. In this model, groupware can be thought of as supporting group interactions occurring at the same/different time/place (see Figure 1). For example, face-to-face interaction can be characterized as "same time, same place" interaction, and could be supported by groupware called meeting room technology. If the members of the group are located in the same place, but they occupy that place at different times, then the interactions will necessarily be *asynchronous*, i.e., information passed from one participant to another will be "posted" for other recipients to access at an arbitrary later time. A physical bulletin board is an example of support for this type of interaction. Same time, different place interactions are referred to as *synchronous, distributed interactions* since information is excehanged among distant participants in real time. A realtime document editor is an example of this kind of interaction (again, see [7]). The final class of interactions are different time, different place interactions, also known as *asynchronous, distributed interactions*; an electronic mail system is an example of groupware supporting such interactions.

The potential pitfalls of groupware have been discussed in conferences, tutorials, and journals concerned with CSCW [12, 19]. As opposed to much of the previous generation of office information systems, this literature addresses the inherently interdisciplinary nature of groupware. To successfully implement groupware in an organization requires well grounded technological development, with careful attention paid to the social and organizational environment into which the technology is being embedded

There are many commercial groupware products [6], although only a few of them capture knowledge of the organizational activity that they are assisting. For example, a group document editor knows nothing about the organizational purpose of the document being edited. Organizationally aware groupware can potentially lead to significantly more powerful and useful systems. One class of organizationally aware groupware is workflow.

1.1 Workflow Systems

Workflow systems are a particular kind of groupware intended to assist groups of people in executing work procedures; they contain knowledge of how work normally flows through the organization. Workflow is defined as "systems that help organizations to specify, execute, monitor, and coordinate the flow of work items within a distributed office environment." [3]

A workflow system contains two basic components: a workflow modeling component and a workflow execution component. The workflow modelling component enables administrators, users, and analysts to define procedures and activities, analyze and simulate them, and allocate them to people. Most commercial workflow products have no model, so this component is reduced to a "specification module", and the execution component is referred to,

	Same time	Different time
Same place	Face-to-face	Asynchronous
Different place	Synchronous Distributed	Asynchronous Distributed

Figure 1: A Time Space Taxonomy of Groupware

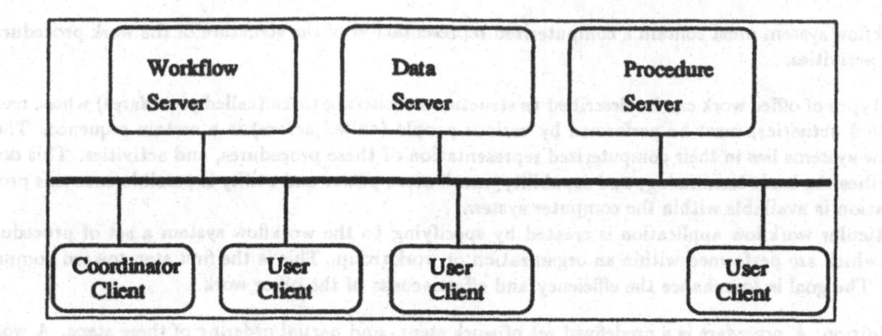

Figure 2: A Workflow Execution Environment

by many authors, as the "workflow system". Usage of this module is typically completed before the flow of work tasks actually begins. We believe that a model of collaboration is a useful entity in all phases of groupware use, so we place high significance upon the modelling component.

The workflow execution component consists of workflow clients and servers which work in combination to assist users in the enactment of their work tasks. The workflow user client consists of the human-computer interface seen by end users with its associated hardware, software, and assesories. The The servers, typically on machines accessible to users over a network, are systems which assists in coordinating and performing the procedures and activities (Figure 2 represents a typical configuration for an execution environment). The workflow server encapsulates the workflow model and the state of its execution; the data server is used to store the application data being processed by the workflow program; and the procedure server stores the individual routines that either makeup the human-computer interface or perform automated processing. The workflow server enables the units of work to flow from one user's workstation to another as the steps of a procedure are completed. Some of these steps may be executed concurrently, and some may be executed automatically by the system. The human-computer interface is utilized for all manual steps; typically it presents forms on the electronic desktop of appropriate workers (users). The user fills in forms with the assistance of an editor. Various databases, servers, and external resources may be accessed in a programmed or ad hoc fashion during the processing of any work step. To enable needed flexibility, we believe that the specification and execution modules need to be tightly interwoven.

In the 1970, workflow products held high expectations, but low success rates in the commercial marketplace [1]. At least part of the reason for this is that they tended to be driven by technological innovation rather than by need. As a consequence, they sometimes solved only (a trivial) part of the task of effectively assisting office work by computer; users tended not to perceive workflow products so much as aids and tools but rather as barriers to accomplishing work. Workflow systems tend to succeed when people creatively violate, augment, or circumvent the standard encoded office procedures when appropriate. It is interesting to note that there has been a recent resurgence of interest in workflow, and the number of vendors offering workflow has in the last few years jumped from a very small number to over 40 (Johansen, private communications.)

In the remainder of this paper, we discuss workflow concepts, workflow modelling, and workflow enactment. Throughout the paper, we emphasize that the intimate interaction of people and computer systems poses a challenge which must be creatively met. Among other requirements, these systems must flexibly accomodate informal work activities, exception handling, and dynamic change.

2 Workflow Concepts and Architecture

2.1 Workflow Definitions and Example

Definition: A *workflow system* is an application level program which helps to define, execute, coordinate and monitor the flow of work within organizations or workgroups. In order to do this, a workflow system must contain a computerized representation of the structure of the work procedures and activities.

Many types of office work can be described as structured recurring tasks (called *procedures*) whose basic work items (called *activities*) must be performed by various people (called *actors*) in a certain sequence. The power of workflow systems lies in their computerized representation of these procedures, and activities. This document only describes the basic terminology and capability; much more power and utility is possible once this procedural representation is available within the computer system.

A particular workflow application is created by specifying to the workflow system a set of procedures and activities which are performed within an organization or workgroup. This is the first step toward computerized workflow. The goal is to enhance the efficiency and effectiveness of the office work.

Definition: A *procedure* is a predefined set of work steps, and partial ordering of these steps. A work step consists of a header (identification, precedence, etc.) and body (the actual work to be done.)

Examples include the "order processing procedure" within an engineering company, and the "claims processing procedure" within an insurance company. Both of these are relatively standardized and structured, and each can be described by a sequence of step. Workflow also attempts to assist in less structured work tasks. Different steps of a procedure may be executed by different people or different groups of people. In some cases several steps of a procedure may be executed at the same time or in any order. In general, we therefore define a procedure to be a partially ordered set of steps rather than a totally ordered set. We also define workflow procedures in such a way that loops are allowed. Procedures typically have attributes, such as name and responsible person, associated with them.

Definition: An *activity* is the body of a work step of a procedure. An activity is either a compound activity, containing another procedure, or an elementary activity.

An elementary activity is a basic unit of work which must be a sequential set of primitive actions executed by a single actor. Alternatively, an elementary activity may be a non-procedural entity whose internals we do not model within our structure. An activity is a reusable unit or work, so one activity may be the body of several work steps. For example, if "order entry" and "credit check" are procedures, then the activity "send out letter" may be an activity in both of these procedures. In this case, these are two distinct steps, but only one activity. An activity instance associated with the body of a particular work step is called a work step activity.

Activities typically have attributes such as description and mode associated with them. An activity has one of three modes. Some work step activities may be automatically executed (automatic mode,) some completely manual (manual mode), and some may require the interaction of people and computers (mixed mode). For example, if the procedure is "order equipment" then there may be work steps of:

1. order entry
2. credit check
3. billing
4. shipping

This level of detail of description is typically adequate for an engineering manager, but is not enough detail for an order administrator . The order administrator would like to look inside of the work step called order entry, and see a procedure that requires logging data and filling out of a form. Thus, the body of this step is itself a procedure with work steps of:

1. log name and arrival time
2. fill out the order form
3. send out acknowledgment letter

Furthermore, step 2 of filling out the order form may itself consist of work steps to fill out the various sections of the form. This example shows that it can be useful to nest procedures within procedures. Thus, a work step body has been defined to possibly contain a procedure. Work steps typically have attributes, such as unique identifier and executor, associated with them.

By definition, a workflow system contains a computerized representation of the structure of procedures and activities. This also implies that there is a means for someone (perhaps a system administrator) to specify and input descriptions of procedures, activities, and orderings into the computer. These specifications are called scripts.

> Definition: A *script* is a specification of a procedure, an activity, or an automatic part of a manual activity. The composition or building of this script from available building blocks is called *scripting*.

Once procedures and activities have been defined, the workflow system can assist in the execution of these procedures. We separate the concept of the static specification of procedure (the template) from its execution.

> Definition: A *job* is the locus of control for a particular execution of a procedure. In some contexts, the job is called a work case; if a procedure is considered a Petri net, then a job is a token flowing through the net. If the procedure is an object Class, then a job is an instance.

In our example, if two customers submit two orders for equipment, then these would represent two different jobs. Each job is a different execution of the procedure. If both jobs are currently being processed by the order entry department, then the state of each job is the order entry state. Jobs typically have parameters such as state, initiator, and history associated with them.

Because of the ever changing and sometimes ad hoc nature of the workplace, it is important for workflow systems to be flexible, and have capabilities to handle exceptions. Many procedures which appear routine and structured, are in reality, highly variable, requiring problem solving and dynamic change. Later in this paper, we describe facilities for the handling of unanticipated events and for ad hoc information access.

One workflow element that helps address these issues is the indirect association of people with activities via the concept of roles.

> Definition: A *role* is a named designator for an actor, or a grouping of actors which conveniently acts as the basis for access control and execution control. The execution of activities is associated with roles rather than end users.

Thus, instead of naming a person as the executor of a step, we can specify that it is to be executed by one or more roles. For example, instead of specifying that Michael executes the order entry activity, we can specify that

1. the order entry activity is executed by the order administrator, and

2. Michael is the order administrator.

There may be a very large number of work steps in which Michael is involved. When Michael goes on vacation, it is not necessary to find and change all procedures and work steps involving Michael. We simply substitute Michael's replacement in the role of order administrator by changing step 2. to

2. Robert is the order administrator.

A role may be associated with a group of actors rather than a single actor. Also, one actor may play many roles within an organization. If there are many order administrators within our example, then these can be defined as a group, and it is easy to send information to all order administrators. In this case, an option may be available to "send to any" administrator, and the system might use some scheduling algorithm to select one. Other flexible scheduling algorithms are possible, including the notification of all members of the group that a job is available, and allowing the first responder to handle the job. In this document, we use the term actor to refer a person, a group, or an automated agent. For example, the credit check activity in our example is really executed by the credit department, not by any single person. And the printing operation is really executed by one of many print servers that might be actors with the role of "printer".

Definition: An *actor* is a person, program, or entity that can fulfill roles to execute, to be responsible for, or to be associated in some way with activities and procedures.

Access attributes or capabilities may be associated with actors and with roles. Other attributes, parameters, and structures can be associated as needed. For example, the role of manager is perhaps only played by Michael within the order entry department. Thus a parameter of the role may be the group within which this role applies.

In summary, we have briefly presented a definition of workflow together with explanations of the concepts of procedure, step, activity, job, script, role, and actor.

2.2 Conceptual Architecture

In the remainder of this section, we present the conceptual architecture of a workflow system using the entity-relationship (E-R) model. The architecture builds upon the general concepts introduced in the previous subsection. It lays out the workflow system conceptual entities and their relationships.

The entity-relationship model is a high level semantic model using nodes and arcs; this model has proven useful as an understandable specification model, has been implemented within E-R databases, directly parallels some object oriented concepts, and has a well known direct mapping into a relational database.

In the E-R model, objects of similar structure are collected into entity sets. The associations among entity sets are represented by named E-R relationships which are either one-to-one, many-to-one, or many-to-many mappings between the sets. The data structures, employing the E-R model, are usually shown pictorially using the E-R diagram. An E-R diagram depicting the conceptual architecture of a workflow system is shown in Figure 3. A labeled rectangle denotes an entity set; a labeled arc connecting rectangles denotes a relationship between the corresponding entity sets.

In Figure 3, the box labeled *procedure* denotes an entity set of procedures that may actually be a table of procedure names and their attributes. Likewise, *activity* may be a table of activity names and their attributes. There is an arc connecting these two boxes because there is a relationship called "part-of" between these two entity sets. Some elements in the activity set are steps of (or parts of) some procedures. This arc is labeled with the relationship name, and a denotation of M and N indicating that this is a many to many relationship. Therefore, a procedure can contain many activities, and an activity can be part of more than one procedure. The arc joining the activity box to itself labeled precedence tells which activities may precede which others.

Since the diagram specifies that this is a many to many relationship, the procedure scripting facility supports the specification of conjunctive and disjunctive precedence relations. For any activity labeled conjunctive, any specification of immediate successors denotes activities which all directly follow the completion of the given activity; specification of immediate predecessors denotes activities which must all complete before the given activitycan begin. Some activities will be labeled disjunctive. OR-out from some activity means that out of the many immediate successor activities, we select only one to actually execute. Similarly, OR-in means that only one of the activities which immediately precede the given activity must complete before it can begin. Thus, any partial ordering of activities using sequencing, and these AND/OR constructs can be specified and supported using workflow.

Other entities shown in Figure 3 are job and data. A job which can be considered to be flowing through a procedure has a state at any instant which is denoted by one or more current activities being executed by the job. The relationship "state-of" captures this state. This relationship gets updated by the system each time that a job moves from one activity to another. This is a many to many relationship, so one job may be executing within several activities in parallel, and one activity may be simultaneously serving several jobs. Similar considerations hold for the data entity which refers to the application data which is accessed by the various activities. People are connected into the system directly if they are listed in the "actor" entity set. Thus, people are players of roles, and roles are designated as the executors of activities.

3 Workflow Modeling

3.1 Petri Net Models of Workflow

Zisman used Petri nets as the base mechanism for representing workflow in 1977 [21]. His model modified the usual Petri net firing rules by adding other rules, (production systems,) to act as guards for transition firing.

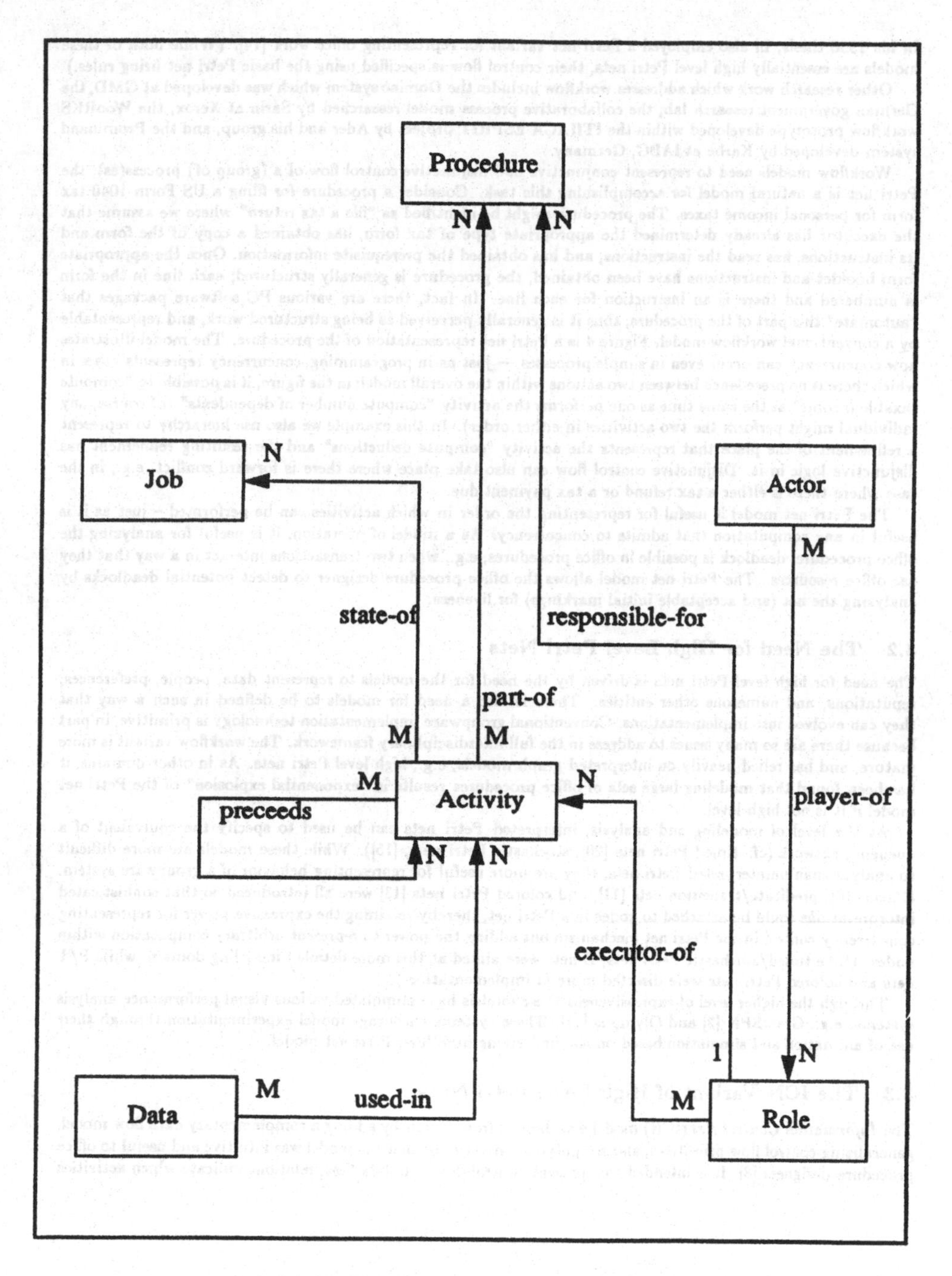

Figure 3: Workflow Conceptual Architecture

In his 1990 thesis, Li also employed a Petri net variant for representing office work [14]. (While both of these models are essentially high level Petri nets, their control flow is specified using the basic Petri net firing rules.)

Other research work which addresses workflow includes the Domino system which was developed at GMD, the German government research lab, the collaborative process model researched by Sarin at Xerox, the WooRKS workflow prototype developed within the ITHACA ESPRIT project by Ader and his group, and the Prominand system developed by Karbe at IABG, Germany.

Workflow models need to represent conjunctive and disjunctive control flow of a (group of) process(es); the Petri net is a natural model for accomplishing this task. Consider a procedure for filing a US Form 1040 tax form for personal income taxes. The procedure might be identified as "file a tax return" where we assume that the executor has already determined the appropriate type of tax form, has obtained a copy of the form and its instructions, has read the instructions, and has obtained the prerequisite information. Once the appropriate form booklet and instructions have been obtained, the procedure is generally structured; each line in the form is numbered and there is an instruction for each line. In fact, there are various PC software packages that "automate" this part of the procedure, thus it is generally perceived as being structured work, and representable by a conventional workflow model. Figure 4 is a Petri net representation of the procedure. The model illustrates how concurrency can occur even in simple processes — just as in programming, concurrency represents cases in which there is no precedence between two actions within the overall model; in the figure, it is possible to "compute taxable income" at the same time as one performs the activity "compute number of dependents" (of course, any individual might perform the two activities in either order). In this example we also use hierarchy to represent a refinement of the place that represents the activity "compute deductions" and the resulting refinement has disjunctive logic in it. Disjunctive control flow can also take place where there is forward conflict, e.g., in the case where there is either a tax refund or a tax payment due.

The Petri net model is useful for representing the order in which activities can be performed – just as it is useful in any computation that admits to concurrency. As a model of operation, it is useful for analyzing the office procedure; deadlock is possible in office procedures, e.g., when two transactions interact in a way that they use office resources. The Petri net model allows the office procedure designer to detect potential deadlocks by analyzing the net (and acceptable initial markings) for liveness.

3.2 The Need for High Level Petri Nets

The need for high level Petri nets is driven by the need for the models to represent data, people, preferences, reputations, and numerous other entities. There is also a need for models to be defined in such a way that they can evolved into implementations. Conventional groupware implementation technology is primitive, in part because there are so many issues to address in the full interdisciplinary framework. The workflow variant is more mature, and has relied heavily on interpreted graph models, e.g., high level Petri nets. As in other domains, it has been found that modeling large sets of office procedures results in "exponential explosion" of the Petri net model if it is not high level.

At the level of modeling and analysis, interpreted Petri nets can be used to specify the equivalent of a queueing network (cf. timed Petri nets [20], stochastic Petri nets, [15]). While these models are more difficult to analyze than uninterpreted Petri nets, they are more useful for representing behavior of a groupware system. E nets [16], predicate/transition nets [11], and colored Petri nets [13] were all introduced so that sophisticated interpretations could be attached to nodes in a Petri net, thereby retaining the expressive power for representing concurrency control in the Petri net mechanism but adding the power to represent arbitrary computation within nodes. (Like timed/stochastic Petri nets, E nets were aimed at this more detailed modeling domain, while P/T nets and colored Petri nets were directed more at implementation.)

Through the higher level of expressiveness these models have stimulated various visual performance analysis systems, e.g., GreatSPN [2] and Olympus [17]. These systems encourage model experimentation through their use of animation and simulation based on the underlying high level Petri net model.

3.3 The ICN Variant of High Level Petri Nets

The *Information Control Net* (ICN) model was derived from E nets by adding a complementary data flow model, generalizing control flow primitives, and simplifying semantics so that the model was intuitive and useful to office procedure designers [8]. It is intended to represent control flow and data flow; relations indicate which activities

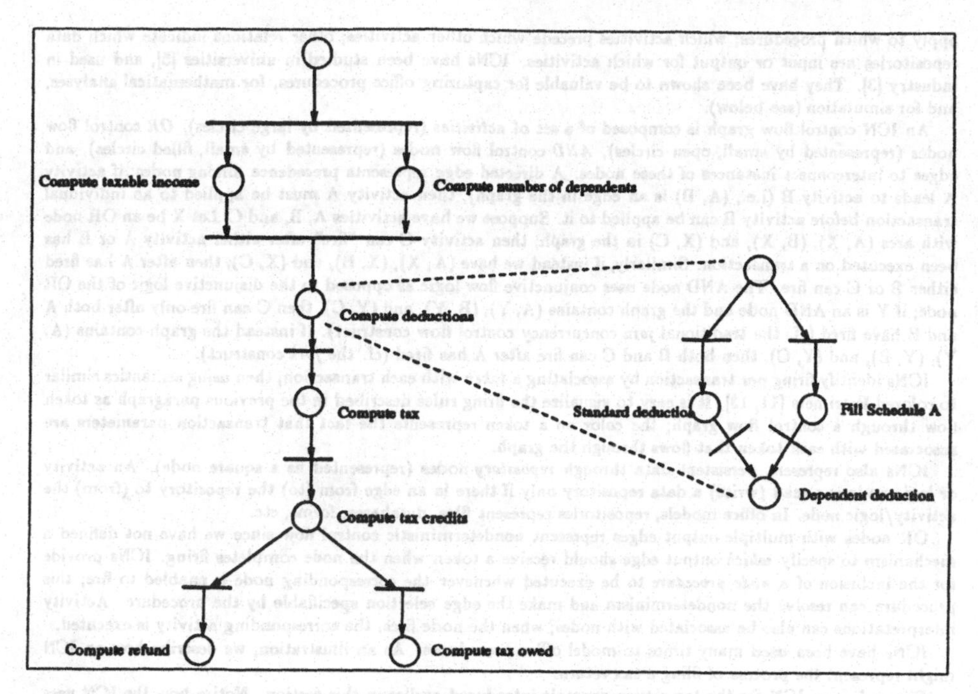

Figure 4: Petri Net for Filing a Tax Return

apply to which procedures; which activities precede which other activities; other relations indicate which data repositories are input or output for which activities. ICNs have been studied in universities [5], and used in industry [3]. They have been shown to be valuable for capturing office procedures, for mathematical analyses, and for simulation (see below).

An ICN control flow graph is composed of a set of *activities* (represented by large circles), *OR* control flow nodes (represented by small, open circles), *AND* control flow nodes (represented by small, filled circles), and edges to interconnect instances of these nodes. A directed edge represents precedence among nodes; if activity A leads to activity B (i.e., (A, B) is an edge in the graph), then activity A must be applied to an individual transaction before activity B can be applied to it. Suppose we have activities A, B, and C. Let X be an OR node with arcs (A, X), (B, X), and (X, C) in the graph; then activity C can "fire" after either activity A or B has been executed on a transaction. Similarly, if instead we have (A, X), (X, B), and (X, C), then after A has fired either B or C can fire. The AND node uses conjunctive flow logic as opposed to the disjunctive logic of the OR node; if Y is an AND node and the graph contains (A, Y), (B, Y), and (Y, C), then C can fire only after both A and B have fired (cf. the traditional *join* concurrency control flow construct). If instead the graph contains (A, Y), (Y, B), and (Y, C), then both B and C can fire after A has fired (cf. the *fork* construct).

ICNs identify firing per transaction by associating a *token* with each transaction, then using semantics similar to colored Petri nets [11, 13]. It is easy to visualize the firing rules described in the previous paragraph as token flow through a control flow graph; the color on a token represents the fact that transaction parameters are associated with each token that flows through the graph.

ICNs also represent persistent data through *repository* nodes (represented as a square node). An activity or logic node can read (write) a data repository only if there is an edge from (to) the repository to (from) the activity/logic node. In office models, repositories represent files, databases, forms, etc.

OR nodes with multiple output edges represent nondeterministic control flow since we have not defined a mechanism to specify *which* output edge should receive a token when the node completes firing. ICNs provide for the inclusion of a *node procedure* to be executed whenever the corresponding node is enabled to fire; this procedure can resolve the nondeterminism and make the edge selection specifiable by the procedure. Activity interpretations can also be associated with nodes; when the node fires, the corresponding activity is executed.

ICNs have been used many times to model office procedures. As an illustration, we describe how an ICN might represent the process of filing a tax return.

Figure 5 is an ICN for the tax return example introduced earlier in this section. Notice how the ICN uses nearly equivalent constructs to represent conjunctive and disjunctive control flow. The ICN workflow model illustrates that the number of dependents can be computed concurrently with (or in either order) computing the total adjusted income. However, before deductions can be computed, the income and number of dependents must have been computed. The OR node is used to differentiate between the cases of overpayment and underpayment of withholding taxes. Also, notice that the activity to compute deductions is refined by the ICN on the right side of the figure. This refined ICN illustrates how to compute a deduction, and implicitly has an activity that is refined ("Fill Schedule A").

3.3.1 Using ICNs for Workflow Implementations

The ICN as described above is useful for *modeling* a procedure independent of how it is actually *implemented*. That is, the tax form ICN could have been completed by one person, a person and a computer, two persons, two persons and a computer, etc. An implementation of the tax form procedure must introduce "engines" to accomplish the work represented by various nodes in the graph, a scheduler for the work, a dispatcher to assign engines to nodes, a mechanism to make data available, a coordinator to oversee the execution, and miscellaneous other tasks. We refine the model to begin to introduce these components.

First, let each individual ICN represent a *procedure* within an enterprise's domain. Various collections of activities and control flow nodes are grouped together to form a *role* within one or more procedures; the idea is that a role represents work characterization for some particular type of "engine," so a role might specify work for an order administrator or a purchasing agent. An *execution environment* for the ICN specifies *actors* that can assume various roles - an actor is a human user or perhaps an automated agent. An ICN is executed by introducing a transaction to a procedure, then by having it pass through the ICN from activity-to-activity — and hence potentially from role-to-role and actor-to-actor.

An execution environment that supports the execution of the workflow specification must schedule and

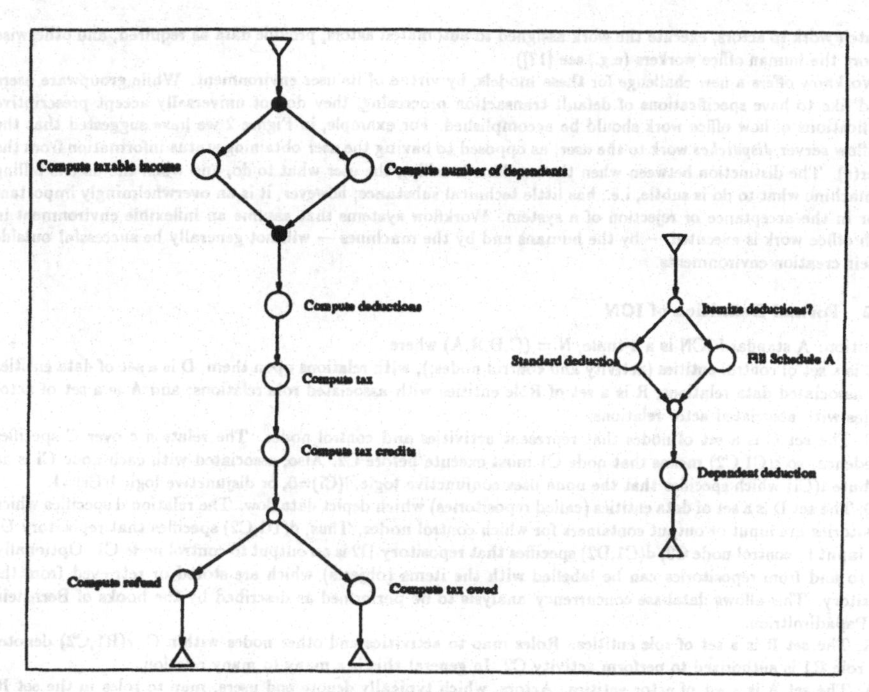

Figure 5: ICN for Filing a Tax Return

dispatch work to actors, execute the work assigned to automated actors, provide data as required, and otherwise support the human office workers (e.g., see [17]).

Workflow offers a new challenge for these models, by virtue of its user environment. While groupware users would like to have specifications of default transaction processing, they do not universally accept prescriptive specifications of how office work should be accomplished. For example, in Figure 2 we have suggested that the workflow server *dispatches* work to the user, as opposed to having the user obtaining status information from the server(s). The distinction between when the machine is telling the user what to do, and when the user is telling the machine what to do is subtle, i.e., has little technical substance; however, it is an overwhelmingly important factor in the acceptance or rejection of a system. Workflow systems that assume an inflexible environment in which office work is executed — by the humans and by the machines — will not generally be successful outside of their creation environments.

3.3.2 Formal Definition of ICN

Definition: A standard ICN is a 4-tuple, N = (C,D,R,A) where

C is a set of control entities (activity and control nodes,), with relations upon them; D is a set of data entities with associated data relations; R is a set of Role entities with associated role relations; and A is a set of actor entities with associated actor relations.

C: The set C is a set of nodes that represent activities and control nodes. The relation c over C specifies precedence, so c(C1,C2) means that node C1 must execute before C2. Also, associated with each node Ci is an attribute l(Ci) which specifies that the node uses conjunctive logic, l(Ci)=0, or disjunctive logic l(Ci)=1.

D: The set D is a set of data entities (called repositories) which depict data flow. The relation d specifies which repositories are input or output containers for which control nodes. Thus, d(D1,C2) specifies that repository D1 is an input to control node C2; d(C1,D2) specifies that repository D2 is an output to control node C1. Optionally, arcs to and from repositories can be labelled with the items (objects) which are stored or retrieved from the repository. This allows database concurrency analysis to be performed as described by the books of Bernstein and Papadimitriou.

R: The set R is a set of role entities. Roles map to activities and other nodes within C. r(R1,C2) denotes that role R1 is authorized to perform activity C2. In general this is a many to many relation.

A: The set A is a set of actor entities. Actors, which typically denote end users, map to roles in the set R. a(A1,R2) specifies that actor A1 qualifies to play the role R2. In general, this is a many to many relation.

Mathematical analyses of workflow have proven useful and tractable. Some of the documented analyses using ICNs include throughput, maximal parallelism, sensitivity analysis, reorganization, and streamlining [4]. Our experience with enactment of prescriptive workflow systems [9, 10, 18] has only met with marginal success, we believe, due to the problem of inflexible environments. While the workflow model must be sufficiently precise to act as a programming language, it must be sufficiently flexible to allow users to operate in the face of exceptional conditions. The workflow system must be able to *assist* the user while he is handling an exception by providing information about the goals of the procedure, by allowing unusual transaction (token) flow, and by generally supporting structural change in the model — even at runtime.

4 Coping with Exceptions

This section describes the SendTo feature which allows any user to send a message to a partial ordering of one or more activities, roles, or users. The message, which may be delayed or immediate, can transport a short informal note or the content of a large job or activity. The feature is extremely flexible, and is intended as a basic mechanism for use in exception handling, in informal communication, in dynamic scheduling, and in notification and reminders.

4.1 Exception Handling

Sometimes, exceptional conditions, which naturally arise in all human endeavor, can be simply handled by a telephone call to the right person. In other cases, activities must be cancelled or undone, events must be rescheduled, or new activities must be dynamically scheduled. To cope with some of these more complex situations, a workflow system can incorporate the SendTo feature which is described here. The description

is functional with parameters. This linear, non-graphic description is NOT the way that the system will present this functionality to the end user. It is not the intent of this document to specify the user interface, nor the implementation details of the SendTo feature.

SendTo (recipients, message-type, activity-state, static routing, receive-time)

This feature causes information to be sent to one or more workflow actors, or system entities. The parameters are:

1. **recipients — activities, roles, or actors to receive message,**
2. **type — this denotes the class of message; it is either a note or else a job with attachments,**
3. **state — the sending activity can be marked (STATE = done) or (STATE = in-progress),**
4. **routing — after sending, continue the steps of the original route or cancel them,**
5. **receive time — immediate (as soon sa possible) or delay (specify later delivery time.)**

Next we explain each of these parameters in greater detail with examples of usage. The examples refer to the order processing procedure described and diagrammed previously in this paper.

The recipient parameter: The receivers of the message are specified by the sender who selects a set of activities, roles and actors. For example, suppose the sales manager is performing the order evaluation activity. If this job is especially complex, then the manager will manually assign the billing task to the most experienced billing clerk (who is named Michael). The manager would do this by using the SendTo with the following parameters:

SendTo (ACTIVITY=billing, ROLE=billing clerk, ACTOR=Michael).

This would override the scheduler's dispatching routine, and select the specific user named Michael. The default data sent would be all of the information associated with the current job, and the manager would select the parameter STATUS=done, to indicate that he has completed the order evaluation step.

Exceptions are one time changes to the normal flow of activities. An exception which frequently occurs is one casually described as "skip over the next step." In our example, if the manager wants to skip over the billing step, and go directly to the shipping step, the he would use the following:

SendTo (ACTIVITY=shipping, ROLE=inventory clerk, ACTOR=null)

The null specification is equivalent to "send to any" and it means that the scheduler can select any appropriate inventory clerk, and send it to that user's workstation. This is more flexible than having a "skip a step" instruction because at times, one may want to skip two or more activities, or return to a previous step. SendTo can be used for all of these types of exceptions.

As another example, suppose that the sales manager needs expert opinion from Marie, an expert who is not normally associated with the procedure before completing the order evaluation. He can send all the job information to Marie with text in the comment field saying "Please take a look, give your opinion, and return to me." The SendTo parameters would be:

SendTo (ACTIVITY=null, ROLE=null, ACTOR=Marie).

4.2 Informal Communications

The type parameter specifies the information, note or activity, to be sent to the recipient. The previous examples were ones in which the manager sent all of the information associated with an activity. If the manager wants to send a reminder after a few days to Marie, then he can select TYPE=note, and just type some text that will then be delivered to Marie. This is useful for many kinds of informal communication.

When the manager sends the order processing job information to Marie, he uses SendTo with TYPE=job. At this stage, he is presented a choice of DATA = all, comment-only, or selected parts. Remember that Marie is not normally associated with the order processing procedure, so there is no specification of what order processing information she should see. The manager can select DATA=all if he thinks that Marie can and needs to see all

of the information in order to formulate an informed opinion. If there are some fields (e.g. salary) which Marie should not see, then the manager would select DATA=selected, and then he would select the fields and parts which are to be sent to Marie. The third option is convenient when only the comment field needs to be sent: DATA=comments.

The default for a SendTo recipient who is not associated with the procedure is to be able to read (see) all information that is sent, and to only be able to write by appending to the document, or adding test to the comment field. Later, the manager can copy the information received from Marie into other fields of the form or document if it is appropriate.

The state of an activity after the SendTo is completed can be STATE=done, indicating that the sender is completed or STATE=suspended, indicating that the sender expects the message to be returned so that the activity can later be completed. The first examples which were provided of exception handling all used the default of STATE=done, since they were forwarding the job to another activity. However, in the last example, the manager will not complete the activity until after he receives a reply from Marie. Thus, the manager sends the job to Marie with parameter STATE = suspended.

4.3 Dynamic Scheduling

The routing parameter determines whether the old (static and normal) route is followed by the job, or a new dynamic route is followed. In our example, the manager is in the midst of the order processing procedure for a complex job, which is automatically monitored and advanced by the workflow system whenever appropriate activities are completed. After a SendTo is completed, the recipient performs some work, and then activates the DONE button indicating completion. At that time, the system can resume the old procedure (if the sender set ROUTING=continue,) or perform steps of the dynamic schedule specified by the sender (if the sender set ROUTING=cancel.) These new steps of the dynamic schedule are a one-time special case, and not part of the static set of procedures. The old procedure is cancelled in this case for this one job only.

These new steps are specified by the sender in the recipients parameter of the SendTo. Thus in the recipient field, one has the option of specifying more than one recipient. If this option is selected then the sender can specify an ICN of ordered recipients.

In all of our previous examples, the "normal procedure" was expected to continue after dealing with the exception, so the manager using SendTo in these examples always chose the default of ROUTING = continue. However, when a job is suddenly cancelled, SendTo can be used to notify all who are currently processing the job. There is no need for the workflow system to continue scheduling and forwarding this job, so ROUTING=cancel is appropriate. Document circulation and mail circulation systems applications frequently do not decide to whom a document should be routed until late in the procedure, and this is constantly subject to change. These situations can all be handled by the SendTo. When a person reads the document and decides that it should be rerouted, she uses the SendTo option, and specifies the new routing within the recipients parameter. Specification of ROUTING=cancel is then used to cancel the old routing.

4.4 Events and Notification

The receive-time parameter can be set to RECEIVE-TIME=immediate, to force delivery as soon as possible, or it can be set to RECEIVE-TIME=delay, in which case the sender is asked to specify date and time of delivery. All of the above examples used the default of RECEIVE-TIME=immediate, but situations in which notifications are needed at a certain time would use RECEIVE-TIME=delay.

As an example, suppose that a new pricing strategy takes effect on February 1. The manager, when notified of this, may want to tell all of the billing clerks immediately, and also send them a reminder on the day before it takes effect. To do this, the manager can send a note of reminder to all billing clerks with a receive date specified as January 31. This would be done as:

SendTo (ACTIVITY=null, ROLE=billing clerk, ACTOR=all, RECEIVE-TIME=delay, DATE=013194, TIME=09:00am, TYPE=note)

This illustrates the SendTo "all" capability allowing the sender to specify all actors who play some role as recipients.

The SendTo with time delay can also be used to initiate periodic automatic activities at a workstation. Since workflow provides programmed access to sending and receiving messages, a script can periodically:

1. test for a new message from itself,
2. perform some activity if a message is received, and
3. send another delayed message to itself if a message is received.

The beginning and ending of activities are examples of "normal" internal events which are expected by the system. This subsection has also discussed how to generate and handle notifications which are considered external events. As with other features, workflow event handling features provide a general mechanism by which different applications can create scripts particular to their needs. Given the huge variety of notifications possibly desired, this is a powerful mechanism because job data, state, and audit information are all available for access within automated scripts.

5 Summary and Future Research

This paper has presented an overview of workflow modelling and enactment based upon the ICN model. An important point emphasized is that workflow systems are people systems. People cannot simply be treat4ed as initial inputters, and consumers of final output. Unlike many other systems, successful utilization is dependent upon social and organizational factors as well as technical factors. Thus, these systems must be flexible, open ended, and support the unstructured work which is prevalent in many organizations.

Workflow modelling was discussed within the framework of ICNs. ICNs, which were formally defined in this paper, have been successfully used in industry and academia for workflow description, simulation, and analysis. Workflow execution environments were discussed and a logical architecture for a generic workflow system was explained. A mechanism was offered to cope with dynamic structural change and exception handling within this environment. Our research group at the University of Colorado, the Collaboration Technology Research Group, is continuing to explore theoretical and applied issues of workflow as well as broader issues of collaboration theory and groupware.

Workflow is a fertile area for research and development. There is high commercial interest, there are formal statements of its properties and issues, and there are significant problems to be solved for its future success. We next mention some of these issues and problems. Much of the work in organizations is goal based rather than procedure based. People do whatever is necessary and expedient to attain these goals. This suggests that there is a need for goal based workflow systems and models. Even without approaching this, our basic ICN definition has only defined one model rather than a family of models. It has become apparent that no one model satisfies all of the varied needs and emphases. Furthermore, our exception handling mechanism allows users to escape the model rather than helping them to analyze and expedite exception handling within the model. Somehow, a mechanism which integrates exception handling into the model needs to be invented.

Besides modelling, there is a need for work on enactment and workflow execution environments. This includes study of issues such as distributed execution, measurement and evaluation of workflow, real-time interaction among actors, and group user interfaces. We need to test solutions to the above in realistic settings, and use this to feedback on other critical research issues. When this is done, one challenging problem which emerges concerns the constant change within organizations. This suggests need for study of dynamic structural change within Petri nets, and questions such as "Given a marked Petri net graph, what structural changes can or cannot be applied while maintaining consistency and correctness." This is an important and tough problem.

The introduction of actors and roles in the ICN model represents a step toward incorporation of an organizational model within workflow, but clearly, much more needs to be done to assimilate appropriate organizational and social sub-models within workflow models. Incorporation of these sub-models is an important ingredient for the success of future generations of high impact workflow systems.

6 Acknowledgements

This work has been supported by a grant from Najah Naffah at Bull Worldwide Information Systems Imaging and Office Solutions.

References

[1] James Bair, editor. *Office Automation Systems: Why Some Work and Others Fail.* Center for Information Technology, 1981. Stanford University conference.

[2] G. Balbo and G. Chiola. Stochastic petri net simulation. In *1989 Winter Simulation Conference Proceedings*, pages 266–276, 1989.

[3] Bull S. A. *FlowPath Functional Specification*, September 1992.

[4] Carolyn Cook. Office streamlining using the icn model and methodology. In *Proceedings of the 1980 National Computer Conference*, 1980.

[5] P. Dumas. *La Methode OSSAD.* Les Editions d'Organization, 1991.

[6] Esther Dyson. *Workflow.* EDventure Holdings, 1992. Release 1.0.

[7] C. Ellis, S. J. Gibbs, and G. L. Rein. Groupware: Some issues and experiences. *Communications of the ACM*, 34(1):38–58, January 1991.

[8] Clarence A. Ellis. Information control nets: A mathematical model of office information flow. In *Proceedings of the 1979 ACM Conference on Simulation, Measurement and Modeling of Computer Systems*, 1979.

[9] Clarence A. Ellis. Officetalk-p: An office information system based upon migrating processes. In Najah Naffah, editor, *Integrated Office Systems*, 1979.

[10] Clarence A. Ellis. Officetalk-d, an experimental office information system. In *Proceedings of the First ACM Conference on Office Information Systems*, 1982.

[11] H. J. Genrich. Predicate/transition nets. In *Advances in Petri Nets 1986*, pages 3–43. Springer Verlag, 1986.

[12] J. Grudin. Why cscw applications fail. In *Proceedings of the CSCW88 Conference*, 1988.

[13] Kurt Jensen. *Coloured Petri Nets: Basic Concepts, Analysis Methods and Practical Use.* Springer Verlag, 1992.

[14] Jianzhong Li. *AMS: A Declarative Formalism for Hierarchical Representation of Procedural Knowledge.* PhD thesis, L'Ecole Nationale Superierure des Telecommunications, 1990.

[15] M. K. Molloy. Performance analysis using stochastic petri nets. *IEEE Transactions on Computers*, C-31(9):913–917, September 1982.

[16] Gary J. Nutt. *The Formulation and Application of Evaluation Nets.* PhD thesis, University of Washington, 1972.

[17] Gary J. Nutt. *A Simulation System Architecture for Graph Models*, pages 417–435. Springer Verlag, 1990.

[18] Gary J. Nutt and Clarence A. Ellis. Backtalk: An office environment simulator. In *ICC 79 Conference Record*, pages 22.3.1–22.3.5, 1979.

[19] S. Poltrock and J. Grudin. Tutorial on computer supported cooperative work and groupware. Presented at the ACM SIGCHI Conference on Human Factors in Computing Systems, April 1980.

[20] C. Ramchandani. *Analysis of Asynchronous Concurrent Systems by Timed Petri Nets.* PhD thesis, MIT, 1974.

[21] Michael D. Zisman. *Representation, Specification and Automation of Office Procedures.* PhD thesis, University of Pennsylvania Wharton School of Business, 1977.

Interleaving Functional and Performance Structural Analysis of Net Models

(Extended Abstract)

Manuel Silva *

Dpto. de Ingeniería Eléctrica e Informática
Centro Politécnico Superior
Universidad de Zaragoza
María de Luna, 3
50015 Zaragoza SPAIN
silva@etsii.unizar.es

Abstract. The design of parallel and distributed systems requires considering both functional and performance specifications and properties. The use of a single family of models to deal with both aspects, plus appropriate analysis techniques, are very important in practice. Petri Nets are one of the formal description techniques able to capture both functional and performance aspects. We illustrate here, after our research experience, how the interleaving of the functional and the performance analysis theories produces a synergetic situation, in which each contributes to the development of the other.

1 Introductory Remarks

A net model consists of two parts: (1) A *net structure*, a weighted-bipartite directed graph, that represents the *static* part of the system; and (2) a *marking*, representing a *distributed state* on the structure. The places are the *state variables* of the system, the transitions their *transformers (actions)*, the weights permit the modelling of *bulk services and arrivals*, while the marking represents the *values* of the state variables.

Modelling with Petri nets may be very convenient for several reasons: true concurrency is naturally captured; states and actions are treated on equal footing; models are essentially state-parametric; there exists a valuable graphic representation of models; top-down and botton-up modelling methodologies can be nicely introduced, etc. A very important point, from the modelling perspective, is that nets allow different *interpretations*, leading to models describing different perspectives of a given system. Uninterpreted net systems are said to be *autonomous*.

* The work of the author in this field is presently supported by the Spanish CICYT ROB 91-0949 and the ESPRIT-BRA Project 7269 (QMIPS) and WG 6067 (CALIBAN). J. Campos, G. Chiola, J.M. Colom and E. Teruel extensively contribute to this research program.

Typically net models are interpreted to deal with the *functional* or with the *performance* behaviour (through a, possibly stochastic, timed interpretation) of a given class of systems. This turns out to be a tremendous advantage, because using Petri nets we can go through the different phases of a design project using models belonging to a *unique family*, what makes easier and more reliable the full process.

Functional analysis of net models deals with properties like deadlock-freeness, or the absence of (store) overflows. Its ultimate goal is proving *correctness* of the modelled system. Properties can be *qualitative* (e.g. the existence of a home space, liveness, boundedness, synchronic relations [SC 88],...), or *quantitative* (e.g. the bound of a place, a synchronic distance,...).

Functional analysis is usually done on the (underlying) *autonomous* system. Based on the separation into a structure and a marking, *structure theory* is a branch of net theory devoted to investigate the relationship between the structure and the behaviour of net system models. Concepts and proof techniques are based on *graph* and on *linear algebra/convex geometry* theories. Let us sketch briefly the evolution of structure theory for autonomous net systems.

Using graph theory, many important results have been obtained in the functional analysis of some autonomous *ordinary* net *subclasses* (see, for instance, [CHEP 71, Hack 72, TV 84, Best 87]).

Linear algebra/convex geometry theories typically allow the study of *necessary* or *sufficient* conditions for functional properties for any net and marking. In this context *invariant analysis* has been developed (see, for example, [Reis 85, Laut 87, Mura 89]). Invariants allow formal reasoning on logical properties, to some extent like "program invariants" do in programs verification; they are suitable for proving (in temporal logic sense) *safety* properties (i.e. those stipulating that some "bad things" do not happen during evolution): mutual exclussions, deadlock-freeness...

Even if invariant analysis is very popular, more efficient algorithms have been developed applying mathematical programming concepts and techniques to the *net system state equation*, which, in fact, contains all the information concerning structural linear invariants. See, for example, [SC 88, CS 91]; in [GL 73] an early application of linear programming to marked graphs is contained.

Recently, graph and linear algebra/convex geometry theories are intensively used in an interleaved way leading to new lights and results in the theory of some already well-known ordinary net subclasses (as *free choice* nets: [CCS 91, ES 91a, ES 91b, Dese 92]), and of some weighted net subclasses as *weighted-T-systems* [TCCS 92], *choice-free* (structurally persistent) systems [TCS 93] and *equal-conflict* (a weighted generalization of extended free choice) systems [TS 93].

Functional properties are also interesting for the analysis of interpreted (eventually timed) net systems. Nevertheless, this is a more difficult problem in general because the *constraints* on the behaviour imposed by the interpretation can be very difficult to be taken into account. Even more, unfortunately the analysis of the underlying autonomous net system may be of *no* help for the analysis of the fully interpreted system (for example, liveness of the autonomous system is neither necessary nor sufficient for liveness of a deterministically timed system [Silv 85]).

Performance analysis of timed models is basically concerned with measures like throughput or utilization rates; that is, with the evaluation of the *efficiency* of the modelled system. Performance properties can be qualitative (e.g. ergodicity, throughput monotonicity wrt. the initial marking,...) or quantitative (e.g. utilization rates, average queue lengths,...).

The next section illustrates the mutual benefits obtained when structure theory is concerned not only with the behaviour of autonomous net systems, but also with performance aspects.

2 Synergy between Functional and Performance Analysis

Structure theory allows the analysis of some autonomous net system models avoiding the *state space explosion* problem. The performance analysis of timed models is usually based on the computation of a state space (the same as for the untimed model if *exponential* pdfs are used), thus suffering also the state space explosion problem. There do not exist too many works in which the structure theory approach has been applied to the performance analysis of timed net models. ([CS 92] is a survey concerning mainly structural techniques for the computation of insensitive performance bounds.)

Performance structural analysis of net models requires some deep understanding of the functional structure analysis of the underlying autonomous net model. Thus, structural analysis theory for the underlying autonomous models allows developing the performance structural analysis theory. But "a priori" this is something easy to expect! What is more surprising is the fact that the development of performance structure theory pushes also the development of the structure theory for the analysis of the underlying autonomous net model. In some sense, these relations are like those between science and technology: Science can lead to improvements in technology, while new technologies may require, or can lead to, improvements in our scientific knowledge.

Besides the "usual" efficiency of the structural analysis algorithms, some performance properties (as throughput bounds, or ergodicity of some throughput monotonic behaviours) can be computed for cases in which the classical Markovian analysis is of no help due to the generality of the pdfs in the timing.

The interleaving of (autonomous) functional and performance properties in the structural analysis of net models has already produced several benefits. Let us mention just a few:

(a) *Introduction of new concepts for the analysis of autonomous net systems.* Liveness and boundedness are two among the most classical properties in functional analysis of net system models. The qualitative concept of boundedness has been classically refined in a quantitative way using the (marking) bound notion: the maximum number of tokens a place or a net system may have. The maximum number of *servers* in steady-state in a station is a crucial point for the performance of a Queueing Network. Its action oriented counterpart in net systems terms is obtained through the so called *liveness bound* (or degree) [CCS 91]. A transition (the analogous of a *station* if timing is associated to transitions) is k-live iff from any reachable marking it can be ultimately k-enabled. Obviously, a transition is not live iff there exists a reachable marking such that it cannot be q-enabled, with any $q \geq 1$, in any successive marking. Thus, the consideration of performance analysis leads to a valued generalization of the liveness concept. The classical liveness non-monotonicity wrt. the initial marking (see, for example, [Reis 85]) can be observed now in more detail: For some net systems, increasing the initial marking decreases the liveness bound of certain transitions.

(b) *Bridges between logical and performance properties.*
Assume, for easy consideration, that the stochastic timing associates only constant rate exponential pdfs to the firing of transitions. In this case it is well-known that the reachability graph of the underlying autonomous net system and the Markov chain of the stochastic net system are *isomorphous.* Therefore if the autonomous net system is bounded and has a *home space* (and this is the case, for example, in the weighted generalization of extended free choice nets [TS 93]), then the stochastic system obtained incorporating the timing is *marking* and *firing ergodic.*
In [CS 90] a class of performance models derived from [Soui 91] is studied. In some sense it can be viewed as an extension of some queueing networks: the servers are modelled with state machines, while some restricted synchronizations among buffers are allowed. The existence of stochastic interpretations leading to ergodic behaviour is characterized by means of two purely structural properties: *consistency* and some *synchronic distance relations.* Even if they are unbounded, exact performance for ergodic systems can be computed in polynomial time.

(c) *Performance analysis suggests new results for checking logical properties.*
The *visit ratio* is a well defined concept in Queueing Networks. The computation of the visit ratio is difficult in general stochastic Petri Net models because it depends not only on the net structure and routing rates, but also on the initial marking and pdfs associated to transitions. Nevertheless, if we reduce our attention to strongly connected and structurally bounded free choice nets, the visit ratio can be efficiently computed for any constant rate resolution policy at conflicts. But this property turns out to be very important because it opens a new algebraic characterization of structural liveness (the *rank theorem* for strongly connected and structurally bounded free choice nets [CCS 91]). A slightly different perspective of this result is given in [ES 91a]. The consequences of the *rank theorem* are enormous for the theory of free choice systems. In particular a *polynomial* characterization of liveness and boundedness and a new, easy, and compact proof of the well-known *duality theorem* [Hack 72] for structurally live and bounded free choice nets are derived from it.

(d) ***Untimed net systems analysis techniques allow developing new performance analysis techniques.***

Let us consider *decomposition* and *reduction* techniques [Reis 85, Silv 85, Mura 89]. It is well known that structurally live and structurally bounded nets can be decomposed into closed subnets generated by the minimal P-semiflows (i.e. minimal non-negative left annullers of the incidence matrix). It has been shown in [CS 93] that under certain conditions, the *structural decomposition* of the net allows to improve the computation of some insensitive bounds of net system models looking at the *embedded queueing networks*. *Reduction* techniques [Bert 87, Silv 85] constitute a well established branch of analysis theory of autonomous net system models. Under certain circumstances some reduction rules allow viewing the reachability space of the reduced systems as an *aggregation* of the original one satisfying interesting properties from the stochastic point of view. This basic idea has been used to derive an iterative technique to approximate the throughput of a stochastic marked graph in [JSS 92]. (Strongly connected stochastic marked graphs are equivalent to the so called *Fork Join Queueing Networks with Blocking* in the performance evaluation field [DLT 90].) In this case the reduction is essentially done in a subnet with one input transition and one output transition (denominated as Single Input/Single Output, SISO, subnets). Even if conceptually this technique is interesting, it is limited from a practical point of view. The consideration of multi-input/multi-output subnets has been possible recently because some new results from the structure theory of autonomous marked graphs have been used [CCJS 93].

Structure theory has been used in the computation of performance properties of manufacturing systems in [CCS 90, DJS 92]. Motivated by the modelling of failure and repair situations on *production flow lines*, a *new* subclass of nets, *Macroplace/Macrotransition (MPMT) nets*, has been introduced [DJS 92]. MPMT-nets extend the modelling power of state machines and marked graphs; they are defined by recursively combining *macroplaces* and *macrotransitions*. This new class of net systems presents a different interplay between choices and synchronizations compared to free choice net systems. The important point is that we for MPMT systems we have a good understanding of their autonomous behaviour, allowing generalizations of the iterative approximation techniques already developed for stochastic strongly connected marked graphs.

Structure theory allows approaching functional and performance analysis of net system models. There is presently a rich body of results, particularly for some net subclasses; but, in the overall, the field can be still considered as relatively inmature, so this is a challenging research area.

References

[Bert 87] G. Berthelot. Transformations and Decompositions of Nets. In *Advance Course on Petri Nets (W. Brauer, W. Reisig, G. Rozenberg, eds.). LNCS 254: 359-376.* Springer-Verlag.

[Best 87] E. Best. Structure theory of Petri Nets: The free choice hiatus. In *Advances Course on Petri Nets. LNCS 254: 168-205.* Springer-Verlag.

[CCCS 92] J. Campos, G. Chiola, J.M. Colom, M. Silva. Properties and performance bounds for timed marked graphs. *IEEE Transactions on Circuits and Systems I: Fundamental theory and applications, 39(5):386-401.*

[CCJS 93] J. Campos, J.M. Colom, H. Jungnitz, M.Silva. A General Iterative Technique for Approximate Throughput Computation of Stochastic Marked Graphs. *Dpto. de Ingeniería Eléctrica e Informática, Universidad de Zaragoza.* RR-93-06. 18 pages.

[CCS 90] J. Campos, J.M. Colom, M. Silva. Performance Evaluation of Repetitive Automated Manufacturing Systems. *Rensselaer's 2nd Int. Conf. on Computer Integrated Manufacturing.* IEEE Computer Society Press: 74-81.

[CCS 91] J. Campos, G. Chiola, M. Silva. Properties and performance bounds for closed free choice synchronized monoclass queueing networks. *IEEE Transactions on Automatic Control (Special Issue on Multidimensional Queueing Networks) 36(12): 1368-1382.*

[CS 90] J. Campos, M. Silva. Steady-state performance evaluation of totally open systems of Markovian sequential processes. In *Decentralized Systems (M. Cosnard and C. Girault, eds.).* North-Holland, Amsterdam, 427-438.

[CS 91] J.M. Colom, M. Silva. Improving the linearly based characterization of P/T nets. In *Advances in Petri Nets 90, LNCS 483:113-145.* Springer-Verlag.

[CS 92] J. Campos, M. Silva. Structural techniques and performance bounds of stochastic Petri net models. *Advances in Petri Nets'92 (G. Rozenberg, ed., LNCS 609:352-391.* Spriger-Verlag.

[CS 93] J. Campos, M. Silva. Embedded queueing networks and the improvement of insensitive performance bounds for Markovian Petri Net systems. To appear in *Performance Evaluation.*

[CHEP 71] F. Commoner, A.W. Holt, S. Even, A. Pnueli. Marked Directed Graphs. *J. Comput. System Sci. 5:72-79.*

[Dese 92] J. Desel. A proof of the Rank theorem for extended free choice nets. In *Applications and Theory of Petri Nets 1992 (Kurt Jensen, ed.). LNCS 616: 134-153.* Springer-Verlag.

[DJS 92] A. Desrochers, H. Jungnitz, M. Silva. An approximation method for the performance analysis of manufacturing systems based on GSPNs. *Procs. of Rensselaer's 3rd. Int. Conference on Computer Integrated Manufacturing, IEEE Computer Society Press, Troy (NY):46-55*

[DLT 90] Y. Dallery, Z. Liu, D. Towsley. Equivalence, Reversibility and Symmetries Properties in Fork Join Queueing Networks with Blocking. In *Technical Report MASI 90-32, Universite Paris 6.*

[ES 91a] J. Esparza, M. Silva. On the analysis and synthesis of free choice systems. *Advances in Petri Nets'90 (G. Rozenberg, ed.), LNCS 483:243-286.* Springer-Verlag.

[ES 91b] J. Esparza, M. Silva. Top-down synthesis of live and bounded free choice nets. *Advances in Petri Nets'91 (G. Rozenberg, ed.), LNCS 524:118-139.* Springer-Verlag.

[GL 73] H.J. Genrich, K. Lautenbach. Synchronizations graphen. In *Acta Informática 2: 143-161.*

[Hack 72] M.H.T. Hack. *Analysis of Production Schemata by Petri Nets.* M.S.Thesis, Project MAC TR-94, MIT, Cambridge, Massachusetts. (Corrections in Computation Structures Note 17, 1974.)

[JSS 92] H. Jungnitz, B. Sanchez, M. Silva. Approximate throughput computation of stochastic marked graphs. *Journal and Parallel and Distributed Computing 15:282-295.*

[Laut 87] K. Lautenbach. Linear Algebraic Techniques for Place/Transition Nets. In *Petri Nets: Central Models and their Properties. Advances in Petri Nets 86, LNCS 254:142-167.* Springer-Verlag.

[Mura 89] T. Murata. Petri Nets: Properties, Analysis and Applications. *Procs. IEEE 77,4:541-580.*

[Reis 85] W. Reisig. *Petri Nets- An Introduction.* Springer-Verlag.

[SC 88] M. Silva, J.M. Colom. On the Computation of Structural Synchronic Invariants in P/T Nets. In *Advances in Petri Nets 88, LNCS 340:387-417.* Springer-Verlag.

[Silv 85] M. Silva. *Las Redes de Petri en la Automática y en la Informática.* Editorial AC, Madrid.

[Soui 91] Y. Souissi. Deterministic Systems of Sequential Processes: a Class of Structured Petri Nets. In *Procs. 12th Int. Conf. on Appl. and Theory of Petri Nets:62-81.* Gjern, Denmark.

[TCCS 92] E. Teruel, P. Chrzastowski-Wachtel, J.M. Colom, M. Silva. On Weighted T-systems. In *Applications and Theory of Petri Nets 1992, LNCS 616:348-367.* Springer-Verlag.

[TCS 93] E. Teruel, J.M. Colom, M. Silva. Modelling and analysis of deterministic concurrent systems with bulk services and arrivals. *Dpto. Ingeniería Eléctrica e Informática, Universidad de Zaragoza, RR-93-09.* 20 pages.

[TS 93] E. Teruel, M. Silva. Liveness and Home States in Equal Conflict Systems. In *Applications and Theory of Petri Nets 1993 (M. Ajmone Marsan, ed.), LNCS.* Springer-Verlag.

[TV 84] P.S. Thiagarajan, K. Voss. A Fresh Look at Free Choice Nets. *Information and Control 61,2:85-113.*

FSPNs: Fluid Stochastic Petri Nets

Kishor S. Trivedi[1] and Vidyadhar G. Kulkarni*
***Department of Operations Research**
University of North Carolina, Chapel Hill, NC 27599, U.S.A.
[1]Department of Electrical Engineering
Duke University, Durham, NC 27708, U.S.A.

Abstract

In this paper we introduce a new class of stochastic Petri nets in which one or more places can hold fluid rather than discrete tokens. After defining the class of fluid stochastic Petri nets, we provide equations for their transient and steady-state behavior. We give two application examples. We hope that this paper will spur further research on this topic.

1 Introduction

One of the difficulties encountered while using Petri nets is that the reachability graph tends to be very large in practical problems. If this is due to an accumulation of large number of tokens in one or more places of the net, we can approximate the number of tokens in such places by continuous quantities. Drawing a parallel with fluid flow approximations in performance analysis of queueing systems, we can define such nets.

Stochastic fluid flow models are increasingly used in the performance analysis of communications [2] and manufacturing systems. On the other hand, stochastic Petri nets with discrete places provide a useful framework for specifying and solving performance and reliability models of discrete event dynamic systems [1, 3, 5, 9]. It is natural to extend the stochastic Petri net framework to Fluid Stochastic Petri Nets (FSPNs) by introducing places with continuous tokens and arcs with fluid flow so as to handle stochastic fluid flow systems.

We consider stochastic Petri nets with two types of places: discrete places containing non-negative integer number of tokens and continuous places containing fluid tokens. All transition firings are determined by discrete places. Fluid flow is controlled by enabled transitions. Associating exponentially distributed or zero firing time with transitions, we can then write the differential equations for the underlying stochastic process.

The paper is organized as follows. In Section 2 we develop the fluid model of a stochastic Petri net and in Section 3 we discuss their analysis. Examples are described in Section 4 while conclusions are given in Section 5.

[1]This research was supported in part by the National Science Foundation under Grant CCR-9108114, and by the Naval Surface Warfare Center under Contract N60921-92-C-0161.

2 The Fluid Model

Following the customary notation [4, 7, 8] for defining Petri nets and their extensions, we define a fluid stochastic Petri net (FSPN) as a 7-tuple $(\mathcal{P}, \mathcal{T}, \mathcal{A}, m_0, \mathcal{F}, \mathcal{W}, \mathcal{R})$. $\mathcal{P}$ is a set of places partitioned into a set of discrete places $\mathcal{P}_d$ and a set of continuous places $\mathcal{P}_c$. In a graphical representation, we shall depict continuous places by means of two concentric circles. The set of transitions $\mathcal{T}$ is partitioned into a set of (exponentially distributed firing) timed transitions $\mathcal{T}_E$ and a set of immediate transitions $\mathcal{T}_I$. The set of directed arcs $\mathcal{A}$ is partitioned into two subsets $\mathcal{A}_c$ and $\mathcal{A}_d$. $\mathcal{A}_c$ is a subset of $(\mathcal{P}_c \times \mathcal{T}_E) \cup (\mathcal{T}_E \times \mathcal{P}_c)$ while $\mathcal{A}_d$ is a subset of $(\mathcal{P}_d \times \mathcal{T}) \cup (\mathcal{T} \times \mathcal{P}_d)$. No arcs are allowed between continuous places and immediate transitions. In a graphical representation, arcs in $\mathcal{A}_c$ are drawn as dotted while those in $\mathcal{A}_d$ are drawn as solid.

Let $m_d = (\#p_i, i \in \mathcal{P}_d)$ be the vector of the number of tokens in discrete places and let $m_c = (x_i, i \in \mathcal{P}_c)$ be the vector of the fluid levels in continuous places. The state (marking) of a fluid Petri net is described by the vector $m = (m_d, m_c)$ where m_d is the marking of the discrete part of the state while m_c keeps track of the fluid levels in the continuous places. The initial marking $m_0 = (m_{d0}, m_{c0})$.

We next discuss the evolution of the FSPN. Evolution of the discrete part of the marking follows the rules of GSPN[2] evolution [1]. Thus a transition is said to be enabled when each of its discrete input places has at least one token in it. (By convention a transition with no discrete input places is always enabled). When a transition fires it removes one token from each of its discrete input places and deposits one token in each of its discrete output places. A discrete marking m'_d is said to be reachable from m_d if there is a sequence of transition firings by which m'_d can be reached, beginning with marking m_d. Let $\mathcal{M}_d$ be the set of all discrete markings reachable from m_{d0}. A marking m_d is said to be a tangible marking if no immediate transition is enabled in m_d; otherwise it is called a vanishing marking.

The firing rate function $\mathcal{F}$ is defined for timed transitions $\mathcal{T}_E$ so that $\mathcal{F} : \mathcal{T}_E \times \mathcal{M}_d \rightarrow \mathbb{R}^+$. Thus if a timed transition j is enabled in (a tangible) marking m_d, it fires after an exponentially distributed time with rate $\mathcal{F}(j, m_d)$.

The weight function $\mathcal{W}$ is defined for immediate transitions $\mathcal{T}_I$ so that $\mathcal{W} : \mathcal{T}_I \times \mathcal{M}_d \rightarrow \mathbb{R}^+$. Thus if an immediate transition j is enabled in (a vanishing) marking m_d, it fires with probability

$$\frac{\mathcal{W}(j, m_d)}{\sum_{i \in \mathcal{T}_I \text{ enabled in } m_d} \mathcal{W}(i, m_d)}.$$

Next we describe the evolution of the continuous part of the marking. The

[2] Arc multiplicities greater than 1 and inhibitor arcs are easily added; we do not consider them here for the sake of simplicity.

flow rate function $\mathcal{R}$ is defined for the arcs connecting a continuous place and a timed transition so that $\mathcal{R} : \mathcal{A}_c \times \mathcal{M}_d \rightarrow \mathbb{R}^+ \cup \{0\}$.Thus when the discrete part of the FSPN marking is m_d, fluid can leave place $i \in \mathcal{P}_c$ along the arc $(i,j) \in \mathcal{A}_c$ at rate $\mathcal{R}((i,j), m_d)$ and enters the continuous place at rate $\mathcal{R}((j,i), m_d)$ along the arc $(j,i) \in \mathcal{A}_c$ for each $j \in \mathcal{T}_E$ that is enabled in m_d. The net rate at which fluid builds in a place $i \in \mathcal{P}_c$ in (a tangible) marking m_d is then given by

$$r_i(m_d) = \sum_{j \in \mathcal{T}_E \text{ enabled in } m_d} \mathcal{R}((j,i), m_d) - \sum_{j \in \mathcal{T}_E \text{ enabled in } m_d} \mathcal{R}((i,j), m_d).$$

Now let $X_i(t)$ be the fluid level at time t in a continuous place $i \in \mathcal{P}_c$. Then the sample path of $X_i(t)$ satisfies the differential equation

$$\frac{dX_i(t)}{dt} = \begin{cases} r_i(m_d) & if \quad X_i(t) > 0 \\ [r_i(m_d)]^+ & if \quad X_i(t) = 0. \end{cases} \tag{1}$$

In the case, $X_i(t) = 0$ and $r_i(m_d) < 0$, we set the actual rate equal to zero (denoted by $[r_i(m_d)]^+$) in order to maintain $X_i(t) \geq 0$.

3 Analysis

Recall that $X_i(t)$ is the fluid level in the ith continuous place at time t. Let the vector $\vec{X}(t) = (X_i(t) : i \in \mathcal{P}_c)$. The reachability graph corresponding to the discrete part of the net can be converted into a continuous-time Markov chain (CTMC) in the usual way [1]. Let S be the state space and let $\mathbf{Q} = [q_{ij}]$ be the infinitesimal generator matrix of the underlying CTMC. S corresponds to the set of tangible discrete markings in $\mathcal{M}_d$. Define the diagonal matrix $\mathbf{R}_i = diag(r_i(m), m \in S)$. Define the distribution function $H(t, \vec{x}, m) = P\left(X_i(t) \leq x_i, i \in \mathcal{P}_c \text{ and } m \in S\right)$and let $\vec{H}(t, \vec{x}) = [H(t, \vec{x}, m), m \in S]$ be a row vector. The next theorem gives the coupled system of partial differential equations satisfied by $\vec{H}(t, \vec{x})$. These equations describe the transient behavior of the FSPN.

<u>Theorem 1:</u>

$$\frac{\partial}{\partial t}\vec{H}(t,\vec{x}) + \sum_i \frac{\partial}{\partial x_i}\vec{H}(t,\vec{x})\mathbf{R}_i = \vec{H}(t,\vec{x})\mathbf{Q} \quad ,\vec{x} > 0 \tag{2}$$

with the boundary condition: $x_i = 0$ and $r_i(m) > 0$ implies that $H(t, \vec{x}, m) = 0$.

<u>Proof:</u>

Considering the evolution of the system in the interval $(t, t+h]$ we obtain:

$$H(t+h, x_1, x_2, \cdots, x_n, m) = \sum_{j \neq m} H(t, x_1 - r_1(j)h, x_2 - r_2(j)h, \cdots, x_n - r_n(j)h, j) q_{jm} h + H(t, x_1 - r_1(m)h, \cdots x_n - r_n(m)h, m)(1 + q_{mm} h) + o(h)$$

where $o(h)$ is such that $o(h)/h \to 0$ as $h \to 0$.

After rearranging terms, dividing by h and taking the limit as h approaches 0, we get

$$\frac{\partial}{\partial t} H(t, \vec{x}, m) + \sum_i \frac{\partial}{\partial x_i} H(t, \vec{x}, m) r_i(m) = \sum_j H(t, \vec{x}, j) q_{jm} .$$

Vector version of this equation yields the required result. The boundary condition follows from the observation that the fluid level in place i cannot remain at 0 for a positive amount of time in CTMC state m if the net rate $r_i(m) > 0$.

□

Now suppose the following limits exist

$$\vec{F}(\vec{x}) = \lim_{t \to \infty} \vec{H}(t, \vec{x}).$$

Then from Theorem 1, we see that the steady-state distribution $\vec{F}(\vec{x})$ obeys the following system of differential equations

$$\sum_i \frac{\partial}{\partial x_i} \vec{F}(\vec{x}) \mathbf{R}_i = \vec{F}(\vec{x}) \mathbf{Q}. \tag{3}$$

3.1 FSPN with a Single Continuous Place

In the special case of an FSPN with a single continuous place, the steady-state distribution $\vec{F}(\vec{x})$ exists provided $\sum_{m \in S} r(m)\pi(m) < 0$ where $\pi(m)$ is the steady-state probability of the CTMC being in state m and $r(m)$ is the net fluid input rate to the continuous place in state m. In this case, Equation (3) reduces to:

$$\frac{d\vec{F}(x)}{dx} \mathbf{R} = \vec{F}(x) \mathbf{Q}. \tag{4}$$

Following [2], solution of such an equation is of the form $\vec{F}(x) = \vec{h} e^{\lambda x}$ where $\vec{h}$ is a row vector and λ is a scalar. Substituting in (4) we have

$$\vec{h}(\lambda \mathbf{R} - \mathbf{Q}) = 0. \tag{5}$$

If a non-zero $\vec{h}$ is to satisfy the above equation, we must have $\det(\lambda \mathbf{R} - \mathbf{Q}) = 0$. The number of solutions of $\det(\lambda \mathbf{R} - \mathbf{Q}) = 0$ equals the number of non-zero diagonal elements of **R**. Let these solutions be denoted by $\lambda_1, \lambda_2 \cdots, \lambda_k$. Let $\vec{h}_i$ be the solution to $\vec{h}_i(\lambda_i \mathbf{R} - \mathbf{Q}) = 0$. Then the general solution to (4) is given by

$$\vec{F}(x) = \sum_{i=1}^{k} a_i \vec{h}_i e^{\lambda_i x} \tag{6}$$

where the scalars a_i need to be determined from the boundary conditions and the boundedness of $\vec{F}(x)$. It is known that the number of λ_i's with positive real part equals the number of negative diagonal entries of **R**. The coefficient a_i corresponding to an eigenvalue λ_i with $Re(\lambda_i) > 0$ must be zero in order to maintain boundedness of $\vec{F}(x)$. The remaining coefficients a_i are uniquely determined by the boundary condition $F(0, m) = 0$ if $r(m) > 0$.

3.2 FSPN with Multiple Continuous Places

When there are two or more continuous places in an FSPN the Equations and (3) are coupled differential equations and they are notoriously hard to solve. Thus computing the transient or steady-state (joint) distribution of $\vec{X}(t)$ is a difficult task. However we can make one observation: the evolution of $X_i(t)(i \in P_c)$ is coupled only through the common discrete marking m_d, which develops independently of the $\vec{X}(t)$ vector. (This is an inherent limitation of our current definition of FSPN and will be removed in future work.) Thus the marginal steady-state distribution $\vec{F}_i(x)$ of $X_i(t)$ satisfies an equation similar to (4) and can be analyzed as described in Section 3.1.

4 Examples

4.1 Machine Breakdown Model

We consider a system with N statistically identical and independent components, each with a failure rate λ and repair rate μ. Work arrives to the system at a constant rate r and it gets completed at rate d per functioning machine. Work here is considered to be a non-negative real quantity. We model this system as an FSPN shown in Figure 1.

The model has two discrete places (p_u and p_d), one continuous place and three timed transitions (t_f, t_r and t_{always}). The number of tokens in p_u models the number of functioning machines (with the initial value N) while the number of tokens in p_d is the number of failed machines undergoing repair. The firing of the transition t_f represents a machine failure; firing rate of this transition is λ times the number of tokens in p_u. Repair of a machine is modeled by the firing of t_r; its firing rate is μ times the number of tokens in p_d. The transition t_{always} is always

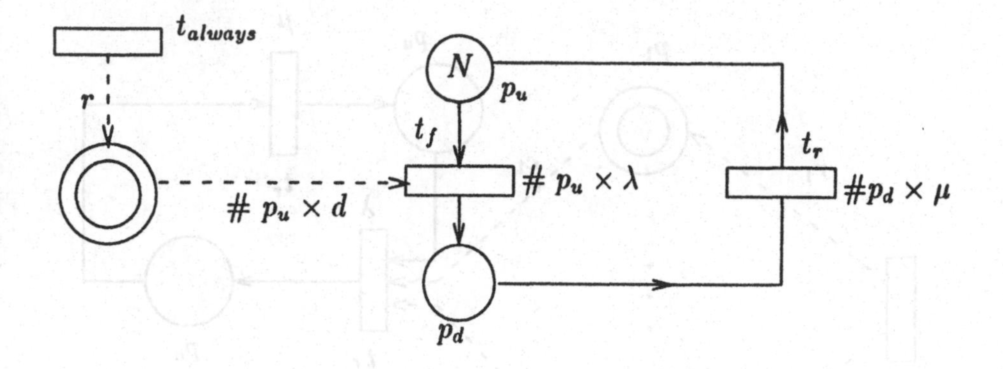

Figure 1: FSPN for the Machine Breakdown Model

enabled and it continuously pumps fluid at rate r into the continuous place. The rate of firing or this transition can be chosen to be any positive value. Whenever transition t_f is enabled, it drains fluid from the continuous place at rate d times the number of tokens in place p_u.

For the steady-state analysis of this FSPN, we can use the development in [2, 6].

4.2 Parallel Buffers Model

For our second example, we consider a system consisting of n buffers in parallel serviced by a single channel subject to breakdown and repairs. Fluid enters buffer i with rate r_i and is emptied at rate c_i whenever the channel is up. The up and down times of the channel are independent exponentially distributed random variables with parameters λ and μ, respectively. The FSPN representation of this system is shown in Figure 2.

The channel is modeled by means of two discrete places (p_u and p_d) and two timed transitions (t_f and t_r) as in the previous example. The initial number of tokens in place p_u is one. Buffers are modeled by n continuous places ($p_1, \cdots, p_n$). Continuous place p_i receives fluid at the rate r_i through the timed transition t_{always}. When the timed transition t_f is enabled (that is, the channel is up), fluid is drained from place p_i at the rate c_i.

The evolution of the discrete part of this FSPN is a two-state CTMC with the infinitesimal generator matrix

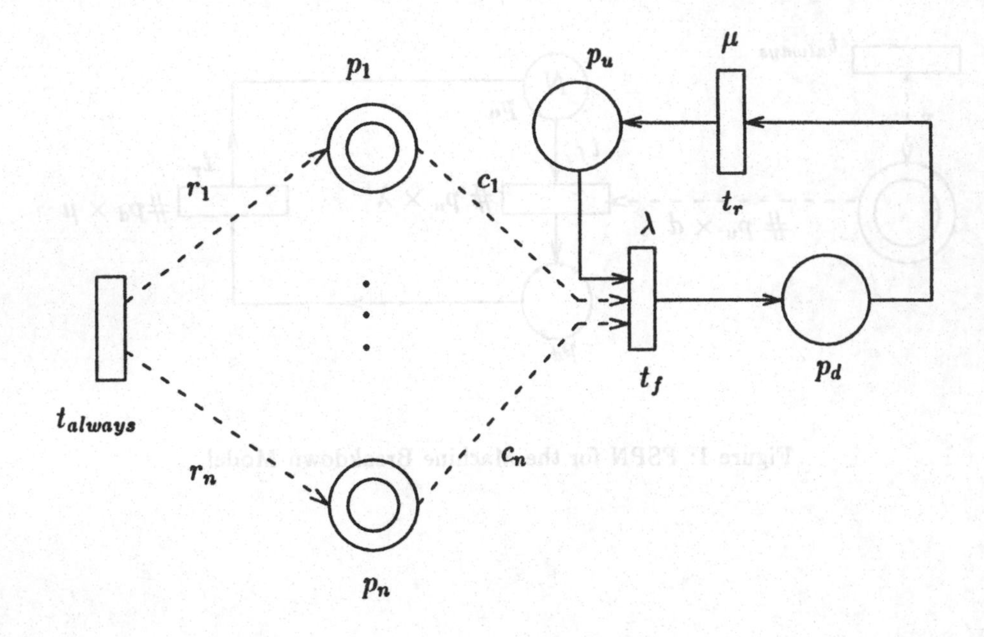

Figure 2: FSPN with Multiple Continuous Places

$$\mathbf{Q} = \begin{bmatrix} -\lambda & \lambda \\ \mu & -\mu \end{bmatrix}.$$

The rate matrix $\mathbf{R}_i$ for $i = 1, \cdots, n$ is

$$\mathbf{R}_i = \begin{bmatrix} r_i - c_i & 0 \\ 0 & r_i \end{bmatrix}.$$

Steady-state marginal distributions $\vec{F}_i(x)$ can be computed by the method discussed in Section 3.1. The steady-state equations for the joint distribution, $\vec{F}(\vec{x}) = (F(\vec{x}, 0), F(\vec{x}, 1))$ are given by Equation (3). The solution of this equation is currently under investigation.

5 Conclusion

We have defined a new class of stochastic Petri nets by introducing places with continuous tokens and arcs with fluid flows. This new class of fluid stochastic Petri nets (FSPNs) should be useful in modeling stochastic fluid flow systems. We have provided formal definition of FSPNs and developed the rules for their dynamic evolution. We have derived coupled systems of partial differential equations for the transient and the steady-state behavior of FSPNs. Spectral representation of the FSPNs with a single continuous place can be adapted from the literature on stochastic fluid flow models. This can also be used to compute the steady-state marginal distributions in an FSPN with multiple continuous places. Solution techniques for the joint distributions in FSPNs with multiple continuous places are currently under investigation.

References

[1] M. Ajmone Marsan, G. Balbo, and G. Conte, *Performance Models of Multiprocessor Systems*, MIT Press, Cambridge, MA, 1986.

[2] D. Anick, D. Mitra and M. Sondhi, "Stochastic Theory of Data-Handling Systems," *The Bell System Technical Journal*, Vol. 61, No. 8, pp. 1871-1894, Oct. 1982.

[3] C. G. Cassandras, *Discrete Event Systems: Modeling and Performance Analysis*, Aksen Associates, Holmwood, IL, 1993.

[4] G. Ciardo, A. Blakemore, P. F. J. Chimento, J. K. Muppala, and K. S. Trivedi, "Automated generation and analysis of Markov reward models using Stochastic Reward Nets", in C. Meyer and R. J. Plemmons, editors, *Linear Algebra, Markov Chains, and Queueing Models*, volume 48 of *IMA Volumes in Mathematics and its Applications*, Springer-Verlag, 1992.

[5] G. Ciardo, J. K. Muppala, and K. S. Trivedi, "Analyzing concurrent and fault-tolerant software using stochastic Petri nets", *J. Par. and Distr. Comp.*, 15(3):255–269, July 1992.

[6] D. Mitra, "Stochastic Theory of Fluid Models of Multiple Failure-Susceptible Producers and Consumers Coupled by a Buffer," *Advances in Applied Probability*, Vol. 20, pp. 646-676, 1988.

[7] T. Murata, "Petri Nets: properties, analysis and applications", *Proceedings of the IEEE*, 77(4):541–579, Apr. 1989.

[8] J. L. Peterson, *Petri Net Theory and the Modeling of Systems*, Prentice-Hall, Englewood Cliffs, NJ, USA, 1981.

[9] N. Viswanadham and Y. Narahari, *Performance Modeling of Automated Manufacturing Systems*, Prentice-Hall, Englewood Cliffs, NJ, 1992.

Taking Advantages of Temporal Redundancy in High Level Petri Nets Implementations

J. A. Bañares, P. R. Muro-Medrano and J. L. Villarroel

Departamento de Ingeniería Eléctrica e Informática
UNIVERSIDAD DE ZARAGOZA
Maria de Luna 3, Zaragoza 50015, Spain**

Abstract. The aim of this paper is to present a software implementation technique for High Level Petri Nets. The proposed technique, implemented for a specialized version of HLPN called KRON, is interpreted and centralized. The approach makes use of the similarities between the inference engine of a rule based system and the interpretation mechanism of a HLPN. It performs an adaptation of the RETE matching algorithm to deal with HLPN implementations. As in RETE, the main objective is to exploit the data temporal redundancy, with this purpose, a RETE-like data structure is implemented. Additionally, our approach benefits from the partition of working memory facilitated by the HLPN. These peculiarities allow the generation of simpler data structures than the ones in more general production systems such as OPS5.
Keywords: Higher-level net models, Rule based systems, Petri net implementation, Matching algorithms.

1 Introduction

A Petri Net software implementation is a computer program created to simulate the firing of transitions following the theoretical behavioral rules imposed by the Petri net semantics. Petri nets implementation techniques can be classified in [CSV86]: *centralized* and *decentralized* implementations. In a centralized implementation, the full net is executed by just one sequential task, commonly called *token player*. Pieces of code associated to transitions (code can be associated to transitions to provide operational capabilities) can be executed as parallel tasks to guarantee the concurrence expressed by the net. In a decentralized implementation, the overall net is decomposed into subnets (normally, sequential subnets) and a task is created to execute each subnet. The other main classification for software implementations of Petri nets distinguishes *interpreted* and *compiled* implementations. An implementation is interpreted if the net structure and the marking are codified as data structures. These data structures are used by one or more tasks called interpreters to make the net evolve. The interpreters do

** This work has been supported in part by project ROB91-0949 from the Comisión Interministerial de Ciencia y Tecnología of Spain and project IT-10/91 from the Diputación General de Aragón. J. A. Bañares is FPU fellow from the Ministerio de Educación y Ciencia of Spain.

not depend on the implemented net. A compiled implementation is based on the generation of one or more tasks whose control flows reproduce the net structure. No data structures are needed in compiled implementations.

The software implementation of High Level Petri Nets (HLPN) has not received much attention so far. Its most representative approaches are briefly stated in the following. [BM86] presents a decentralized and compiled implementation of a subclass of HLPN (PROT nets). A PROT net is partially unfolded and decomposed into sequential processes that are implemented as a set of communicating Ada tasks. Other representative approach has been stablished in [CSV86], where a centralized and interpreted implementation of Colored Petri Nets is proposed. One of the main ideas of this work is the extension of representing place concept to HLPN. Another centralized and interpreted implementation of HLPN is proposed in [Har87], in this case, several interpreters compete to make the net evolve. The net is codified in just one data structure. A highly decentralized implementation is proposed in [BBM89], each place and transition are implemented as an OCCAM process and the implementation runs on a transputer based architecture. [VB90], the closest related work, will be explained later.

The aim of this paper is to present a software implementation approach for HLPN developed for the KRON knowledge representation language ([MM90] and [Vil90]). The proposed technique is interpreted and, in principle, centralized. However, the basis of this technique can also be used in decentralized implementations. This technique is based on the similarities between the inference engine of a rule based system and the interpretation mechanism of a HLPN. In both, efficiency is an important consideration since rule based systems and HLPN based systems, may be expected to exhibit high performance in interactive or real-time domains. The proposed HLPN implementation technique makes use of an adaptation of the RETE matching algorithm [For82] which is used in the OPS5 [BFKM85] rule based language. As in RETE, the main idea is to exploit temporal data redundancies (comming from the marking that is not changed during transition firing).

The adequacy of the RETE match algorithm for software implementations of HLPN is getting greater interest within the research groups working on PN/AI. Thus, [BE86] and [DB88] adopt OPS5 as the implementation language whereas [Vil90] and [VB90] provide a more in-depth knowledge about the use of the OPS5 matching strategy (RETE) to deal with HLPN software implementations. The paper of Valette and col. provides a pretty nice conceptual and pedagogical view for the specific use of RETE for the matching process. The approach presented here goes further in these ideas, which have been refined with the feedback of our implementation experience. We propose a strategy to specialize the RETE match algorithm to be more suitable for HLPN software implementations. A specific implementation is proposed and some aspects, as the reduction of test tree, are treated with more deep. Some details, which are interesting from the software point of view (mainly related with data structures), are also considered.

The rest of the paper is structured as follows. In section 2 rule based systems and its interpretation mechanism are explained. Additionally, the matching process and the RETE match algorithm are illustrated in more detail. In section 3 a special class of high level Petri nets (KRON nets) and its relationship with rule based systems are briefly presented. Software issues involved in the KRON interpreter and the RETE influences are considered in section 4. The advantages and limitations of the proposed approach are analyzed in section 5. Finally, section 6 collects some conclusions of the work.

2 Rule based systems components and rule interpreter mechanism

Rule based systems can be decomposed into three parts: rule memory, working memory and inference engine. The rule memory contains declarative information based in precondition/consequence sentences. The working memory (data memory) contains dynamic data that is compared with the precondition part of the rules. The individual elements of the working memory are referred to as the working memory elements. The rule interpretation mechanism is materialized by the inference engine. The inference engine executes what is called a recognize-act cycle [BFKM85] which is itself composed by three main activities:

1. Match: Performs the comparison of the dynamically changing working memory elements to the precondition part of the rules. If a precondition is satisfied, the rule is included in the conflict set (the set of executable rules for the present working memory state). Currently, the match phase problem is often solved by the RETE match algorithm [For82].
2. Select: Selects one rule from the conflict set. Two main strategies are usually used depending on the recency of individual condition elements and the specificity of the precondition (LEX and MEA strategies).
3. Act: Executes the selected rule according to its consequence part.

The inference engine cycles until no rules are satisfied or the system is explicitly halted.

2.1 Matching process

The match phase of recognize-act cycle consumes the most time of the rule interpreter mechanism (using conventional approaches the interpreter can spend more than 90% of this time). To have a better understanding of the computational complexity involved in this phase we will have a look at the different steps of the matching process in rule based systems.

Rule preconditions are composed by condition elements. In a condition element, an attribute name is followed by an expression specifying restrictions on its value. If the expression specifies a constant value, the condition element will match only the working memory elements with that value. The expression can also be a variable, in this case it will match every working memory element having that attribute. For every match, a binding between the variable and the attribute value of the matched element is created (the variable is bound to the value it matches). When a specific variable can be bound to the same value for each occurrence in the precondition part of the rule, we say that a consistent binding has been found [BFKM85].

From the computational point of view, the set of variables of the rule can be seen as a pattern that must be specified in a consistent way according with the restrictions imposed by the preconditions. The calculation of these patterns can be made in a step by step manner. At the first level, the possible bindings related to the first and second preconditions, are establishing resulting in two sets of patterns partially specified. Next, the consistency of each of the patterns of the first set must be contrasted for consistency against each one of the second set. When two patterns are consistent, a new pattern is generated that will hold the bindings or specifications of both (in general, more specific than its predecessors). Next, the possible bindings with respect the third condition element must be computed and its consistency must be contrasted against the set of patterns computed in the previous level. The result will be again another set of patterns still more specifics. This process goes on until restrictions of the last precondition are integrated and the set of consistent bindings represented by completely filled patterns are obtained. The calculation strategy is graphically illustrated by the binary tree shown in figure 1.

This general framework is instanciated for different match algorithms. Their main differences are the calculations they make each cycle and the information stored to remember previous calculations. In straightforward implementations there are many redundancies in the matching process which decrease efficiency. This execution efficiency problem led to the development of the RETE algorithm [For82], the TREAT algorithm [Mir86] and other related algorithms (see [Pas92] or [SPA92] for review). In the following paragraphs we explain RETE, a well-recognized algorithm which exploits these redundancies, and then we will see how these ideas can be used to improve the efficiency in high level Petri nets software implementations.

2.2 The RETE algoritm

The RETE match algorithm [For82], originally implemented in the OPS5 rule-based programming tool [BFKM85] avoids the brute force approach taking advantage of *temporal redundancy* (persistence of information in the working memory across the recognize-act cycle is called temporal redundancy). This is accomplished by matching only the changed data elements against the rules rather than repeatedly matching the rules against all the data. With this purpose, the information about previous matchings is recorded in a graph structure called *network*.

The network is composed by a global *test tree*, common for all rules, and a *join tree* specific for each rule. The test tree is composed by *one-input* nodes, each of them representing a test over some attribute. A path between the root node and a leaf node represents the sequence of tests required by a rule precondition. The consistent bindings are established following the computation strategy explained

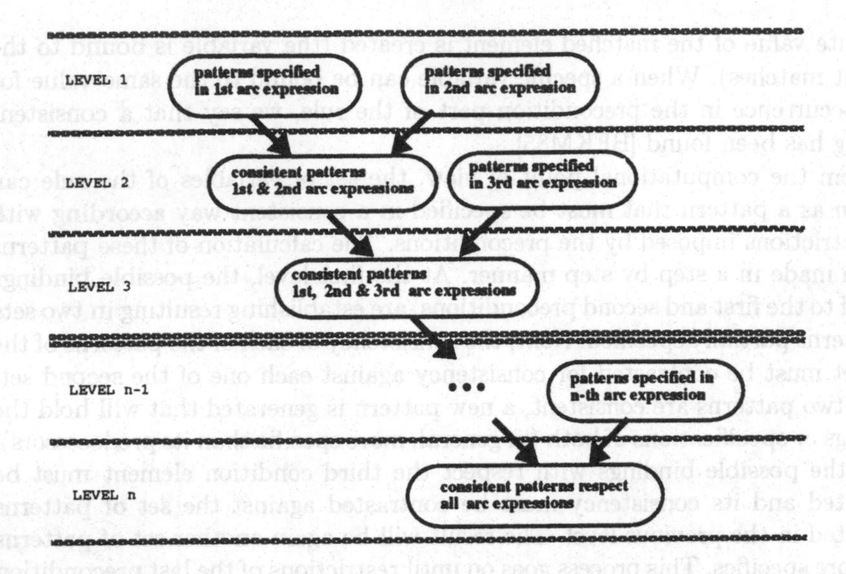

Fig. 1. Graphical view of the general computation process used during the match phase.

above (see figure 1). The information of this process is represented by the join tree associated to the rule (join tree nodes are called *two-input* nodes).

When a working memory element is added to the working memory, a pointer to the element is entered in the root of the test tree and propagated in case the test is successful. The working memory element pointers coming out of the test tree leaves are entered in the join tree. Then, the pointers are combined in tuples and stored in each two-input node. Each tuple collects the working memory elements allowing a consistent binding at the corresponding level. A tuple of the root node points to the working memory elements with which the associated rule can be applied.

On the other hand, when working memory element is removed, the corresponding pointer must be removed from the test tree root. The tuples having this pointer must be also removed from the two-input nodes. When many working memory elements match the same condition element, removing a working memory element is expansive. It takes a linear search to find the particular element to remove in the list of pointers for a condition.

3 High level Petri net/rule based model equivalence

Different versions of high level Petri nets have been proposed in the scientific literature (see the book of Jensen and Rozenberg [JR91]), we restrict ourselves here to the approach used in our software implementation, which is called KRON. KRON (Knowledge Representation Oriented Nets) is a frame based knowledge representation language for discrete event systems with concurrent behavior. It

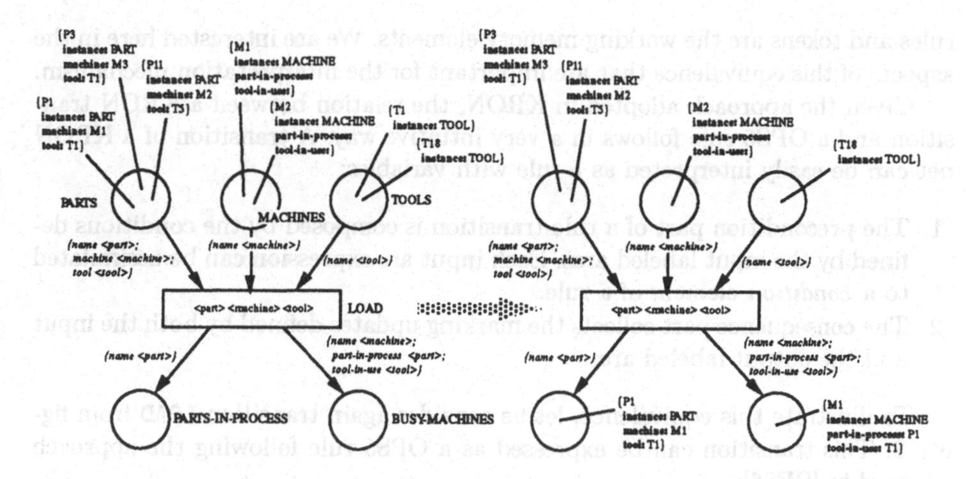

Fig. 2. A simple KRON net example.

belongs to the object oriented programming paradigm and incorporates a specialization of the color Petri net formalism for the representation of the dynamic aspects. The reader is referred to [MM90] and [Vil90] for more theoretical bases.

KRON provides a set of classes of specialized objects and primitives for the construction of a system model where the dynamic behavior is defined by an underlying HLPN that we will call in the sequel KRON nets. Given the scope of this paper, we will briefly introduce KRON through an illustrative example of a very simple KRON net, which is shown in figure 2. The net, composed by five places and one transition, is representing the load for a set of machines. The arcs are labeled by expressions that are lists of attribute-variable pairs. The set of all variables included in the arc expressions is associated to each transition. Each pair defines possible bindings between a variable and attribute values of tokens in the related place. For example, arc expression `{name <part>; machine <machine>; tool <tool>}` in transition `LOAD` allows three different bindings for variables `<part>`, `<machine>` and `<tool>` regarding to the actual marking:
`(<part> = P1; <machine> = M1; <tool> = T1)`
`(<part> = P3; <machine> = M3; <tool> = T1)` and
`(<part> = P11; <machine> = M2; <tool> = T5)`.

A transition is enabled if there exists a *consistent binding* for its associated variables. This means that all transition variables are bound, the bindings from all arc expressions are the same and the variable values verify the restrictions imposed by the predicate associate to the transition. Each of these consistent bindings defines a different *firing mode*. In the example shown in figure 2, there exists only one firing mode that is defined by the following consistent binding:
`(<part> = P1; <machine> = M1; <tool> = T1)`.

The similarities between Petri nets and rule based systems have been pointed out in several papers [BE86], [DB88], [VB90], [HLMMM91], [MMEV92]. A HLPN can be interpreted as a rule based system in which transitions have the role of rules and tokens are the working memory elements. We are interested here in the aspects of this equivalence that are important for the interpretation mechanism.

Given the approach adopted in KRON, the relation between a KRON transition and a OPS5 rule follows in a very intuitive way. A transition of a KRON net can be easily interpreted as a rule with variables:

1. The precondition part of a rule-transition is composed by the conditions defined by the input labeled arcs. Each input arc expression can be assimilated to a *condition element* of a rule.
2. The consequence part collects the marking updates defined by both the input and the output labeled arcs.

To illustrate this equivalence, let us consider again transition **LOAD** from figure 2. This transition can be expressed as a OPS5 rule following the approach adopted in [BE86]:

```
(p LOAD
    (PART ↑name <name> ↑machine <machine> ↑tool <tool> ↑place PARTS)
    (MACHINE ↑name <machine> ↑place MACHINES)
    (TOOL ↑name <tool> ↑place TOOLS)
  →
    (modify <part> ↑place PARTS-IN-PROCESS)
    (modify <machine> ↑part-in-process <part> ↑tool-in-use <tool>
        ↑place BUSY-MACHINES)
    (modify <tool> ↑place NIL))
```

In this approach, tokens (data elements) have a special attribute called (**↑place**) used to specify the place where the tokens are. The RETE test tree and join tree associated to the rule **LOAD** is shown in figure 3.

On the other hand, HLPN provide some features that make its implementation more specific than the one in general rule-based systems. The main differences are related with the *working memory.* A rule based system has just one global working memory for all rules, whereas a HLPN has its working memory splited in places. The preconditions of a transition only must match against the tokens (data elements) of its input places. This fact allows a place to be seen as a working memory partition. The main effects on a RETE network produced by this partition can be established as follows:

– The root node is splited into several nodes, one for each input place, each one defining itself a local working memory.
– The test tree of the network is reduced because no class or place tests are needed. Consider the OPS5 rule of transition **LOAD**. Each precondition element of this rule has an implicit test over the token class and an explicit test over the place where the token is located. Place tests can be avoided making use of the HLPN structure because each condition element has implicitly associated an input arc. The class test (**PART, MACHINE** or **TOOL**) can also be avoided because each place has associated a set/class of possible tokens. All tokens in a place belong to the same class.

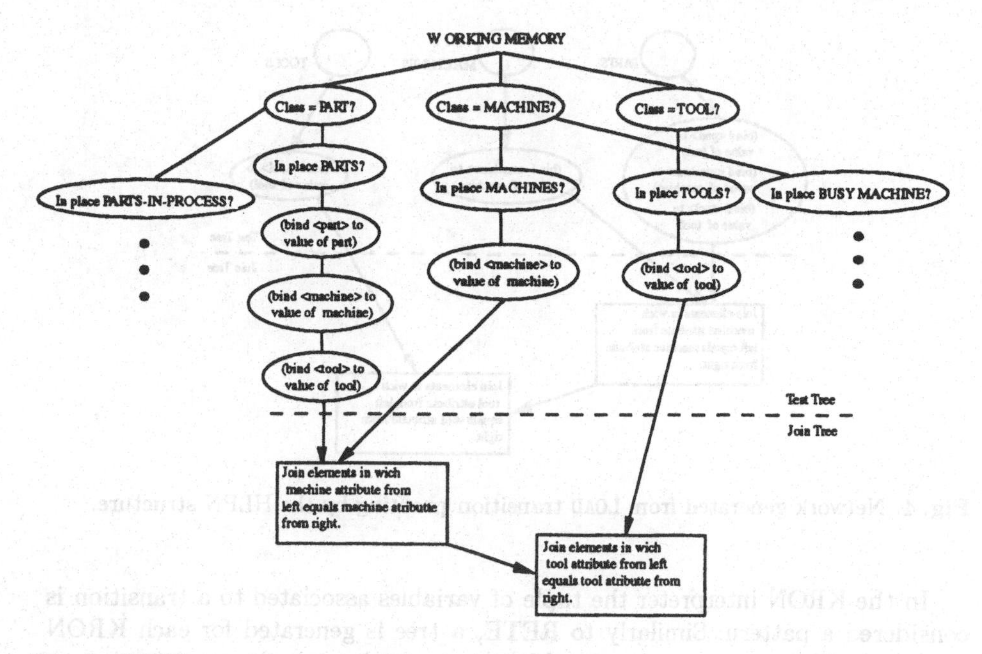

Fig. 3. RETE network generated from transition LOAD.

- It can not make use of the *structural similarity* [BFKM85]. The structural similarity allows the sharing of partial branches of the test tree by different, but similar, condition elements. Except in places that are shared by several transitions, the branches of the test tree come out from separate root nodes. This fact makes impossible for nodes to be shared.

If the structural similarity reminded are not used, the test tree obtained is a set of separated chains coming out from root nodes. Thus, each chain can be reduced to only one node representing all test and bindings of the corresponding condition element. It is graphically illustrated by the network shown in figure 4. It can also be compared with the RETE network in figure 3.

4 Interpretation mechanism in KRON

4.1 Matching phase

Our approach for the interpretation mechanism makes use of the temporal redundancy features pointed out in the RETE matching algorithm. With this purpose, a similar data structure to the RETE network is implemented. Additionally, our approach benefits from the partition of working memory provided by the HLPN that implies the generation of simpler networks than the ones in more general production systems such as OPS5.

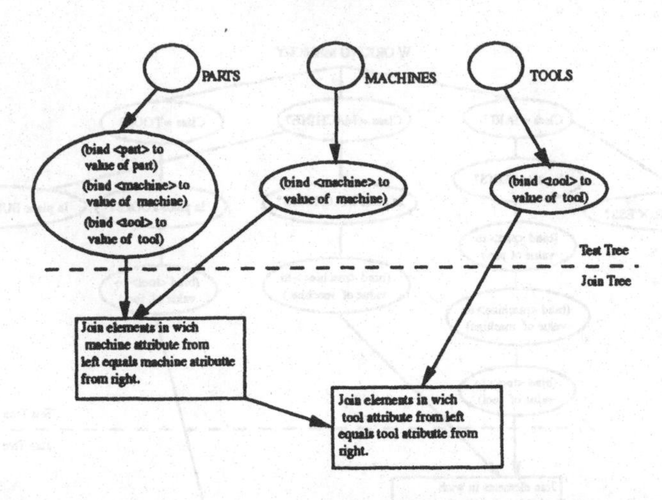

Fig. 4. Network generated from LOAD transition profiting by the HLPN structure.

In the KRON interpreter the tuple of variables associated to a transition is considered a pattern. Similarly to RETE, a tree is generated for each KRON transition. These trees are composed by two node classes:

1. Entry Nodes. Do not have predecessors and correspond to patterns with bindings generated by a single input arc. Each entry node has a link to the corresponding input place which is, in fact, a pointer to its local working memory. These nodes correspond to folded branches of the test tree in a RETE network.
2. Join Nodes. Correspond to consistent patterns with respect to several input arcs. These nodes are equivalent to the join nodes in a RETE network.

The tree is organized into levels:

- LEVEL 1: Two entry nodes corresponding to the patterns of the first and second input arcs (any order is accepted).
- LEVEL i ($1< i < n$): An entry node corresponding to the i+1 input arc and a join node created upon the nodes of level i-1.
- LEVEL n: A join node, which represents the consistent bindings of the transition.

Calculation is performed in an increasing direction of levels and, inside each level, from left to right. As in a RETE network the intermediate results of the matching process are stored and recalculation can be avoided meanwhile no changes in the allocation of the tokens of the places involved are made. To do that, the information about the set of patterns partially specified is stored in a special data structure. To provide fast accessibility each of these patterns is linked to its predecessors and successors. The patterns corresponding to the entry nodes do not have predecessors. In this case, a link is set between the pattern and the token that generated it. Figure 5 shows a graphical representation of

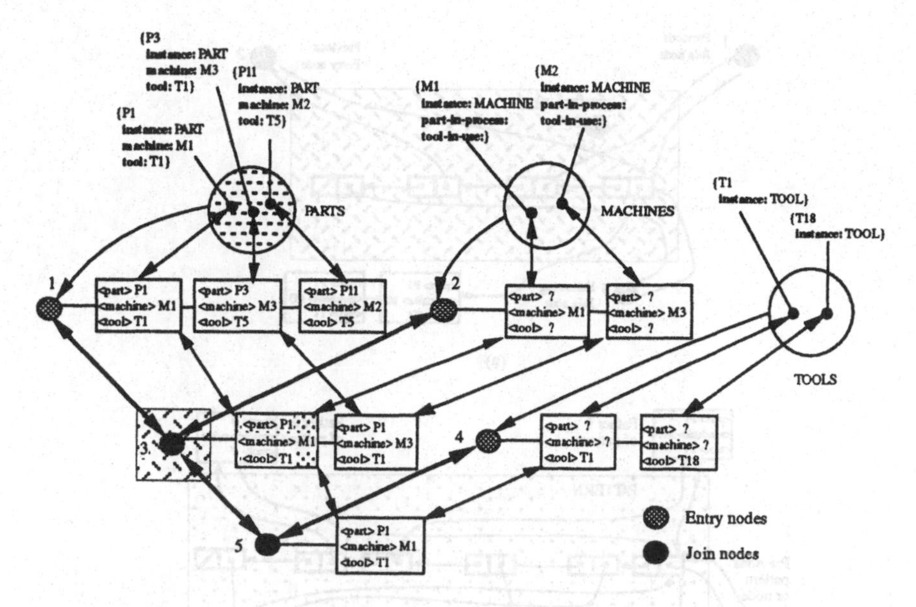

Fig. 5. Link structure for the LOAD network.

the network, this highly related data structure allows a fast tree update because it avoids computational searches.

Figure 6 shows more details about the link structure developed in our implementation. To facilitate software development, KRON has been implemented in KEE (Knowledge Engineering Environment from Intellicorp) which runs on top of Common Lisp. The interpreter itself has been implemented exclusively using Common Lisp primitives (Common Lisp is specially attractive to deal with pointers). The typical Lisp *cons cells* are used in the figure to describe more graphically the link structure. A cons cell is stood for a box with two pointers. The cons cells are linked by arrowlike pointers emerging from their right and left sides. Figure 6.a shows a prototypic tree node that is composed by a list of pointers: the first one points to the list of patterns, the rest are pointers to the previous and successor nodes. Figure 6.b shows the data structure for a pattern composed by: a list of pointers to the previous and following patterns at the same and different levels, another pointer points to an association list of pairs' variable-value (the tuple of partially specified variables of the transition). Finally, figure 6.c shows the data structure used in the place: the first element is a list of pointers to the entry nodes that pick up the patterns with bindings, the rest is a list with the links that are needed for each token in the place, such as the pointers to its generated patterns.

From the data structure point of view, there are two aspects that make differences between the RETE algorithm and the KRON interpreter. The first one is that partially specified patterns are stored instead of tuples of pointers as happens in the RETE algorithm. Another difference is related with the link

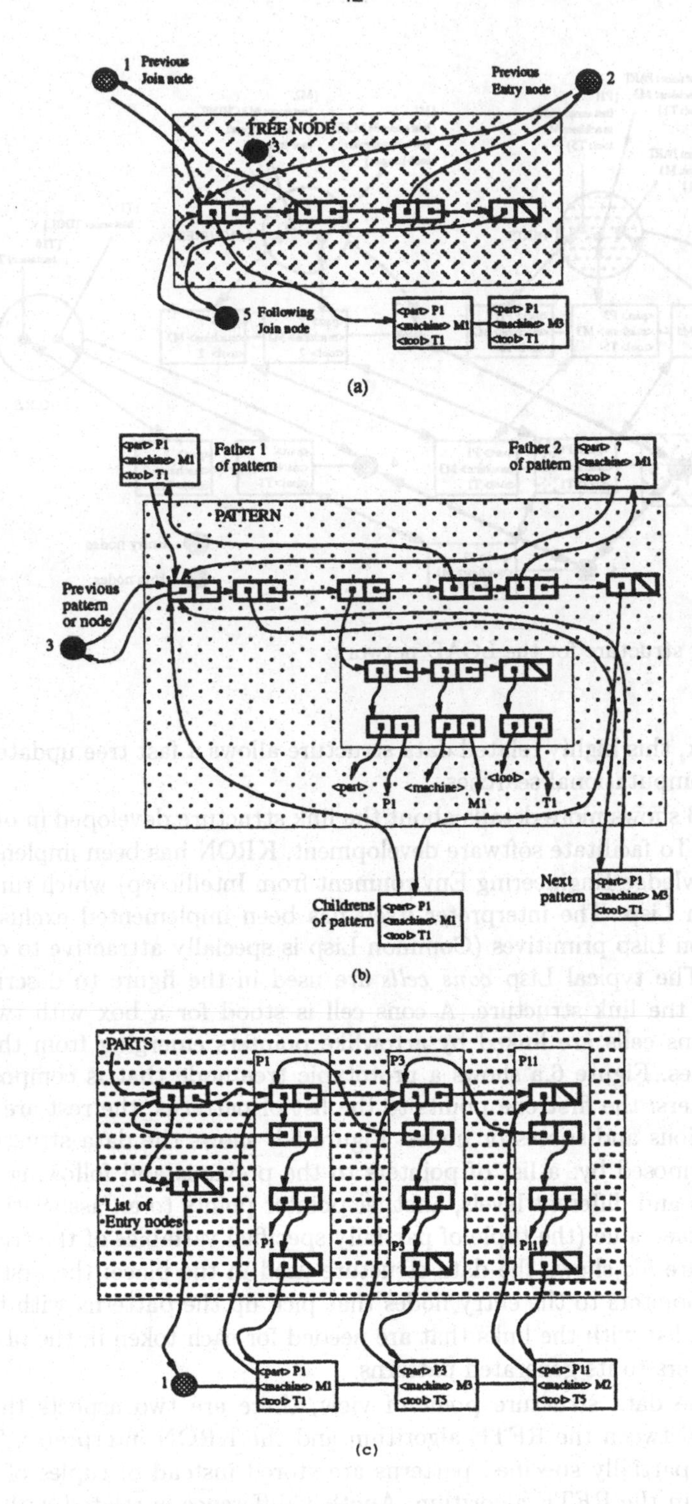

Fig. 6. Some more details of the link structure.

structure between patterns that avoids searches in a node pattern list when an element must be eliminated.

The data structure created during the matching process is not modified until a change in the token distribution happens. Two cases must be distinguished:

1. **Deleting a token**. Carries on the elimination of the patterns specified by this token. The calculation trees of all descending transitions of the place containing the token are examined, and its successors' patterns deleted.
2. **Adding a token**. Originates new bindings of the transition variables. If the token has been added in the place corresponding to the i-th input arc, a new pattern or patterns are created in the entry node of the level i-1. The new pattern causes the verification of its consistency with respect the node of the same level. For each detected consistency a new pattern is created in the immediate following level. Actualization is propagated following this strategy.

4.2 Selection phase

Select one rule from the conflict set is accomplished in rule-based systems applying concepts such as recency, refraction or specificity (see strategies LEX and MEA from OPS5 language [BFKM85]). However, there are other important issues to be considered in implementing HLPN:

1. A HLPN can model concurrence. In each execution cycle, more than one enabled transition can be fired.
2. The strategies to solve conflicts depend on the application, and the resolution objectives can be different in each conflict.

In the KRON interpreter, transitions are grouped by conflicts, each one having its resolution strategy that we call control policy. During selection, all enabled transitions in a conflict are considered together, the conflict control policy is the responsible to provide a solution (transitions and firing modes must be chosen). The KRON interpreter offers a control policy library but the user can design new control policies according to the application domain.

In each execution cycle the set of all enabled transitions are known due to the incremental operation of the KRON interpreter [VB90]. It is an important difference from other centralized implementations, as the ones based on representing places [CSV86], where only a list of likely enabled transitions is known. This list can contain transitions that are not enabled. The consideration of these additional transitions makes the efficiency of the interpreter to decrease.

4.3 Firing phase

A transition firing with respect to a firing mode (pattern of the n level) carries on the following steps:

1. The calculation tree is traversed backwards from the consistent binding (firing mode) to find out which are the tokens used to generate it. Each of these

tokens is then removed from the input places and a tree update is performed in all the transitions sharing these input places.
2. The piece of code associated to the firing mode is executed.
3. The tokens specified by the output arcs of the transition are added to the output places and a tree updates is performed in all the transitions having these places as inputs.

5 Efficiency issues

The efficiency of a Petri net implementation can be measured by a function of the number of operations performed by transition firing. In the proposed implementation a transition firing carries on the elimination of the tokens produced by the firing mode and the addition of tokens in the output places. In these activities tree elemental operations are involved: creating a pattern, delete a pattern and make new pattern matching.

In a transition firing, the number of performed operations depends on two main factors: the number of tokens removed from the input places and the number of tokens added to the output places. The cost of adding a token in a place depends on the number of descending transitions, that is the number of trees to be updated. The operations involved in adding a token are: pattern creation and pattern matching. These operations depend on:

1. The number of input places. A large number of places means a deep join tree with many entry nodes. The average number of pattern matching operations grows with the deep of the join tree.
2. The patterns corresponding to entry nodes. The number of patterns in entry nodes depends on the number of tokens in places, the expressions labeling arcs and the cardinality of token attributes. Thus, if a token's attribute holds many values, one token may generate many patterns in an entry node. The number of pattern matching operations also increase if the number of patterns is bigger.
3. The arcs order. The interpretation mechanism evaluates the expressions labeling the arcs sequential`y, stopping when there are not consistent matches with an initial sequence of arc expressions. Locating earlier the arc labeled with the most restrictive expression, or the one who has its place normally empty reduce the number of consistent matches that are passed down in the join tree. It makes sense to locate at the end the arcs whose matches are changed frequently (arcs whose places support frequently operations of adding and deleting tokens).

Pattern deletion is the operation involved on removing tokens from places. The number of these operations depends on the former factors in a similar way.

Unfortunately, the application of these ideas for improving efficiency and getting down the cost of calculate the firing modes are sometimes contradictory. For example, the principle of putting frequently changing marking places later may be in conflict with the principle of putting arcs with restrictive expressions earlier if an arc is both restrictive and matched by the tokens that are frequently put and deleted from their place.

For centralized implementations, the main emphasis has been put on the selective scanning of transitions. These implementations avoid the checking of all transitions, and the idea is to make a quick selection of a subset of transitions likely to be fired. In non incremental implementations, the number of operations in an execution cycle can be considered as: the number of created patterns and pattern matching operations coming from completely recomputing the firing modes of the fired transition and the transitions likely to be fired.

The adopted approach can be more efficient than non incremental implementation because avoids recalculations and considers only totally enabled transitions. However, there are situations where a non incremental technique may be more efficient as in the HLPN of figure 7. Let us show the number of operations by the firing of transition T_1 using both, the KRON interpreter approach and another non incremental technique. When transition T_1 is fired, token A_1 must be deleted from place P_1 and the same token must be added to place P_1 again.

Using the KRON interpreter approach, on the one hand deleting A_1 from P_1 implies that 7 patterns must be removed from the tree attached to T_2, and another one from tree attached to T_1. On the other hand, adding A_1 in P_1 implies that 7 patterns must be created and 6 pattern matching operations must be performed in the tree attached to T_2 and 1 pattern is created in the tree attached to T_2.

Nevertheless, using a non incremental implementation such as *representing places* the number of operations is reduced. Let us take P_1 as representing place of T_1 and P_4 as representing place of T_2. In the place P_4 there are no tokens, therefore only the transition T_1 is in the list of likely enabled transitions. The completely recompute of firing modes of transition T_1 generates 2 patterns. The difference from KRON approach is obvious.

Since the RETE algorithm maintains state between cycles, KRON interpreter is efficient in situations where there are many tokens and most of them do not change on a cycle. On the other hand, the number of operations in a transition firing depends on the output transitions of the input and output places. That is, the average number of operations by transition firing depends on the average number of output transitions. There are two situations where the KRON interpreter does not take advantage of temporal redundancy: when the number of tokens that are involved in a transition firing is large with respect the total number of tokens; and when the average number of descending transitions from places is high. In both cases the number of operations required to update the stored information can be greater than the number required for a recalculation.

6 Conclusion.

This paper provides inside knowledge of a software implementation for HLPN. To be more specific we use a specialized version of HLPN called KRON. KRON

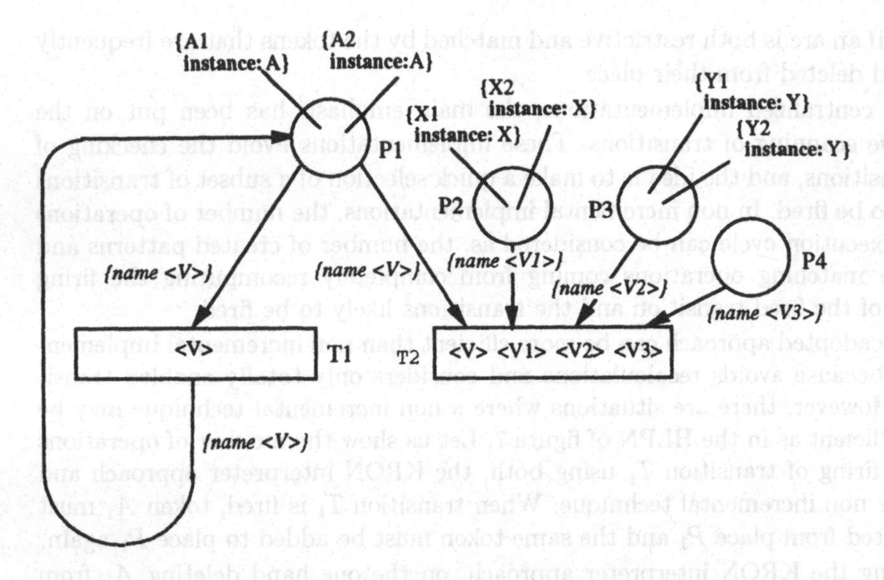

Fig. 7. A sample where the KRON interpreter shows lower efficiency.

(Knowledge Representation Oriented Nets) is a frame based knowledge representation language for discrete event systems. In this representation schema, the system dynamic behavior is represented by a HLPN that we call KRON net.

We follow an interpreted and centralized approach. This interpreter makes use of the similarities between the inference engine of a rule based system and the interpretation mechanism of a HLPN, specifically it uses an adaptation of the RETE matching algorithm used in the OPS5 rule based language. As in RETE, the main idea is to exploit temporal data redundancies.

A strategy to specialize the RETE match algorithm to be more suitable for HLPN software implementations has been proposed. Our approach for the interpretation mechanism makes use of a similar data structure to the RETE network. Additionally, our approach benefits from the partition of working memory in places provided by the HLPN that implies the generation of simpler networks than the ones in more general production systems such as OPS5

From the data structure point of view, there are two aspects that make differences between the RETE algorithm and the KRON interpreter. The first one is that partially specified patterns are stored instead of tuples of pointers as happens in the RETE algorithm. Another difference is related with the link structure between patterns that avoids searches in a node pattern list when an element must be eliminated.

An important difference from other centralized implementations is that the operation of the KRON interpreter is incremental. On each execution cycle, the set of all enabled transitions are known instead of a list of likely enabled transitions. The consideration of these not enabled transitions makes the efficiency of the interpreter to decrease.

The KRON interpreter has been designed to take advantage of temporal redundancy. However, there are two situations where the KRON interpreter fails in this objective: when the number of tokens that are involved in a transition firing is large with respect the total number of tokens; and when the average number of descending transitions from places is high. In both cases the number of operations required for update the stored information can be greater than the number required for a recalculation.

We can conclude that the implementation technique proposed is efficient in situations where there is a large marking and it is relatively stable. KRON has been implemented in a SUN workstation running KEE. The interpreter has been written using Common Lisp primitives, whereas KEE primitives have been use to access to the objects' information.

References

[BBM89] R. Esser B. B utler and R. Mattmann. A distributed simulator for high order petri nets. In *Proc. of International Conference on Applications and Theory of Petri Nets*, pages 22–34, Bonn, 1989.

[BE86] G. Bruno and A. Elia. Operational specification of process control systems: Execution of prot nets using ops5. In *Proc. of IFIC'86, Dublin*, 1986.

[BFKM85] L. Browston, R. Farrell, E. Kant, and N. Martin. *Programming Expert Systems in OPS5: An Introduction to Rule-Based Programming*. Adisson-Wesley, 1985.

[BM86] G. Bruno and G. Marchetto. Process-translatable petri nets for the rapid prototyping of process control systems. *IEEE transactions on Software Engineering*, 12(2):346–357, February 1986.

[CSV86] J.M. Colom, M. Silva, and J.L. Villarroel. On software implementation of petri nets and colored petri nets using high-level concurrent languages. In *Proc of 7th European Workshop on Application and Theory of Petri Nets*, pages 207–241, Oxford, July 1986.

[DB88] J. Duggan and J. Browne. Espnet: expert-system-based simulator of petri nets. *IEEE Proceedings*, 135(4):239–247, July 1988.

[For82] C. Forgy. A fast algorithm for many pattern / many object pattern match problem. *Artificial Intelligence*, 19:17–37, 1982.

[Har87] G. Hartung. Programming a closely coupled multiprocessor system with high level petri nets. In *Proc. of 8th European Workshop on Application and Theory of Petri Nets*, pages 489–508, June 1987.

[HLMMM91] G. Harhalakis, C.P. Lin, L. Mark, and P.R. Muro-Medrano. Information systems for integrated manufacturing (insim) - a design methodology. *International Journal of Computer Integrated Manufacturing*, 4(6), 1991.

[JR91] K. Jensen and G. Rozenberg, editors. *High-level Petri Nets*. Springer-Verlag, Berlin, 1991.

[Mir86] A. Miranker. *TREAT: A new and efficient match algorithm for AI production systems*. PhD thesis, Dep. Comput, Sci., Columbia University, 1986.

[MM90] P.R. Muro-Medrano. *Aplicación de Técnicas de Inteligencia Artificial al Diseño de Sistemas Informáticos de Control de Sistemas de Producción.*

PhD thesis, Dpto. de Ingeniería Eléctrica e Informática, University of Zaragoza, June 1990.

[MMEV92] P.R. Muro-Medrano, J. Ezpeleta, and J.L. Villarroel. *Aceptado en IMACS Transactions*, chapter Knowledge Based Manufacturing Modeling and Analysis by Integrating Petri Nets, 1992.

[Pas92] A. Pasik. A source-to-source transformation for increasing rule-based system paralellism. *IEEE Tran. on Knowledge and Data Engineering*, 4(4):336–343, August 1992.

[SPA92] M. Sartori, K. Passino, and P. Antsaklis. A multilayer perceptron solution to the match phase problem in rule-based artificial intelligence systems. *IEEE Tran. on Knowledge and Data Engineering*, 4(3):290–297, June 1992.

[VB90] R. Valette and B.: Bako. Software implementation of petri nets and compilation of rule-based systems. In *11th International Conference on Application and Theory of Petri Nets*, Paris, 1990.

[Vil90] J.L. Villarroel. *Integración Informática del Control de Sistemas Flexibles de Fabricación.* PhD thesis, Dpto. de Ingeniería Eléctrica e Informática, University of Zaragoza, September 1990.

A Subset of Lotos with the Computational Power of Place/Transition-Nets [1]

Michel Barbeau, Gregor v. Bochmann[2]

Abstract

In this paper, we define a subset of Lotos that can be modelled by finite Place/Transition-nets (P/T-nets). That means that specifications in that Lotos subset can be translated into finite P/T-nets and validated using P/T-net verification techniques. An important aspect of our work is that we show that conversely P/T-nets can be simulated in our Lotos subset. It means that the constraints we put on Lotos in order to obtain finite nets are minimally restrictive. We may also conclude that our Lotos subset and P/T-nets have equivalent computational power. To the best of our knowledge, no such bidirectional translation scheme has been published before.

Topics:

Relationships between net theory and other approaches.

1. Introduction

In this paper, we define a subset of Basic Lotos [Bolo 87, ISO 88] that can be modelled by finite Place/Transition-nets (P/T-nets). That means that specifications in that Lotos subset can be represented and translated into finite P/T-nets and validated using P/T-net verification techniques. An important aspect of our work is that we show that conversely P/T-nets can be simulated in our Lotos subset. It means that the constraints we put on Lotos in order to obtain finite nets are minimally restrictive. We may also conclude that our Lotos subset and P/T-nets have equivalent computational power. To the best of our knowledge, no such bidirectional translation scheme has been published before.

The problem of modelling process-oriented languages, and more specifically CCS and CSP like languages, by Petri nets has been tackled by several authors. Cindio et al. [Cind 83], Degano et al. [Dega 88], Glabbeek [Glab 87], Goltz [Golt 84a, 84b, 88], Nielsen [Niel 86], Olderog [Olde 91] and Taubner [Taub 89] considered CCS or CSP, or both. Lotos has been worked by Marchena and Leon [Marc 89], and Garavel and Sifakis [Gara 90]. The approaches may be

[1]This work was performed within a research project on object-oriented specifications funded by Bell-Northern Research (BNR) and the Computer Research Institute of Montréal (CRIM). Funding from the Natural Sciences and Engineering Research Council of Canada is also acknowledged.

[2]First author's address: Université de Sherbrooke, Département de mathématiques et d'informatique, Sherbrooke (Québec), Canada, J1K 2R1. Second author's address: Université de Montréal, Département d'IRO, C.P. 128, Succ. "A", Montréal (Québec), Canada, H3C 3J7.

categorized based on the following criteria: i) style of definition, ii) finiteness of the representation, and iii) distinction of concurrency and nondeterminism.

One of two definition styles may be adopted, namely denotational or operational. A denotational style is used in: [Cind 83], [Gara 90], [Glab 87], [Golt 84a, 84b, 88], [Niel 86], [Marc 89] and [Taub 89], whereas an operational style, à la Plotkin, is used in: [Dega 88], [Olde 91] and in the present paper. In opposition to the operational approach, the denotational style is constructive. It means that the definition yields directly to a procedure for translating terms of the process-oriented language to Petri nets. However, we shown in [Barb 91a, b] that thanks to our operational definition an important P/T-net verification method can be adapted to Lotos without even translating the latter to the former.

Another important matter is whether or not the Petri net representation of the process-oriented language is finite. It is well known that an unbounded number of Petri net places and transitions is required to represent a process-oriented language when recursion is combined with parallel composition, sequential composition, hiding and disabling operators. This difficulty means that it is impossible to transfer to the process-oriented language several important verification techniques elaborated for Petri nets, since they require finite nets. Note that in our mind, finite nets does not mean finite state systems. Finite representations can be obtained by restricting the process-oriented language or using high-level Petri net models. Finite representations for subsets of CCS are proposed in [Golt 88], using P/T-nets, and in [Taub 89], using Predicate/Transition-nets, which is a high-level model. Finite extended Petri nets are generated from Lotos, with the finite control property, in [Gara 90], this work is also interesting because the data part of Lotos is also handled. In this paper we define a subset of Basic Lotos, with syntactical constraints, that can be modelled by finite P/T-nets.

Non distinction of concurrency and nondeterminism means that Lotos expressions such as $a; stop|||b; stop$ and $a; b; stop[]b; a; stop$ have the same semantic interpretation. Distinction of concurrency and nondeterminism allows accurate representation of behaviors by partial orders. It is a representation that shows just natural dependencies between actions. Multi-sets of actions are possible in a single transition. This has an impact on treatment of fairness problems [Reis 84]. Our Place/Transition-net semantics is less attractive, than definitions described in Refs. [Dega 88], [Golt 88], [Niel 86] and [Olde 91], with respect to distinction of concurrency and nondeterminism.

An important feature in our approach is that we show that P/T-nets can be simulated in our Lotos subset. Other authors have proposed simulations of Petri nets in languages such as Prolog, Azema et al. [Azem 84], or Meije, Boudol et al. [Boud 85]. These simulations are not in languages that have been shown translatable into finite Peo nets. The goal of Azema et al. is to use Prolog as a simulation tool for Petri nets whereas the aim of Boudol et al. is to provide a textual representation for Petri nets. Translation into Lotos of another graphical representation for behaviors, called *Process-Gate Network*, is described [Bolo 90].

In Section 2, we introduce the P/T-net model. Our Basic Lotos subset that can be translated into finite P/T-nets is called PLotos and is defined in Section

3. In Section 4, we discuss modelling of PLotos by P/T-nets. The converse simulation is presented in Section 5. We conclude in Section 6.

2. P/T-nets

We represent a P/T-net [Pete 81] as a tuple (P, T, Act, M_0) where:

- P, is a set of places $\{p_1, ..., p_n\}$,
- $T \subseteq \mathcal{N}^P \times Act \times \mathcal{N}^P$, is a transition relation,
- Act, is a set of transition labels, and
- $M_0 \in \mathcal{N}^P$, is the initial marking.

A P/T-net is **finite** if the sets P, T and Act are finite.

$\mathcal{N}$ is the set of non-negative integers. $\mathcal{N}^P$ denotes the set of multi-sets over the set P. An element $t = (X, a, Y) \in T$ is also denoted as $X - a \rightarrow Y$. Its **preset** $pre(t)$ is X, **postset** $post(t)$ is Y and **action** $act(t)$ is a. The multi-sets X and Y are also called respectively the input and output places of t. We denote as $pre(t)(p)$ $(post(t)(p))$ the number of instances of the element p in the preset (postset) of t.

The operators $\leq$, $+$ and $-$ denote respectively multi-set inclusion, summation and difference. A multi-set X can also be seen as the formal sum: $X = \sum_{p \in P} pre(t)(p)p$.

A Petri net marking is also a multi-set. We denote by $M(p_i)$ the number of instances of the element p_i in the multi-set M. Instances of the element p_i are also called tokens inside place p_i.

$pre(t)(p)$ is the number of tokens that place p must contain to enable transition t. A transition $t \in T$ is **enabled** in marking M if $pre(t) \leq M$. This is denoted as $M(t >$. An enabled transition can be **fired** and the successor marking M' is defined as:

$$M' = M - pre(t) + post(t)$$

this is represented as $M(t > M'$.

We define the **reachability graph** of a P/T-net $N = (P, T, Act, M_0)$ as a graph $RG(N) = (RS, E, M_0)$ where:

1. RS is the reachability set, i.e. a set of markings of N,
2. $E \subseteq RS \times Act \times RS$, is a transition relation, and
3. for all $M \in RS, t \in T$, if $M(t > M'$ then $M' \in RS$ and $(M, act(t), M') \in E$.

3. Definition of the Syntax of PLotos

In this section we define the syntax of a subset of Basic Lotos, namely PLotos which is equivalent, in terms of computational power, to finite P/T-nets (to be shown formally in Section 4). First, we discuss Basic Lotos. Then, we define PLotos as Basic Lotos along with syntactical constraints. The syntax of Basic Lotos is given in Ref. [Bolo 87] and in Appendix A.

It is well known that Basic Lotos has the computational power of Turing machines. Our aim is to reduce the power of Basic Lotos to the one of P/T-nets. Before we state the syntactical constraints that make PLotos equivalent to P/T-nets, we define preliminary concepts.

The "calls" relation

Let p_1 be a process and B_{p_1} its defining behavior-expression. We say that p_1 **calls** p_2 if B_{p_1} has one or more occurrences of p_2. This relation is denoted as:

$$C = \{(p_1, p_2) : p_1 \text{ calls } p_2\}$$

The mutual recursion relation

Let C^+ be the transitive closure of C. We define in terms of C^+ the **mutual recursion** relation Φ as follows:

$$\Phi = \{(p_1, p_2) : (p_1, p_2) \in C^+ \wedge (p_2, p_1) \in C^+\}$$

Recursive process

The process p is recursive if $(p, p) \in \Phi$.

Functionality

The **functionality** of a behavior B is equal to *exit iff* every alternative in B terminates with the successful termination action δ, otherwise it is equal to *noexit* [Bolo 87].

Context

A Lotos **context** $C[\,]$ is a Lotos behavior-expression with a formal "behavior-expression" parameter denoted as "[ ]". If $C[\,]$ is a context and B is a behavior-expression then $C[B]$ is the behavior-expression that is the result of replacing all occurrences of "[ ]" in $C[\,]$ by B. For example, let $C[\,]$ be the Lotos context $g; [\,]$. The behavior-expression $C[stop]$ is defined as $g; stop$.

Guarded process

A process instantiation term p is guarded if it occurs in any of the following forms:

- $C_1[a; C_2[p]]$
- $C_1[B >> C_2[p]]$
- $C_1[B[> C_2[p]]$

where $C_1[\ ]$ and $C_2[\ ]$ are any contexts, a is any gate identifier and B is any behavior-expression.

PLotos

PLotos is defined as the subset of Basic Lotos that satisfies the following syntactical constraints:

1. Terms that instantiate recursive processes must be guarded.
2. Operands B_1 and B_2 in a parallel composition $B_1|||B_2$ must have the *noexit* functionality.
3. Let $C_1[\ ]$ and $C_2[\ ]$ denote two contexts and B denote a behavior-expression. For any pair $(p_1, p_2) \in \Phi$ the defining behavior-expression of p_1 may not have the following patterns:

   3.1 $C_1[C_2[p_2] * B]$, where the operator "$*$" is either "$|[g_1, ..., g_n]|$" or "$>>$" or "$[>$"

   3.2. $C_1[B|[g_1, ..., g_n]|C_2[p_2]]$

   3.3 $C_1[\mathit{hide}\ g_1, ..., g_n\ \mathit{in}\ C_2[p_2]]$
4. The behavior-expression B_1 must have the *exit* functionality in behavior-expressions of the forms: $B_1 >> B_2$ or $B_1[> B_2$.

Mutual recursion is possible in sub-terms of the form "$B_1|||B_2$", with operands of functionality *noexit* (i.e. constraint 2). The control is not finite state but can be represented by a finite P/T-net. It is possible to simulate an arbitrarily large stack if the constraint 3.1 is unsatisfied (e.g. [Gotz 86]). Arbitrarily large stacks cannot be simulated by finite P/T-nets.

PLotos has the computational power of finite P/T-nets. That is, every PLotos specification can be modelled by an equivalent finite P/T-net. Conversely, every finite P/T-net can be modelled by an equivalent PLotos specification. In Section 4, we show how a PLotos specification can be modelled by a finite P/T-net. The converse is demonstrated in Section 5.

4. P/T-net Semantics for PLotos

4.1. General Idea

Our PLotos to P/T-nets mapping is inspired by the work of Olderog [Olde 91] for CSP. In general, a Lotos behavior-expression B represents the composition of several concurrent components. In our simulation of PLotos by P/T-nets, the expression B is explicitly decomposed into its components which become tokens when this behavior is activated. More precisely, parallel components and states of parallel components are respectively modelled by Petri net tokens and places. The place in which a token is contained denotes the state of the corresponding component. Every Lotos gate occurrence is modelled by a Petri net transition. Tokens, contained in the transition input places, represent components synchronized on the gate. Tokens deposited into the transition output places represent the successor components after the transition has occurred. Several tokens, contained in the same place, represent several identical components. This models unbounded process instantiation with finite P/T-nets.

For example, the Lotos expression $u; v; stop|[u]|u; stop$ represents two concurrent components. The first component executes actions u and v and then stops. The second component executes action u and becomes inactive. Both components are coupled on gate u and are therefore dependent on each other with respect to the occurrence of u. The decomposition of $u; v; stop|[u]|u; stop$ into its components is denoted as the multi-set $\{u; v; stop|[u]|, |[u]|u; stop\}$. In this syntax, we represent explicitly the fact that the components are coupled on gate u by concatenating the symbol $|[u]|$ to the right of $u; v; stop$ and to the left of $u; stop$.

Places modelling states of components are labelled by the corresponding component-expressions. Transitions are labelled by gate names. The "stop" expression represents inaction and does not appear in the P/T-net. In our construction, edges from places to transitions are always one valued (i.e. $(\forall t, p)[pre(t)(p)$ equals 0 or 1]) and every place has a distinct label. **We unambiguously denote a place by its label**. ¿From the above multi-set of components, it is possible to derive the transition represented as the triple:

$$\{u; v; stop|[u]|, |[u]|u; stop\} - u \rightarrow \{v; stop|[u]|\}$$

To derive such triples, we define: i) a function decomposing PLotos behavior-expressions into component-expressions, and ii) a system of inference rules. The head of each rule matches a term of the form:

$$\{p_1, ..., p_m\} - a \rightarrow \{q_1, ..., q_n\}$$

Such a rule can be applied to infer, as a function of the component-expressions, a transition with preset $\{p_1, ..., p_m\}$, action a and postset $\{q_1, ..., q_n\}$. For instance, the rule:

if $M_1 - a \rightarrow M_1'$ and $a \notin \{S, \delta\}$
then $M_1.|[S]| - a \rightarrow M_1'.|[S]|$

is used to infer the transition:

$$\{v; stop|[u]|\} - v \rightarrow \{\}$$

We substituted $\{v; stop\}$, u and v to respectively M_1, S and a. M_1' is empty because the decomposition of "stop" is defined as the empty set.

We introduce the decomposition function in Section 4.3 then we present, in Section 4.4, the inference rules. But first, in Section 4.2 we translate Lotos specifications in a form that makes easier development of consistency proofs.

4.2. Normal Form Specifications

PLotos specifications are rewritten into simpler forms, called normal form specifications. Sub-terms in which mutual recursion does not occur are expanded, that is, process definitions are substituted for process calls. Then we distinguish every parallel composition $B_1|[g_1, ..., g_n]|B_2$ by labelling the operator with an unique value k. This is represented as $|[g_1, ..., g_n]|_k$. For example:

process $p_1[a, b, c]$: *noexit* :=
$(p_2[a, b]|||p_2[a, b])[]c; p_1[a, b, c]$
endproc
process $p_2[a, b]$: *noexit* :=
$a; b; stop|[a]|a; stop$
endproc

is rewritten as:

process $p_1[a, b, c]$: *noexit* :=
$((a; b; stop\ |[a]|_1\ a; stop)|||(a; b; stop\ |[a]|_2\ a; stop))[]c; p_1[a, b, c]$
endproc

Static relabelling instead of dynamic relabelling is performed when process instantiation terms are substituted by the corresponding defining behavior-expressions. This issue is further discussed in Section 4.3.

As discussed in Section 4.1, with the small example: $\{u; v; stop|[u]|, |[u]|u; stop\}$, every general parallel composition is decomposed into two or more component-expressions during the PLotos to the P/T-net modelling process. Labelling of general parallel operators with a unique value is required to preserve important contextual information of component-expressions. This information is required to unambiguously determine which component-expressions need to be synchronized together.

4.3. Decomposition Function

The decomposition function is denoted as *dec*. Its domain is the set of well-formed PLotos behavior-expressions. Its range is the set of all possible finite multi-sets of component-expressions.

Let B_1, B_2 denote syntactically correct PLotos behavior-expressions, a denote an action name and $S = g_1, ..., g_n$ a list of synchronization gates, the decomposition function *dec* is defined as follows:

$$
\begin{array}{lrl}
(d1) & dec(stop) & := \{\} \\
(d2) & dec(a; B_1) & := \{a; B_1\} \\
(d3) & dec(B_1 [] B_2) & := \{B_1 [] B_2\} \\
(d4) & dec(p[g_1, ..., g_n]) & := dec(B_p[g_1/h_1, ..., g_n/h_n]) \\
(d5) & dec(B_1 ||| B_2) & := dec(B_1) + dec(B_2) \\
(d6) & dec(B_1 |[S]|_k B_2) & := dec(B_1).|[S]|_k + |[S]|_k.dec(B_2) \\
(d7) & dec(B_1 >> B_2) & := \{B_1 >> B_2\} \\
(d8) & dec(B_1 [> B_2) & := \{B_1 [> B_2\} \\
(d9) & dec(hide\ S\ in\ B_1) & := hide\ S\ in.dec(B_1) \\
(d10) & dec(exit) & := \{exit\}
\end{array}
$$

where:

- in (d4), B_p represents the body of process definition p,
- $g_1, ..., g_n$ is a list of actual gates,
- $h_1, ..., h_n$ is a list of formal gates,
- $[g_1/h_1, ..., g_n/h_n]$ is the relabelling postfix operator, gate h_i becomes gate g_i $(i = 1, ..., n)$, and
- the expression $dec(B_1).|[S]|_k$ denotes $\{x|[S]|_k : x \in dec(B_1)\}$, similarly for $|[S]|_k.dec(B_2)$ and the expression $hide\ S\ in.dec(B_1)$ denotes $\{hide\ S\ in\ x : x \in dec(B_1)\}$.

The *dec* function is deterministic, taking into account operator precedences given in [ISO 88]. The restriction to guarded recursive processes (see Section 3) is required to stop recursion in the *dec* function.

The relabelling operator is not user accessible and exists for the semantic description of process instantiation. In Lotos, relabelling is dynamic; gates are renamed at the execution time. For instance, let us consider this process definition:

process $p[a, b]$: **noexit** :=
$\quad a; stop|[a, b]|b; stop$
endproc

Instantiating p with $p[a, a]$ yields an inactive process with dynamic relabelling, since the expression $a; stop|[a, b]|b; stop$ is inactive. Nevertheless, with static renaming $p[a, a]$ yields the expression $a; stop|[a, a]|a; stop$ which may perform the action a and becomes inactive.

It can be shown easily that for injective relabelling operators, static and dynamic relabelling are equivalent. For the sake of simplicity, hereafter we consider solely injective relabellings and perform static renaming, that is syntactical substitution. We believe that this restriction is not significant, at least from a computational point of view, and it is fulfilled in many applications.

4.4. Inference Rules

This section exposes the inference rules of our PLotos to P/T-nets mapping. The P/T-net $N = (P, T, Act, M_0)$, with reachability set RS, associated to a PLotos behavior-expression B is defined as:

1. • $M_0 = dec(B)$
   • $M_0 \in RS$
   • $(\forall p)[M_0(p) > 0 \Rightarrow p \in P]$

2. if $M \in RS$ and $X \leq M$ and $X - a \rightarrow Y$ then
   • $(\forall p)[Y(p) > 0 \Rightarrow p \in P]$
   • $(X, a, Y) \in T$
   • $a \in Act$
   • $M' = M - X + Y$
   • $M' \in RS$

3. only the elements that can be obtained from items 1 or 2 are in P, T and Act

The transition instances are inferred from the rules below. For all PLotos behavior-expressions B_1, B_1', B_2, B_2', action name a, list $S = g_1, ..., g_n$ of synchronization gates and component-expression multi-sets M_1, M_2, M_1', M_2':

(r1) $\{a; B_1\} - a \rightarrow dec(B_1)$
(r2) if $B_1 - a \rightarrow B_1'$
then $\{B_1 [] B_2\} - a \rightarrow dec(B_1')$
(r3) if $B_2 - a \rightarrow B_2'$
then $\{B_1 [] B_2\} - a \rightarrow dec(B_2')$
(r4) if $M_1 - a \rightarrow M_1'$ and $a \notin \{S, \delta\}$
then $M_1 . \|[S]\|_k - a \rightarrow M_1' . \|[S]\|_k$
(r5) if $M_2 - a \rightarrow M_2'$ and $a \notin \{S, \delta\}$
then $\|[S]\|_k . M_2 - a \rightarrow \|[S]\|_k . M_2'$
(r6) if $M_1 - a \rightarrow M_1'$ and $M_2 - a \rightarrow M_2'$ and $a \in \{S, \delta\}$
then $M_1 . \|[S]\|_k + \|[S]\|_k . M_2 - a \rightarrow M_1' . \|[S]\|_k + \|[S]\|_k . M_2'$
(r7) if $B_1 - a \rightarrow B_1'$ and $a \neq \delta$
then $\{B_1 >> B_2\} - a \rightarrow \{B_1' >> B_2\}$
(r8) if $B_1 - \delta \rightarrow B_1'$
then $\{B_1 >> B_2\} - i \rightarrow dec(B_2)$
(r9) if $B_1 - a \rightarrow B_1'$ and $a \neq \delta$
then $\{B_1 [> B_2\} - a \rightarrow \{B_1' [> B_2\}$
(r10) if $B_1 - \delta \rightarrow B_1'$
then $\{B_1 [> B_2\} - \delta \rightarrow dec(B_1')$
(r11) if $B_2 - a \rightarrow B_2'$
then $\{B_1 [> B_2\} - a \rightarrow dec(B_2')$

(r12) if $M_1 - a \rightarrow M'_1$ and $a \notin \{S\}$
then *hide S in*.$M_1 - a \rightarrow$ *hide S in*.M'_1
(r13) if $M_1 - a \rightarrow M'_1$ and $a \in \{S\}$
then *hide S in*.$M_1 - i \rightarrow$ *hide S in*.M'_1
(r14) $\{exit\} - \delta \rightarrow \{stop\}$

In the "if part" of inference rules (r2), (r3), (r7) (r8), (r9), (r10) and (r11) behavior B_1 (B_2) makes a transition to behavior B'_1 (B'_2) on action a or δ in accordance with the original Basic Lotos semantics.

Theorem 1 *(Boundedness theorem) PLotos can be modelled by a finite P/T-net.*

We must show [3] that any PLotos normal form specification:

specification ... **behavior**B_0 ... **endspec**

can be modelled by a P/T-net $N = (P, T, Act, M_0)$ whose sets P, T and Act are finite (note that the associated reachability set RS is not necessarily finite).

In the sequel, the operators in component-expressions are classified as follows:

- **stop**, **exit**, and $p[g_1, ..., g_n]$ are nullary operators.
- $|[S]|_k$, and **hide** S **in** are unary operators.
- ";", ">>" and "[>" are binary operators.

Note that the operator "|||" never appears in a component-expression.

(*The set Act is finite*). In a normal form PLotos specification there is a finite number of gates. Lotos gates are translated to P/T-net transition labels, i.e. elements of *Act*. Consequently, the set *Act* is finite.

(*The set P is finite*). The statement "The set P is finite" is equivalent to the statement:

> *S1*: There exists a K such that for all $p \in P$, the number of nullary operators in p is less than K.

This equivalence is a consequence of the conjunction of the following facts (let us suppose that we distinguish, in the normal form specification, every operator from the others): i) The set of gates and nullary, unary and binary operators, that can possibly be used in a component-expression is finite. ii) Every unary or binary operator is used at most once in a component-expression. iii) Using a finite number of gates and nullary, unary and binary operators, and zero or one occurrence of every unary or binary operator, there is a finite number of syntactically different component-expressions that can be constructed.

The negation of statement *S1* is the following statement:

[3] The proof technique is similar to the one used in [Gara 89].

S2: It is possible to infer from $dec(B_0)$ a component-expression p in which there is an unbounded number of nullary operators.

Statement *S2* implies that there exist processes p_1, p_2 with:

$$(p_1, p_2) \in \Phi$$

a marking:

$$M_n \in RS$$

and a component-expression p with:

$$p \in M_n$$

where there is a nullary operator who occurs an unbounded number of times in p. This unbounded number of occurrences is due to the substitution of recursive instantiation terms of p_1 by its defining behavior-expression B_{p_1} of p_1. Nonetheless, this true solely if B_{p_1} has one of the following patterns:

- $C_1[C_2[p_2] * B]$, where the operator "$*$" is either "$|[g_1, ..., g_n]|$" or "$>>$" or "$[>$".
- $C_1[B|[g_1, ..., g_n]|C_2[p_2]]$.
- $C_1[hide\ g_1, ..., g_n\ in\ C_2[p_2]]$

where $C_1[\ ]$ and $C_2[\cdot]$ denote two contexts and B denote any behavior-expression. However, these patterns ar disallowed in PLotos (see Section 3).

(*The set T is finite*). This follows from the fact that from a finite set of syntactically different component-expressions, application of the inference rules can derive a finite number of transitions.

The next theorem states that the P/T-net semantics is in accordance with the original semantics of Lotos.

Theorem 2 *(Consistency theorem) The Petri net semantics of Lotos is consistent with the standard Lotos semantics. That is, for all PLotos behavior expression B, marking M with $dec(B) := M$:*

1. $[B - a \rightarrow B'] \Rightarrow (\exists M')(\exists t)[M(t > M' \wedge act(t) = a \wedge dec(B') := M']$

2. $[M(t > M'] \Rightarrow (\exists B')[B - act(t) \rightarrow B' \wedge dec(B') := M']$

The proof is by induction on the number of operators in a behavior-expression B and refers to the standard Lotos semantics in Refs. [Bolo87] and [ISO88].

Definition 1 *Two graphs $A_1 = (S_1, E_1, n_1)$ and $A_2 = (S_2, E_2, n_2)$ are bisimular [Park 81] if there exists a relation $R \subseteq S_1 \times S_2$, called a bisimulation relation, with:*

1. $(n_1, n_2) \in R$, *and for all* $(n, m) \in R$

2. $[(n,a,n') \in E_1] \Rightarrow (\exists m')[(m,a,m') \in E_2 \wedge (n',m') \in R]$, *and*

3. $[(m,a,m') \in E_2] \Rightarrow (\exists n')[(n,a,n') \in E_1 \wedge (n',m') \in R]$.

Corollary 1 *Let B be a PLotos behavior-expression, with transition graph TG, and let $N = (P, T, Act, M_0)$ be the associated P/T-net with reachability graph $RG(N)$. The dec function is a homomorphism from TG reachability set to $RG(N)$ reachability set. TG and $RG(N)$ are bisimular under the bisimulation relation R defined as:*

1. $(B, M_0) \in R$, *and*

2. For all B' in the TG reachability set and for all M in $RG(N)$ reachability set:

$$(B', M) \in R \Longleftrightarrow dec(B') := M$$

The *dec* function is a graph homomorphism because it identifies equivalent Lotos behavior-expressions $B_1|||B_2$ with $B_2|||B_1$, and $B_1|||(B_2|||B_3)$ with $(B_1|||B_2)|||B_3$. These equivalences are in accordance with the commutativity and the associativity laws in [ISO 88]. Solely syntactic nature information is lost, "*dec*" preserves all semantic properties. This can be illustrated by the following commutative diagram:

$$\begin{array}{ccc} B & -dec \rightarrow & M \\ | & & | \\ a & & a \\ \downarrow & & \downarrow \\ B' & -dec \rightarrow & M' \end{array}$$

5. Simulation of P/T-nets in PLotos

In Section 4, we identified a subset of Lotos, PLotos, that can be modelled by finite P/T-nets. In this section, we show that conversely P/T-nets can be simulated by PLotos. These two facts lead to the conclusion that PLotos and P/T-nets are equivalent models, that is models with equivalent computational power.

We make two reasonable hypotheses. First, we simulate in PLotos, P/T-nets whose place to transition edges are one valued, i.e.:

$$(\forall t, p)[pre(t)(p) \text{ equals } 0 \text{ or } 1]$$

This restriction is not a handicap because it has been proved [Kasa 82] that P/T-nets of arbitrary edge valuation can be simulated by P/T-nets whose edges are all valued to one, with language equality equivalence.

Second, we assume that no place is simultaneously in the preset and the postset of a single transition. This restriction is not significant. P/T-nets with circuits made of one place and one transition can be simulated by circuit free (*pure*) P/T-nets [Bram 83].

Before we go into detail, we give a brief overview of the simulation. Given a P/T-net $N = (P, T, Act, M)$ with set of places $P = \{p_1, ..., p_n\}$, we define a PLotos process $N_{1,n}$ with equivalent behavior. That is, the reachability graph of N and the transition graph of $N_{1,n}$ are bisimular. The process $N_{1,n}$ is defined inductively on the number n of places.

Every transition $t \in T$ is mapped to a Lotos gate, also named t. Every place $p_i \in P$ is mapped to three PLotos processes, namely $token_i$, p_i and $P_i(k)$. The process $token_i$ models a token inside the place p_i. It participates in actions that occur at gates corresponding to outgoing transitions of the place p_i. Instances of $token_i$ are created by the process p_i when the place p_i incoming transitions are fired.

The process $P_i(k)$ models place p_i containing k tokens and is defined as the independent parallel composition of one instance of process p_i and k instances of process $token_i$. Simulation of a place p_i in PLotos is further discussed in Section 5.1 (with an example in App. B).

The whole PLotos model of the P/T-net N, with current (or initial) marking M is defined inductively. The PLotos model $N_{1,1}$ of N, restricted to place p_1, is defined as the process $P_1(M(1))$.

The model $N_{1,i}$ of N, restricted to places $p_1, ..., p_i$, is defined as the parallel composition of the process $N_{1,i-1}$ that models N restricted to places $p_1, ..., p_{i-1}$ and the process $P_i(M(i))$. These two processes are synchronized on the set of transitions that place p_i shares with places $p_1, ..., p_{i-1}$. This construction is presented formally in Section 5.2 (with an example in App. B).

5.1 Modelling of Places

Let $N = (P, T, Act, M)$ be a P/T-net with $P = \{p_1, ..., p_n\}$ and $T = \{t_1, ..., t_m\}$. We first discuss how tokens inside places are represented by Lotos processes.

Given place $p_i \in P$, let:

- $\Gamma^{-1}(p_i) = \{t_I : post(t_I)(p_i) > 0\}$, transitions that deposit tokens into place p_i
- $\Gamma(p_i) = \{t_O : pre(t_O)(p_i) > 0\}$, transitions that extract tokens from place p_i
- $T(X) = \bigcup_{p_i \in X}(\Gamma(p_i) \cup \Gamma^{-1}(p_i))$, transitions connected to places in X

A token inside place p_i can participate in the firing of a transition in $\Gamma(p_i)$.

Definition 2 *Let $\Gamma(p_i) = \{t_{O1}, ..., t_{Ov}\}$, a token inside place p_i is represented as the following PLotos process:*

process $token_i[t_{O1}, ..., t_{Ov}]$:***noexit***:=
$t_{O1}; stop [] ... [] t_{Ov}; stop$
endproc

If $\Gamma(p_i)$ is empty, then the body of $token_i$ is stop. Let $\Gamma^{-1}(p_i) = \{t_{I1}, ..., t_{Iu}\}$, the place p_i is modelled by the following process:

$$
\begin{array}{l}
\textbf{\textit{process}}\ p_i[t_{I1},...,t_{Iu},t_{O1},...,t_{Ov}]\textbf{\textit{:noexit:=}}\\
\quad t_{I1};\ (\underbrace{(token_i[t_{O1},...,t_{Ov}]|||\cdots|||token_i[t_{O1},...,t_{Ov}])}_{(*\ post(t_{I1})(p_i)\ times\ *)}\\
\quad\quad |||\\
\quad\quad p_i[t_{I1},...,t_{Iu},t_{O1},...,t_{Ov}])\\
\quad [] \cdots []\\
\quad t_{Iu};\ (\underbrace{(token_i[t_{O1},...,t_{Ov}]|||\cdots|||token_i[t_{O1},...,t_{Ov}])}_{(*\ post(t_{Iu})(p_i)\ times\ *)}\\
\quad\quad |||\\
\quad\quad p_i[t_{I1},...,t_{Iu},t_{O1},...,t_{Ov}])\\
\textbf{\textit{endproc}}
\end{array}
$$

Note that recursive calls to p_i are guarded and allowed in a pure interleaving. Informally, this says that when an input transition of place p_i is fired, either t_{I1} or ... or t_{Iu}, then tokens are deposited inside place p_i (instances of process $token_i$ are created). These new tokens can enable and fire transitions in $\Gamma(p_i)$. If $\Gamma^{-1}(p_i)$ is empty, then the body of p_i is stop.

The next lemma demonstrates the consistency of the PLotos model of a place.

Lemma 1 *Let, for $k \in \mathcal{N}$, $P_i(k)$ denote the place p_i containing k tokens, modelled as the PLotos processes*[4]*:*

$$
\begin{array}{ll}
P_i(0) & = p_i\\
P_i(k) & = token_i|||P_i(k-1) \quad \textit{if } k > 0
\end{array}
$$

For all $k \in \mathcal{N}$:

$$
\begin{array}{lll}
P_i(k) - t \rightarrow p & \Leftrightarrow & [(t \in \Gamma^{-1}(p_i) \wedge p = P_i(k + post(t)(p_i)))\\
 & \vee & (k > 0 \wedge t \in \Gamma(p_i) \wedge p = P_i(k-1))].
\end{array}
$$

The proof is by induction on k.

5.2 Modelling of P/T-nets

The model of a P/T-net in PLotos is also defined inductively. We first consider unlabelled P/T-nets. For $1 \leq i \leq n$, we denote by:

$$N_{1,i} = (P_{1,i}, T_{1,i}, M_{1,i})$$

the subnet of $N = (P, T, M)$ restricted to places $\{p_1, ..., p_i\}$ where:

- $P_{1,i} = \{p_1, ..., p_i\}$
- $T_{1,i} = \{(X, act(t), Y) : t \in T \wedge X = \sum_{p \in P_{1,i}} pre(t)(p)p$ $\wedge Y = \sum_{p \in P_{1,i}} post(t)(p)p \wedge (X \neq \{\} \vee Y \neq \{\})\}$

[4] *For the sake or readability, we omit gate-tuples.*

- $M_{1,i}$, the marking M restricted to places in $P_{1,i}$

Note that $N_{1,n} = N$. We denote by $M_{1,i}(i)$ the number of tokens inside place p_i for the marking $M_{1,i}$.

Definition 3 *For $1 \leq i \leq n$, the subnet $N_{1,i}$ is modelled by a PLotos process named $N_{1,i}(M_{1,i})$ defined as:*

process $N_{1,1}(M_{1,1})[t_1, ..., t_m]$***:noexit:=***
$P_1(M_{1,1}(1))$
endproc

For $i > 1$, $N_{1,i}(M_{1,i})$ is defined as:

process $N_{1,i}(M_{1,i})[t_1, ..., t_m]$***:noexit:=***
$P_i(M_{1,i}(i)) \; |[T(\{p_i\}) \cap T(\{p_1, ..., p_{i-1}\})]| \; N_{1,i-1}(M_{1,i-1})[t_1, ..., t_m]$
endproc

Note that, for $i = 1, ..., n$, $N_{1,i}(M_{1,i})$ is not recursive (i.e constraint 3.1 and 3.2 are not violated).

The model of a P/T-net N in PLotos is the process $N_{1,n}(M_{1,n})$. The next lemma demonstrates the consistency of the PLotos model of P/T-nets.

Lemma 2 *Let $N = (P, T, M)$ be a P/T-net and $M' \in \mathcal{N}^P$ be a marking. For every $i = 1, ..., n$, let $N_{1,i} = (P_{1,i}, T_{1,i}, M_{1,i})$ be the subnet defined as above, then for all $t \in T_{1,i}$:*

$$M_{1,i}(t > M'_{1,i} \Leftrightarrow N_{1,i}(M_{1,i}) - t \rightarrow N_{1,i}(M'_{1,i})$$

The proof is by induction on i.

Let $N = (P, T, Act, M)$ be a labelled P/T-net and let $N_{1,n}(M_{1,n})$ be the PLotos model of the corresponding unlabelled P/T-net. We may add labels $a_1, ..., a_l$ in Act, of transitions in T, to the PLotos model as follows:

process $LabelledN_{1,n}(M_{1,n})[a_1, ..., a_l]$**:noexit:=**
hide $t_1, ..., t_m$ **in** $A[a_1, ..., a_l, t_1, ..., t_m] \; |[t_1, ..., t_m]| \; N_{1,n}(M_{1,n})[t_1, ..., t_m]$
where
process $A[a_1, ..., a_l, t_1, ..., t_m]$**:noexit:=**
$t_1; act(t_1); A[a_1, ..., a_l, t_1, ..., t_m] \; [] \cdots [] \; t_m; act(t_m); A[a_1, ..., a_l, t_1, ..., t_m]$
endproc
endproc

Note that in the process $LabelledN_{1,n}$, the constraint 3.3 is not violated.

6. Conclusion

The fact that PLotos has the computational power of P/T-nets, with bisimulation equivalence, means that:

1. properties that are decidable for P/T-nets are decidable as well for PLotos, and
2. algorithms for deciding properties of P/T-nets can be adapted to PLotos.

Furthermore, the aforementioned items are obtained by minimally restricting Lotos, since P/T-nets can be modelled by PLotos. We have investigated adaptation of P/T-nets verification techniques to PLotos in Refs. [Barb 91a] and [Barb 91b].

Acknowledgements

The authors thank the members of the CRIM/BNR project for many fruitful discussions. We wish to thank Prof. Alain Finkel from École normale supérieure de Cachan who contributed to the proofs in Section 5.

References

[Azem 84] P. Azema, G. Juanole, E. Sanchis, M. Montbernard, *Specification and Verification of Distributed Systems Using Prolog Interpreted Petri Nets*, 7th International Conference on Software Engineering, 1984.

[Barb 91a] M. Barbeau, G. v. Bochmann, *Extension of the Karp and Miller Procedure to Lotos Specifications*, Computer Aided Verification'90, ACM/AMS DIMACS Series in Discrete Mathematics and Theoretical Computer Science, Vol. 3, 1991, pp. 103-119; and Springer- Verlag, LNCS 531, pp. 333-342.

[Barb 91b] M. Barbeau, G. v. Bochmann, *The Lotos Model of a Fault Protected System and its Verification Using a Petri Net Based Approach*, Workshop on Computer-aided verification, Aalborg, Danemark, 1991; and Springer-Verlag, LNCS 575.

[Bolo 87] T. Bolognesi, E. Brinksma, *Introduction to the ISO Specification Language Lotos*, Computer Networks and ISDN Systems, Vol. 14, No. 1, 1987, pp. 25-59.

[Bolo 90] T. Bolognesi, *A Graphical Composition Theorem for Networks of Lotos Processes*, Proceedings of Distributed Computing Systems, Paris, May-June 1990, pp. 88-95.

[Boud 85] G. Boudol, G. Roucairol, R. de Simone, *Petri Nets and Algebraic Calculi of Processes*, Advances in Petri Nets, 1985, pp. 41-58.

[Bram 83] G. W. Brams, *Réseaux de Petri: Théorie et Pratique - T.1. Théorie et analyse*, Masson, Paris, 1983.

[Cind 83] F. de Cindio, G. de Michelis, L. Pomello, C. Simone, *Milner's Communicating Systems and Petri Nets*, in: A. Pagnoni, G. Rozenberg (Eds.), Application and Theory of Petri Nets, Springer-Verlag, IFB 66, 1983, pp. 40-59.

[Dega 88] P. Degano, R. de Nicola, U. Montanari, *A Distributed Operational Semantics for CCS Based on Condition/Event Systems*, Acta Informatica, Vol. 26, 1988, pp. 59-91.

[Gara 89] H. Garavel, E. Najm, *Tilt: From Lotos to Labelled Transition Systems*, in: P. H. J. van Eijk, C. A. Vissers, M. Diaz (Eds.), The Formal Description Technique Lotos, North-Holland, 1989, pp. 327-336.

[Gara 90] H. Garavel, J. Sifakis, *Compilation and Verification of Lotos Specifications*, PSTV X, Ottawa, 1990, pp. 359-376.

[Glab 87] R. J. van Glabbeek, F. W. Vaandrager, *Petri Net Models for Algebraic Theories of Concurrency*, Proceedings of PARLE, Vol. II, LNCS 259, Springer-Verlag, 1987.

[Golt 84a] U. Goltz, A. Mycroft, *On the Relationship of CCS and Petri Nets*, in: J. Paredaens (Ed.), Proceedings of ICALP 84, LNCS 172, Springer-Verlag, 1984, pp. 196-208.

[Golt 84b] U. Goltz, W. Reisig, *CSP-Programs as Nets with Individual Tokens*, in: G. Rozenberg (Ed.), Advances in Petri Nets 1984, LNCS 188, Springer-Verlag, 1985, pp. 169-196.

[Golt 88] U. Goltz, *On Representing CCS Programs by Finite Petri Nets*, in: M. Chytil et al. (Eds.), Mathematical Foundations of Computer Science 1988, LNCS 324, Springer-Verlag, 1988, pp. 339-350.

[Gotz 86] R. Gotzhein, *Specifying Abstract Data Types with Lotos*, Proc. of PSTV VI, Montréal, 1986.

[ISO 88] ISO, *Lotos - A Formal Description Technique Based on the Temporal Ordering of Observational Behavior*, IS 8807, E. Brinksma (Ed.), 1988.

[Kasa 82] T. Kasai, R. E. Miller, *Homomorphisms Between Models of Parallel Computation*, J.C.S.S., Vol. 25, 1982, pp. 285-331.

[Marc 89] S. Marchena, G. Leon, *Transformation from Lotos Specs to Galileo Nets*, in: K. J. Turner (Ed.), Formal Description Techniques, North-Holland, 1989.

[Niel 86] M. Nielsen, *CCS and its Relationship to Net Theory*, in: W. Brauer, Advances in Petri Nets 1986, Part II, LNCS 255, Springer-Verlag, 1986.

[Olde 91] E.-R. Olderog, *Nets, Terms and Formulas: Three Views of Concurrent Processes and their Relationships*, Cambridge Tracts in Theoretical Computer Science 23, Cambridge University Press, 1991.

[Park 81] D. M. R. Park, *Concurrency and Automata on Infinite Sequences*, Proceedings of 5th GI Conf. on Theoretical Computer Science, LNCS 104, Springer-Verlag, 1981, pp. 167-183.

[Pete 81] J. L. Peterson, *Petri Net Theory and the Modelling of Systems*, Prentice Hall, 1981.

[Reis 84] W. Reisig, *Partial Order Semantics Versus Interleaving Semantics for CSP-like Languages and Its Impact on Fairness*, in: G. Goos, J. Hartmanis, 11th Colloquium on Automata, Languages and Programming, LNCS 172, Springer-Verlag, 1984, pp. 403-413.

[Taub 89] D. Taubner, *Finite Representation of CCS and TCSP Programs by Automata and Petri Nets*, LNCS 369, Springer-Verlag, 1989.

Appendix A: Basic Lotos

A.1. Syntax of Basic Lotos

We assume that Basic Lotos specifications are constructed as follows:

specification ::= **specification** specification-identifier formal-parameter-list
behavior
behavior-expression
[local-definitions]
endspec

formal-parameter-list ::= [gate-tuple] ":" functionality

gate-tuple ::= "[" gate-identifier-list "]"

gate-identifier-list ::= gate-identifier { "," gate-identifier }

functionality ::= **exit** | **noexit**

behavior-expression ::=

stop
gate-identifier ";" behavior-expression
behavior-expression "[]" behavior-expression
process-identifier [gate-tuple]
behavior-expression "|||" behavior-expression
behavior-expression "|[" gate-identifier-list "]|" behavior-expression
exit
behavior-expression ">>" behavior-expression
behavior-expression "[>" behavior-expression
hide gate-identifier-list **in** behavior-expression

local-definitions ::= **where** process-definition { process-definition }

process-definition ::=
process process-identifier formal-parameter-list ":="
behavior-expression
endproc

specification-identifier ::= identifier

process-identifier ::= identifier

gate-identifier ::= identifier

identifier ::= letter [{ normal-character | "_" } normal-character]

normal-character ::= letter | digit

In a "process-definition", the term "behavior-expression" is called the defining behavior-expression of the process named "process-identifier".

Appendix B: The simulation of a P/T-net in PLotos

P/T-net

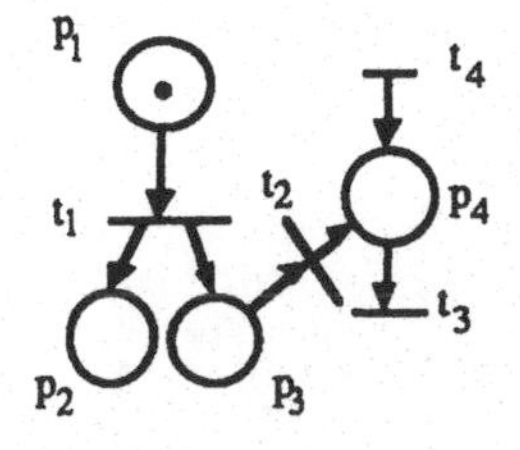

Translation of places

```
process p1[t1]:noexit:=
    stop
endproc

process p2[t1]:noexit:=
    t1; token2[ ]|||p2[t1]
endproc

process p3[t1, t2]:noexit:=
    t1; token3[t2]|||p3[t1, t2]
endproc

process p4[t2, t3, t4]:noexit:=
    t2; token4[t3]|||p4[t2, t3, t4]
[ ]
    t4;token4[t3]|||p4[t2, t3, t4]
endproc

process token1[t1]:noexit:=
    t1; stop
endproc

process token2[ ]:noexit:=
    stop
endproc

process token3[t2]:noexit:=
    t2;stop
endproc

process token4[t3]:noexit:=
    t3;stop
endproc
```

Lotos model of the P/T-net

process $N_{1,1}((1))[t_1, t_2, t_3, t_4]$:noexit:= $P_1(1)$ endproc

process $N_{1,2}((1,0))[t_1, t_2, t_3, t_4]$:noexit:= $P_2(0)|[t_1]|N_{1,1}((1))[t_1, t_2, t_3, t_4]$ endproc

process $N_{1,3}((1,0,0))[t_1, t_2, t_3, t_4]$:noexit:= $P_3(0)|[t_1]|N_{1,2}((1,0))[t_1, t_2, t_3, t_4]$
endproc

process $N_{1,4}((1,0,0,0))[t_1, t_2, t_3, t_4]$:noexit:= $P_4(0)|[t_2]|N_{1,3}((1,0,0))[t_1, t_2, t_3, t_4]$
endproc

where

$P_1(1) = \text{token}_1[t_1]|||p_1[t_1]$
$P_2(0) = p_2[t_1]$
$P_3(0) = p_3[t_1, t_2]$
$p_4(0) = p_4[t_2, t_3, t_4]$

An Efficient Algorithm for Finding Structural Deadlocks in Colored Petri Nets

K. BARKAOUI * C. DUTHEILLET # S. HADDAD #

* Laboratoire CEDRIC
Conservatoire National des Arts et Métiers
292 rue Saint-Martin
75003 Paris - FRANCE
e-mail : barkaoui@cnam.cnam.fr

IBP - Laboratoire MASI
Université P. & M. Curie
4 , Place Jussieu
75252 Paris Cedex 05 - FRANCE
e-mail : haddad@masi.ibp.fr

Abstract In this paper, we present an algorithm to compute structural deadlocks in colored nets under specified conditions. Instead of applying the ordinary algorithm on the unfolded Petri net, our algorithm takes advantage of the structure of the color functions. It is obtained by iterative optimizations of the ordinary algorithm. Each optimization is specified by a meta-rule, whose application is detected during the computation of the algorithm. The application of such meta-rules speeds up a step of the algorithm with a factor proportional to the size of a color domain. We illustrate the efficiency of this algorithm compared to the classical approach on a colored net modelling the dining philosophers problem.

1 Introduction

Analysis techniques of Place-Transition nets [1] can be grouped in two broad classes. The first approach consists in building the reachability graph, which gives full information but is usually very expensive [2].

Another approach - *the structural analysis* - consists in getting information about the behavior of the model directly from the structure of its underlying bipartite and valuated digraph, the initial marking being considered as a parameter [3]. Two kinds of structural analysis can be distinguished:

- *The algebraic analysis*, where the structure of the net is represented by the incidence matrix associated with its underlying digraph. It provides results such as conditions for liveness and boundedness of the net, or linear invariants [4] [5].

- *The graph theoretical analysis*, where the behavior of the net is related to the flow relation of subnets generated by remarkable subsets of places, such as structural deadlocks and traps [6]. With such techniques, liveness can be decided in polynomial time for different classes of nets [7] [8] [9].

However, modelling real complex systems yields so large models that it is difficult to handle the complexity of both the structural and the reachability analyses. Thus High-Level nets - Predicate/Transition nets [10] or Colored nets [11] have been introduced to concisely model large systems. In order to take advantage of this high-level description, the analysis of the model must be performed without refering to its equivalent unfolded net, i.e., a place-transition net with the same behavior. Thus, a lot of research is being done to directly apply to High-Level nets analysis techniques similar to those existing for Place-Transition nets.

The present paper is a contribution to graph theoretical analysis for colored nets. On Place-Transition nets, this approach uses the flow relation of subnets generated by structural deadlocks and traps, which are remarkable subsets of places. For these nets, the problem of finding structural deadlocks was attacked from different points of view. The technique of computing strongly connected deadlocks via a positive flow calculus on an expanded net [8] was optimized and extended to the class of Unary Predicate-Transition nets without guards [12]. The application of this method to less restrictive classes requires a generalization of positive flow computation [13], which seems to be a very difficult task [14]. In any case, the problem of obtaining the set of minimal deadlocks by means of flow computation is NP-complete.

In [15], the problem is expressed as a logic programming problem, leading to identify structural deadlocks (traps) with the set of solutions of a Horn-clause satisfiability problem. In colored nets, the flow relation is defined by both the underlying bipartite digraph and the functions labelling the arcs. Hence, a deadlock of a colored net may not correspond to a deadlock of the net's skeleton in the sense of [16], while the converse is true. Yet, for a restricted class of colored nets, the problem of finding deadlocks can be reduced to the problem of finding deadlocks in the skeleton [17], and the technique proposed in [15] could be exploited. But an efficient extension of the method to general colored nets is still an open problem.

In this paper, we develop a method for solving efficiently the following problem `Pb` by reasoning directly on the color functions, i.e., without an effective unfolding.

`Pb`:	For any pair of disjoints subsets of places P_{exc} and P_{inc}, find a structural deadlock D such that $D \cap P_{exc} = \emptyset$ and $P_{inc} \subseteq D$,
or	decide that no such deadlock exists.

The paper is organized as follows. Section 2 introduces a property of deadlocks which yields propositional deduction rules. These rules are used in an algorithm that solves problem `Pb` for Place-Transition nets. We then extend this approach to colored Petri nets, rewriting the algorithm in terms of deduction rules of first order logic.

In Section 3, we optimize the applications of the previous rules by defining meta-rules. These meta-rules transform a set of rules so that one application of a transformed rule corresponds to the application of several initial rules.

In Section 4, we present the optimized algorithm that exploits the effects of these meta-rules, and we apply it on a model of the dining philosophers. Then we compare its execution on a colored net with the execution of the ordinary algorithm on the equivalent unfolded net. Section 5 contains the perspectives of this work.

2 Structural Deadlocks in Colored Nets

In the first part of this section, we recall the basic notations of Petri nets, together with the definition of a deadlock. We then extend this definition to colored Petri nets. Finally, we show on an example how the color functions of a model could be exploited to improve the efficiency of deadlock characterization algorithms.

2.1 Structural Deadlocks in Ordinary Petri Nets

Definition 2.1 *A Petri net N is a 4-tuple* $<P, T, W^-, W^+>$ *where*

P *is a finite set of places,*

T *is a finite set of transitions,*

W^- *(resp.* W^+*)* $: P \times T \rightarrow \mathbb{N}$ *is the input (resp. output) function.*

We also define the input set and output set of a subset of places, which contain respectively the input and output transitions of the places under consideration.

Definition 2.2 *Let* $p \in P$*. The input (resp. output) set of p, denoted by* $\bullet p$ *(resp.* $p\bullet$*) is defined as:*

$$\bullet p = \{t \in T \,/\, W^+(p, t) \neq 0\}$$

$$(resp. \quad p\bullet = \{t \in T \,/\, W^-(p, t) \neq 0\})$$

This definition can be extended to a subset D of places:

$$\bullet D = \bigcup_{p \in D} \bullet p \qquad (resp. \quad D\bullet = \bigcup_{p \in D} p\bullet)$$

A similar definition exists for transitions.

Definition 2.3 *Let* $t \in T$*. The input (resp. output) set of t, denoted by* $\bullet t$ *(resp.* $t\bullet$*) is defined as :*

$$\bullet t = \{p \in P \,/\, W^-(p, t) \neq 0\} \qquad (resp. \quad t\bullet = \{p \in P \,/\, W^+(p, t) \neq 0\})$$

We now give the definition of a structural deadlock:

Definition 2.4 *Let D be a non-empty subset of places. D is a structural deadlock iff* $\bullet D \subseteq D\bullet$*.*

The following property states that if no input place of a transition belongs to a deadlock, then the output places of this transition neither belong to the deadlock. Such a property is introduced since it leads to the construction of maximal deadlocks.

Property 2.1 *Let D be a non-empty subset of places. We have:*

$$[\, \forall t \in T, \quad \bullet t \subseteq (P \setminus D) \Rightarrow t\bullet \subseteq (P \setminus D)\,] \Leftrightarrow \text{D is a structural deadlock}$$

Proof: If we consider the negation of the left-hand part of the property, we can write:

$[\forall t \in T, \ \exists p \in D, p \in t\bullet \Rightarrow \exists p' \in D, p' \in \bullet t\,] \Leftrightarrow$ D is a deadlock.

but we also have $\quad \exists p \in D, \ p \in t\bullet \Leftrightarrow t \in \bullet D$

and $\quad \exists p' \in D, \ p' \in \bullet t \Leftrightarrow t \in D\bullet$

Hence, the left-hand part of the property is equivalent to:

$$\forall t \in T, \quad t \in \bullet D \Rightarrow t \in D\bullet$$

which is the definition of a structural deadlock.

Actually, Property 2.1 defines removal rules according to which places that do not belong to a deadlock can be eliminated from the net. We know that a place cannot belong to a deadlock if it is an output place of a transition such that no input place of this transition belongs to the deadlock.

Definition 2.5

- *Let $t \in T$. Then $R(t)$ is the removal rule associated with t and is written as:*

$$R(t) \; : \; \bullet t \;\Rightarrow\; t\bullet$$

- *Let E be a subset of P. Then $R(t)$ is applicable on E iff $\bullet t \subseteq E$. The application of t on E is given by $E := E \cup t\bullet$.*

We will call hypothesis the left-hand part of the rule, whereas the right-hand part will be called conclusion. Using Definition 2.5, we can write a generic algorithm for finding a structural deadlock satisfying two constraints:

- places contained in a set called P_{exc} must not belong to the deadlock,
- places contained in a set called P_{inc} must belong to the deadlock.

The principle of the algorithm is simple: a set R is initialized with all the rules of the net, a set Removed is initialized with the set P_{exc} and we try to apply on Removed the different rules of R, i.e., to remove the places that do not belong to the deadlock. When no more rule is applicable, we verify that P_{inc} is included in the complementary of Removed. If so, the algorithm has produced the maximal deadlock satisfying the constraints. Else, there is no deadlock satisfying these constraints.

Abstract Algorithm 1

```
Removed := Pexc
While ∃ t such that R(t) is applicable on Removed do
    Apply R(t) on Removed
    Delete R(t) in R
done;
Deadlock := P \ Removed
If Pinc ⊆ Deadlock and Deadlock ≠ Ø then return (success,Deadlock)
else return(failure)
```

An optimized implementation of this algorithm [15] requires an execution time proportional to the size of the net (expressed as the sum of the number of nodes and arcs). In the next section we extend the definitions and the algorithm to colored nets.

2.2 Structural Deadlocks in Colored Nets

Before giving a formal definition of a structural deadlock in a colored net, we recall the notations associated with this model.

Definition 2.6 *A colored Petri net N is a 5-tuple $<P, T, C, W^-, W^+>$ where*

- P — *is a finite set of places,*
- T — *is a finite set of transitions,*
- C — *is the color function, mapping $P \cup T$ onto Ω, where Ω is some finite set of finite and non-empty sets. $C(s)$ is called the color set of s.*
- W^- — *(resp. W^+) is the input (resp. output) function defined on $P \times T$, where $W^-(p, t)$ (resp. $W^+(p, t)$) is a function from $C(t)_{MS}$ to $C(p)_{MS}$.*

In this definition, E_{MS} denotes the set of multisets over a set E. Informally, an element of E_{MS} is a subset of E where the same element can appear several times.

Notation:
In the rest of this section, we will use PC to denote the set $[\bigcup_{p \in P} \{p\} \times C(p)]$, i.e., the set containing couples of places and associated color instances.

Definitions 2.7 and 2.8 are the extension to colored nets of Definitions 2.2 and 2.3.

Definition 2.7 *Let p be a place, and $c \in C(p)$ be a color instance of p. The input (resp. output) set of (p, c) is defined by the set $\bullet(p, c)$ (resp. $(p, c)\bullet$)*

$$\bullet(p, c) = \{(t, c') \mid W^+(p, t)(c')(c) > 0\}$$

$$(\text{resp.} \quad (p, c)\bullet = \{(t, c') \mid W^-(p, t)(c')(c) > 0\})$$

This definition can be extended to a subset $D \subseteq PC$, i.e., a subset D of places with associated color instances:

$$\bullet D = \bigcup_{(p, c) \in D} \bullet(p, c) \qquad (\text{resp.} \quad D\bullet = \bigcup_{(p, c) \in D} (p, c)\bullet\,)$$

Definition 2.8 *Let t be a transition, and $c \in C(t)$ be a color instance of t. The input (resp. output) set of (t, c) is defined by the set $\bullet(t, c)$ (resp. $(t, c)\bullet$)*

$$\bullet(t, c) = \{(p, c') \mid W^-(p, t)(c)(c') > 0\}$$

$$(\text{resp.} \quad (t, c)\bullet = \{(p, c') \mid W^+(p, t)(c)(c') > 0\})$$

Now the formal expression of the definition of a structural deadlock is quite similar to the definition of a deadlock in an ordinary Petri net.

Definition 2.9 *Let D be a non-empty subset of PC. D is a structural deadlock iff* $\bullet D \subseteq D\bullet$.

In the same way as we did for ordinary Petri nets, we now give a property that is equivalent to the definition of a deadlock.

Property 2.2 *Let D be a non-empty subset of PC. We have:*

$$[\,\forall (t, c) \in \bigcup_{t \in T} \{t\} \times C(t), \quad \bullet(t, c) \subseteq (PC \setminus D) \;\Rightarrow\; (t, c)\bullet \subseteq (PC \setminus D)\,]$$

$$\Leftrightarrow \; D \text{ is a structural deadlock}$$

The proof of this property is quite similar to the proof of Property 2.1.

In order to extend the algorithm computing the maximal structural deadlock that verifies some conditions, we extend the definition of removal rules to colored nets. Considering a transition t and an associated color c, we know from Property 2.2 that if no input place of (t, c) in the unfolded net belongs to a deadlock, then also none of the output places belongs to this deadlock. Hence, we introduce a function that gives for each place p the set of colors of p that are input (resp. output) of (t, c). These functions map the color domain of t onto the powerset of the color domain of p, i.e., the set of subsets of C(p), that we denote by $\mathcal{P}[C(p)]$.

Definition 2.10 *Let p be a place and t be a transition. The function $\Gamma^-(p, t) : C(t) \to \mathcal{P}[C(p)]$ defines for every color instance c of t the corresponding input colors in p:*

$$\Gamma^-(p, t)(c) = \{c' \mid W^-(p, t)(c)(c') > 0\}$$

The output colors are defined by $\Gamma^+(p, t) : C(t) \to \mathcal{P}[C(p)]$

$$\Gamma^+(p, t)(c) = \{c' \mid W^+(p, t)(c)(c') > 0\}$$

However, we already know that if there is no arc between p and t (resp. t and p), the set $\Gamma^-(p, t)(c)$ (resp. $\Gamma^+(p, t)(c)$) will be empty for any value of c. Hence, we need to determine which places and transitions are connected, disregarding the color functions. The sets •t and t• that we define now are the sets that would be calculated on the ordinary Petri net obtained when ignoring the color functions on the arcs.

Definition 2.11 *Let t be a transition. The input (resp. output) places of t are defined by the set •t (resp. t•):*

$$\bullet t = \{p \in P \mid \exists c \in C(t),\ \Gamma^-(p, t)(c) \neq \emptyset\}$$

$$\text{(resp.} \quad t\bullet = \{p \in P \mid \exists c \in C(t),\ \Gamma^+(p, t)(c) \neq \emptyset\})$$

The following definition introduces new notations that allow a more compact expression of set of places and associated colors.

Definition 2.12 *Let p and q be two places, E and F be subsets of C(p) and C(q) respectively. We define:*

- $[p, E] = \{(p, c) \mid c \in E\}$
- $[p, E] \wedge [q, F] = [p, E] \cup [q, F]$

Now we have all the elements that allow us to define the rules, whose application will determine if a place belongs to a deadlock or not.

Definition 2.13

- *Let $t \in T$. Then R(t) is the removal rule associated to t and is written as:*

$$R(t) \;:\; \bigwedge_{p \in \bullet t} [p, \Gamma^-(p, t)] \Rightarrow \bigwedge_{p \in t\bullet} [p, \Gamma^+(p, t)]$$

- *The color domain C[R(t)] of the rule is the color domain of the transition.*
- *Let E be a subset of PC, and $c \in C[R(t)]$. R(t) is applicable for c on E iff*

$$\bigwedge_{p \in \bullet t} [p, \Gamma^-(p, t)(c)] \subseteq E$$

- *The application of R(t) for c on E is given by*

$$E := E \cup \bigwedge_{p \in t\bullet} [p, \Gamma^+(p, t)(c)]$$

We use removal rules to write an algorithm that computes structural deadlocks. We first define a set P_{exc} (resp. P_{inc}) of places and associated color instances that are excluded from the deadlock (resp. that must belong to the deadlock). We initialize a set R with the rules of the net, a set Removed with P_{exc} and we try to apply the removal rules. If a rule R(t) is applicable for a color c, we add the output places of (t, c) to Removed, i.e., we remove these places because they cannot belong to the deadlock, and we also remove c from the color domain of the rule. When this color domain is empty, we remove the rule from R. When no more rule is applicable, places that have not been included in Removed form the maximal deadlock under the constraints of P_{exc}. If P_{inc} is included in the complementary of Removed, then the research has been successful.

```
Abstract   Algorithm   2
Removed := Pexc;
While ∃ t and c such that R(t) is applicable for c on Removed do
      Apply R(t) for c on Removed ;
      C[R(t)] := C[R(t)] \ {c};
      If C[R(t)] = Ø then Delete R(t) in R;
done;
Deadlock := PC \ Removed;
If Pinc ⊆ Deadlock and Deadlock ≠ Ø then return (success, Deadlock)
else return (failure)
```

Clearly, the complexity of this algorithm depends on the size of the unfolded net, even if we extend the optimized version that exists for ordinary Petri nets. But such an algorithm does not exploit at all the structure of the color functions, which corresponds to particular structures of the graph of the unfolded net. Actually, these functions often have a regular structure that can be used to optimize the application of removal rules. We present two examples of this optimization in the next section.

2.3 Two Introductory Examples

In this section, we present two structures of ordinary Petri nets that correspond to usual color functions, and we show how these color functions can be exploited to improve the efficiency of algorithms for finding structural deadlocks.

The first case where we can benefit from the color functions is the case where several places of the unfolded net can be eliminated at the same time, because their input transitions have the same input places. This is what we call "parallel deduction". We show on an example how it works. Let us consider the colored Petri net in Figure 1. The removal rule associated with the transition of the colored net is:

$$[P1, \varepsilon] \Rightarrow [P2, X]$$

But when considering the unfolded net, we immediately remark that places P2a, P2b and P2c have the same set of input places, namely place P1. Hence, instead of trying the possible assignments of variable X, if we already know that P1 does not belong to a deadlock, we could eliminate all of the instances of P2 at the same time.

To do so, we should rewrite the former removal rule as:

$$[P1, \varepsilon] \Rightarrow [P2, C1.all]$$

where function C1.all is defined by : $\quad C1.all(x) = \bigcup_{x \in C1} \{x\}$

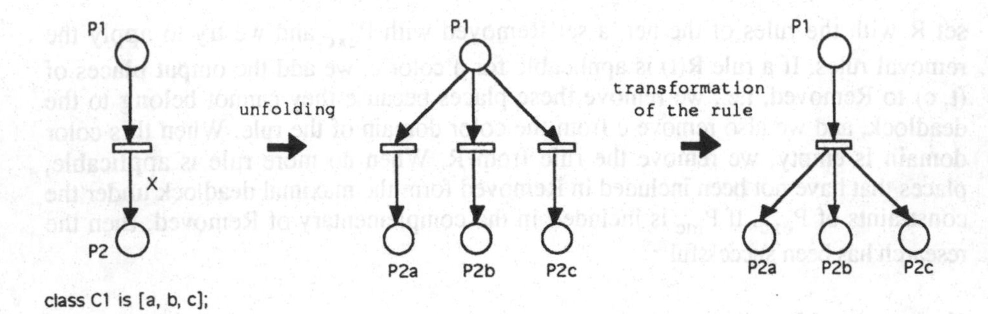

Fig. 1: Parallel deduction

The applicability of this new rule does not depend on any variable, whereas the original rule required the assignment of X. Hence, the number of applications of the rule is divided by the cardinality of the class on which X is defined.

We consider now the colored net in Figure 2, whose graph contains a loop. According to Property 2.2, we know that (P, a) cannot belong to a deadlock, and then according to Definition 2.13, the removal rule associated with T1 is:

$$\texttt{True} \Rightarrow \texttt{[P, a]}$$

If we consider now transition T2, the associated removal rule is:

$$\texttt{[P, X]} \Rightarrow \texttt{[P, X++1]}$$

where `X++1` stands for the successor function, modulo the color class. By trying successively the different assignments of this rule, we deduce that if (P, a) does not belong to a deadlock, then neither (P, b) nor (P, c) belongs to the deadlock. In fact, the successive applications of the rule are equivalent to finding a deadlock in the unfolded net. These successive applications are what we call "iterative deduction".

However, if we try to iterate the application of the rule before any assignment, then we immediately find that this rule can be replaced by the following rule:

$$\texttt{[P, X]} \Rightarrow \texttt{[P, C.all]}$$

for which only one application is necessary in order to obtain the wanted information. The application of this new rule, combined with the application of the rule associated with T1 ensures that no occurrence of P belongs to a deadlock. With this rule, the number of steps of the removal algorithm is divided by the cardinality of the color domain of T2.

In fact, the transformation of the rule is based on the cyclic structure of the graph of the colored net. One lap in the colored cycle provides information, through the color functions, on the color instances of places that must be excluded from the deadlock. This information is immediately reused to try to obtain more results.

Hence, by a combination of the color functions that appear in the cycle, we obtain directly the information given by a repeated application of the original rule.

These two examples show that for particular structures of the unfolded net, represented by specific structures of both the color functions and the graph of the net, the deadlock computation algorithm can be improved.

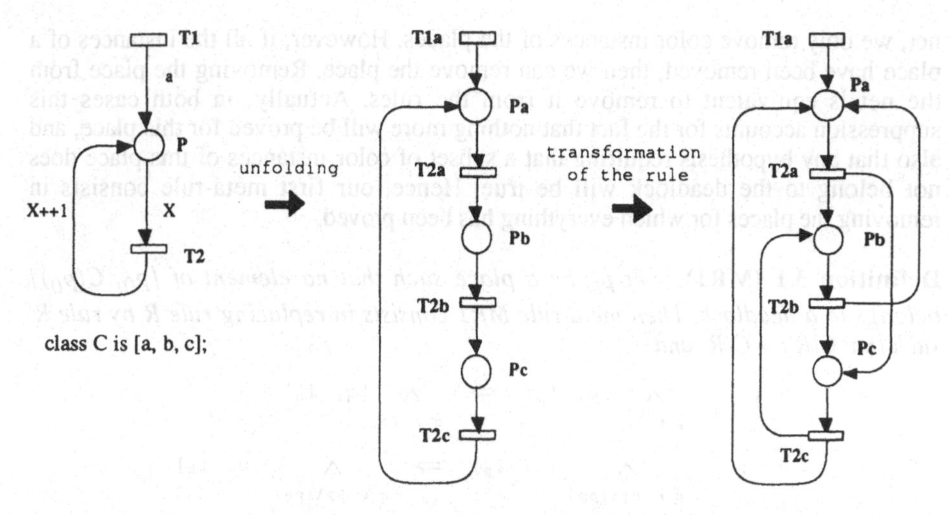

Fig. 2: Iterative deduction

What we have presented on the examples is in fact a very general approach, namely transforming the removal rules of the unfolded net by applying meta-rules on them. In order to obtain automatic transformations, we develop and formalize the approach in the next section.

3 Removal Rules and Meta-Rules

The aim of the meta-rules is to rewrite removal rules in order to improve their application. At every step of the computation of a deadlock, one application of a rule must provide as much information as possible. The computation of a deadlock relies on the initial creation of one rule per transition of the net. A meta-rule is a transformation that applies to a rule R, or to a set of rules, and that produces a new rule R', or a set of new rules. A meta-rule can modify not only the rule, but also its color domain, i.e., the possible assignments of the variables that appear in the rule. The application of a meta-rule transforms a rule in such a way that one application of the new rule corresponds to a series of applications of the original rule in the unfolded net. The efficiency of the approach is thus linked to the number of ordinary rules that have been replaced by this transformation.

As removal rules correspond to transitions, we can identify a set of rules with the colored net from which they have been written. The modification of a rule that results from the application of a meta-rule can be seen as a transformation of the associated Petri net. Based on this identification, we extend to rules notions that exist for colored nets. The skeleton of a rule will be the set of places that appear in the rule, disregarding the color instances that are associated with these places. We will also use p• to denote the set of rules for which p appears in the hypothesis.

3.1 Basic Meta-Rules

In an ordinary Petri net, the algorithm for computing a structural deadlock consists in removing the places that cannot belong to the deadlock. In a colored Petri net, we only remove color instances of the places. However, if all the instances of a place have been removed, then we can remove the place. Removing the place from the net is equivalent to remove it from the rules. Actually, in both cases this suppression accounts for the fact that nothing more will be proved for this place, and also that any hypothesis requiring that a subset of color instances of this place does not belong to the deadlock will be true. Hence, our first meta-rule consists in removing the places for which everything has been proved.

Definition 3.1 (MR1) *Let p_0 be a place such that no element of $[p_0, C(p_0)]$ belongs to a deadlock. Then meta-rule MR1 consists in replacing rule R by rule R' such that C(R') = C(R) and:*

$$R : \bigwedge_{p \in P_1} [p, f_p] \Rightarrow \bigwedge_{q \in P_2} [q, f_q]$$

$$R' : \bigwedge_{p \in P_1 \setminus \{p_0\}} [p, f_p] \Rightarrow \bigwedge_{q \in P_2 \setminus \{p_0\}} [q, f_q]$$

Once MR1 has been applied, some rules may no longer have a conclusion. These rules can be removed because they will give no further information.

Definition 3.2 (MR2) *Let R be a rule whose conclusion is empty. Then meta-rule MR2 consists in removing R.*

The two former meta-rules do not depend on the color functions that appear around the transition corresponding to rule R. Hence, they can be applied on the skeleton of the colored net as well. We present now two meta-rules that improve the removal process by using the structure of the colored net through the color functions.

3.2 Parallel Deduction

The following meta-rule corresponds to the parallel deduction process. As we have shown on the example of Section 2.3, we can define a meta-rule exploiting the case where the hypothesis of a rule is partially independent of the conclusion. The meaning of this meta-rule can be explained as follows. We recognize on the color functions around a transition that, in the unfolded net, some transitions have the same input places. The semantics of the meta-rule is then to substitute all these transitions by a unique transition, whose input places are those common input places and whose output places are the union of the output places of the transitions. And the substitution is performed at the colored net level.

In order to give a formal expression of this meta-rule, we introduce two functions:

Definition 3.3 *Let C1 and C2 be two sets.*

- *The projection π of $C1 \times C2$ on C1 is defined by:* $\pi(\langle x, y\rangle) = x$
- *Σ is a function on C1 that defines the sum over the elements of C2:*

$$\Sigma(x) = \bigcup_{y \in C2} \{\langle x, y\rangle\}$$

Consider now a rule R with a domain $C(R) = C1 \times C2$, where C1 and C2 can in turn be Cartesian products of color sets. For a partial independence of the hypothesis and the conclusion, we want a condition on the color functions in the hypothesis, such that the hypothesis is verified independently of the assignment of the variable in C2.

A necessary and sufficient condition is that the functions of the hypothesis can be written as $g_p = f_p \circ \pi$, where f_p is a function defined on C1. After the application of the meta-rule, the composition of function f_p with a projection is no longer necessary, as the domain of the rule has been reduced to C1.

In the conclusion of rule R, a function f_q that applies to $C1 \times C2$ is associated to each place q. But the aim of the meta-rule is to obtain a new rule, whose domain is only C1 and whose conclusion is the union of the conclusions obtained for all the assignments of the variable in C2. Hence, once an assignment has been done for the variable in C1, we first compute all the couples that associate this assignment to an element of C2. This is done by applying function Σ. Then we apply f_q to all these couples.

Definition 3.4 (MR3) *Let R be a rule whose domain can be written as $C(R) = C1 \times C2$, and whose expression is:*

$$R : \bigwedge_{p \in P1} [p,\ f_p \circ \pi] \Rightarrow \bigwedge_{q \in P2} [q,\ f_q]$$

Then the meta-rule MR3 produces a new rule R' whose domain is $C(R') = C1$, and whose expression is :

$$R' : \bigwedge_{p \in P1} [p,\ f_p] \Rightarrow \bigwedge_{q \in P2} [q,\ f_q \circ \Sigma]$$

However, the problem is to detect in which cases this meta-rule can be applied. The easiest to detect and also the most frequent case is when the expression of a rule contains in the conclusion a variable that does not appear in the hypothesis. The application of the meta-rule then consists in replacing this variable by the constant representing all the elements of the domain of the variable.

We now present the last meta-rule that corresponds to the iterative deduction process.

3.3 Iterative Deduction

The following meta-rule corresponds to the iterative deduction process. As we have shown on the example of Section 2.3, we can define a meta-rule exploiting the case where there exists a circuit such that each of its transition has only one input place in the graph of the Petri net. We call external places w.r.t. a circuit places that are connected to a transition of the circuit but do not belong to the circuit. We are thus interested in circuits without external input places.

Moreover, if the functions valuating the input arcs of the transitions of the circuit are all identity functions, then all the transitions of the circuit only have one input place in the unfolded net. Hence, if a place of the unfolded circuit does not belong to the deadlock, then no descendant of this place obtained by a path meeting only transitions of the circuit can belong to the deadlock.

The meaning of the meta-rule can be explained as follows. We substitute to the output places of a transition the set of descendant places of this transition. This set of places can be obtained by computing the transitive closure of the subnet under consideration. An algorithm for this computation can be found in [18]. It uses the transitive closure of a color function that we recall now.

Definition 3.5 *Let E and F be two sets, f be a function $\mathcal{P}(E) \rightarrow \mathcal{P}(F)$. The transitive closure f^* of f is defined by:*

$$c' \in f^*(c) \iff \exists\, n \geq 0 \text{ such that } c' \in f^n(c)$$

where $f^n = f \circ \ldots \circ f$ (n times).

For the sake of clarity, we first present a simplified version of the meta-rule. In this version, we only consider a circuit without external output places.

Let $p_0, \ldots, p_{n-1}$ be the places belonging to the circuit, and let f_i be the color function valuating the input arc of place p_{i+1}. Starting from an instance c of p_i, we can reach the descendant instances c' of p_j that are such that:

$$c' \in f_{j-1} \circ \ldots \circ f_i(c)$$

Hence, as the graph is cyclic, from an instance c of p_i, we may reach all the instances c' of p_i that are such that

$$c' \in f_{i-1} \circ \ldots \circ f_i(c)$$

But these instances may in turn reach instances c" such that

$$c'' \in f_{i-1} \circ \ldots \circ f_i(c'), \text{ i.e., } c'' \in f_{i-1} \circ \ldots \circ f_i \circ f_{i-1} \circ \ldots \circ f_i(c)$$

By repeating the process, from an instance c of p_i, we can reach the descendant instances c' of p_i that are such that

$$c' \in (f_{i-1} \circ \ldots \circ f_i)^*(c)$$

As know how to compute the colors that can be reached from these colors of p_i for any place p_j belonging to the circuit , we can now give the general expression of the meta-rule for a circuit without external output places.

Definition 3.6 (simplified MR4)
Let $p_0, \ldots, p_{n-1}$ be n places belonging to a cycle of the colored net. Let t_i be the output transition of p_i in the cycle, and R_i be the rule associated with t_i. If R_i has the following expression:

$$R_i : \quad [p_i, id] \Rightarrow [p_{i+1}, f_i]$$

the application of the meta-rule transforms R_i in R'_i such that:

$$R'_i : \quad [p_i, id] \Rightarrow \bigwedge_{j=0}^{n-1} [p_j, f_{j-1} \circ \ldots \circ f_i \circ (f_{i-1} \circ \ldots \circ f_i)^*]$$

where the operations on the indices are performed modulo n.

The introduction of external output places does not change the expression of the color functions, but these places must now appear in all the rules. The meta-rule becomes:

Definition 3.7 (MR4)
Let $p_0, \ldots, p_{n-1}$ be n places belonging to a cycle of the colored net. Let t_i be the output transition of p_i in the cycle, and P_i be the set of output places of t_i that do not belong to the cycle. Let R_i be the rule associated to t_i. If R_i has the following expression:

$$R_i : \quad [p_i, id] \Rightarrow [p_{i+1}, f_i] \quad \bigwedge_{q \in P_i} [q, g_q]$$

the application of the meta-rule transforms R_i in R'_i such that:

$$R'_i : \quad [p_i, id] \Rightarrow \bigwedge_{j=0}^{n-1} \Big\{ [p_j, f_{j-1} \circ \ldots \circ f_i \circ (f_{i-1} \circ \ldots \circ f_i)^*] \bigwedge_{q \in P_j} [q, g_q \circ f_{j-1} \circ \ldots \circ f_i \circ (f_{i-1} \circ \ldots \circ f_i)^*] \Big\}$$

where the operations on the indices are performed modulo n.

We have considered only identity functions on input arcs. The result can be easily extended to the case where the functions on input arcs are bijective, by using the same kind of transformation as for reductions in colored nets [19].

4 An Algorithm to Compute Structural Deadlocks

4.1 Presentation and Definition of the Algorithm

This algorithm differs from Algorithm 2 on the two following points:

- The occurrence of places in rules is used only at the "skeleton" level.
- The structures of both the color functions and the skeleton are exploited by meta-rules MR3 and MR4.

More precisely, the algorithm works as follows. The first step (instructions 1-7) initializes the color instances of places that are excluded from the deadlock, the rules to be examined - those with a hypothesis containing a place for which at least one color is excluded from the deadlock - and tries to apply the meta-rules once.

The main loop (instructions 8-20) then applies one rule at a time with all the possible assignments, updating the excluded color instances of places and the rules to be examined. The main optimization (instructions 15-17) is the detection of the application of the meta-rules; due to the test in (15), the execution of instruction 16 always applies at least one meta-rule (see below). In this loop, all the control instructions work at the skeleton level and colors only appear in the application of an ordinary rule. The end of the algorithm (instructions 21-22) is identical to that of Algorithm 2. The algorithm is developed below.

Notation Given E, some subset of PC, we need to know the subset of colors of a place included in E. So we introduce the following notation: $E.p = \{c \mid (p, c) \in E\}$

Abstract Algorithm 3

```
(1)  Compute the elementary circuits of the skeleton
(2)  Removed := P_exc
(3)  To_Examine := Ø
(4)  For all place p such that Removed.p ≠ Ø do
(5)       Insert p• in To_Examine
(6)  done
(7)  Apply the Meta-rules on Removed
(8)  While To_Examine ≠ Ø Do
(9)       Extract R from To_Examine
(10)      A := {c | R is applicable for c on Removed}
(11)      For all c in A do
(12)           Apply R for c on Removed ;
(13)           For all p such that Removed.p has increased do
(14)                Insert p• in To_Examine;
(15)                If Removed.p = C(p) then
(16)                     Apply the Meta-rules
(17)                endif
(18)           done
(19)      done
(20) done
(21) Deadlock := PC \ Removed
(22) If P_inc ⊆ Deadlock and Deadlock ≠ Ø then
          return (success, Deadlock)
     else return (failure).
```

Explanations and details of implementation

(1) In order to apply meta-rule MR4, we compute all the elementary circuits of the skeleton. We choose to compute the circuits before eliminating the places in P_{exc} because the result of this computation can be used for different problems on the same net, such as the research of deadlocks and traps, or the research of deadlocks with different conditions, namely different sets P_{exc}.

The computation of the circuits can be done in a time proportional to the product of the size of the skeleton and the number of circuits [18]. Anyway, this time is independent of the size of the color domains, which is the relevant criterion of complexity for colored nets. To test efficiently if a circuit fulfils the condition of MR4, we associate to each circuit the number of external input places and we link each place to the circuits of which it is an external input place. Each time we apply MR1, we update these numbers and see if new applications of MR4 are possible.

(3),(5),(8),(9),(14) `To_Examine` is a set-type variable which provides a termination test to the algorithm. Again the updating of this variable is related to the skeleton of the set of rules. Each time a color of a place is added to Removed, the rules with this place appearing in the hypothesis are selected for examination. With a link between a place and each rule where the place appears in the hypothesis, the updating of `To_Examine` is quick.

(10),(12) These steps are the most time-consuming ones and they depend on the color domains. Many heuristics, used in simulation techniques, (and already applicable to Algorithm 2) optimize this step but the complexity remains color domain dependent. *But the aim of Algorithm 3 is to reduce the color domain of the rules and to transform the conclusions of the rules in such a way that the cost of testing is minimal and the information brought by the application of the rule is maximal.*

(13),(15) The evolution of Removed.p can easily be memorised with two counters, one for the current cardinality and one for the preceding one. Hence these two tests are a comparison of integers. If the test of instruction (15) is successful, we already know that we can apply the deletion meta-rule MR1 on place p, and possibly any of the three other meta-rules:

- MR2 becomes applicable if p was the last element of the conclusion of a rule
- MR3 becomes applicable if p appeared in the hypothesis of a rule and determined part of the assignment. This can be detected by a very simple syntactic analysis of the color functions, such as for instance the vanishing of a variable in the hypothesis of a rule. The next section will give an illustration of this.
- MR4 becomes applicable if p was the last external input place of a circuit.

One can see that the overhead time added by the test of the meta-rules is minimal and once again independent of the color domain.

4.2 Application: The Dining Philosophers

The net in Figure 3 models the dining philosophers, in the case where they do not take both forks at the same time. Initially all the philosophers are `Thinking` and all `Forks` are free. When a philosopher `X` wants to eat, he takes his right fork (`X++1`) and `Waits` for his left fork (`X`). If his left fork is free, then he `Eats`. When he stops eating, he releases the two forks and starts thinking again. We will use C to denote the set of philosophers (and of forks).

We are looking for possibly deficient, i.e., insufficiently marked deadlocks. For the philosophers net, such a deadlock may appear with waiting philosophers.

So we look for deadlocks that do not contain place Wait, and we start our algorithm with P_{exc} = [Wait, C(Wait)] and $P_{inc} = \emptyset$. The rules associated with transitions T1, T2 and T3 respectively are:

```
R1    [Think, X] ∧ [Forks, X++1] ⇒ [Wait, X]
R2    [Wait, X] ∧ [Forks, X] ⇒ [Eat, X]
R3    [Eat, X] ⇒ [Forks, X + X++1] ∧ [Think, X]
```

We first (instruction 1) compute the elementary circuits and we find three ones:

(Think, Wait, Eat), (Forks, Wait, Eat), (Forks, Eat)

Then we initialize (instruction 2) Removed with [Wait, C] and (instructions 3-6) we update To_Examine with rule R2. We now apply the meta-rules (instruction 7): MR1 is applicable on Wait, and the rules become:

```
R1    [Think, X] ∧ [Forks, X++1] ⇒ True
R2    [Forks, X] ⇒ [Eat, X]
R3    [Eat, X] ⇒ [Forks, X + X++1] ∧ [Think, X]
```

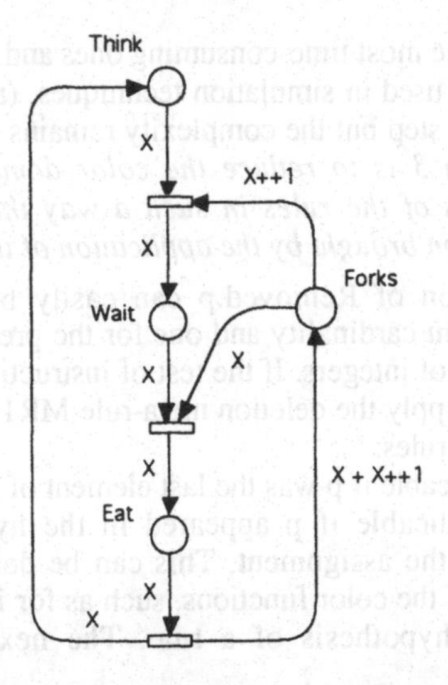

Fig. 3: The dining philosophers with forks taken separately.

MR2 is applicable on R1 and the rules become:

```
R2    [Forks, X] ⇒ [Eat, X]
R3    [Eat, X] ⇒ [Forks, X + X++1] ∧ [Think, X]
```

MR4 is applicable on the circuit (Forks, Eat). As X^* is X and $(X + X{+}{+}1)^*$ is `C.All`, the rules become:

```
R2    [Forks, X] ⇒ [Forks,C.all] ∧[Eat,C.all] ∧ [Think,C.all]
R3    [Eat, X] ⇒[Eat,C.all] ∧ [Forks,C.all] ∧ [Think,C.all]
```

The net in Figure 4 graphically describes the two rules. At instruction 7, To_examine contains R2, so we extract it and try to apply it.

However, as Removed.forks is empty, no assignment is possible and A is empty (instruction 10). Thus we exit from the main loop and return success with the following deadlock:

```
[Eat, C.all] ∧ [Forks, C.all] ∧ [Think, C.all]
```

This deadlock contains all the forks. Is there a deadlock that contains a strict subset of forks ? To answer this question, we initialize P_{exc} = [Wait, C(Wait)] ∪ [Forks, c] and $P_{inc} = \varnothing$ where c is an arbitrary color. All the steps until instruction 10 are identical to the former ones.

We find now that rule R2 is applicable for c and its application updates Removed to PC. Then instructions 15-16 delete all places and rules. The algorithm finishes with an empty deadlock, hence returning failure.

It must be noted that the most efficient implementation of Algorithm 2 would have required 2.n (n is the number of philosophers) applications of rules, whereas Algorithm 3 requires only one application !

For a last illustration of Algorithm 3, we test if our first deadlock contains traps.

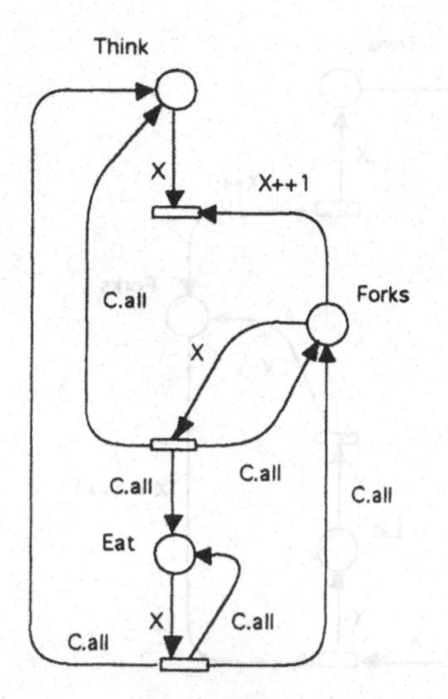

Fig. 4: Transformed net after one application of MR1, MR2 and MR4

So we start our algorithm with the reversed net and with P_{exc} = [Wait, C(Wait)] and $P_{inc} = \emptyset$. The model is presented in Figure 5. The rules become:

```
R1    [Wait, X] ⇒ [Think, X] ∧ [Forks, X++1]
R2    [Eat, X] ⇒ [Wait, X] ∧ [Forks, X]
R3    [Forks, X + X++1] ∧ [Think, X] ⇒ [Eat, X]
```

The circuits are the reverse of the preceding circuits. To_Examine is initialized with R1. Let us look at the application of the meta-rules: MR1 deletes Wait and the rules become:

```
R1    True ⇒ [Think, X] ∧ [Forks, X++1]
R2    [Eat, X] ⇒ [Forks, X]
R3    [Forks, X + X++1] ∧ [Think, X] ⇒ [Eat, X]
```

Now MR3 is applicable on rule R1, the decomposition of the domain is a particular case C = {ε} x C and is detected as variable X appears in the conclusion and does not appear in the hypothesis. Thus the rules become:

```
R1    True ⇒ [Think, C.all] ∧ [Forks, C.all]
R2    [Eat, X] ⇒  [Forks, X]
R3    [Forks, X + X++1] ∧ [Think, X] ⇒ [Eat, X]
```

The loop is executed once with an application of R1; Removed is updated with [Think, C] and [Forks, C], so we test the application of the meta-rules. MR1 deletes places Think and Forks; MR2 deletes rules R1 and R2; MR3 transforms rule R3 in:

```
R3    True ⇒ [Eat, C.all]
```

The second execution of the loop applies rule R3 which updates Removed to PC. The application of meta-rules is detected again and place Eat and rule R3 are deleted. Then the algorithm exits from the loop and returns failure since Deadlock is empty. Hence, the deadlock excluding place Wait we had obtained does not contain a trap.

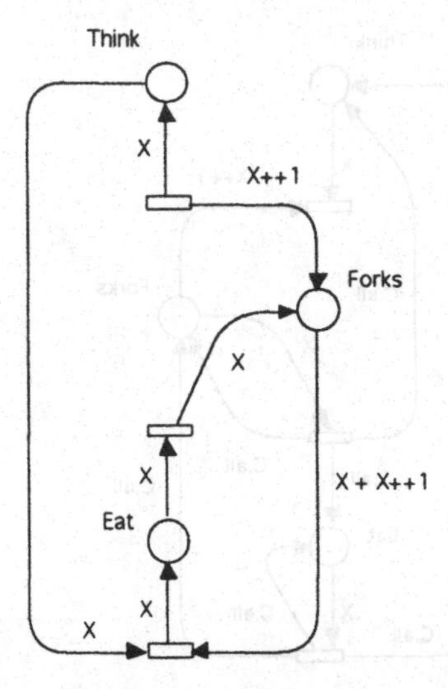

Fig. 5: Search for traps, Wait excluded

It must be noted that the most efficient implementation of Algorithm 2 would have required 2.n (n being the number of philosophers) applications of rules, whereas Algorithm 3 requires only two applications !

4.3 Comparison Between Algorithms 2 and 3

The efficiency of Algorithm 3 compared with Algorithm 2 depends on the number of applications of meta-rules MR3 and MR4 along its execution. So it is difficult to give any theoretical measure of the complexity. In the worst case, the overhead time added to Algorithm 3 is negligible compared with the assignment of the rules, and thus the complexity is in the same order.

However, to estimate the average complexity, we can observe that MR3 becomes applicable as soon as a place conditioning the assignment of a transition disappears, and this happens frequently when the algorithm is applied to colored nets modelling real systems. Moreover MR4 is based on the existence of circuits, which are numerous in the skeleton of a colored net, especially when liveness and boundedness are required. The additional constraints are usually not fulfilled initially. Nevertheless the deletion of places increases the possibility of satisfying the structural condition (no external input places) and the transformation of the color domains by meta-rule MR3 yields the occurrence of the functional condition (existence of identities).

One application of MR3 followed by the application of the transformed rule corresponds to n applications of the initial rule where n is the size of the vanishing color domain. One application of MR4 followed by the application of the transformed rule corresponds to at least one application of all the rules of the circuit. In fact as soon as the output functions of the circuit are different from identity, the reduction factor is proportional to the product of a color domain by the length of the circuit.

We point out that in many cases the computation is parameterized: it remains valid for a family of models where only the size of the color classes changes. Thus the deadlock characterization we perform on the model of the philosophers is independent of the number of philosophers. Such a characterization would have been impossible with Algorithm 2. We now plan to develop a parameterized version of Algorithm 3 for well-formed nets [20].

Last but not least, the results are easier to interpret. Their expression is only a function of the high-level description, and thus uses the same notations that have been given by the designer to describe the model. Unlike our algorithm, Algorithm 2 provides an extensive representation of deadlocks.

All the results can be easily transposed for the detection of traps, as we have shown in the example of the philosophers.

5 Conclusions

In this paper we have presented an efficient algorithm for finding deadlocks and traps in colored nets. The algorithm exploits both the structure of the net and the structure of the color functions. We have shown on an example how the meta-rules speed up the computation of colored deadlocks. The efficiency of the algorithm strongly depends on the number of times meta-rules can be applied. The first experiments have shown that in most cases, the conditions of application are fulfilled. We are now working on the integration of the algorithm in the CASE AMI [21] in order to obtain statistical results on the efficiency of the algorithm.

A forthcoming work is the specialization of this algorithm for syntactically well defined nets, with a complexity almost independent of the size of the color domains. After characterizing classes of colored nets for which some structural property is a necessary or sufficient liveness condition, this algorithm will allow us to decide liveness for such nets.

References

[1] W.Reisig: *Place-Transition Systems.* In Petri Nets: Central models and their properties, W.Brauer, W.Reisig and G.Rozenberg eds., LNCS n° 254, Springer-Verlag, 1987, pp 117-141.

[2] E.W. Mayr: *An Algorithm for the General Petri Net Reachability Problem.* In SIAM. Journal of Computing n° 13, 1984.

[3] E. Best: *Structure Theory of Petri Nets: the Free Choice Hiatus.* In Petri Nets: Central models and their properties, W.Brauer, W.Reisig and G.Rozenberg eds., LNCS n° 254, Springer- Verlag, 1986, pp 168-205.

[4] J. Martinez, M. Silva: *A Simple and Fast Algorithm to Obtain all Invariants of a Generalized Petri Net.* In Informatik Fachberichte n° 52, C.Girault and W.Reisig eds., Springer-Verlag, 1982, pp 301-310.

[5] K.Lautenbach: *Linear Algebraic Techniques for Place / Transition Nets.* In Petri Nets: Central models and their properties, W.Brauer, W.Reisig and G.Rozenberg eds., LNCS n° 254, Springer- Verlag, 1986, pp 142-167.

[6] F. Commoner: *Deadlocks in Petri nets.* In Applied Data Res. Inc., Wakefield, MA, 1972.

[7] J. Esparza, M. Silva: *A Polynomial-Time Algorithm to Decide Liveness of Bounded Free-Choice Nets.* Hildesheimer Informatikberichte 12/90, Institut für Informatik, Univ. Hildesheim.

[8] K.Lautenbach: *Linear Algebraic Calculation of Deadlocks and Traps.* In Concurrency and Nets, K.Voss, H.Genrich and G.Rozenberg eds., Springer Verlag, 1987, pp 315-336.

[9] K.Barkaoui, M.Minoux: *A Polynomial-Time Graph Algorithm to Decide Liveness of Some Basic Classes of Bounded Petri Nets.* In Application and Theory of Petri Nets 92, Proc. of the 13th Conference, K. Jensen ed., LNCS n° 616, Springer-Verlag, Sheffield, UK, 1992, pp 62-74.

[10] H.J. Genrich: *Predicate / Transition Nets.* In High-level Petri Nets. Theory and Application, K. Jensen and G. Rozenberg eds., Springer-Verlag, 1991, pp 3-43.

[11] K. Jensen: *Coloured Petri Nets: A High Level Language for System Design and Analysis.* In High-level Petri Nets. Theory and Application, K. Jensen and G. Rozenberg eds., Springer-Verlag, 1991, pp 44-119.

[12] J. Ezpeleta, J.M. Couvreur: *A New Technique for Finding a Generating Family of Siphons, Traps and ST-Components. Application to Colored Petri Nets.* In proc. of the 12th International Conference on Application and Theory of Petri Nets, Gjern, Denmark, June 1991, pp 145-164.

[13] J.M. Couvreur, S. Haddad, J.F. Peyre: *Computation of Generative Families of Positive Flows in Coloured Nets.* In proc. of the 12th International Conference on Application and Theory of Petri Nets, Gjern, Denmark, June 1991.

[14] J.F. Peyre: *Résolution Paramétrée de Systèmes Linéaires. Application au Calcul d'Invariants et à la Génération de Code Parallèle.* Thèse de l'Université Paris 6, March 1993 (in French).

[15] M. Minoux, K. Barkaoui: *Deadlocks and Traps in Petri Nets as Horn-Satisfiability Solutions and some Related Polynomially Solvable Problems.* Discrete Applied Mathematics n° 29, 1990.

[16] J. Vautherin: *Parallel Systems Specifications with Coloured Petri Nets and Algebraic Specifications.* In Advances in Petri Nets 1987, Springer-Verlag, 1987, pp 293-308.

[17] G. Findlow: *Obtaining Deadlock-Preserving Skeletons for Coloured Nets,* in Application and Theory of Petri Nets 92, Proc. of the 13th Conference, K. Jensen ed., LNCS n° 616, Springer-Verlag, Sheffield, UK, 1992, pp 173-192.

[18] D.B. Johnson: *Finding all Elementary Circuits of a Directed Graph,* SIAM J.Computer, vol.4, n° 1, 1975.

[19] S. Haddad: *A Reduction Theory for Coloured Nets.* In High-level Petri Nets. Theory and Application, K. Jensen and G. Rozenberg eds., Springer-Verlag, 1991, pp 399-425.

[20] G. Chiola, C. Dutheillet, G. Franceschinis, S. Haddad: *Stochastic Well-Formed Nets and Symmetric Modeling Applications,* to appear in IEEE Transactions on Computers.

[21] J.M. Bernard, J.L. Mounier, N. Beldiceanu, S. Haddad: *AMI an Extensible Petri Net Interactive Workshop,* Proc. of the 9th European Workshop on Application and Theory of Petri Nets, Venice, Italy, June 1988.

Synthesis of Net Systems

Luca Bernardinello
Università degli Studi di Milano
Dipartimento di Scienze dell'Informazione
via Comelico 39, I-20135 Milano
e-mail: bernardinello@hermes.unimi.it

Abstract. Transition Systems are among the most general mathematical structures used to describe the behaviour of systems, both sequential and concurrent. Ehrenfeucht and Rozenberg identified a basic notion for this structures, and called it *region*. They used it to define a construction yielding a Net System whose behaviour can be described by a given Transition System. In this paper, properties of the set of regions of Transition Systems are investigated and an alternative construction is defined.

Topics: theory of concurrency, analysis and synthesis, structure and behaviour of nets.

1 Introduction

The behaviour of a concurrent system is often described operationally as a transition system, that is a set of states connected by a state-transition relation. Each transition is labelled by a symbol denoting the event whose occurrence causes the change of state. On the other hand, net theory gives a formalism to describe the structure of a distributed system. One of the main mathematical tools to describe the behaviour of a net is its sequential case graph, which is a transition system. Elementary Net (EN) Systems, first defined in [5] are one of the basic net-based models. They allow to formulate in a neat and rigorous way the fundamental situations arising in concurrent systems, such as concurrency, conflict and sequence. Elementary Transition Systems were introduced in [1] and shown to correspond in a very strong sense to sequential case graphs of Elementary Net Systems. In the same paper the authors defined a procedure to synthesize an Elementary Net System exhibiting a behaviour given in abstract form as an Elementary Transition System. The net systems built in this way are called saturated. The construction makes use of the notion of region. Regions are sets of states of a transition system characterized by a property which relates them to event occurrences. They can be interpreted as local states of the system whose behaviour is described by the transition system, hence they correspond to conditions in net systems. The saturated net version of a transition system owes its name to the fact that it is 'saturated' with respect to conditions: adding a condition to this net system changes its behaviour. From certain points of view some of the conditions in a saturated net system can be considered as redundant, since one can remove them without changing the behaviour. Hence it is interesting to look for different ways of constructing net systems from a given transition system.

In this paper properties of the set of regions of a transition system are investigated. Results of this analysis suggest an alternative synthesis procedure, based on the set of minimal (with respect to set inclusion) regions. The net systems constructed in this way have some nice properties: they are contact-free and decomposable in state-machine components.

The paper is organized as follows. In section 2 we recall the basic definitions related to Elementary Transition Systems and prove some properties of the set of regions of an Elementary Transition System. In section 3 the definition of Elementary Net Systems is recalled. Section 4 presents the synthesis of the saturated net system associated to an Elementary Transition System and the synthesis of the corresponding minimal-regions version. Finally, section 5 discusses the results obtained and sketches future developments.

2 Transition Systems

Within the general context of transition systems it is possible to characterize specific classes of structures corresponding to physical or conceptual system models. In this section we focus attention on two models of transition systems, namely Elementary and Condition/Event Transition Systems, which were introduced by Ehrenfeucht and Rozenberg (see [1] and [2]), and present some of their basic properties.

Definition 2.1 *Transition System*
A Transition System is a structure $A=(S,E,T)$, where S is a set of states, E is a set of events, $T \subseteq S \times E \times S$ is the set of transitions.

A Transition System is *finite* if S and E are finite.

We consider finite Transition Systems satisfying the following properties:

1. $\forall (s,e,s') \in T \quad s \neq s'$
2. $\forall (s,e_1,s_1),(s,e_2,s_2) \in T \quad [s_1 = s_2 \Rightarrow e_1 = e_2]$
3. $\forall e \in E \; \exists (s,e,s') \in T$

The notion of region is basic in studying properties of transition systems. It was introduced in [1]. A region is a set of states such that every occurrence of a given event must have the same crossing relation (entering, leaving or non crossing) with respect to the region itself, and this property holds for all events. For the rest of this section let $A = (S,E,T)$ be a fixed Transition System.

Definition 2.2 *region*
A set of states $r \subseteq S$ is said to be a region iff:
$\forall (s_1,e,s_{1'})(s_2,e,s_{2'}) \in T : (s_1 \in r$ and $s_{1'} \notin r) \Leftrightarrow (s_2 \in r$ and $s_{2'} \notin r)$ and $(s_1 \notin r$ and $s_{1'} \in r) \Leftrightarrow (s_2 \notin r$ and $s_{2'} \in r)$

Remark. It is easy to verify that, for each Transition System, both the empty set and the set of all states are regions. They are called the trivial regions. We are interested in the set of non-trivial regions, which will be denoted by R_A. For each $s \in S$, R_s will denote the set of non-trivial regions containing s.

Example 1. Consider the transition system in figure 1. The set $\{3,4\}$ is a region. On the other hand, $r = \{1\}$ is not a region, since an occurence of $e1$ "leaves" r, while another occurrence of the same event neither "leaves" nor "enters" r.

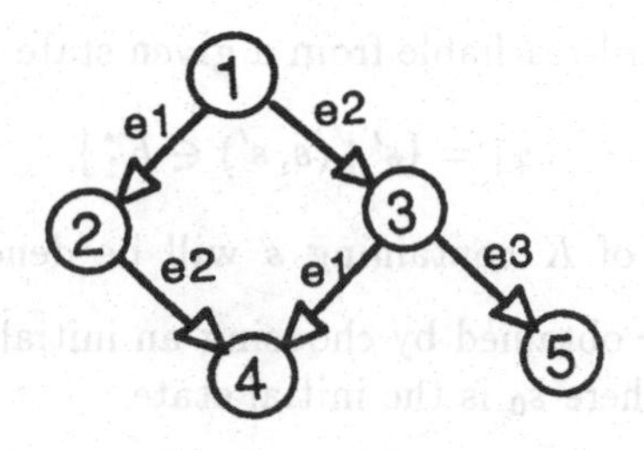

Fig. 1. An Elementary Transition System

The condition defining regions allows us to define a relation between events and regions formalizing the crossing relation. This is captured by the notions of pre- and post-sets of regions and of pre- and post-regions of events.

Definition 2.3 *pre- and post-sets, pre- and post-regions*
Let r be a region of A. Then the pre-set of r, denoted by $^\bullet r$, and the post-set of r, denoted by $r^\bullet$, are defined by:

$$^\bullet r = \{e \in E \mid \exists (s,e,s') \in T : s \notin r \text{ and } s' \in r\}$$
$$r^\bullet = \{e \in E \mid \exists (s,e,s') \in T : s \in r \text{ and } s' \notin r\}$$

Let $e \in E$. Then the set of pre-regions and the set of post-regions of e, denoted by, respectively, $^\bullet e$ and $e^\bullet$, are defined by:

$$^\bullet e = \{r \mid r \in R_A \text{ and } \exists (s,e,s') \in T \quad s \in r \text{ and } s' \notin r\}$$
$$e^\bullet = \{r \mid r \in R_A \text{ and } \exists (s,e,s') \in T \quad s \notin r \text{ and } s' \in r\}$$

Example 2. Consider again the transition system in figure 1. Its regions are:

$$r_1 = \{1,2\} \; r_6 = \{3,4,5\}$$
$$r_2 = \{1,3\} \; r_7 = \{2,4,5\}$$
$$r_3 = \{2,4\} \; r_8 = \{1,3,5\}$$
$$r_4 = \{3,4\} \; r_9 = \{1,2,5\}$$
$$r_5 = \{5\} \quad r_{10} = \{1,2,3,4\}$$

Some examples of pre- and post-sets of regions:

$$^\bullet r_4 = \{e_2\} \quad ^\bullet r_7 = \{e_1, e_3\} \quad r_7^\bullet = \emptyset$$

In order to distinguish different classes of Transition Systems, we introduce some 'reachability' relations.

Definition 2.4 *Reachability relations*

1. $(s_1, s_2) \in K_1 \Leftrightarrow \exists (s_1, e, s_2) \in T$
2. $K := (K_1 \cup K_1^{-1})^*$

The set of states forward-reachable from a given state s will be denoted by $s\uparrow$:

$$s\uparrow := \{s' \mid (s, s') \in K_1^*\}$$

The equivalence class of K containing s will be denoted by $[s]_K$. Initialized Transition Systems are obtained by choosing an initial state. They are denoted by $A = (S, E, T, s_0)$, where s_0 is the initial state.

Definition 2.5 *isomorphism*
Let $A_1 = (S_1, E_1, T_1, s_{01})$ and $A_2 = (S_2, E_2, T_2, s_{02})$ be two Initialized Transition Systems. They are said isomorphic iff there exist two bijections, $\beta : S_1 \rightarrow S_2$ and $\eta : E_1 \rightarrow E_2$ such that:

1. $\beta(s_{01}) = s_{02}$
2. $\forall (s, e, s') \in T_1 \quad [(\beta(s), \eta(e), \beta(s') \in T_2]$
3. $\forall (s, e, s') \in T_2 \quad [(\beta^{-1}(s), \eta^{-1}(e), \beta^{-1}(s') \in T_1]$

Now we are ready to give the definition of Elementary Transition (ET) System.

Definition 2.6 *Elementary Transition System*
An Initialized Transition System $A = (S, E, T, s_0)$ is an Elementary Transition System, shortly ET System, iff it satisfies the following axioms:

1. $s\uparrow = S$
2. $\forall s, s' \in S[R_s = R_{s'} \Rightarrow s = s']$
3. $\forall s \in S \forall e \in E[^\bullet e \subseteq R_s \Rightarrow \exists s' \in S \quad (s, e, s') \in T]$

Another type of Transition System can be defined by dropping axiom 1, which requires reachability of all states from an initial state, and adding a 'backward closure' axiom, symmetric to axiom 3. This class was introduced and discussed in [1]. Here we call its elements *Condition Event Transition Systems* because it corresponds, in a sense to be precised later, to Condition Event Net Systems.

Definition 2.7 *Condition Event Transition System*
A Transition System $A = (S, E, T)$ is a Condition Event Transition System, denoted CET System, iff it satisfies the following axioms:

1. $\forall s \in S \;\; [s] = S$
2. $\forall s, s' \in S \quad [R_s = R_{s'} \Rightarrow s = s']$
3. $\forall s \in S \forall e \in E \quad [{}^\bullet e \subseteq R_s \Rightarrow \exists s' \in S \quad (s, e, s') \in T]$
4. $\forall s \in S \forall e \in E \quad [e^\bullet \subseteq R_s \Rightarrow \exists s' \in S \quad (s', e, s) \in T]$

In order to clarify the difference between the two models, consider the transition systems in figure 2. A_1 is an Elementary Transition System if we choose 1 as initial state. However, it is not a Condition Event Transition System. In fact, it is easy to see that $F^\bullet \subseteq R_9$, while there is no F-labelled transition leading into 9. If we add a new state, like in A_2, then the backward-closure axiom is satisfied, hence A_2 is a CET System. Notice that it is not Elementary, since the axiom requiring reachability from an initial state can be satisfied by no choice of initial state.

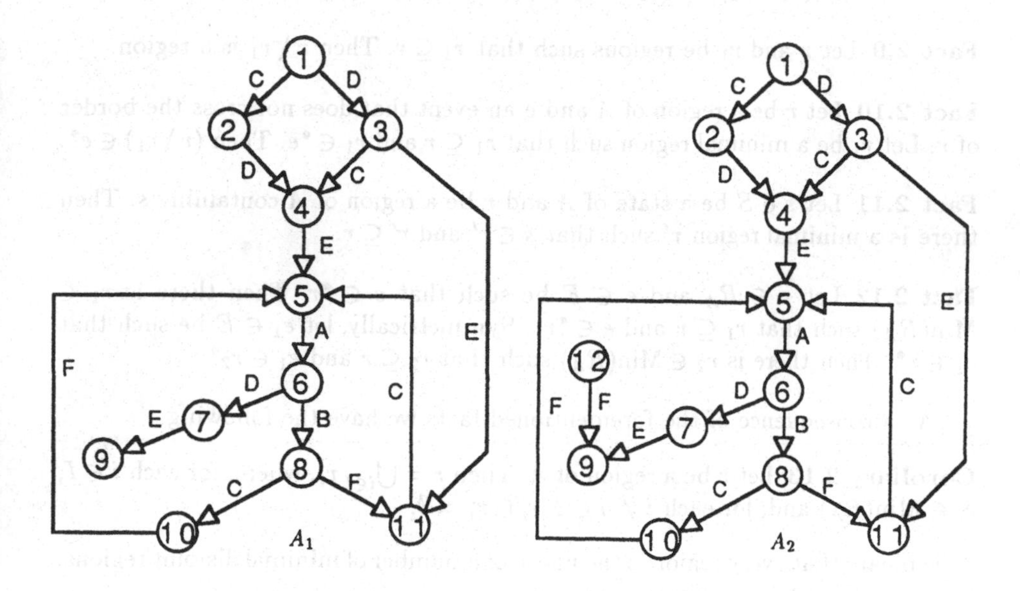

Fig. 2. ET Systems vs CET Systems

2.1 On the Algebraic Structure of the Set of Regions

The set of regions of a Transition System is a partially ordered set, with respect to set inclusion. We will now state some simple algebraic properties of this set. They provide a background for the results in the next section.

Throughout this section we consider a fixed ET system $A = (S, E, T, s_0)$.

Remark. All the results presented in this section hold also for CET Systems and can be proved in the same way, since proofs do not use the axiom of reachability from the initial state.

In the following $\mathrm{Min}(R_A)$ will denote the set of minimal regions with respect to set inclusion, that is:

$$\mathrm{Min}(R_A) = \{r \mid r \in R_A \text{ and } [\forall r' \in R_A[r' \neq r \Rightarrow r' \not\subseteq r]]\}$$

The following properties can be easily proved by verification.

Fact 2.8 Let r_1 and r_2 be two disjoint regions of A. Then $r_1 \cup r_2$ is a region. Moreover,

$$^\bullet(r_1 \cup r_2) = ({}^\bullet r_1 \cup {}^\bullet r_2) \setminus (({}^\bullet r_1 \cap r_2{}^\bullet) \cup ({}^\bullet r_2 \cap r_1{}^\bullet))$$

$$(r_1 \cup r_2)^\bullet = (r_1{}^\bullet \cup r_2{}^\bullet) \setminus (({}^\bullet r_1 \cap r_2{}^\bullet) \cup ({}^\bullet r_2 \cap r_1{}^\bullet))$$

Fact 2.9 Let r and r_1 be regions such that $r_1 \subseteq r$. Then $r \setminus r_1$ is a region.

Fact 2.10 Let r be a region of A and e an event that does not cross the border of r. Let r_1 be a minimal region such that $r_1 \subseteq r$ and $r_1 \in {}^\bullet e$. Then $(r \setminus r_1) \in e^\bullet$.

Fact 2.11 Let $s \in S$ be a state of A and r be a region of A containing s. Then there is a minimal region r' such that $s \in r'$ and $r' \subseteq r$.

Fact 2.12 Let $r \in R_A$ and $e \in E$ be such that $e \in {}^\bullet r$. Then there is $r_1 \in \mathrm{Min}(R_A)$ such that $r_1 \subseteq r$ and $e \in {}^\bullet r_1$. Symmetrically, let $e_1 \in E$ be such that $e_1 \in r^\bullet$. Then there is $r_2 \in \mathrm{Min}(R_A)$ such that $r_2 \subseteq r$ and $e_1 \in r_2{}^\bullet$.

As a consequence of the forementioned facts we have the following

Corollary 2.13 Let r be a region of A. Then $r = \bigcup_{i \in I} r_i$, where, for each $i \in I$, $r_i \in \mathrm{Min}(R_A)$ and, for each $i \neq j \in I, r_i \cap r_j = \emptyset$.

This means that every region is the union of a number of minimal disjoint regions. In general the decomposition is not unique.

3 Net Systems

In this section we briefly recall the basic definitions related to Elementary Net Systems (shortly EN Systems) and Condition/Event Systems (shortly CE Systems). The main reference for the former class is [5]; for CE Systems see [3].

Definition 3.1 *net*
A net is a triple $N = (B, E, F)$ where B and E are finite sets such that: $B \cap E = \emptyset, B \cup E \neq \emptyset$, $F \subseteq (B \times E) \cup (E \times B)$ is the flow relation, $\mathrm{dom}(F) \cup \mathrm{ran}(F) = B \cup E$.

B is the set of *conditions*, E is the set of *events*, F is the *flow relation*.

We will use the standard graphical notation for nets.

Definition 3.2 *pre- and post-sets*
Given a net $N = (B, E, F)$ and an element $x \in B \cup E$ we define the pre- and post-set of x, denoted by, respectively, ${}^\bullet x$ and $x^\bullet$, and the neighbourhood of x, denoted by ${}^\bullet x^\bullet$:

$$ {}^\bullet x = \{y \mid (y, x) \in F\} $$
$$ x^\bullet = \{y \mid (x, y) \in F\} $$
$$ {}^\bullet x^\bullet = {}^\bullet x \cup x^\bullet $$

This notation is extended to subsets of $B \cup E$ in a standard way.

Definition 3.3 *simple, connected, strongly connected net*
A net $N = (B, E, F)$ is:

- *simple* iff $\forall x, y \in B \cup E \quad {}^\bullet x = {}^\bullet y \text{ and } x^\bullet = y^\bullet \Rightarrow x = y$
- *connected* iff $\forall x, y \in B \cup E \quad (x, y) \in (F \cup F^{-1})^*$
- *strongly connected* iff $\forall x, y \in B \cup E \quad (x, y) \in F^*$

Definition 3.4 *subnet*
Let $N_i = (B_i, E_i, F_i)$ be a net for $i = 1, 2$. Then N_1 is a subnet of N_2 iff:

- $B_1 \subseteq B_2, E_1 \subseteq E_2$
- $F_1 = F_2 \cap ((B_1 \times E_1) \cup (E_1 \times B_1))$

Let $X \subseteq B_2$. Then the subnet of N_2 generated by X is

$$ N_X = (X, {}^\bullet X^\bullet, F_2 \cap ((X \times {}^\bullet X^\bullet) \cup ({}^\bullet X^\bullet \times X))) $$

The global state of a net system is described by the set of conditions holding in that state. The occurrence of an event changes the state according to the transition rule defined as follows.

Definition 3.5 *transition rule, transition relation*
Let $N = (B, E, F)$ be a net, $e \in E$ and $c \subseteq B$.

1. e is said to be enabled at c, denoted $c[e>$, iff ${}^\bullet e \subseteq c$ and $e^\bullet \cap c = \emptyset$.
2. If e is enabled at c, then the occurrence of e leads from c to c', denoted $c[e>c'$, iff $c' = (c \setminus {}^\bullet e) \cup e^\bullet$.
3. The transition relation of N, $\longrightarrow_N \subseteq 2^B \times E \times 2^B$ is given by: $\longrightarrow_N = \{(c, e, c') \mid c[e > c'\}$.

Now we can define, in the context of net systems, the reachability relations defined for Transition Systems.

Definition 3.6 *reachability relations for net systems*

1. $(c_1, c_2) \in Q_1 \Leftrightarrow c_1[e>c_2$ for some $e \in E$.
2. $Q_2 := Q_1^*$.
3. $Q := (Q_1 \cup Q_1^{-1})^*$

Note that Q is an equivalence relation (usually it is called the *full reachability* relation); $[c]_Q$ will denote the class of equivalence of c.

3.1 Elementary Net Systems

Definition 3.7 *Elementary Net System*
An Elementary Net (EN) System is a quadruple $\Sigma = (B, E, F, c_0)$ where

1. (B, E, F) is a simple net
2. $c_0 \subseteq B$ is the initial case.

The definition of subnet can be easily raised to the level of net systems by taking into account the initial case.

Definition 3.8 *subsystem*
Let $\Sigma_i = (B_i, E_i, F_i, c_{0i})$ be an EN system for $i = 1, 2$. Then Σ_1 is a subsystem of Σ_2 iff (B_1, E_1, F_1) is a subnet of (B_2, E_2, F_2) and $c_{01} = c_{02} \cap B_1$

Definition 3.9 *set of cases of an EN System*
Let $\Sigma = (B, E, F, c_0)$ be an EN System.
The set of cases of Σ, denoted C_Σ, is the smallest subset of 2^B such that:

1. $c_{in} \in C_\Sigma$
2. $\forall c \in C_\Sigma, \forall c' \subseteq B, \forall e \in E[\ c[e\!>\!c' \Rightarrow c' \in C_\Sigma]$.

The behaviour of an EN system Σ can be operationally described by an initialized transition system whose nodes are the reachable cases of Σ and whose arcs are labelled by events.

Definition 3.10 *sequential case graph*
Let $\Sigma = (B, E, F, c_0)$ be an EN system. The sequential case graph of Σ is the transition system $\mathrm{SCG}(\Sigma) = (C_\Sigma, E, \rightarrow_\Sigma, c_0)$, where $\rightarrow_\Sigma$ is $\rightarrow_N$ restricted to $C_\Sigma \times E \times C_\Sigma$.

Remark. In [1] it is proved that $\mathrm{SCG}(\Sigma)$ is an Elementary Transition System.

Example 3. Figure 3 shows an EN system and its sequential case graph.

The following definition collects properties of Elementary Net Systems.

Definition 3.11 *dynamic properties*
Let $\Sigma = (B, E, F, c_0)$ be an EN System with $\mathrm{SCG}(\Sigma) = (C_\Sigma, E, \rightarrow_\Sigma, c_{in})$.

1. Σ is contact-free iff $\forall e \in E, \forall c \in C_\Sigma[{}^\bullet e \subseteq c \Rightarrow e^\bullet \cap c = \emptyset]$
2. An event $e \in E$ is dead in Σ iff $\nexists c \in C_\Sigma \quad c[e>$
3. Σ is reduced (1-live) iff $\forall e \in E \exists c, c' \in C_\Sigma \quad (c, e, c') \in \rightarrow_\Sigma$

From now on we will consider only reduced EN systems.

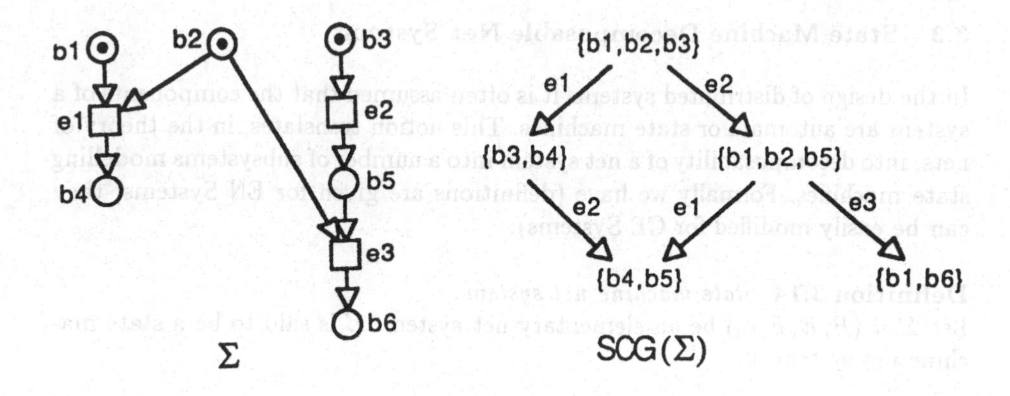

Fig. 3. An EN System and its sequential case graph

3.2 Condition Event Systems

The class of reachable cases of an EN System is defined as the class of subsets of conditions which are forward-reachable from a distinguished initial case. If, instead, we define the reachable cases as an equivalence class of the full reachability relation associated to a net, we get a different model, namely Condition Event Net Systems.

The notions of subsystem, sequential case graph and the properties defined for EN System can be easily rephrased for CE Systems. The only property which requires a little attention is contact-freeness, since now we must take into account the possibility of "backward-enabled" events.

Definition 3.12 *Condition Event System*
A Condition Event (CE) System is a tuple $\Sigma = (B, E, F, C)$, where (B, E, F) is a simple, pure net and C is an equivalence class of the full reachability relation associated to (B, E, F). A CE System can also be denoted in the form $\Sigma = (B, E, F, [c]_Q)$, where c is a representative case.

Definition 3.13 *contact-free CE System*
Let $\Sigma = (B, E, F, C)$ be a CE System with $\mathrm{SCG}(\Sigma) = (C, E, \rightarrow_\Sigma)$. Σ is contact-free iff $\forall e \in E, \forall c \in C$

- $^\bullet e \subseteq c \Rightarrow e^\bullet \cap c = \emptyset$
- $e^\bullet \subseteq c \Rightarrow {}^\bullet e \cap c = \emptyset$

Ehrenfeucht and Rozenberg have proved that the sequential case graph of a CE System is a CE Transition System.

3.3 State Machine Decomposable Net Systems

In the design of distributed systems it is often assumed that the components of a system are automata or state machines. This notion translates, in the theory of nets, into decomposability of a net system into a number of subsystems modelling state machines. Formally we have (definitions are given for EN Systems; they can be easily modified for CE Systems):

Definition 3.14 *State machine net system*
Let $\Sigma = (B, E, F, c_0)$ be an elementary net system. Σ is said to be a state machine net system iff:

1. $\forall e \in E \quad |{}^\bullet e| = 1 = |e^\bullet|$
2. $|c_0| = 1$

Definition 3.15 *State machine decomposable net system*
Let $\Sigma = (B, E, F, c_0)$ be an Elementary Net System. Σ is said to be decomposable into state machine components iff there exists a set $\Sigma_i = (B_i, E_i, F_i, c_{0i}), i = 1, \ldots, m$ of connected subsystems of Σ generated by subsets of conditions such that:

1. $\forall i \in \{1, \ldots, m\} \quad \Sigma_i$ is a state machine net system
2. $B = \bigcup_i B_i, E = \bigcup_i E_i$

The Σ_i are called state machine components (or sequential components) of Σ.

Example 4. Figure 4 shows the state machine components of the net system of Figure 3.

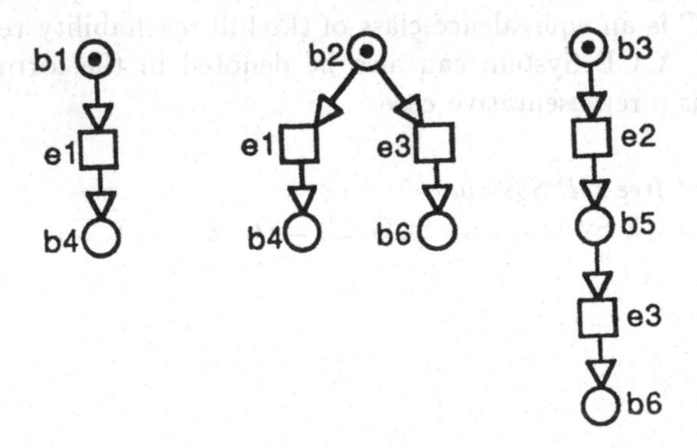

Fig. 4. Three state machine net systems

4 Synthesis

In this section we face the synthesis problem, expressed in the following form: given a transition system A, find a net system Σ such that the sequential case graph of Σ is isomorphic to A. A first solution, applicable to both Elementary and Condition Event Transition Systems, was given in [1]; this solution was further discussed in [2]. The key idea behind it is that regions are actually local states. The net systems constructed according to this procedure have a condition for each region in the Transition System. The flow relation is determined by pre- and post-sets of regions. We will call a net system obtained in this way from a transition system A, the ***saturated net version*** of A. It has a very peculiar property: it is maximal with respect to the number of conditions among all the net systems whose sequential case graph is isomorphic to A. This observation leads to look for alternative synthesis procedures yielding "more economic" nets (see [4]). The section is organized as follows: first we briefly recall the definition of the saturated net version of a transition system. Then we present an alternative synthesis procedure and analyze its properties. In the next section a conjecture concerning a characterization of net systems constructed with this new method will be discussed.

Definition 4.1 *saturated net version*
Let $A = (S, E, T, s_0)$ be an ET system. The saturated net version of A is the structure $\Sigma_A = (R_A, E, F, c_0)$, where
- $F = \{(r, e) \mid r \in {}^\bullet e\} \cup \{(e, r) \mid r \in e^\bullet\}$
- $c_0 = R_{s_0}$

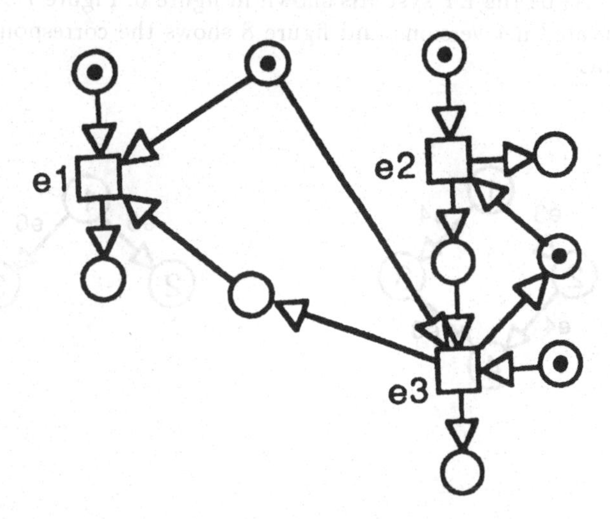

Fig. 5. A saturated net system

Example 5. Let $A = (S, E, T, s_0)$ be the ET system shown in figure 1. Then its saturated net version is the net system in figure 5.

Ehrenfeucht and Rozenberg have proved that A and SCG Σ_A are isomorphic.

Now we define a new operator which associates an EN system to an ET system. Corollary 2.13 suggests that the minimal regions of an ET system contain enough information to construct all regions, hence they implicitly contain all the information about behavioural relations among events. On these grounds, we build the net associated to the ET system A in such a way that the conditions correspond to the minimal regions of A and the flow relation is coherent with pre- and post-sets of regions in A.

From now on the dot notation refers only to regions of transition systems. To denote pre- and post-sets of an element x of a net Σ_N we will use respectively $\text{pre}_N(x)$ and $\text{post}_N(x)$.

Definition 4.2 ***minimal regions version***
Let $A = (S, E, T, s_0)$ be an Elementary Transition System. Then the minimal-regions version of A, denoted by $M(A)$, is the structure $M(A) = (\text{Min}(R_A), E, F, c_0)$, where

$$\forall r \in \text{Min}(R_A) \forall e \in E \ [\ (r, e) \in F \Leftrightarrow e \in r^\bullet \text{ and } (e, r) \in F \Leftrightarrow e \in {}^\bullet r]$$

$$c_0 = \{r \mid r \in \text{Min}(R_A) \text{ and } s_0 \in r\}$$

Remark. $M(A)$ is a subsystem of the saturated net version of A.

Example 6. Let A_1, A_2, A_3 be the ET systems shown in figure 6. Figure 7 shows the corresponding saturated net versions and figure 8 shows the corresponding minimal-regions versions.

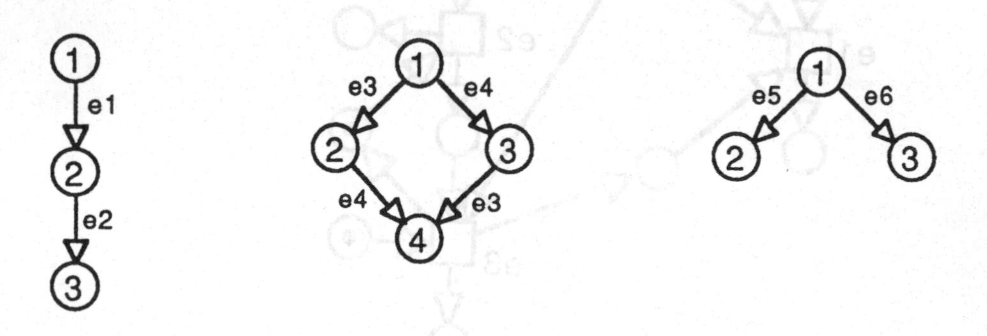

Fig. 6.

The following results are related to the structure of $M(A)$.

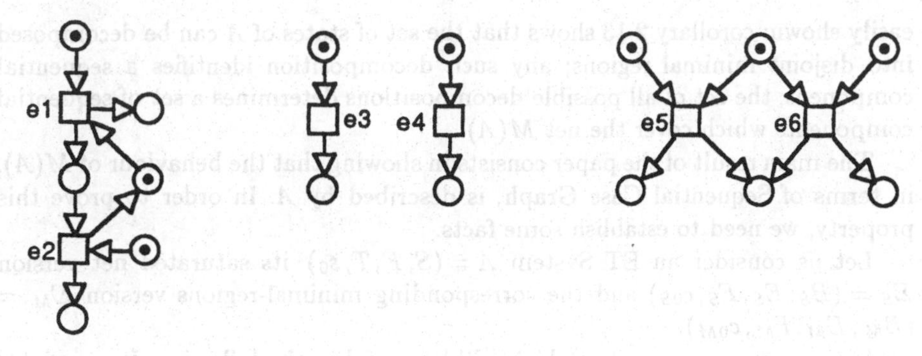

Fig. 7.

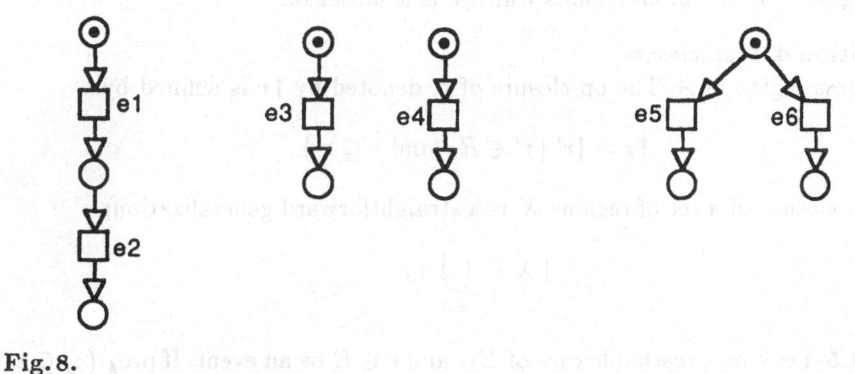

Fig. 8.

Fact 4.3 Let $Z = \{r_1, \ldots, r_k\}$ be a family of regions of A with the following properties:
$\forall i, j \in \{1, \ldots, k\}: \quad r_i \cap r_j = \emptyset$
$\forall r \in R_A: \quad r \notin Z \Rightarrow [\exists r_i \in Z: \quad r \cap r_i \neq \emptyset]$
Then:

1. $\bigcup r_i = S$
2. $\forall e \in E: \mid {}^\bullet e \cap Z \mid \leq 1$ and $\mid e^\bullet \cap Z \mid \leq 1$
3. $\forall e \in E: e \in {}^\bullet r_i \Leftrightarrow \exists j: e \in r_j{}^\bullet$

Proof. 1. Suppose there exists a state $s \in S \setminus \bigcup r_i$; since $\bigcup r_i$ is a region (by Fact 2.8), $S \setminus \bigcup r_i$ is a region too; moreover r_i and $S \setminus \bigcup r_i$ are (obviously) disjoint, hence the hypothesis is contradicted.
2. By contradiction: suppose that $e \in {}^\bullet r_i \cap {}^\bullet r_j$, then for all $(s_1, e, s_2) \in T$ we have $s_2 \in r_i$ and $s_2 \in r_j$, which contradicts the hypothesis.
3. Suppose $e \in {}^\bullet r_i$. Then there is a transition (s_1, e, s_2) and $s_1 \notin r_1, s_2 \in r_i$. By Fact 4.3 there is r_j containing s_1. Obviously $e \in r_j{}^\bullet$. The converse property can be proved analogously.

□

Properties 2 and 3 state that Z, interpreted as a set of conditions in $M(A)$, generates a state machine component. Now decomposability of $M(A)$ can be easily shown: corollary 2.13 shows that the set of states of A can be decomposed into disjoint minimal regions; any such decomposition identifies a sequential component; the set of all possible decompositions determines a set of sequential components which cover the net $M(A)$.

The main result of the paper consists in showing that the behaviour of $M(A)$, in terms of Sequential Case Graph, is described by A. In order to prove this property, we need to establish some facts.

Let us consider an ET System $A = (S, E, T, s_0)$ its saturated net version $\Sigma_S = (B_S, E_S, F_S, c_{0S})$ and the corresponding minimal-regions version $\Sigma_M = (B_M, E_M, F_M, c_{0M})$.

Let us define an operator which will be useful in the following. It associates to a region r the set of all regions which r is a subset of.

Definition 4.4 *up-closure*
Let r be a region of A. The up-closure of r, denoted by $\uparrow r$ is defined by:

$$\uparrow r = \{r' \mid r' \in R_A \text{ and } r \subseteq r'\}$$

The up-closure of a set of regions X is a straightforward generalization:

$$\uparrow X = \bigcup_{r \in X} \uparrow r$$

Fact 4.5 Let c be a reachable case of Σ_M and $e \in E$ be an event. If $\text{pre}_M(e) \subseteq c$ then $\text{pre}_S(e) \subseteq \uparrow c$. If $\text{post}_M(e) \subseteq c$ then $\text{post}_S(e) \subseteq \uparrow c$

Proof. Let $r \in \text{pre}_S(e)$. By Fact 2.12 there is a minimal region $r' \subseteq r$ such that $r' \in {}^\bullet e$. Then $r' \in \text{pre}_M(e)$, hence $r' \in c$ and $r \in \uparrow c$. The second thesis can be proved analogously. □

Fact 4.6 $c_{0S} = \uparrow c_{0M}$.

Proof. Take $r \in c_{0S}$. If r is minimal, then $r \in c_{0M}$, hence $r \in \uparrow c_{0M}$. If r is not minimal, then, by Fact 2.11, there is a minimal region $r\prime$ such that $s_0 \in r\prime$ and $r\prime \subset r$, so $r\prime \in c_{0M}$ and $r \in \uparrow r\prime$.
Take $r \in \uparrow c_{0M}$; $s_0 \in r$, so $r \in c_{0S}$. □

Fact 4.7 Let b be a minimal region of A and c be a reachable case of Σ_S containing b. Then $\uparrow b \subseteq c$.

Proof. Every reachable case of Σ_S corresponds to a state of A. Let $s \in S$ be the state of A corresponding to c. In [2] it is proved that c is the set of the regions of A containing s, hence $s \in b$. Every region r such that $b \subseteq r$ ($r \in \uparrow b$) contains s, hence $r \in c$. □

Corollary 4.8 For each reachable case c of Σ_S we have $c = \uparrow \text{Min}(c)$, where $\text{Min}(c) := c \cap \text{Min}(R_A)$.

Fact 4.9 Let c be a reachable case of Σ_M such that $\uparrow c$ is reachable in Σ_S. If $c \xrightarrow{e} c_1$ in Σ_M, then $\uparrow c \xrightarrow{e} \uparrow c_1$ in Σ_S.

Proof. The proof is divided into two parts. First we show that, if e is enabled at c in Σ_M, then it is enabled at $\uparrow c$ in Σ_S. Take $b \in \mathrm{pre}_S(e)$. If $b \in \mathrm{Min}(R_A)$, then $b \in c$, hence $b \in \uparrow c$. If b is not minimal, then, by Fact 2.12, there is a minimal region r such that $r \subseteq b$ and $e \in {}^\bullet r$, hence $b \in \uparrow r$; since $r \in c$ we have $b \in \uparrow c$. This proves that $\mathrm{pre}_S(e) \subseteq \uparrow c$. Since Σ_S is contact-free and $\uparrow c$ is reachable in Σ_S by hypothesis, we can conclude that e is enabled at $\uparrow c$.
Suppose that $\uparrow c \xrightarrow{e} c_2$. It is easy to verify that $\uparrow c_1 \subseteq c_2$. We will prove the inclusion in the other direction. We have

$$c_2 = (\uparrow c \setminus \mathrm{pre}_S(e)) \cup \mathrm{post}_S(e)$$

Take $b \in \mathrm{post}_S(e)$. By Fact 2.12 there is a minimal region r such that $r \subseteq b$ and $e \in {}^\bullet r$. Then $r \in c_1$, hence $b \in \uparrow c_1$. Take $b \in (\uparrow c \setminus \mathrm{pre}_S(e))$. If b is minimal then $b \in c_1$, hence $b \in \uparrow c_1$. If b is not minimal, then there is a minimal region r such that $r \in c$ and $r \subseteq b$. Consider separately two cases:

- $r \notin {}^\bullet e$. In this case $r \in c_1$, hence $b \in \uparrow c_1$
- $r \in {}^\bullet e$. By Fact 2.10 $(b \setminus r) \in e^\bullet$, hence, by Fact 2.12, there is a minimal region r_1 such that $r_1 \subseteq b$ and $r_1 \in e^\bullet$, so $r_1 \in c_1$ and $b \in \uparrow c_1$.

□

Remark. From Facts 4.5 and 4.9 and from contact-freeness of Σ_S it follows that $M(A)$ is contact-free.

Fact 4.10 If $c_{OM} \xrightarrow{w} c$ in Σ_M, then $c_{OS} \xrightarrow{w} \uparrow c$ in Σ_S.

Proof. The thesis follows from Fact 4.6 and from Fact 4.9. □

Fact 4.11 For each reachable case c of Σ_S, the following holds: if $c \xrightarrow{e} c_1$ in Σ_S, then $\mathrm{Min}(c) \xrightarrow{e} \mathrm{Min}(c_1)$ in Σ_M.

Proof. By definition $\mathrm{pre}_M(e) = \mathrm{Min}(\mathrm{pre}_S(e))$ and $\mathrm{post}_M(e) = \mathrm{Min}(\mathrm{post}_S(e))$. □

Fact 4.12 If $c_{OS} \xrightarrow{w} c$ in Σ_S, then $c_{OM} \xrightarrow{w} \mathrm{Min}(c)$ in Σ_M.

Proof. Obviously $c_{OM} = \mathrm{Min}(c_{OS})$. Then the thesis can be proved by induction using Fact 4.11. □

The results of the last part of this section can be summarized in the following

Theorem 4.13 Let $A = (S, E, T, s_0)$ be an Elementary Transition System. Let Σ_S and Σ_M be, respectively, its saturated net and minimal-regions version. Then $\mathrm{SCG}(\Sigma_S), \mathrm{SCG}(\Sigma_M)$ and A are all isomorphic.

4.1 Condition Event Systems

Theorem 4.13 has been stated in the context of Elementary Transition Systems and, correspondingly, Elementary Net Systems; however the minimal-regions version can be defined also for Condition Event Transition Systems. It is not hard to see that all the results, suitably reformulated, hold also in this case. Since we have a bijective correspondence between states of the transition system and reachable cases of the net, it suffices to substitute an arbitrarily selected state for the initial state and consider both backward and forward transition in proofs carried out by induction.

5 Discussion

We have mentioned the fact that the saturated net version of an ET, or CE, System can be characterized by a "maximality" property. We would like to have a similar, possibly dual, characterization for the minimal-regions version. It is easy to see that we cannot simply state the dual property. Consider, for instance, the first net in figure 8. If we remove the output place of e_2, then the behaviour does not change. However, the new net system is no more contact-free, while we could be interested in net systems satisfying certain given properties, for instance contact-freeness. Let us call a net system *optimal* with respect to a certain set of properties if removing a condition from the underlying net causes loss of some of the properties.

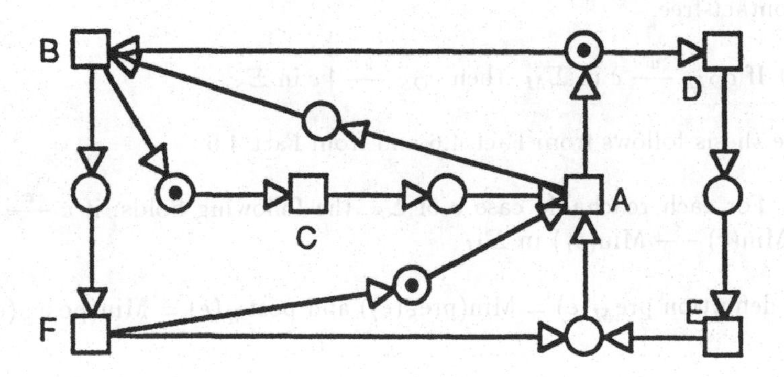

Fig. 9.

Following [4], let us call *admissible* a set of regions of a transition system A if the net system generated by that set of regions has a sequential case graph isomorphic to A. Furthermore, let us call *CF-admissible* a set of regions if it is admissible and the net system it generates is contact-free. One could conjecture that the minimal regions version of an ETS is optimal with respect to contact-freeness. The ET System A_1 in figure 2 is a counterexample; its minimal-regions version is shown in figure 9. If we remove the condition b with ${}^\bullet b = \{F\}$ and

$b^\bullet = \{A\}$, the behaviour does not change and the new net system is still contact-free; however, if we "backward-close" the Transition System, transforming it into a CET System, and then compute the minimal-regions version, we get the net system of figure 9 without the condition b with $^\bullet b = \{F\}$ and $b^\bullet = \{A\}$, and this is an optimal Net System. Hence it is still an open problem whether, in the context of CET Systems and CE Net Systems, the minimal-regions version is optimal with respect to CF-admissibility.

Acknowledgments
This work has been supported by the Esprit Basic Research Working Group 6067 CALIBAN.
The author wishes to thank Lucia Pomello, Giorgio De Michelis and Giovanni Guariglia for fruitful discussions and four anonymous referees for their valuable comments.

References

1. A.Ehrenfeucht, G.Rozenberg: *Partial (Set) 2-Structures. I and II*, in: Acta Informatica, Vol. 27, No. 4, pp. 315-368 (1990).
2. M.Nielsen, G.Rozenberg, P.S.Thiagarajan: *Elementary Transition Systems*, in: Theoretical Computer Science, vol. 96, No. 1, pp. 3-33 (1992).
3. W.Reisig: *Petri Nets*, EATCS Monographs, Springer-Verlag, (1985).
4. W.Reisig: *Towards "handier" solutions of the synthesis problem*, invited talk at the REX Concurrencydag, Leiden, 2 October 1992.
5. G.Rozenberg, P.S.Thiagarajan: *Petri nets: basic notions, structure, behaviour*, in: J.W.de Bakker, W.P.de Roever, G.Rozenberg (eds), Current trends in concurrency, pp 585-668, Springer-Verlag, 1986.

Hierarchies in Colored GSPNs

Peter Buchholz

Informatik IV, Universität Dortmund
P.O. Box 50 05 00, 4600 Dortmund 50
Germany

Abstract. Hierarchies are integrated in the class of Generalized Colored Stochastic Petri Nets (GCSPNs). The use of hierarchies supports the specification of nets describing large real-world systems. Moreover, the new model class can be analysed extremely efficient according to qualitative and quantitative results. Techniques for quantitative analysis, qualitative analysis and subnet aggregation are introduced. The usability of the approach is shown by means of a non-trivial example from literature.

1 Introduction

Stochastic Petri Nets have become very popular for performance analysis of complex systems. Especially the class of Generalized Stochastic Petri Nets (GSPNs) is proposed in the literature as a good modeling tool [1]. GSPNs have been subsequently extended in particular by the introduction of token colors to Generalized Colored Stochastic Petri Nets (GCSPNs) [9]. Additionally the class of Stochastic Well-formed colored Nets (SWNs) allowing a parametrized specification and possibly a more efficient analysis due to model symmetries has been introduced [8].

The use of colors in GSPNs allows a more compact representation of models including symmetric parts. GCSPNs and SWNs are a first substantial step toward an adequate model world for complex systems. However, colored nets still describe a flat view of a model, the possibility of hiding details is restricted to the folding of symmetric parts of the model. A second step toward a tool for the handling of complex systems would be the integration of hierarchies as stated in [16]. The use of hierarchies to handle complexity is common in system modeling and design. The advantages of hierarchical models from a specification point are obvious, since they

- hide details in a well defined way.
- separate the model into well defined components.
- allow to reuse components.
- support top-down and bottom-up design.
- can be integrated naturally in a graphical interface.
- allow the modification of a model without starting from scratch.

Hierarchies in untimed Petri Nets have been investigated in various papers (e.g., [10,16]). These approaches consider mainly the static net structure and not the dynamic behavior. In the area of GSPNs there are several attempts to integrate hierarchies (e.g., [2,3,11]). All these approaches are restricted to specific model classes or analysis techniques and are not formalized. Recently a class of hierarchical GCSPNs (HGCSPNs) has been introduced by the author [7]. It has been shown that the hierarchical structure of the net is reflected in the underlying Markov chain and that the state space and generator matrix can be computed by combining state spaces and matrices of isolated system parts. Using this approach beside the specification also the model analysis is supported, since

- qualitative analysis can be made more efficient by first analyzing and reducing subnets and afterward generating a reduced marking space for the overall net.
- vanishing markings can be eliminated efficiently inside isolated subnets.
- iterative numerical solution techniques become more efficient and allow to extend the dimension of an analyzable Markov chain by a factor 10 compared with the conventional solution approach (or around 2 million states on contemporary workstations).
- several aggregation techniques based on loosely coupled subnets or symmetric model parts can be integrated.
- identical subnets in a model need to be handled only once.

Therefore the use of hierarchies is not only useful for specification convenience, it is also essential for an efficient analysis. In this paper we extend the class of HGCSPNs significantly by allowing complicated interactions between subnets and embedding higher level net and vice versa. Although a formal introduction of HGCSPNs is given, our main goal is to provide a framework for a modular specification and analysis of complex models.

The paper is organized as follows: First we describe an abstract view of a hierarchical net and extend the class of GCSPNs to include an abstract view of subnets and an environment for a subnet. In the third section the generation and structure of the marking space and Markov chain underlying a HGCSPN is introduced. Section 4 investigates very briefly analysis algorithms. In Sect. 5 an examples is given.

2 HGCSPNs

Before we introduce hierarchies, flat GCSPNs should be defined. The class of GCSPNs includes beside colors for places and transitions also priorities among immediate transitions and inhibitor arcs. We assume that the reader is familiar with colored Petri nets and colored GSPNs as defined elsewhere [9,15] and restrict the introduction to the pure definition.

Definition 1. A generalized colored stochastic Petri net is defined as a 9-tuple: $GCSPN = (P, T, C, W^+, W^-, W^h, \pi, m_0, Y)$, where

- P is a finite set of places,
- T is a finite set of transitions, such that $P \cap T = \emptyset$, $P \cup T \neq \emptyset$,
- $C : P \cup T \rightarrow \mathcal{C}, \mathcal{C} \cap P = \emptyset, \mathcal{C} \cap T = \emptyset$ is a function mapping $P \cup T$ on a non-empty set of colors,
- $W^-, W^+, W^h : P \times T \rightarrow F$ are the input, output and inhibition functions, where F is the union of the null function with the set of functions $[C(P) \times C(T) \rightarrow \mathbb{N}]$,
- $\pi : T \rightarrow \mathbb{N}$ is the priority function, mapping transitions into natural numbers,
- $m_0 : [P \rightarrow [C(P) \rightarrow \mathbb{N}]]$ is the initial marking,

– $Y : T \rightarrow [C(T) \times M \rightarrow \mathbb{R}^+]$ is a possibly marking dependent weight function associating a positive real number with each transition, where M is a set of markings.

$C(p)$ (resp. $C(t)$) is the color domain of p (resp. t).

In contrast to most other papers markings are denoted by small letters. We only consider bounded nets in this paper. Thus, a marking m can be interpreted as a $\|P\| \times \|C\|$ dimensional vector including the number of color c token on place p in position $m(p, c)$. The lexicographical ordering of markings is well defined and used for the comparison of different markings. Capital letters M ($M(j)$) specify the set of all marking (on place j).

2.1 An Abstract View of the Hierarchy

The goal of any kind of hierarchy is to divide a complex model into several smaller parts and to include in each part only detailed local information and, possibly, a coarse view of the rest of the model. We assume here a hierarchy which is structured like a tree and covers only two levels. The unique *top level net* is numbered with 0, subnets are numbered from 1 to J. The restriction to two levels is mainly caused by the lack of space. Using more than two levels provides no particular problems apart from a more complex notation, the underlying ideas are very similar.

The idea of hierarchical modeling proposed here requires that the marking space and transitions matrices of parts of a model (i.e., a subnet or the *top level net*) are generated in isolation and are latter combined. Since the marking space of a subnet (or the *top level net*) depends on the environment (or the subnets included), an aggregated view of the behavior of the environment (subnets) is needed which covers influences on the part of the system (subnet or *top level net*) that is considered in detail. We will come back to assumption after extending GCSPNs for the description of an aggregated view on non-local model parts.

2.2 Extension to GCSPNs for Hierarchical Modeling

The class of GCSPNs has to be extended to describe an abstract view of a subnet in the *top level net* and to specify an abstract view of an environment in a subnet. In a hierarchically structured system subsystems are rather autonomous in their behavior, therefore it is not necessary to allow arbitrarily complex interactions between subnets and their environment. We assume here that a subnet is influenced by its environment only through tokens entering from the environment and the environment is affected by the subnet only through leaving tokens.

Since both, environment and lower level subnets, describe a coarse view on non-local model parts, they can be specified by similar constructs. In both cases we need a virtual place to include a marking that describes an abstract view of the marking of non-local parts and a virtual transition specifying variations in the visible marking of non-local parts and the departure of tokens from the virtual place. Obviously a virtual transition has to be specified according to its qualitative behavior, the quantitative behavior depends on the whole system and should not be included in the specification of an isolated model part. Since we have assumed that subnets and environment are autonomous, the only input and inhibitition place of a virtual transition is the virtual place belonging to this virtual transition and each virtual place belongs to exactly one virtual transition and is not input or inhibitition place for any other transition. Let p_j and t_j be a virtual place and transition, respectively. The qualitative behavior of the virtual transition is similar to the behavior of an ordinary transition. In marking $m(p_j)$ t_j can fire for color $c \in C(t_j)$ removing $W^-(p_j,t_j)(c',c)$ color c' token from p_j and adding $W^+(p_i,t_j)(c'',c)$ color c'' token to place p_i. p_i can be either an ordinary place or another virtual place. We often use the shorthand notation $m(j)$ for $m(p_j)$ and $W^-(p_j,t_i)(c)$ $(W^+(p_j,t_i)(c))$ for the multiset of tokens that is added to (removed from) p_j by firing t_i in c.

To describe the qualitative behavior of a virtual transition we have to assign a priority to it. We assume that a virtual transition has priority 0 if it describes a subnet in the *top level net* and a priority lower than any local immediate transition if it describes the environment of a subnet. It follows that in a subnet tokens cannot depart and arrive immediately and the priority of immediate transitions in the *top level net* is lower than the priority of immediate transitions in the subnets. Both assumptions are reasonable from the modeling viewpoint.

Another extension of the behavior of virtual places has shown to be useful during the specification of example nets. With the definition up to now each token that is fired onto p_j from the outside is visible in the marking of a virtual place p_j since $W^+(p_j,t_i)(c',c)$ is added to $m(p_j,c')$. However, often different token colors need not to be distinguished inside the subnet and therefore also not in the marking of the virtual place. Suppose for example that the subnet models a part of a distributed system and entering tokens describe messages from other components which answer previous requests. Let c_{succ} and c_{miss} be two colors describing successful or missed requests. If we further assume that successful and missed requests inside the subnet are only distinguished by the point where they start (i.e., a successful request starts with a new cycle and a missed request with a new attempt to call the remote component), then it is sufficient that the marking of the virtual place includes only the number of remaining requests which can be sent. Therefore the arrival of color c_{succ} and color c_{miss} tokens cause the same change in the marking of place p_j. To specify this behavior a function $\mathcal{W}_j(W^+(p_j,t_i)(c'),m(j))$ is defined for each virtual place p_j that has as result a multiset including the number of tokens of colors from $C(p_j)$ to be added to the marking $m(j)$. After this informal description of virtual transitions and places, we give a formal definition.

Definition 2. A virtual transition t_j and a virtual place p_j in a GCSPN are defined by

- $C(t_j)$ and $C(p_j)$ the color domains,
- $\forall c \in C(t_j), c' \in C(p_i)$, if $W^-(p_i,t_j)(c',c) \neq 0$ or $W^h(p_i,t_j)(c',c) \neq 0$, then $p_i = p_j$,
- $\forall c \in C(p_j), c' \in C(t_i)$, if $W^-(p_j,t_i)(c,c') \neq 0$ or $W^h(p_j,t_i)(c,c') \neq 0$, then $t_i = t_j$,
- a marking dependent priority function $\pi_j : C(t_j) \times M(j) \rightarrow \mathbf{N}$, $\pi_j(c,m(j)) = 0$ if p_j, t_j describe a subnet in the top level net, $\pi_j(c,m(j)) < \pi(t_i)$ for all ordinary transitions t_i, if p_j, t_j describe the environment of a subnet,

– and a function $W_j(n, m(j))$ specifying the change in the marking of p_j after the arrival of a multiset n of tokens in marking $m(j)$.

We further assume that the firing of tokens on a virtual place is an atomic operation in the sense that at a given time only one timed or immediate transition adds tokens to a virtual place. This restriction is only for notational convenience. Often several immediate transitions firing at the same time can be substituted by a place and an immediate transition collecting first the tokens and firing then (i.e., immediately) tokens in one step. However, this does not mean that only one token can enter or leave a virtual place at a given time (also only this behavior is used in the example below), arbitrarily complex interactions can be defined. In the graphical representation virtual places are denoted by double circles, virtual transitions by double bars.

Before we consider the introduced restrictions and assumptions in terms of modeling power, *top level net* and subnets are introduced formally. Then a small example model is given to clarify the concepts.

2.3 Top Level Net

The *top level net* specifies the most abstract view of the model, it includes a virtual place and transition for each subnet $1 \ldots J$ and a number of ordinary places and (timed/immediate) transitions. We assume that virtual transitions and places are numbered from 1 to J. Ordinary places and transitions receive numbers larger than J.

If virtual transitions are handled like timed transitions, the marking space of the *top level net* can be generated by standard means starting from the initial marking m_0. The initial marking is defined by the distribution of tokens over the places, including virtual places. Since the weights of immediate transitions are completely known, all transition probabilities starting in vanishing markings are defined. Therefore vanishing markings can be eliminated by standard means [7] resulting in a marking space and transition matrix on the set of tangible markings. In contrast to a flat HGCSPN, the transition matrix contains rates of timed transitions and colors for which virtual transitions can fire. Virtual transitions specify an abstract view of a more complicated behavior in the lower level.

Let M_0 be the set of tangible markings and $\underline{Q}_0$ be the transition matrix. $\underline{Q}_0$ is a $\|M_0\| \times \|M_0\|$ matrix, which includes in position m, m' the possible transitions between marking m and m' in the *top level net*. The change of the marking results from the firing of a timed or virtual transition, possibly in combination with the firing of several immediate transitions. The transition from m to m' possibly results from the firing of more than one transition in the net. An element $Q_0(m, m')$ is given by a list of elements of the form (j, c, p, n), each realizing a transition from m to m', where

- j is the virtual or timed transition which fires,
- c is the color for which j fires,
- if j is a virtual transition, then p is the probability of the subsequent firing of immediate transitions to yield marking m' ($p = 1.0$ if no immediate transitions

can fire after j), if j is a timed transition, then p includes the product of the firing rate and the probability of subsequent firing of immediate transitions[1],
- n is a vector which includes in component $n(i, c')$ the number of arriving color c' tokens on each virtual place i[2].

In flat GCSPNs transitions that fire without changing the marking of the net can simply be eliminated. In the *top level net* this is not as easy, since the firing of a transition might yield a situation where tokens arrive on a virtual place without changing its marking observable from the higher level, but, nevertheless, resulting in another internal state of the subnet represented by the virtual place. Therefore diagonal elements $Q_0(m, m)$ have the same structure as non-diagonal elements and additionally include the sum of rates of all activated timed transitions in marking m denoted by $\lambda(m)$.

Under the condition that virtual transitions describe the correct behavior of the subnets as seen in the *top level net*, the *top level net* includes an abstract view of the behavior of the overall net and the behavior of the environment for each subnet is defined in the *top level net*. To specify the behavior of the environment with respect to subnet j, a first step is the representation of all places from the *top level net* except p_j by a single virtual place p_{0j}. The possible markings $m \in M_0$ excluding $m(j) \in M(j)$ have to be coded in the marking of p_{0j} by introducing an appropriate color domain[3]. The virtual transition t_{0j} belonging to p_{0j} has to specify all transitions in the *top level net* that are not originated by t_j. The resulting net consists of two virtual places/transitions p_{0j}/t_{0j}, p_j/t_j and includes the complete specification of the *top level net*. However, with respect to subnet j different markings of the environment are only interesting if they describe different qualitative influences on j. Thus it is desirable to aggregate markings of p_{0j} to get a more compact representation. Two markings $m, m' \in M(0)$ can be aggregated if m' is reachable from m without firing t_j and without firing tokens on p_j. The aggregation is performed by choosing m as representative for m and m' and changing t_{0j} such that a transition from m' to m'' is substituted by a transition m to m'' and a transition m'' to m' is substituted by a transition m'' to m. The informally specified algorithm is rather straightforward to use and implement and yields a minimal representation of the set of markings of the environment for subnet j.

Before subnets are specified, a small example for a *top level net* is given, which will be continued and extended in the sequel of this paper. The example describes a very abstract view of a client-server model and is, of course, mainly for the introduction of concepts rather than specifying a real world application. In Fig. 1 a first and very simple *top level net* is shown including only two virtual places and transitions. The first place p_1 describes the clients, the second p_2 the server. The color sets of the

[1] The probabilities of firing a sequence of immediate transitions are calculated during the elimination of vanishing markings (see [7]).

[2] Since i is a virtual places $m'(i) = m(i) - W^-(p_i, t_j)(c) + \mathcal{W}_i(n(i), m(i))$. Ordinary places need not to be considered here, since the number of tokens arriving on an ordinary place can be computed from m and m'.

[3] This can be done straightforward by introducing a unique color c_i for each place $p_i \neq p_j$ in the *top level net* and extending the color of each token on p_i from $c \in C(p_i)$ to $< c_i, c >$.

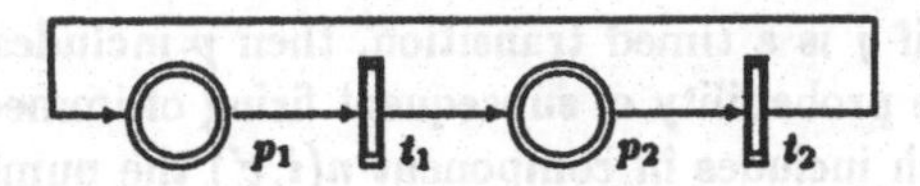

Fig. 1. Top level net

places include both one color c_r. Tokens of this color specify the number of requests served in the clients or the server, respectively. Clients send requests to the server. There is only one kind of request, therefore t_1 can only fire in color c_r, removing one token from p_1 and adding it to p_2. However, a request can be successful (i.e., t_2 fire in color c_s) or can miss (i.e., t_2 fire in color c_m). In both cases one token is removed from p_2 and a token of color c_s or c_m, enters p_1. The functions $\mathcal{W}_1(..)$ are defined that the arrival of a token, either of color c_s or c_m, yields a new marking for p_1 with one additional token of color c_r. The initial marking is given by locating all tokens in p_1.

For the example we assume that the *top level net* includes three tokens, yielding the following four markings in M_0:
1) (3,0) 2) (2,1) 3) (1,2) 4) (0,3)
Markings will be identified by their number. $\underline{Q}_0$ contains the following non-zero elements[4]:
$Q_0(1,2) = Q_0(2,3) = Q_0(3,4) = (1, c_r, 1.0, n(2, c_r) = 1)$
$Q_0(2,1) = Q_0(3,2) = Q_0(4,3) = (2, c_s, 1.0, n(1, c_s) = 1), (2, c_m, 1.0, n(1, c_m) = 1)$.

Since the environment of each subnet consists of a single virtual place, the representation need not to be modified. Furthermore, the marking of the environment is minimal since every transition results either in the firing of a token from the environment in the subnet or vice versa.

2.4 Low Level Subnets

A subnet j is a flat GCSPN with an environment realized by a virtual place p_{0j} and a virtual transition t_{0j}. The behavior of the environment is known in the sense that p_{0j} and t_{0j} are completely specified.

The local marking of subnet j is visible in the *top level net* only through an abstract view described by a marking $n \in M_0(j)$ of the virtual place p_j. Each marking of the virtual place covers a set of markings of subnet j. The marking space M_j of j in combination with the virtual place for the environment can be decomposed into disjoint subsets $M_j(n)$ including all markings in the subset belonging to marking n of p_j. The initial marking of a subnet is influenced by the initial marking of the *top level net* specifying the marking of p_{0j} and the initial marking of p_j which has to be chosen from the subset $M_j(m_0(j))$, where $m_0(j)$ is the initial marking of place p_j in the *top level net*. The main question is how to assign markings of the subnet to the subsets $M_j(n)$. Often the marking of the virtual place describes a state of the subnet which allows to assign a marking to a subset in a natural way (e.g., the

[4] Elements are described in the form (i, c, p, n) given on the previous page. For notational convenience we only specify the non-zero elements of the vector n.

marking of the virtual place includes the population in the subnet). More formalized the following conditions have to hold for a marking m to belong to subset $M_j(n)$:

- For each color c for which t_j can fire in marking n, there has to be a marking $m' \in M_j(n)$ such that the next tangible marking in the subnet belongs to $M_j(n - W^-(p_j,t_j)(c))$ and the new marking of p_{0j} is given by $m'(p_{0j}) + \mathcal{W}_{0j}(W^+(p_{0j},t_j)(c), m'(p_{0j}))$, where $W^+(p_{0j},t_j)(c)$ is given by the sum of $W^+(p_i,t_j)(c)$ (for all $i \neq j$).
- From a marking $m \in M_j(n)$ and a color c for which t_j can fire in marking n, a marking m' fulfilling the above condition is reachable.
- For each transition from a tangible marking $m \in M(n)$ to another tangible marking $m' \in M(n')$ with $n' \neq n$ or $m(p_{0j}) \neq m'(p_{0j})$ exists a color c for which t_j can fire, such that $n' = n - W^-(p_j,t_j)(c)$ and $m'(p_{0j}) = m(p_{0j}) + \mathcal{W}_{0j}(W^+(p_{0j},t_j)(c), m(p_{0j}))$.

The above conditions are sufficient, if the behavior of the environment is correctly specified by p_{0j}, t_{0j}. Otherwise the second condition has to include the condition that m' is reachable from m without firing t_{0j} and without firing tokens on p_{0j}. This condition, of course, is rather strict, but is necessary for a correct specification of a HGCSPN, if the behavior of non-local system parts is not exactly known (see Sect. 2.5).

Due to the above assumptions it is guaranteed that after the arrival of tokens in the subnet or the departure from the subnet a tangible marking of the subnet is reached before the next arrival or departure[5]. Each subset of markings $M_j(n)$ includes at least one tangible marking and vanishing markings can be eliminated with standard techniques inside the subsets $M_j(n)$, for details see [7]. For the sequel we assume that $M_j(n)$ contains only tangible markings.

The next step is the determination of transitions on the marking space M_j. We can distinguish transitions originated in the subnet and transitions originated in the environment. The former are fully specified by the subnet and the transition rates are known, the latter are only qualitatively specified, no firing rates are known. Nevertheless, transitions from the environment into subnet j can be quantified by conditional probabilities with the condition that j is in marking $m \in M_j$ and a multiset of tokens $m'(j)$ arrives. Three different types of matrices specify the behavior of a subnet.

The first type includes transition rates among markings from a set $M_j(n)$ and in the main diagonal the negative sum of all transition rates activated in the actual marking. The matrices are denoted by $\underline{Q}_j^n$. Each $\underline{Q}_j^n$ is a $\|M_j(n)\| \times \|M_j(n)\|$ matrix with non-negative non-diagonal elements and negative diagonal elements, the row sum of each row is less or equal zero.

The second type of matrices describes the departure of tokens from j to the environment, realized by firing the virtual transition t_j for color c in the *top level net*. If j is in a marking from $M_j(n)$, then the successor marking is from $M_j(n')$

[5] It is, of course, possible that tokens depart and arrive immediately in a subnet, but since we have assumed that immediate transitions in the subnet have priority over immediate transitions in the *top level net*, the subnet reaches a tangible marking before new tokens arrive.

$(n' = n - W^{-}(t_j, p_j)(c))$. The rates of these transitions are completely defined in j and are collected in a non-negative $\|M_j(n)\| \times \|M_j(n')\|$ matrix $\underline{S}_j^n(c)$. $\underline{S}_j^n(c)$ is a zero matrix if t_j cannot fire for color c in marking n.

The last type of matrices describes the arrival of tokens in j. Assume that j is in a marking from the set $M_j(n)$ and a multiset of tokens m arrives simultaneously, then the subsequent marking is from $M_j(n')$ $(n' = n + W_j(m, n))$. Transitions are described by conditional probabilities in a stochastic matrix $\underline{U}_j^{n'}(m)$ of order $\|M_j(n)\| \times \|M_j(n')\|$. The matrices $\underline{U}_j^{n'}(m)$ have to be specified for every possible multiset of tokens m that can arrive in a marking from $M_j(n)$. Of course, to each m belongs a firing of the virtual transition t_{0j} with $W^{+}(t_{0j}, p_j)(c') = m$. The above matrices specify the behavior of a subnet completely and can be generated from the subnet specification including p_{0j} and t_{0j}.

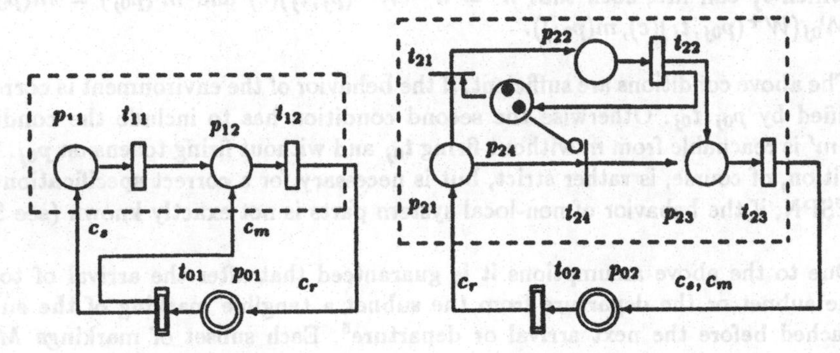

Fig. 2. Client and server subnet

We now continue the small example by specifying the subnets *clients* and *server* (see Fig. 2). The client subnet consists of two places p_{11}, p_{12} and two timed transitions t_{11}, t_{12}. Each subset $M_1(m(1))$ includes $m(1) + 1$ markings specifying the distribution of tokens over the places. Tokens on p_{11} describe clients that are locally computing without outstanding requests. p_{12} includes requests which are transferred to the server. If a successful request (a token of color c_s) enters the subnet, it starts in p_{11}. A missed request (a token of color c_m) starts immediately in p_{12} describing a new attempt to send the request to the server. The server subnet includes four places $p_{21} \ldots p_{24}$, two timed and two immediate transitions. Arriving tokens enter place p_{21}. Place p_{24} describes the availability of the server. If p_{24} contains tokens, a new arriving request is immediately routed to p_{22} and the number of available server resources is decreased by one, otherwise the request is rejected by firing transition t_{24} and changing the color of the token to c_m. p_{22} includes requests in service, after ending the service tokens are fired on p_{23} changing the color to c_s and increasing the number of available servers on p_{24} by one. Tokens on p_{23} are transferred back to the clients. The initial marking of the client subnet is given by putting 3 tokens in place p_{11}, the initial marking of the server subnet includes 2 resource tokens in p_{24} and 3 tokens on the environment place p_{02}. This initial marking is consistent with the initial marking of the *top level net* with 3 tokens in subnet place p_1 specifying the clients.

The complete marking space M_1 is given below in the form $(\sharp p_{11}, \sharp p_{12})$ ($\sharp p$ is the number of tokens on place p, $\sharp p(c)$ is the number of color c tokens on p).

$M_1(0)$ (0,0) $M_1(2)$ (0,2) (1,1) (2,0)
$M_1(1)$ (0,1) (1,0) $M_1(3)$ (0,3) (1,2) (2,1) (3,0)

The marking space of subnet 2 is denoted by $(\sharp p_{22}, \sharp p_{23}(c_s), \sharp p_{23}(c_m))$. Notice that the number of tokens of color c_m is restricted to one, since only the third (and last) token can miss its request to the server.

$M_1(0)$ (0,0,0) $M_1(1)$ (0,0,1) (0,1,0) (1,0,0)
$M_1(2)$ (0,1,1) (0,2,0) (1,0,1) (1,1,0) (2,0,0)
$M_1(3)$ (0,2,1) (0,3,0) (1,1,1) (1,2,0) (2,0,1) (2,1,0) (3,0,0)

2.5 Specification of HGCSPNs

In the previous subsections we have introduced a hierarchical extension to the class of GCSPNs. The goal of this subsection is more practical oriented, since the specification of hierarchical nets and the practical applicability of the approach will be considered briefly. The first and, of course, fundamental question is, if the assumptions made about the knowledge of subnet behavior are realistic. Taking an arbitrary GCSPN as a subnet, it is impossible to specify a priori the qualitative behavior in terms of a virtual transition and a virtual place. However, in realistic system modeling the situation is not as hopeless, since in hierarchical models the input and output behavior of submodels is usually well defined. If the behavior of a subnet is not known exactly, we can choose an iterative approach by first assuming a specific behavior, then generate the *top-level net* marking space M_0 and the matrix $\underline{Q}_0$. From M_0 and $\underline{Q}_0$ the environment for every subnet can be generated and during marking space generation of a subnet it can be tested, if the behavior of the subnet as visible in the *top-level net* matches the behavior of the corresponding virtual transition. If virtual transitions are not correctly specified in the *top-level net*, the specification is changed and M_0, $\underline{Q}_0$ are recomputed yielding possibly new environments for all or some subnets. This approach is iterated until the virtual transitions and places describe the correct behavior of all subnets in the *top-level net*. If aggregated views are refined in each iteration step, convergence can be assured. Of course, the hierarchical structure of a net is not unique and depends strongly on the autonomy of different model parts.

After specification only a qualitative analysis to detect specification errors is necessary. In the small example introduced above, the structure of the marking space for subnet 1 is valid for every environment which generates successful requests. For the second subnet assumptions about the population are necessary (e.g., in an environment with only two outstanding requests, no request will ever miss and therefore also the subsets $M_2(1)$ and $M_2(2)$ do not include markings where a missed request is transferred back to the clients).

3 Marking Space and Transition Matrices

Up to now isolated marking spaces and transition matrices have been generated for the different parts of the hierarchical net. In this section it is shown how the isolated parts are combined to form the marking space and generator matrix of the complete HGCSPN.

3.1 Marking Space Structures

A single marking of the HGCSPN describes implicitly J markings of the subnets and a marking of the *top level net*. Let M be the marking space of the HGCSPN including only tangible markings. For each $m \in M$ let $m_0 \in M_0$ be the marking of the *top level net* and m_j the marking of subnet j. According to the marking of the *top level net*, M can be decomposed into disjoint subsets of markings $M(m_0)$ including all markings that are indistinguishable in the *top level net*. This yields the following structure of M.

$$M = \cup_{m_0 \in M_0} M(m_0) = \cup_{m_0 \in M_0} \times_{j=1}^{J} M_j(m_0(j)) \tag{1}$$

Let $\|M\|$ be the cardinality of a marking space, then

$$\|M\| = \sum_{m_0 \in M_0} \|M(m_0)\| = \sum_{m_0 \in M_0} \prod_{j=1}^{J} \|M_j(m_0(j))\| \quad . \tag{2}$$

Markings from M are described by $J+1$ components specifying the marking of each subnet and the *top level net*. We will now transform the specification in a two dimensional description and assume that markings from each subset $M_j(n)$ are numbered consecutive from 1 to $\|M_j(n)\|$. Markings from M_0 are numbered from 1 to $\|M_0\|$. Let $\sharp m_j$ be the number of marking $m_j \in M_j(m_0(j))$ and $\sharp m_0$ be the number of $m_0 \in M_0$. Each marking $m \in M$ gets a two dimensional number.

$$\sharp m = (\sharp m_0, (\sum_{j=1}^{J} (\sharp m_j - 1) \prod_{i=j+1}^{J} \|M_i(m_0(i))\|) + 1) \tag{3}$$

3.2 Markov Chain

In the previous subsection the marking space description has been transformed from $J+1$ to 2 dimensions. Since M contains only tangible markings it is isomorph to the state space of the Markov chain underlying the HGCSPN [1]. For notational convenience we will not distinguish between marking space and state space of the Markov chain.

Let $\underline{Q}$ be the $\|M\| \times \|M\|$ generator matrix. $\underline{Q}$ can be decomposed into $\|M_0\|^2$ submatrices $Q^{m,m'}$ $(m, m' \in M_0)$, which include transitions between markings from $M(m)$ and $\overline{M}(m')$. This mapping specifies the first part of the number of a marking (i.e., all transitions between markings belonging to the same marking in the *top level net* are collected in one submatrix). The second part of the marking number specifies the position in the submatrices. Before we introduce the computation of submatrices from subnet matrices and transitions in the *top level net*, tensor multiplication has to be defined as a basic operation for matrix generation [12].

Definition 3. The tensor product of two matrices $\underline{A}_1 \in \mathbb{R}^{r_1 \times c_1}$ and $\underline{A}_2 \in \mathbb{R}^{r_2 \times c_2}$ is defined as $\underline{C} = \underline{A}_1 \otimes \underline{A}_2$, $\underline{C} \in \mathbb{R}^{r_1 r_2 \times c_1 c_2}$,
where $c((i_1 - 1) * r_2 + i_2, (j_1 - 1) * c_2 + j_2) = a_1(i_1, j_1) a_2(i_2, j_2)$
$(1 \le i_x \le r_x, 1 \le j_x \le c_x, x \in \{1,2\})$.
The generalized tensor product is given by $\otimes_{j=1}^{J} \underline{A}_j = \underline{A}_1 \otimes (\otimes_{j=2}^{J} \underline{A}_j)$ and $\otimes_{j=J}^{J} \underline{A}_j = \underline{A}_J$.

To clarify the tensor product, we consider the following example.

$$\begin{pmatrix} a(1,1) & a(1,2) \\ a(2,1) & a(2,2) \end{pmatrix} \otimes \underline{B} = \begin{pmatrix} a(1,1)\underline{B} & a(1,2)\underline{B} \\ a(2,1)\underline{B} & a(2,2)\underline{B} \end{pmatrix}$$

Tensor multiplication implicitly realizes a linearization according to the lexicographical ordering of indices in the involved matrices, which is consistent with the linearization introduced for markings from subsets $M(m)$ in the previous subsection. The following functions are needed for matrix generation.

$$l_j(m) = \prod_{i=1}^{j-1} \|M_i(m(i))\| \quad u_j(m) = \prod_{i=j+1}^{J} \|M_i(m(i))\| \quad \text{where } m \in M_0,\ 1 \le j \le J \tag{4}$$

We have to distinguish transitions observable in the *top level net* and transitions local in one subnet without affecting another subnet or the *top level net*. Transitions that are observable in the *top level net* are specified in the matrix $\underline{Q}_0$. For notational convenience we define matrices $\underline{V}^m(j,c,n)$ describing transitions observable in the *top level net* on the level of the involved subnets as follows

$$\underline{V}^m(j,c,n) = \begin{cases} \underline{I}_{l_j(m)} \otimes \underline{S}_j^{m(j)}(c) \otimes \underline{I}_{u_j(m')} \prod_{n(i) \ne \underline{0}} \underline{I}_{l_i(m)} \otimes \underline{U}_i^{m'(i)}(n(i)) \otimes \underline{I}_{u_i(m')} & \text{if } 1 \le j \le J \\ \prod_{n(i) \ne \underline{0}} \underline{I}_{l_i(m)} \otimes \underline{U}_i^{m'(i)}(n(i)) \otimes \underline{I}_{u_i(m')} & \text{else} \end{cases}$$

where $\underline{I}_x$ is the identity matrix of order x and $m'(i) = m(i) + \mathcal{W}_i(n(i), m(i))$

(5)

If virtual transition t_j fires, the marking of subnet j is modified as specified in the $\underline{S}$-matrix. Tokens arriving in a virtual place p_i (i.e., $n(i) \ne \underline{0}$) change the marking of subnet i as specified in the $\underline{U}$-matrix. Caused by the firing of one transition in the *top level net* tokens can only leave from one subnet, but can arrive to several subnets.

With the matrices $\underline{V}^m(j,c,n)$ submatrices $\underline{Q}^{m,m'}$ of the generator $\underline{Q}$ are computed. We use the notation (j,c,p,n) to specify all elements in position $\underline{Q}_0(m,m')$. If $\underline{Q}_0(m,m') = \emptyset$, then $\underline{Q}^{m,m'} = \underline{0}$.

$$\underline{Q}^{m,m'} = \begin{cases} \sum_{(j,c,p,n)} p\underline{V}^m(j,c,n) & \text{if } m \neq m' \\ -\lambda(m)\underline{I}_{l_{J+1}(m)} + \sum_{j=1}^{J} \underline{I}_{l_i(m)} \otimes \underline{Q}_j^{m(j)} \otimes \underline{I}_{u_i(m)} + \sum_{(j,c,p,n)} p\underline{V}^m(j,c,n) & \text{if } m = m' \end{cases} \quad (6)$$

The non-diagonal blocks are given by a $\underline{V}$-matrix and the value p including the firing rate or probability, respectively. In the diagonal-blocks we have additionally the negative sum of firing rates of activated timed transitions in the *top level net* and the $\underline{Q}$-matrices specifying transitions inside subnets without affecting the *top level net*.

Equation (5) and (6) describe a way to compute the generator of the complete HGCSPN using matrices of subnets and the *top level net*. Moreover, the structure of these matrices is not complicated, since they result from the sum or product of matrices of the form $\underline{I}_l \otimes \underline{A} \otimes \underline{I}_u$. Such a matrix contains l non-zero diagonal blocks, and each diagonal block consists of a modified matrix $\underline{A}$, where an element $A(m, m')$ is substituted by a $u \times u$ diagonal matrix with $A(m, m')$ in the main diagonal.

4 Analysis Algorithms

We have considered the specification of HGCSPNs and have introduced a method to compute the generator matrix of the underlying Markov chain. The approach allows to perform several steps in parallel and the elimination of vanishing markings locally in subnets and the *top level net* is very efficient. However, the hierarchical structure of a net can also be used for a very efficient analysis. Appropriate techniques have been published recently (see [4,5,6,7]) and are briefly reviewed here.

There are two different goals of the analysis of a HGCSPN. The first is related to qualitative analysis of the net yielding results like "liveness" or the existence of "home states". The second is related to the performance of the net yielding results like throughputs or mean populations. Our purpose is on the quantitative analysis. Nevertheless, we cannot neglect qualitative aspects, since they might indicate specification errors in the net.

4.1 Qualitative Analysis

Qualitative analysis of a model is necessary to ensure that the model is correct in a rather restricted sense. In this paper we introduce the influence of the hierarchy on a qualitative analysis rather than specifying new analysis techniques or results. Qualitative analysis is supported by the hierarchical net structure since it allows to analyse several isolated parts sequentially or in parallel which is much more efficient than the analysis of the flat overall net. A first approach for qualitative analysis has been already presented during the generation of the marking space. If virtual transitions describe the behavior of non-local model parts, then a subnet or the *top level net* can be analysed locally. Using the iterative approach explained in Subsect.

2.5 the qualitative behavior can be analysed implicitly during the generation and interpretation of subnet marking spaces.

A second approach has been introduced in [7] and will be only very briefly reflected here. The idea is to reduce the marking space of the subnet and to generate afterward a reduced marking space for the complete net which can be analysed much more efficient. Marking spaces of subnets are reduced by aggregating markings which are equivalent according to reachability (i.e., if one marking is reached all equivalently reachable markings can be reached by internal transitions in the subnets). The reduction can be performed on net level using behavior preserving reduction rules as defined in [15] or directly on the marking spaces $M_j(n)$ which covers more general situations but needs also more effort. Using the approach on the marking space the subspaces $M_j(n)$ often can be reduced significantly and, of course, the reduced marking space of the complete net becomes also much smaller. It should be mentioned that the reduction preserves only the qualitative behavior of the net and not the quantitative, therefore it can only be used for qualitative not for quantitative analysis. If we use this approach for the server subnet of our example net, the marking spaces $M_2(2)$ and $M_2(1)$ of the server subnet can be reduced to two markings distinguishing if a missed request is transferred back or not. We have noticed before that a request can only miss if more than two requests arrive and this is, of course, not visible locally in the server subnet.

4.2 Numerical Quantitative Analysis

Quantitative analysis of GCSPNs is preformed either by numerical analysis of the underlying Markov chain or by simulation. Numerical analysis is often preferable, but is faced with the problem of state explosion yielding Markov chains of an extremely high dimension. However, the hierarchical structure of a net can be integrated in numerical solution techniques allowing the analysis of much larger nets.

Quantitative analysis distinguishes steady state and transient analysis. In the former the interest lies in the steady state vector $\underline{p}$ given as the solution of $\underline{p}\underline{Q} = \underline{0}$ and the additional condition $\underline{p}\underline{e}^T = 1.0$. Transient analysis requires the determination of a vector $\underline{\pi}(t)$ including the state probabilities at time $t > 0$ which depend on the initial distribution $\underline{\pi}(0)$.

We start with steady state analysis. Since the matrix $\underline{Q}$ is normally huge and sparse, the solution is usually computed by an iterative solution technique preserving the non-zero structure of $\underline{Q}$. One often used (although not very efficient) iterative solution technique is the power method, yielding the following iteration scheme.

$$\underline{p}^k = \underline{p}^{k-1} + \frac{1.0}{\alpha}\underline{p}^{k-1}\underline{Q}$$

where $\alpha > \max(|Q(i,i)|)$ and $\underline{p}^0$ is the initial vector with $\underline{p}^0 \geq \underline{0}$ and $\underline{p}\underline{e}^T = 1.0$

(7)

Taking the structure of the HGCSPN into account a vector $\underline{p}$ can be decomposed into subvector $\underline{p}_n$ $(n \in M_0)$ including state probabilities for states from the subset $M(n)$. With this decomposition the iteration scheme of the power method becomes

$$\underline{p}_n^k = \underline{p}_n^{k-1} + \frac{1.0}{\alpha} \sum_{n' \in M_0} \underline{p}_{n'}^{k-1} \underline{Q}^{n',n} \quad (8)$$

The above iteration is based on the (repeated) multiplication of a vector with a matrix of the form $\underline{I}_l \otimes \underline{A} \otimes \underline{I}_u$. Such a multiplication can be performed without generating the matrix using only the values l, u and the matrix $\underline{A}$. A procedure in pseudocode is defined in [5] and needs approximately 20 lines of code. The approach has the advantage that we need not store the huge matrix $\underline{Q}$, only the subnet matrices and $\underline{Q}_0$ are used. The difference in memory requirements between the conventional approach using $\underline{Q}$ and the new approach presented here is analysed in [5]. Very roughly the size of $\underline{Q}$ is the limiting factor of the conventional analysis[6]. For the new techniques the vector $\underline{p}$ becomes the limiting factor. In practice this has the effect that the size of solvable models on a given hardware is extended approximately by the order of a magnitude. On a standard workstation with $32MB$ virtual memory around $2.0e+6$ instead of $2.0e+5$ states can be handled. Additionally the analysis of larger nets becomes much more efficient since paging rates are reduced due to the reduction in memory requirements. Examples for the savings are given in [5,7].

The structured description of an iteration step allows a parallel implementation based on the net structure and the tensor products (see [12] for the parallel handling of tensor multiplication). This approach is very well suited for parallel architectures without shared memory since there is a large amount of data locality as already noticed in [13] for a different class of Petri nets with a generator matrix built by tensor products. Nevertheless, a parallel implementation is subject of further research.

We have only considered the power method for stationary solution, however, more efficient iteration techniques like Jacobi, Gauss-Seidel or SOR can be handled in similar way [5,7]. Furthermore aggregation/disaggregation steps [17] can be integrated naturally to improve the convergence rate of iterative methods for many models. For details of such an integration we refer to the cited papers.

Transient analysis is handled very similar by using the so called "randomization" method which has proofed to be the most efficient method for numerical transient analysis [18]. The determination of the vector $\underline{\pi}(t)$ using randomization is given in the following equation. The vectors $\underline{p}^k$ are determined as shown in (8) starting with $\underline{p}^0 = \underline{\pi}(0)$. In practice only a finite number of steps in the sum below will be evaluated which can be used to determine bounds for the resulting vector (see [18]).

$$\underline{\pi}(t) = \sum_{k=0}^{\infty} \underline{p}^k e^{-\alpha t} \frac{(\alpha t)^k}{k!} \quad (9)$$

4.3 Aggregation Approaches

With the methods presented in the previous subsection rather large models can be analysed, however, the size of realistic models often exceeds the size of solvable models by orders of a magnitude. Numerical analysis of such models is only possible using exact or approximative aggregation methods. A wide variety of aggregation

[6] This holds although the matrix is sparse and sparse storage schemes are used.

methods has been proposed during the last decade in the literature. In this paper we will concentrate on methods based on the decomposition on net level rather than decomposing the generator matrix. For really large nets only this approach is feasible. We can only give here a very brief overview of the use of aggregation techniques and describe the approaches mainly by means of small examples. More detailed results will be published in a forthcoming paper.

There are two main approaches for aggregation in HGCSPNs; the first is based on symmetries in the net, the second on the loose coupling between subnets. The latter approach will not be further investigated since it has been considered in [7] and extends earlier work from [2]. Aggregation based on symmetries is exact for steady state and several transient results. A method to generate directly an aggregated Markov chain from the net specification has been proposed recently for the class of Stochastic Well-formed Colored Nets [8]. Other approaches for hierarchical models generate reduced Markov chains by exploiting symmetries in a submodel [4] and between identical submodels [6]. The latter two approaches are adopted for the class of HGCSPNs.

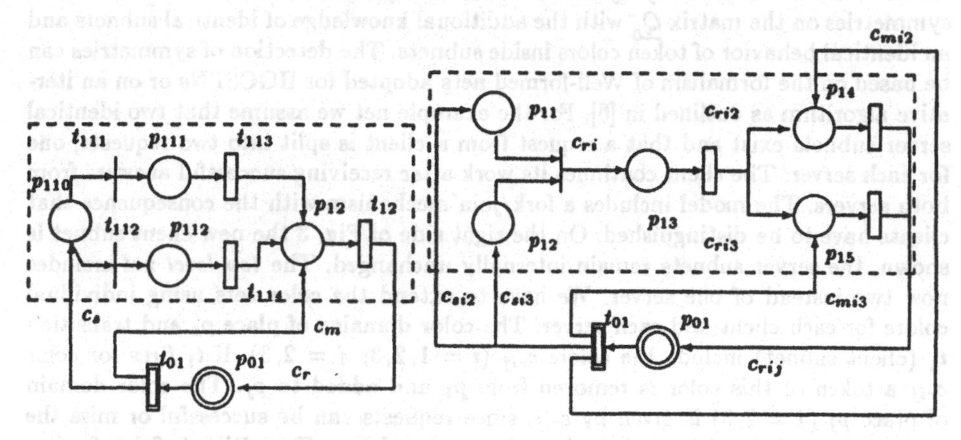

Fig. 3. Second and third client subnet

We start with symmetries inside an isolated subnet and modify the client subnet as shown on the left side of Fig. 3. In the entire model all clients run on the same CPU, in the modified net two CPUs are available to run the client processes. The place p_{11} is substituted by three places p_{110}, p_{111} and p_{112}. Successful requests arrive in place p_{110} and choose one of the CPUs by firing immediate transition t_{111} or t_{112}. The weights of these transitions are allowed to depend on the marking of the subnet, but this dependency should be symmetric. Assume that the weight of t_{111} is 1.0 if $\sharp p_{111} \leq \sharp p_{112}$ and 0 otherwise. The weight of t_{112} is 1.0 if $\sharp p_{112} \leq \sharp p_{111}$ and 0 otherwise. An arriving request chooses the CPU with the lowest load (smallest number of tokens) or one of the CPUs with equal probabilities if both are equally loaded. Furthermore let the firing rates of t_{113} and t_{114} be identical. Obviously both CPUs are completely symmetric and a marking with one token in place p_{111} and none in p_{112} is symmetric to a marking with one token in place p_{112} and none in

p_{111}. Both markings can be lumped together yielding a reduced marking space and Markov chain which, nevertheless, allows the computation of the exact result for steady state and several transient measures [4,6,8]. It is interesting to note that the proposed reduction can be performed locally on the subnet completely independent from the embedding environment and the reduction of the subnet marking space by a factor r yields a reduction of the marking space of a complete HGCSPN including the reduced instead of the original subnet by a factor of approximately r. To generate the reduced marking space there are two different ways. The first is to use the formalism of well-formed nets which, of course, has to be extended to HGCSPNs. An alternative is to use an iterative algorithm for the reduction of the subnet matrices as proposed in [4]. The latter approach is independent from the subnet specification and allows, additionally, to extend the notation of symmetric states to nearly symmetric states yielding further reductions and approximative aggregates.

The second aggregation technique is based on symmetries which are specified in the whole HGCSPN rather than locally in a subnet. There are several levels of symmetries and reduction in hierarchical models as described in [6]. We will present the results here very briefly by means of an example. The central idea is to detect symmetries on the matrix $\underline{Q}_0$ with the additional knowledge of identical subnets and an identical behavior of token colors inside subnets. The detection of symmetries can be based on the formalism of Well-formed nets adopted for HGCSPNs or on an iterative algorithm as outlined in [6]. For the example net we assume that two identical server subnets exist and that a request from a client is split into two requests, one for each server. The client continues its work after receiving successful answers from both servers. The model includes a fork/join mechanism with the consequence that clients have to be distinguished. On the right side of Fig. 3 the new client subnet is shown, the server subnets remain internally unchanged. The *top level net* includes now two instead of one server. We have to extend the color sets using individual colors for each client and each server. The color domains of place p_1 and transition t_1 (client subnet) include the colors c_{rij} ($i = 1,2,3$; $j = 2,3$). If t_1 fires for color c_{rij} a token of this color is removed from p_1 and added to p_j. The color domain of place p_j ($j = 2,3$) is given by c_{rij}, since requests can be successful or miss the color domain of transition t_j includes colors c_{sij} and c_{mij}. Transition t_j firing for c_{sij} (c_{mij}) removes a token of color c_{rij} from p_j and fires a color c_{sij} (c_{mij}) token on p_1. $\mathcal{W}_1(\ldots)$ defines that the arrival of a color c_{sij} or c_{mij} token change the marking of p_1 by adding one token of color c_{rij}. The client subnet consists of five places. Place p_{13} includes locally computing clients, place p_{14} and p_{15} include requests that are transferred to server 2 and 3 (tokens of color c_{ri2} and c_{ri3}), respectively. Requests that have missed (tokens of color c_{mij}) immediately enter place p_{1k} ($k = j+2$) as color c_{rij} tokens for a new attempt. The places p_{11} and p_{12} include tokens waiting for their counterpart to complete. A token of color c_{sij} enters place p_{1k} ($k = j+1$), if its counterpart is available both are joined and the request has completed. If we assume that all clients and servers are identical, the model includes two symmetric levels which are independent from the concrete realization of the subnets. On the first level we need not to distinguish the concrete identity of a client, on the second level we need not to distinguish the concrete identity of a server. Reduction is performed in the *top level net* yielding a reduced marking space M_0 which is used to generate efficiently a reduced marking space and generator matrix of the HGCSPN.

For details of the technique the reader is referred to [6].

5 An Example

We show the usability of HGCSPNs by means of an examples borrowed from literature [1]. The main goal of this section is to specify a realistic example with HGCSPNs rather than describing applications in detail or presenting analysis results. The model specifies a flexible manufacturing system (FMS) with faulty machines and a repair station. The *top level net* (see Fig. 4) specifies the mechanism machines get faulty and are repaired in the repair facility. The virtual place and transition specify the FMS. The failure of a machine is realized by the virtual transition firing a token onto the place *w_rep*. If a repair resource is available on place *r_res* the faulty machine immediately gets in repair on place *p_rep*. After ending repair the resource returns to *r_res* and the repaired machine returns to the virtual place describing the FMS. We assume that three distinguishable machines exist in the FMS specified by three tokens with color c_1, c_2 and c_3, respectively. The state of the FMS visible in the *top level net* is given by the number of available machines. Machines can fail in different modes, machine 1 in mode 0, 1 and 2, machine 2 in mode 0 or 2 and machine 3 in mode 0 or 1 (see the FMS specification below). If machine i fails in mode j, this is realized by firing a token of color c_{ij} onto place *w_rep*. The marking of the virtual place is afterward given by removing one token of color c_i. If a token of color c_{ij} enters the virtual place, its marking is modified by adding one token of color c_i. The possibility of failures in different modes is an extension of the original model where idle machines cannot fail.

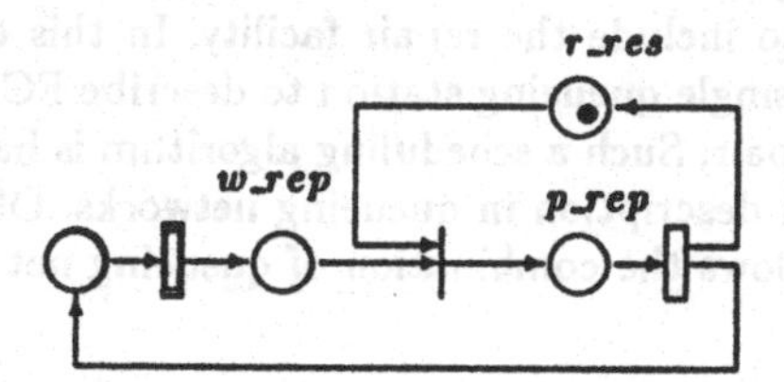

Fig. 4. FMS top level net

In the FMS subnet two parts of material are handled. Each part has to be first loaded on a pallet and is afterward moved to the machines. Parts of type 1 have to be processed by machine 1 and afterward by machine 3. Parts of type 2 are processed by machine 1 or 2. The failure of machine i working on part j yields to the firing of a color c_{ij} token on the environment place p_{01}. The failure of an idle machine i is characterized by the firing of a token of color c_{i0} on p_{01}. If a machine returns from repair, it continues in its old state. It should be noted that the state of a failed machine is defined by the color of the token in the repair station and needs not be stored in the marking of the FMS subnet.

The hierarchical structure of the HGCSPN and the underlying generator matrix in this example conforms with a time scale decomposition since failure and

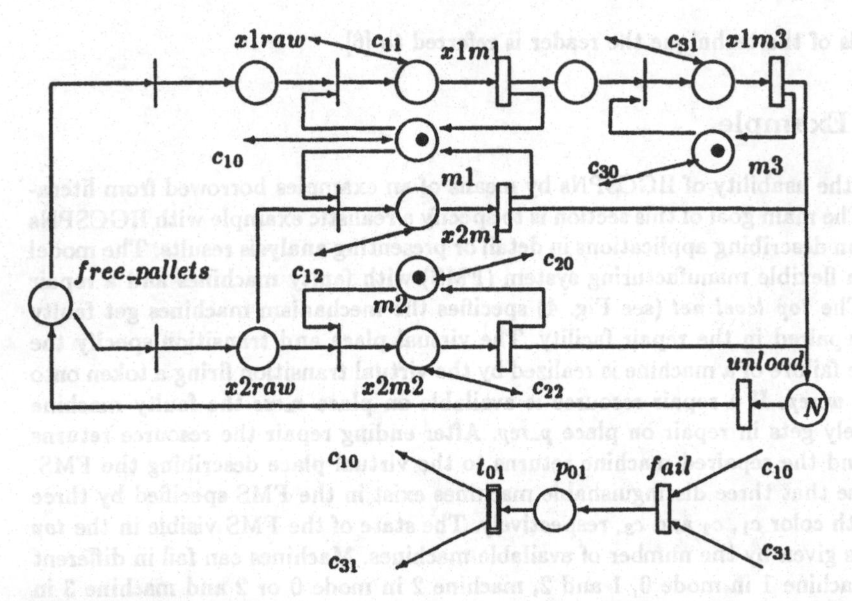

Fig. 5. FMS subnet

repair rates are normally much smaller than working rates of the machines. Therefore aggregation techniques based on loose coupling or the integration of aggregation/disaggregation steps in the iterative solution techniques increases the speed of the solution significantly. The model can be modified in various ways. One possibility is to use a subnet to include the repair facility. In this case the repair facility might be specified as a single queueing station to describe FCFS scheduling of failed machines waiting for repair. Such a scheduling algorithm is hard to integrate in GCSPNs but has a natural description in queueing networks. Of course, this approach can be extended and allows the combination of queueing networks with GSPNs.

6 Conclusions

We have presented a class of hierarchical GCSPNs by extending earlier work in [7]. The hierarchical net structure allows a modular specification of complex models. The general approach presented in this paper is obviously not restricted to the class of GCSPNs. It can be integrated in the same way in other classes of timed Petri Nets with an underlying Markov chain and it allows to specify and analyse real multi-paradigm models.

There are, of course, several topics for future research. The use of aggregation techniques has to be formalized in the given framework, in particular the notation of well-formed nets should be extended to hierarchical nets. Furthermore the efficient use of the approach on parallel architectures will be investigated.

References

1. Ajmone-Marsan, M.A., Balbo, G., Chiola, G., Conte, G., Donatelli, S., Franceschinis, G.: An Introduction to Generalized Stochastic Petri Nets. *Microelectron. Reliab. 31, 4 (1991), 699-725.*
2. Ammar, H.H., Islam, S.M.: Time Scale Decomposition of a Class of Generalized Stochastic Petri Net Models. *IEEE Trans. on Software Eng.*, 15, *6 (1989), 809-820.*
3. Balbo, G., Bruell, S.C., Ghanta, S.: Combining Queueing Networks and Generalized Stochastic Petri Nets for the Solution of Complex System Behavior. *IEEE Trans. on Comp.* 37, *10 (1988) 1251-1268.*
4. Buchholz, P.: The Aggregation of Markovian Submodels in Isolation. *Universität Dortmund, Fachbereich Informatik, Forschungsbericht 369, 1990.*
5. Buchholz, P.: Numerical Solution Methods based on Structured Descriptions of Markovian Models. *In G. Balbo, G. Serazzi (eds.): Computer Performance Evaluation, Modelling Techniques and Tools, North Holland 1992, pp. 251-267.*
6. Buchholz, P.: Hierarchical Markovian Models -Symmetries and Aggregation-. *Proc. of the Sixth International Conference on Modelling Tools and Techniques for Computer Performance Evaluation, Edinburgh, Scotland, UK, September 15-18, 1992, pp. 305-319.*
7. Buchholz, P.: A Hierarchical View of GCSPNs and its Impact on Qualtitative and Quantitative Analysis. *Journal of Parallel and Distributed Computing* 15, *2 (1992), 207-224.*
8. Chiola, G., Dutheillet, C., Franceschinis, G., Haddad, S.: Stochastic Well-Formed colored Nets and Multiprocessor modeling Application. *In K. Jensen, G. Rozenberg (eds.): High-Level Petri Nets. Theory and Application, Springer 1991.*
9. Chiola, G., Bruno, G., Demaria, T.: Introducing a color Formalism into Generalized Stochastic Petri Nets. *Proc. of the 9th European Workshop on Application and Theory of Petri Nets, Venezia, Italy, 1988.*
10. Chehaibar, G.: Use of Reentrant Nets in Modular Analysis of Colored Nets. *In K. Jensen, G. Rozenberg (eds.): High-Level Petri Nets. Theory and Application, Springer 1991.*
11. Ciardo, G., Trivedi, K.S.: Solution of Large GSPN Models. *In G. Stewart (ed.): Numerical Solution of Markov Chains, Marcel Dekker 1991.*
12. Davio, M.: Kronecker Products and Shuffle Algebra. *IEEE Trans. on Comp.* 30, *2 (1981), 116-125.*
13. Donatelli, S.: Superposed Stochastic Automata: a class of Stochastic Petri nets amenable to parallel solution. *Proc. of the Fourth Int. Workshop on Petri Nets and Performance Models, Melbourne, 1991.*
14. Fehling, R.: A Concept of Hierarchical Petri Nets with Building Blocks. *Proc. of the 12th. Int. Conf. on Application and Theory of Petri Nets, Gjern, Denmark, June 1991.*
15. Haddad, M.S.: A Reduction Theory for colored Nets. *In G. Rozenberg (ed.): Advances in Petri Nets, Lecture Notes in Computer Sciences, vol. 424, Springer 1990, pp. 209-235.*
16. Huber, P., Jensen, K., Shapiro, R.M.: Hierarchies in colored Petri Nets. *In G. Rozenberg (ed.): Advances in Petri Nets, Lecture Notes in Computer Sciences, vol. 483, Springer 1990, pp. 313-341.*
17. Krieger, U., Müller-Clostermann, B., Sczittnick, M.: Modeling and Analysis of Communication Systems Based on Computational Methods for Markov Chains. *IEEE J. on Selected Areas in Communication 8, 9 (1990), 1630-1648.*
18. Reibman, A., Trivedi, K.S.: Numerical Transient Analysis of Markov Models. *Comput. Opns. Res.* 15, *1 (1988), 19-36.*

Variable Reasoning and Analysis about Uncertainty with Fuzzy Petri Nets*

Tiehua Cao and Arthur C. Sanderson

Electrical, Computer, and Systems Engineering Department
Rensselaer Polytechnic Institute
Troy, NY 12180-3590, USA

Abstract. Ordinary Petri nets often do not have sufficient power to represent and handle approximate and uncertain information. Fuzzy Petri nets are defined in this paper using three types of fuzzy variables: local fuzzy variables, fuzzy marking variables, and global fuzzy variables. These three types of variables are used to model uncertainty based on different aspects of fuzzy information. Several basic types of fuzzy Petri nets are analyzed, and the necessary and/or sufficient conditions of boundedness, liveness, and reversibility are given. An example of modeling sensory transitions in a robotic system is discussed to illustrate reasoning about input local fuzzy variables to obtain mutually exclusive tokens in the output places.

1 Introduction

Petri nets[20, 21] have been widely used to model computer systems[18, 19, 22], manufacturing systems[1], robotic systems[3-8], knowledge-based systems[2, 14], and other kinds of engineering applications. An operation and its preconditions and postconditions in a manufacturing system can be represented by a transition and its input places and output places in a Petri net model. While modeling and representation of lower level operations and objects in a robot system are necessary[6, 8], ordinary Petri nets are found not sufficient to represent uncertainty and approximate information. The uncertainty in a robotic system may occur due to many factors during the execution of a planned sequence or program. When an operation such as 'the gripper A grasps the object B', is modeled by a transition t_i in a Petri net N, one kind of uncertainty within this grasping operation is the geometric uncertainty in the coordinate of the grasp point or the contacting point of the gripper with the object. Because this point is important for the succeeding operations of the robot gripper such as motion, force control, and assembly, we need to represent uncertain information in the output place which shows the result of the robotic grasp operation. Another kind of uncertainty within this operation is the uncertainty of the success of this operation,

* This research was supported in part by the NASA Center for Intelligent Robotic Systems for Space Exploration at Rensselaer Polytechnic Institute.

and the uncertainty of degree completion for the whole task of the robotic system, such as assembling a complete set of objects in a certain configuration or reaching a final system state.

Based on the above discussions, a fuzzy Petri net may be used to describe the operations and conditions with uncertainty. Uncertain states are associated with objects, and transitions are used to model fuzzy operations such that the input variables of transitions will be reasoned approximately and efficiently rather than precisely. This approximation or uncertainty is propagated along the net so that a predefined transition sequence may be followed as desired, or, if errors accumulate, the operations sequence should be stopped so that the error can be detected by the system monitor and the correct recovery sequence may be followed to recover to a correct system state. In this paper, we use local fuzzy variables to model the information which locally affects an operation, and we use fuzzy marking variables to represent the state of the system. The main differences between an ordinary Petri net and a fuzzy Petri net are the fuzzy values associated with places and tokens, and the reasoning rules which govern the firing of transitions. For an ordinary Petri net, a transition is fired if all input places contain at least one token. For a fuzzy Petri net, the condition of firing is also based on the local fuzzy variables associated with input places. Using a fuzzy Petri net to represent a robotic or manufacturing system, we are able to handle approximate information or uncertainty in the system. The reasoning about this information is incorporated into the firing rules of transitions.

Fuzzy Petri nets have been used to model robotic assembly systems[4, 5, 8]. A generalized definition was given in [8]. In this definition, local fuzzy variables, fuzzy marking variables, and global fuzzy variables were defined as different fuzzy information carried through the net. In [4], global fuzzy variables were used to model the degree of completion of the robotic task so that an operations sequence could be planned off-line while searching in the fuzzy Petri net model. Compared with the strategy used in ordinary Petri nets, computational time and space are saved because when the order of *key transitions* are given, all planned sequences should imply this order. Therefore, a correct sequence is defined as a feasible, complete, and correctly ordered sequence. Any sequence which fails in satisfying this definition is discarded during the off-line searching process. In [5], local fuzzy variables and global fuzzy variables are used simultaneously to model sensor-based robot task sequence planning and the execution of operations involving the use of sensory data. During the execution of a robot operations sequence, sensory operations can detect the partial result of some *key operations* on-line and the accumulation of errors will cause a local error recovery sequence or a global error recovery sequence. The local fuzzy variables decide the choice of error recovery strategy.

Previous work on predicate/transition nets has described approaches to handling predicate related expressions[12, 13]. Tokens in predicate/transition nets can be structured objects carrying values, and transition firing can be controlled by imposing conditions on the token values. Predicate/transition nets have been used for the management of expert systems, analysis in database systems, and many other applications. Research on colored Petri nets[16] reports related results though enabling limited reasoning capacity. The major difference between the predicate/transition net or the colored Petri net and the fuzzy Petri net is that the fuzzy Petri net can represent more generalized data using fuzzy numbers and fuzzy reasoning functions. The results of firing on some transition will depend not only on the input values, but also on a reasoning process built in the transition. The transition firing may depend on the local fuzzy variable or the global marking based on different applications. The strategy of property analysis on predicate/transition nets thus cannot be directly applied to the fuzzy Petri net.

Investigation of the properties of fuzzy Petri nets is very important for performance evaluation of a system being modeled. Reachability is a fundamental problem in the research on ordinary Petri nets. The reachability problem on fuzzy Petri nets is also defined on a feasible reachable set from an initial state. The reachability problem can influence other properties such as liveness, boundedness, and reversibility. Boundedness of fuzzy Petri nets is defined under the assumption that no more than one copy of a single object appears in the system. Liveness of a fuzzy Petri net implies that the reasoning process or the execution process can continue when the accumulation of errors is still within a range of tolerance, and when errors go over a threshold, an alternative sequence can be chosen in place of the original sequence. Reversibility implies that at any time if errors are too large, a home state is reachable.

The discussions on fuzzy Petri nets in this paper are arranged as follows: the next section presents the definition of the fuzzy Petri net. The concepts of fuzzy Petri nets and some related definitions will be given. Section 3 gives analysis for some basic cases. In Section 4, sensing operations are modeled as mutually exclusive transitions in a fuzzy Petri net and therefore reasoning on sensory data can be performed in these transitions.

2 Fuzzy Petri Nets

Some types of fuzzy Petri nets have been proposed to handle problems in different applications. When the Petri net model is applied to fuzzy rule-based reasoning using propositional logic[17], the tokens which represent *conditions* or *truths* are marked by crisp values between 0 and 1. The resulting net is considered as a new type of neural network where the transitions serve as the *neurons*, and the places serve as the *conditions*, so that fuzzy reasoning for knowledge could be performed. Based on the specific characteristic in logic reasoning that the truths are not consumed (or the tokens do not disappear) after the corresponding transitions are fired, the algorithm to reason about the fuzzy truth state vector using the Petri net model is based on the iterative conjunctive and disjunctive operations to update the truth tokens and neuron(transition) state components until a convergence is reached.

Using an augmented Petri net model, Chen, Ke, and Chang[10] developed an algorithm to determine whether there exists an antecedent-consequence relationship from a proposition d_s to another proposition d_j, and the relationship of the degree of truth of d_s and d_j may be devised. A production rule such as R_i : $IF\ d_j\ THEN\ d_k(CF = \mu_i)$, is modeled by a transition with its input place and output place. All possible paths are found from d_s to d_j following different types of reasoning rules represented by fuzzy Petri nets, and the number of paths is verified to be finite. Similarly, Garg, Ahson, and Gupta[11] proposed a set of reduction rules for the verification of consistency of a fuzzy knowledge base. An algorithm based on these reduction rules is introduced using Petri nets. In this approach, crisp values between 0 and 1 are associated with transitions for the truth values of fuzzy formulas, rather than tokens. In the process of reduction of an FPN, if a "Null Transition", which is equivalent to a "Null Clause", obtains a truth value greater than 0.5, then the knowledge base is inconsistent. Otherwise, it is consistent. This approach is useful for theorem proving using refutation.

The above approaches introduced real numbers in $[0,1]$ to describe the token values in places and/or the weights of transitions rather than conventional fuzzy numbers. Thus, their applications are very limited and all are used in reasoning about propositions in rule-based systems, or more generally, possibly in expert systems. However, the uncertainty and approximation in manufacturing or robotic systems are more complicated than those in rule-based systems, and crisp values in $[0,1]$ are not sufficient to describe them. In our approach, we propose a generalized definition of fuzzy Petri nets which incorporate reasoning and computational capability using fuzzy sets as well as non-fuzzy values such as those in the above discussions.

In monitoring and control of a flexible manufacturing system using Petri nets with objects[23], it may happen that the existence of an object modeled by a token is known but its localization is imprecise in terms of a set of alternative places. In [9, 24], Valette *et al.* modeled an event transforming an imprecise marking into another one using fuzzy dates, and represented the imprecision of the possible location of an object by possibility distributions. In a Petri net with objects, tokens are tuples of physical objects such as parts, machines or tools, and places are predicates representing the state of some objects. When an event is deduced by ***pseudo-firing*** a transition, tokens are added in output places of this transition, and the tokens in its input places are not removed. Thus the imprecision with the state of objects is augmented. When a message is received about the actual state of the system, the imprecision may be decreased, i.e., the certainty of the tokens may be restored. In [9], an algorithm is given for recomputing possibility distributions of the objects when a message arrives. Because the enabling duration associated with a transition is modeled by a fuzzy interval, a fuzzy firing date can be computed for reasoning about imprecision. In their recent article[25], the relations between Petri nets and logic and the implementation of Petri nets as rules are discussed.

Cao and Sanderson[4] applied global fuzzy variables in a fuzzy Petri net to the representation and planning of operations sequences using fuzzy reasoning strategies. The uncertainty of the accomplishment of subgoals was proposed for planning and reasoning about feasible, complete, and correctly ordered operations sequences with a robotic assembly system. Using prime numbers to mark the key transitions and the real numbers between 0 and 1 to mark the tokens, we are able to efficiently search all correct sequences and the time and space for searching would be saved. In a later approach[5], local fuzzy variables of the fuzzy Petri net were used for reasoning about and recovering from errors using a local or global error recovery sequence in the execution of a task sequence. A checking and verification procedure which may use sensing operations for the quality of the fulfillment of subgoals was investigated. Using this method, we reduce the propagation of errors for subgoals to a range of tolerance so that the correctness of the final goal is guaranteed. In [8], state representation corresponding to three different fuzzy variables, and the fuzzy reasoning rules are discussed.

The definition of the generalized fuzzy Petri net is shown as follows[8]:

Definition 1 *A fuzzy Petri net* is formally defined as an 8-tuple:

$$FPN = (P,\ T,\ Q_t,\ \alpha,\ \beta,\ m_f,\ m_t,\ \mu_f),$$

where

1) $P = \{p_1,\ p_2,\ \ldots,\ p_n\}$ is a finite set of *places*, $n \geq 0$.

2) $T = \{t_1,\ t_2,\ \ldots,\ t_m\}$ is a finite set of *transitions*, $m \geq 0$. $P \bigcap T = \emptyset$.

3) $Q_t = \{q_1,\ q_2,\ \ldots,\ q_l\}$ is a finite set of *state tokens*, $l \geq 0$.

4) $\alpha \subseteq \{P \times T\}$ is the input function, a set of directed arcs from places to transitions. We call each p_i where $(p_i, t_j) \in \alpha$ as an *input place* of t_j.

5) $\beta \subseteq \{T \times P\}$ is the output function, a set of directed arcs from transitions to places. We call each p_i where $(t_j, p_i) \in \beta$ as an *output place* of t_j.

6) $m_f : P \rightarrow \{(\rho, \varrho)\}$ assigns p_i the value of a 2-tuple, (ρ, ϱ), where ρ represents the *local fuzzy variable* and ϱ represents the *fuzzy marking variable*.

7) $m_t : Q_t \rightarrow \{\bigcup_i(k_i, \sigma_{k_i}),\ C\}$ is a mapping from a token to a union of 2-tuples of k_i and the k_ith *global fuzzy variable*, σ_{k_i}, or, to a constant, C, which indicates no global fuzzy variable is attached to the token. σ_{k_i} is a membership function in a universe of discourse.

8) $\mu_f : T \rightarrow \{f_1, f_2, \ldots, f_m\}$ is an association function, a mapping from transitions to corresponding *reasoning functions*. A reasoning function f_i maps variables associated with input places and a set of tokens to variables associated with output places and another set of tokens.

From the above definition, we know that three different types of variables are operated on or carried along through the net. A *local fuzzy variable* is attached to a place. Its value indicates uncertainty of the local variable or object which is attached to the place. A *fuzzy marking variable* is attached to a place and denotes uncertainty that a token exists in the given place. A *global fuzzy variable* is attached to a token and this is a characteristic variable of a global task. For instance, when we model a manufacturing system where object A would follow a sequence of state changes: $A_1, A_2, \ldots, A_k$. In this case, a fuzzy marking variable is used to indicate whether A_i exists; a local fuzzy variable is used to describe the local or geometric configuration of A_i; and a global fuzzy variable is used to represent the degree of completion of A_i compared with the global task which is a state sequence in this case for A.

If a place p_i represents an object O_i, then the local fuzzy variable associated with it can be represented as $\rho(p_i)$; the fuzzy marking variable associated with it can be represented as $\varrho(p_i)$; and the k_ith global fuzzy variable within this place can be represented as $\sigma_{k_i}(q_j)$, where q_j occupies p_i in the current state. If we consider the three types of variables for a set of places, $p_{j_1}, p_{j_2}, \ldots, p_{j_w}$, we can write them as

$$\rho(p_{j_1}, p_{j_2}, \ldots, p_{j_w}) = (\rho(p_{j_1}), \rho(p_{j_2}), \ldots, \rho(p_{j_w})) ;$$

$$\varrho(p_{j_1}, p_{j_2}, \ldots, p_{j_w}) = (\varrho(p_{j_1}), \varrho(p_{j_2}), \ldots, \varrho(p_{j_w})) ;$$

$$\sigma_{k_i}(q_{j'_1}, q_{j'_2}, \ldots, q_{j'_w}) = (\sigma_{k_i}(q_{j'_1}), \sigma_{k_i}(q_{j'_2}), \ldots, \sigma_{k_i}(q_{j'_w})) ;$$

where $q_{j'_1}, q_{j'_2}, \ldots, q_{j'_w}$ occupy $p_{j_1}, p_{j_2}, \ldots, p_{j_w}$. In the above statements, the values of ρ, ϱ, and σ may be either fuzzy membership functions or non-fuzzy crisp values.

Figure 1 shows an example of a robot assembly task which illustrates these three types of fuzzy variables. The robot gripper(R) moves to and grasps a strut(S) on the table, and the grasping state is described as RS. R has 6 degrees of freedom for the position and orientation of the gripper, $(x, y, z, \phi, \omega, \psi)$. S is defined by the position of its center and the angle between S and the x' axis, (x', y', θ). RS is described by x'', the distance between the grasping point and O'', the center point of S, under the assumption that the grasp position will be on the strut.

This assembly task can be represented by a fuzzy Petri net shown in Fig.2. In this representation, R_o and R_c means the robot gripper is open or closed, respectively. $\rho(R_o)$ and $\rho(R_c)$ are 6-D membership functions representing uncertainty in the robot position. $\rho(S)$ is a 3-D membership function representing uncertainty in the position of the strut on the table. $\rho(R_cS)$ is a 1-D membership function with parameter x'' which represents the uncertainty in the grasp point along the strut. We assume $\varrho(p_i)$ for this example is a fuzzy singleton, which describes the uncertainty of event completion. For example, in Fig.1(a), $\varrho(R_o, S, R_cS, R_c) = (1,1,0,0)$, and in Fig.1(b), $\varrho(R_o, S, R_cS, R_c) = (0, 0.1, 0.9, 0.1)$. If the token which represents an entity containing R carries a global fuzzy variable indicating the degree of completion of the task and under the assumption of $m_t(q_j)$ as a fuzzy singleton, then in Fig.2(a), $\sigma(R_o) = 0$, and in Fig.2(b), $\sigma(R_cS) = 1$, $\sigma(R_c) = 0$.

As with other Petri net models, we would like to use the fuzzy Petri net to analyze many properties such as reachability, liveness, boundedness, and reversibility. We define the reachable set of a fuzzy Petri net as the set of markings(consisting of fuzzy marking variables) reachable from the initial marking after all feasible transition sequences are fired. While it is complicated to analyze the properties of a fuzzy Petri net with all three types of fuzzy variables, some basic cases can be treated to yield useful properties. For an ordinary Petri net, all useful properties are defined based upon the markings on the net. The

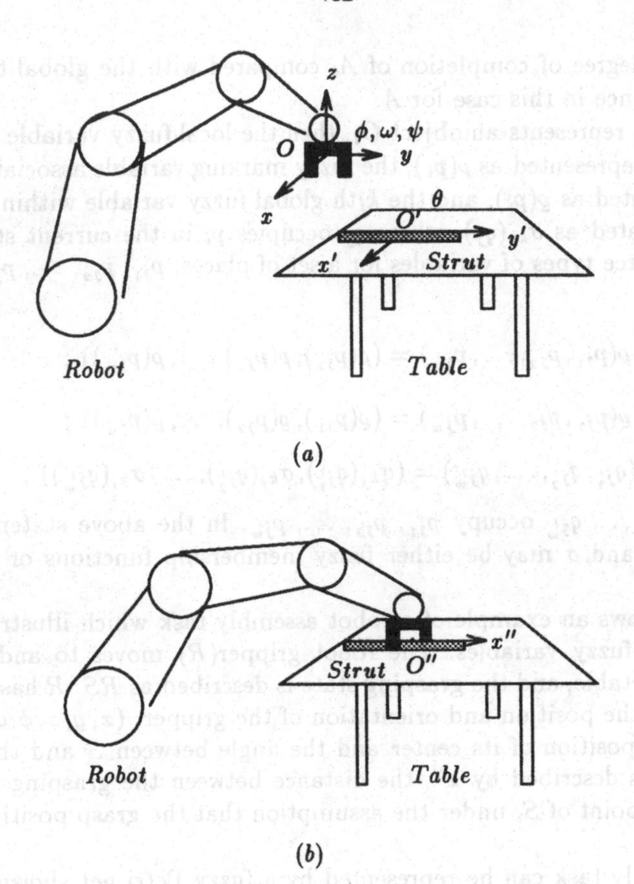

Fig. 1. A robotic task in an assembly system. (a) shows the initial state before "grasping". (b) shows the final state after "grasping".

fuzzy Petri net model adds the complexity of local and global variables which must be considered. In this analysis, we will consider first the problem where only local fuzzy variables exist in the fuzzy Petri net model.

There may be three different cases according to the above assumptions. First, firing rules follow from the ordinary Petri net and local fuzzy variables are unchanged. Second, firing rules follow from the ordinary Petri net and local fuzzy variables are changed. Third, firing rules are conditional upon input variables and local fuzzy variables are changed. These three cases are discussed in the following section.

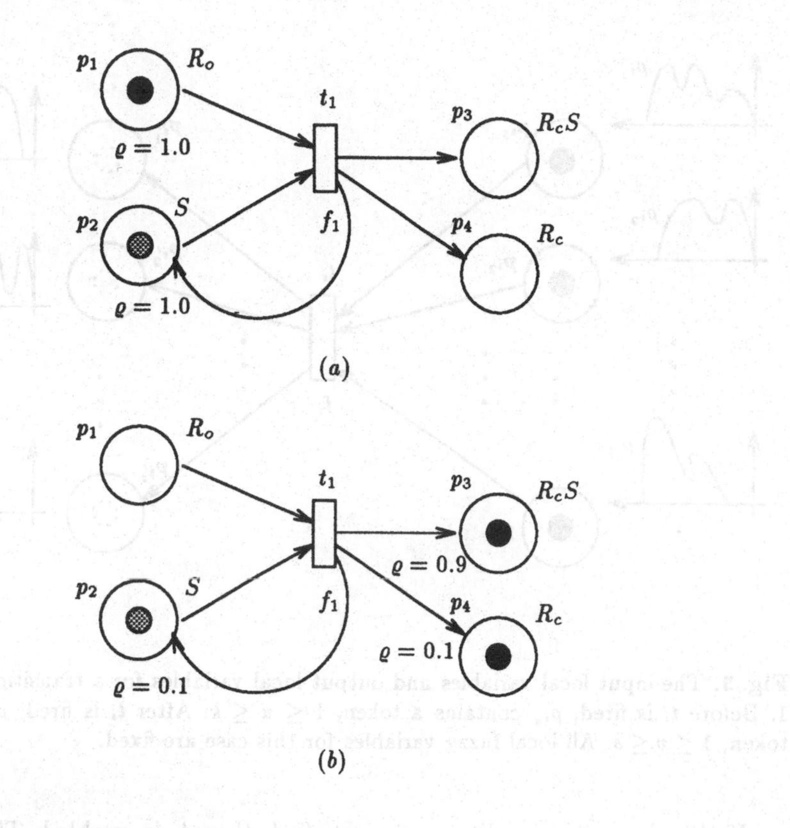

Fig. 2. The fuzzy Petri net representation for the robotic assembly task shown in Fig.1. (a) shows the initial state. (b) shows the final state.

3 Property Analysis for Several Basic Cases with Local Fuzzy Variables

Figure 3 shows a transition t_i in the net which has k input places which correspond to k variables, $p_{i_1}, p_{i_2}, \ldots, p_{i_k}$. t_i has s output places which correspond to s variables, $p_{i'_1}, p_{i'_2}, \ldots, p_{i'_s}$. The corresponding local fuzzy variables are represented as $\rho_{i_1}, \rho_{i_2}, \ldots, \rho_{i_k}$ and $\rho_{i'_1}, \rho_{i'_2}, \ldots, \rho_{i'_s}$. For the sake of simplicity, we draw one-dimensional membership functions to represent all these variables. Transition t_i has a reasoning function f_i. Two types of reasoning rules for f_i, $r_\rho^{t_i}$ and $r_\varrho^{t_i}$, are used for the following discussions.

3.1 Case 1: Local Fuzzy Variables Unmodified by Transitions

Case 1 can be described by two conditions: (1) membership function; (2) for some transition t_i, if all input places have at least one token at time N, i.e., $\varrho(p_{i_u}^{(N)}) \geq 1$, $1 \leq u \leq k$.

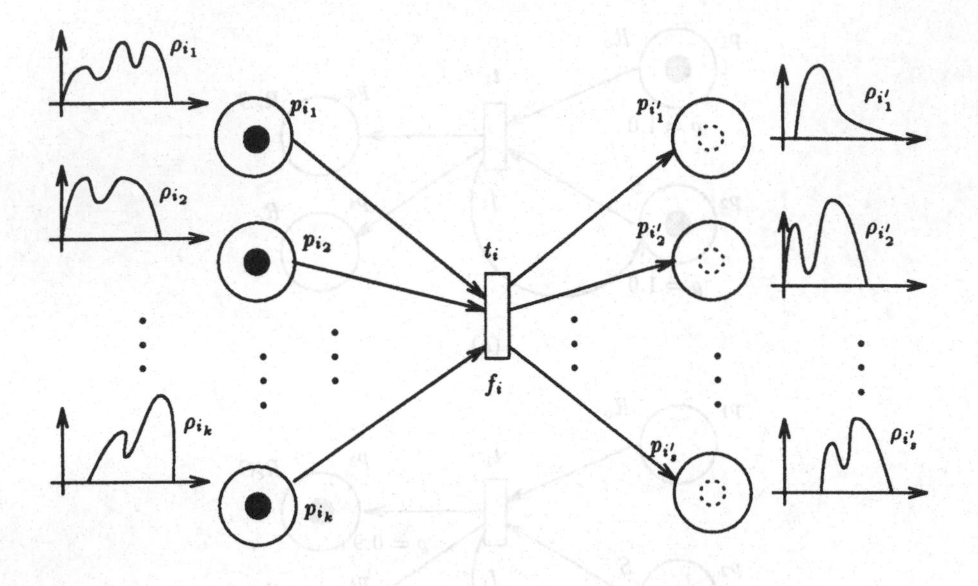

Fig. 3. The input local variables and output local variables for a transition t_i of Case 1. Before t_i is fired, p_{i_u} contains a token, $1 \leq u \leq k$. After t_i is fired, p_{i_v} obtains a token, $1 \leq v \leq s$. All local fuzzy variables for this case are fixed.

If the above two conditions are satisfied, then t_i is enabled. The resulting rule $r_{\varrho}^{t_i}$ can be written as

$$r_{\varrho}^{t_i}(\varrho(p_{i_1}^{(N)}), \varrho(p_{i_2}^{(N)}), \ldots, \varrho(p_{i_k}^{(N)})) = (\Delta(p_{i'_1}), \Delta(p_{i'_2}), \ldots, \Delta(p_{i'_s})) ,$$

and

$$\Delta(p_{i'_v}) = 1 , \quad 1 \leq v \leq s .$$

Then we can obtain ϱ for each output place as

$$\varrho(p_{i'_v}^{(N+1)}) = \varrho(p_{i'_v}^{(N)}) + \Delta(p_{i'_v}) , \quad 1 \leq v \leq s .$$

and

$$\varrho(p_{i_u}^{(N+1)}) = \varrho(p_{i_u}^{(N)}) - 1 , \quad 1 \leq u \leq k .$$

The membership functions for all local fuzzy variables are represented by the solid curves in Fig.3, and these curves are unchangeable when the system is executed.

Therefore, each time when we reach and fire transition t_i, we expect the same input local variables and output local variables and the output marking only depend on the input marking of t_i. This case can be considered as the same as the ordinary Petri net firing mechanism. The following properties can be seen for this type of FPN.

Theorem 1 Assume a fuzzy Petri net of Case 1 is mapped from an ordinary Petri net by assigning to each place a fixed local fuzzy variable, and assume the reasoning function is the same firing rule as in ordinary Petri nets. The resulting fuzzy Petri net is live, bounded, and/or reversible if and only if the original Petri net is live, bounded, and/or reversible.

Proof. All these properties are dependent upon the firing mechanisms of transitions in the net. Because the enabling conditions of transitions only depend on the markings of input places, the properties of liveness, boundedness, and reversibility are unchanged while the fuzzy Petri net is mapped from an ordinary Petri net or the fuzzy Petri net is mapped back to the ordinary Petri net. □

3.2 Case 2: Local Fuzzy Variables Modified by Transitions

Case 2 can be described by two conditions: (1) The output local variables of a transition t_i may be changed after t_i is fired, i.e.,

$$r_\rho^{t_i}(\rho(p_{i_1}^{(N)}), \rho(p_{i_2}^{(N)}), \ldots, \rho(p_{i_k}^{(N)})) = (\rho(p_{i'_1}^{(N+1)}), \rho(p_{i'_2}^{(N+1)}), \ldots, \rho(p_{i'_s}^{(N+1)})) ,$$

where N and $N+1$ represent the different time slots before and after t_i is fired; (2) for some transition t_i, if all input places have at least one token at time N, i.e., $\varrho(p_{i_u}^{(N)}) \geq 1$, $1 \leq u \leq k$.

If the above two conditions are satisfied, then t_i is enabled. The resulting rule $r_\varrho^{t_i}$ can be written as

$$r_\varrho^{t_i}(\varrho(p_{i_1}^{(N)}), \varrho(p_{i_2}^{(N)}), \ldots, \varrho(p_{i_k}^{(N)})) = (\Delta(p_{i'_1}), \Delta(p_{i'_2}), \ldots, \Delta(p_{i'_s})) ,$$

and

$$\varrho(p_{i'_v}^{(N+1)}) = \varrho(p_{i'_v}^{(N)}) + \Delta(p_{i'_v}) , \quad \Delta(p_{i'_v}) = 1 , \text{ and } 1 \leq v \leq s .$$

and

$$\varrho(p_{i_u}^{(N+1)}) = \varrho(p_{i_u}^{(N)}) - 1 , \quad 1 \leq u \leq k.$$

The membership functions of all local fuzzy variables are represented by dotted curves in Fig.4. These curves may be changed during the execution on the system. The generation of the output local variables are dependent on the input local variables and the rule $r_\rho^{t_i}$.

Case 2 is more general than Case 1. For the robotic assembly task in Fig.1, Case 1 fixes the uncertainty associated with each place, and thus the errors this case can model are very limited. For case 2, the configuration R_cS depends not only on transition t_1 but also the initial configuration $\rho(R_o)$ and $\rho(S)$. The following properties can be found for this type of fuzzy Petri net.

Theorem 2 Suppose a fuzzy Petri net is mapped from an ordinary Petri net by assigning to each place a changeable local fuzzy variable and the reasoning functions are the same as the firing strategy as defined in ordinary Petri nets. The resulting fuzzy Petri net is live, bounded, and/or reversible if and only if the original Petri net is live, bounded, and/or reversible.

Proof. The proof is similar to Theorem 1 because the changeable local fuzzy variables still have no influence on deciding the marking of the net. □

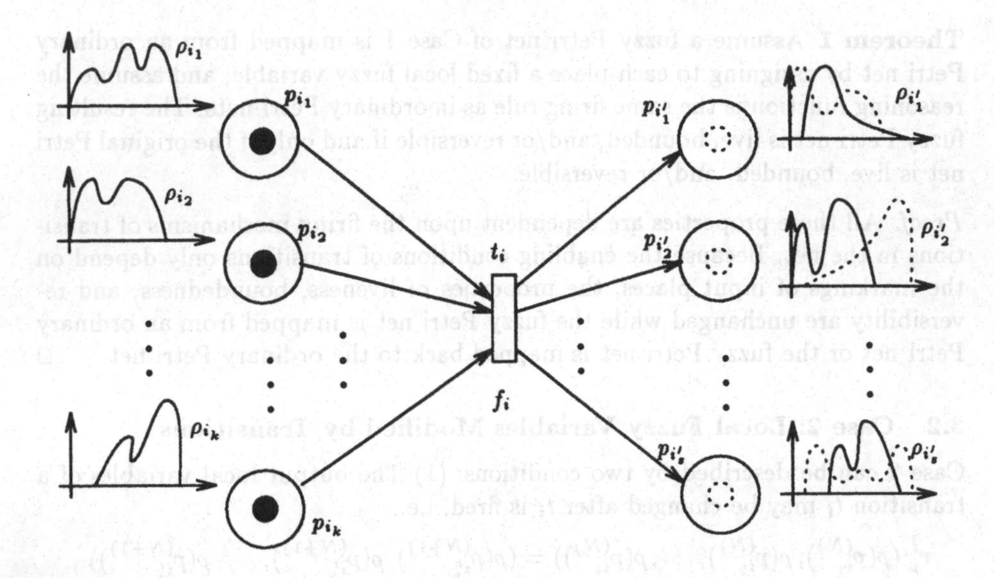

Fig. 4. The input local variables and output local variables for a transition t_i of Case 2. Before t_i is fired, p_{i_u} contains a token, $1 \leq u \leq k$. After t_i is fired, p_{i_v} obtains a token, $1 \leq v \leq s$. The output local fuzzy variables for this case are changed.

3.3 Case 3: Transition Firing Depends on Input Local Fuzzy Variables

Case 3 can be described by two conditions: (1) The output local variables of a transition t_i may be changed after t_i is fired, i.e.,

$$r_\rho^{t_i}(\rho(p_{i_1}^{(N)}), \rho(p_{i_2}^{(N)}), \ldots, \rho(p_{i_k}^{(N)})) = (\rho(p_{i'_1}^{(N+1)}), \rho(p_{i'_2}^{(N+1)}), \ldots, \rho(p_{i'_s}^{(N+1)})) ,$$

where N and $N+1$ represent the different time slots before and after t_i is fired; (2) for some transition t_i, if all input places have at least one token at time N, i.e., $\varrho(p_{i_u}^{(N)}) \geq 1$, $1 \leq u \leq k$.

If the above two conditions are satisfied, then t_i is enabled. If we fire t_i, then

$$r_\varrho^{t_i}(\varrho(p_{i_1}^{(N)}), \varrho(p_{i_2}^{(N)}), \ldots, \varrho(p_{i_k}^{(N)}), \rho(p_{i_1}^{(N)}), \rho(p_{i_2}^{(N)}), \ldots, \rho(p_{i_k}^{(N)}))$$
$$= (\Delta(p_{i'_1}), \Delta(p_{i'_2}), \ldots, \Delta(p_{i'_s})) ,$$

and

$$\Delta(p_{i'_v}) \in \{0,1\} , \quad 1 \leq v \leq s .$$

Then we can obtain ϱ for each output place as

$$\varrho(p_{i'_v}^{(N+1)}) = \varrho(p_{i'_v}^{(N)}) + \Delta(p_{i'_v}) , \quad 1 \leq v \leq s ,$$

and

$$\varrho(p_{i_u}^{(N+1)}) = \varrho(p_{i_u}^{(N)}) - 1 \; , \quad 1 \leq u \leq k \; .$$

The membership functions of all local fuzzy variables are represented by dotted curves in Fig.5. These curves may change during the execution of the system.

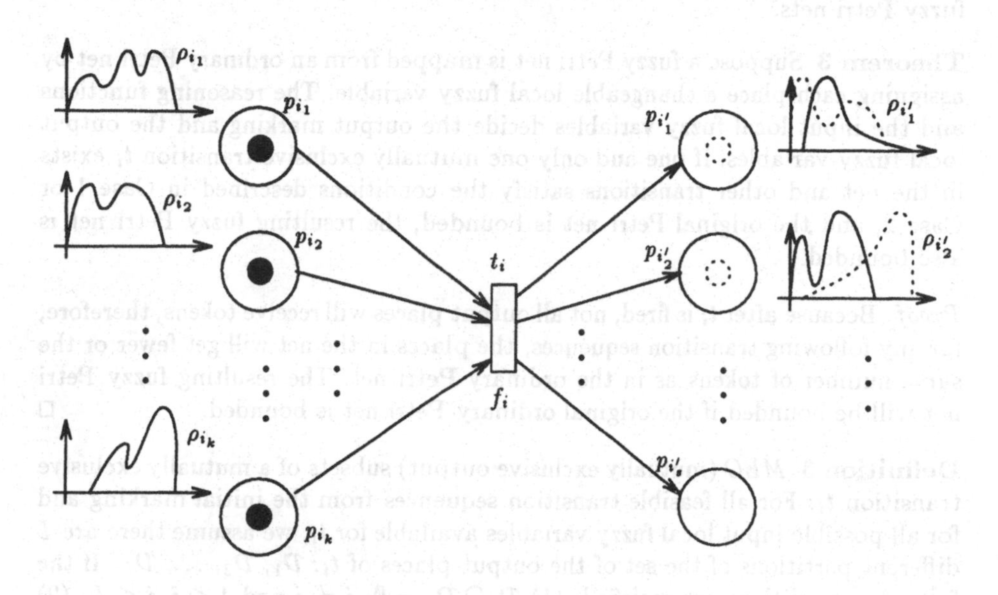

Fig. 5. The input local variables and output local variables for a transition t_i of Case 3. Before t_i is fired, p_{i_u} contains a token, $1 \leq u \leq k$. After t_i is fired, some p_{i_v} obtain a token, $1 \leq v < s$. The output local fuzzy variables for this case are changed.

For this case, the output local fuzzy variables are dependent on input local fuzzy variables and the fuzzy reasoning rules, and the output markings after the t_i is fired are dependent upon the input local fuzzy variables, the input marking, and the firing functions. Therefore, even if all input places have tokens, after t_i is fired, it does not necessarily guarantee that all output places get more tokens. One application of this type of fuzzy Petri net is sensor-based selection for on-line robotic operations. After a sensor is used to verify a system state, the following operation may be local error recovery, global error recovery, or continuation of the execution of the planned task sequence, all of which depend on the token in one output place of a sensing transition.

The properties of this type of fuzzy Petri net are not provable in general because of different reasoning strategies which decide the availability of tokens in output places dependent upon input local fuzzy variables and the rule $r_{\varrho}^{t_i}$. Some useful subclasses of this case are worthwhile to investigate. We first give the following definition:

Definition 2 A mutually exclusive transition t_i: (1) if $p_{i'_1}$, $p_{i'_2}$, ..., and $p_{i'_s}$ are s output places *of* t_i; (2) After t_i is fired, $\Delta(\varrho(p_{i'_j})) = 1$ or $\Delta(\varrho(p_{i'_j})) = 0$, $1 \leq j \leq s$, and $\mathcal{D} = \bigcup_j \{p_{i'_j} : \Delta(\varrho(p_{i'_j})) = 1\}$, $\mathcal{D}' = \bigcup_j \{p_{i'_j} : \Delta(\varrho(p_{i'_j})) = 0\}$; (3) $\mathcal{D} \neq \emptyset$, $\mathcal{D}' \neq \emptyset$, $\mathcal{D} \bigcap \mathcal{D}' = \emptyset$, and $\mathcal{D} \bigcup \mathcal{D}' = \bigcup_{j=1}^{s} \{p_{i'_j}\}$.

The following theorem provides the condition for boundedness for a class of fuzzy Petri nets.

Theorem 3 Suppose a fuzzy Petri net is mapped from an ordinary Petri net by assigning each place a changeable local fuzzy variable. The reasoning functions and the input local fuzzy variables decide the output marking and the output local fuzzy variables. If one and only one mutually exclusive transition t_i exists in the net and other transitions satisfy the conditions described in Case 1 or Case 2, and the original Petri net is bounded, the resulting fuzzy Petri net is also bounded.

Proof. Because after t_i is fired, not all output places will receive tokens, therefore, for any following transition sequences, the places in the net will get fewer or the same number of tokens as in the ordinary Petri net. The resulting fuzzy Petri net will be bounded if the original ordinary Petri net is bounded. □

Definition 3 *MEO* (mutually exclusive output) subsets of a mutually exclusive transition t_i: For all feasible transition sequences from the initial marking and for all possible input local fuzzy variables available for t_i, we assume there are L different partitions of the set of the output places of t_i: $\mathcal{D}_1$, $\mathcal{D}_2$, ..., $\mathcal{D}_L$. If the following conditions are satisfied: (1) $\mathcal{D}_i \bigcap \mathcal{D}_j = \emptyset$, $i \neq j$ and $1 \leq i, j \leq L$; (2) $\bigcup_{l=1}^{L} \mathcal{D}_l = \bigcup_{j=1}^{s} p_{i'_j}$; (3) At any time after t_i is fired, $\Delta(p_{i'_v}) = 1$, for all $p_{i'_v} \in \mathcal{D}_j$ and $\Delta(p_{i'_v}) = 0$ for all $p_{i'_v} \notin \mathcal{D}_j$, $1 \leq v \leq s$ and $1 \leq j \leq L$, then $\mathcal{D}_1$, $\mathcal{D}_2$, ..., $\mathcal{D}_L$ are called *MEO* subsets of t_i.

Figure 6 shows a mutually exclusive transition with four output places and four examples of possible output markings. Figure 6(a), 6(b), and 6(c) show *MEO* subsets. In Fig.6(a), $L = 2$, $\mathcal{D}_1 = \{p_3, p_4\}$, $\mathcal{D}_2 = \{p_5, p_6\}$, which indicates that sometimes after t_1 is fired, $\varrho(p_3) = \varrho(p_4) = 1$, and $\varrho(p_5) = \varrho(p_6) = 0$, and sometimes $\varrho(p_5) = \varrho(p_6) = 1$ and $\varrho(p_3) = \varrho(p_4) = 0$. No other possibilities may appear for ϱ. In Fig.6(b), $L = 3$, $\mathcal{D}_1 = \{p_3\}$, $\mathcal{D}_2 = \{p_4, p_5\}$, and $\mathcal{D}_3 = \{p_6\}$, and in Fig.6(c), $L = 2$, $\mathcal{D}_1 = \{p_3, p_6\}$, *and* $\mathcal{D}_2 = \{p_4, p_5\}$. Fig.6(d) shows non-*MEO* subsets where $\varrho(p_4) = 1$, $\varrho(p_5) = 1$ and $\varrho(p_3) = 1$, $\varrho(p_4) = 1$ are both possible after firing t_i. The following theorems define a class of fuzzy Petri nets which guarantee the properties of liveness, boundedness, and reversibility.

Before we discuss the following theorem, we give definitions of the addition of two Petri nets and subtraction of a subnet from a Petri net. Similar operations on Petri nets are used in synthesis techniques of Petri nets[15].

Definition 4 The addition("+") of a net $N_1 = (P_1, T_1, \alpha_1, \beta_1)$ and a net $N_2 = (P_2, T_2, \alpha_2, \beta_2)$: $N_1 + N_2 = (P_1 \bigcup P_2, T_1 \bigcup T_2, \alpha, \beta)$ where $\alpha = \bigcup_{ij} \{(p_i, t_j)\}$, $(p_i, t_j) \in \alpha_1 \bigcup \alpha_2$, and $\beta = \bigcup_{ij} \{(t_j, p_i)\}$, $(t_j, p_i) \in \beta_1 \bigcup \beta_2$.

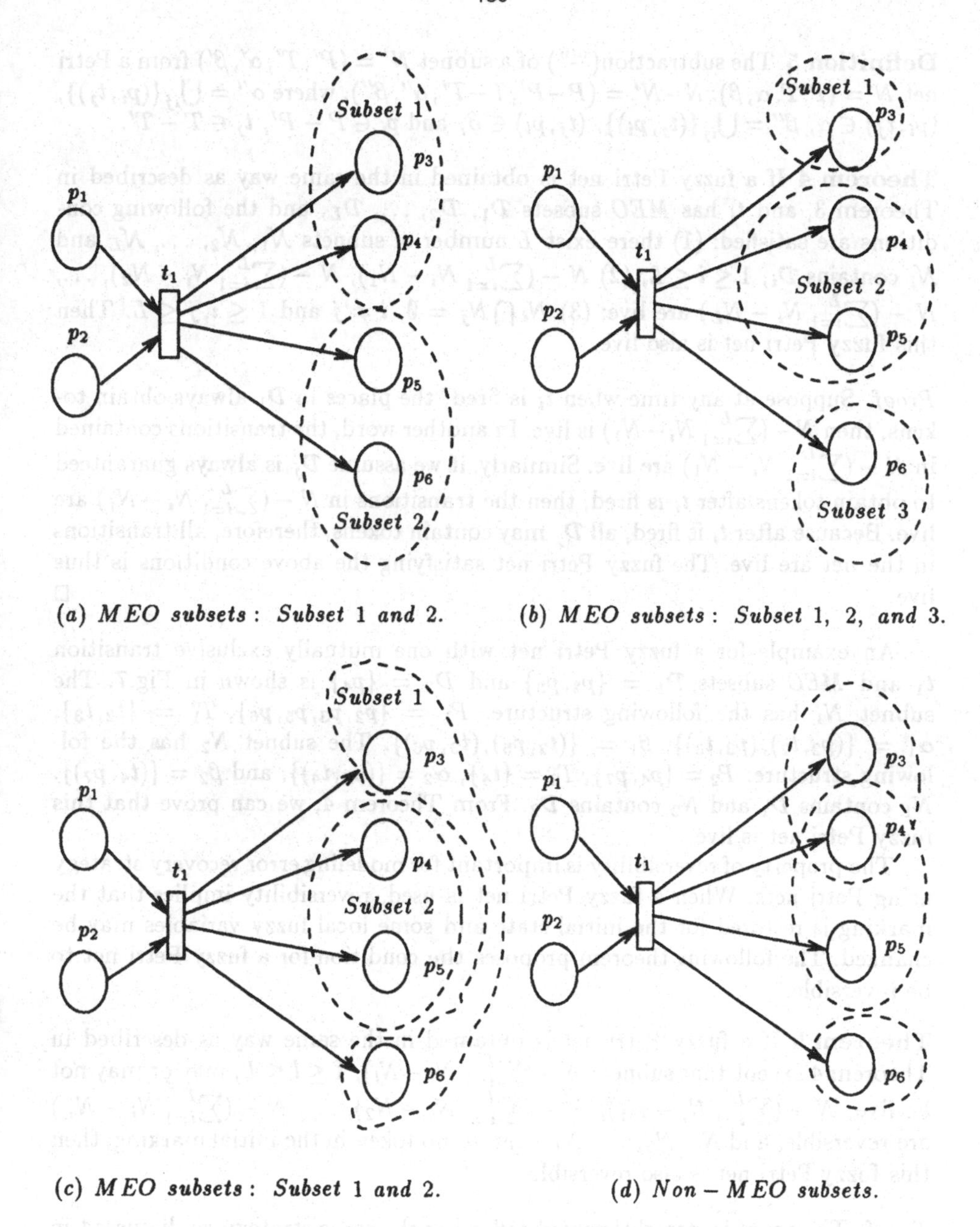

(a) *MEO subsets : Subset* 1 *and* 2. (b) *MEO subsets : Subset* 1, 2, *and* 3.

(c) *MEO subsets : Subset* 1 *and* 2. (d) *Non − MEO subsets.*

Fig. 6. Some examples of *MEO* subsets for a mutually exclusive transition with four output places.

Definition 5 The subtraction("-") of a subnet $N' = (P', T', \alpha', \beta')$ from a Petri net $N = (P, T, \alpha, \beta)$: $N - N' = (P - P', T - T', \alpha'', \beta'')$, where $\alpha'' = \bigcup_{ij}\{(p_i, t_j)\}$, $(p_i, t_j) \in \alpha$, $\beta'' = \bigcup_{ij}\{(t_j, p_i)\}$, $(t_j, p_i) \in \beta$, and $p_i \in P - P'$, $t_j \in T - T'$.

Theorem 4 If a fuzzy Petri net is obtained in the same way as described in Theorem 3, and t_i has *MEO* subsets $\mathcal{D}_1, \mathcal{D}_2, \ldots, \mathcal{D}_L$, and the following conditions are satisfied: (1) there exist L number of subnets $\mathcal{N}_1, \mathcal{N}_2, \ldots, \mathcal{N}_L$ and $\mathcal{N}_i$ contains $\mathcal{D}_i$, $1 \le i \le L$; (2) $N - (\sum_{i=1}^{L} N_i - N_1)$, $N - (\sum_{i=1}^{L} N_i - N_2)$, ..., $N - (\sum_{i=1}^{L} N_i - N_L)$ are live; (3) $N_i \bigcap N_j = \emptyset$, $i \neq j$ and $1 \le i, j \le L$. Then this fuzzy Petri net is also live.

Proof. Suppose at any time when t_i is fired, the places in $\mathcal{D}_1$ always obtain tokens, then $N - (\sum_{i=1}^{L} N_i - N_1)$ is live. In another word, the transitions contained in $N - (\sum_{i=1}^{L} N_i - N_1)$ are live. Similarly, if we assume $\mathcal{D}_j$ is always guaranteed to obtain tokens after t_i is fired, then the transitions in $N - (\sum_{i=1}^{L} N_i - N_j)$ are live. Because after t_i is fired, all $\mathcal{D}_j$ may contain tokens, therefore, all transitions in the net are live. The fuzzy Petri net satisfying the above conditions is thus live. □

An example for a fuzzy Petri net with one mutually exclusive transition t_1 and *MEO* subsets $\mathcal{D}_1 = \{p_2, p_3\}$ and $\mathcal{D}_2 = \{p_4\}$ is shown in Fig.7. The subnet N_1 has the following structure: $P_1 = \{p_2, p_3, p_5, p_6\}$, $T_1 = \{t_2, t_3\}$, $\alpha_1 = \{(p_2, t_2), (p_3, t_3)\}$, $\beta_1 = \{(t_2, p_5), (t_3, p_6)\}$. The subnet N_2 has the following structure: $P_2 = \{p_4, p_7\}$, $T_2 = \{t_4\}$, $\alpha_2 = \{(p_4, t_4)\}$, and $\beta_2 = \{(t_4, p_7)\}$. N_1 contains $\mathcal{D}_1$ and N_2 contains $\mathcal{D}_2$. From Theorem 4, we can prove that this fuzzy Petri net is live.

The property of reversibility is important for modeling error recovery strategy using Petri nets. When a fuzzy Petri net is used, reversibility implies that the marking is restored for the initial state and some local fuzzy variables may be changed. The following theorem proposes the condition for a fuzzy Petri net to be reversible.

Theorem 5 If a fuzzy Petri net is obtained in the same way as described in Theorem 4 except that subnets $N - (\sum_{i=1}^{L} N_i - N_l)$, $1 \le l \le L$, may or may not be live, $N - (\sum_{i=1}^{L} N_i - N_1)$, $N - (\sum_{i=1}^{L} N_i - N_2)$, ..., $N - (\sum_{i=1}^{L} N_i - N_L)$ are reversible, and $N_1, N_2, \ldots, N_L$ contain no token in the initial marking, then this fuzzy Petri net is also reversible.

Proof. The proof is straightforward following the same strategy as discussed in Theorem 4. □

If there are more than one mutually exclusive transition, $t_{i_1}, t_{i_2}, \ldots, t_{i_r}$, existing in the fuzzy Petri net, we can generalize the above discussions to the following corollaries:

Corollary 1 Suppose a fuzzy Petri net is mapped from an ordinary Petri net by assigning each place a changeable local fuzzy variable. The reasoning functions

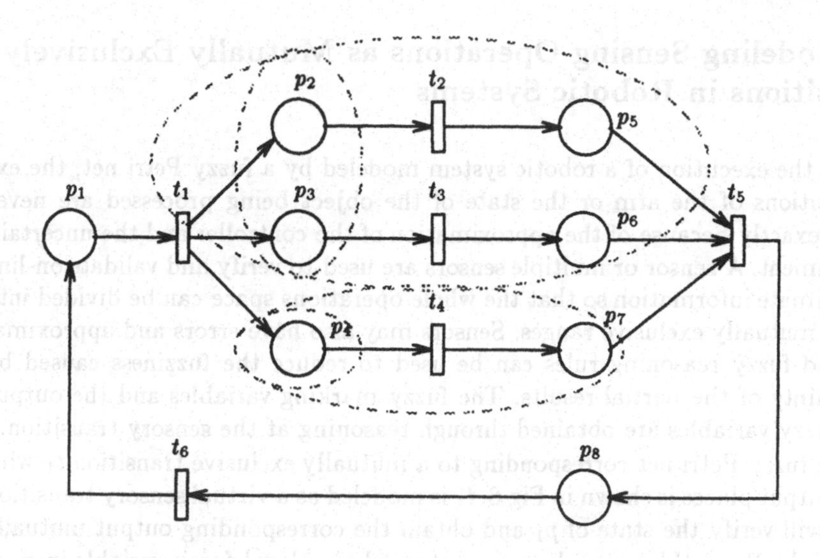

Fig. 7. A fuzzy Petri net with one mutually exclusive transition t_1. After t_1 is fired, p_2 and p_3 or p_4 will receive the tokens based on the local variable available in p_1 and $r_{\varrho}^{t_1}$.

and the input local fuzzy variables decide the output marking and the output local fuzzy variables. If more than one mutually exclusive transition, $t_{i_1}, t_{i_2}, \ldots, t_{i_r}$, exists in the net and the original Petri net is bounded, the resulting fuzzy Petri net is also bounded.

Corollary 2 If a fuzzy Petri net is obtained in the same way as described in Corollary 1, and $t_{i_u} (1 \leq u \leq r)$ has *MEO* subsets $D_{i_{u_1}}, D_{i_{u_2}}, \ldots, D_{i_{u_L}}$. And the following conditions are satisfied: (1) there exist u_L number of subnets $N_{i_{u_1}}, N_{i_{u_2}}, \ldots, N_{i_{u_L}}$ and N_{i_j} contain $\mathcal{D}_{i_j}$, $u_1 \leq j \leq u_L$; (2) $N - (\sum_{j=u_1}^{u_L} N_{i_j} - N_{i_{u_1}})$, $N - (\sum_{j=u_1}^{u_L} N_{i_j} - N_{i_{u_2}}), \ldots, N - (\sum_{j=u_1}^{u_L} N_{i_j} - N_{i_{u_L}})$ are live; (3) $N_{i_p} \bigcap N_{i_q} = \emptyset$, $p \neq q$ and $u_1 \leq p, q \leq u_L$, then this fuzzy Petri net is also live.

Corollary 3 If a fuzzy Petri net is obtained in the same way as described in Corollary 2 except that the subnets $N - (\sum_{j=u_1}^{u_L} N_{i_j} - N_{i_l})$, $u_1 \leq l \leq u_L$ may or may not be live. $N - (\sum_{j=u_1}^{u_L} N_j - N_{i_{u_1}})$, $N - (\sum_{j=u_1}^{u_L} N_j - N_{i_{u_2}}), \ldots, N - (\sum_{j=u_1}^{u_L} - N_{i_{u_L}})$ are reversible, and $N_{i_1}, N_{i_2}, \ldots, N_{L_u}$ contain no token in the initial marking, then this fuzzy Petri net is also reversible.

In the above discussions, Case 3 may be considered as the generalization of Case 1 and Case 2, and Case 1 or Case 2 are special examples of Case 3. Thus, The results obtained for Case 3 can also be used for Case 1 and Case 2.

4 Modeling Sensing Operations as Mutually Exclusively Transitions in Robotic Systems

During the execution of a robotic system modeled by a fuzzy Petri net, the exact positions of the arm or the state of the object being processed are never known exactly because of the approximation of the controller and the uncertain environment. A sensor or multiple sensors are used to verify and validate on-line approximate information so that the whole operations space can be divided into several mutually exclusive ranges. Sensors may also have errors and approximation and fuzzy reasoning rules can be used to reduce the fuzziness caused by uncertainty of the partial results. The fuzzy marking variables and the output local fuzzy variables are obtained through reasoning at the sensory transition.

The fuzzy Petri net corresponding to a mutually exclusive transition t_1 with three output places is shown in Fig.8. t_1 is modeled as a virtual sensory transition which will verify the state of p_1 and obtain the corresponding output mutually exclusively. From this example, we see that when a local fuzzy variable in p_1 is obtained, it is input into t_1 for reasoning about the token in an output place. The reasoning function $r_e^{t_1}$ consists of three steps: (1) intersect input fuzzy variables with the expected membership function residing in p_2 and obtain the intersection area or the highest membership degree, (2) intersect input fuzzy variables with the expected membership function residing in p_3 and obtain the intersection area or the highest membership degree, (3) intersect input fuzzy variables with the expected membership function residing in p_4 and obtain the intersection area or the highest membership degree. Then, from these three partial results, we can get a maximum value corresponding to a certain place p_i, $2 \leq p \leq 4$. In this example, p_3 has the maximum intersection area as shown by the shaded area in Fig.8(a). Therefore, a token is put in p_3 after the reasoning process ends as shown in Fig.8(b).

In a real system, a sensing operation may direct the following transition sequences to continue the execution, locally recover from an error if that error is locally recoverable, or globally recover from an error if that error is not locally recoverable. Each sensing operation is a mutually exclusive transition, and its output places constitute *MEO* subsets.

5 Conclusions

In this paper, we present a generalized definition of fuzzy Petri nets using three different types of fuzzy variables: local fuzzy variables, fuzzy marking variables, and global fuzzy variables. Fuzzy local variables are examined in detail, and are used to reason about parameters associated with places. Properties of these nets have been defined for specific cases of interest. An example of the application of these properties is modeling and analysis of a sensor-based robotic system. Uncertain sensory input data can be handled and sensory transitions may be modeled as mutually exclusive transitions. Only a subset of the output places

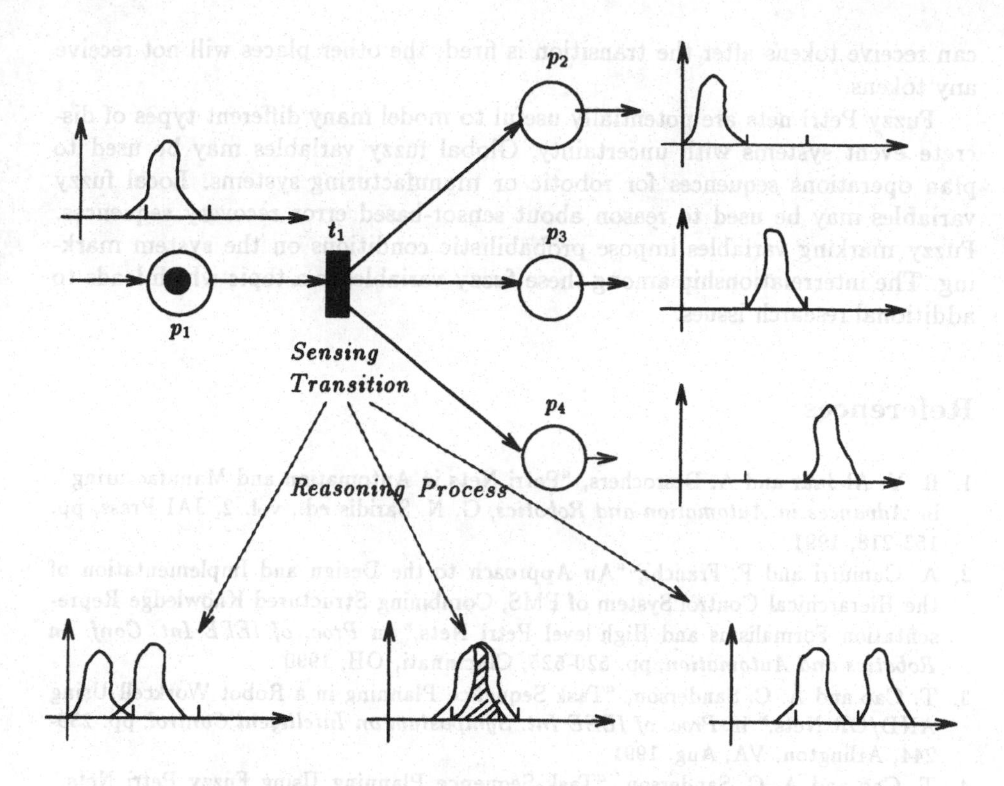

(a) Fuzzy Reasoning of a mutually exclusively sensing transition.

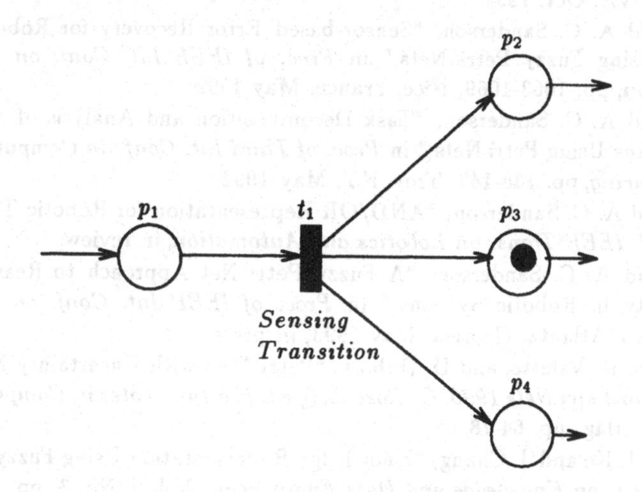

(b) The resulting token in place p_3.

Fig. 8. Fuzzy reasoning in a fuzzy Petri net for obtaining a token in an output place mutually exclusively.

can receive tokens after the transition is fired, the other places will not receive any tokens.

Fuzzy Petri nets are potentially useful to model many different types of discrete event systems with uncertainty. Global fuzzy variables may be used to plan operations sequences for robotic or manufacturing systems. Local fuzzy variables may be used to reason about sensor-based error recovery sequences. Fuzzy marking variables impose probabilistic conditions on the system marking. The interrelationship among these fuzzy variables is a topic which leads to additional research issues.

References

1. R. Y. Al-Jaar and A. Desrochers, "Petri Nets in Automation and Manufacturing," in *Advances in Automation and Robotics*, G. N. Saridis ed., vol. 2, JAI Press, pp. 153-218, 1991
2. A. Camurri and P. Franchi, "An Approach to the Design and Implementation of the Hierarchical Control System of FMS, Combining Structured Knowledge Representation Formalisms and High-level Petri Nets," in *Proc. of IEEE Int. Conf. on Robotics and Automation*, pp. 520-525, Cincinnati, OH, 1990
3. T. Cao and A. C. Sanderson, "Task Sequence Planning in a Robot Workcell Using AND/OR Nets," in *Proc. of IEEE Int. Symposium on Intelligent Control*, pp. 239-244, Arlington, VA, Aug. 1991
4. T. Cao and A. C. Sanderson, "Task Sequence Planning Using Fuzzy Petri Nets," in *Proc. of IEEE Int. Conf. on Systems, Man, and Cybernetics*, pp. 349-354, Charlottesville, VA, Oct. 1991
5. T. Cao and A. C. Sanderson, "Sensor-based Error Recovery for Robotic Task Sequences Using Fuzzy Petri Nets," in *Proc. of IEEE Int. Conf. on Robotics and Automation*, pp. 1063-1069, Nice, France, May 1992
6. T. Cao and A. C. Sanderson, "Task Decomposition and Analysis of Assembly Sequence Plans Using Petri Nets," in *Proc. of Third Int. Conf. on Computer Integrated Manufacturing*, pp. 138-147, Troy, NY, May 1992
7. T. Cao and A. C. Sanderson, "AND/OR Representation for Robotic Task Sequence Planning," *IEEE Trans. on Robotics and Automation*, in review
8. T. Cao and A. C. Sanderson, "A Fuzzy Petri Net Approach to Reasoning about Uncertainty in Robotic Systems," in *Proc. of IEEE Int. Conf. on Robotics and Automation*, Atlanta, Georgia, May 1993, in press
9. J. Cardoso, R. Valette, and D. Dubois, "Petri Nets with Uncertainty Markings," in *Advances in Petri Nets 1990, G. Rozenberg ed. (Lecture Notes in Computer Science)*, Springer-Verlag, pp. 64-78
10. S. Chen, J. Ke and J. Chang, "Knowledge Representation Using Fuzzy Petri Nets," *IEEE Trans. on Knowledge and Data Engineering*, Vol. 2, No. 3, pp. 311-319, Sep. 1990
11. M. L. Garg, S. I. Ahson, and P. V. Gupta, "A Fuzzy Petri Net for Knowledge Representation and Reasoning," *Information Processing Letters*, Vol. 39, pp. 165-171, 1991
12. H. J. Genrich and K. Lautenbach, "Systems Modeling with High-Level Petri Nets," *Theoretical Computer Sciences*, 13, pp. 109-136, 1981

13. A. Giordana and L. Saitta, "Modeling Production Rules by Means of Predicate Transition Networks," *Information Sciences*, 35, pp. 1-41, 1985
14. M. Jantzen, "Structured Representation of Knowledge by Petri Nets as an aid for Teaching and Research," in *Net Theory and Applications (Lecture Notes in Computer Science 84)*, Springer-Verlag, pp. 507-516, 1980
15. M. D. Jeng, *Theory and Applications of Resource Control Petri Nets for Automated Manufacturing Systems*, Ph.D. Thesis, Rensselaer Polytechnic Institute, Aug. 1992
16. K. Jensen, "Coloured Petri Nets," in *Advances in Petri Nets 1986*, Vol. 254, Springer-Verlag, pp. 288-299, 1988
17. C. G. Looney, "Fuzzy Petri Nets for Rule-Based Decisionmaking," *IEEE Trans. on Systems, Man, and Cybernetics*, Vol. 18, No. 1, pp. 178-183, Jan./Feb. 1988
18. R. E. Miller, "A Comparison of Some Theoretical Models of Parallel Computations," *IEEE Trans. on Computers*, Vol. C-22, No. 8, pp. 710-717, Aug. 1973
19. S. Moriguchi and G. S. Shedler, "Diagnosis of Computer Systems By Stochastic Petri Nets," *Ieice Trans. on Fundamentals of Electronics Communications and Computer Sciences*, Vol. E75A, pp. 1369-1377, Oct. 1992
20. T. Murata, "Petri Nets: Properties, Analysis and Applications," *Proc. of the IEEE*, Vol. 77, No. 4, pp. 541-580, Apr. 1989
21. J. L. Peterson, *Petri Net Theory and the Modeling of Systems*, Prentice Hall, 1981
22. C. V. Ramamoorthy and G. S. Ho, "Performance Evaluation of Asynchronous Concurrent Systems Using Petri Nets," *IEEE Trans. on Software Engineering*, Vol. SE-6, No. 5, pp. 440-449, 1980
23. C. Sibertin-Blanc, "High-level Petri Nets with Data Structures," in *6th European Workshop on Applications and Theory of Petri Nets*, Helsinki, Finland, June 1985.
24. R. Valette, J. Cardoso and D. Dubois, "Monitoring Manufacturing Systems by Means of Petri Nets with Imprecise Markings," in *Proc. of IEEE Int. Symposium on Intelligent Control*, pp. 233-237,1989
25. R. Valette and M. Courvoisier, "Petri Nets and Artificial Intelligence," *International Workshop on Emerging Technologies for Factory Automation*, North Queensland, Australia, Aug. 17-19, 1992

Distributed Simulation of Timed Petri Nets: Exploiting the Net Structure to Obtain Efficiency*

Giovanni Chiola[1] and Alois Ferscha[2]

[1] Università di Torino, Dipartimento di Informatica
I-10149 Torino, Italy
[2] Universität Wien, Institut für Statistik und Informatik
A-1080 Vienna, Austria

Abstract. The conservative and the optimistic approaches of distributed discrete event simulation (DDES) are used as the starting point to develop an optimized simulation framework for studying the behaviour of large and complex timed transition Petri net (TTPN) models. This work systematically investigates the interdependencies among the DDES strategy (conservative, Time Warp) and the spatial decomposition of TTPNs into logical processes to be run concurrently on individual processing nodes in a message passing and shared memory multiprocessor environment. Partitioning heuristics are developed taking into account the structural properties of the TTPN model, and the simulation strategy is tuned accordingly in order to attain the maximum computational speedup. Implementations of the simulation framework have been undertaken for the Intel iPSC/860 hypercube, the Sequent Balance and a Transputer based multiprocessor. The simulation results show that the use of the Petri net formalism allows an automatic extraction of the parallelism and causality relations inherent to the model.

1 Introduction

The popularity of the timed Petri net modelling and analysis framework for the quantitative (and simultaneously qualitative) study of asynchronous concurrent systems is mainly due to the availability of software packages automating the analysis algorithms and often providing a powerful, graphic user interface. Although the use of Petri net tools shows that complex simulation models are hard and costly to evaluate, simulation is often the only practical analysis means. Parallel and distributed simulation techniques have in some practical applications, and at least in theory, improved the performance of large systems simulation, but are still far from a general acceptance in science and engineering. The main reasons for this pessimism are the lack of automated tools and simulation languages

* The work at the University of Torino was performed in the framework of the Esprit BRA Project No.7269, QMIPS. The work at the University of Vienna was financially supported by the Austrian Ministry for Science and Research under Grant GZ 613.525/2-26/90.

suited to parallel and distributed simulation, as well as performance limitations inherent to the simulation strategies.

This paper points out the special suitability of the Petri net formalism for parallel and distributed simulation: it allows an automatic exploitation of useful model parallelism by structural Petri net analysis; moreover, based on some Petri net properties, simulation strategy can be designed in order to optimize overall execution time. Causal relations among events are explicitly described by the Petri net structure so that a proper exploitation of this information may potentially improve the performance of general purpose, distributed simulation engines. A TTPN is decomposed into a set of spatial regions that bear potentially concurrent events, based on a preanalysis of the net topology and initial marking. The decomposed TTPN is assigned to multiple processors with distributed memory, working asynchronously in parallel to perform the global simulation task. Every processor hence simulates events according to a local (virtual) time and synchronizes with other processors by rollbacks of its local time caused by messages with time stamp in the local past — thus no synchronous communication is required at the software level (neither the send nor the receive message operations are blocking to the processors). Input buffers are assumed to be available for every processor to keep arriving messages until they are withdrawn by an input process between the simulation of two consecutive events.

1.1 Timed Transition Petri Nets

In the sequel of this work we consider the class of timed Petri net models where timing information is associated to transitions [1]:

Definition 1.1 *A timed Petri net is a tuple* $TTPN = (P, T, F, W, \Pi, \tau, M_0)$ *where:*

(i) $P = \{p_1, p_2, \ldots, p_{n_P}\}$ *is a finite set of P-elements (places),*
$T = \{t_1, t_2, \ldots, t_{n_T}\}$ *is a finite set of T-elements (transitions)*
with $P \cap T = \emptyset$ *and a nonempty set of nodes* $(P \cup T \neq \emptyset)$.

(ii) $F \subseteq (P \times T) \cup (T \times P)$ *is a finite set of arcs between P-elements and T-elements denoting input flows* $(P \times T)$ *to and output flows* $(T \times P)$ *from transitions.*

(iii) $W : F \mapsto \mathbf{N}$ *assigns weights* $w(f)$ *to elements of* $f \in F$ *denoting the multiplicity of unary arcs between the connected nodes.*

(iv) $\Pi : T \mapsto \mathbf{N}$ *assigns priorities* π_i *to T-elements* $t_i \in T$.

(v) $\tau : T \mapsto \mathbf{R}$ *assigns firing delays* τ_i *to T-elements* $t_i \in T$.

(vi) $M_0 : P \mapsto \mathbf{N}$ *is the marking* $m_i = m(p_i)$ *of P-elements* $p_i \in P$ *with tokens in the initial state of TTPN (initial marking).*

A transition t_i with $\pi_i = \Pi(t_i)$ is said to be *enabled with degree* $k > 0$ *at priority level* π_i in the marking M, iff $k > 0$ is the maximum integer such that $\forall p_j \in {}^\bullet t_i$, $m(p_j) \geq k\ w(p_j, t_i)$ in M (${}^\bullet t_i = \{p_j \mid (p_j, t_i) \in F\}$ and $t_i^\bullet = \{p_j \mid (t_i, p_j) \in F\}$). t_i is said to be *enabled with degree* $k > 0$, iff there

are no transitions enabled with positive degree on a priority level higher than π_i. The multiset of all transitions enabled in M (with positive enabling degree representing the multiplicity of the transition) is denoted $E(M)$. A transition instance t_i that has been continuously enabled during the time τ_i must *fire*, in that it removes $w(p_j, t_i)$ tokens from every $p_j \in {}^\bullet t_i$ by at the same time placing $w(t_i, p_k)$ tokens into every $p_k \in t_i^\bullet$. This firing of a transition takes zero time and is denoted by $M[t_i\rangle M'$.

1.2 Approaches to DDES in the Literature

Discrete event simulations of TTPNs exploit a natural correspondence between transition firings and event occurrences: an event in the simulated system relates to the firing of a transition in the TTPN model. In general, a simulation engine (SE) maintains an ordered list of time stamped events (transitions) scheduled for their occurrence (firing) time and repeatedly executes the first event in the list by simulating the behaviour caused by the event. The *virtual time* of the simulation is changed to the event's timestamp; new events caused are scheduled and events having lost their cause are descheduled. In terms of the TTPN model the marking is changed according to the transition firing, and newly enabled (disabled) transitions are scheduled (descheduled).

Distributed simulation strategies generally aim at dividing a global simulation task into a set of communicating *logical processes* (LPs) trying and exploiting the parallelism inherent in the concurrent execution of these processes. With respect to the target architecture we distinguish among *parallel discrete event simulation* if the implementation is for a tightly coupled multiprocessor allowing simultaneous access to shared data (shared memory machines), and *distributed discrete event simulation* (DDES) if loosely coupled multiprocessors with communication based on message passing are addressed. In this work we only restrict to DDES. Notice however that the simulation algorithms developed for DDES trivially adapt to the execution on shared memory multiprocessors without performance loss.

Indeed event processing must respect causality constraints even in parallel and distributed simulation: classical DDES approaches preserve causality by simply enforcing timestamp ordering in event processing. Two main protocols have been proposed for the implementation of DDES: *conservative* that enforces time stamp order processing by restricting the parallelism among LPs and *optimistic* that instead allows out of timestamp order to occur and then correct the situation by "undoing" part of the simulation. Both, the *conservative* [2] and the *optimistic* strategy [3] have led to implementations of general purpose discrete event simulators on parallel and distributed computing facilities (for a survey see [4]). The main potential contribution of the use of Petri net model descriptions for DDES is to define causality relations explicitly through the graph structure, thus allowing one to relax constraints on timestamp ordering for the processing of events that are not causally related. From this consideration we can thus expect to be able to increase the parallelism of a DDES engine with respect to the

standard approaches by just taking the structure of the PN model into proper account.

In conservative DDES strategies all interactions among LPs are based on timestamped event messages sent between the corresponding LPs. Out-of-time-order processing of events in LPs is prevented by forcing LPs to block as long as there is the possibility for receiving messages with lower timestamp, which in turn has a severe impact on simulation speed. Optimistic DDES strategies weaken this strict synchronization rule by allowing a preliminary causality violation, which is corrected by a (costly) detection and recovery mechanism usually called rollback (in simulated virtual time).

In order to guarantee a proper synchronization among LPs in optimistic simulation, two rollback mechanisms have been studied. In the original Time Warp [3] an LP receiving a message with timestamp smaller than the LPs locally simulated time (*straggler message*) starts a *rollback* procedure right away. Rollback is initiated by revoking all effects of messages that have been processed prematurely by restoring to the nearest safe state, and by neutralizing, i.e. sending "antimessages" for messages sent since that state (*aggressive cancellation, ac*). The impact of the erroneously sent messages is that also succeeding LPs might be forced to rollback, thus generating a rollback chain that eventually terminates. Reducing the size of the rollback chain is attempted by "filtering" messages with preempted effects, by postponing erroneous message annihilation until it turns out that they are not reproduced in the repeated simulation (*lazy cancellation, lc*) [5], or by maintaining a causality record for each event to support *direct cancellation* of messages that will definitely not be reproduced. The performance of rollback mechanisms is investigated in [6] and [7].

1.3 Paper Organization

The balance of the rest of this paper is the following. Section 2 introduces classical concepts of DDES and adapts them to the particular case of TTPN models. Different kinds of customizations of DDES strategies and partitioning in concurrent LPs are discussed in a general framework. Section 3 presents empirical results derived from the implementation of several variations of DDES strategies for TTPN models on 3 different multiprocessor architectures. Results are analyzed to identify weaknesses and potentialities of the different alternatives and to validate some of the proposed net specific optimizations. Section 4 outlines additional optimizations that may be introduced in the most promising DDES strategy to take into account particularly favourable net topologies. Section 5 contains concluding remarks and perspectives for future developments of the research.

2 A Distributed Simulation Framework for TTPNs

2.1 Logical Processes

In the most general case of distributed simulation of TTPNs we consider a decomposition of the "sequential" discrete event simulation task into a set of communicating LPs to be assigned to individual processing elements working autonomously and communicating with each other for synchronization. Each and only one LP is assigned to a dedicated processor and is residing there for the whole real simulation time. A message transportation system is assumed to be available (either implemented fully in software or by hardware routing facilities) that implements directed, reliable, FIFO communication channels among LPs. Every LP applies one and the same discrete event strategy to the simulation of the local subnet. We identify the work partition assigned to LPs, the LPs' communication behaviour and the LPs' simulation strategy as the constituent parts of a distributed simulation framework, and will in further treat these three components in a very abstract form.

Definition 2.1 *A* Logical Process *is a tuple* $LP_k = (TTPNR_k, I_k, SE)$ *where*

(*i*) **region**
$TTPNR_k = (P_k, T_k, F_k \subseteq (P_k \times T_k) \cup (T_k \times P_k), W, \Pi, \tau, M_0)$ *is a (spatial)* region *of some TTPN such that* $P = \bigcup_{k=1}^{N_{LP}} P_k$, $T = \bigcup_{k=1}^{N_{LP}} T_k$, $F = \bigcup_{k=1}^{N_{LP}} F_k \bigcup_{i \in neighborhood(TTPNR_k)}((P_k \times T_i) \cup (T_k \times P_i))$, *which are again TTPNs in the sense of Definition 1.1,*

(*ii*) **communication interface**
$I_k = (CH, m)$, *the* communication interface *of* LP_k *is a set of communication channels* $CH = \bigcup_{k,i} ch_{k,i}$ *where* $ch_{k,i} = (LP_k, LP_i)$ *is a directed channel from* LP_k *to* LP_i, *corresponding to an arc* $f \in F$ *and carrying messages* $m = \langle w(f), D, TT \rangle$ *where* $w(f)$ *(*w_f *for short) is the number of tokens transferred,* D *is an identifier of the destination (place or transition) and* TT *(token time) is the timestamp of the local virtual time of the sending* LP_k *at send time.*

(*iii*) **simulation engine**
SE *is a* simulation engine *implementing the simulation strategy.*

The set of all LPs $LP = \bigcup_k LP_k$ *together with directed communication channels* $CH = \bigcup_{k,i} ch_{k,i} = \bigcup_{k,i}(LP_k, LP_i)$ *constitute the* Graph of Logical Processes $GLP = (LP, CH)$.

Within the frame of this general definition of a logical simulation process a variety of different distributed simulations are possible. Not all combinations of decompositions into regions, communication interfaces and simulation engines make sense in practical situations: most of the combinations will not conform to the expectations of increasing simulation speed over sequential simulation. Moreover they are highly interrelated: e.g. the simulation engine affects the communication interface and the appropriateness of decompositions.

To the best of our knowledge three approaches have been developed so far in the literature ([8], [9], [10]), which would be categorized according to our framework as follows.

In [8] an LP is created for every $p \in P$ and every $t \in T$, resulting in a very large number of LPs ($| LP | = | P | + | T |$) of smallest possible grain size, where the maximum number of arcs between any pair of LPs is 1. The idea behind this partition choice is to maximize the "potential parallelism" of the DDES in terms of number of LPs. For every arc an unidirectional channel with the same orientation as the arc is introduced. For every input arc from a place's LP and a transition's LP another unidirectional channels with the opposite orientation is required (control channel) in order to solve conflicts. The I_k proposed is based on a protocol invoking four messages ("activate," "request," "grant," and "confirm") among a transition's LP and all its input places' LPs in order to fire the transition. Obviously the (possibly tremendous) amount of messages inherent to this approach prevents efficiency if simulated in a distributed memory environment (their protocol has been simulated on a shared memory multiprocessor). The proposed simulation engine is *conservative*.

Another conservative approach has been introduced by [9], allowing *"completely general"* decompositions into regions and employing a simulation engine that handles three kinds of events: the arrival of a token, the start and the end of transition firing in order to exploit lookahead. The generality of the partition is limited in such a way that conflicting transitions together with all their input places are always put together in the same LP, in order to avoid distributed conflict resolution.

Recently [10] a Time Warp simulation (i.e., a simulation engine following the optimistic approach) of SPNs has been proposed also allowing any partitioning in the $TTPNR_k$ sense, but with a redundant representation of places (and the corresponding arcs) in adjacent LPs. Their I_k maintains time consistency of markings among LPs by exchanging five different types of messages. The freedom of allowing any partitioning that stems from arbitrary arc cutting makes the message exchange protocol and the simulation engine unnecessarily complex.

In the following we systematically work out *useful*, (and then *improved*) combinations of TTPN regions, communication interfaces and simulation engines.

2.2 TTPN Regions

The most natural decomposition of work is the spatial partitioning of a TTPN into disjoint *regions* TTPNR_k representing small(er) sized TTPNs. This also supports the application of one and the same simulation engine to all regions.

Apparently the decomposition of a TTPN has strong impact on the DDES performance. Small sized regions naturally raise high communication demands, whereas large scale regions can help to clamp local TTPN behaviour inside one LP. The common drawback of the intuitive partitionings in [8] and [10] is the necessity of developing a proper communication protocol among LPs in order to implement a distributed conflict resolution policy for transitions sharing input places. In a message passing environment for inter process communication, such a conflict resolution protocol may induce substantial overhead in the distributed simulation, thus nullifying the advantages of the potential model parallelism on the simulation time. "Arbitrary arc cutting" for partitioning can burden performance since arcs $f \in (P \times T)$ have a (severe) consequence on the efficient computation of enabling and firing, whereas arcs out of $(T \times P)$ have not such consequence. The latter is the reason why the partitioning has to be related to the TTPN firing rule: an LP is a minimum set of transitions T_k along with its input set ($^\bullet(T_k) = \bigcup_{i|t_i \in T_k} {}^\bullet t_i$) such that local information is sufficient to decide upon the enabling and firing of any $t_i \in T_k$. This is in order to minimize conflict resolution overhead, since in this way conflict resolution always occurs internally to an LP, thus involving no communication overhead.

Minimum Regions Let $CS(t_i)$ be the set of all transitions $t_j \in T$ in structural conflict with t_i ($t_i, t_j \in T$ are said to be in ***structural conflict*** denoted by (t_i SC t_j), iff $^\bullet t_i \cup {}^\bullet t_j \neq \emptyset$. In the case of GSPNs with priority structures [11] ***extended conflict set*** $ECS(t_i)$ have been defined as the sets of all transitions $t_j \in T$ in indirect structural conflict with t_i. To relate the region partitioning to the firing semantics definition of TTPNs, all transitions within the same CS (or ECS) have to be included in the same TTPNR$_k$, since a distribution over several regions would require distributed conflict resolution (which would in any case degrade performance). This consideration leads directly to a partitioning into ***minimum regions***, i.e. TTPNR$_k$s that should not be subdivided further in order to be practically ***useful***:

Definition 2.2 ***TTPNR$_k$ = $(P_k, T_k, F_k, W, \Pi, \tau, M_0)$ of some TTPN, $\bigcup_{k=1}^{N_{LP}} T_k = T$, where $T_k = \{t_i\}$, $t_i \in T$ is a single transition iff $t_i \in T$ does not participate in any ECS or $T_k = ECS(t_i) \subseteq T$, $t_i \in T$ is the whole ECS iff $t_i \in T$ participates in some ECS, $P_k = \bigcup_{t_i \in T_k} {}^\bullet(t_i)$ and $F_k \subseteq (P_k \times T_k)$ is called a*** **minimum region** ***of TTPN.***

Any general TTPN can always (and uniquely) be partitioned into ***minimum regions*** according to Definition 2.2.

2.3 The Communication Interface

Of course, the decomposition of TTPNs requires an interface among TTPNRs preserving the behavioural semantics of the whole TTPN. Such an interface has to be implemented by an appropriate communication protocol among the regions, according to the strategy chosen for the simulation engines.

Since we have forced the regions not to distribute $p \in P_i$ and $t \in T_k$ over different regions TTPNR$_i$ and TTPNR$_k$ if there exists an arc $f = (p,t) \in (P_i \times T_k)$ (remember that for each $TTPNR_k$, $t_i \in T_k \Rightarrow {}^\bullet(t_i) \in P_k$), we can map the sets of arcs $(T_k \times P_j) \subseteq F$ interconnecting different TTPN regions to the channels of the communication interface. In other words we define $I_k = (CH, m)$ of TTPNR$_k$ to be the communication interface with channels $CH = \bigcup_{k,i} ch_{k,i}$, where $ch_{k,i} = (LP_k, LP_i)$ is a set of directed channel from LP$_k$ to LP$_i$, corresponding to all the arcs $(t_l, p_m) \in (T_k \times P_i) \subseteq F$, $t_l \in T_k$, $p_m \in P_i$, carrying messages m of identical type. A message m represents tokens to be deposited in places of adjacent LPs due to the firing of local transitions.

The description (and implementation) of the interaction among LPs is also simplified in the following way. Consider two TTPN regions TTPNR$_i$ and TTPNR$_j$ and let $T_i^{\triangleright j} = \bigcup_l t_l \mid (t_l, p) \in (T_i \times P_j)$ be the set of transitions in TTPNR$_i$ incident to arcs pointing to places in TTPNR$_j$, and analogously $P_j^{\triangleleft i} = \bigcup_l p_l \mid$

$(t, p_l) \in (T_i \times P_j)$ the set of places in TTPNR$_j$ incident to arcs originating from transitions in TTPNR$_j$, then the interaction of LP$_i$ and LP$_j$ only concerns $T_i^{\triangleright j}$ and $P_j^{\triangleleft i}$. We call $T_i^{\triangleright j}$ ($P_j^{\triangleleft i}$) the *output border* (*input border*) of TTPNR$_i$ *towards* TTPNR$_j$ (of TTPNR$_j$ *from* TTPNR$_i$). $T_k^{\triangleright} = \bigcup_i T_k^{\triangleright i}$ is the *output border* of TTPNR$_k$, $P_k^{\triangleleft} = \bigcup_i P_k^{\triangleleft i}$ is the *input border*. (Naturally, $P_k^{\triangleleft} = P_k$ and $T_k^{\triangleright} = T_k$ for a minimum region.)

As a static measure of the potential communication induced by LP$_i$ we can define its *arc-degree* (*of connectivity*) as the number of arcs $f \in F$ originating in TTPNR$_i$ and pointing to places in adjacent TTPNRs and vice-versa:

$$\mathrm{AD}(TTPNR_i) = \left| \bigcup_{k \in \mathrm{neighborhood}(TTPNR_i)} (P_i \times T_k) \cup (T_k \times P_i) \right|$$

Similarly we define the *channel-degree*

$$\mathrm{CD}(TTPNR_i) = \left| \bigcup_k ch_{i,k} \right| + \left| \bigcup_k ch_{k,i} \right|$$

as the number of channels connected to a region. Note that these measures can be evaluated in a static preanalysis of the decomposed TTPN.

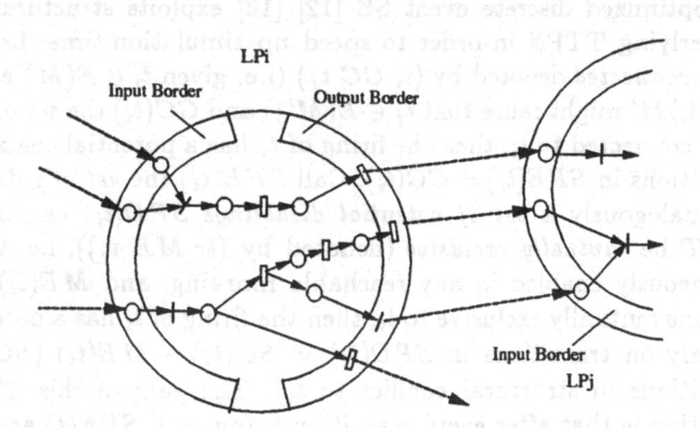

Fig. 1. TTPN Region for a Logical Process.

Figure 1 shows a TTPN region to become the "local simulation task" of some LP. The set of dashed TTPN arcs from LP$_i$ to LP$_j$ represent the directed communication channel among them. The arc-degree of connectivity of LP$_i$ in Fig. 1 is 7, while a single channel is needed to connect LP$_i$ to LP$_j$, thus inducing a channel-degree ≤ 5 for LP$_i$. Figure 2 shows the minimum region partitioning — Definition 2.2 — of a GSPN model of the reader-writer problem [11].

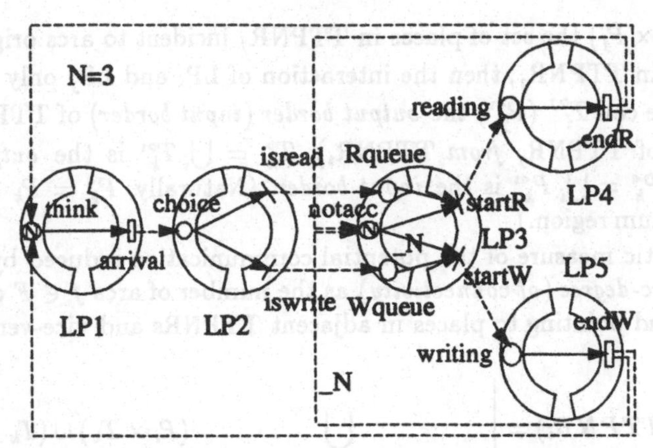

Fig. 2. Reader/Writer Example: Minimum Region Partitioning.

2.4 The Simulation Engine

A general *simulation engine* (SE) implements the simulation of the occurrence of events in virtual time according to their causality; at the same time it collects a trace of event occurrences over the whole simulation period. Data structures of a general SE include: the TTPN representation (in some internal form); an event list (EVL) with entries $e_k = \langle t_i @ FT \rangle$ (fire transition t_i at time FT); the virtual time (VT); the event stack (ES) with entries of the form $\langle t_i, VT, M \rangle$ (t_i is the transition that has fired at virtual time VT, yielding a new marking M).

An optimized discrete event SE [12] [13] exploits structural properties of the underlying TTPN in order to speed up simulation time: Let $t_i, t_j \in T$ be *causally connected* denoted by $(t_i\ CC\ t_j)$ (i.e. given $t_i \in E(M)$ and $t_j \notin E(M)$, then $M[t_i\rangle M'$ might cause that $t_j \in E(M')$) and $CC(t_i)$ the set of all transitions causally connected to t_i, then the firing of t_i has a potential enabling effect only on transitions in $SPE(t_i) = CC(t_i)$. Call $SPE(t_i)$ the *set of potential enablings* of t_i. Analogously a *set of potential disablings* $SPD(t_i)$ can be defined: Let $t_i, t_j \in T$ be *mutually exclusive* (denoted by $(t_i\ ME\ t_j)$), i.e. they cannot be simultaneously enabled in any reachable marking, and $ME(t_i)$ the set of all transitions mutually exclusive to t_i, then the firing of t_i has a potential disabling effect only on transitions in $SPD(t_i) = SC(t_i) - ME(t_i)$ ($SC(t_i)$ is the set of transitions in structural conflict to t_i). The gain of this TTPN structure exploitation is that after every transition firing only $SPE(t)$ and $SPD(t)$ have to be investigated rather than the whole T, which is significant for $\mid SPE(t_i) \mid$, $\mid SPD(t_i) \mid \ll \mid T \mid$.

The (optimized) SE behaves as follows. After an initialization and a preliminary scheduling of events caused by the initial marking the algorithm performs — if there are events to be simulated — the following steps until some end time is reached. First all simultaneously enabled transitions are generated and conflict is resolved identifying one transition per actual conflict set to be fired; the first scheduled transition is fired in that it is removed from EVL, the marking is changed accordingly, possible new (obsolete) transition instances are scheduled (descheduled) by insertion (deletion) of new (old) events into (from) EVL (investigating only SPE and SPD !); the occurrence of the event at its virtual simulated time and the new marking are logged into ES, and the VT is updated to the occurrence time of the event processed. This (sequential) general SE is adapted to DDES strategies as described in the following sections.

2.5 The Conservative Strategy

The SE following the conservative approach allows only the processing of *safe* events, i.e. the firing of transitions up to a *local* virtual time LVT for which the LP has been guaranteed not to receive token messages with timestamp "in the past." The causality of events is preserved over all LPs by sending timestamped token messages of type $m = \langle w, D, TT \rangle$ (with $w > 0$) in non decreasing order, or at least a promise $m = \langle 0, D, TT \rangle$ (*null message*) not to send a new message time-stamped earlier than TT, and by processing the corresponding events in nondecreasing time stamp order.

One basic practical problem is the determination of when it is safe to process an event, since the degree to which LPs can *look ahead* and predict future events plays a critical role in the performance of the DDES. For conservative DDES so called "lookahead" coming directly from the TTPN structure can be exploited. Given some transition t_i located in the output border of some LP$_k$ ($t_i \in T_k^{\triangleright}$). t_i is said to be *persistent* if there is no $t_j \in T_k$ such that if $t_i, t_j \in E(M)$, and $M[t_j\rangle M'$ causes that $t_i \notin E(M')$ for all reachable markings M. A sufficient condition for persistence of t_i is that $\forall t_j,\ {}^\bullet t_i \cap {}^\bullet t_j = \emptyset$. Call t_j the persistent predecessor of t_i if ${}^\bullet t_i = t_j^\bullet$ and $\forall k \neq j,\ {}^\bullet t_i \cap t_k^\bullet = \emptyset$; let $T_k^{pers}(t_i)$ be the set of all persistent predecessors ahead of t_i, i.e. the transitive closure of the persistent predecessor relation. Then a lower bound for the degree of lookahead exposed by LP$_k$ via t_i is $\sum_{j|t_j \in T_k^{pers}(t_i)} \tau_j$, i.e., the sum over all firing times of transitions in $T_k^{pers}(t_i)$.

Definition 2.3 *Let $t_i \in T_k^{\triangleright}$ and $t_j \in T_k$ be persistent $(({}^\bullet t_i)^\bullet = t_i, ({}^\bullet t_j)^\bullet = t_j)$. t_i and t_j are said to be in the same persistence chain denoted by the set $(t_j \leadsto t_i)$, iff there exists a sequence $S = \{t_r, t_s \dots t_t\}$ of persistent predecessors $(t_r, t_s \dots t_t \in T_k)$ such that $t_j \in E(M) \Rightarrow t_i \in E(M')$ $M[S\rangle M'$.*

Since every $t \in S$ is persistent, we can state the following:

Corollary 2.1 *Given that $t_j \in T_k \leadsto_i \in T_k^{\triangleright}$ and M with $t_j \in E(M)$ at time τ, then M' with $t_i \in E(M')$ is reached at time $\tau + \sum_{t \in S} \tau(t)$ at the earliest.*

For two transitions $t_j \in T_k \leadsto t_i \in T_k^{\triangleright}$ in the same persistence chain we can define the amount of lookahead of t_j on t_i by $la((t_j \leadsto t_i)) = \sum_{t \in (t_j \leadsto t_i)} \tau(t)$ (with the particular case $la((t_j \leadsto t_i)) = 0$ if $(t_j \leadsto t_i) = \emptyset$). The value of la can be established for all pairs of transitions in the TTPN region (one in the output border, the other not in the output border) by a static preanalysis of the region structure. This can improve the simulation performance since upon firing of any transition within some $TTPNR_k$ the timestamp of output messages caused by transitions in the output border which are in the same persistence chain can be increased (and thus improved) by la, thus relaxing the synchronization constraints on the adjacent LPs.

Assuming that the communication requirements of $GLP = (LP, CH)$ are supported by the multiprocessor hardware (i.e. there exist communication media for all $ch_{k,i} = (LP_k, LP_i)$), then we derive the following conservative simulation engine SE^{co} to be applied in every $LP_i \in LP$. Two types of messages are necessary to implement the communication interface: ***token messages*** $m = \langle w_f, D, TT\rangle$ carrying a specific number of tokens (w_f or #) to some destination place D (which uniquely defines the destination LP), and ***null messages*** $m = \langle 0, D, TT\rangle$. SE^{co} holds the net data of $TTPNR_k$ and simulates the local behaviour of the region by holding transitions to be fired in a local EVL, recording event occurrences in a local ES and incrementing a local time LVT. For every input channel an input queue collects incoming messages, the first element of which (that defines the ***channel clock***, CC) is used for synchronization. Output buffers OB keep messages to be sent to other LPs, one per output channel. The behaviour of SE^{co} is to process the first event of EVL if there is no token message in one of the CC_is with smaller timestamp (***process first event***), or to process the token message with the minimum token time (***process first message***). Processing the first event (i.e., firing transition t) is similar to the general SE (change marking, schedule/deschedule events, increment LVT), but also invokes the sending of messages: If $t \in T_k^{\triangleright}$ then a message $\langle \#, D, LVT\rangle$ is generated and deposited in the corresponding OB; if $t \notin T_k^{\triangleright}$ then a null message $\langle 0, D, LVT + la((t \rightsquigarrow t_i))\rangle$ is deposited for every $t_i \in T_k^{\triangleright}$ in the corresponding OBs – thus giving maximum lookahead to all the following LPs. After processing the first event the contents of all OBs is transmitted, except for null messages with token times that have already been distributed in a previous step (to reduce the number of null messages). The processing of the first message invokes removing the message with minimum timestamp over all CC_is from CC_i (the head of IQ_i), while leaving a "null message copy" (change # to 0 in the message head) of it in CC_i if it has been the last message in IQ_i. M and LVT are changed accordingly; also scheduling/descheduling of events might become necessary.

2.6 Time Warp Strategies

In order to simulate a $TTPNR_k$ according to the Time Warp (optimistic) strategy [14] one ***input queue*** IQ and one ***output queue*** OQ with time sequenced message entries are maintained. Either ***positive (token) messages*** ($m = \langle w_f, D, TT,$ '+'$\rangle$) or ***negative (annihilation) messages*** ($m = \langle w_f, D, TT,$ '−'$\rangle$) are received from other LPs out of $GLP = (LP, CH)$ not necessarily in time stamp order, indicating either the transfer of tokens or the request to annihilate a previously received positive message. Messages are assumed to be buffered in some input buffer IB upon their arrival, to be taken over into IQ eventually. Messages generated during a simulation step are held in an ***output buffer*** OB to be sent upon completion of the step itself, all at once. The data structure of an LP with an optimistic SE (SE^{opt}) is depicted in Fig. 3; a simulation engine is shown in Fig. 4.

SE^{opt} behaves as follows: first all $t_i \in T_k$ enabled in M_0 are scheduled. Messages received are processed according to their timestamp and sign; messages with timestamp in the local future (*tokentime*(m) > LVT) are inserted into IQ (if the sign of m is '+' then it is inserted in timestamp order, a message with sign(m) = '−' annihilates its positive counterpart in IQ. ***Straggler messages*** (*tokentime*(m) < LVT) force the LP to ***roll back*** in local simulated time by restoring the most recent *valid* state. Processing the first event (as in SE^{co}) simulates the firing of a transition and the generation of output messages, while processing the first message changes M and possibly schedules/deschedules new/old events. Whenever an event is processed the ***event stack*** ES additionally records all state variables such that a past state can be reconstructed on occasion. The algorithm requires the knowledge of the global virtual time (GVT), a lower bound for the timestamp of any unprocessed message in $GLP = (LP, CH)$, in order to reduce bookkeeping efforts for past states (ES).

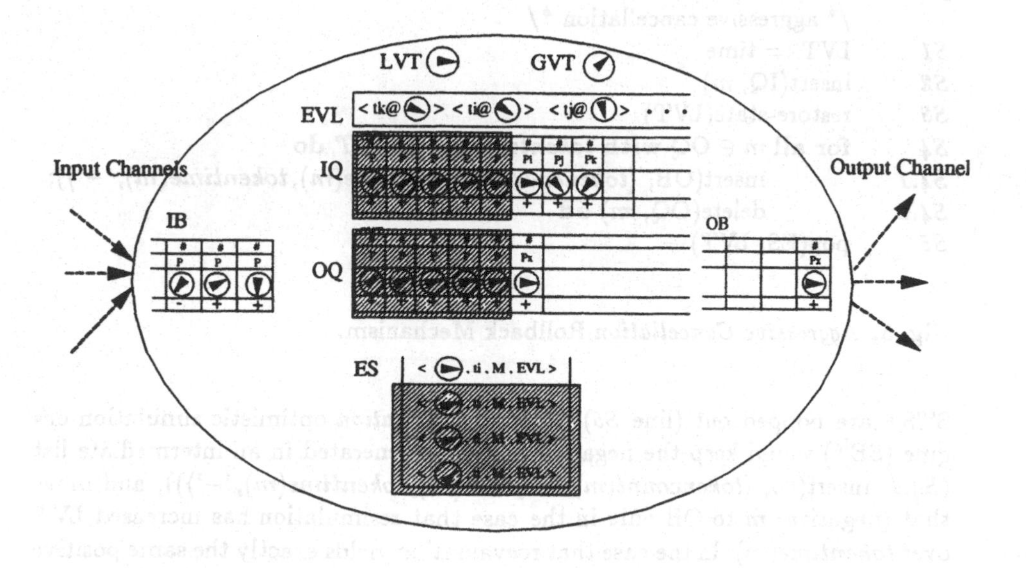

Fig. 3. Logical Process for Optimistic Strategies

Rollback can be applied in two different ways. An optimistic simulation engine with *aggressive cancellation* (SE^{ac}, see Fig. 5) first inserts the straggler message into IQ, and sets LVT to *tokentime*(m). Then the state at time LVT is restored, which is the marking at time of the already processed event closest but not exceeding LVT in SQ. All messages in OQ with token time larger than the rolled back LVT are annihilated by removing them from OQ and sending corresponding antimessages. Finally, all records prematurely pushed in the

```
program SE^opt(TTPN_k)
S1        GVT:=0; LVT := 0; EVL := {}; M := M_0; stop := false;
S2        for all t_i ∈ E(M_0) do schedule(⟨t_i@τ_i⟩) od
S3        while not stop do
S3.1          for all m ∈ IB do
S3.1.1            if tokentime(m) < LVT
S3.1.1.1              then rollback(tokentime(m), m) else insert(IQ, m) od
S3.2          if EVL = {} ∧ IQ = {} then goto S3.1
S3.3          if (GVT ≤ endtime) ∧ (not empty(EVL)) do
S3.3.1            if tokentime(first(EVL)) < tokentime(first(IQ))
S3.3.1.1              then process(first(EVL)) else process(first(IQ)) od
S3.4          for all m ∈ OB do send(m) od
S3.5          if (GVT ≥ endtime) then stop := true;
S3.6      od while
```

Fig. 4. Aggressive Cancellation Simulation Engine for TTPN_k.

```
procedure rollback(time, m)
          /* aggressive cancellation */
S1        LVT := time
S2        insert(IQ, m)
S3        restore-state(LVT)
S4        for all m ∈ OQ with tokentime(m) ≥ LVT do
S4.1          insert(OB, ⟨tokencount(m), outplace(m), tokentime(m), '−'⟩);
S4.2          delete(OQ, m) od
S5        pop(ES, LVT)
```

Fig. 5. *Aggressive Cancellation* Rollback Mechanism.

STST are popped out (line *S5*). A *lazy cancellation* optimistic simulation engine (SE^{lc}) would keep the negative messages generated in an intermediate list (*S4.1:* insert(IL, $\langle tokencount(m), outplace(m), tokentime(m), `-'\rangle$)), and move that (negative) m to OB only in the case that resimulation has increased LVT over $tokentime(m)$. In the case that reevaluation yields exactly the same positive message as already sent before, the new positive message is not resent, but used to compensate the corresponding negative message from IL — thus annihilating the (obsolete) causality correction locally, and preventing from unnecessary message transfers as well as possibly new rollbacks in other LPs.

3 Performance Influences

Due to the variety of degrees of freedom in composing a DDES for TTPNs as offered by the framework developed above, the question for optimizing the performance of a distributed simulation on a physical multiprocessor naturally arises. We have implemented the SEs and communication interfaces for a 30 node Sequent Balance, a 16 node Intel iPSC/860 and a 16 node T800 Transputer multiprocessor. Case studies showed that there are influences on the performance of the distributed simulation which are inherent to a proper combination of the *partitioning in regions* ($TTPNR_k$s), the *communication interface* (I_ks) and the *simulation engine*, but also caused by the physical hardware as such. In the following we will evaluate the main observations and use the empirical results for an improved distributed simulator for TTPNs.

We consider the small net depicted in Fig. 2 as a test case since the automatic partitioning of arbitrary nets is still under development and for the time being the partitioning has to be hand coded in our prototypes. In practice one would apply DDES techniques to substantially larger models. However the test case is significant since it exhibits most of the characteristic features of larger models and the results may be extrapolated for larger nets.

3.1 Partitioning

We already proposed a partitioning such that several transitions together with all their input places are simulated by a single logical process. Subnets are constructed from the topologic description of the TTPN so that conflict resolution always occurs internally to a logical process. Depending on the particular multiprocessor architecture on which the DDES is run, such minimal region partitioning may however turn out to be too fine grain to efficiently adapt to the interprocessor communication overhead. One should thus be willing to partition the DDES in a lower number of LPs in order to reduce the communication overhead and attain better performance. In practice a trade-off must be sought for different target architectures based on empirical cases in order to achieve speedup over sequential simulation.

Figure 6 compares (in the 5 LP columns) absolute execution time of the minimum region partitioning with a SE^{lc} of the reader writer net in Fig. 2 in a balanced parametrization: $\lambda_{arrival} = 1.0$, $\lambda_{endR} = 2.0$, $\lambda_{endW} = 0.5$, $prob(isread) := 0.8$, $prob(iswrite) := 0.2$ and $N = 4$ processes. The results show that the Sequent Balance takes physically more communication time, although the work profile is the same as for the iPSC, which uses a faster processor.

From the communication overheads encountered for the minimum region partitioning it is obvious to see that the communication/computation ratio has to be increased to use this kind of multiprocessor hardware more efficiently. Moreover, additional aggregation of conflict sets into larger logical simulation processes may increase the balancing of the distributed simulation without decreasing its inherent parallelism if some conditions are verified on the TTPN model structure (grain packing). For example transitions *endR* and *endW* are never simultaneously enabled, thus LP_4 and LP_5 in Fig. 2 can be aggregated to a single LP without loss of potential model parallelism.

The following rules can be applied to pack grains starting from minimum $TTPNR_k$s:

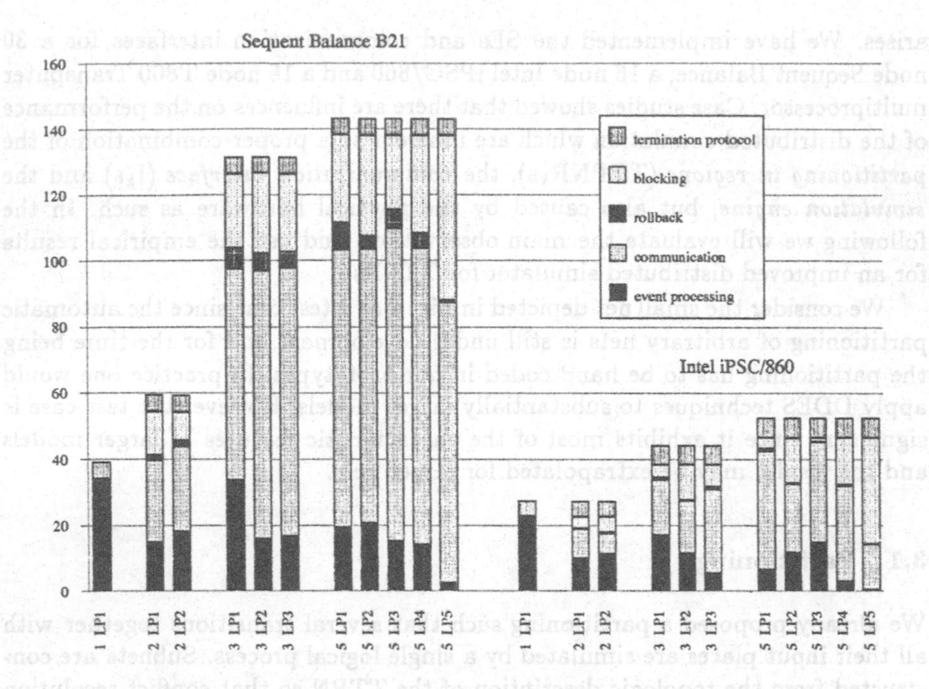

Fig. 6. Comparison of Partitionings on Sequent Balance and Intel iPSC/860

Rule 1 **mutually exclusive (ME) transitions go into one LP since they bear no potential parallelism. Two transitions $t_i, t_j \in T$ are said to be *mutually exclusive*, denoted by $(t_i\ ME\ t_j)$, if and only if they cannot be simultaneously enabled in any reachable marking. A sufficient condition for $(t_i\ ME\ t_j)$ is that the number of tokens in a P-invariant out of which t_i and t_j share places prohibits a simultaneous enabling.**
Another sufficient condition for $(t_i\ ME\ t_j)$ is that $\exists t_k : \pi_k > \pi_j$ such that $\forall p \in {}^\bullet t_k, p \in {}^\bullet t_i \cup {}^\bullet t_j \wedge w(t_k, p) \leq \max(w(t_i, p), w(t_j, p))$.

Rule 2 **endogenous simulation speed is balanced to prevent from rollbacks, i.e. the probability of receiving straggler messages is reduced by balanced virtual time increases in all LPs**

Rule 3 **LPs with high message traffic intensity are clamped to save message transfer costs**

Rule 4 **persistent net parts and free choice conflicts are always placed to the output border to allow sending out messages ahead the LVT (lookahead) without possibility of rollback, (i.e. sending ahead messages that will be inevitably generated by future events — unless local rollback occurs)**

Rule 5 **transitions having a single input place can also be connected to the input border, since the enabling test can be avoided for these transitions (firing can be scheduled immediately upon receipt of the positive token message without additional overhead)**

One may realize by looking at the connectivity of LP_3 (Fig. 2) that this logical process needs to treat 3 events out of 4 per access cycle (the marking of the input border of LP_3 is affected by the firing of transitions "isread,", "iswrite," "startR," "startW," "endR," and "endW," and is not affected only by the firing of "arrival"). This structural characteristics limits the maximum speedup of a distributed simulation of the model to the value 4/3 (since each marking update accounts for one simulation step of LP_3). As LP_4 and LP_5 are persistent supporting the exploitation of lookahead both of them will be aggregated to LP_3 to form the output border of a new LP_{3+4+5} (Rule 4). (Moreover LP_3 and LP_5 are mutually exclusive (Rule 1), and the aggregation reduces the external communications (*endW*, *notacc*) and (*endR*, *notacc*) (Rule 3).)

Since we observe an overlap of real simulation work only among LP_1, LP_2 and LP_3 in the minimum region decomposition we can use the following arguments to merge LP_1 and LP_2 to a new LP_{1+2}: LP_{1+2} contains an TTPNR with single input transition connected to the input border and free-choice conflicting transitions in the output border, so that 1) the enabling test for transition *arrival* can be avoided on receipt of a token message (Rule 5), and 2) the output message can be sent ahead of simulation due to the free choice conflict in the output border (Rule 4). We finally end up with a two LPs (LP_{1+2}, LP_{3+4+5}) partitioning which is optimum in the particular model.

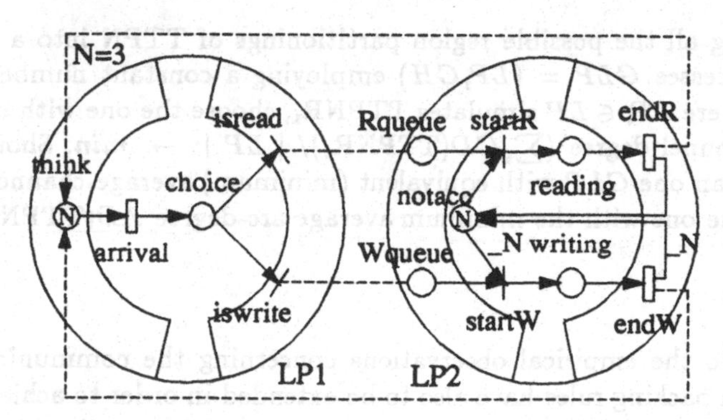

Fig. 7. Reader/Writer Example: Optimum Partitioning.

Figures 7 and 8 show the optimum partitioning and the performance (total simulation time) of its Transputer implementation (Part. 1 refers to the partitioning (LP_1, $LP_{2+3+4+5}$), Part. 2 to (LP_{1+2}, LP_{3+4+5}) and Part. 3 to (LP_{1+2+3}, LP_{4+5}). Figure 6 compares the Balance and iPSC performance for the minimum region partitioning, a decomposition into 3 regions (LP_{1+2}, LP_3, LP_{4+5}), the optimum partitioning and the engine simulating the whole TTPN (one LP, i.e. sequentially) according to the lazy cancellation strategy.

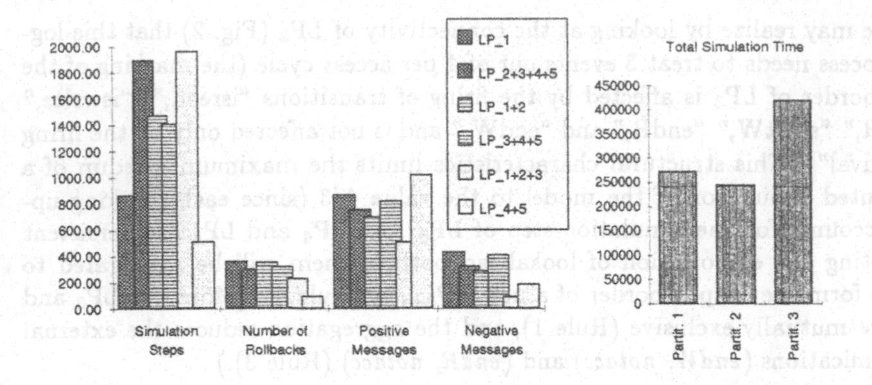

Fig. 8. Reader/Writer Example: Optimum Partitioning Performance (Transputer)

3.2 Communication

Since communication latency is rather high compared to the raw processing power for the multiprocessor systems under investigation, communication is the dominating performance influence factor (see e.g. Fig. 6) for all simulation engines. The most promising tuning of a DDES of TTPNs is by making arc-degree and channel-degree reduction the main principles of the partitioning process. Additionally to the grain packing rules we can state:

Rule 7 **Among all the possible region partitionings of TTPN into a graph of logical processes $GLP = (LP, CH)$ employing a constant number of LPs ($| LP |$) where $LP_i \in LP$ simulates TTPNR$_i$, choose the one with minimum average channel degree $(\sum_i CD(\text{TTPNR}_i)/ \mid LP \mid \; \rightarrow \; min$. Should there be more than one GLP with equivalent (minimum) average-channel degree, then use the one with the minimum average arc-degree $AD(\text{TTPNR}_i)$.**

3.3 Load

With respect to the empirical observations concerning the communication latency the grain packing rules have also to be extended in order to achieve *actual* speedup when simulating the TTPN regions in parallel. Naturally this can happen as soon as the efforts for real (local) simulation work exceed the communication efforts:

Rule 8 **Cluster TTPNR$_k$s such that the local simulation work in terms of physical processor cycles exceeds a certain (hardware specific) computation/communication threshold in order to observe real speedup.**

3.4 The Strategy

Obviously the simulation strategy has a strong influence on the performance of a DDES. Since SE^{co} strictly adheres the local causality constraint by processing

events only in non decreasing timestamp order, the SE^{co} cannot fully exploit the parallelism available in the simulation application. In the case where one transition firing *might* affect (directly or indirectly) the firing of another transition SE^{co} must execute the firings sequentially — hence it forces sequential execution even if it is not necessary. In all cases where causality violations due to interference among transition are logically possible but occurs seldomly in the simulation run, SE^{co} is overly pessimistic in the majority of cases. SE^{ac} and SE^{lc} gain from a proper partitioning and the placement of net parts in the input or output border of the LP (as described), but suffer from tremendous memory requirements and memory accesses. Although empirical observations give raise for best speedup attainable by the use of SE^{lc} and large $TTPNR_k$s with minimum average channel degree, a general rule cannot be stated on the SE to be applied.

4 Optimizing the SE

With SE^{co} we have seen how to exploit the $TTPNR_k$ structure to improve the standard simulation engine by introducing persistence chains for transitions in the output border. Based on particular model structures the optimistic simulation engines can also be optimized in different respects. One of these optimizations have already been undertaken with the local annihilation of negative messages in SE^{lc}.

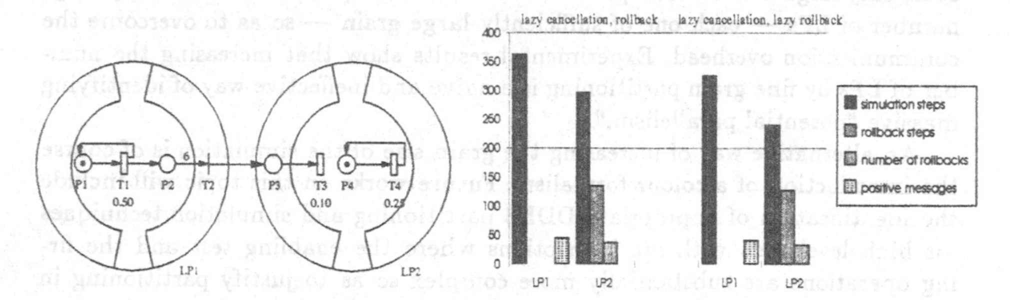

Fig. 9. Performance improvement by Lazy Rollback

In addition to Rules 4 and 5 already stated, we may identify the case, where a straggler positive message changes the marking but does not cause the enabling of any new event in the past (but e.g. in the future), a simplified rollback mechanism (*lazy rollback*) is sufficient to recover from the causality error. This is best explained by a ceteris paribus analysis of a simple example. The two LPs in Fig. 9) simulate (approx.) at the same load, whereas LP_2 ($\lambda(T4) = 0.5$) increments LVT twice as fast as LP_1 ($\lambda(T1) = 0.25$) does. After every 6th step in LP_1 $T2$ generates a straggler message for LP_2 (i.e. time stamped in the past of LP_2, and with effect possibly ($\lambda(T3) = 0.1$) in the future of LP_2) which potentially does not violate any causality in LP_2. In this case no rollback would be invoked. Should the effect of the message received however be in the past of LP_2, then only an appropriate insertion of the firing of $T3$ is made on ES, and the top of ES (entries with time stamps in between the occurrence time of $T3$ and LVT) is copied considering a potential change in the marking (not necessary in the example). The effect of lazy rollback is also shown in Fig. 9 (Transputer implementation of SE^{lc} with lazy rollback).

5 Conclusions

In this paper we have described the implementation of various adaptions of classical DDES strategies to the simulation of TTPN models. Several prototypes have been developed and run on three different distributed architectures, allowing the collection of many empirical data from the measurement of the performance of such prototypes. Additional results are reported in [13]. These results have been used to validate the different variations of the techniques on some case study as well as to identify problems and potentialities of the approach.

The main performance problem that we found (not surprisingly) was related to the interprocessor communication latency inherent to distributed architectures. The conclusion that can be drawn from our preliminary results is that DDES has no hope to attain real speedup over sequential simulation unless the intrinsic properties of parallelism and causality of the simulated model are properly identified and exploited to optimize the parallel execution of LPs. Moreover, only large TTPN descriptions have a chance to produce a sufficiently large number of LPs — each one of sufficiently large grain — so as to overcome the communication overhead. Experimental results show that increasing the number of LPs by fine grain partitioning is a naive and ineffective way of identifying massive "potential parallelism."

An alternative way of increasing the grain size of the simulation is of course the introduction of a colour formalism. Future works on this topic will include the identification of appropriate DDES partitioning and simulation techniques for high level nets with arc inscriptions where the enabling test and the firing operations are substantially more complex so as to justify partitioning in distributed processes also for structurally small net structures.

Each model may be characterized by its inherent parallelism independently of the number of places and transitions of the TTPN description, and it is this inherent parallelism that we should try and capture in order to achieve speedup over sequential simulation. In this sense, the use of a TTPN formalism may provide a substantial contribution to the implementation of efficient, general purpose DDES engines. Indeed most of the relevant characteristics that have to be taken into account to produce efficient LPs are determined by the Petri net structure. Appropriate phases of structural analysis may be implemented in order to capture these relevant characteristics automatically from the model structure.

The work presented in this paper should be considered only as a first step in the direction of exploiting Petri net structural analysis for the efficient implementation of DDES techniques. We already identified some net patterns that yield particularly efficient simulation strategies. We believe however that several other net dependent optimizations may be developed in order to obtain practical advantages from the application of DDES techniques to real models.

References

1. T. Murata. Petri nets: properties, analysis, and applications. *Proceedings of the IEEE*, 77(4):541–580, April 1989.
2. K.M. Chandy and J. Misra. Distributed simulation: A case study in design and verification of distributed programs. *IEEE Transactions on Software Engineering*, 5(11):440–452, September 1979.
3. D.A. Jefferson. Virtual time. *ACM Transactions on Programming Languages and Systems*, 7(3):404–425, July 1985.
4. R.M. Fujimoto. Parallel discrete event simulation. *Communications of the ACM*, 33(10):30–53, October 1990.
5. A. Gafni. Rollback mechanisms for optimistic distributed simulation systems. In *Proc. Conference on Distributed Simulation 1988*, pages 61–67, California, 1988. Society for Computer Simulation.
6. Y.B. Lin and E.D. Lazowska. A study of the time warp rollback mechanism. *ACM Transactions on Modeling and Computer Simulation*, 1(1):51–72, January 1991.
7. B. Lubachevsky, A. Weiss, and A. Shwartz. An analysis of rollback based simulation. *ACM Transactions on Modeling and Computer Simulation*, 1(2):154–193, April 1991.
8. G.S. Thomas and J. Zahorjan. Parallel simulation of performance Petri nets: Extending the domain of parallel simulation. In B. Nelson, D. Kelton, and G. Clark, editors, *Proc. 1991 Winter Simulation Conference*, 1991.
9. D.M. Nicol and S. Roy. Parallel simulation of timed Petri nets. In B. Nelson, D. Kelton, and G. Clark, editors, *Proc. 1991 Winter Simulation Conference*, pages 574–583, 1991.
10. H.H. Ammar and S. Deng. Time warp simulation of stochastic Petri nets. In *Proc. 4th Intern. Workshop on Petri Nets and Performance Models*, pages 186–195, Melbourne, Australia, December 1991. IEEE-CS Press.
11. M. Ajmone Marsan, G. Balbo, G. Chiola, G. Conte, S. Donatelli, and G. Franceschinis. An introduction to Generalized Stochastic Petri Nets. *Microelectronics and Reliability*, 31(4):699–725, 1991. Special issue on Petri nets and related graph models.
12. G. Balbo and G. Chiola. Stochastic Petri net simulation. In *Proc. 1989 Winter Simulation Conference*, Washington D.C., December 1989.
13. G. Chiola and A. Ferscha. Distributed discrete event simulation of timed Petri nets. Technical report, Austrian Center for Parallel Computation, Technical University of Vienna, 1993. to appear.
14. D. Jefferson and H. Sowizral. Fast concurrent simulation using the time warp mechanism. In P. Reynolds, editor, *Proc. Conference on Distributed Simulation 1985*, pages 63–69, La Jolla, California, 1985. Society for Computer Simulation.

Transient Analysis of Deterministic and Stochastic Petri Nets

Hoon Choi,[1] Vidyadhar G. Kulkarni* and Kishor S. Trivedi[2]

[1]Department of Computer Science
[2]Department of Electrical Engineering
Duke University, Durham, NC 27708, U.S.A.
*Department of Operations Research
University of North Carolina, Chapel Hill, NC 27514, U.S.A.

Abstract

Deterministic and stochastic Petri nets (DSPNs) are recognized as a useful modeling technique because of their capability to represent constant delays which appear in many practical systems. If at most one deterministic transition is allowed to be enabled in each marking, the state probabilities of a DSPN can be obtained analytically rather than by simulation. We show that the continuous time stochastic process underlying the DSPN with this condition is a Markov regenerative process and develop a method for computing the transient (time dependent) behavior. We also provide a steady state solution method using Markov regenerative process theory and show that it is consistent with the method of Ajmone Marsan and Chiola.

1 Introduction

Several classes of timed transition Petri nets have been proposed in order to provide a unified framework for performance and reliability analysis of computer and communication systems. This includes stochastic Petri nets (SPNs) [14], generalized stochastic Petri nets (GSPNs) [2], extended stochastic Petri nets (ESPNs) [7], and deterministic and stochastic Petri nets (DSPNs) [1]. In an SPN, a transition fires after an exponentially distributed amount of time (firing time) when it is enabled. A GSPN allows transitions with zero firing time or exponentially distributed firing time. The stochastic process underlying an SPN or a GSPN is a continuous-time Markov chain. An ESPN allows generally distributed firing times. Under some restrictions, the underlying stochastic process of an ESPN is a semi-Markov process. A DSPN allows transitions with zero firing time or exponentially distributed or deterministic firing time. *The underlying stochastic process of a DSPN is neither a continuous-time Markov chain nor a semi-Markov process.*

DSPNs are suited to represent system features such as time-outs, propagation delays which are naturally associated with constant delays. DSPNs have been

---

[1]This research was supported in part by the IBM Graduate Fellowship and by the Overseas Study Program of ETRI.

[2]This research was supported in part by the National Science Foundation under Grant CCR-9108114, and by the Naval Surface Warfare Center under contract N60921-92-C-0161.

applied for modeling an Ethernet bus LAN, a fault-tolerant clocking system, and CSMA/CD protocol with deterministic collision resolution time. A steady state solution method for DSPNs was introduced by Ajmone Marsan and Chiola [1], and Lindemann [12] improved the numerical algorithm of the steady state analysis. Algorithms for parametric sensitivity analysis of the steady state solution are presented in [6].

In this paper, we show that ***the underlying stochastic process of a DSPN is a Markov regenerative process. We then develop a solution method for the transient analysis of DSPNs.*** We also provide the steady state solution method of DSPNs using Markov regenerative theory and show that it is consistent with the method in [1]. Following [1], we assume that at most one deterministic transition is allowed to be enabled in each marking and the firing time of a deterministic transition is marking independent. The state space of the underlying stochastic process is also assumed to be finite.

The paper is organized as follows. After defining basic terminology and concepts in Section 2, we show in Section 3 that the underlying stochastic process of the DSPN is a Markov regenerative process. The transient analysis and the steady state analysis methods are described in Section 4 and 5, respectively. The computational aspect of the method is studied in Section 6. As an example, the M/D/1/2/2 queueing system is analyzed in Section 7.

2 Definitions

The state of a Petri net is defined by the number of tokens in each place, and is represented by a vector $m = (\#(p1), \#(p2), \cdots, \#(pk))$, called a *marking* of the Petri net, where $\#(pi)$ is the number of tokens in place pi and k is the number of places in the net. The firing of a transition may change the allocation of tokens to the places and, thus, may create a new marking. A marking m_j is said to be ***reachable*** from a marking m_i if, starting from m_i, there exists a sequence of transitions whose firings generate m_j. The ***reachability graph*** can be constructed by connecting a marking m_j from a marking m_i with a directed arc if the marking m_j can result from the firing of some transition enabled in m_i.

Transitions which have zero firing time are called ***immediate transitions*** and those with nonzero firing time are called ***timed transitions***. We will call a timed transition whose firing time is exponentially (deterministically) distributed an *EXP* (*DET*) transition. The markings of timed Petri nets can be classified into two types: ***vanishing markings*** and ***tangible markings***. A vanishing marking is one in which at least one immediate transition is enabled and a tangible marking is one in which no immediate transition is enabled.

The ***reduced reachability graph*** is obtained from the reachability graph of a DSPN by merging the vanishing markings into their successor tangible markings.

Suppose a timed transition t fires in a tangible marking m_i and the successor marking is a vanishing marking m_k. Then some immediate transition enabled in m_k will fire and the next successor marking may be either vanishing or tangible. The probability to visit a tangible marking m_j after visiting an arbitrary number (including 0) of vanishing markings given that the current marking is vanishing marking m_k is X_{kj} of the system of linear equations [15]:

$$(I - B^{VV})X = B^{VT}, \tag{1}$$

where B^{VV} (B^{VT}) is the one-step transition probability matrix from vanishing to vanishing (tangible) markings in the reachability graph. All the sequence of vanishing markings between m_i and m_j are substituted by the branching probability X_{kj} in the following way:

1. If t is an EXP transition with the firing rate λ, each of the successor tangible marking m_j is connected from m_i directly with the firing rate $X_{kj} \cdot \lambda$.
2. If t is a DET transition with the firing time τ, each of the successor tangible marking m_j is connected from m_i directly with the firing time τ. The branching probabilities in this case is kept in the *branching-probability matrix* Δ where $\Delta(i,j) = X_{kj}$. (The probability to reach m_k from m_i conditioned on firing of t is 1.) When a tangible marking m_j is reached directly from m_i then $\Delta(i,j)$ is set to 1. $\Delta(i,j)$ will be explained again in Equation (10).

A timed transition t is said[3] to be *exclusive* in a marking m if t is the only transition enabled in m. When a timed transition t is enabled together with another transition t', t is called a *competitive* transition in the marking m if the firing of t' disables the transition t. If the firing of t' does not disable transition t, then t is called a *concurrent* transition in the marking m. Figure 1 shows two example DSPNs (a) and (c) with their (reduced) reachability graphs (b) and (d), respectively. In the DSPNs, filled rectangles denote DET transitions and empty rectangles denote EXP transitions. In the reachability graphs, solid arcs denote state transitions by DET transitions and dotted arcs denote state transitions by EXP transitions. A 4-tuple in each marking of Figure 1(b) denotes $(\#(p1), \#(p2), \#(p3), \#(p4))$. Similarly, a 3-tuple in each marking of Figure 1(d) denotes $(\#(p5), \#(p6), \#(p7))$. Transition *tj* represents the submission of a job by one of two clients. The rate of this event is *marking dependent*, i.e., the firing time of this transition is exponentially distributed with rate $\#(p1) \cdot \lambda$. Transition *ts* takes an exponentially distributed amount of time to fire, and *tv* takes a deterministic time to fire. Markings m_1 and m_2 of Figure 1(b) are the ones in which a DET transition and an EXP transition are enabled concurrently. Marking m_3 of Figure 1(b) is the one in which a DET transition is enabled exclusively. Marking m_7 of Figure 1(d) is the one in which a DET transition and an EXP transition are enabled competitively. When the EXP transition *te* fires, the DET transition *td* is

[3] This definition of exclusive, competitive, and concurrent transitions is slightly different from the original definition given in [7].

disabled. Whenever *te* fires, one token is put back in place *p5* making *td* enabled again. This represents the *reset* of a deterministic timer by a preemptable event which is modeled by *te*.

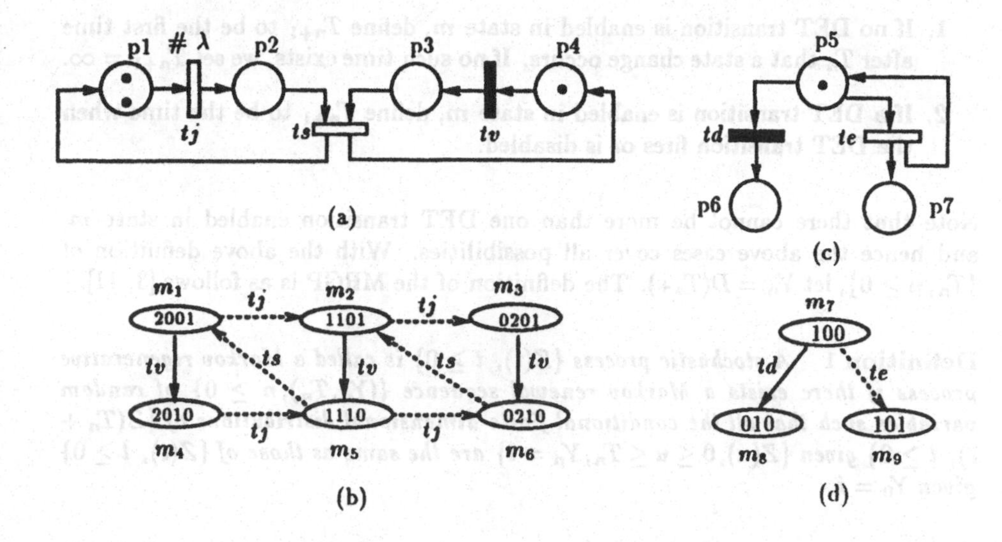

Figure 1: Examples of DSPNs and Their Markings

3 DSPNs and Markov Regenerative Processes

The piecewise constant, right-continuous, continuous-time stochastic process underlying the DSPN is formed by the changes of the tangible markings over the time domain. The tangible markings of the reduced reachability graph at time t are in one-to-one correspondence with the states of the stochastic process underlying the DSPN at time t. We hereafter call the stochastic process underlying the DSPN the *marking-process* $\{D(t), t \geq 0\}$ of the DSPN, where $D(t)$ is the tangible marking of the DSPN at time t [9].

We assume that *at most one DET transition is enabled in each marking of the DSPN and the deterministic firing time is marking-independent* as done in [1]. We shall show that, under this assumption, the marking-process $\{D(t), t \geq 0\}$ is a Markov regenerative process (MRGP), also known as the semi-regenerative process [3, 11]. We need to introduce the following notation in order to prove this result.

Let Ω be the set of all tangible markings of the reduced reachability graph of the DSPN. Then Ω is also the state space of the marking-process of the DSPN. For example, $\Omega = \{m_1, m_2, m_3, m_4, m_5, m_6\}$ is the state space of the marking-process of the DSPN shown in Figure 1(a). Consider a sequence $\{T_n, n \geq 0\}$, of epochs when the DSPN is observed. Let $T_0 = 0$ and define $\{T_n, n \geq 0\}$ recursively as follows: Suppose $D(T_n+) = m$.

1. If no DET transition is enabled in state m, define T_{n+1} to be the first time after T_n that a state change occurs. If no such time exists, we set $T_{n+1} = \infty$.
2. If a DET transition is enabled in state m, define T_{n+1} to be the time when the DET transition fires or is disabled.

Note that there cannot be more than one DET transition enabled in state m, and hence the above cases cover all possibilities. With the above definition of $\{T_n, n \geq 0\}$, let $Y_n = D(T_n+)$. The definition of the MRGP is as follows [3, 11].

Definition 1 *A stochastic process $\{Z(t),\ t \geq 0\}$ is called a Markov regenerative process if there exists a Markov renewal sequence $\{(Y_n, T_n), n \geq 0\}$ of random variables such that all the conditional finite dimensional distributions of $\{Z(T_n + t),\ t \geq 0\}$ given $\{Z(u), 0 \leq u \leq T_n, Y_n = i\}$ are the same as those of $\{Z(t),\ t \geq 0\}$ given $Y_0 = i$.*

This definition implies that

$$P\{Z(T_n + t) = j \mid Z(u), 0 \leq u \leq T_n, Y_n = i\} = P\{Z(t) = j \mid Y_0 = i\}.$$

Theorem 1 *The marking-process $\{D(t),\ t \geq 0\}$ of a DSPN is a Markov regenerative process.*

(Proof) First we shall show that $\{(Y_n, T_n), n \geq 0\}$ embedded in the marking-process of the DSPN is a Markov renewal sequence, i.e.,

$$\begin{aligned} &\forall\, i, j \in \Omega, \\ &\quad P\{Y_{n+1} = j,\ T_{n+1} - T_n \leq t \mid Y_n = i, T_n, Y_{n-1}, T_{n-1}, \ldots, Y_0, T_0\} \\ &\quad = P\{Y_{n+1} = j,\ T_{n+1} - T_n \leq t \mid Y_n = i\} \quad \text{(Markov Property)} \\ &\quad = P\{Y_1 = j,\ T_1 \leq t \mid Y_0 = i\} \qquad \text{(Time Homogeneity).} \qquad (2) \end{aligned}$$

Suppose the past history $Y_0, T_0, \ldots, Y_{n-1}, T_{n-1}, Y_n, T_n$ is given and $Y_n = i$. Consider two cases.

1. No DET transition is enabled, i.e., all the transitions that are enabled in state i at time T_n are EXP transitions. In this case, due to the memoryless property of the exponential random variable, the future of the marking-process depends only on the current state i and does not depend upon the past history or the time index n.

2. **Exactly one DET transition is enabled in state i. There may be other EXP transitions enabled in state i. In this case, T_{n+1} is the next time when the DET transition fires or is disabled. The joint distribution of Y_{n+1} and $(T_{n+1} - T_n)$ will depend only on the state at time T_n. (That is, the next transition time depends on the destination state Y_{n+1} and the state Y_{n+1} will be decided from the current state Y_n. Therefore, where to and when to move next is only dependent on the current state at time T_n.) It is thus independent of the past history and the time index n.**

This proves that $\{(Y_n, T_n), n \geq 0\}$ is a Markov renewal sequence.

Now consider the marking-process from time T_n onwards, namely $\{D(T_n + t),\ t \geq 0\}$. Given the history $\{D(u), 0 \leq u \leq T_n, D(T_n+) = i\}$, it is clear from the above argument that the stochastic behavior of $\{D(T_n + t),\ t \geq 0\}$ depends only on $D(T_n) = i$. Thus,

$$\begin{aligned}
&\{D(T_n + t),\ t \geq 0 \mid D(u), 0 \leq u \leq T_n, D(T_n) = i\} \\
&\stackrel{d}{=} \{D(T_n + t), t \geq 0 \mid D(T_n) = i\} \\
&\stackrel{d}{=} \{D(t), t \geq 0 \mid D(0) = i\} \qquad \forall i \in \Omega
\end{aligned}$$

where $\stackrel{d}{=}$ denotes equality in distribution. This proves that $\{D(t), t \geq 0\}$ is an MRGP. □

From Theorem 1, we see that the following conditional probabilities play an important role:

$$K_{ij}(t) = P\{Y_1 = j,\ T_1 \leq t \mid Y_0 = i\} \qquad i, j \in \Omega. \tag{3}$$

The matrix $K(t) = [K_{ij}(t)]$ is called the ***kernel***. With the kernel, the distribution function of T_1 starting from state i is defined as:

$$H_i(t) = \sum_{j \in \Omega} K_{ij}(t) = P\{T_1 \leq t \mid Y_0 = i\}, \quad t \geq 0, \ \ i \in \Omega. \tag{4}$$

Corollary 1 *$\{Y_n, n \geq 0\}$ is a DTMC with transition probability matrix $K(\infty)$.*

(Proof) Follows from Equation (2) by letting $t \to \infty$. □

Remark: $\{Y_n, n \geq 0\}$ is called the ***embedded Markov chain*** (EMC) for the DSPN.

Let $N(t) = sup\{n \geq 0 : T_n \leq t\}$. The process $\{X(t),\ t \geq 0\}$ such that:

$$X(t) = Y_{N(t)}, \quad t \geq 0 \tag{5}$$

is a *semi-Markov process* (SMP) of the DSPN with kernel $K(\cdot)$.

Using the theory of MRGP, we carry out the transient analysis and the steady state analysis in the following sections.

4 Transient Analysis of DSPN

In this section, we develop a method of computing the transient distribution of the marking-process $\{D(t),\ t \geq 0\}$. Define the transition probability

$$V_{ij}(t) = P\{D(t) = j \mid D(0) = Y_0 = i\} \tag{6}$$

and let $V(t) = [V_{ij}(t)]$. Let

$$K_{iu} * V_{uj}(t) = \int_0^t dK_{iu}(x)\ V_{uj}(t-x), \tag{7}$$

and $K * V(t)$ be a matrix whose (i,j) element is $\sum_u K_{iu} * V_{uj}(t)$.

The transient analysis of the marking-process of the DSPN is based on the following general theorem on MRGPs with kernel $K(\cdot)$. We include the proof for completeness.

Theorem 2 *The transition matrix $V(t)$ satisfies the following generalized Markov renewal equation:*

$$V(t) = E(t) + K * V(t) \tag{8}$$

where $E_{ij}(t) = P\{D(t) = j, T_1 > t \mid Y_0 = i\}$.

(Proof) Conditioning on Y_1 and T_1 and using the Markov regenerative properties of $\{D(t),\ t \geq 0\}$, we obtain:

$$P\{D(t) = j \mid Y_0 = i, Y_1 = k, T_1 = x\} = \begin{cases} P\{D(t) = j \mid Y_0 = i, T_1 = x\} & t < x \\ V_{kj}(t-x) & t \geq x. \end{cases}$$

Unconditioning, we get:

$$\begin{aligned} V_{ij}(t) &= \sum_{k \in \Omega} \int_0^\infty P\{D(t) = j \mid Y_0 = i, Y_1 = k, T_1 = x\}\ dK_{ik}(x) \\ &= \sum_{k \in \Omega} \int_t^\infty P\{D(t) = j \mid Y_0 = i, T_1 = x\}\ dK_{ik}(x) \\ &\quad + \sum_{k \in \Omega} \int_0^t V_{kj}(t-x)\ dK_{ik}(x) \end{aligned}$$

$$= P\{D(t) = j, T_1 > t \mid Y_0 = i\} + \sum_{k \in \Omega} K_{ik} * V_{kj}(t)$$

$$= E_{ij}(t) + [K * V]_{ij}(t)$$

which in matrix form is Equation (8). □

Note that $E_{ij}(t)$ describes the behavior of the marking-process between two transition epochs of the EMC, i.e., over the time interval $[0, T_1)$. We will call the matrix $E(t)$ the *local kernel*, opposed to the kernel $K(t)$ which is global in the marking-process.

In order to use the above theorem, we need to specify $E(t) = [E_{ij}(t)]$ and $K(t) = [K_{ij}(t)]$ matrices for the DSPN. We give the following definitions for this purpose.

Consider a state m in Ω. Let $\mathcal{D}(m)$ be the set of DET transitions and $\mathcal{E}(m)$ be the set of EXP transitions enabled in state m. For example, $\mathcal{D}(m_1) = \{tv\}, \mathcal{E}(m_1) = \{tj\}$ for state m_1 of Figure 1(b). Consider the following cases.

Case 1 : $\mathcal{D}(m) = \emptyset$, i.e., no DET transition is enabled in m.
In this case, define:

$$\Lambda_m = \sum_{n \in \Omega} \lambda(m, n) \quad (9)$$

where $\lambda(m, n)$ is the transition rate from marking m to n in the reduced reachability graph. Note that we are dealing with the reduced reachability graph of the DSPN, therefore the transition rates between states of the marking-process (tangible markings of the reduced reachability graph) are not necessarily the same as the original firing rates of EXP transitions assigned in the DSPN. As described in Section 2, the transition rates may have been altered by branching probabilities of the immediate transitions. If more than one EXP transitions in $\mathcal{E}(m)$ lead to marking n after the firing, then $\lambda(m, n)$ is the sum of the transition rates of these EXP transitions. T_1, in this case, is exponentially distributed with rate Λ_m and $D(t) = Y_0$ for $0 \le t < T_1$.

Case 2 : $\mathcal{D}(m) = \{d\}$, i.e., exactly one DET transition d is enabled.
Suppose $Y_0 = m$. In this case T_1 is the time when d fires or is disabled due to the firing of a competitive EXP transition. Define $\Omega(m)$ to be the set of all states reachable from m in which the marking-process can spend a non-zero time before the next EMC transition occurs (before the firing of the DET transition d or competitive EXP transitions enabled in state m). For example, $\Omega(m_1) = \{m_1, m_2, m_3\}$ for m_1 of Figure 1(b), and $\Omega(m_7) = \{m_7\}$ for m_7 of Figure 1(d). The marking-process during $[0, T_1)$ is a CTMC, which is called the *subordinated CTMC* [12], on state space Ω with the generator matrix $Q(m)$. The generator matrix $Q(m)$ is formed as follows: for any $n \in \Omega(m)$, the rate from n to $n' \in \Omega$ is given by $\lambda(n, n')$, and if $n \notin \Omega(m)$, the rates out of a marking n are zeros[4].

---

[4]Even though $Q(m)$ is defined for each m, it does not have to be distinct. For instance,

Next, define $\Omega_\mathcal{E}(m)$ to be the set of states which can be reached starting from m (not necessarily directly) by firing of a competitive EXP transition. Similarly, define $\Omega_\mathcal{D}(m)$ to be the set of states reached by firing of the DET transition d. For example, $\Omega_\mathcal{D}(m_7) = \{m_8\}, \Omega_\mathcal{E}(m_7) = \{m_9\}$ for state m_7 of Figure 1(d) and $\Omega_\mathcal{D}(m_1) = \{m_4, m_5, m_6\}$, $\Omega_\mathcal{E}(m_1) = \emptyset$, for state m_1 of Figure 1(b).

Next, define the branching-probability matrix $\Delta = [\Delta(n, n')]$ for $\forall\ n \in \Omega(m)$ and $n' \notin \Omega(m)$ by:

$$\Delta(n, n') = P\{\text{next marking is } n' \mid \text{current marking is } n \text{ and the transition } d \text{ fires}\}. \tag{10}$$

This accounts for the branching probabilities after the DET transition d fires. The branching-probability matrix is obtained when the reachability graph is generated. This completes the description of the process in case 2.

Using the above cases, we now describe the kernel $K(t)$.

Theorem 3 *The kernel $K(t) = [K_{m,n}(t)]$ $(m, n \in \Omega)$ of the marking-process of the DSPN is given by:*

1. for state m s.t. $\mathcal{D}(m) = \emptyset$,

$$K_{m,n}(t) = \begin{cases} 0 & \Lambda_m = 0 \\ \frac{\lambda(m,n)}{\Lambda_m}(1 - e^{-\Lambda_m t}) & \Lambda_m > 0 \end{cases} \tag{11}$$

2. for state m s.t. $\mathcal{D}(m) = \{d\}$ with τ being the firing time of transition d, if $n \in \Omega_\mathcal{E}(m)$ but $n \notin \Omega_\mathcal{D}(m)$:

$$K_{m,n}(t) = \begin{cases} [e^{Q(m)t}]_{m,n} & t < \tau \\ [e^{Q(m)\tau}]_{m,n} & t \geq \tau \end{cases} \tag{12}$$

if $n \notin \Omega_\mathcal{E}(m)$ but $n \in \Omega_\mathcal{D}(m)$:

$$K_{m,n}(t) = \begin{cases} 0 & t < \tau \\ \sum_{m' \in \Omega(m)} [e^{Q(m)\tau}]_{m,m'}\, \Delta(m', n) & t \geq \tau \end{cases} \tag{13}$$

if $n \in \Omega_\mathcal{E}(m)$ and also $n \in \Omega_\mathcal{D}(m)$:

$$K_{m,n}(t) = \begin{cases} [e^{Q(m)t}]_{m,n} & t < \tau \\ [e^{Q(m)\tau}]_{m,n} + \sum_{m' \in \Omega(m)} [e^{Q(m)\tau}]_{m,m'}\, \Delta(m', n) & t \geq \tau \end{cases} \tag{14}$$

$Q(m_1) = Q(m_2) = Q(m_3)$ in Figure 1(b).

if $n \notin \Omega_{\mathcal{E}}(m)$ *and also* $n \notin \Omega_{\mathcal{D}}(m)$:

$$K_{m,n}(t) = 0, \qquad t \geq 0 \,. \tag{15}$$

(Proof) **As given in Equation (3), $K_{m,n}(t)$ is defined by:**

$$K_{m,n}(t) = P\{Y_1 = n,\ T_1 \leq t \mid Y_0 = m\} \qquad m,n \in \Omega.$$

We consider each of the above cases.

1. **When no DET transition is enabled in state m, the firing of any EXP transition in $\mathcal{E}(m)$ triggers the state change of the EMC. Hence it is clear that:**

$$\begin{aligned} & P\{Y_1 = n,\ T_1 \leq t \mid Y_0 = m\} \\ = \;& P\{\text{state transition occurs until } t \mid Y_0 = m\} \times \\ & P\{n \text{ is reached by the transition} \mid Y_0 = m\} \\ = \;& (1 - e^{-\Lambda_m t}) \times \frac{\lambda(m,n)}{\Lambda_m} \,. \end{aligned}$$

   If $\mathcal{E}(m) = \emptyset$, then $\lambda(m,n) = \Lambda_m = 0$. In this case, further state transitions are not possible from the state m, i.e., m is an absorbing state and $P\{Y_1 = n,\ T_1 \leq t \mid Y_0 = m\}$ is computed to be 0 for all $n \in \Omega$.

2. **When a DET transition d is enabled, the EMC state change is triggered either at the time of firing of d or when d is disabled by a competitive EXP transition. Recall that the set $\Omega_{\mathcal{E}}(m)$ defines the states which are reachable from m by firing of the competitive EXP transitions. Depending on the type of state n, we have the following cases.**

   (a) **If n is in $\Omega_{\mathcal{E}}(m)$ but not in $\Omega_{\mathcal{D}}(m)$, then n is only reachable by firing of a competitive EXP transition.**

   - **When $t < \tau$,**

$$\begin{aligned} & P\{Y_1 = n,\ T_1 \leq t \mid Y_0 = m\} \\ = \;& P\{\text{state of the subordinated CTMC is } n \text{ at time } t \mid Y_0 = m\} \\ = \;& [e^{Q(m)t}]_{m,n} \quad n \in \Omega_{\mathcal{E}}(m) \,. \end{aligned}$$

   - **When $t \geq \tau$, $n \in \Omega_{\mathcal{E}}(m)$ may have been reached during $[0,\tau)$. The probability of this event is $[e^{Q(m)\tau}]_{m,n}$.**

   (b) **If n is in $\Omega_{\mathcal{D}}(m)$ but not in $\Omega_{\mathcal{E}}(m)$, then n is reachable by firing of d.**

   - **When $t < \tau$, the DET transition d will not fire by time t, thus the probability of EMC state change is 0.**

- **When $t \geq \tau$, the DET transition d will fire and $n \in \Omega_{\mathcal{D}}(m)$ will be reached at time τ. The probability of this event is $\sum_{m' \in \Omega(m)} [e^{Q(m)\tau}]_{m,m'} \Delta(m', n)$ which means that the marking-process is in the states $m' \in \Omega(m)$ by $[0, \tau)$ and then jump to $n \in \Omega_{\mathcal{D}}(m)$ at time τ.**

(c) If n is both in $\Omega_{\mathcal{E}}(m)$ and in $\Omega_{\mathcal{D}}(m)$, n can be reached either by firing of a competitive EXP transition or by firing of d. Equation (14) is obtained by combining the previous two cases.

(d) If n is neither in $\Omega_{\mathcal{E}}(m)$ nor in $\Omega_{\mathcal{D}}(m)$, n is not reachable from m at the next EMC transition. Therefore $K_{m,n}(t) = 0$.

The above cases cover all the possibilities. □

Corollary 2 ***The one-step transition probability matrix $P = [P_{m,n}]$ of the EMC is given by:***

1. for state m s.t. $\mathcal{D}(m) = \emptyset$,

$$P_{m,n} = \begin{cases} 0 & \Lambda_m = 0 \\ \frac{\lambda(m,n)}{\Lambda_m} & \Lambda_m > 0 \end{cases} \tag{16}$$

2. for state m s.t. $\mathcal{D}(m) = \{d\}$ with τ being the firing time of transition d, if $n \in \Omega_{\mathcal{E}}(m)$ but $n \notin \Omega_{\mathcal{D}}(m)$:

$$P_{m,n} = [e^{Q(m)\tau}]_{m,n} \tag{17}$$

if $n \notin \Omega_{\mathcal{E}}(m)$ but $n \in \Omega_{\mathcal{D}}(m)$:

$$P_{m,n} = \sum_{m' \in \Omega(m)} [e^{Q(m)\tau}]_{m,m'} \Delta(m', n) \tag{18}$$

if $n \in \Omega_{\mathcal{E}}(m)$ and also $n \in \Omega_{\mathcal{D}}(m)$:

$$P_{m,n} = [e^{Q(m)\tau}]_{m,n} + \sum_{m' \in \Omega(m)} [e^{Q(m)\tau}]_{m,m'} \Delta(m', n) \tag{19}$$

if $n \notin \Omega_{\mathcal{E}}(m)$ and also $n \notin \Omega_{\mathcal{D}}(m)$:

$$P_{m,n} = 0. \tag{20}$$

(Proof) **These results follow from Theorem 3 and the fact $P = K(\infty)$.** □

Now we derive expression for the local kernel $E(t)$.

Theorem 4 *The local kernel $E(t) = [E_{m,n}(t)]$ $(m, n \in \Omega)$ of the DSPN is given by:*

1. *when $\mathcal{D}(m) = \emptyset$:*
$$E_{m,n}(t) = \delta_{m,n}\ e^{-\Lambda_m t} \tag{21}$$
where $\delta_{m,n}$ is the Kronecker δ defined by $\delta_{m,n} = 1$ if $m = n$ and 0 otherwise,
2. *when $\mathcal{D}(m) = \{d\}$:*
for $n \in \Omega(m)$,
$$E_{m,n}(t) = \begin{cases} [e^{Q(m)t}]_{m,n} & t < \tau \\ 0 & t \geq \tau \end{cases} \tag{22}$$
for $n \notin \Omega(m)$, $\qquad E_{m,n}(t) = 0.$

(Proof) As defined in Theorem 2:
$$E_{m,n}(t) = P\{D(t) = n, T_1 > t \mid Y_0 = m\}$$
which is the state transition probability of the marking-process between two transition epochs of the EMC. Starting with $D(0) = m$ and knowing that $T_1 > t$, the marking-process $\{D(u),\ 0 \leq u \leq t\}$ is a CTMC whose infinitesimal generator matrix is $Q(m)$.

1. Suppose $\mathcal{D}(m) = \emptyset$. Then the firing of any EXP transition triggers the state change of the EMC. The probability of marking-process being in state n at time t (before the EMC state change occurs) given that it entered state m at time 0 is the probability that the marking-process stays at the initial state m until time t, i.e., $D(u) = D(0)$ for all $0 \leq u < T_1$. The $\{D(u),\ 0 \leq u < T_1 \mid D(0) = m\}$ in this case is a *degenerative CTMC* that stays in state m. Therefore:
$$P\{D(t) = n, T_1 > t \mid Y_0 = m\} = \delta_{m,n}\ (1 - H_m(t)) = \delta_{m,n}\ e^{-\Lambda_m t}$$
where $H_m(t)$ is defined in Equation (4).
2. Suppose $\mathcal{D}(m) = \{d\}$. Then the EMC state change is triggered either at the time of firing of d or when d is disabled. The set $\Omega(m)$ consists of the states that marking-process can visit starting from m until the next EMC state change occurs. Since the EMC state change occurs after time t, the marking-process at time t will remain in $\Omega(m)$. Therefore, the probability to move a state outside of $\Omega(m)$ is 0.
$$\forall\, n \notin \Omega(m), \qquad E_{m,n}(t) = 0\,.$$
For state $n \in \Omega(m)$, the marking-process is captured by the subordinated CTMC with generator matrix $Q(m)$ which describes the behavior of the

marking-process over the time interval $[0, T_1)$. The transition probability of this CTMC at time t is:

$$P\{D(t) = n, T_1 > t \mid Y_0 = m\} = \begin{cases} [e^{Q(m)t}]_{m,n} & t < \tau \\ 0 & t \geq \tau \end{cases}$$

The above two cases cover all the possibilities. □

Corollary 3 *Given the state transition probability matrix $V(t)$ and the initial probability distribution $p(0) = (p_j(0))$, the state probability at time t of the DSPN is computed by:*

$$p_j(t) = \sum_{i \in \Omega} V_{ij}(t)\, p_i(0) \qquad i, j \in \Omega. \tag{23}$$

(Proof) Follows by conditioning on the initial state Y_0. □

5 Steady State Analysis of the DSPNs

In this section, we consider the steady state analysis of a DSPN whose underlying SMP is finite and ergodic (irreducible, aperiodic and positive recurrent) so that the limiting probability distributions exist. These can be computed using the general theory of MRGP. We briefly review the theory below. Define

$$\mu_m = E[T_1 \mid Y_0 = m]$$

$$\alpha_{mn} = E[\text{time spent by the marking-process in state } n \text{ during } [0, T_1) \mid Y_0 = m]$$

and the steady state probability vector $v = (v_j)$ of the EMC:

$$v = vP, \qquad \sum_{j \in \Omega} v_j = 1$$

where $P = K(\infty)$ is the one-step transition probability matrix of the EMC defined in Corollary 2. When $\mathcal{D}(m) = \emptyset$, T_1 is the time until any of EXP transitions fires. Hence, $\mu_m = E[T_1 \mid Y_0 = m] = 1/\Lambda_m$ where Λ_m is as defined in Equation (9). When $\mathcal{D}(m) = \{d\}$, μ_m is computed by:

$$\mu_m = E[T_1 \mid Y_0 = m] = \sum_{n \in \Omega(m)} \int_0^{\tau} [e^{Q(m)t}]_{m,n}\, dt\ . \tag{24}$$

α_{mn} is computed by:

$$\alpha_{mn} = \int_0^{\infty} P\{D(t) = n, T_1 > t \mid Y_0 = m\} dt = \int_0^{\infty} E_{m,n}(t)\, dt\ . \tag{25}$$

Note that $\mu_m = \sum_{n \in \Omega} \alpha_{mn}$.

The following theorem describes the steady state probability distribution of the DSPN.

Theorem 5 *Let $\{D(t),\ t \geq 0\}$ be an MRGP with embedded Markov renewal sequence $\{(Y_n, T_n), n \geq 0\}$. Suppose $v = (v_j)$ is a positive solution to $v = vP$ and the SMP of the MRGP is finite and ergodic. The limiting distribution $p = (p_j)$ of the state probabilities of the MRGP is given by:*

$$p_j = \lim_{t \to \infty} P\{D(t) = j \mid Y_0 = m\} = \frac{\sum_{k \in \Omega} v_k \alpha_{kj}}{\sum_{k \in \Omega} v_k \mu_k} = \sum_{k \in \Omega} \beta_k \frac{\alpha_{kj}}{\mu_k} \qquad (26)$$

where

$$\beta_k = \frac{v_k \mu_k}{\sum_{r \in \Omega} v_r \mu_r}.$$

See [11] for the proof of the theorem. Intuitively p_j is the fraction of time the marking-process spends in state j and is given by:

$$\begin{aligned} p_j &= \sum_{k \in \Omega} (\text{fraction of time the SMP of the DSPN spends in state } k) \times \\ &\quad (\text{time the marking-process spends in state } j \text{ per unit time of the SMP} \\ &\quad \text{spent in state } k) \\ &= \sum_{k \in \Omega} \beta_k \frac{\alpha_{kj}}{\mu_k}. \end{aligned}$$

We thus see that the steady state analysis by Ajmone Marsan and Chiola [1] agrees with the solution derived using Markov regenerative theory. The one-step transition probabilities shown in Equation (3.c) and (5) in [1] agree with Equation (17) and Equation (18) of this paper. The mean sojourn time in states of the SMP shown in Equation (8.a) of their work also agrees with Equation (24) of this paper. The expression for conversion factor $C_d(k, j)$ of Equation (9.a) in [1] is indeed (α_{kj}/μ_k) here.

6 Transient Analysis in Transform Domain

We discuss methods of computing Equation (8) in this section.

A direct approach may be to solve the system of integral equations:

$$\forall\, i,j \in \Omega, \qquad V_{ij}(t) = E_{ij}(t) + \sum_{k\in\Omega} \int_0^t V_{kj}(t-x)\, dK_{ik}(x) \tag{27}$$

which is discussed in detail in [8].

Laplace-Stieltjes transformation can also be employed. Define the Laplace transform (LT) of a function $f(t)$ to be $f^*(s) = \int_0^\infty e^{-st} f(t)\, dt$ and the Laplace-Stieltjes transform (LST) of a function $f(t)$ to be $f^\sim(s) = \int_0^\infty e^{-st}\, df(t)$.

If we take the LST's on both sides of Equation (8), we get:

$$V^\sim(s) = E^\sim(s) + K^\sim(s)V^\sim(s)\,,$$
$$V^\sim(s) = [I - K^\sim(s)]^{-1} E^\sim(s)\,. \tag{28}$$

Therefore, if we are able to obtain expressions for $K^\sim(s)$ and $E^\sim(s)$, we can get the LST of the state transition probability $V(t)$, and then invert $V^\sim(s)$ into $V(t)$ either analytically or numerically. We now show that $V(t)$ can be computed this way by deriving the expressions of the LST's of $K(t)$ and $E(t)$ of the DSPN.

Theorem 6 *The LST of $K(t) = [K_{m,n}(t)]$ $(m,n \in \Omega)$ of the DSPN is given by:*

1. *for state m s.t. $\mathcal{D}(m) = \emptyset$:*

$$K^\sim_{m,n}(s) = \frac{\lambda(m,n)}{s + \Lambda_m} \tag{29}$$

2. *for state m s.t. $\mathcal{D}(m) = \{d\}$ with τ being the firing time of transition d, if $n \in \Omega_\mathcal{E}(m)$ but $n \notin \Omega_\mathcal{D}(m)$:*

$$K^\sim_{m,n}(s) = \left[s(sI - Q(m))^{-1} \left\{ I - e^{-(sI-Q(m))\tau} \right\} + e^{-s\tau}\, e^{Q(m)\tau} \right]_{m,n} \tag{30}$$

if $n \notin \Omega_\mathcal{E}(m)$ but $n \in \Omega_\mathcal{D}(m)$:

$$K^\sim_{m,n}(s) = \left[e^{-s\tau} \sum_{m'\in\Omega(m)} [e^{Q(m)\tau}]_{m,m'}\, \Delta(m',n) \right]_{m,n} \tag{31}$$

if $n \in \Omega_\mathcal{E}(m)$ and also $n \in \Omega_\mathcal{D}(m)$:

$$K^\sim_{m,n}(s) = \Bigg[s(sI - Q(m))^{-1} \left\{ I - e^{-(sI-Q(m))\tau} \right\} + \left\{ e^{Q(m)\tau} + \sum_{m'\in\Omega(m)} [e^{Q(m)\tau}]_{m,m'}\, \Delta(m',n) \right\} e^{-s\tau} \Bigg]_{m,n} \tag{32}$$

if $n \notin \Omega_\mathcal{E}(m)$ and also $n \notin \Omega_\mathcal{D}(m)$:

$$K^\sim_{m,n}(s) = 0. \tag{33}$$

(Proof) Equation (29), (31), (33) are obtained by directly taking LST's of Equation (11), (13), (15) respectively. Equation (30) may be obtained as follows: The LT of $K(t)$ is given by:

$$
\begin{aligned}
K^*_{m,n}(s) &= \left[\int_0^{\tau} e^{-st} e^{Q(m)t}\, dt + \int_{\tau}^{\infty} e^{-st} e^{Q(m)\tau}\, dt\right]_{m,n} \\
&= \left[\sum_{r=0}^{\infty} \frac{(Q(m))^r}{r!} \int_0^{\tau} e^{-st}\, t^r\, dt + \frac{e^{-s\tau}}{s}\, e^{Q(m)\tau}\right]_{m,n} \\
&= \left[\sum_{r=0}^{\infty} \frac{(Q(m))^r}{r!} \frac{r!}{s^{r+1}} \times \{1 - \sum_{k=0}^{r} \frac{e^{-s\tau}(s\tau)^k}{k!}\} + \frac{e^{-s\tau}}{s}\, e^{Q(m)\tau}\right]_{m,n} \\
&= \left[\sum_{k=1}^{\infty} \frac{e^{-s\tau}(s\tau)^k}{k!s} \sum_{r=0}^{k-1} \frac{(Q(m))^r}{s^r} + \frac{e^{-s\tau}}{s}\, e^{Q(m)\tau}\right]_{m,n} .
\end{aligned}
$$

The third line is obtained using the Erlang distribution function. After some algebra, we get:

$$
K^*_{m,n}(s) = [\,(sI - Q(m))^{-1}\{I - e^{-(sI-Q(m))\tau}\} + \frac{e^{-s\tau}}{s}\, e^{Q(m)\tau}\,]_{m,n} .
$$

Converting LT into LST, we get Equation (30). Equation (32) is obtained by the same technique. □

Theorem 7 *The LST of $E(t) = [E_{m,n}(t)]$ $(m, n \in \Omega)$ of the DSPN is given by:*

1. *for state m s.t. $\mathcal{D}(m) = \emptyset$:*

$$
E^{\sim}_{m,n}(s) = \delta_{m,n} \frac{s}{s + \Lambda_m} \tag{34}
$$

2. *for state m s.t. $\mathcal{D}(m) = \{d\}$,*
   for $n \in \Omega(m)$:

$$
E^{\sim}_{m,n}(s) = [\, s(sI - Q(m))^{-1}\, \{I - e^{-(sI-Q(m))\tau}\}\,]_{m,n} \tag{35}
$$

   for $n \notin \Omega(m)$: $\quad E^{\sim}_{m,n}(s) = 0.$

(Proof) Equation (34) is obtained by directly taking LST of Equation (21):

$$
E^{\sim}_{m,n}(s) = \delta_{m,n}\, \{1 + \int_{0+}^{\infty} e^{-st}\, (-\Lambda_m) e^{-\Lambda_m t}\, dt\} = \delta_{m,n} \frac{s}{s + \Lambda_m}.
$$

Equation (35) is obtained from Equation (22) by:

$$
\begin{aligned}
E^{\sim}_{m,n}(s) &= [\int_0^{\tau} e^{-st}\, d(e^{Q(m)t}) - e^{-s\tau}\, e^{Q(m)\tau}\,]_{m,n} \\
&= [\, s(sI - Q(m))^{-1}\{I - e^{-(sI-Q(m))\tau}\}\,]_{m,n} .
\end{aligned}
$$

The integration in the above equation is the LST of $e^{Q(m)t}$ over the time interval $[0, \tau)$ which we already computed for Equation (30). □

7 Numerical Example

We illustrate the transient analysis through an example of the M/D/1/2/2 queueing system. Figure 2 shows the DSPN, its reachability graph, and its reduced reachability graph [1]. Each client is assumed to submit a job with an interval

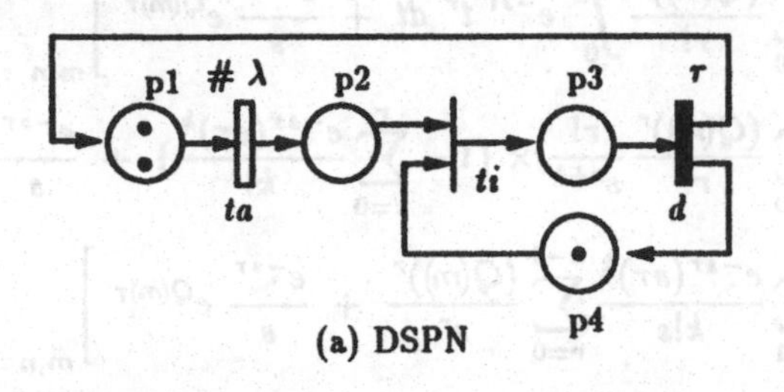

(a) DSPN

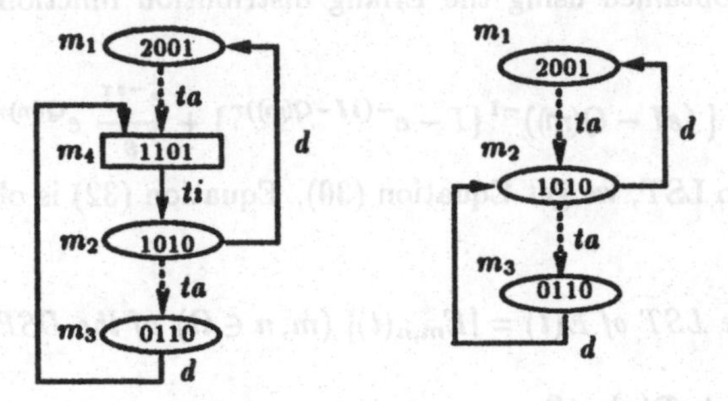

(b) Reachability Graph (c) Reduced Reachability Graph

Figure 2: DSPN and Reachability Graphs of M/D/1/2/2 System

that is exponentially distributed with rate $\lambda = 0.5$ job/hour, and the service time of a job is a constant $\tau = 1.0$ hour. The # sign above the transition *ta* indicates the marking dependent firing rate as explained in Section 2. The transition *ti* is an immediate transition. Dotted (thick solid) arcs represent the state transitions by EXP (DET) transitions and thin solid arcs out of rectangles represent the state transitions by immediate transitions. We compute the state probability vector $p(t) = (p_j(t))$, $(j = 1, 2, 3)$ at time t with given initial marking $m_1 = (2001)$ so that $p_j(t) = V_{1j}(t)$.

The kernel $K(t) = [K_{ij}(t)]$ $(i, j = 1, 2, 3)$ of this model is given as:

$$K_{21}(t) = \begin{cases} 0 & t < \tau \\ e^{-\lambda\tau} & t \geq \tau \end{cases} \quad K_{22}(t) = \begin{cases} 0 & t < \tau \\ 1 - e^{-\lambda\tau} & t \geq \tau \end{cases} \quad K_{32}(t) = \begin{cases} 0 & t < \tau \\ 1 & t \geq \tau \end{cases}$$

$K_{12}(t) = 1 - e^{-2\lambda t}$, and $K_{ij}(t) = 0$ for all other i, j.

$E(t) = [E_{ij}(t)]$ $(i, j = 1, 2, 3)$ is given as:

$$E_{22}(t) = \begin{cases} e^{-\lambda t} & t < \tau \\ 0 & t \geq \tau \end{cases} \quad E_{23}(t) = \begin{cases} 1 - e^{-\lambda t} & t < \tau \\ 0 & t \geq \tau \end{cases} \quad E_{33}(t) = \begin{cases} 1 & t < \tau \\ 0 & t \geq \tau \end{cases}$$

$E_{11}(t) = e^{-2\lambda t}$, and $E_{ij}(t) = 0$ for all other i, j.

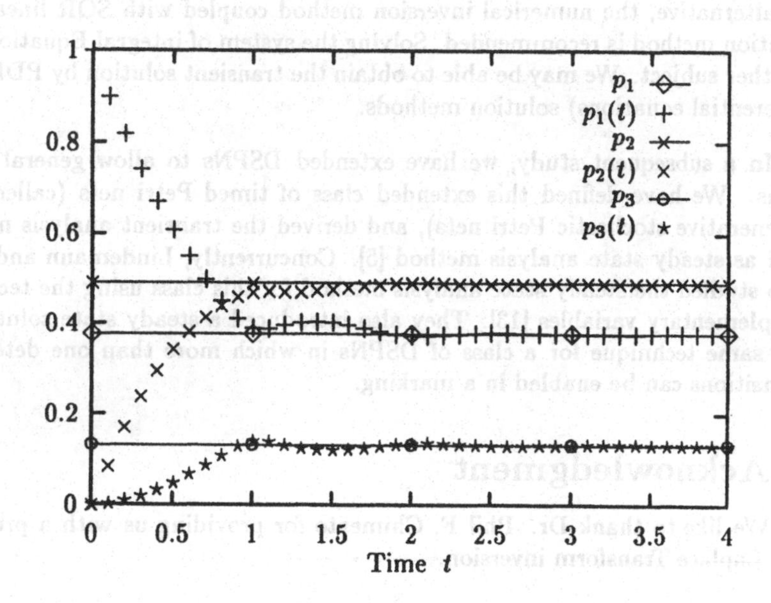

Figure 3: Transient and Steady State Probabilities of M/D/1/2/2 System

We computed LST's of $K(t)$ and $E(t)$ according to Theorem 6 and Theorem 7 respectively and computed the LST of $V(t)$ using Equation (28). Then we computed the LT of $V(t)$ by $V^*(s) = V^{\sim}(s)/s$. We obtained $V(t)$ for a fixed t by inverting the LT of $V(t)$ numerically using Jagerman's method [10] as adapted by Chimento and Trivedi [4]. The plot in Figure 3 shows the transient (time dependent) state probabilities $p_j(t)$ over a time interval [0,4] along with the steady state probabilities p_j which are computed analytically from Theorem 5. As expected, the transient state probabilities approach the steady state probabilities as time approaches infinity.

8 Conclusions

We have discussed analytical solution techniques for DSPNs in which the underlying stochastic process is not Markovian. With the condition that at most one deterministic transition is enabled in each marking, a DSPN can be solved analytically. We have shown that the stochastic process of DSPNs with this condition is a Markov regenerative process and derived a method for their transient solution. We illustrated our method by means a simple example.

Efficient numerical methods are crucial for the computation of transient and steady state solutions for large models. The computational method described in this paper needs a matrix inversion which is costly for a large size matrix. As an alternative, the numerical inversion method coupled with SOR linear system solution method is recommended. Solving the system of integral Equations (27) is another subject. We may be able to obtain the transient solution by PDE (partial differential equations) solution methods.

In a subsequent study, we have extended DSPNs to allow general distributions. We have defined this extended class of timed Petri nets (called Markov regenerative stochastic Petri nets), and derived the transient analysis method as well as steady state analysis method [5]. Concurrently, Lindemann and German also studied the steady state analysis method for this class using the technique of supplementary variables [13]. They also introduced a steady state solution using the same technique for a class of DSPNs in which more than one deterministic transitions can be enabled in a marking.

Acknowledgment

We like to thank Dr. Phil F. Chimento for providing us with a program for the Laplace Transform inversion.

References

[1] M. Ajmone-Marsan and G. Chiola. On Petri nets with deterministic and exponentially distributed firing times. In *Lecture Notes in Computer Science*, volume 266, pages 132–145. Springer-Verlag, 1987.

[2] M. Ajmone-Marsan, G. Conte, and G. Balbo. A class of generalized stochastic Petri nets for the performance evaluation of multiprocessor systems. *ACM Transactions on Computer Systems*, 2(2):93–122, May 1984.

[3] E. Çinlar. *Introduction to Stochastic Processes*. Prentice-Hall, Englewood Cliffs, U.S.A., 1975.

[4] P. F. Chimento and K. S. Trivedi. The completion time of programs on processors subject to failure and repair. *IEEE Transactions on Computers*, To appear.

[5] H. Choi, V. G. Kulkarni, and K. S. Trivedi. Markov Regenerative Stochastic Petri Nets. Technical Report DUKE-CCSR-93-001, Center for Computer Systems Research, Duke University, 1993.

[6] H. Choi, V. Mainkar, and K. S. Trivedi. Sensitivity analysis of deterministic and stochastic Petri nets. In *Proceedings of MASCOTS'93, the International Workshop on Modeling, Analysis and Simulation of Computer and Telecommunication Systems*, San Diego, USA, Jan 1993.

[7] J. B. Dugan, K. S. Trivedi, R. M. Geist, and V. F. Nicola. Extended stochastic Petri nets: Applications and analysis. In E. Gelenbe, editor, *Performance '84*, pages 507–519. Elsevier Science Publishers B. V. (North-Holland), Amsterdam, Netherlands, 1985.

[8] R. M. Geist, M. K. Smotherman, K. S. Trivedi, and J. B. Dugan. The reliability of life-critical systems. *Acta Informatica*, 23:621–642, 1986.

[9] P. J. Haas and G. S Shedler. Stochastic Petri nets with timed and immediate transitions. *Communications in Statistics- Stochastic Models*, 5(4):563–600, 1989.

[10] D. L. Jagerman. An inversion technique for the Laplace Transforms. *Bell System Technical Journal*, 61(8):1995–2002, Sep. 1982.

[11] V. G. Kulkarni. *Lecture Notes on Stochastic Models in Operations Research*. University of North Carolina, Chapel Hill, U.S.A., 1990.

[12] C. Lindemann. An improved numerical algorithm for calculating steady-state solutions of deterministic and stochastic Petri net models. In *Proceedings of the 4th International Workshop on Petri Nets and Performance Models*, pages 176–185, Melbourne, Australia, Dec. 3-5 1991.

[13] C. Lindemann and R. German. Analysis of stochastic Petri nets by the method of supplementary variables. Technical Report 1992-43, Technical University of Berlin, 1992.

[14] M. K. Molloy. Performance analysis using stochastic Petri nets. *IEEE Transactions on Computers*, C-31(9):913–917, Sep. 1982.

[15] J. K. Muppala and K. S. Trivedi. GSPN models: Sensitivity analysis and applications. In *Proceedings of the 28th ACM Southeast Region Conference*, pages 24–33, Apr. 1990.

Coloured Petri Nets Extended with Place Capacities, Test Arcs and Inhibitor Arcs

Søren Christensen
Niels Damgaard Hansen
Computer Science Department, Aarhus University
Ny Munkegade, Bldg. 540
DK-8000 Aarhus C, Denmark
Phone: +45 86 12 71 88
Telefax: +45 86 13 57 25
Telex: 64767 aausci dk
E-mail: schristensen@daimi.aau.dk, ndh@daimi.aau.dk

Abstract

In this paper we show how to extend Coloured Petri Nets (CP-nets), with three new modelling primitives—place capacities, test arcs and inhibitor arcs. The new modelling primitives are introduced to improve the possibilities of creating models that are on the one hand compact and comprehensive and on the other hand easy to develop, understand and analyse. A number of different place capacity and inhibitor concepts have been suggested earlier, e.g., integer and multi-set capacities and zero-testing and threshold inhibitors. These concepts can all be described as special cases of the more general place capacity and inhibitor concepts defined in this paper.

We give an informal description of the new concepts and show how the concepts can be formally defined and integrated in the Petri net framework—keeping the basic properties of CP-nets. In contrast to a number of the previously suggested extensions to CP-nets the new modelling primitives preserve the concurrency properties of CP-nets. We show how CP-nets with place capacities, test arcs and inhibitor arcs can be transformed into behaviourally equivalent CP-nets without these primitives. From this we conclude that the basic properties of CP-nets are preserved and that the theory developed for CP-nets can be applied to the extended CP-nets. This means that, e.g., the generalisations of analysis methods of CP-nets to cover the new modelling primitives are straightforward.

The reader is assumed to be familiar with the notion of CP-nets.

The work presented in this paper has been supported by a grant from the Danish Research Programme for Informatics (grant number 5.26.18.19).

Introduction

The development of high-level Petri Nets, CP-nets [Jen91], [Jen92] and Predicate/Transition nets [Gen86], during the late 70's and the 80's has been a substantial step in the direction of breaking the limitations of Petri Nets for modelling large and complex systems. We have, nevertheless, during our CP-net modelling, recognised that there exist a number of problems that are hard to capture by means of the existing modelling primitives. This often results in unnecessarily complex models or models which do not satisfactorily describe the modelled systems.

This is the motivation for the introduction of three new modelling primitives: place capacities, test arcs and inhibitor arcs. The primitives are general in the sense that they are equally useful for modelling in a number of different application areas—not just tailored for use in one specific area. Place capacities, test arcs and inhibitor arcs facilitate the creation of compact, comprehensive but still easily understandable models. It should be noted that the inclusion of these new modelling primitives does not, in theory, increase the computational power of CP-nets. All CP-nets including place capacities, test arcs and inhibitor arcs can be transformed into CP-nets without these primitives, but the nets will often be more difficult to read. We show how CP-nets with place capacities, test arcs and inhibitor arcs can be transformed into behaviourally equivalent CP-nets without these primitives. It is hereby ensured that the theory developed for CP-nets can be applied to the extended CP-nets. This is especially valuable when the analysis methods of CP-nets are generalised to cover extended CP-nets.

The discussion of place capacities and inhibitor arcs is based on work presented in [Bil88]. We show how the work presented in [Bil88] can be generalised and improved to overcome a number of the problems noted there.

Our paper is organised as follows. First place capacities, test arcs and inhibitor arcs are described one at a time. Each modelling primitive is presented informally before it is formally defined. It is possible to read the informal descriptions without reading the formal definitions, but it is not recommended to read the formal definitions without reading the informal introductions.

In this paper we have left out the technical details of some of the proofs. The detailed proofs can be found in [CD92].

1. Informal Introduction to Place Capacities

In this section we propose to extend CP-nets with a general place capacity construct. The approach is based on ideas originating from Kurt Jensen and briefly mentioned in [Bil88]. First we informally describe the generalised place capacity concept. Then we discuss how it relates to a number of previously proposed concepts. The formal definition of the generalised place capacity concept can be found in section 2.

Previously proposed place capacity concepts are different generalisations of the place capacity construct of P/T-nets to cover high level nets, see, e.g., [GLT79], [Res85] or [Dev89]. As they have all proved to be useful from a practical point of view for different modelling purposes, we want to define a place capacity concept for CP-nets general enough to cover all of them.

The basic idea behind our approach is described in the following. A small example illustrating our place capacity concept is given below the description. We propose to express place capacity as a combination of:

- A ***capacity colour set*** associated with each capacity place.

- A ***capacity projection***, which is a linear function mapping markings of the capacity place into multi-sets over the capacity colour set. We use the term projection to indicate that this function is often used to disregard some of the information associated with the marking of the capacity place. The result of mapping a multi-set, ms, over the colour set into a multi-set over the capacity colour set by means of the capacity projection is called the ***volume*** of ms. Throughout this paper we use the term volume of an arc expression to denote the volume of the multi-set of tokens determined by the arc expression.
- A ***capacity multi-set*** restricting the acceptable markings of the capacity place. The capacity multi-set is a multi-set over the capacity colour set. A marking of a capacity place is legal if the volume of the marking does not exceed the capacity multi-set.

In Fig. 1.1 we show a capacity stating that a place p, holding pairs, can be marked with at most two tokens with identical first components. The colour set of p is the product of two colour sets, U and V. The place capacity of p is specified as the triple: (U, Proj1, 2*U), where U is the capacity colour set, Proj1 is the capacity projection mapping pairs onto their first component and 2*U is the capacity multi-set consisting of 2 elements of each colour in U.

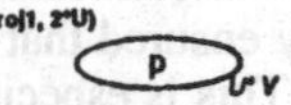

(U, Proj1, 2*U)
p
U*V

Fig 1.1: Place capacity expressing an upper limit on identical first components

In the following we show how three earlier proposed capacity concepts: ***integer capacity***, ***multi-set capacity*** and ***multi-set capacity with unbounded colours*** [Bil88], can be expressed in terms of our general capacity concept.

Integer capacity, sometimes also called ***total capacity***, can be represented as a single integer specifying the upper limit of the number of tokens on the place regardless of token colours. In our general framework we handle this as follows: the capacity colour set is a singleton set, e.g., E = {e}. The capacity projection is the function, Ig, ignoring the colour of the tokens. This means that Ig maps any colour into e, e.g., Ig(3\`a + 2\`c) = 5\`e, where a and c are colours of the colour set U of the capacity place. Finally the capacity multi-set is k\`e where k is the integer determining the upper limit of the number of tokens.

From a modelling point of view, the new place capacity concept may seem to create an unnecessary overhead: you now have to specify both the capacity colour set and the capacity projection in addition to the original integer capacity. To reduce this overhead we suggest to use a shorthand notation for some of the most commonly used capacity restrictions. In the following figures the general notation is illustrated in the left-hand side whereas a shorthand notation is shown in the right-hand side.

Fig 1.2 illustrates how the integer capacity of a place p can be expressed in terms of the general place capacity concept.

(E, Ig, 10\`e) (10)
p p
U U

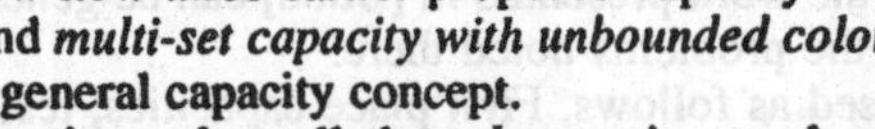

Figure 1.2: Integer capacity (left: general notation, right: shorthand notation)

Multi-set capacity expresses place capacity in terms of a multi-set over the colour set of the place. This multi-set denotes the upper limit of tokens of specific colours. Multi-set capacity can, in the same way as integer capacities, easily be expressed by means of the generalised place capacity concept. The capacity colour set of a place is identical to the colour set of the place. The capacity projection is the identity function, and the capacity multi-set is identical to the original multi-set capacity. This is illustrated in Fig 1.3 together with a shorthand notation of it.

Figure 1.3: Multi-set capacity (left: general notation, right: shorthand notation)

Multi-set capacities with unbounded colours are identical to multi-set capacities except that the multi-set is allowed to contain infinite elements [Bil88]. In terms of our general capacity concept multi-set capacities with unbounded colours can be expressed as follows. The capacity colour set of a place is identical to the colour set of the place. The capacity projection is the identity function except that unbounded colours are discarded, i.e., mapped into the empty multi-set. In this paper we denote by Id* the identity function omitting the colours having a zero coefficient in the associated multi-set parameter (in [] brackets), e.g., Id*[2`a+7`b](4`b + 3`c) = 4`b. Finally, the capacity multi-set is the multi-set, where unbounded colours are discarded. Please note that the functions Id and Id* are identical when the parameter multi-set of Id* contains all members of the colour set. Fig. 1.4 shows a multi-set capacity with unbounded colours.

Figure 1.4: Multi-set capacity with unbounded colours (left: general notation, right: shorthand notation)

Above we have only considered simple place capacity restrictions. The general concept of place capacity also allows us to specify more elaborated capacities, e.g., only allowing lists of a certain length or pairs where the first component differs from the second.

In order to formally define the behaviour of CP-nets with place capacities, we extend the enabling rule of CP-nets to handle place capacities, but we do not change the occurrence rule. Adding a capacity restriction to a place constrains the enabling of the input transitions of the place. Informally, a step is enabled in CP-nets with place capacities if the following conditions are satisfied:

(i) The step is enabled in the traditional CP-net sense—that is, there are enough tokens of the right colours on the input places.

(ii) The marking of capacity places does not violate the specified capacity restrictions when the tokens produced in the step are added.

Intuitively, the second condition seems too restrictive, because you would have expected to be allowed to subtract the tokens consumed in the step before testing the capacity restriction. This would, however, violate the so-called diamond rule of Petri nets stating that steps can be split into an arbitrary sequence of substeps. The diamond rule is illustrated in Fig. 1.5: transitions A and B are concurrently enabled in marking M with bindings b_A and b_B. If A occurs with binding b_A, we arrive at marking Ma where B is enabled with binding b_B, and vice versa. If A and B occur concurrently with bindings b_A and b_B respectively, we arrive immediately at marking Mab.

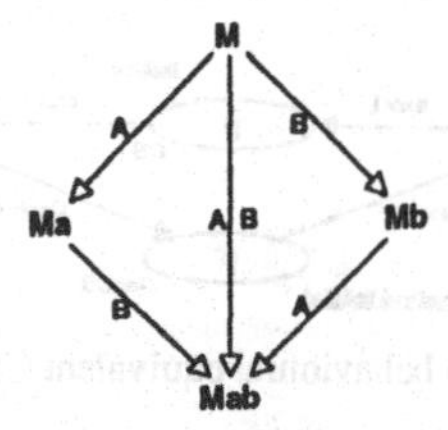

Figure 1.5. The diamond rule of Petri nets

The small net in Fig. 1.6 illustrates why it is necessary to use the enabling rule outlined above. Let us assume that the place p has integer capacity 1 and it is initially marked with one token. Transition A adds a token to p and B consumes a token from p.

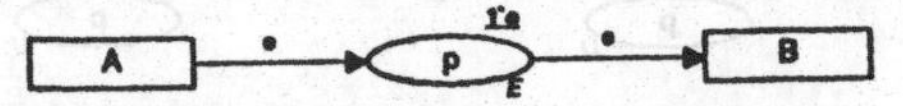

Figure 1.6: Net illustrating the enabling rule

If we use the obvious enabling rule, where we subtract the tokens consumed in the step when we calculate the enabling, this would mean that A and B are concurrently enabled, B can occur before A, but A cannot occur before B, thus violating the diamond rule as illustrated in Fig. 1.7.

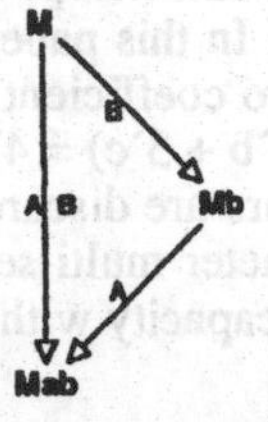

Figure 1.7: Diamond rule diagram when using obvious enabling rule

If we instead use the enabling rule (ii), outlined above, the diamond rule is preserved since our enabling rule prevents A and B from being concurrently enabled without each of them being enabled alone. We then obtain the diagram shown in Fig. 1.8.

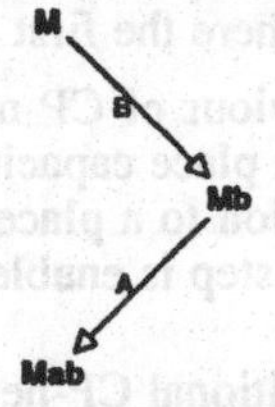

Figure 1.8: Diamond rule diagram when using our restricted enabling rule

As mentioned in the introduction, it is valuable to know that a CP-net with the new modelling primitives can be transformed into a behaviourally equivalent CP-net without the primitives. The idea behind the transformation of capacities is to add a complementary place for each capacity place and then omit the capacity restriction without hereby changing the behaviour of the model. The complementary place p' of a capacity place p can be viewed as modelling the "free space" available on the capacity place. Fig. 1.9 illustrates the transformation.

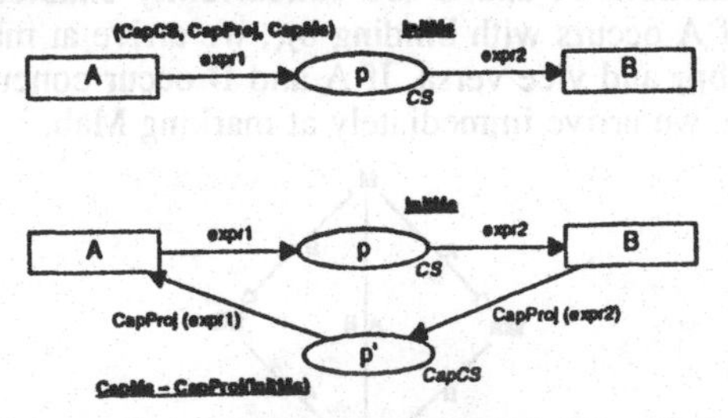

Figure 1.9: Transformation into behavioural equivalent CP-net without place capacities

The colour set of p' is equal to the capacity colour set of p. The initial marking of p' is the multi-set capacity of p subtracted the volume of the initial marking of p. For each arc, a, connected to p we have an arc a' connecting the transition of a to p'. If a is an input arc of p, a' is an output arc of p' and vice versa. The arc expression of a' is the volume of the arc expression of a, i.e., if a multi-set m is added to p by means of the arc a, the volume of m is removed from p', and vice versa.

The behaviour is not changed by the transformation because the arcs to the complementary place specify the same restrictions to the ordinary enabling rule as the enabling condition (ii) above. In Fig. 1.10 we have shown the CP-net obtained by transforming the net of Fig. 1.6.

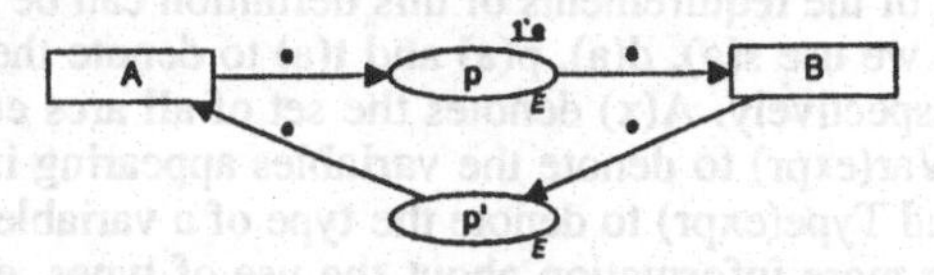

Figure 1.10: Behaviourally equivalent CP-net of the net shown in Fig. 1.6

The formal definition of CP-nets with capacities and the transformation of CP-nets with place capacity into behaviourally equivalent CP-nets can be found in section 2.

2. Formal Definition of CP-nets with Place Capacities

In this section we formalise how the place capacity concepts introduced in the previous section can be added to CP-nets. The notation used in this chapter is consistent with the one used in [Jen91] and [Jen92]. First we recall the definition and notation for multi-sets.

A **multi-set** m, over a non-empty set S, is a function $m \in [S \rightarrow N]$. The non-negative integer $m(s) \in N$ is the number of appearances of the element s in the multi-set m. We usually represent the multi-set m by a formal sum:

$$\sum_{s \in S} m(s)`s$$

By S_{MS} we denote the set of all multi-sets over S. The non-negative integers $\{m(s) \mid s \in S\}$ are called the coefficients of the multi-set m, and the number of appearances of s, m(s), is called the **coefficient** of s. An element $s \in S$ is said to **belong** to the multi-set m iff $m(s) \neq 0$ and we then write $s \in m$.

2.1 The Definition of CP-nets with Place Capacities

We recall the definition of non-hierarchical CP-nets:

Definition 2.1 ([Jen92], Def. 2.5)
A non-hierarchical **CP-net** is a tuple CPN = (Σ, P, T, A, N, C, G, E, I) satisfying the requirements below:

(i) Σ is a finite set of non-empty types, called **colour sets**.
(ii) P is a finite set of **places**.
(iii) T is a finite set of **transitions**.
(iv) A is a finite set of **arcs** such that:
• $P \cap T = P \cap A = T \cap A = \emptyset$.
(v) N is a **node** function. It is defined from A into $P \times T \cup T \times P$.
(vi) C is a **colour** function. It is defined from P into Σ. *Continues*

(vii) **G is a guard function.** It is defined from T into expressions such that:
- $\forall t \in T$: $[Type(G(t)) = \mathbf{B} \wedge Type(Var(G(t))) \subseteq \Sigma]$.

(viii) E is an **arc expression function.** It is defined from A into expressions such that:
- $\forall a \in A$: $[Type(E(a)) = C(p(a))_{MS} \wedge Type(Var(E(a))) \subseteq \Sigma]$ where p(a) is the place of N(a).

(ix) I is an **initialization function.** It is defined from P into closed expressions such that:
- $\forall p \in P$: $[Type(I(p)) = C(p)_{MS}]$.

A detailed explanation of the requirements of this definition can be found in [Jen91] and [Jen92]. For an arc, a, we use s(a), d(a), p(a) and t(a) to denote the source, destination, place and transition respectively. A(x) denotes the set of all arcs connected to the node x. Moreover, we use Var(expr) to denote the variables appearing in an expression expr and we use Type(v) and Type(expr) to denote the type of a variable v and an expression expr, respectively. For more information about the use of types, expressions and variables in CP-nets see [Jen91] or [Jen92].

We now formally define CP-nets with place capacities. Some explanation is given below the formal definition. It is recommended to read this explanation in parallel with the definition.

Definition 2.2

A CP-net with place capacities is a tuple $CPN_C = (CPN, CS)$ satisfying the requirements below:

(i) CPN is a non-hierarchical CP-net: (Σ, P, T, A, N, C, G, E, I).

(ii) CS is a **capacity specification:** (CN, CC, CP, CM).

(iii) $CN \subseteq P$ is a set of **capacity nodes.**

(iv) CC is a **capacity colour** function. It is defined from CN into Σ.

(v) CP is a **capacity projection** function. It is defined from CN into linear functions such that:
- $\forall p \in CN$: $CP(p) \in [C(p)_{MS} \rightarrow CC(p)_{MS}]_L$.

(vi) CM is a **capacity multi-set function.** It is defined from CN into closed expressions such that:
- $\forall p \in CP$: $[Type(CM(p)) = CC(p)_{MS} \wedge CP(p)(I(p)) \le CM(p)]$.

(i) CPN is a CP-net, cf. Def. 2.1.

(ii) A capacity specification is defined as a four tuple satisfying (iii) - (vi).

(iii) CN denotes the subset of places with capacity restriction. If a place p belongs to CN we say that p is a capacity place. In Fig. 1.1 we have a single capacity place p.

(iv) For each capacity place we specify a capacity colour set. The capacity colour set of p in Fig. 1.1 is U.

(v) The capacity projection maps multi-sets over the colour set of the place to multi-sets over the capacity colour set. We demand that CP is linear. We say that a multi-set ms over the colour set of a capacity place p has **volume** CP(p)(ms). Analogously for an arc a we say that the volume of the arc expression of a is the volume of the multi-set determined by the arc expression, i.e., CP(p(a))(E(a)). The projection function of p in Fig. 1.1 is *Proj1*. If the marking of p is 1\`(a,x) + 1\`(a,y) + 2\`(b,x) the volume of p is 2\`a + 2\`b.

(vi) The capacity multi-set of a place is a multi-set over the capacity colour set of the place. We demand that the volume of the initial marking of p is less than or equal to the capacity multi-set of p. In Fig. 1.1 we have a capacity multi-set: 2*U.

2.2 Behaviour of CP-nets with Place Capacities

Having defined the static structure of CP-nets with place capacities we are now ready to consider their behaviour. Adding a capacity restriction to a given place constrains the enabling of the input transitions of the place, but does not change the effect of an occurrence.

Our definitions of bindings, B(t), token elements, TE, binding elements, BE, markings, M, and steps, Y, are identical to the definitions given in [Jen92], which except for technical details are identical to those of [Jen91]. However, the enabling rule of CP-nets must be extended to take place capacities into account.

Definition 2.3

A step $Y \in \mathbf{Y}$ is enabled in a marking $M \in \mathbf{M}$ iff the following properties are satisfied:

(i) $\forall p \in P\colon \sum_{(t,b) \in Y} E(p,t)<b> \le M(p)$.

(ii) $\forall p \in CN\colon CP(p)\big(M(p) + \sum_{(t,b) \in Y} E(t,p)<b>\big) \le CM(p)$.

If Y fulfils (ii) it is said to be capacity enabled.

(i) This is the enabling rule of CP-nets, See [Jen91] Def. 2.6 or [Jen92] Def. 2.8.

(ii) The capacity enabling rule ensures that all reachable markings fulfil the restriction imposed by the place capacities, i.e., the volume of the marking of each capacity place must be less than or equal to the capacity multi-set.
This rule is comparable to the usual enabling rule ensuring that the marking of a place exceeds the multi-set of tokens consumed by an enabled step without taking advantage of the tokens added in the step. In the capacity enabling we also include the volume of the tokens added, but we do not subtract the volume of the tokens removed as discussed earlier.

2.3 Behaviourally Equivalent CP-net

All CP-nets with place capacities can be transformed into behaviourally equivalent CP-nets. The transformation is informally described in section 1.

Definition 2.4

Let a CP-net with place capacities $CPN_C = (CPN, CS)$ be given with $CPN = (\Sigma, P, T, A, N, C, G, E, I)$ and $CS = (CN, CC, CP, CM)$. Then we define the equivalent CP-net to be $CPN^* = (\Sigma^*, P^*, T^*, A^*, N^*, C^*, G^*, E^*, I^*)$ where:

(i) $\Sigma^* = \Sigma$.

(ii) $P^* = P \cup P'$ where $P' = \{p' \mid p \in CN\}$.

(iii) $T^* = T$.

(iv) $A^* = A \cup A'$ where $A' = \{a' \mid a \in A(CN)\}$.

(v) $N^*(a) = \begin{cases} N(a) & \text{iff } a \in A \\ (p',t) & \text{iff } a = b' \in A' \wedge N(b) = (t,p) \\ (t,p') & \text{iff } a = b' \in A' \wedge N(b) = (p,t). \end{cases}$

(vi) $C^*(p) = \begin{cases} C(p) & \text{iff } p \in P \\ CC(q) & \text{iff } p = q' \in P'. \end{cases}$

(vii) $G^* = G$.

(viii) $E^*(a) = \begin{cases} E(a) & \text{iff } a \in A \\ CP(p(b))(E(b)) & \text{iff } a = b' \in A'. \end{cases}$

Continues

(ix) $I^*(p) = \begin{cases} I(p) & \text{iff } p \in P \\ CM(q) - CP(q)(I(q)) & \text{iff } p = q' \in P'. \end{cases}$

(i) The set of colour sets is unchanged.

(ii) We extend the set of places to include a complementary place of each capacity place. We use single prime to indicate a complementary place, i.e., the complementary place of p is denoted p'. We assume P' to be disjoint from other net elements.

(iii) The set of transitions is unchanged.

(iv) The set of arcs is extended to include arcs connected to the complementary places. For each arc a connected to a capacity place we have an arc, a', connected to the complementary place. We assume A' to be disjoint from all other net elements.

(v) The node function is changed to match (iv). A complementary arc a' connects a complementary place with the transition t(a). If a is an input arc of p, a' will be an output arc of p' and vice versa.

(vi) The colour function is extended to handle the complementary places.

(vii) The guard function is unchanged.

(viii) The arc expressions are changed to match (iv). The arc expression of an arc connected to a complementary place is the volume of the arc expression of the arc connected to the original place.

(ix) The initialisation expressions are extended to handle the complementary places. The initial marking of a complementary place is the multi-set difference between the capacity multi-set and the volume of the initial marking of the original place.

Before we show that the transformation specified in Def. 2.4 preserves the behaviour of the CP-nets, we show a lemma that allows us to restrict the set of markings without neglecting any reachable marking.

Lemma 2.5
Let CPN be a CP-net with complementary places corresponding to the capacity specification CS = (CN, CC, CP, CM).
Then we have the following property:
$\forall p \in CN \; \forall M \in [M_0\rangle : M(p') = CM(p) - CP(p)(M(p)).$

Proof: The proof can be derived from the previous definitions in the following way: the property is true in the initial marking due to the transformation in Def. 2.4 (ix). If a binding changes the marking of a capacity place the arcs to the complementary place ensure that the marking of the complementary place is updated accordingly: from Def. 2.4(iv) we know that an arc a, connected to a capacity place will have a corresponding arc a'. From Def. 2.4(v) we know that if an input arc of p then a' is an output arc of p' and vice versa. Finally Def. 2.4(viii) specifies that the arc expression of a' is the volume of the arc expression of a. ♦

We are now ready to show that the transformation specified in Def. 2.4 preserves the behaviour of the CP-nets. We show that there is a correspondence between markings and steps of the original CP-net with place capacities and markings and steps of the transformed CP-net.

M denotes the set of all markings and Y denotes the set of all bindings, see ([Jen92], Def. 2.13). We use M | Q to denote function restriction of M to the subset of places specified by Q. All concepts with a star refer to CPN*, while those without refer to CPN_C. We denote the subset of all markings M* fulfilling the capacity restriction of lemma 2.5 by M^+.

Theorem 2.6

Let CPN_C be a CP-net with place capacities and let CPN* be the equivalent CP-net. Then we have the following properties:

(i) $\mathbb{M} = (\mathbb{M}^* \mid P) \;\wedge\; M_0 = (M_0^* \mid P)$.

(ii) $\mathbb{Y} = \mathbb{Y}^*$.

(iii) $\forall M_1, M_2 \in \mathbb{M}^+, \forall Y \in \mathbb{Y}: (M_1 \mid P)\,[Y\rangle_{CPN_C} (M_2 \mid P) \Leftrightarrow M_1\,[Y\rangle_{CPN^*} M_2$.

Proof: A detailed proof can be found in [CD92].

3. Informal Introduction to Test Arcs

When modelling with CP-nets you often have to model a situation similar to a traditional database system where multiple users share some common data. A straightforward CP-net description of this situation is shown in Fig. 3.1, where the actual database is modelled as a token on the place Data Base.

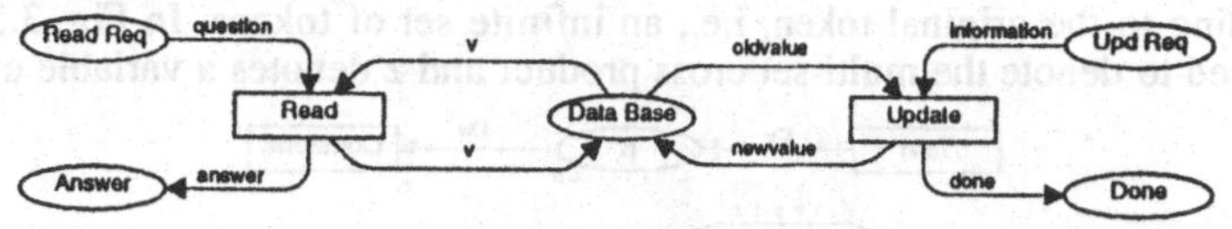

Figure 3.1: Small database example

Users are often allowed to read the data simultaneously, but only one at a time is allowed to update the data—often called multiple-read/single-write. Hereby data integrity is ensured.

Unfortunately, the model of Fig. 3.1 suffers from the lack of concurrency between simultaneous occurrences of the Read transition. Due to the enabling rule of CP-nets, an occurrence step can only contain one instance of either the Read transition or the Update transition, because each occurrence of the transitions monopolises the token, hereby preventing simultaneous access. In the case of the Update transition, this is exactly what we want to model, but the Read transitions are not able to access the data simultaneously. We are not able to model the wanted concurrency—at least not in the straightforward way shown above. If we have an upper limit for the number of simultaneous reads we could, of course, create sufficient tokens for all of these and then let each Read access only a single token, while Update changes all tokens. This solution unfortunately increases the complexity of the net inscriptions and the model contains a limit to the number of simultaneous reads.

As illustrated above, this lack of appropriate modelling primitives has often resulted in descriptions with either reduced concurrency or increased complexity of the net structure and/or the net inscriptions. We therefore propose to extend CP-nets with a new kind of arcs called test arcs. In contrast to ordinary arcs, several test arcs are allowed to access the same tokens in the same occurrence step, but not in the same step as ordinary arcs access the tokens. Test arcs cannot change the marking of a place.

Besides being a convenient modelling primitive, test arcs also simplify the transformation of CP-nets with inhibitor arcs into CP-nets, as we shall see in sections 5 and 6.

We have chosen to draw test arcs as connectors with a cross bar at both ends of the connector instead of arrowheads, because they are easy to draw and they can easily be distinguished from both source and destination of ordinary arcs. The small database example of Fig. 3.1 is then changed into the model shown in Fig. 3.2. The model in Fig 3.2 now expresses the wanted functionality and concurrency. Several instances of the Read transition may occur in the same occurrence step, but not concurrent with the Update transition since, due to the ordinary arc, it still needs to get exclusive access to the token.

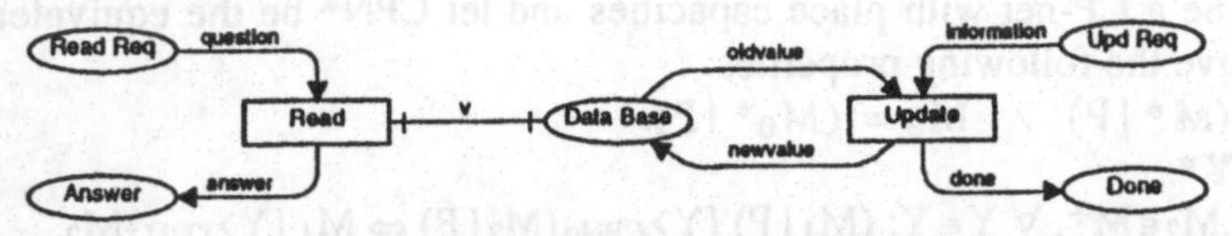

Figure 3.2: Small database example using test arcs

In the following section we formally define how CP-nets can be extended with test arcs. We also describe how a model with test arcs can be formally transformed into a behaviourally equivalent CP-net without test arcs. The idea behind the transformation is illustrated in Fig. 3.3. For each token in the original net we generate an infinite set of tokens in the equivalent CP-net by using the cross product with an infinite colour set Z. A test arc is then transformed into two ordinary arcs with opposite directions, which produce and consume only one single token, while ordinary arcs consume all tokens corresponding to the original token, i.e., an infinite set of tokens. In Fig. 3.3 the × operator is used to denote the multi-set cross product and z denotes a variable of type Z.

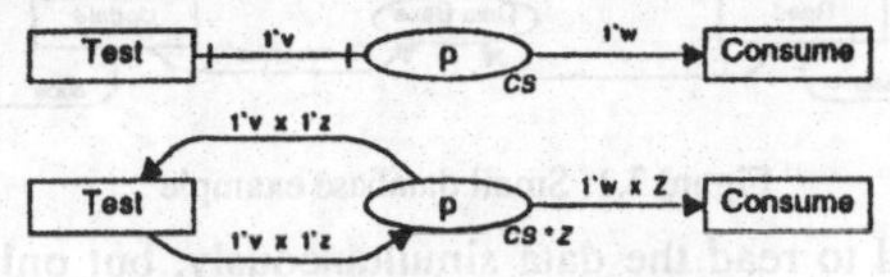

Figure 3.3: Transformation of test arcs

This means, that if the place p of the original net with test arcs is marked with, e.g., 2\`a, the corresponding place in the transformed net is marked with $2`(a,z_1) + 2`(a,z_2) + 2`(a,z_3) + \cdots$, where $z_1, z_2, z_3, \cdots$ are the colours of the infinite colour set Z. The Test transition of the transformed net consumes and produces only one token, e.g., $1`(a,z_7)$, whereas Consume removes $1`(a,z_1) + 1`(a,z_2) + 1`(a,z_3) + \cdots$. Having infinitely many different tokens in the marking of a place is not a theoretical problem, it corresponds to a PTnet with infinitely many places each having a finite marking.

4. Formal Definition of CP-nets with Test Arcs

In this section we show how test arcs are formally defined. In Def. 4.1 we give the structural definition of test arcs. In Def. 4.2 we define the additions to the enabling rule of CP-nets. Finally, in Def. 4.3, Lemma 4.4 and Theorem 4.5 we define the transformation of CP-nets with test arcs into behaviourally equivalent CP-nets.

4.1 The Definition of CP-nets with Test Arcs

Definition 4.1
A CP-net with test arcs is a tuple CPN_T = (CPN, TS) satisfying the requirements below:
(i) CPN is a non-hierarchical CP-net: (Σ, P, T, A, N, C, G, E, I).
(ii) TS is a **test arc specification**: (A_T, N_T, E_T).
(iii) A_T is a set of **test arcs**, such that: $A_T \cap (P \cup T \cup A) = \emptyset$. *Continues*

(iv) N_T is a **test node function**. It is defined from A_T into P×T.
(v) E_T is the **test expression function**. It is defined from A_T into expressions such that:
• $\forall a \in A_T$: $[Type(E_T(a)) = C(p(a))_{MS} \wedge Type(Var(E_T(a))) \subseteq \Sigma]$
where p(a) is the place of $N_T(a)$.

(i) CPN is a CP-net, see Def. 2.1.
(ii) TS is a triple, defined as follows:
(iii) A_T is the set of test arcs. All test arcs are disjoint from other net elements.
(iv) The test node function maps each test arc into a pair, where the first element is a place and the second a transition. Even though we use P×T it is important to remember that test arcs have no direction.
(v) The type of the test expression of a test arc, a, must match the colour set of the place of a. All variables of a test expression must be of known colour sets. This is exactly the same demand as for ordinary arc expressions, Def. 2.1 (viii).

We extend the p and t functions so that they also handle test arcs. This can be done by means of the N_T function.

4.2 Behaviour of CP-nets with Test Arcs

Test arcs only influence the enabling of bindings, i.e., not the actual tokens moved when an enabled binding occurs. This means that we only need to specify the new enabling rule, and keep the occurrence rule of CP-nets unchanged.

Definition 4.2
A step $Y \in \mathbf{Y}$ is **enabled** in a marking $M \in \mathbf{M}$ iff the following properties are satisfied:

(i) $\forall p \in P: \sum_{(t,b) \in Y} E(p,t)\langle b\rangle \leq M(p)$.

(ii) $\forall a \in A_T: \forall (t(a),b) \in Y: E_T(a)\langle b\rangle \leq M(p(a)) - \sum_{(t',b') \in Y} E(p(a),t')\langle b'\rangle$.

If Y fulfils (ii) it is said to be **test arc enabled.**

(i) This is the enabling rule of CP-nets, See [Jen91] Def. 2.6 or [Jen92] Def. 2.8.
(ii) For each test arc we demand that the required multi-set of tokens can be accessed, after the tokens used in the step have been consumed but before the tokens produced in the step have been added.

4.3 Behaviourally Equivalent CP-net

All CP-nets with test arcs can be transformed into behaviourally equivalent CP-nets. The idea behind the transformation was illustrated in section 3.

Definition 4.3
Let a CP-net with test arcs CPN_T = (CPN, TS) be given with CPN = (Σ, P, T, A, N, C, G, E, I) and TS = (A_T, N_T, E_T). Then we define the **equivalent CP-net** to be CPN* = (Σ*, P*, T*, A*, N*, C*, G*, E*, I*) where:
(i) $\Sigma^* = \Sigma \cup \{U \times Z \mid U \in \Sigma\} \cup \{Z\}$, where $Z \notin \Sigma$ is an infinite colour set.
(ii) P* = P.
(iii) T* = T.
(iv) $A^* = A \cup (A_T \times Dir)$, where Dir = {In,Out}. *Continues*

(v) $N^*(a) = \begin{cases} N(a) & \text{iff } a \in A \\ (p(a'),t(a')) & \text{iff } a=(a',In) \in A_T \times Dir \\ (t(a'),p(a')) & \text{iff } a=(a',Out) \in A_T \times Dir. \end{cases}$

(vi) $C^*(p) = C(p) \times Z.$

(vii) $G^* = G.$

(viii) $E^*(a) = \begin{cases} E(a) * Z & \text{iff } a \in A \\ E_T(a') * 1`v_Z & \text{iff } a=(a',d) \in (A_T \times Dir) \end{cases}$

where v_Z is a variable of type Z. v_Z must be unique for each test arc a.

(ix) $I^*(p) = I(p) \times Z.$

(i) The set of colour sets is extended to include the colour sets determined by the cross product of each original colour set and a new infinite colour set called Z. Z itself is also included in the set of colour sets.

(ii) The set of places is unchanged.

(iii) The set of transitions is unchanged.

(iv) The set of arcs is extended by arcs corresponding to the test arcs. Each test arc is substituted by a pair of ordinary arcs.

(v) The node function is extended to handle the arcs corresponding to the test arcs. For each test arc, a', we have an arc (a,In) having p(a) as the source and t(a) as the destination, and an arc (a,Out) having t(a) as the source and p(a) as the destination.

(vi) The colour set of each place is the cross product of the old colour set and the new infinite colour set Z.

(vii) The guard function is unchanged.

(viii) The arc expressions are changed to match (vi). For the ordinary arcs we specify that the expressions must take a token for each colour in Z, i.e., an infinite set of tokens. The expressions corresponding to the test arcs will take only a single token. The tokens consumed by an arc (a,In) will have the same value as the tokens produced by (a,Out).

(ix) The initialization expression is changed to fit (vi).

Before we show that the transformation specified in Def. 4.3 preserves the behaviour of the CP-nets, we show a lemma that allows us to restrict the set of markings without neglecting any reachable marking. We show that for any reachable marking of a place we have a full set of Z tokens for each colour c, where (c,z) is a member of the marking.

Lemma 4.4
Let CPN = (Σ, P, T, A, N, C, G, E, I) be a CP-net corresponding to the translation of a test arc specification.
Then we have the following property:

$$\forall p \in P\ \forall M \in [M_0\rangle\ \forall (c,z) \in M(p)\text{: } M(p)(c,z) * (1`c \times Z) \leq M(p).$$

Proof: From Def. 4.3(vi) we know that the colour set of a place p has the structure CS×Z. The lemma holds for M_0 due to the transformation specified in Def. 4.3(ix). From Def. 4.3(viii) we have that each arc corresponding to an ordinary arc moves a full set of Z tokens, while the arcs corresponding to a test arc do not change the marking. This means that all reachable markings will fulfil lemma 4.4. ♦

We are now ready to show that the transformation specified in Def. 4.3 preserves the behaviour. M denotes the set of all markings and Y denotes the set of all bindings, see ([Jen92], Def. 2.13). We use M×Z to denote the marking, which for each place p, has the tokens: $M(p)\times Z$. We use $Y\backslash V_Z$ to denote the restriction of a binding element $Y=(t,b)$ where variables of colour set Z are discarded, i.e., $Y\backslash V_Z = (t,b\backslash V_Z)$. All concepts with a * refer to CPN*, while those without refer to CPN_T. $\mathbb{M}^+$ denotes the subset of all markings, $\mathbb{M}^*$, fulfilling the restriction of lemma 4.4.

Theorem 4.5

Let CPN_T be a CP-net with test arcs and let CPN* be the equivalent CP-net. Then we have the following properties:

(i) $\mathbb{M}^+ = \{M\times Z \mid M\in\mathbb{M}\} \wedge M_0^* = M_0\times Z$.

(ii) $\mathbb{Y} = \{Y^*\backslash V_Z \mid Y^*\in\mathbb{Y}^*\}$.

(iii) $\forall M_1,M_2\in\mathbb{M}, \forall Y^*\in\mathbb{Y}^*$:

$$M_1\times Z\ [Y^*\rangle_{CPN^*} M_2\times Z \Rightarrow M_1\ [Y^*\backslash V_Z\rangle_{CPN_T} M_2.$$

$\forall M_1,M_2\in\mathbb{M}, \forall Y\in\mathbb{Y}$:

$$M_1\ [Y\rangle_{CPN_T} M_2 \Rightarrow \exists\, Y^*\in\mathbb{Y}^*: Y = Y^*\backslash V_Z \wedge M_1\times Z\ [Y^*\rangle_{CPN^*} M_2\times Z.$$

Proof: A detailed proof can be found in [CD92].

5. Informal Introduction to Inhibitor Arcs

The convenience of inhibitor arcs providing the ability to test for the absence of tokens has been known in the Petri net area for a long time. As in the case of place capacities, a number of different inhibitor concepts have been proposed, reflecting the fact that they are useful for different modelling purposes—see, e.g., [Bil88] or [CDF91]. Nevertheless, the concepts seem to be generalisations of the P/T inhibitor concepts to cover high-level Petri nets, e.g., zero-testing inhibitor arcs and threshold inhibitor arcs. A transition t connected to a place p by a *zero-testing inhibitor arc* can only occur if the marking of p is empty. If t and p are connected by a *threshold inhibitor arc* t can only occur if the marking of p is less than or equal to a specified multi-set called the threshold.

In this paper we improve and generalise the inhibitor arcs defined in [Bil88] in several ways. This section informally describes by means of small examples both the reasons why we find it necessary to change the concepts and the actual changes we propose. The generalised inhibitor arcs are formally defined in section 6.

Please note that we throughout this paper have chosen to draw inhibitor arcs as connectors with a circle at both ends because this notation ensures that inhibitor arcs are easily distinguished from both source and destination of ordinary arcs. Furthermore, this notation indicates that inhibitor arcs define an undirected relationship between places and transitions.

In [Bil88] the enabling rule defined to cover inhibitor arcs states that a multi-set of bindings is enabled if there are enough tokens on the input places, there is enough volume left in the capacity places and the inhibitor thresholds are not exceeded. Unfortunately this enabling rule does not preserve the diamond rule.

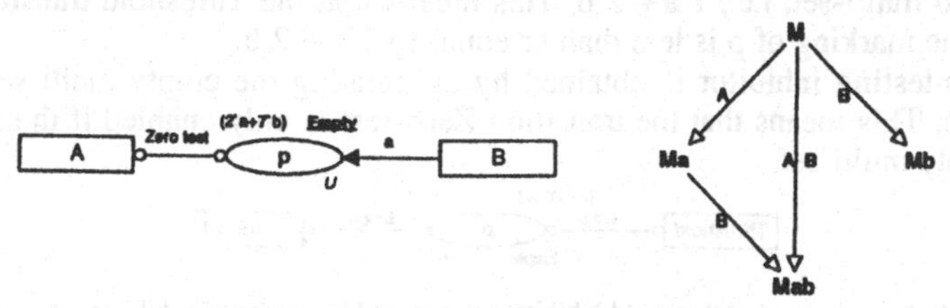

Figure 5.1: CP-net with inhibitor arcs violating the diamond rule

Let us assume that the inhibitor arc of Fig 5.1 is a zero-testing inhibitor arc, and the place p is unmarked. Then A and B are concurrently enabled according to the above enabling rule. If B occurs before A, A is no longer enabled, and the diamond rule is not fulfilled—an edge of the diamond is missing as illustrated in the diamond rule diagram of Fig. 5.1. To ensure that the diamond rule is preserved, we strengthen the enabling rule concerning inhibitors arcs. We demand that the threshold restrictions are valid for the current marking plus the tokens produced in the step. Hereby the diamond rule is satisfied since the concurrency causing the problem is transformed into a conflict—see Fig. 5.2.

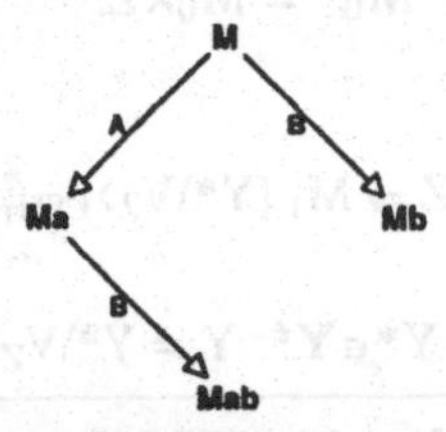

Figure 5.2: Diamond rule diagram for changed enabling rule

One may, of course, argue that this enabling rule is too restrictive from a modelling point of view, because models are hereby left out. It is, however, our belief that the diamond rule is such an essential and integral part of the Petri net framework and necessary in order to be able to understand the behaviour of CP-nets, that one must not deviate from it.

The next change we propose concerns the generality of the inhibitor concept. From our point of view there is no reason why the class of inhibitor arcs should be limited to zero-testing inhibitors and threshold inhibitors. We therefore propose to generalise the inhibitor concept along the line of the generalisation of the place capacity concept. We introduce a new generalised concept covering both zero-testing and threshold inhibitors. In order to be able to transform a CP-net with inhibitor arcs to a behaviourally equivalent CP-net in a straightforward way, we demand—like [Bil88]—that the place of an inhibitor arc must be a capacity place. The reason for this demand will be clear later when we discuss the transformation.

We suggest to define the inhibitor condition as an ordinary arc expression, except it must evaluate to a multi-set over the capacity colour set of the place the inhibitor arc is connected to. The inhibitor arc expression then expresses the threshold that must not be exceeded by the volume of the marking of the place.

In Fig. 5.3 we show how zero-testing and threshold inhibitor arcs are easily expressed by means of the new inhibitor concept. The place p of Fig. 5.3 is a capacity place with multi-set capacity 2\`a + 7\`b + 1\`c. Since the capacity projection is equal to the identity function the volume of the marking of p is equal to the marking of p. The threshold inhibitor is then expressed by defining the inhibitor arc expression equal to the threshold multi-set, i.e., 1\`a + 2\`b. This means that the Threshold transition is only enabled if the marking of p is less than or equal to 1\`a + 2\`b.

The zero-testing inhibitor is obtained by associating the empty multi-set, with the inhibitor arc. This means that the transition Zero-test is only enabled if the marking of p is the empty multi-set.

Threshold — 1\`a+2\`b — p (2\`a+7\`b+1\`c) Empty U — Empty — Zero-test

Figure 5.3: Zero-testing and threshold inhibitors expressed by means of inhibitor arc expressions

In [Bil88] it is pointed out several times that it is important to be able to transform proposed extensions of CP-nets into behaviourally equivalent CP-nets without the extensions. This is an obvious way of ensuring that the basic properties of CP-nets are preserved. Unfortunately, the transformation of inhibitor arcs proposed in [Bil88] does not preserve the entire behaviour. Only the single step behaviour—also called interleaving behaviour is preserved. This means that the behaviours of the two models are only equivalent as long as we consider steps without concurrency. The author is well aware of this problem, and it is suggested to be subject for further work.

There are two reasons for the difference in the behaviour of the two models in [Bil88]: the problem with the enabling rule illustrated above and the selected transformation. In [Bil88] the basic idea is to transform an inhibitor arc a, connected to a place p, into two ordinary arcs with opposite directions connected to the complementary place p'. The arc expressions of the arcs connected to p' are the capacity of p minus the arc expression of a. The transformation of the zero-testing inhibitor of Fig 5.1 is illustrated in Fig. 5.4.

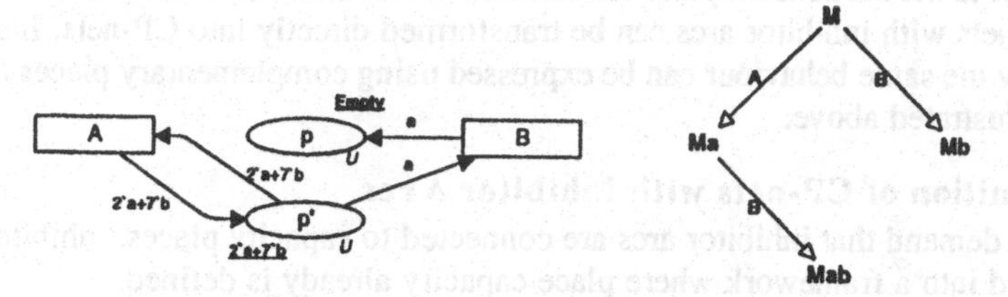

Figure 5.4: Transformation of zero-testing inhibitor arc and associated diamond rule

It is obvious from Fig. 5.1 and 5.4 that this transformation may limit the amount of concurrency. If p is unmarked the transitions A and B are concurrently enabled in the model of Fig. 5.1, but not in the transformed model in Fig 5.4. Since A and B concurrently access the same tokens a conflict arises between the transitions—a conflict that did not exist in the original model because of the enabling rule chosen in [Bil88]. If two transitions are connected to the same place by means of inhibitor arcs, the transformation may also cause the two transitions to be in conflict, because they—when transformed to ordinary arcs—try to access the same tokens in the same step. Fig. 5.1 and 5.4 also illustrate that the difference in behaviour results in different diamond rule diagrams.

Our solution to this problem is straightforward. We propose to use our strengthened enabling rule and to use test arcs in the transformation instead of two ordinary arcs with opposite directions. Hereby we avoid the problems described above. We propose to transform an inhibitor arc between a transition A and a place p into a test arc between A and the complementary place p' hereby ensuring that the concurrency of the model is preserved. Note that the arc expression of an inhibitor arc evaluates to a multi-set over the capacity colour set. The transformation of inhibitor arcs is illustrated in Fig. 5.5.

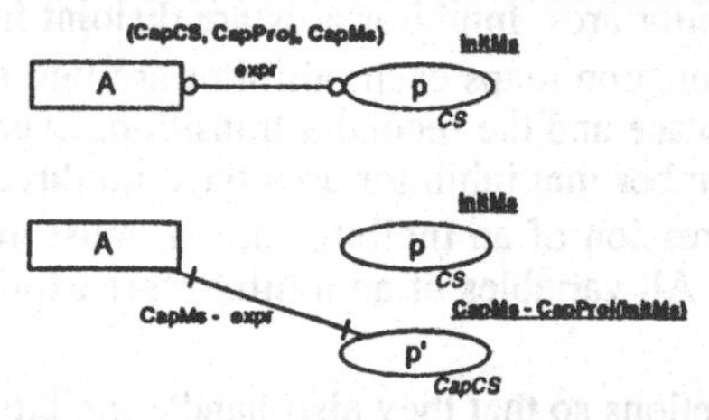

Figure 5.5: Transformation of an inhibitor arc to test arc connected to the complementary place

Fig 5.5 also illustrates why we demand that the place of an inhibitor arc must be a capacity place—otherwise it would be impossible to compute the initial marking of the complementary place and the test arc inscription.

Please note that our transformation—due to the use of test arcs—preserves the concurrency between two transitions connected to the same place by inhibitor arcs. This is illustrated in Fig. 5.6 showing the transformation of Fig. 5.3.

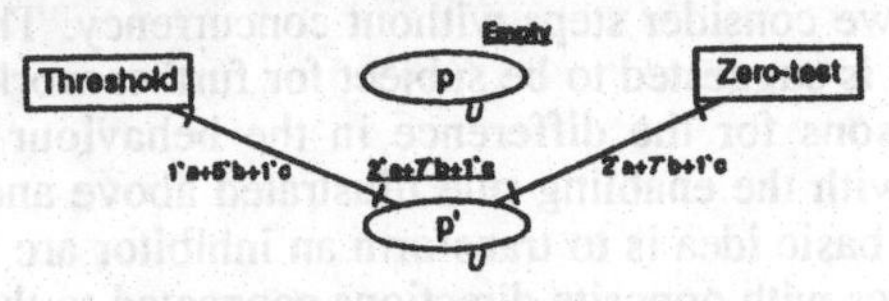

Figure 5.6: Transformation of Fig. 5.3

6. Formal Definition of CP-nets with Inhibitor Arcs

In contrast to the sections on place capacities and test arcs this section does not show how CP-nets with inhibitor arcs can be transformed directly into CP-nets. Instead we show how the same behaviour can be expressed using complementary places and test arcs as illustrated above.

6.1 Definition of CP-nets with Inhibitor Arcs

Since we demand that inhibitor arcs are connected to capacity places, inhibitor arcs are introduced into a framework where place capacity already is defined.

Definition 6.1
A CP-net with inhibitor arcs is a tuple CPN_{CI} = (CPN, CS, IS) satisfying the requirements below:

(i) CPN is a non-hierarchical CP-net: (Σ, P, T, A, N, C, G, E, I).
(ii) CS is a capacity specification: (CN, CC, CP, CM).
(iii) IS is an inhibitor arc specification: (A_I, N_I, E_I).
(iv) A_I is a set of inhibitor arcs, such that: $A_I \cap (P \cup T \cup A) = \emptyset$.
(v) N_I is the inhibitor node function. It is defined from A_I into CN×T.
(vi) E_I is the inhibitor expression function. It is defined from A_I into expressions such that:
- $\forall a \in A_I : [Type(E_I(a)) = CC(p(a))_{MS} \wedge Type(Var(E_I(a))) \subseteq \Sigma]$

where p(a) is the capacity place of $N_I(a)$.

(i) CPN is a CP-net, see Def. 2.1.
(ii) CS is a capacity specification, see Def. 2.2.
(iii) IS is an inhibitor arc specification, defined as a triple with the following elements:
(iv) A_I is the set of inhibitor arcs. Inhibitor arcs are disjoint from other net elements.
(v) The inhibitor node function maps each inhibitor arc into a pair, where the first element is a capacity place and the second a transition. Even though we use CN×T it is important to remember that inhibitor arcs have no direction.
(vi) The type of the expression of an inhibitor arc, a, must match the capacity colour set of the place of a. All variables of an inhibitor arc expression must be of known colour sets.

We extend the p and t functions so that they also handle inhibitor arcs. This can be done by means of the N_I function.

6.2 Behaviour of CP-nets with Inhibitor Arcs

The behaviour of CP-nets with inhibitor arcs is described as follows:

Definition 6.2
A step $Y \in \mathbb{Y}$ is enabled in a marking $M \in \mathbb{M}$ iff the following properties are satisfied:

(i) $\forall p \in P: \quad \sum_{(t,b) \in Y} E(p,t)\langle b\rangle \leq M(p).$

(ii) $\forall p \in CN: \quad CP(p)\big(M(p) + \sum_{(t,b) \in Y} E(t,p)\langle b\rangle\big) \leq CM(p).$

(iii) $\forall a \in A_I: \forall (t(a),b) \in Y: E_I(a)\langle b\rangle \geq CP(p(a))\big(M(p(a)) + \sum_{(t',b') \in Y} E(t',p(a))\langle b'\rangle\big).$

If Y fulfils (iii) it is said to be **inhibitor arc enabled.**

(i) This is the enabling rule of CP-nets, See [Jen91] Def. 2.6 or [Jen92] Def. 2.8.
(ii) The capacity enabling rule is defined in Def. 2.3.
(iii) For each inhibitor arc we demand that the threshold expression evaluates to a multi-set which is larger than the volume of the marking together with the volume of the added tokens.

6.3 Behaviourally Equivalent CP-net

In sections 2.3 and 4.3 we have shown how to transform CP-nets with place capacities and CP-nets with test arcs into behaviourally equivalent CP-nets. We could specify a similar transformation for CP-nets with inhibitor arcs, but instead we take the easy way out. We show how to transform inhibitor arcs into test arcs connected to complementary places. This means that we must do the following: by means of Def. 2.3, we transform a CP-net with place capacities and inhibitor arcs into a CP-net with complementary places and inhibitor arcs. Then we use Def. 6.3, see below, to transform the inhibitor arcs into test arcs connected to the complementary places. Finally, we use Def. 4.3 to transform a CP-net with test arcs into an ordinary CP-net.

Definition 6.3
Let a CP-net with inhibitor arcs and complementary places $CPN_I = (CPN, IS)$ be the result of the transformation of a CP-net having the capacity specification $CS = (CN, CC, CP, CM)$. We use the following notation: $CPN = (\Sigma, P \cup P', T, A \cup A', N, C, G, E, I)$ and $IS = (A_I, N_I, E_I)$. We define the equivalent CP-net with test arcs to be $CPN_T{}^* = (CPN^*, TS)$ where $TS = (A_T, N_T, E_T)$ where:

(i) $CPN^* = CPN.$
(ii) $A_T = A_I.$
(iii) $N_T(a) = (p',t)$ where $N_I(a) = (p,t).$
(iv) $E_T(a) = CM(p(a)) - E_I(a).$

(i) The basic net definition is unchanged.
(ii) Each inhibitor arc is substituted by a test arc.
(iii) The test node function maps a test arc into the pair consisting of the complementary place and the transition of the inhibitor arc.
(iv) The expression of the test arc is the multi-set difference between the capacity multi-set and the inhibitor expression of the original arc.

Theorem 6.4
Let CPN_I be a CP-net with inhibitor arcs and complementary places and let $CPN_T{}^*$ be the equivalent CP-net with test arcs. Then we have the following properties:
(i) $M = M^* \wedge M_0 = M_0{}^*$
(ii) $Y = Y^*$.
(iii) $\forall M_1, M_2 \in M, \forall Y \in Y: M_1 [Y\rangle_{CPN_I} M_2 \Leftrightarrow M_1 [Y\rangle_{CPN_T{}^*} M_2$.

Proof: A detailed proof can be found in [CD92].

In the previous sections we have given an informal introduction to the notion of place capacity, test arcs and inhibitor arcs. We have also shown how these concepts can be formally defined in such a way that the extended CP-nets can be transformed into behaviourally equivalent CP-nets without extensions. For reasons of simplicity we have introduced the primitives one at a time. When modelling you would probably often want to use all three primitives. In the following section we have therefore shown how the three concepts can be combined into a single net formalism called CP-nets with place capacities, test arcs and inhibitor arcs.

7. CP-nets with place capacities, test arcs and inhibitor arcs

The three extensions described in the previous sections can easily be combined into a single model making it possible to model systems using place capacities, test arcs as well as inhibitor arcs.

Using the transformation defined in Def. 2.4, 4.3 and 6.3 we have shown, that a CP-net with place capacities, test arcs or inhibitor arcs can be transformed into a behaviourally equivalent CP-net. Since we from section 6.3 know that the transformation of inhibitor arcs uses both complementary places and test arcs, the straightforward order of transformation of a CP-net with both place capacities, test arcs and inhibitor arcs is: first place capacities are transformed into complementary places using Def. 4.2. Then we use Def. 6.3 to transform the inhibitor arcs into test arcs connected to the complementary places. Finally we use Def. 4.3 to transform the test arcs either remaining from the original net or resulting from the transformation of the inhibitor arcs into a CP-net.

8. Conclusion

In this paper we have extended CP-nets to include three new modelling primitives: place capacities, test arcs and inhibitor arcs, hereby improving the possibilities to create compact and understandable descriptions reflecting the modelled system in a straightforward way.

Our place capacity concept may be viewed as a generalisation of the previously proposed place capacity concepts: integer capacity, multi-set capacity and multi-set capacity with unbounded colours. In addition to these a number of other useful capacity restrictions can be formulated by means of our generalised concept.

Test arcs are—in contrast to ordinary arcs—able to access the same tokens simultaneously, but test arcs are not allowed to change the marking. Often the modelled system allows concurrent access to common data provided that the data is not changed. Without test arcs or similar modelling primitives this situation is difficult to model and therefore often results in models with either reduced concurrency or increased complexity in net structure and/or net inscriptions.

The definition of inhibitor arcs is also a generalisation of previously proposed inhibitor arc primitives. We associate an inhibitor arc expression with each inhibitor arc. The inhibitor arc expression is written like an ordinary arc expression. The result of the evaluation of the inhibitor arc expression is denoted the threshold value and the associated transition is disabled if the threshold value is exceeded. A number of useful inhibitor conditions including zero-testing inhibitors and threshold inhibitors are easily expressed by means of our generalised inhibitor concept. To preserve the diamond rule—stating that steps can be split up into substeps—we need to strengthen the enabling rule known from previously proposed inhibitor concepts. We demand that the thresholds are not exceeded for the current marking and with the tokens produced in the step.

We have given an informal description of the new concepts, a formal definition and shown how they fit nicely into the Petri net framework. Both the static and dynamic properties of CP-nets with place capacities, test arcs and inhibitor arcs are formally defined. In order to ensure that the extensions do not violate the basic properties of CP-nets we have shown how CP-nets with place capacities, test arcs and inhibitor arcs can be transformed into behaviourally equivalent CP-nets. The possibility to transform a CP-net with the three new primitives to a behaviourally equivalent CP-net allows us to relate properties of the extended CP-nets to properties of the CP-nets. This is, e.g., useful when the analysis methods of CP-nets are generalised to cover extended CP-nets.

It should be noted, that place capacities, test arcs and inhibitor arcs from a theoretical point of view do not increase the computational power of CP-nets. It is, however, widely accepted that place capacities and inhibitor arcs are valuable modelling primitives. We furthermore believe that the inclusion of test arcs results in models reflecting the concurrency of the modelled system to a larger extent.

Acknowledgement

We would like to thank Kurt Jensen for valuable discussions and Peter Huber, Kjeld Høyer Mortensen, Mogens Nielsen and Laura Petrucci for many helpful comments on earlier versions of this paper.

References

[Bil88] J. Billington: **Extending Coloured Petri Nets.** University of Cambridge, Computer Laboratory, Technical Report No. 148, October 1988. (Ph.D. Thesis).

[CD92] S. Christensen, N. Damgaard Hansen: **Coloured Petri nets extended with place capacities, test arcs and inhibitor arcs.** Daimi PB-398, ISSN 0105-8517, May 1992. Available as: Daimi PB-398, ISSN 0105-8517, May 1992.

[CDF91] G. Chiola, S. Donatelli and G. Franceschinis. **Priorities, Inhibitor Arcs and Concurrency in P/T Nets.** Proceedings of the 12th International Conference on Application and Theory of Petri Nets, Aarhus 1991, 182-205.

[Dev89] R. Devillers: **The Semantics of Capacities in P/T Nets.** In: G. Rozenberg (ed.): Advances in Petri Nets 1989. Lecture Notes in Computer Science, vol. 424. Springer-Verlag, 1989, pp. 128-150.

[Gen86] H.J. Genrich: **Predicate/Transition nets.** In: W. Brauer, W. Reisig and G. Rozenberg (eds.): Petri Nets: Central Models and Their Properties, Advances in Petri Nets 1986 Part I, Lecture Notes in Computer Science vol. 254, Springer-Verlag 1987, 207-247.

[GLT79] H.J. Genrich, K. Lautenbach and P.S. Thiagarajan: **Elements of general net theory.** In: W. Brauer (ed.): Net theory and applications. Proceedings of the Advanced Course on General Net Theory of Processes and Systems, Hamburg 1979, Lecture Notes in Computer Science vol. 84, Springer-Verlag 1980, 21-163.

[Jen91] K. Jensen: **Coloured Petri nets: A high level language for system design and analysis.** In: G. Rozenberg (ed.): Advances in Petri Nets 1990. Lecture Notes in Computer Science, vol. 383. Springer-Verlag, 1990, pp. 342-416.

[Jen92] K. Jensen: **Coloured Petri nets. Basic concepts, analysis methods and practical use. Volume 1: Basic concepts.** EATCS monographs on Theoretical Computer Science, Springer-Verlag 1992.

[Res85] W. Reisig: **Petri Nets. An introduction.** EATCS Monographs on Theoretical Computer Science, Vol. 4, ISBN 3-540-13723-8 or 0-387-13723-8, Springer-Verlag, 1985.

Integrating Software Engineering Methods and Petri Nets for the Specification and Prototyping of Complex Information Systems*

Yi Deng[1], S.K. Chang[2], Jorge C.A. de Figueired[2]* and Angelo Perkusich[2]*

[1] School of Computer Science
Florida International University - Miami, FL 33199 - USA
[2] Department of Computer Science
University of Pittsburgh - Pittsburgh, PA 15260 - USA

Abstract. We present a Petri net based framework called *G-Net* for the modular design of complex information systems. The motivation of this framework is to integrate Petri net theory with a modular, object-oriented approach for the specification and prototyping of complex software systems. We use the client/server example to illustrate the *G-Net* specification of distributed systems, and how such a specification can be translated into a *Predicate/Transition net* for formal analysis. The differences between *G-Net* and hierarchical Petri net, as well as some limitations of the transformation technique, are then discussed.

1 Introduction

First introduced by C.A. Petri in the early sixties, Petri nets have been actively studied and applied to many areas of computer science. Different variations or extensions of Petri nets have been proposed. The early Condition/Event (CE) nets and their extension, Place/Transition nets, serve as the basis for other higher level or more specialized Petri net models. However, when modeling systems that consist of many interrelated identical components, these basic Petri net models exhibit great redundancy, since the only way to differentiate two identical elements is to specify an identical subnet for each one. An important development in Petri net theory is the introduction of *individual tokens*, that is, to allow individual objects as tokens. With this extension, we can model the common component only once and assign one distinguishable token to each identical component. This development leads to a class of Petri-Nets called *high-level Petri-Nets*, including *Predicate/Transition nets (PrT-nets)* [10, 9] *Colored Petri-Nets (CP-Nets)* [12] *Relation Nets (Rel-Nets)* [17], and *Petri Nets with Individual Tokens* [18]. However, even with high level Petri nets, system modeling can still be difficult because no mechanisms or notations are provided in these

* Also with Departamento de Engenharia Elétrica, Universidade Federal da Paraíba, 58100, Campina Grande, PB - Brazil. The authors would like to acknowledge the financial support of CNPq/Brazil in the form of scholarship 201463/91-1 and 201462/91-5

models to specify the *structure* of a system. For example, there is no concept of hierarchy. To remedy the problem, another extension has been proposed to introduce the concept of hierarchy to high-level Petri nets. The resulting Petri net models are called hierarchical Petri nets, including *Hierarchical CP-Nets* [11, 12] and *Hierarchical PrT-nets* [15]. As indicated in [11] although the hierarchical constructs introduced do not extend the modeling power of CP-Nets, they do provide structuring tools to construct large system models in practice.

In this paper, we present a Petri net based framework called *G-Net* for the modular design and specification of distributed information systems. The framework [4, 5, 6] is an integration of Petri net theory with the software engineering approach for system design. The motivation of this integration is to bridge the gap between the formal treatment of Petri nets and a modular, object-oriented approach for the specification and prototyping of complex software systems.

The *G-Net* notation incorporates the notions of module and system structure into Petri nets; and promotes abstraction, encapsulation and loose coupling among the modules. The former feature makes *G-Net* a more suitable tool for specifying complex software systems. The latter feature supports incremental modification of the specification of a complex system.

A specification based on *G-Nets* (called a *G-Net specification*) consists of a set of independent and loosely-coupled modules (*G-Nets*) organized in terms of various system structures. Recursive structures or functions can be easily and naturally specified by *G-Nets*. A *G-Net* is encapsulated in such a way that a module can only access another module through a well defined mechanism called *G-Net abstraction*, and no *G-Net* can directly affect the internal structure of another module. As indicated by Booch [1] we can hardly make a complex design right at the first try, thus complex software design is an evolutionary process, where repeated changes are necessary. Also, as argued by Luqi [13], the potential benefits of prototyping depend critically on the ability to modify the prototype's behavior with substantially less effort than required to modify the production software. The modular features of *G-Nets* provide the necessary support for incremental design and successive modification.

A *G-Net* specification can be directly executed in a distributed environment, where a G-Net is a natural unit for distribution and execution [3]. The execution mechanism allows multiple instances of invocation on the same *G-Net* to be executed simultaneously. The execution of a *G-Net* handled by multiple distributed computation agents enables the execution to be conducted in a parallel fashion.

The paper is organized as follows. In Section 2 we introduce the basic notations and concepts. The unique modularization features of the *G-Nets* and their impact on incremental design and continuous modification are discussed. We also show that recursive structures can be easily and naturally specified using *G-Nets*. A prototype *G-Net* system has been implemented on SUN workstations connected by a LAN. Some examples are used to show the execution of the system. The modeling of a client/server interaction using *G-Nets* is presented in Section 3. A formal transformation technique has been developed, which is capable of translating a *G-Net* specification to a semantically equivalent *Predi-*

cate/Transition nets (PrT-nets). With this transformation, the formal analysis techniques for *PrT-nets* can then be applied to *G-Net* specifications. We will present the transformation procedure in Section 4. Finally, the conclusion of this paper is given in Section 5, where the differences between *G-Net* and hierarchical Petri net are also discussed.

2 G-Net Abstraction and Instantiation

A widely accepted software engineering principle of design is that a system should be composed of a set of independent modules, where each module in the system hides the internal details of its processing activities and modules communicate through well-defined interfaces [7]. The *G-Net* notations provide a strong support to this principle. As mentioned earlier, a system modeled in terms of a *G-Net* specification is composed of a set of self-contained modules called *G-Nets*. A *G-Net*, G, is composed of two parts: a special place called *Generic Switch Place (GSP)* and an *Internal Structure (IS)*. The *GSP* provides the abstraction of the module, and serves as the only interface between the *G-Net* and other modules. The *IS*, a modified Petri Net, represents the detailed internal realization of the modeled application. An example is shown in Figure 1.

The *GSP* of a *G-Net*, G, (denoted by $G.GSP$ in the ellipse of Figure 1) uniquely identifies the module. The *GSP* contains one or more executable methods (denoted by $G.MS$ in the round-cornered rectangle in Figure 1) specifying the functions, operations or services defined by the net, and a set of attributes (denoted by $G.AS$) specifying the passive properties of the module (if any). The detailed structures and information flows of each method in $G.MS$ are defined by a piece of Petri net in $G.IS$. More specifically, a method $mtd \in G.MS$ defines the input parameters, the initial marking of the corresponding internal Petri net (the initial state of the execution), a set of special places called *Goal Places* (denoted by double circle in Figure 1) represents the final state of the execution, and the results (if any) to be returned. The collection of the methods and the attributes (if any) provides the abstraction or the external view of the module.

In $G.IS$, Petri Net places represent *primitives*; and transitions, together with arcs, represent connections or relations among the primitives. These primitives may be actions, predicates, data entities, and *Instantiated Switch Place (ISP)*. A primitive is said to be enabled if it receives a token, and an enabled primitive can be executed. A transition, together with arcs, defines the synchronization and coordinates the information transfer between its input and output places (primitives). More specifically, an input arc between a transition t and an input place p may carry an inscription denoted by $I(p,t)$. $I(p,t)$ is composed of a set of items, each of which is either a variable or a logical expression of variables. $I(p,t)$, $p \in I(t)$, defines the conditions when t is enabled and defines the values to be exported by the primitive p. We assume that the inscriptions on the input arcs of t do not share variables to avoid ambiguity. An output arc may also have an inscription, which is a list of variables defining the content of the input token to each output place.

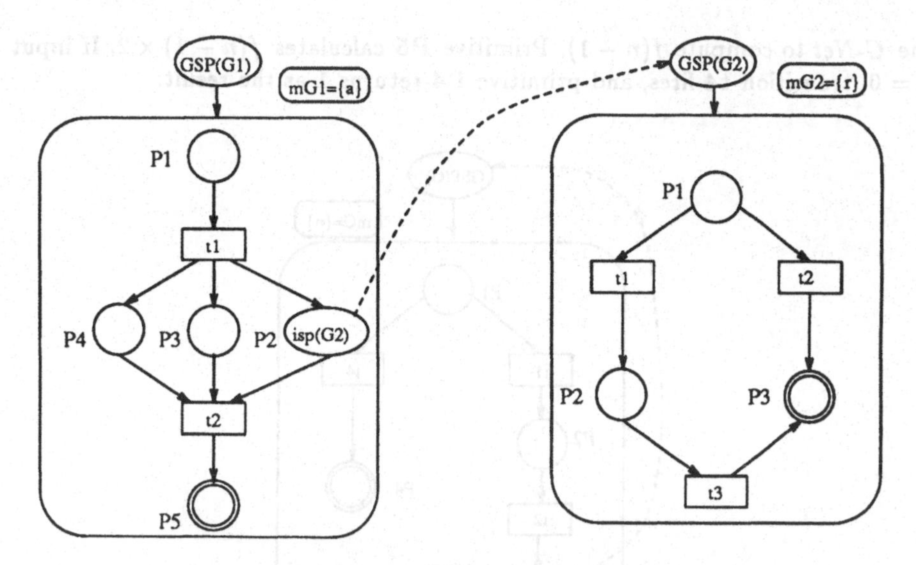

Figure 1. Two *G-Nets* connected through an isp

Given a *G-Net* G, an ISP of G is a tuple $(G.Nid, mtd)$ which is denoted by $isp(G.Nid, mtd)$ (or simply $isp(G)$ if no ambiguity occurs), where $G.Nid$ is the unique identification of G, and $mtd \in G.MS$. Each $isp(G.Nid, mtd)$ denotes an instantiation of the *G-Net* G, i.e. an instance of invocation of G based on the method $mtd \in G.MS$. Therefore, executing the *ISP* primitive implies invoking G (by sending a token to G) based on the specified method using the content of the token (received at the *ISP*) as input (if any). On the other hand, using an *ISP* of one *G-Net* (say $isp(G.Nid, mtd)$) as a building block (primitive) in the Internal Structure of another *G-Net*, $G'.IS$, specifies a connection between the two *G-Nets*. Since the *ISP* is treated as a primitive of $G'.IS$, the precise meaning of the connection between G' and G is defined by the connections between the primitives of $G'.IS$. From the view point of G', invoking G from its internal structure is just like executing another primitive.

The *ISP* notation serves as the primary mechanism for specifying the connections or relationships between different *G-Nets* (modules). Embedding an *ISP* of a lower level *G-Net* into the *IS* of a higher *G-Net* specifies a hierarchical configuration. Embedding an *ISP* of a *G-Net* into its own *IS* specifies a recursive configuration. Furthermore, embedding an *ISP* of a *G-Net* specifying a server into a *G-Net* specifying a client results in a client-server relation. Figure 1 shows an example of two *G-Nets* connected by an *isp*, where an *isp* of G_2 is used as a primitive of G_1. For net G_1 it is defined one method, m_{G1}, that receives an integer a, and for net G_2 it is defined one method, m_{G2}, that receives a integer r. Figure 2 shows a *G-Net* specifying the recursive function $f(n) = 2^{f(n-1)}$; $f(0) = 1$, where $n \in \mathbb{N}$. The *G-Net* has one method, m_1, with input n and initial marking **P1**. If $n > 0$, transition **t1** becomes enabled and fires. Primitive **P2** decreases the value of n by 1. Primitive **P3** is an *ISP* which recursively invokes the *G-Net* to compute $f(n-1)$. Primitive **P5** calculates $f(n-1) \times 2$. If input $n = 0$, transition **t4** fires, and primitive **P4** returns 1 as the result.

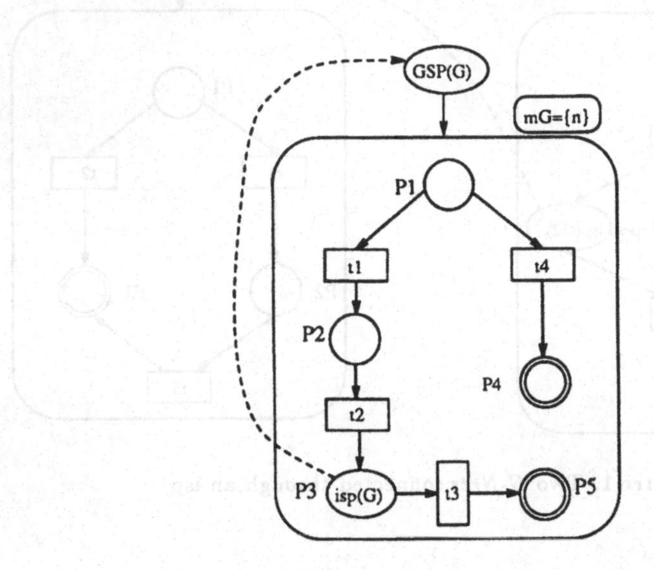

Figure 2. *G-Net* specification of a recursive function

The *G-Nets* in a *G-Net system* are loosely coupled because they don't share variables, and they interact with each other only through their *GSPs* (by the means of token passing). The internal structure of a *G-Net* is transparent to the others. Therefore, any change made to the internal structure, *IS*, of a *G-Net* is unlikely to affect the structures of the other *G-Nets* as long as the interface defined by its *GSP* remains unchanged. Furthermore, the entire *G-Net* specification can be built incrementally in a *G-Net* by *G-Net* basis, and a *G-Net* can be referenced by the other *G-Nets* even before the design of its internal structure, because the interface of the net is solely determined by its *GSP*.

An interesting feature of the *G-Net* framework is that it allows more than one invocation of a *G-Net* to be executed simultaneously. This feature is due to the unique token structure in the *G-Net* framework. A token, *tkn*, is defined as a triple, i.e. $tkn = (seq, sc, msg)$, where *tkn.seq* is called *propagation sequence*, *tkn.sc* is called the status color, and *tkn.msg* is called the message of the token.

Tkn.seq is a sequence of $\langle Nid, isp, Pid \rangle$, where *Nid* is the identifier of a *G-Net*, *isp* is the name of an *ISP* and *Pid* is a process identification. The propagation sequence of a token is only changed in *ISP* and *GSP*. When a *G-Net*, *G*, is invoked from an *ISP*, isp_i, in another *G-Net*, G', (when isp_i receives a token), a triple $\langle G', isp_i, Pid_{G'} \rangle$, where $Pid_{G'}$ is the identifier of the process executing G', is appended to the propagation sequence of the token before it is sent to *G-Net G*. This triple indicates that when the execution of *G* is over, the resulting token should be returned to the place identified by isp_i in *G-Net* G'. The process identifier is needed to distinguish which instance of the execution on G' the returning token belongs to. When the input token received at the *GSP* G, a triple $\langle 0, 0, Pid_G \rangle$ is appended to the end of its propagation sequence indicating that the agent responsible for executing the invocation is identified by Pid_G. Because Pid_G is unique, the propagation sequence is also unique. The token structure in the *G-Net* framework thus not only guarantees that all tokens belong to an instance of a *G-Net* execution have the same and unique propagation sequence, but also contains the complete propagation history of the tokens, which governs the interactions between the processes executing a *G-Net* specification.

The status color of a token has two possible values, either before or after. A token is said to be *available* if its sc = after, otherwise, it is said to be *unavailable*. When a new token is received at a place, its status color is set to be before, and after the primitive action at the place is taken, the status color of the token is set to be after, indicating that the token is ready to be used in subsequent transition firing. The message field of a token is a list of application specific values.

The *G-Net* transition firing mechanism is defined by the following *transition (firing) rules*:

1. A transition t is said *enabled* iff: (a) every place $p \in I(t)$ holds at least one token and its content satisfies the condition defined by $I(p,t)$; (b) all the tokens stated in (a) have the same propagation sequence; (c) all the tokens stated in (a) and (b) have their sc = after; and (d) the number of tokens in $O(t)$ is smaller than the capacities of the places.
2. An enabled transition t may *fire*. When firing t: (a) a token which satisfies $I(p,t)$ is removed from every $p \in I(t)$; and (b) a token whose propagation sequence is the same as the one in tokens removed from $I(t)$, whose sc = before, and whose message part is determined by $O(t,p)$ is deposited to every place $p \in O(t)$.

An invocation of a *G-Net*, G, based on a method $mtd \in G.MS$ is carried out in the following way:

1. Determine the initial marking of G based on the definition of *mtd* (the content of the tokens depends on the content of the input token);
2. Fire enabled transitions, if any;
3. Invoke enabled primitives, if any;
4. Repeat (2)-(3) until a *Goal Place* is reached;
5. Send the result of the execution (if defined by *mtd*) to the invoker.

The execution of a *G-Net* specification is carried out by a set of concurrent processes called agents. A *G-Net* can be associated with one of the two execution modes, namely, *instantiation mode* or *server mode*. In the instantiation mode, the function of an agent is solely to execute *one* instance of a *G-Net* invocation based on a specified method. A new agent is created for every instantiation of the *G-Net*; and the agent is terminated when the instance of the *G-Net* execution is done. In the server mode, only one agent is associated with a *G-Net* specifying a server. The agent is called a server agent which handles all client request sent

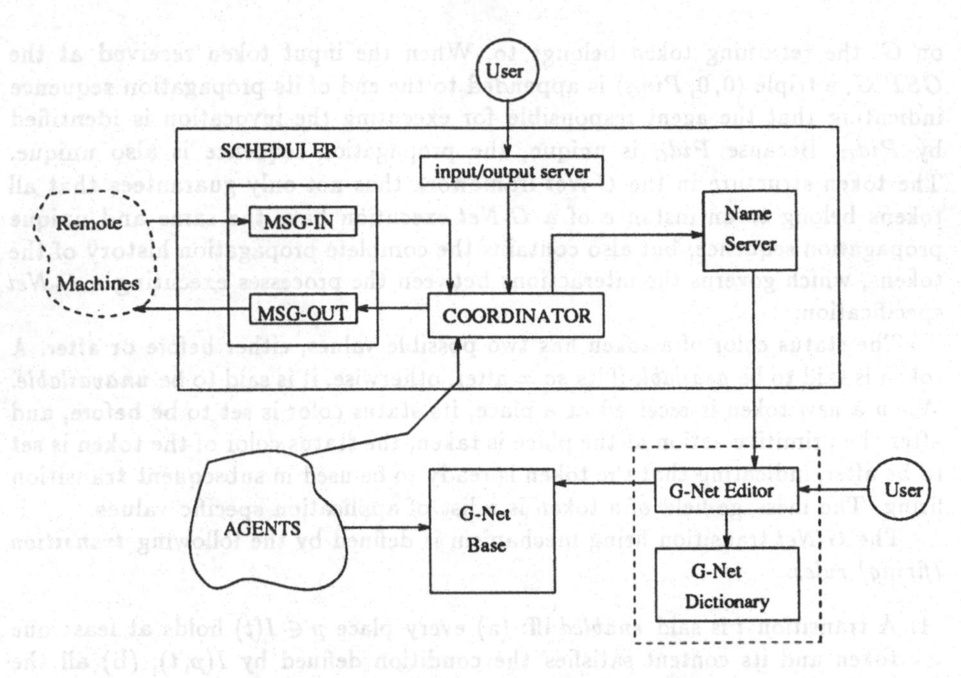

Figure 3. Block diagram of the *G-Net* simulator

to the server. The agents communicate to each other through token passing as described above. The *G-Net* framework maximize the concurrency among the agents because (1) the agents are independent in the sense that no agent has direct control over another; and (2) while an agent is waiting for the result of an invocation of another *G-Net*, it can still execute other primitives which do not depend on the result of the communication. Such concurrency among the primitives is explicitly specified by Petri net notations. For instance, in Figure 1, while the agent executing G_1 is waiting for the reply from G_2 (at **P2**), it can still execute **P3** and **P4**. Thus the execution of the both *G-Nets* are conducted in parallel.

A prototype of a simulator for *G-Net* systems has been implemented on SUN workstations connected by a LAN. The structure of the system and the details of the implementation can be found in [3, 4]. In Figure 3 the block diagram of the simulator is shown. A copy of the simulator resides in each machine of a distributed system executing a *G-Net* specification. The SCHEDULER has four basic modules. The SHELL module implements the interface between the user and the module COORD, named coordinator. The module COORD is the main part of the SCHEDULER, its main functions are: creation of computation agents to execute a *G-Net* specification; and provision of an interface between the agents and the message server. The message server, named MSG SERVER, handles inter-process comunication for local and remote agents. The module NAME-SERVER provides a mapping from a given logical *G-Net* identifier to a phisycal address for the system network. As shown in Figure 3, there is also an

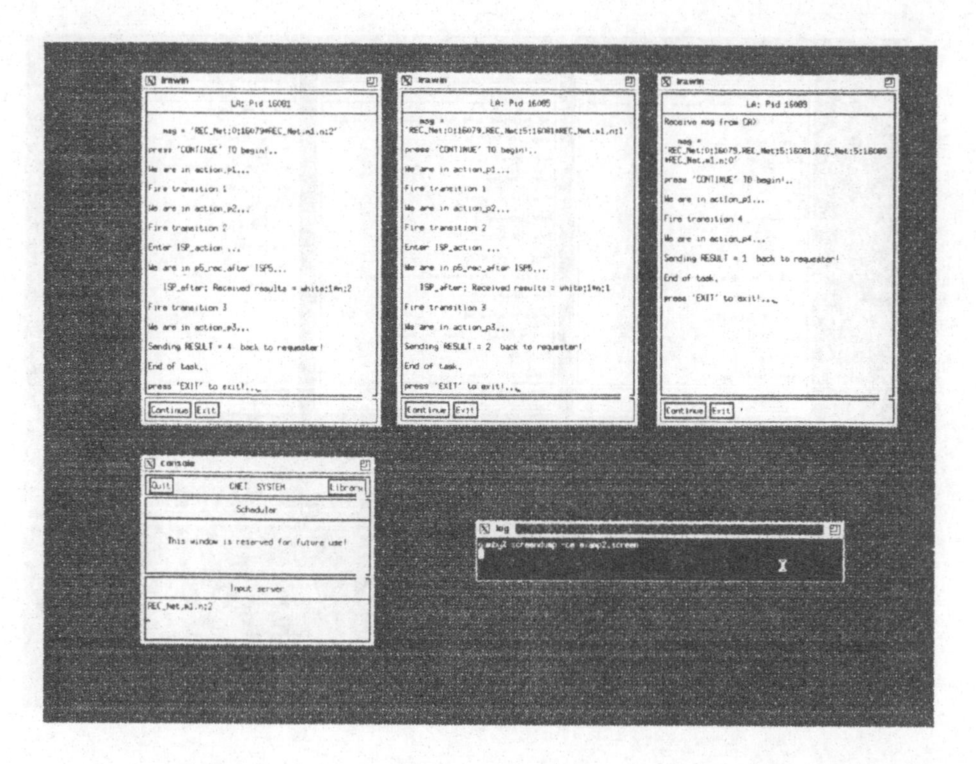

Figure 4. The execution of the *G-Nets* of Figure 1

off-line user inteface. The main modules are the *G-Net Editor* and the *G-Net Dictionary*. The *G-Net Editor* is a prompt-driven editor that allows the user to input or change a *G-Net* specification. The *G-Net Dictionary* is a set of predefined high level *G-Net* constructs. Finally, the *G-Net Base* contains the text files specifying each *G-Net* alocated to a machine.

The execution of the *G-Nets* of Figure 1 is shown in Figure 4, where the *G-Net* G_1 is allocated in a machine called "gumby", and *G-Net* G_2 is allocated in a machine called "elvis". Two middle windows in the figure are for the SCHEDULER subsystem in the two machines respectively. These subsystems run on every involved machines to control the execution and to coordinate the communications between the computation agents executing the *G-Nets*. The two bottom windows show the traces of the execution of the two *G-Nets* G_1 and G_2, respectively. The execution of the recursive *G-Net* G of Figure 2 is shown in Figure 5, where the *G-Net* G is invoked with input $n = 2$. The three windows on the top of the figure are associated with three recursive instantiations of the net with input n being 2, 1 and 0, respectively.

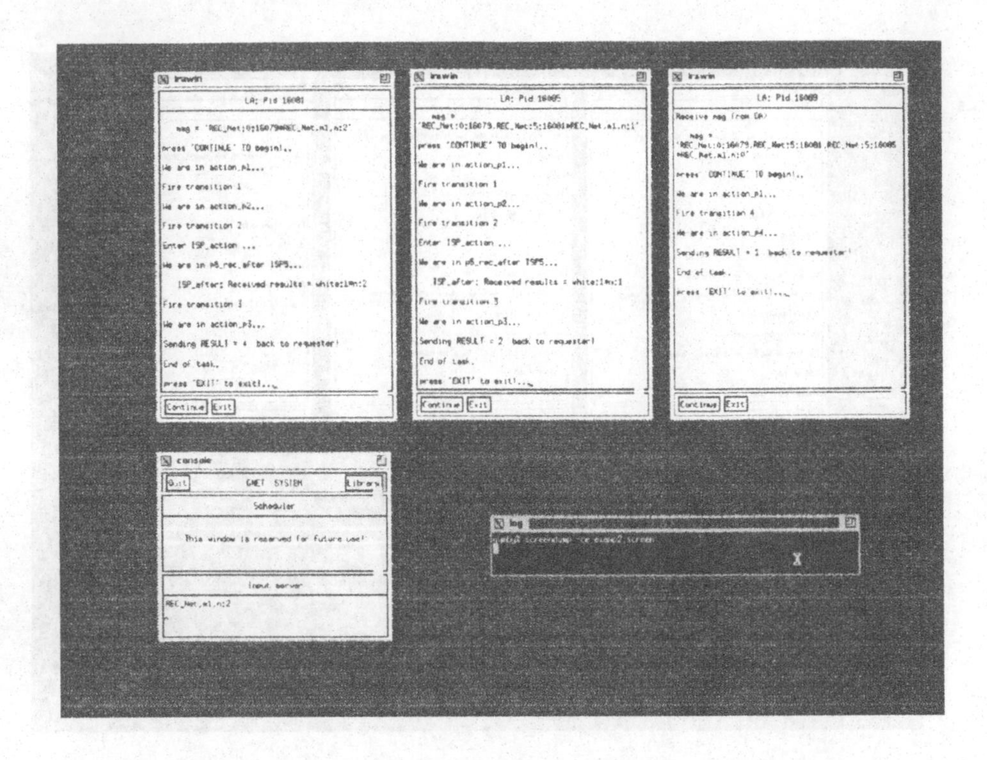

Figure 5. The execution of the recursive *G-Net* of Figure 2

3 Client-Server Example

In this section we present the modeling of a client/server interaction using *G-Nets.* The server may attend requests from clients and it is provided a buffer for pending requests. The client may issue various requests to the server, depending on the service that is to be executed. After a successful request, the client sends the data to the server to trigger the execution of the pending request. When the server receives a request it checks if it can either accept the request in case there is space in the buffer, or reject the request otherwise. If the pending requests buffer is not full it puts the requested information, in this case the service needed, in a buffer, to be processed when the data arrives. The availability of buffer space is modeled by an attribute, *BS*, associated with the server. Upon starting the execution of request the server updates the attribute related to the buffer and starts the execution of the service. After executing a request the server sends a positive acknowledge to the client.

The client/sever interaction as described above is modeled for the i-th client by *G-Net* $G(C)$, presented in Figure 6, and for the server by *G-Net* $G(S)$ modeling the client and presented in Figure 7. The process starts when the client has a request to send to the server. In this case, a token is deposited in place

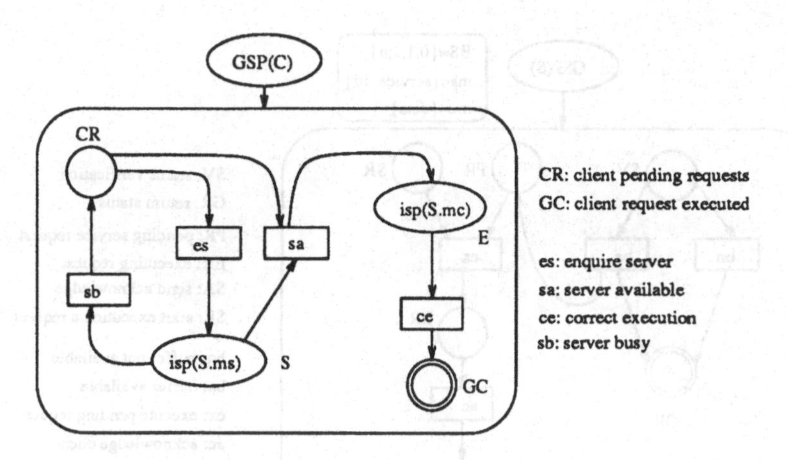

Figure 6. *G-Net* modeling the Client

CR in the *G-Net* $G(C)$. The transition **es** fires and a token is put in place **isp(S.ms)**, with the service identification, service_id, associated to it. Then, the *G-Net* $G(S)$ is invoked with method *ms* to verify if the server is available. A token is deposited in place **SV** in the *G-Net* $G(S)$. The availability of buffer space will determine which transition will fire, i.e., **bn** or **ba**. If no space is available in the buffer, transition **bn** fires and a token is deposited in place **GD** and the invocation resumes. In this case, after the invocation, the transition **sb** in *G-Net* $G(C)$ fires, a token is deposited in place **CR** and the process is repeated until the pending request is executed. If space is available in the buffer, the transition **ba** in $G(S)$ fires and tokens are deposited in places **PR** and **GR**. The buffer capacity attribute is decremented by one and the invocation resumes. In this case, transition **sa** fires, and the data inforamation is obtained from place **CR**, a token is deposited in place **isp(S.mc)** that invokes the *G-Net* $G(S)$ with method *mc*. Then, a token is put in place **TR** in *G-Net* $G(S)$ and transition **ex** fires and the pending request is processed, and the buffer capacity attribute, *BS*, is incremented by one. After processing the request (token in place **ER**), transition **ac** fires, an acknowledgment is sent to the client and the invocation resumes. Transition **ce** in *G-net* $G(C)$ fires, a token is deposited in goal place **GC** and the interaction finishes.

4 A Formal Transformation Technique

A formal transformation technique has been proposed in [4]. The technique translates a *G-Net* specification to a set of *PrT-nets*, where each *PrT-net* corresponds to a method in the *G-Net* specification, and has the equivalent semantics as the latter. By this transformation, the formal analysis techniques developed for *PrT-nets* can be used to analyze *G-Net* specifications.

There are some limitations associated with this formal transformation technique. The transformation is, at this time, restricted to *G-Nets* without net-wide

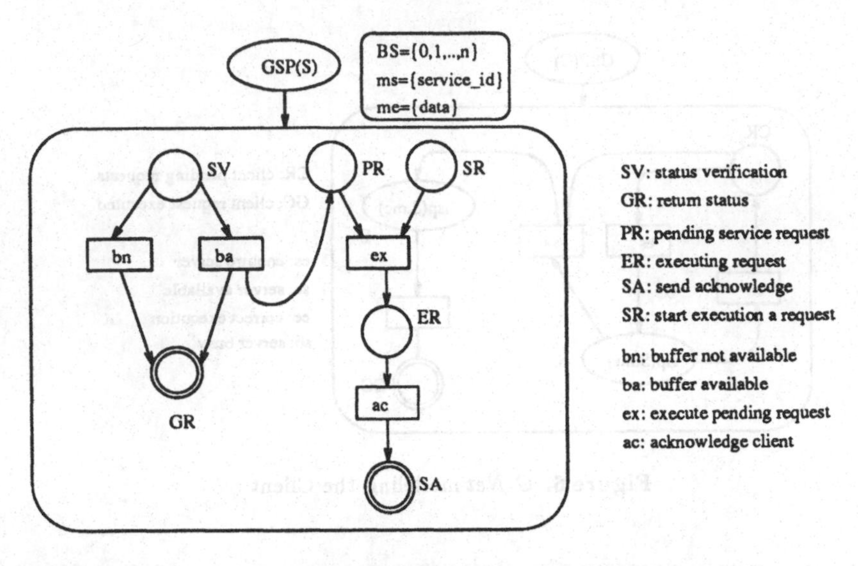

Figure 7. *G-Net* modeling the Server

attribute set. A more serious weakness is that the technique does not take advantage of the abstraction and modularization features of *G-Nets*. Therefore, when it is applied to analyze complex *G-Net* specifications, we are likely to face the so-called *state explosion* problem, which is considered to be a major weakness of Petri nets [14]. Nevertheless, this formal transformation does provide a theoretical foundation for the *G-Net* framework.

In this section, we use the *G-Nets* in Figure 1 as an example for illustrate the transformation procedure. A more formal presentation of the technique and associated proofs can be found in [4].

The transformation algorithm accepts a *G-Net* specification defined by $GS = G_1, G_2, \cdots, G_k$ and a method $mtd \in G.MS$, $G \in GS$, as inputs. It produces a semantically equivalent *PrT-Net* corresponding to the method as output. The key of this transformation is the treatment of GSPs and ISPs. The basic idea of the algorithm can be described as follows: For each regular place (primitive action) and each transition, we create an equivalent segment in *PrT-net* notation. As the result of the translation of an ISP, $isp(G')$, the *PrT-net* representation of $G'.IS$ is also created, recursively. Because multiple ISPs corresponding to G' may be used in the *G-Net* specification, we must be careful that $G'.IS$ is only translated once. The definition of the method *mtd* will determine the initial marking and final marking of the *PrT-net*, these markings become the interfaces to other part of the specification. Another issue is that, since the concept of propagation sequence is not defined in *PrT-nets*, we have to use regular attributes to mimic of the effect of *G-Net* token propagation sequence.

We start with elementary transformations. The transformation of regular places and transitions is quite intuitive. A regular place in *G-Net* can be viewed as a mapping, denoted by $A(p)$, between a set of input parameters, $I(p) =$

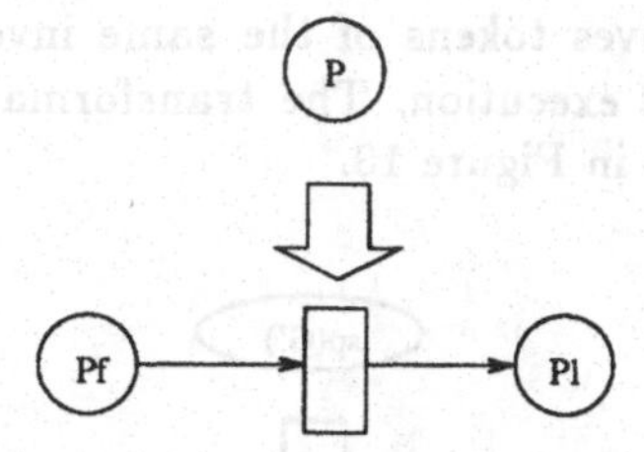

Figure 8. The transformation of a primitive

$\langle seq, x_1, x_2, \ldots, x_q\rangle$, (carried by the input token, an input token is one whose $sc =$ before), and a set of output parameters, $O(p) = \langle seq, y_1, y_2, \ldots, y_r\rangle$, which constitute the content of the output token from the place. Such a place can be easily transformed to *PrT-net* representation shown in Figure 8. Notice that the propagation sequence of the token remains unchanged. A transition, t, in a *G-Net* can also be easily transformed to *PrT-net* notation as shown in Figure 9.

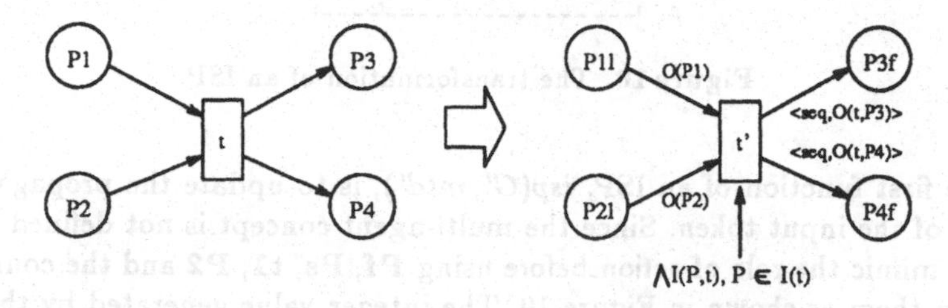

Figure 9. The transformation of a transition and its surrounding arcs

Recall that an ISP denotes a *G-Net* instantiation. Every time the ISP receives an input token, the *G-Net* corresponding to the ISP is invoked. More specifically, the semantics of an ISP, $Isp(G'.Nid, mtd')$, is defined in the following procedure, where Step 1 and 4 are to update the propagation sequence of the token.

1. action_before : $tkn.seq \leftarrow tkn.seq + \langle Nid, NAME(p), Pid\rangle$;
2. Invoking *G-Net* G' based on method *mtd'* using *tkn* as input (by sending the token to G');
3. Waiting for the result of step (2)
4. action_after : $tkn.seq \leftarrow tkn.seq - LAST(tkn.seq)$; $tkn.sc \leftarrow$ after,

There are two basic issues need to be addressed in the transformation of an ISP. The first is how to interface the *G-Net*, *G*, which contains the ISP, and the *G-Net* G'. The second is how to simulate the semantics of the *G-Net* token propagation sequence, so as to maintain the uniqueness of the tokens in a *G-Net* invocation. This is important because in the resulting *PrT-net* we must be able to distinguish the tokens belonging to one invocation of a *G-Net* from the tokens belonging to other invocations. Such an ability will ensure that the firing

of a transition only involves tokens of the same invocation, thus maintaining the semantics of a *G-Net* execution. The transformation procedure of an ISP, *isp(G′, mtd′)*, is illustrate in Figure 10.

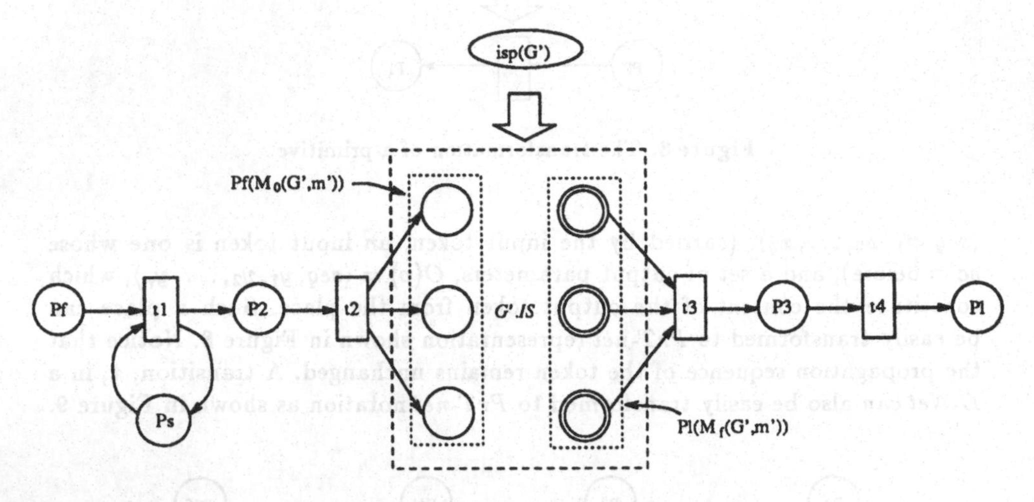

Figure 10. The transformation of an ISP

The first function of an ISP, *isp(G′, mtd′)*, is to update the propagation sequence of the input token. Since the multi-agent concept is not defined in *PrT-net*, we mimic the role of action_before using Pf, Ps, t1, P2 and the connections between them as shown in Figure 10. The integer value generated by the marking of Ps, which is increased by one each time t1 fires, is used to mimic the unique process identifications in *G-Nets*. The value is then combined with the propagation sequence of the input token of the ISP serving as the propagation sequence of the tokens involved in the execution of *G′*.

In the transformation shown in Figure 10, transition t2 serves as the connection point of *G* and *G′*. The output places of t2 are the places defined in the initial marking of *mtd′* in *G′*. The inscriptions on the arcs between t2 and its output places can also be determined from *mtd* definition. Therefore, t2 serves as a gate in the resulting *PrT-net*, and the firing of t2 moves tokens from one *G-Net* to another. The part of *PrT-net* corresponding to *G′* is produced by recursively calling the *G-Net* transformation procedure, which guarantees that each *G-Net* is only transformed at most once.

Similarly, transition t3 serves as another connection point between *G* and *G′* in the resulting *PrT-net*. The input place of t3 is the goal place in *G′* defined by *mtd*. The firing of t3 indicates the end of execution of *G′*. Finally, transition t4 represent event (defined by action_after) which recovers the propagation sequence of the tokens back from *G′* to the original one in *G*.

A final note is that the status color of tokens is not used in the transformation, because the structure of the transformation sequence has reflected the order of processing, thus status color is no longer necessary.

In summary, the following description provides the framework of the transformation algorithm, where $GS = G_1, G_2, \ldots G_k$ is a *G-Net* specification, $G \in GS$, and $mtd \in G.MS$.

1. IF $G \notin GS$ THEN
   return the initial marking and the goal place of mtd of G;
   exit;
2. $GS \leftarrow GS - G$
3. FOR every place $p \in G$ DO
   If p is an regular place THEN
      Place_Transformation (p);
   ELSE
      ISP_Transformation (p, GS);
4. FOR every transition t in G DO
   T_transformation (t);
5. return the initial marking and the goal place of mtd of G; exit;

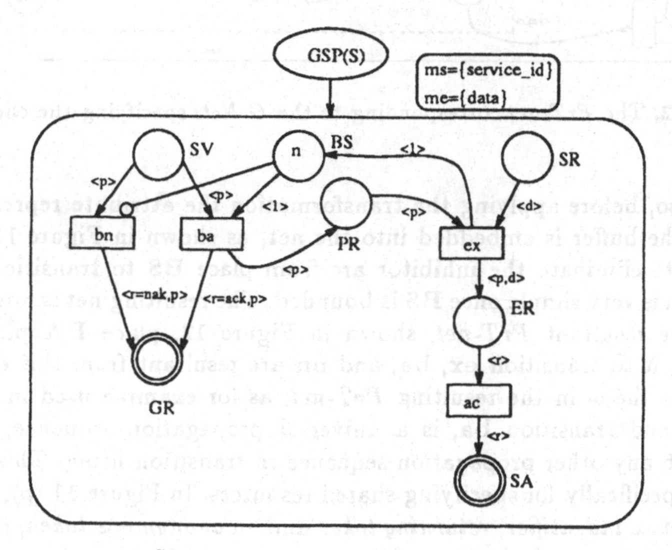

Figure 11. *G-Net* modeling the Server with a place modeling the buffer state

It has been shown that the time complexity of the transformation algorithm and the size of the resulting *PrT-net* are both linear to the size of the *G-Net* specification *GS*.

As an example, the transformation of the client/server interaction specified as in Section 3 will be shown. Before applying the transformation to the G-Nets specifying the client/server interaction, the G-Nets are redrawn with annotations about the inscriptions on arcs. In Figure 11 the revised *G-Net* for the server is

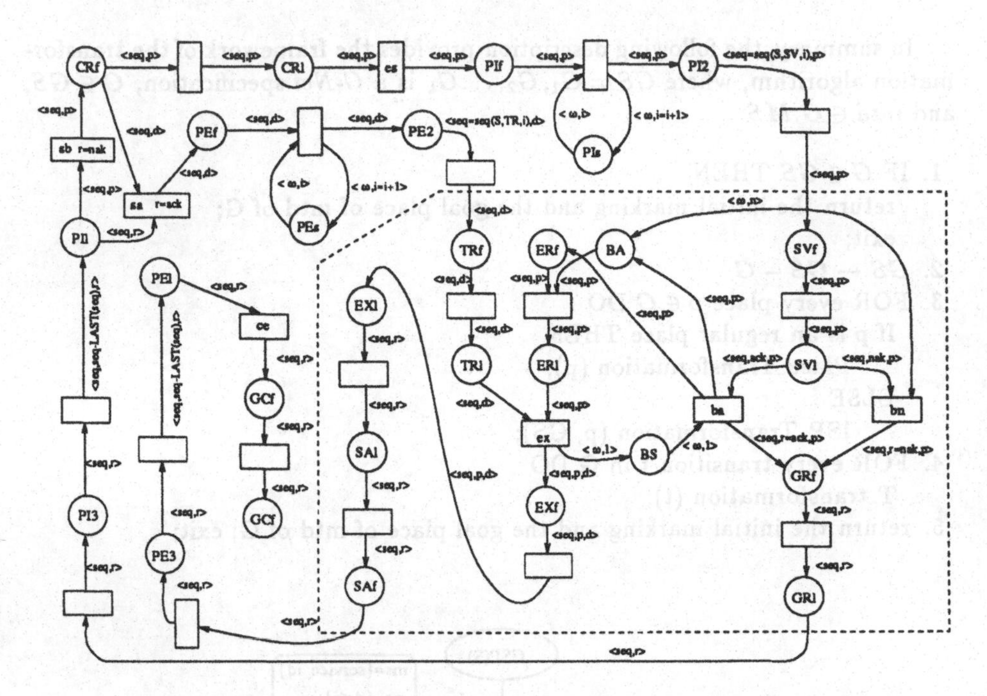

Figure 12. The *PrT-net* corresponding to the *G-Nets* specifying the client/server

shown. Also, before applying the transformation the attribute representing the status of the buffer is embedded into the net, as shown in Figure 11. It is also necessary to eliminate the inhibitor arc from place **BS** to transition **bn**. This elimination is very simple once **BS** is bounded. The resulting net is not presented. But on the resultant *PrT-net*, shown in Figure 12, place **BA** plus the arcs connecting it to transition **ex**, **ba**, and **bn** are resultant from this elimination. Notice that the ω in the resulting *PrT-net*, as for example used in connecting place **BS** and transition **ba**, is a universal propagation sequence, which can match with any other propagation sequence in transition firing. This concept is designed specifically for specifying shared resources. In Figure 11 $\langle p \rangle$, $\langle r \rangle$ and $\langle 1 \rangle$ are the *service indentifier*, *returning token* and an *anonymous token*, respectively.

5 Discussion and Conclusion

In the preceding sections, we presented a modular, object-oriented approach for the specification and prototyping of complex software systems. The client/server example was used to illustrate the *G-net* specification of distributed systems, and how such a specification can be translated into a PrT net for formal analysis. *G-Nets* differ from hierarchical Petri nets in both technical and paradigmatic aspects. These differences will now be discussed.

System modeling based on hierarchical Petri nets follows the stepwise refinement approach, in which high-level specifications will be eventually replaced by more detailed lower level specifications. This is clearly shown by the rationale behind the substitution of transitions and places [11]. On the other hand, *G-Nets* are a multi-level modeling and specification tool, in which each higher level *G-Net* imposes constraints and requirements on lower level ones so that all levels remain part of the final specification [8].

The fundamental concepts behind *G-Nets*, namely Logical Entity Abstraction and Instantiation (LEAI) [4] are independent of, and have no counterpart in the Petri net theory. These concepts are the cornerstones of the *G-Net* framework. The notion of *G-Net* is fundamentally different from the notion of a subnet. A subnet in hierarchical Petri nets is a specification of some functional unit with single entry and exit. In addition, conventional Petri nets emphasize control information and provide little opportunity for representing data [19]. In contrast, *G-Net* is a unified notion for specifying not only functional modules but also objects and data.

In hierarchical Petri nets, subnets are tightly coupled, where socket places (transitions) must match exactly to the corresponding port places (transitions), because socket places or transitions are the interface of a higher-level net to a lower-level net whose interface to the high-level net is in turn represented by the corresponding port places or transitions [11]. In a *G-Net* specification, each *G-Net* is a self-contained and independent module. The *G-Net* abstraction separates the external properties of the module from its realization or internal structure. No *G-Net* can directly access the internal structure of another *G-Net*. Under the concept of *G-Net* instantiation, the *G-Nets* in a system specification are loosely coupled through a well defined and encapsulated mechanism, named *G-Net abstraction*. Any change made to the internal structure of a *G-Net* are transparent to other *G-Nets*. The self-contained nature of *G-Nets* also provides the means to allocate a system specification in a distributed environment and to concurrently execute the specification.

In the development of the theory of *G-Nets* the most powerful way to analyze a *G-Net* specification is through simulation. Since *a G-Net* system is an executable specification and a simulator has already been implemented, analysis by simulation is feasible. This gives the designer a powerful tool to validate a specification.

Currently the transformation procedure presented in Section 4 can only handle functional or control modules and restricted types of attributes, e.g. finite integers. Another limitation of the technique is that it does not take advantage of the abstraction and modularization features of the *G-Nets*. Therefore, when it is applied to the analysis of complex *G-Net* specifications, we are likely to face the so-called *state explosion* problem, a major problem of Petri nets [14]. A more desirable approach would be to develop a localized analysis technique, which allows us to analyze each *G-Net* individually, and to express and verify the properties of a *G-Net* specification by composing the results of analysis of the individual *G-Nets*. Since the size of each *G-Net* can be kept small with proper design, the state explosion problem can be significantly reduced. This localized analysis technique is currently under investigation. Also, the timing analysis of

G-Net has been addressed. An extension to *G-Net* is being investigated in order to permit the specification of timing requirements and the performance analysis of systems. In [16] these two aspects together with an approach to embed fault-tolerant properties in the design of complex software systems are described.

We are currently investigating two problems based on the *G-Net* framework. First, we are using *G-Net* as a specification and simulation tool for the design of distributed multimedia systems [20]. Second, we are cooperating with an industrial partner to apply the *G-Net* system tool for the fault-tolerant design and performance analysis of real time control systems [2].

References

1. G. Booch. *Object Oriented Design with Applications.* The Benjamin/Cummings Publishing Company Inc., Redwood City, CA, 1991.
2. S.K. Chang, A. Perkusich, B. de Figueiredo, J.C.A. Yu, and M.J. Ehrenberger. The Design of Real-Time Distributed Information Systems with Object-Oriented and Fault-Tolerant Characteristics. To appear in Proc. of The Fifth International Conference on Software Engineering and Knowledge Engineering, June 1993.
3. T.C. Chen, Y. Deng, and S.K. Chang. A Simulator for Distributed Systems Using G-Nets. In *Proc. of 1992 Pittsburgh Simulation Conference*, Pittsburgh, PA, USA, May 1992.
4. Y. Deng. *A Unified Framework for the Modeling, Prototyping and Design of Distributed Information Systems.* PhD thesis, Department of Computer Science, University of Pittsburgh, 1992.
5. Y. Deng and S.K. Chang. A Framework for the Modeling and Prototyping of Distributed Information Systems. *International Journal of Software Engineering and Knowledge Engineering, SEKE93*, 2(3):203–226, September 1991.
6. Y. Deng and S.K. Chang. Unifying Multi-Paradigms in Software System Design. In *Proc. of the 4th Int. Conf. on Software Engineering and Knowledge Engineering*, Capri, Italy, June 1992.
7. R. Fairley. *Software Engineering Concepts.* MacGraw-Hill, New York, NJ, 1985.
8. A. Gabrielian and M.K. Franklin. Multilevel Specification of Real Time Systems. *Com. of ACM*, 34(5):50–60, May 1991.
9. H.J. Genrich. Predicate/Transition Nets. In W. Brauer, W. Reisig, and G. Rozemberg, editors, *Lecture Notes in Computer Science, Petri Nets: Central Models and Their Properties*, volume 254, pages 207–247. Springer Verlag, 1987.
10. H.J. Genrich and K. Lautenbach. System Modeling with High Level-Petri Nets. *Theorical Computer Science*, 13:109–136, 1981.
11. P. Huber, K. Jensen, and R.M. Shapiro. Hierarchies in Coloured Petri Nets. In Jensen. K. and Rozenberg, G., editor, *High-Level Petri Nets: Theory and Application*, pages 313–341. Springer-Verlag, 1991.
12. K. Jensen. Coloured Petri Nets: A High Level Language for System Design and Analysis. In *Lecture Notes in Computer Science 483, Advances in Petri Nets 1990.* Springer-Verlag, 1990.
13. Luqi. Software Evolution Through Rapid Prototyping. *IEEE Transactions on Computers*, pages 13–25, May 1989.
14. T. Murata. Petri Nets: Properties, Analysis and Applications. *Proc. of the IEEE*, 77(4):541–580, April 1989.

15. H. Oswald, R. Esser, and R. Mattmann. An Environment for Specifying and Executing Hierarchical Petri Nets. In *Proc. of the 12th International Conference on Software Engineering*, pages 164–172, 1990.
16. A. Perkusich, J.C.A. de Figueiredo, and S.K Chang. Embedding Fault-Tolerant Properties in the Design of Complex Systems. To appear in Journal of Systems and Software, May 1993.
17. W. Reisig. *Petri Nets*. Springer-Verlag, 1985.
18. W Reisig. Petri Nets with Individual Tokens. *Theoritical Computer Science*, 41:185–213, 1985.
19. P Zave. An Insider's Evaluation of PAISLey. *IEEE Transactions on Software Engineering*, 17(3):212–225, March 1991.
20. T. Znati, Y. Deng, B. Field, and S.K. Chang. Multi-Level Specification and Protocol Design for Distributed Multimedia Communication. In *Proc. of Conference on Organizational Computing Systems*, pages 255–268, Atlanta, GA, USA, November 1991.

Shortest Paths in Reachability Graphs[1]

Jörg Desel

Institut für Informatik, Technische Universität München
Arcisstraße 21, D-8000 München 2

Javier Esparza

Institut für Informatik, Universität Hildesheim
Samelsonplatz 1, D-3200 Hildesheim

Abstract. We prove the following property for safe conflict-free Petri nets and live and safe extended free-choice Petri nets:
Given two markings M_1, M_2 of the reachability graph, if some path leads from M_1 to M_2, then some path of polynomial length in the number of transitions of the net leads from M_1 to M_2.

1 Introduction

Let M_1, M_2 be two markings of the reachability graph of a safe Petri net such that M_2 is reachable from M_1. What can be said about the length of the shortest path of the graph leading from M_1 to M_2 ?
Since a safe Petri net with n places has less than 2^n markings, this length is smaller than 2^n. However, in some situations we would like to have a better bound. A first example is a system with a home state[2] which should be reached after a system failure in order to start a recovery action: if the home state can only be reached after an exponential number of steps, then the system cannot recover in reasonable time. It has also been recently observed that the length of shortest paths between pairs of markings is related to the complexity of the model checker developed in [3,7] for arbitrary safe Petri nets. This model checker (based on the unfolding technique developed in [13]) does not construct the reachability graph, but an unfolding of the Petri net. It happens that the size of the unfolding – and, with it, the complexity of the model checker – is strongly related to the length of the shortest paths between markings. Therefore, a good bound on this parameter can be used to derive a good bound on the complexity of verifying all the properties expressible in a temporal logic.

---

[1]Work partly done within the Esprit Basic Research WG 6067: CALIBAN and within SFB 342, WG A3: SEMAFOR

[2]A marking reachable from any other reachable marking

We prove in this paper the following two results:

- If the Petri net is conflict-free [12,11], then the length of the shortest path is at most
$$\frac{|T| \cdot (|T|+1)}{2}$$
- If the Petri net is live and extended free-choice [10], then the length of the shortest path is at most
$$\frac{|T| \cdot (|T|+1) \cdot (|T|+2)}{6}$$

where T is the set of transitions of the net.
The first of these two results has already been used in [7] to prove that the complexity of the model checking technique developed there is polynomial in the size of the system for conflict-free Petri nets. Our second result complements the result of [5], namely that live and safe extended free choice nets have home states: not only they exist, they are also reachable from any other reachable marking in a short number of steps.

The paper is organised as follows. Section 2 contains basic definitions and results. Section 3 studies so-called biased sequences. Using the results of Section 3, our two results are proved in Section 4 and Section 5. Finally, Section 6 shows that for safe persistent systems there exist no polynomial bounds for the length of shortest paths.

2 Definitions and Preliminaries

Let S and T be finite and nonempty disjoint sets and let $F \subseteq (S \times T) \cup (T \times S)$. Assume that for each $x \in (S \cup T)$ there exists a $y \in (S \cup T)$ satisfying $(x, y) \in F$ or $(y, x) \in F$. Then $N = (S, T, F)$ is called a *net*. S is the set of *places* and T the set of *transitions* of N.

N is *connected* if for every two elements x, y of N, the pair (x, y) belongs to the reflexive and transitive closure of $F \cup F^{-1}$. N is *strongly connected* if for every two elements x, y of N, the pair (x, y) belongs to the reflexive and transitive closure of F.

A *path* of N is a nonempty sequence $x_1 \ldots x_k$ of elements (places and transitions) of N satisfying $(x_1, x_2), \ldots, (x_{k-1}, x_k) \in F$. Such a sequence is a *circuit* if, moreover, $(x_k, x_1) \in F$.

Pre- and *post-sets* of elements are denoted by the dot-notation: ${}^\bullet x = \{y \mid (y, x) \in F\}$ and $x^\bullet = \{y \mid (x, y) \in F\}$. This notion is extended to sets of elements: ${}^\bullet X$ is the union of the pre-sets of elements of X and $X^\bullet$ is the union of the post-sets of elements of X.

A set $c \subseteq T$ is a *conflict set* if either $c = s^\bullet$ for some place s or $c = \{t\}$ for some transition satisfying ${}^\bullet t = \emptyset$.

A *marking* of N is a mapping $M: S \to \mathbb{N}$. A place s is called *marked* by a marking M if $M(s) > 0$.

A marking M *enables* a transition t if it marks every place of $^\bullet t$. The *occurrence* of an enabled transition t leads to the *successor marking* M' (written $M \xrightarrow{t} M'$) which is defined for every place s by

$$M'(s) = \begin{cases} M(s) - 1 & \text{if} \quad s \in {}^\bullet t \setminus t^\bullet \\ M(s) + 1 & \text{if} \quad s \in t^\bullet \setminus {}^\bullet t \\ M(s) & \text{if} \quad s \notin {}^\bullet t \cup t^\bullet \text{ or } s \in {}^\bullet t \cap t^\bullet \end{cases}$$

If $M_0 \xrightarrow{t_1} M_1 \xrightarrow{t_2} \cdots \xrightarrow{t_n} M_n$, then $\sigma = t_1\, t_2 \ldots t_n$ is called *occurrence sequence* and we write $M_0 \xrightarrow{\sigma} M_n$ (sometimes we say that $M_0 \xrightarrow{\sigma} M_n$ is an occurrence sequence, meaning that σ is an occurrence sequence leading from M_0 to M_n). This notion includes the empty sequence ϵ: $M \xrightarrow{\epsilon} M$ for each marking M. We call M' *reachable* from M if $M \xrightarrow{\sigma} M'$ for some occurrence sequence σ. $[M\rangle$ denotes the set of all markings reachable from M.

For a sequence σ of transitions and a transition t, $\#(t, \sigma)$ denotes the *number of occurrences* of t in σ. For a set of transitions T', $\#(T', \sigma)$ is the sum of all $\#(t, \sigma)$ for $t \in T'$. If T' is the set of all transitions T, then $\#(T', \sigma)$ is called the *length* of σ.

A sequence σ of transitions is a *permutation* of a sequence τ if $\#(t, \sigma) = \#(t, \tau)$ for every transition t.

A *net system* (or just a *system*) is a pair (N, M_0), where N is a net and M_0 a marking of N. If N is (strongly) connected, we call the system (N, M_0) (strongly) connected.

A *reachable marking* of (N, M_0) is a marking reachable from M_0.

(N, M_0) is called *live* if for every reachable marking M and every transition t there exists a marking $M' \in [M\rangle$ that enables t. (N, M_0) is called *safe* if every reachable marking M satisfies $M(s) \leq 1$ for every place s.

The *reachability graph* (V, E) of (N, M_0) is the directed graph defined by $V = [M_0\rangle$ and $E = \{(M_1, M_2) \in V \times V \mid M_1 \xrightarrow{t} M_2 \text{ for some transition } t\}$.

We use the two following results, which are well known:

Lemma 2.1

(1) *Let $M_1 \xrightarrow{\sigma} M_2$ be an occurrence sequence of a net N. Then, for every place s,*

$$M_2(s) = M_1(s) + \#({}^\bullet s, \sigma) - \#(s^\bullet, \sigma)$$

(2) *Let $M_1 \xrightarrow{\sigma} M_2$ and $M_1 \xrightarrow{\tau} M_3$ be occurrence sequences of a net N. If τ is a permutation of σ then $M_2 = M_3$.* ∎

3 Biased Occurrence Sequences

The purpose of this section is to prove Theorem 3.5, which yields un upper bound for the shortest paths between two markings M_1 and M_2 when M_2 can be reached from M_1 by means of a so called biased occurrence sequence. This theorem will easily lead to our first result concerning conflict-free systems, and will be used as lemma in the proof of our second result on extended free-choice systems.

The results of this section are a reformulation and small extension of results of [15].

Definition 3.1

Let N be a net. A sequence σ of transitions of N is called *biased* if for every conflict set c of N at most one transition of c occurs in σ.

Lemma 3.2

Let (N, M_0) be a safe system and M_1 a reachable marking.
Let σ be a biased sequence of transitions of N such that $M_1 \xrightarrow{\sigma} M_2$. Let t be a transition occurring in σ and u a transition satisfying $u^\bullet \cap {}^\bullet t \neq \emptyset$.
Then $\#(u,\sigma) - \#(t,\sigma) \leq 1$.

Proof:
Let $s \in u^\bullet \cap {}^\bullet t$. Since σ is biased and $t \in s^\bullet$ occurs in σ, no other transition of $s^\bullet$ occurs in σ. So $\#(t,\sigma) = \#(s^\bullet,\sigma)$. We have then:

$$\begin{array}{rcll} \#(u,\sigma) - \#(t,\sigma) & = & \#(u,\sigma) - \#(s^\bullet,\sigma) & \\ & \leq & \#({}^\bullet s,\sigma) - \#(s^\bullet,\sigma) & (\, u \in {}^\bullet s \,) \\ & = & M_2(s) - M_1(s) & (\text{ Lemma 2.1(1) }) \\ & \leq & M_2(s) & \\ & \leq & 1 & (\,(N, M_0) \text{ is safe }) \end{array}$$

■

Lemma 3.3

Let (N, M_0) be a safe system and M_1 a reachable marking.
Let $\sigma_1\, \sigma_2\, t$ be a biased sequence of transitions of N such that

(i) *t does not occur in σ_1 and*

(ii) *every transition occurring in σ_2 also occurs in σ_1*

If $M_1 \xrightarrow{\sigma_1\, \sigma_2\, t} M_2$ is an occurrence sequence then $M_1 \xrightarrow{\sigma_1\, t\, \sigma_2} M_2$ is also an occurrence sequence.

Proof:
By induction on the length of σ_2.
Base: If σ_2 is the empty sequence then $\sigma_1\, \sigma_2\, t = \sigma_1\, t = \sigma_1\, t\, \sigma_2$.
Step: Assume that σ_2 is not empty and define $\sigma_2 = \sigma_2'\, u$, where u is a transition.
Let $M_1 \xrightarrow{\sigma_1} M_3 \xrightarrow{\sigma_2'} M_4 \xrightarrow{u} M_5 \xrightarrow{t} M_2$.
By (ii), u also occurs in σ_1. So u occurs at least twice in $\sigma_1\, \sigma_2$.
By (i) and (ii), t does not occur in $\sigma_1\, \sigma_2$. So, by Lemma 3.2, $u^\bullet \cap {}^\bullet t = \emptyset$.
Hence t is already enabled at M_4. Let $M_4 \xrightarrow{t} M_6$.
Since $\sigma_1\, \sigma_2\, t$ is biased, ${}^\bullet t \cap {}^\bullet u = \emptyset$. Hence the occurrence of t does not disable u, and so u is enabled at M_6. Since $u\, t$ and $t\, u$ are permutations, we get $M_6 \xrightarrow{u} M_2$.
The application of the induction hypothesis to $\sigma_1\, \sigma_2'\, t$ (taking σ_2' for σ_2) yields an occurrence sequence $M_1 \xrightarrow{\sigma_1\, t\, \sigma_2'} M_6$. The result follows from $M_6 \xrightarrow{u} M_2$ and $\sigma_2'\, u = \sigma_2$.

■

Lemma 3.4

Let (N, M_0) be a safe system and M_1 a reachable marking.
Let $M_1 \xrightarrow{\sigma} M_2$ be a biased occurrence sequence.
Then there exists a permutation $\sigma_1\,\sigma_2$ of σ such that $M_1 \xrightarrow{\sigma_1\,\sigma_2} M_2$, no transition occurs more than once in σ_1 and every transition occurring in σ_2 also occurs in σ_1.

Proof:
By induction on the length of σ.
Base: If $\sigma = \epsilon$, then take $\sigma_1 = \sigma_2 = \epsilon$.
Step: Assume that σ is not the empty sequence. Let $\sigma = \tau\, t$.
By the induction hypothesis, there is a permutation $\tau_1\,\tau_2$ of τ such that no transition occurs more than once in τ_1 and every transition occurring in τ_2 also occurs in τ_1.
If t occurs in τ_1 then $\sigma_1 = \tau_1$ and $\sigma_2 = \tau_2\, t$ satisfy the requirements.
If t does not occur in τ_1 then $\tau_1\,\tau_2\, t$ satisfies the conditions of Lemma 3.3, and so $M_1 \xrightarrow{\tau_1\, t\, \tau_2} M_2$ is an occurrence sequence. Take then $\sigma_1 = \tau_1\, t$ and $\sigma_2 = \tau_2$. ■

Theorem 3.5

Let (N, M_0) be a safe system and M_1 a reachable marking.
Let $M_1 \xrightarrow{\sigma} M_2$ be a biased occurrence sequence. Let k be the number of distinct transitions occurring in σ.
Then there exists a sequence τ of transitions satisfying

(i) *$M_1 \xrightarrow{\tau} M_2$, and*

(ii) *the length of τ is at most* $\dfrac{k \cdot (k+1)}{2}$

Proof:
By induction on the length of σ.
Base: If $\sigma = \epsilon$ then choose $\tau = \epsilon$.
Step: Assume that σ is not the empty sequence.
By Lemma 3.4, there exists a permutation $\tau_1\,\tau_2$ of σ such that $M_1 \xrightarrow{\tau_1\,\tau_2} M_2$, every transition occurring in τ_2 occurs in τ_1, and no transition occurs in τ_1 more than once. Since σ is not the empty sequence, τ_1 is not empty, and therefore τ_2 is shorter than σ. Let $M_1 \xrightarrow{\tau_1} M_3 \xrightarrow{\tau_2} M_2$. We distinguish two cases:
Case 1: Every transition occurring in τ_1 occurs in τ_2.
Again by Lemma 3.4, there exists a permutation $\rho_1\,\rho_2$ of τ_2 such that $M_3 \xrightarrow{\rho_1\,\rho_2} M_2$, every transition occurring in ρ_2 occurs in ρ_1, and no transition occurs in ρ_1 more than once. Then a transition occurs in τ_1 if and only if it occurs in ρ_1. Moreover, no transition occurs more than once in either sequence. So every transition t satisfies $\#(t, \tau_1) = \#(t, \rho_1)$. Let $M_1 \xrightarrow{\tau_1} M_3 \xrightarrow{\rho_1} M_4$. Then, for each place s,

$$M_4(s) = M_1(s) + \#({}^\bullet s, \tau_1) - \#(s^\bullet, \tau_1) + \#({}^\bullet s, \rho_1) - \#(s^\bullet, \rho_1)$$

and hence

$$M_4(s) = M_1(s) + 2 \cdot (\#({}^\bullet s, \tau_1) - \#(s^\bullet, \tau_1))$$

Since (N, M_0) is safe and $M_1, M_4 \in [M_0\rangle$, $M_1(s)$ and $M_4(s)$ are both either 0 or 1. Therefore, $\#({}^\bullet s, \tau_1) - \#(s^\bullet, \tau_1) = 0$ and hence $M_1(s) = M_4(s)$.
So $M_1 = M_4$ and $M_1 \xrightarrow{\rho_2} M_2$. Since ρ_2 is shorter than σ, we can apply the induction hypothesis to it, which yields an occurrence sequence τ satisfying (i) and (ii).

Case 2: There exists a transition which occurs in τ_1 but does not occur in τ_2.
We apply the induction hypothesis to $M_3 \xrightarrow{\tau_2} M_2$.
Since the number of distinct transitions occurring in τ_2 is at most $k-1$, we get a sequence $M_3 \xrightarrow{\rho} M_2$ such that the length of ρ is at most $\dfrac{(k-1)\cdot k}{2}$.
Since each transition occurs in τ_1 at most once, the length of τ_1 is bounded by k.
The sequence $\tau = \tau_1\,\rho$ satisfies (i). Its length is at most $\dfrac{(k-1)\cdot k}{2} + k = \dfrac{k\cdot(k+1)}{2}$, so it also satisfies (ii). ∎

4 T-Systems and Conflict-Free Systems

If a system has no forward branching places (i.e., $|s^\bullet| \leq 1$ for every place) then all its occurrence sequences are biased. Hence Theorem 3.5 applies to every occurrence sequence, and we get the following result:

Theorem 4.1

Let (N, M_0) be a safe system where $N = (S, T, F)$ and $|s^\bullet| \leq 1$ for every $s \in S$, and let M_1 be a reachable marking. Let M_2 be a marking reachable from M_1. Then there exists an occurrence sequence $M_1 \xrightarrow{\tau} M_2$ such that the length of τ is at most

$$\frac{|T|\cdot(|T|+1)}{2}$$

Proof:
Since M_2 is reachable from M_1, there exists an occurrence sequence $M_1 \xrightarrow{\sigma} M_2$. σ is biased because every conflict set of N contains exactly one transition. The number of distinct transitions occurring in σ is at most $|T|$. The result follows from Theorem 3.5. ∎

This theorem applies in particular to T-systems, in which $|s^\bullet| \leq 1$ and $|{}^\bullet s| \leq 1$ for every place s (T-systems are also called *marked graphs* [6] and *synchronisation graphs* [9]). The bound of Theorem 4.1 is reachable for T-systems, i.e., there exist T-systems and pairs of markings M_1, M_2 for which the bound above is the exact value of the length of the shortest path leading from M_1 to M_2. Consider the family of T-systems of Fig. 1. The marking M_{odd} that puts a token in all places with odd indices (shown in the figure) is safe. It is not difficult to see that the marking M_{even} that puts a token in all places with even indices is reachable from M_{odd}. Moreover, the shortest path leading from M_{odd} to M_{even} has length $\dfrac{n\cdot(n+1)}{2}$.

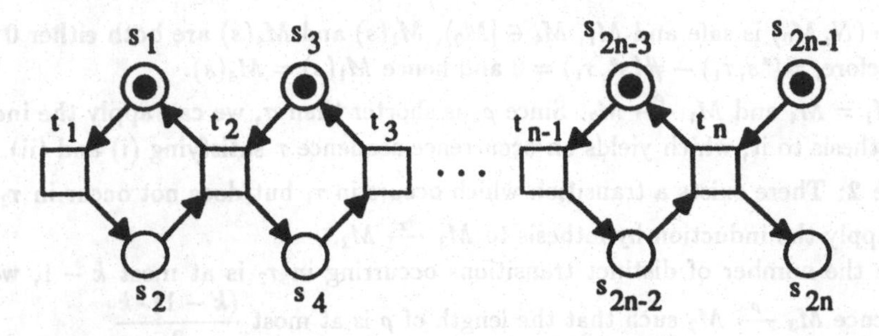

Fig. 1 **A family of T-systems for which the bound of Theorem 4.1 is tight**

Therefore, if the only available information is the number of transitions of the net, then the bound of Theorem 4.1 cannot be improved.

Theorem 4.1 can be easily generalised to conflict-free nets, a well-known class of nets studied e.g. in [12,11,15].

Definition 4.2

A net N is called *conflict-free* if every place s of N satisfies either $|s^\bullet| \leq 1$ or $s^\bullet \subseteq {}^\bullet s$.

A system (N, M_0) is conflict-free if N is conflict-free.

Theorem 4.3

Let (N, M_0) be a safe conflict free system where $N = (S, T, F)$, and let M_1 be a reachable marking. Let M_2 be a marking reachable from M_1.
Then there exists an occurrence sequence $M_1 \xrightarrow{\tau} M_2$ such that the length of τ is at most

$$\frac{|T| \cdot (|T| + 1)}{2}$$

Proof:
Since M_2 is reachable from M_1, there exists an occurrence sequence $M_1 \xrightarrow{\sigma} M_2$.
Let S' be the set of places of N with more than one output transition. We proceed by induction on $|S'|$.
Base: If $S' = \emptyset$ then the result follows by Theorem 4.1.
Step: Assume that $S' \neq \emptyset$ and let $s \in S'$.
We show that the behaviour of N can be simulated by some conflict-free net N' which has less forward branched places than N. N' is obtained from N by the following transformation (note that by the conflict-freeness of N, $s^\bullet \setminus {}^\bullet s$ is empty):

- For each $t \in s^\bullet \cap {}^\bullet s$, define a new place s_t and arcs (s_t, t) and (t, s_t).
- For each $t' \in {}^\bullet s \setminus s^\bullet$ and each $t \in s^\bullet \cap {}^\bullet s$, define an arc (t', s_t).
- Delete s and adjacent arcs.

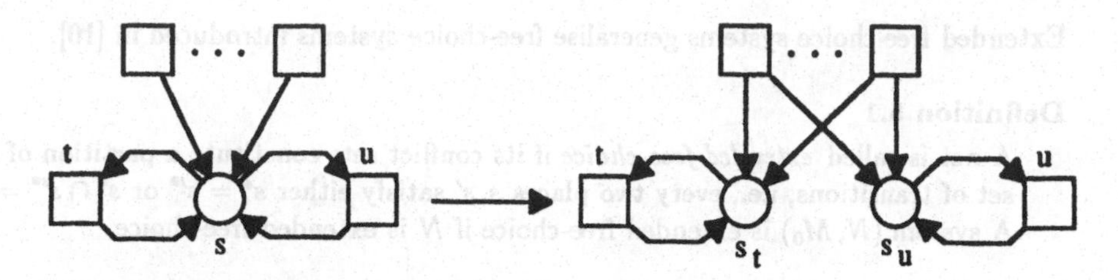

Fig. 2 Transformation of a conflict-free net into a net without forward branching places.

This transformation is shown in Fig. 2

For every marking M of N, we define a marking M' of N' as follows:

$$M'(s') = \begin{cases} M(s') & \text{if } s' \text{ is a place of } N \\ M(s) & \text{if } s' = s_t \end{cases}$$

We claim that $M_1 \xrightarrow{\rho} M_2$ is an occurrence sequence of N iff $M_1' \xrightarrow{\rho} M_2'$ is an occurrence sequence of N'.

Clearly, it suffices to prove the claim for sequences ρ having the length one; the general case follows by induction. So let $\rho = t$ for some transition t. We distinguish four cases (where in the sequel the •-notion is used for pre- and post-sets in N and the symbol o is used for pre- and post-sets in N'):

(i) $t \notin {}^\bullet s \cup s^\bullet$. Then ${}^\bullet t = {}^\circ t$ and $t^\bullet = t^\circ$, and the result follows.

(ii) $t \in {}^\bullet s \setminus s^\bullet$. Then, in N', $t \in {}^\circ s_u \setminus s_u^\circ$ for each transition $u \in s^\bullet$, and the result follows.

(iii) $t \in s^\bullet \setminus {}^\bullet s$. This case is impossible since N is conflict-free.

(iv) $t \in {}^\bullet s \cap s^\bullet$. Then $t \in {}^\circ s_u \cap s_u^\circ$ for each transition $u \in s^\bullet$, and the result follows.

By this claim, $M_1' \xrightarrow{\sigma} M_2'$ is an occurrence sequence of N'.

By construction, N' is conflict-free. Moreover, the set of places of N' with more than one output transition is $S' \setminus \{s\}$. Hence, we can apply the induction hypothesis; there exists an occurrence sequence $M_1' \xrightarrow{\tau} M_2'$ such that the length of τ is at most $\frac{|T| \cdot (|T|+1)}{2}$.

Again by the above claim, $M_1 \xrightarrow{\tau} M_2$ is an occurrence sequence of N. ∎

5 Extended Free-Choice Systems

In this section we obtain an upper bound for the length of the shortest path between two reachable markings of live and safe extended free-choice systems: it is never longer as

$$\frac{|T| \cdot (|T|+1) \cdot (|T|+2)}{6}$$

Extended free-choice systems generalise free-choice systems introduced in [10].

Definition 5.1

A net is called *extended free-choice* if its conflict sets constitute a partition of its set of transitions, i.e., every two places s, s' satisfy either $s^\bullet = s'^\bullet$ or $s^\bullet \cap s'^\bullet = \emptyset$. A system (N, M_0) is extended free-choice if N is extended free-choice.

Note that every net without forward branching places is extended free-choice.
The proof of our result is based on the notions of conflict order and sorted sequence. They are introduced in the next definition.

Definition 5.2

Let N be an extended free-choice net and let T be the set of transitions of N.
A *conflict order* $\preceq \subseteq T \times T$ is a partial order such that two transitions t and u are comparable (i.e., $t \preceq u$ or $u \preceq t$) if and only if they belong to the same conflict set. $u \prec t$ denotes $u \preceq t$ and $u \neq t$.
Let σ be a sequence of transitions of N.
A conflict order $\preceq$ is said to *agree* with σ if for every conflict set c, either no transition of c occurs in σ, or the last transition of c occurring in σ is maximal, i.e., the greatest transition of c with respect to $\preceq$.

The sequence σ is called *sorted* with respect to a conflict order $\preceq$ if for every two transitions t, u satisfying $t \prec u$, t does not occur after u in σ.

We outline the proof of the result. Let (N, M_1) be a live and safe extended free-choice system and $M_1 \xrightarrow{\sigma} M_2$ an occurrence sequence. We shall show:

(1) There exists a conflict order $\preceq$ that agrees with σ and a sorted permutation τ of σ such that $M_1 \xrightarrow{\tau} M_2$.

(2) $\tau = \tau_1 \tau_2 \ldots \tau_k$, where τ_i is a biased sequence for every i, and k is less or equal than the number of transitions of N.

Using (2) and Theorem 3.5, we shall prove that there exist sequences $\rho_1, \rho_2, \ldots, \rho_k$ of bounded length such that, for every i, if $M_i \xrightarrow{\tau_i} M_{i+1}$ then $M_i \xrightarrow{\rho_i} M_{i+1}$.
We define $\rho = \rho_1 \rho_2 \ldots \rho_k$. Then $M_1 \xrightarrow{\rho} M_2$. Some arithmetic will yield the upper bound on the length of ρ we are looking for.

Of these two steps, (1) is more involved (step (2) shall follow easily from the definition of sorted sequence). To prove (1), we shall make use of the well-known decomposition theorem of the theory of free-choice nets, which states that every live and safe extended free-choice system can be decomposed into S-components carrying one token. Let us recall both the definition of S-component and the decomposition theorem.

Definition 5.3

An *S-net* is a net satisfying $|{}^\bullet t| = |t^\bullet| = 1$ for each transition t.
(N, M_0) is an *S-system* if N is an S-net.

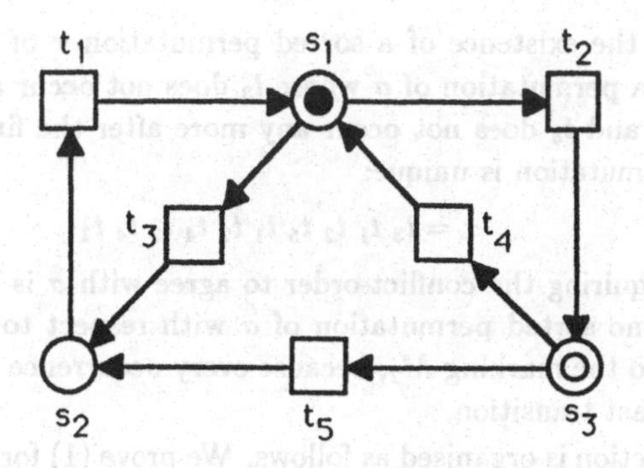

Fig. 3 An S-net and two live and safe markings

Definition 5.4

A strongly connected S-net N_1 is an *S-component* of a net N if for every place s of N_1 holds:

- s is a place of N,
- the pre-set of s in N_1 equals the pre-set of s in N, and
- the post-set of s in N_1 equals the post-set of s in N.

A net N is covered by a set of S-components $\{N_1, \ldots, N_n\}$ if every place of N is contained in some S-component N_i of this set.

Theorem 5.5 [10,2]

Let (N, M_0) be a live and safe extended free-choice system.
Then N is covered by a set of S-components $\{N_1, \ldots, N_n\}$ such that each N_i has exactly one marked place (which contains only one token because (N, M_0) is safe).

■

We shall prove (1) in two steps. First, we shall show that the statement holds for connected live and safe S-systems (notice that every S-system is extended free-choice). Then, using this result and Theorem 5.5, we shall extend the result to arbitrary live and safe extended free-choice systems.

Let us illustrate the meaning of (1) with an example. Since (1) is already non-trivial for the special case of S-systems, we choose as example the connected live and safe S-system (N, M_1) of Fig. 3, where M_1 is the marking that puts one token in s_1 (black token), and M_2 is the marking that puts one token in s_3 (white token).

We have $M_1 \xrightarrow{\sigma} M_2$ for the sequence

$$\sigma = t_2\, t_4\, t_3\, t_1\, t_2\, t_5\, t_1\, t_2\, t_4\, t_2$$

The conflict sets of the net are $\{t_1\}$, $\{t_2, t_3\}$ and $\{t_4, t_5\}$. The last transition of $\{t_2, t_3\}$ occurring in σ is t_2; the last transition of $\{t_4, t_5\}$ occurring in σ is t_4. Therefore, the only conflict order that agrees with σ is the one given by $t_3 \prec t_2$ and $t_5 \prec t_4$.

Now, (1) asserts the existence of a sorted permutation τ of σ that also leads from M_1 to M_2 – i.e., a permutation of σ where t_3 does not occur any more after the first occurrence of t_2, and t_5 does not occur any more after the first occurrence of t_4. In this case, the permutation is unique:

$$\tau = t_3\, t_1\, t_2\, t_5\, t_1\, t_2\, t_4\, t_2\, t_4\, t_2$$

The condition requiring the conflict-order to agree with σ is essential for the result. In our example, no sorted permutation of σ with respect to a conflict order where $t_2 \prec t_3$ can lead to the marking M_2, because every occurrence sequence leading to M_2 must have t_2 as last transition.

The rest of the section is organised as follows. We prove (1) for live and safe connected S-systems – actually, we prove a stronger result – in Proposition 5.8. We generalise the result to live and safe extended free choice systems in Theorem 5.10. Finally, we obtain the upper bound in Theorem 5.11.

5.1 Sorted Occurrence Sequences of S-Systems

The result we wish to prove has a strong graph theoretical flavour, because the occurrence sequences of live and safe S-systems correspond to paths of S-nets. In fact, the main idea of our proof is taken from the proof of the BEST-theorem [8] of graph theory, which gives the number of Eulerian trails of a directed graph. In [8], [1] is cited as the original reference.

The following result is well-known:

Lemma 5.6 [4]

A connected S-system (N, M_0) is live and safe if and only if it is strongly connected and exactly one place is marked with one token at M_0. ■

Lemma 5.7

Let (N, M_0) be a live and safe connected S-system and let M_1 be a reachable marking. Let $M_1 \xrightarrow{\sigma} M_2$ be an occurrence sequence.

Then (N, M_2) is still live and safe. Let s be the unique place satisfying $M_2(s) = 1$. Let $\preceq$ be a conflict order which agrees with σ and let T_m be the set of maximal transitions (with respect to $\preceq$) occurring in σ.

Then every circuit of N containing only transitions of T_m contains the place s.

Proof:

Assume there exists a circuit of N which contains only transitions of T_m but does not contain the place s.

Let t, r, u be three consecutive nodes of the circuit, where t, u are transitions and r is a place. Since $t \in T_m$, t occurs in σ. Let $\sigma = \tau\, t\, \rho$ such that t does not occur in ρ. We have $r \neq s$, because the place s is not contained in the circuit. Since r is marked after the occurrence of t, some transition which removes the token from r – i.e., some transition of the conflict set $r^\bullet$ – occurs in ρ. In particular, the maximal transition of $r^\bullet$ (with respect to $\preceq$) occurring in σ occurs in ρ; by the definition of the set T_m, this transition is u.

So, for every pair of consecutive transitions t and u of the circuit, u occurs after t in σ. This contradicts the finiteness of σ. ■

Proposition 5.8

Let (N, M_0) be a live and safe connected S-system and let M_1 be a reachable marking. Let $M_1 \xrightarrow{\sigma} M_2$ be an occurrence sequence.
Let $\preceq$ be a conflict order which agrees with σ.
Then there exists a sequence τ of transitions of N such that

(i) *τ is sorted with respect to $\preceq$,*

(ii) *$M_1 \xrightarrow{\tau} M_2$, and*

(iii) *τ is a permutation of σ.*

Proof:
Construct an occurrence sequence τ as follows:
Start with M_1. At every reached marking, choose an enabled transition according to the following rule:

> **Take the least enabled transition (with respect to $\preceq$) which occurs more often in σ than in the sequence obtained so far.**

τ is the sequence obtained after applying this rule as long as possible. Notice that the procedure eventually stops, because the rule can only be applied if the length of the sequence constructed so far is less than the length of σ.
Let $M_1 \xrightarrow{\tau} M_3$. Then (N, M_3) is still live and safe; let s be the unique place of N marked by M_3 (Lemma 5.6). By construction, τ satisfies the following two properties:

- **For every transition t of N, $\#(t,\tau) \leq \#(t,\sigma)$, and**
- **For every transition t of $s^\bullet$, $\#(t,\tau) = \#(t,\sigma)$.**
  (since every transition of $s^\bullet$ is enabled at M_3, if for some transition $t \in s^\bullet$ we have $\#(t,\tau) < \#(t,\sigma)$, then τ can be extended to $\tau\, t$ using the rule, which contradicts the definition of τ.)

We claim that τ satisfies (i) to (iii).

(i) **τ is sorted by construction.**

(ii) **We show $M_3 = M_2$. By Lemma 5.6, and since (N, M_2) as well as (N, M_3) are live and safe, both markings mark exactly one place with one token. Since $M_3(s) = 1$, it suffices to prove $M_2(s) \geq M_3(s)$.**

$$\begin{aligned} M_2(s) &= M_1(s) + \#({}^\bullet s,\sigma) - \#(s^\bullet,\sigma) && (\, M_1 \xrightarrow{\sigma} M_2 \,) \\ &\geq M_1(s) + \#({}^\bullet s,\tau) - \#(s^\bullet,\tau) && (\text{ properties of } \tau \,) \\ &= M_3(s) && (\, M_1 \xrightarrow{\tau} M_3 \,) \end{aligned}$$

(iii) **Assume that τ is not a permutation of σ.**
Then there are transitions occurring in σ more often than in τ. By construction of τ, there are maximal transitions (with respect to $\preceq$) with the same property.

Let T_m be the set of maximal transitions t satisfying $\#(t,\tau) < \#(t,\sigma)$.
Let $s \in T_m^\bullet$. By (ii), $M_2 = M_3$ and therefore

$$\#({}^\bullet s,\sigma) - \#(s^\bullet,\sigma) = \#({}^\bullet s,\tau) - \#(s^\bullet,\tau)$$

By the first property of τ, $\#(t,\tau) \leq \#(t,\sigma)$ for every $t \in {}^\bullet s$. Since $s \in T_m^\bullet$, we have $\#({}^\bullet s,\tau) < \#({}^\bullet s,\sigma)$. So $\#(s^\bullet,\tau) < \#(s^\bullet,\sigma)$. Let t be the maximal transition in $s^\bullet$. As τ is sorted, $\#(t,\tau) < \#(t,\sigma)$. So $t \in T_m$.
Therefore $T_m^\bullet \subseteq {}^\bullet T_m$.
Since $T_m \neq \emptyset$ and by the finiteness of N, we find a circuit of N containing only (places and) transitions of T_m. Since all transitions of T_m are maximal, we can apply Lemma 5.7: the circuit contains the unique place s marked at M_3. Let t be the unique transition of $s^\bullet$ contained in the circuit. Then t is enabled at M_3. Since $t \in T_m$, we have $\#(t,\sigma) > \#(t,\tau)$ – contradicting the second property of τ. ∎

Our goal (1) was to prove the existence of a conflict order and a sorted permutation τ of σ leading to the same marking as σ. Proposition 5.8 proves a stronger result, namely that the conflict order can be arbitrarily chosen among those that agree with σ (notice that there always exist some conflict order that agrees with σ).

5.2 Sorted Occurrence Sequences of Extended Free-Choice Systems

Theorem 5.5 suggests to look at extended free-choice systems as a set of sequential systems (corresponding to the S-components carrying one token) which communicate by means of shared transitions. The following lemma states that the projection of an occurrence sequence of the system on one of its S-components yields a 'local' occurrence sequence of the component. The proof is simple (see e.g. [14]).

Lemma 5.9

Let (N, M_0) be a system and let M_1 be a reachable marking. Let $M_1 \xrightarrow{\sigma} M_2$ be an occurrence sequence.
Let N_i be an S-component of N. Let M_1^i, M_2^i be the restriction of the markings M_1, M_2 to the places of N_i. Let σ_i denote the sequence obtained from σ by deletion of all transitions which do not belong to N_i.
Then $M_1^i \xrightarrow{\sigma_i} M_2^i$ is an occurrence sequence of N_i. ∎

Using this lemma, we now generalise Proposition 5.8 to live and safe extended free-choice systems.

Proposition 5.10

Let (N, M_0) be a live and safe extended free-choice system and let M_1 be a reachable marking. Let $M_1 \xrightarrow{\sigma} M_2$ be an occurrence sequence.
Let $\preceq$ be a conflict order which agrees with σ.
Then there exists a sequence τ of transitions of N such that

(i) τ is sorted with respect to $\preceq$,

(ii) $M_1 \xrightarrow{\tau} M_2$, *and*

(iii) τ *is a permutation of* σ.

Proof:

By Theorem 5.5, N is covered by a set $\{N_1, \ldots, N_n\}$ of S-components with exactly one place marked. In the sequel, we call these S-components *state-machines* of N.

Let N_i be a state-machine of N. For each marking M of N, we define M^i as the restriction of N to the set of places of N_i. For a sequence of transitions α, α_i denotes the sequence obtained from α by deletion of all transitions which do not belong to N_i.

By Lemma 5.9, for every state-machine N_i, $M_1^i \xrightarrow{\sigma_i} M_2^i$ is an occurrence sequence of N_i. Since every conflict set of a state-machine N_i is a conflict set of N, the restriction of $\preceq$ to transitions of N_i agrees with σ_i.

By Lemma 5.6, (N_i, M_0^i) is live and safe. By Proposition 5.8, we find for every state-machine N_i a sorted permutation ρ_i of σ_i satisfying $M_1^i \xrightarrow{\rho_i} M_2^i$.

Now we define τ to be a maximal sequence (with respect to prefix ordering) satisfying

(a) τ is an occurrence sequence

(b) For every state-machine N_i, τ_i is a prefix of ρ_i

Since the empty sequence enjoys (a) and (b), such a maximal sequence τ exists.

τ is sorted because every conflict set is contained in some state-machine N_i, τ_i is a prefix of ρ_i, and ρ_i is sorted.

It remains to prove that $M_1 \xrightarrow{\tau} M_2$ and that τ is a permutation of σ. Since τ is an occurrence sequence by construction, it suffices to prove the second part, i.e., that $\#(t,\tau) = \#(t,\sigma)$ for every transition t of N.

Let t be a transition of N and let N_i be a state-machine containing t. We have:

$$\begin{aligned} \#(t,\tau) &= \#(t,\tau_i) \\ &\leq \#(t,\rho_i) && (\ \tau_i \text{ is a prefix of } \rho_i\ ) \\ &= \#(t,\sigma_i) && (\ \rho_i \text{ is a permutation of } \sigma_i\ ) \\ &= \#(t,\sigma) \end{aligned}$$

Let T_l be the set of transitions t satisfying $\#(t,\tau) < \#(t,\sigma)$. We prove $T_l = \emptyset$.

Let S' (T') be the set of places (transitions) of the state-machines that contain some transition of T_l. For each state-machine N_i define $\rho_i = \tau_i \tau_i'$ (which is possible because τ_i is a prefix of ρ_i).

Let $M_1 \xrightarrow{\tau} M_3$. We show first that every transition $t \in T'$ has an input place in the set S' which is moreover unmarked at M_3.

Case 1: t is in the conflict set of some transition in T_l.

Since N is an extended free-choice net, every two transitions of this conflict set have the same presets. Hence we can assume without loss of generality that t is the least transition in the conflict set which belongs to T_l, i.e., $t \in T_l$ and $\#(t',\rho) = \#(t',\sigma)$ for every $t' \prec t$.

Let $s \in {}^\bullet t$. Every state-machine containing s also contains t. Since $t \in T_l$, $s \in S'$. So ${}^\bullet t \subseteq S'$. It remains to show that t has an unmarked input place.
Assume that every place $s \in {}^\bullet t$ is marked at M_3. Then t is enabled at M_3.
Let N_i be an arbitrary state-machine containing t. By assumption, the unique place s marked at M_3^i is in ${}^\bullet t$.
We claim the following:

(1) t occurs in τ_i'.
We have:

$$\begin{aligned} \#(t,\sigma_i) &= \#(t,\rho_i) && (\ \rho_i \text{ is a permutation of } \sigma_i\ ) \\ &= \#(t,\tau_i\,\tau_i') && (\text{ definition of } \tau_i'\ ) \\ &= \#(t,\tau_i) + \#(t,\tau_i') \end{aligned}$$

Since $t \in T_l$, $\#(t,\tau) < \#(t,\sigma)$, and therefore $\#(t,\tau_i) < \#(t,\sigma_i)$. So $\#(t,\tau_i') > 0$, and therefore t occurs in τ_i'.

(2) For every $t' \prec t$, t' does not occur in τ_i'.
Using the same arguments as in (1), we have $\#(t',\sigma_i) = \#(t',\tau_i) + \#(t',\tau_i')$. Since t' does not belong to T_l, $\#(t',\tau) = \#(t',\sigma)$, and therefore $\#(t',\tau_i) = \#(t',\sigma_i)$. So $\#(t',\tau_i') = 0$.

Since $M_1^i \xrightarrow{\tau_i} M_3^i$ is an occurrence sequence of N_i, τ_i' starts with some transition of $s^\bullet$, the conflict set containing t. τ_i' does not start with a transition less than t by (2). τ_i' does not start with a transition greater than t because τ_i is sorted, and t is the least transition in the conflict set that belongs to T_l. Hence τ_i' starts with t.
Since this holds for all state-machines N_i containing t, the sequence $\tau' = \tau\, t$ satisfies (a) and (b) – contradicting the definition of τ.

Case 2: t is not in the conflict set of any transition in T_l.
Since $t \in T'$, there exists a state-machine N_i containing t and some transition of T_l.
Let s be the unique place marked at M_3^i.
Since N_i contains a transition of T_l, τ_i' is not empty (use the same argument of (1) in Case 1). Let t' be the first transition of τ_i'. Then $t' \in T_l$. Since $M_1^i \xrightarrow{\tau_i} M_3^i$ is an occurrence sequence of N_i, $t' \in s^\bullet$.
Since t and t' do not belong to the same conflict set, $t \notin s^\bullet$.
Hence the unique place of N_i in ${}^\bullet t$ is unmarked at M_3^i. This place is in S' by definition of S'.

So every transition $t \in T'$ has an input place in the set S' which is moreover unmarked at M_3. Assume $T_l \neq \emptyset$. Then $T' \neq \emptyset$.
Since every transition in T' has an unmarked input place, no transition in T' is enabled at M_3. Since M_1 is a live marking, we find an occurrence sequence $M_1 \xrightarrow{\chi} M$ such that M enables a transition t of T'. Assume without loss of generality that χ is minimal, i.e., no transition occurring in χ belongs to T'.
Let s be an input-place of t such that $s \in S'$ and s is not marked at M_3. Since t is enabled at M, χ contains a transition $t' \in {}^\bullet s$. Every state-machine containing s contains t'; hence $t' \in T'$ – contradicting the minimality of χ. ∎

5.3 An Upper Bound on the Length of Shortest Paths

We are finally ready to prove the result stated in the introduction.

Theorem 5.11

Let (N, M_0) be a live and safe extended free-choice system where $N = (S,T,F)$, and let M_1 be a reachable marking. Let M_2 be a marking reachable from M_1. Then there is an occurrence sequence $M_1 \xrightarrow{\rho} M_2$ such that the length of ρ is at most

$$\frac{|T| \cdot (|T|+1) \cdot (|T|+2)}{6}$$

Proof:

Since M_2 is reachable from M_1, there exists an occurrence sequence $M_1 \xrightarrow{\sigma} M_2$. By Proposition 5.10, there is a conflict order $\preceq$ and an occurrence sequence $M_1 \xrightarrow{\tau} M_2$ such that τ is sorted with respect to $\preceq$.

Let k be the number of distinct transitions occurring in τ. Then $k \leq |T|$. We show that there exists an occurrence sequence $M_1 \xrightarrow{\rho} M_2$ such that the length of ρ is at most $\dfrac{k \cdot (k+1) \cdot (k+2)}{6}$.

We proceed by induction on k.

Base: If $k = 0$ then there is nothing to be shown.

Step: Assume that $k > 0$.

Decompose $\tau = \tau_1\,\tau_2$ such that τ_1 is the maximal prefix of ρ that contains at most one transition of each conflict set. Then τ_1 is biased. Let $M_1 \xrightarrow{\tau_1} M_3 \xrightarrow{\tau_2} M_2$. By Theorem 3.5, there is an occurrence sequence $M_1 \xrightarrow{\rho_1} M_3$ such that the length of ρ_1 is at most $\dfrac{k \cdot (k+1)}{2}$.

If $M_3 = M_2$, then we are finished because

$$\frac{k \cdot (k+1)}{2} \leq \frac{k \cdot (k+1) \cdot (k+2)}{6}$$

So assume that $M_3 \neq M_2$. Then τ_2 is not empty and starts with a transition t. Since τ_1 is maximal, τ_1 contains a transition t' in the conflict set of t.

Since τ is sorted, $t' \prec t$ and t' does not occur in τ_2.

So the number of distinct transitions occurring in τ_2 is at most $k-1$.

By the induction hypothesis, there exists an occurrence sequence $M_3 \xrightarrow{\rho_2} M_2$ such that the length of ρ_2 is at most $\dfrac{(k-1) \cdot k \cdot (k+1)}{6}$.

Define $\rho = \rho_1\,\rho_2$. Then $M_1 \xrightarrow{\rho} M_2$ and the length of ρ is at most

$$\frac{k \cdot (k+1)}{2} + \frac{(k-1) \cdot k \cdot (k+1)}{6} = \frac{k \cdot (k+1) \cdot (k+2)}{6}$$

■

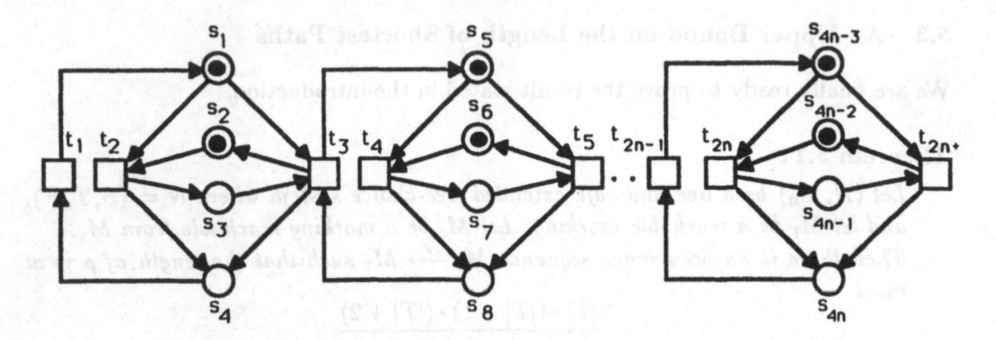

Fig. 4 A family of systems with exponential shortest paths

6 A Family of Systems with Exponential Shortest Paths

We exhibit in this section a family of systems for which there exists no polynomial upper bound in the length of the shortest paths. The family is shown in Fig. 4. All the systems of the family are live and safe. They are even persistent, i.e, a transition can only cease to be enabled by its own firing.

The shortest path that, from the marking shown in the figure, reaches the marking that puts a token in the places of the set

$$\{s_1, s_3, s_5, s_7, \ldots, s_{4n-3}, s_{4n-1}\}$$

has exponential length in the number of transitions of the net. This can be easily proved by showing that, in order to reach this marking, transition t_{2n-1} has to occur at least once and, for every $1 \leq i \leq n$, transition t_{2i-1} has to occur twice as often as transition t_{2i+1}.

7 Conclusions

We have obtained polynomial bounds for the length of the shortest paths connecting two given markings for two classes of net systems: safe conflict-free systems and live and safe extended free-choice systems. Furthermore, we have shown that in the case of safe conflict-free systems the bound is reachable, and that the length of shortest paths in safe persistent systems can be exponential in the number of transitions. In the proofs we have made strong use of results of Yen [15] on conflict-free systems and of graph theoretical results on Eulerian trails [1,8].

Using the results of this paper, we have been able to prove in [7] that the model checker described there has polynomial complexity in the size of the system for safe conflict-free systems.

We do not know at the moment if the bound for live and safe free-choice nets is reachable. In fact, we believe that a reachable bound should be quadratic in the number of transitions. We are also working in the generalisation of our results to the bounded case.

Acknowledgments. We thank Eike Best, Klaus-Jörn Lange and Walter Vogler for helpful comments and suggestions.

References

[1] T. van Aardenne-Ehrenfest and N.G. de Bruijn: Circuits and Trees in Oriented Linear Graphs, Simon Stevin 28, 203-217 (1951).

[2] E. Best and J. Desel: Partial Order Behaviour and Structure of Petri Nets. Formal Aspects of Computing Vol.2 No.2, 123–138 (1990).

[3] E. Best and J. Esparza: Model Checking of Persistent Petri Nets. Computer Science Logic 91, E. Börger, G. Jäger, H. Kleine Büning and M.M. Richter (eds.), LNCS 626, 35–53 (1992).

[4] E. Best and P.S. Thiagarajan: Some Classes of Live and Save Petri Nets. Concurrency and Nets, K. Voss, H.J. Genrich, G. Rozenberg, G. (eds.), Advances in Petri Nets. — Berlin: Springer-Verlag, 71–94 (1987).

[5] E. Best and K. Voss: Free Choice Systems have Home States. Acta Informatica 21, 89–100 (1984).

[6] F. Commoner, A.W. Holt, S. Even and A. Pnueli: Marked Directed Graphs. Journal of Computer and System Science Vol.5, 511–523 (1971).

[7] J. Esparza: Model Checking Using net Unfoldings. Hildesheimer Informatik Fachbericht 14/92 (October 1992). To appear in the Proceedings of TAPSOFT'93.

[8] H. Fleischner: Eulerian Graphs and Related Topics, Part 1, Volume 1. Annals of Discrete Mathematics Vol.45. North-Holland (1990).

[9] H.J. Genrich and K. Lautenbach: Synchronisationsgraphen. Acta Informatica Vol.2, 143–161 (1973).

[10] M. Hack: Analysis of Production Schemata by Petri Nets. TR-94, MIT-MAC (1972). Corrections (1974).

[11] R. Howell and L. Rosier: On questions of fairness and temporal logic for conflict-free Petri nets. Advances in Petri Nets 1988, G. Rozenberg (ed.), LNCS 340, 200–226 (1988).

[12] L. Landweber and E. Robertson: Properties of Conflict-Free and Persistent Petri Nets. JACM, Vol.25, No.3, 352-364 (1978).

[13] K.L. McMillan: Using unfoldings to avoid the state explosion problem in the verification of asynchronous circuits. Proceedings of the 4th Workshop on Computer Aided Verification, Montreal, pp. 164–174 (1992).

[14] P.S. Thiagarajan and K. Voss: A Fresh look at free-choice Nets. Information and Control, Vol. 61, No. 2, 85-113 (1984).

[15] H. Yen: A polynomial time algorithm to decide pairwise concurrency of transitions for 1-bounded conflict-free Petri nets. Information Processing Letters 38, 71–76 (1991).

Construction of S-invariants and S-components for Refined Petri Boxes [1]

Raymond Devillers[2]

Abstract

The paper shows how to synthesize S-invariants and S-components for a refined Petri Box, from the characteristics of its constituents. The construction is based on the tree structure of the interface places gluing the refining fragments to the remaining part of the Box to be refined.

1 Introduction

In previous papers [3, 1, 2, 9], together with E. Best, J. Hall and J. Esparza, we defined the basis of a general Petri Box Calculus (PBC). This calculus, which has been developed in the Esprit Basic Research Action DEMON, is based on a Petri net semantics and aims at easing the compositional definition of the semantics of various concurrent programming languages such as occam [13, 6]. With respect to Milner's CCS [14] from which it is largely inspired, the PBC features a different synchronisation operator, a refinement operator and a true sequence operator; as a consequence the recursion operator is much more general and not limited to tail-end recursion.

Moreover, the refinement operator plays a central rôle in the theory since not only recursion but also sequence, choice and iteration may be synthesized from it. Many equational properties of these operators were derived [1, 2] but there is still the need for a structural and behavioural analysis of the Boxes so constructed from other, more elementary (in general) ones. As various behavioural properties, like safeness and self-concurrency freeness, are connected to the knowledge of structural invariants, we will here be concerned with the construction of S-invariants and S-components for a refined Box, from corresponding characteristics of the constituent Boxes. We will see that this is driven by the structure of the interface places introduced to glue the refining fragments to the Box undergoing the refinement. Important consequences will then arise for whole families of Petri Boxes, especially the one introduced to model an algebra of Box expressions [3].

The next section will recall the main features of the PBC we will need in the following. The third section gives the central results we have derived on the construction of S-invariants and S-components for finite Boxes, the fourth section shows the 2-safeness of large families of refined Boxes and the last one is devoted to the extension of our results to recursion and infinite Boxes.

[1] Work done within the Esprit Basic Research Working Group 6067 CALIBAN (CAusal calcuLI BAsed on Nets).

[2] Laboratoire d'Informatique Théorique, Université Libre de Bruxelles, Boulevard du Triomphe, B-1050 Bruxelles, Belgium; e-mail: rdevil@ulb.ac.be.

2 Petri Boxes and the refinement operator

Petri Boxes, or Boxes for short, are defined [3] as equivalence classes of labelled Petri Nets. However, as Box operators are first defined on nets and then lifted to the Box level (after verifying that the result does not depend on the choice of the representatives), in order to simplify the presentation we shall in a first stage work at the net level; the impact of the introduction of equivalence classes will be considered in section 5.

Our labelled nets will be denoted as $\Sigma = (S, T, W, \lambda)$, where S is the place set, T the transition set, W the arc weights and λ the labelling function; the labels for the places may be: e (entry places, where tokens may *enter* the net), x (exit places, from where tokens may *leave* the net) and $\emptyset$ (internal places); the transition labels may be *communication* labels (elementary actions or finite multisets of them) or *hierarchical* labels (of the form[3] X, where X is a variable name). The nets are unmarked but the e-places define a natural initial marking M_e (one token on each e-place and no tokens elsewhere), and the x-places a natural terminating marking M_x (one token on each x-place and no tokens elsewhere), thus allowing to speak about the behaviour of the net; they may be weighted (but for the sake of simplicity none of our examples will exhibit this feature) and fulfill some contraints:

- each transition has input and output places;
- there are entry and exit places;
- there are no arcs to entry places or from exit places.

${}^\bullet\Sigma$ ($\Sigma^\bullet$) will denote the set of entry (exit) places of Σ.

Figure 1 shows examples of such labelled nets (the strange names used in the third net will be explained soon).

Those nets are the base of a translation from a process algebra of *Box expressions* into Petri Boxes [3]. Various operations are defined on those nets, which match the syntactic Box expression operators.

The operation we shall mainly be concerned with here is the refinement $\Sigma[X_i \leftarrow \Sigma_i | i \in I]$, meaning 'net Σ where all X_i-labelled transitions are refined into (i.e., replaced by a copy of) net Σ_i, for each $i \in I$'. This operation was formally defined in [1] in terms of a sequences set device, and more informally in terms of labelled trees. We shall formalize here the tree approach, and mention how to translate it into the sequence set one.

Let us first consider an illustrative example of a single refinement $\Sigma[X \leftarrow \Sigma_1]$, as shown by Figure 1. The intuitive idea is to replace each X-labelled transition in Σ (here χ, ψ) by a separate copy of the interior of Σ_1 (i.e., everything but the entry and exit places; here this means transitions γ, α and place 6) and to replace the places (here $0, 1$) of Σ connected to the refined transitions by adequate interface

[3] More exactly, the general theory allows for labels of the form (f, X), or $f(X)$, where f is a pending relabelling function (or relation [12]), in order to compositionally incorporate a relabelling operator; however, as relabellings have no impact on the structure of the nets and thus on the S-invariant analysis but complicate the definition of the labels in refined nets, for the sake of simplicity we shall omit them here.

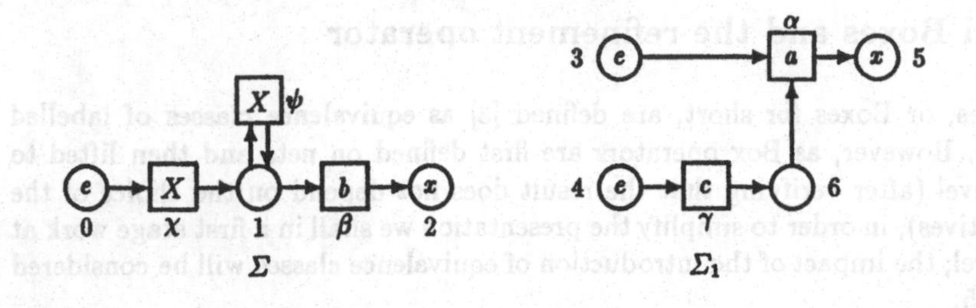

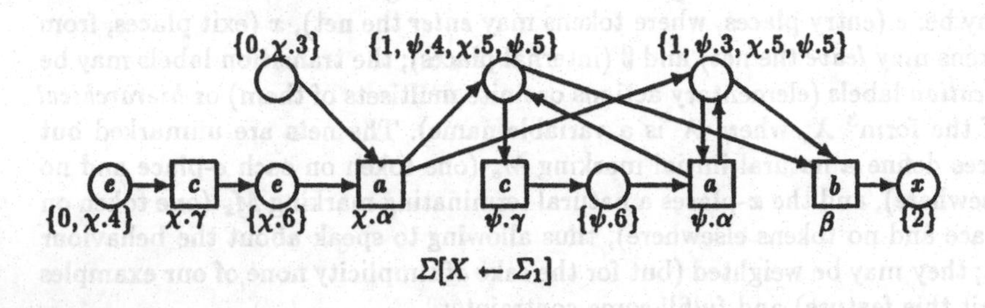

Fig. 1. A refinement with a side loop.

places. The (relative) complexity in defining the right interface places arises from the generality we want to reach, allowing any number of refined transitions, any type of connectivity (including side loops, like $1-\psi$), and any number of entry/exit places. Those interface places will thus have to mix together the places of Σ and, depending on their connectivity to X-labelled transitions, the entry/exit places of the refining net; it has turned out most convenient to define them as labelled trees of the kind

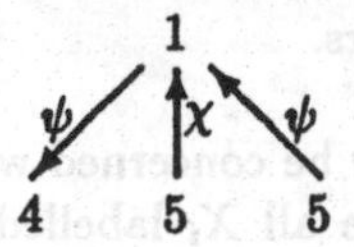

meaning that place 1 of Σ may be melted through its output transition ψ with the entry place 4 of Σ_1, through its input transition χ with the exit place 5 of Σ_1 and finally through its input transition ψ with the exit place 5 of Σ_1 again.

This tree may also be represented by a sequence set $\{1, \psi.4, \chi.5, \psi.5\}$, giving the root and the various children together with their associated arcs. There are other trees of that kind, like the ones represented here by $\{0, \chi.3\}$, $\{0, \chi.4\}$ and $\{1, \psi.3, \chi.5, \psi.5\}$.

The various copies will be distinguished by a prefixing device (e.g. $\psi.\alpha$ will be the copy of α corresponding to the refinement of transition ψ).

The label of a tree-place like the one exhibited above will be the same as its root (here $\emptyset$) and its connectivity will be driven by its structure: it will be connected to an non-refined transition (like β) like the root (1) is connected to it in Σ, it will be connected to a copied transition like $\psi.\gamma$ like 1 is connected to ψ in Σ and 4 to γ in Σ_1.

In order to get a more uniform presentation, the places of Σ (like 2) which are not replaced by interface places and the copied internal places (like ψ.6) may themselves be considered as labelled trees reduced to their root, with that very same label.

It may be verified that, if we start from the natural initial marking, $\Sigma[X \leftarrow \Sigma_1]$ may first perform a c followed by an a (like Σ_1) corresponding to the execution of χ, then any series of c-followed-by-a corresponding to the iterative execution of ψ, and finally a b like in Σ. That is, the refined net indeed exhibits the expected behaviour.

If we generalise these ideas to simultaneous refinements, and formalise them, this leads us to the following.

If X is a variable name and $\mathcal{X} = \{X_i \mid i \in I\}$ is a family of (distinct) such names, let us define $T^X = \{t \in T \mid \lambda(t) = X\}$ $(= \lambda^{-1}(X))$ and $T^{\mathcal{X}} = \bigcup_{X \in \mathcal{X}} T^X$.

Definition 1. *The Refinement Operator*

Let $\Sigma = (S, T, W, \lambda)$ and $\Sigma_i = (S_i, T_i, W_i, \lambda_i)$ (for $i \in I$) be labelled nets of the kind explained above.

$\Sigma[X_i \leftarrow \Sigma_i \mid i \in I]$ is defined as the labelled net $\tilde{\Sigma} = (\tilde{S}, \tilde{T}, \widetilde{W}, \tilde{\lambda})$ with

$\tilde{T} = (T \backslash T^{\mathcal{X}}) \cup \bigcup_{i \in I} T^i$

with $T^i = \{t.t_i \mid t \in T^{X_i}, t_i \in T_i\}$

$\tilde{S} = \bigcup_{i \in I} S^i \cup \bigcup_{s \in S} S^s$

with $S^i = \{t.s_i \mid t \in T^{X_i}, s_i \in S_i \backslash ({}^\bullet\Sigma_i \cup \Sigma_i^\bullet)\}$

and S^s is the set of all (up to isomorphism) the labelled trees of the following form:

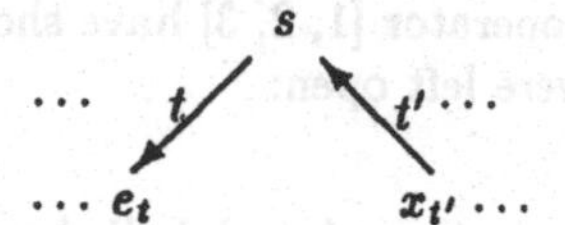

i.e., there is an arc labelled t for each (if any) $t \in s^\bullet$ with a label of the form X_i, going to any entry place e_t of Σ_i and there is an arc labelled t' for each (if any) $t' \in {}^\bullet s$ with a label of the form X_i, coming from any exit place $x_{t'}$ of Σ_i (for each $i \in I$).

$$\widetilde{W}(\tilde{t}, \tilde{s}) = \begin{cases} w & \text{if } \tilde{t} = t \in (T \backslash T^{\mathcal{X}}), \tilde{s} \in S^s \\ & \quad \text{and } W(t, s) = w \\ w \cdot w_i & \text{if } \tilde{t} = t.t_i \in T^i, \quad \stackrel{t}{\leftarrow} x_i \text{ occurs in } \tilde{s} \in S^s \\ & \quad W(t, s) = w \text{ and } W_i(t_i, x_i) = w_i \\ w_i & \text{if } \tilde{t} = t.t_i \in T^i, \quad \tilde{s} = t.s_i \in S^i \\ & \quad \text{and } W_i(t_i, s_i) = w_i \\ 0 & \text{otherwise} \end{cases}$$

$\widetilde{W}(\tilde{s}, \tilde{t})$ is defined analogously.

$$\tilde{\lambda}(\tilde{t}) = \begin{cases} \lambda(t) & \text{if } \tilde{t} = t \in (T \backslash T^{\mathcal{X}}) \\ \lambda_i(t_i) & \text{if } \tilde{t} = t.t_i, t_i \in T_i \text{ and } \lambda(t) = X_i \end{cases}$$

$$\tilde{\lambda}(\tilde{s}) = \begin{cases} \lambda(s) & \text{if } \tilde{s} \in S^s \\ \emptyset & \text{otherwise} \end{cases}$$

■ 1

As mentioned previously, we will often represent equivalently a tree in S^s by a set of sequences $\{s, t.e_t, \cdots, t'.x_{t'}, \cdots\}$, describing the root and all the children together with the arc label. It may be verified that Figure 1 indeed conforms to this formalisation and convention.

Let us notice that S^s may be a single tree reduced to its root (if s has no input or output transition labelled by one of the variables X_i); similarly, as was said before, in order to get a more uniform description, one may consider the components of S^i as labelled trees reduced to their root.

S^i and T^i give the fragments of the refined net corresponding to disjoint copies of the interior of Σ_i, one for each transition in Σ with a label X_i.

$T \backslash T^X$ gives the transitions of Σ which do not have to be refined.

S^s gives the interface places originating from s; this may be s itself (i.e., a single root) if s is not connected to any X_i-labelled transition; this may be a true tree-set, as shown above, if s is connected in any way to some X_i-labelled transitions; they all have the same label as s.

A transition t is connected to a place in S^s in $\tilde{\Sigma}$ like it is connected to s in Σ; a transition $t \cdot t_i$ is connected to a place $t \cdot s_i$ like t_i is connected to s_i in Σ_i; it is connected to a place $\tilde{s}$ in S^s like t_i is connected to e_t or x_t in Σ_i (and like t is connected to s in Σ).

Finally, we may notice that if s is an entry (exit) place of Σ, all the children (if any) of the tree-places in S^s are entry (exit) places of the Σ_i's.

Previous papers on this operator [1, 2, 3] have shown that it behaves nicely, but some important questions were left open:

- when translating Box expressions into labelled nets, the resulting nets are generally 1-safe, but not always: some expressions mixing iterations and parallel compositions lead to 2-safe nets; the question then arises to know if things could still be worse, i.e., if some more complicated expressions could lead to 3- or 4-safe nets, or even unsafe ones, and when we may be certain that the resulting nets will always be 1-safe;
- even in the (known) 2-safe cases, it has been observed that no transition is self-concurrent in the nets modelling Box expressions; again, the question arises to know if this is general or not;
- finally, it has been observed that, in the nets modelling Box expressions, when from the natural initial marking a marking is reached where all the exit places have a token, then this is the natural final marking, so that that there is exactly one token in each exit place and nothing elsewhere; again we would like to know if this is general.

In order to solve these problems, at least for the finite nets modelling Box expressions, in such a way that our techniques may be extended to infinite nets, to the recursion operator, and to other operators if the need arises in the future to extend the theory, we shall conduct an S-invariant analysis and synthesise S-invariants for refined nets from the similar characteristics of their constituents.

3 S-invariants and S-components

It may be observed that, if Σ and each Σ_i ($i \in I$) are finite, then so is $\tilde{\Sigma} = \Sigma[X_i \leftarrow \Sigma_i | i \leftarrow I]$.

Classically [5, 15], an S-invariant ν of a finite (labelled) net $\Sigma = (S, T, W, \lambda)$ may be defined as a function $\nu : S \rightarrow \mathcal{N}$ such that

$$\forall t \in T : \sum_{s \in S} W(s,t) \cdot \nu(s) = \sum_{s \in S} W(t,s) \cdot \nu(s) \tag{1}$$

i.e., for any transition t, the amount of tokens, weighted by ν, absorbed by t is the same as the amount of tokens, still weighted by ν, produced by t.

$(\mathcal{N}, +)$ may be any commutative monoid, depending on the type of usage which is considered; generally, it is the set of natural numbers, relative integers, (nonnegative) rationals or (nonnegative) reals, but it may also be the set of complex numbers, vectors or arrays of them, etc. Here we will essentially consider the real case ($\mathcal{N} = \Re$).

For instance, it may be verified that the net Σ_1 in Figure 1 has the S-invariant $\nu = \{3 \rightarrow \frac{1}{2}, 4 \rightarrow \frac{1}{2}, 5 \rightarrow 1, 6 \rightarrow \frac{1}{2}\}$.

ν may be extended to T, by defining

$\nu(t) = \sum_{s \in S} W(s,t) \cdot \nu(s) \qquad (= \sum_{s \in S} W(t,s) \cdot \nu(s)$ from (1)),

and to markings, by defining

$\nu(M) = \sum_{s \in S} M(s) \cdot \nu(s)$.

The same example as before leads to the extension $\nu \cup \{\alpha \rightarrow 1, \gamma \rightarrow \frac{1}{2}\}$.

The invariance property relates to the observation that $\nu(M) = \nu(M_0)$ for any marking M reachable from M_0, or from which M_0 is reachable.

Nonnegative S-invariants are especially important in order to get structural bounds on the possible evolutions of a net; indeed, it may be observed that if ν is nonnegative and if the initial marking is M_0:

for any $s \in S$, if $\nu(s) > 0$ then s is $\lfloor \frac{\nu(M_0)}{\nu(s)} \rfloor$-safe,
and is thus structurally bounded;
s is then said to be covered by ν;

for any $t \in T$, if $\nu(t) > 0$ then t may only occur $\lfloor \frac{\nu(M_0)}{\nu(t)} \rfloor$ times concurrently
with itself, which is also a kind of structural boundedness.

In particular,

if $\nu(t) > \nu(M_0), t$ is dead;
if $\nu(M_0) \geq \nu(t) > \frac{\nu(M_0)}{2}, t$ has no self-concurrency;
if $\nu(s) > \nu(M_0)$, s will always be empty
and all its surrounding transitions are dead;
if $\nu(M_0) \geq \nu(s) > \frac{\nu(M_0)}{2}$, s is 1-safe
and none of its surrounding transitions exhibits concurrency.

A net is S-covered by S-invariants if for each $s \in S$ there is a nonnegative S-invariant ν such that $\nu(s) > 0$; since (nonnegative) linear combinations of (nonnegative) S-invariants are (nonnegative) S-invariants, this is equivalent to the condition that there is a nonnegative S-invariant ν such that for each $s \in S$: $\nu(s) > 0$; the net is then structurally bounded.

A net is T-covered by S-invariants if for each $t \in T$ there is a nonnegative S-invariant ν such that $\nu(t) > 0$; this is again equivalent to the condition that there is a nonnegative S-invariant ν such that for each $t \in T$: $\nu(t) > 0$; the self-concurrency of the net is then structurally bounded. It may be observed that, since we only consider nets without isolated transitions, S-coveredness implies T-coveredness, but the converse is not true.

Since our nets have a natural initial marking M_e but also a natural final marking M_x, a special interest will be devoted to *conservative* S-invariants, where, by definition, $\nu(M_e) = \nu(M_x)$. Since the application of a scaling factor to a conservative S-invariant still gives a conservative S-invariant, we may restrict our attention to 1-conservative S-invariants, where by definition $\nu(M_e) = 1 = \nu(M_x)$, and to 0-conservative S-invariants, where by definition $\nu(M_e) = 0 = \nu(M_x)$ (but the latter are generally less interesting).

We may also observe that a net S-covered by nonnegative conservative S-invariants satisfies a *generalised emptiness property*, characterising the fact that from the natural initial marking (where only the entry places are marked, by 1), if a marking is reached where all the exit places are marked, then the reached marking is the natural final one (each exit place has exactly one token and all the other places are empty). This is indeed a generalised version of the emptiness property used for instance in [4, 7], where there was only a single entry place and a single exit place.

S-components can be viewed as a special class of S-invariants, where $\forall s \in S : \nu(s) \in \{0,1\}$ and $\forall t \in T : \nu(t) \in \{0,1\}$; this partitions S and T into two parts: $S_0^\nu = \nu^{-1}(0) \cap S$, $S_1^\nu = \nu^{-1}(1) \cap S$, $T_0^\nu = \nu^{-1}(0) \cap T$ and $T_1^\nu = \nu^{-1}(1) \cap T$. Each $t \in T_1^\nu$ has exactly one input place and one output place in S_1^ν, with arc weight 1, and each $s \in S_1^\nu$ has all its connected transitions in T_1^ν, with arc weights 1. This thus corresponds to the classical definition found in [5].

Again, 1-conservative S-components will be especially interesting since then, from what we said previously, if we start from the natural initial marking, each place in S_1^ν is 1-safe and each transition in T_1^ν is self-concurrency free. Also, there is exactly one entry place $e \in S_1^\nu$ and one exit place $x \in S_1^\nu$; initially, e is the only marked place in S_1^ν, and if a marking is reached where x is marked then it is the only one which is marked in S_1^ν, and it has one token.

A net will be said to be S-covered by S-components if each $s \in S$ belongs to S_1^ν for some 1-conservative S-component: it is then 1-safe (from the natural initial marking); it will be said to be T-covered by S-components if each $t \in T$ belongs to T_1^ν for some 1-conservative S-component: it is then self-concurrency free (from the natural initial marking). Again, S-coveredness implies T-coveredness, but not the converse.

The basic property of this paper is then:

Theorem 2. *S-invariant Synthesis*

If ν is an S-invariant of Σ and $\forall i \in I$, $\forall t \in \lambda^{-1}(X_i)$: ν_t is a 1-conservative S-invariant of Σ_i, then $\tilde{\nu}$ defined as follows

$$\tilde{\nu}(t \cdot s_i) = \nu(t) \cdot \nu_t(s_i)$$
$$\tilde{\nu}(\tilde{s}) = \nu(s) \cdot \prod_t \nu_t(e_t) \cdot \prod_{t'} \nu_{t'}(x_{t'})$$

if $\tilde{s} = \cdots$ [diagram: place s with arcs t to $\ldots e_t$ and t' to $x_{t'} \ldots$]

is an S-invariant of $\tilde{\Sigma} = \Sigma[X_i \leftarrow \Sigma_i | i \in I]$;
moreover

$$\tilde{\nu}(\widetilde{M_e}) = \nu(M_e)\ ,\ \tilde{\nu}(\widetilde{M_x}) = \nu(M_x)$$
$$\tilde{\nu}(t) = \nu(t)$$
$$\tilde{\nu}(t \cdot t_i) = \nu(t) \cdot \nu_t(t_i)\ .$$

Proof: 1) From the connectivity of transitions t in $\tilde{\Sigma}$, we get (by denoting Σ_t for Σ_i if $\lambda(t) = X_i$):

$$\begin{aligned}\sum_{\tilde{s}\in\tilde{S}} \widetilde{W}(\tilde{s},t)\cdot\tilde{\nu}(\tilde{s}) &= \sum_{s\in S}\sum_{e_{t''}\in{}^\bullet\Sigma_{t''}} \cdots \sum_{x_{t'}\in\Sigma_{t'}^\bullet} \cdots W(s,t)\cdot\nu(s)\cdot \\ &\qquad \prod_{t''\in s^\bullet\cap\lambda^{-1}(\mathcal{X})} \nu_{t''}(e_{t''}) \cdot \prod_{t'\in{}^\bullet s\cap\lambda^{-1}(\mathcal{X})} \nu_{t'}(x_{t'}) \\ &= \sum_{s\in S} W(s,t)\cdot\nu(s)\cdot\prod_{t''}[\sum_{e_{t''}}\nu_{t''}(e_{t''})]\cdot\prod_{t'}[\sum_{x_{t'}}\nu_{t'}(x_{t'})] \\ &= \sum_{s\in S} W(s,t)\cdot\nu(s) \\ &= \nu(t)\end{aligned}$$

since, due to the 1-conservativeness of each $\nu_{t''}$ or $\nu_{t'}$, $\sum_{e_{t''}} \nu_{t''}(e_{t''}) = 1$ and $\sum_{x_{t'}} \nu_{t'}(x_{t'}) = 1$;
symmetrically, $\sum_{\tilde{s}\in\tilde{S}} \widetilde{W}(t,s)\cdot\tilde{\nu}(\tilde{s}) = \nu(t)$.

2) From the connectivity of transitions $t \cdot t_i$ in $\tilde{\Sigma}$, we get

$$\begin{aligned}\sum_{\tilde{s}\in\tilde{S}} \widetilde{W}(\tilde{s},t.t_i)\cdot\tilde{\nu}(\tilde{s}) &= \sum_{s_i\in\Sigma_t\setminus({}^\bullet\Sigma_t\cup\Sigma_t^\bullet)} W_t(s_i,t_i)\cdot\nu(t)\cdot\nu_t(s_i) \\ &\quad + \sum_{s\in S}\sum_{e_{t''}}\cdots\sum_{x_{t'}}\cdots W(s,t)\cdot W_t(e_t,t_i)\cdot\nu(s)\cdot \\ &\qquad \prod_{t''}\nu_{t''}(e_{t''})\cdot\prod_{t'}\nu_{t'}(x_{t'}) \\ &= \nu(t)\cdot[\sum_{s_i\in\Sigma_t\setminus({}^\bullet\Sigma_t\cup\Sigma_t^\bullet)} W_t(s_i,t_i)\cdot\nu_t(s_i)]\end{aligned}$$

$$+[\sum_{s\in S} W(s,t)\cdot\nu(s)]\cdot[\sum_{e_t\in{}^\bullet\Sigma_t} W_t(e_t,t_i)\cdot\nu_t(e_t)]\cdot$$
$$\prod_{t''\neq t}[\sum_{e_{t''}\in{}^\bullet\Sigma_{t''}} \nu_{t''}(e_{t''})]\cdot\prod_{t'}[\sum_{x_{t'}\in\Sigma_{t'}^\bullet} \nu_{t'}(x_{t'})]$$
$$=\nu(t)\cdot[\sum_{s_i\in\Sigma_t\setminus({}^\bullet\Sigma_t\cup\Sigma_t^\bullet)} W_t(s_i,t_i)\cdot\nu_t(s_i)]$$
$$+\nu(t)\cdot[\sum_{e_t\in{}^\bullet\Sigma_t} W_t(e_t,t_i)\cdot\nu_t(e_t)]$$
$$=\nu(t)\cdot\nu_t(t_i)$$

since $\forall x\in\Sigma_t^\bullet\ :\ W_t(x,t_i)=0$, and again $\sum_{e_{t''}}\nu_{t''}(e_{t''})=1$ and $\sum_{x_{t'}}\nu_{t'}(x_{t'})=1$;
symmetrically, $\sum_{\tilde{s}\in\tilde{S}}\widetilde{W}(t\cdot t_i,\tilde{s})\cdot\tilde{\nu}(\tilde{s})=\nu(t)\cdot\nu_t(t_i)$.

3) finally

$$\sum_{\tilde{s}\in{}^\bullet\tilde{\Sigma}}\tilde{\nu}(\tilde{s})=\sum_{s\in{}^\bullet\Sigma}\sum_{\tilde{s}\in S^\bullet}\tilde{\nu}(\tilde{s})=\sum_{s\in{}^\bullet\Sigma}\sum_{e_t}\cdots\nu(s)\cdot\prod_t\nu_t(e_t)$$
$$=\sum_{s\in{}^\bullet\Sigma}\nu(s)\cdot\prod_t[\sum_{e_t}\nu_t(e_t)]=\nu(M_e)$$

let us notice that here we do not have to consider x-children since an entry place s of Σ may only have output transitions;
symmetrically, $\tilde{\nu}(\widetilde{M_x})=\sum_{\tilde{s}\in\tilde{\Sigma}^\bullet}\tilde{\nu}(\tilde{s})=\nu(M_x)$. ■ 2

For instance, if we apply the construction to the example considered in Figure 1, with $\nu=\{0\rightarrow 1,1\rightarrow 1,2\rightarrow 1;\chi\rightarrow 1,\psi\rightarrow 1,\beta\rightarrow 1\}$ and $\nu_\chi=\nu_\psi=\{3\rightarrow\frac{1}{2},4\rightarrow\frac{1}{2},6\rightarrow\frac{1}{2},5\rightarrow 1;\gamma\rightarrow\frac{1}{2},\alpha\rightarrow 1\}$, we get $\tilde{\nu}$ associating a weight 1 to place $\{2\}$ and to transitions $\chi.\alpha$, $\psi.\alpha$ and β, and a weight $\frac{1}{2}$ to all the other places and transitions; it may be verified that this is indeed an S-invariant of $\Sigma[X\leftarrow\Sigma_1]$.

Despite its simplicity, this property is very rich and leads to many applications:

Corollary 3. ***Various Special Cases***

With the hypotheses of Theorem 2,

1. *if moreover, ν is conservative, so is $\tilde{\nu}$;*
2. *if moreover, ν is 1-conservative, so is $\tilde{\nu}$;*
3. *if moreover, ν is 0-conservative, so is $\tilde{\nu}$;*
4. *if ν and each ν_t is nonnegative, so is $\tilde{\nu}$;*
5. *if Σ and each Σ_i is T-covered by S-invariants, so is $\tilde{\Sigma}$;*
6. *if Σ and each Σ_i is S-covered by S-invariants, so is $\tilde{\Sigma}$;*
7. *if ν and each ν_t is a (1-conservative) S-component, so is $\tilde{\nu}$;*
8. *if Σ and each Σ_i is T-covered by S-components, so is $\tilde{\Sigma}$;*
9. *if Σ and each Σ_i is S-covered by S-components and $\tilde{s}$ is such that, for each side loop t of its root (if any), e_t and x_t are covered by ν_t, then $\tilde{s}$ is covered by $\tilde{\nu}$. In particular, this is true for all the entry and exit places of $\tilde{\Sigma}$;*

10. *if no X_i-labelled transition (for $i \in I$) has a side loop in Σ, then if Σ and each Σ_i is S-covered by S-components, so is $\tilde{\Sigma}$.*

Proof: Immediate from 2 ■ 3

The interest of those properties stems from the fact that, in the Petri Box theory, we are especially interested (see [1, 2]) in the nets obtained through refinements from the basic nets (Net(α) and Net(X)) and from the operative nets ($\Sigma_;$, $\Sigma_{\Box}$, $\Sigma_{\|}$ and Σ_*) exhibited in Figure 2: it may be observed that they are all T- and S-covered by (1-conservative) S-components.

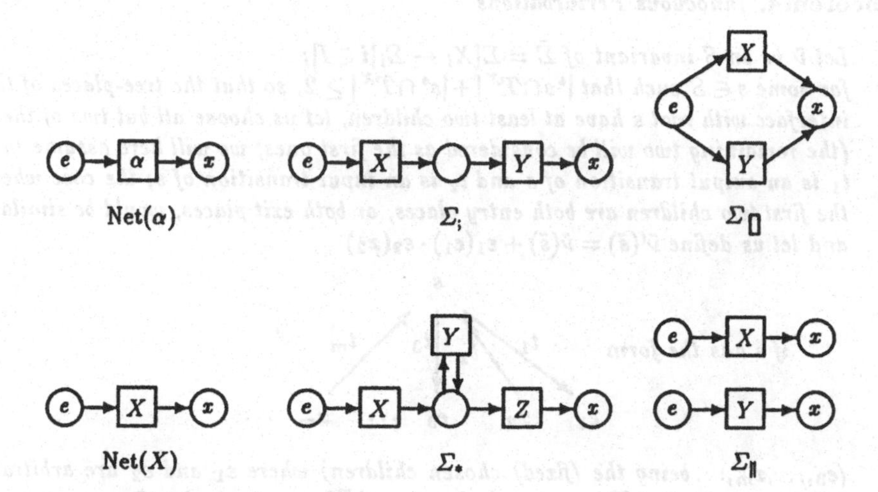

Fig. 2. The basic and operative nets

Let us notice that the side-loop freeness condition of 3(10) is essential, since for some choice of e_t and x_t, which may occur simultaneously in some $\tilde{s} \in S'$ if s is a side condition of t, it may happen that there is no 1-conservative S-component ν_t of Σ_t for which $\nu_t(e_t) = 1 = \nu_t(x_t)$ simultaneously.

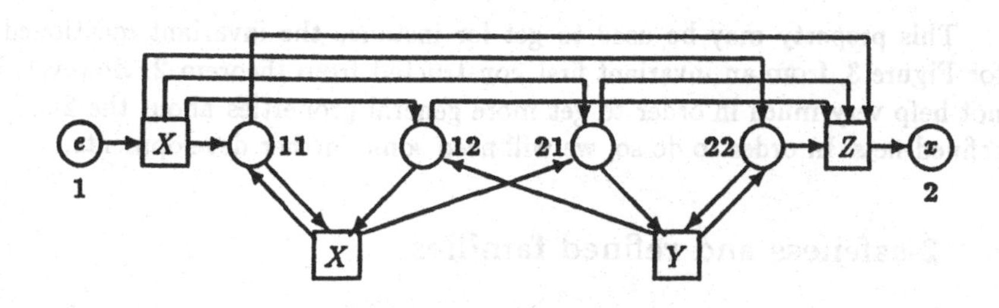

Fig. 3. $\Sigma_*[Y \leftarrow \Sigma_{\|}]$

For example, it may be checked that the net $\Sigma_*[Y \leftarrow \Sigma_{||}]$, exhibited (up to isomorphism) in Figure 3, is not S-covered by S-components; indeed, from the natural initial marking, the places 12 and 21 are not 1-safe, but only 2-safe.

This is related to the fact that

$$\tilde{\nu} = \{1 \to 1, 11 \to 0, 12 \to \tfrac{1}{2}, 21 \to \tfrac{1}{2}, 22 \to 0, 2 \to 1\}$$

is a 1-conservative S-invariant; it may also be observed that this invariant may not be obtained through the application of theorem 2 (since if $11 \to 0$ then either 12 or 21 should also be weighted by 0), so that this theorem is simply a way to construct S-invariants for a refined net, but by no means the only one. For instance, we may mention

Theorem 4. *Innocuous Perturbations*

Let $\tilde{\nu}$ be an S-invariant of $\tilde{\Sigma} = \Sigma[X_i \leftarrow \Sigma_i | i \in I]$;
for some $s \in S$ such that $|^{\bullet}s \cap T^{\mathcal{X}}| + |s^{\bullet} \cap T^{\mathcal{X}}| \geq 2$, so that the tree-places of the interface with root s have at least two children, let us choose all but two of them (the remaining two will be considered as the first ones; we will here assume that t_1 is an output transition of s and t_2 is an input transition of s; the case where the first two children are both entry places, or both exit places, would be similar) and let us define $\tilde{\nu}'(\tilde{s}) = \tilde{\nu}(\tilde{s}) + \varepsilon_1(e_1) \cdot \varepsilon_2(x_2)$

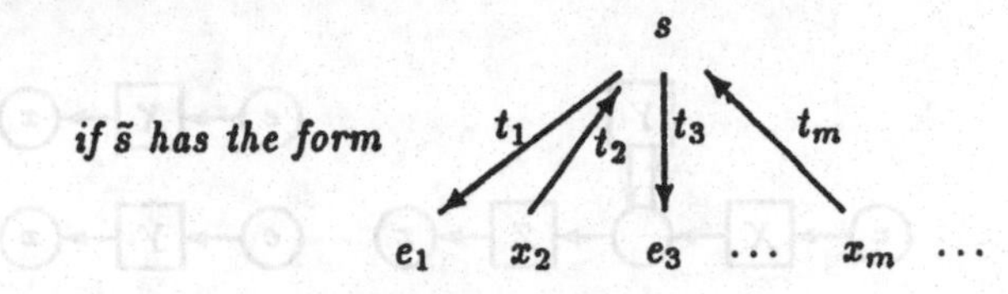

($e_3, \ldots, x_m, \ldots$ being the (fixed) chosen children) where ε_1 and ε_2 are arbitrary functions such that $\sum_{e_1 \in {}^{\bullet}\Sigma_{t_1}} \varepsilon_1(e_1) = 0$ and $\sum_{x_2 \in \Sigma^{\bullet}_{t_2}} \varepsilon_2(x_2) = 0$;
if $\tilde{\nu}' = \tilde{\nu}$ in all the other cases, $\tilde{\nu}'$ is still an S-invariant of $\tilde{\Sigma}$.

Proof: This results from the observations that in the sums we have to perform to check the invariance (cf. the proof of theorem 2), the perturbating terms may always be grouped, with a common factor $\sum_{e_1} \varepsilon_1(e_1)$ or $\sum_{x_2} \varepsilon_2(x_2)$ (or both), and thus vanish. Notice however that the perturbation is only interesting if $|^{\bullet}\Sigma_{t_1}| \geq 2 \leq |\Sigma^{\bullet}_{t_2}|$. ■ 4

This property may be used to get for instance the invariant mentioned above for Figure 3, from an invariant first constructed from theorem 2. However, it does not help very much in order to get more general properties about the 2-safeness of refined nets. In order to do so, we will need some further developments.

4 2-safeness and refined families

Let us first notice that from what we said about S-invariants, 2-safeness is clearly connected to nonnegative 1-conservative S-invariants with weights $\frac{1}{2}$. This leads us to introduce the following property:

Definition 5. *1/2-Property*

A system Σ will be said to have the 1/2-property if

1. for any $s \in S$ there is a nonnegative 1-conservative S-invariant ν such that $\nu(s) \in \{\frac{1}{2}, 1\}$;
2. for any $e \in {}^\bullet\Sigma$ there is a place $x \in \Sigma^\bullet$ and a nonnegative 1-conservative S-invariant ν such that $\nu(e) = 1 = \nu(x)$;
3. for any $x \in \Sigma^\bullet$ there is a place $e \in {}^\bullet\Sigma$ and a nonnegative 1-conservative S-invariant ν such that $\nu(e) = 1 = \nu(x)$. ■ 5

A basic stability feature about the 1/2-property is then

Theorem 6. *Stability of the 1/2-Property*

If Σ is S-covered by S-components and, for each $i \in I$, Σ_i has the 1/2-property, then $\tilde{\Sigma} = \Sigma[X_i \leftarrow \Sigma_i \mid i \in I]$ also has the 1/2-property.

Proof: – if $\tilde{s} \in \tilde{S}$ is of the form $t \cdot s_i$, then from theorem 2, corollaries 3(2 and 4), the hypotheses and the fact that S-coveredness implies T-coveredness, it is always possible to choose ν and ν_t in such a way that $\tilde{\nu}$ is a nonnegative 1-conservative S-invariant and $\tilde{\nu}(\tilde{s}) = \nu(t) \cdot \nu_t(s_i) = \nu_t(s_i) \in \{\frac{1}{2}, 1\}$; hence 5(1) is fulfilled for the copied internal places;

– if $\tilde{s} \in {}^\bullet\tilde{\Sigma}$, $\tilde{s}$ is of the form

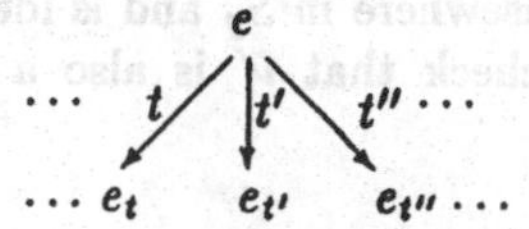

and again it is possible to choose the composing S-invariants in such a way that $\tilde{\nu}$ is nonnegative and 1-conservative, $\tilde{\nu}(\tilde{s}) = 1$ and, on $\tilde{\Sigma}^\bullet$, $\tilde{\nu} \in \{0, 1\}$ so that exactly one exit place of $\tilde{\Sigma}$ will have a weight 1; hence 5(2);

– the symmetrical argument shows that 5(3) is fulfilled;

– we now have to prove 5(1) for the interface tree-places; the main problem here rests with the side loops: it may be impossible to assign a weight 1 simultaneously to the corresponding two children through a nonnegative 1-conservative S-invariant ν_t. Figure 4 shows a tree-place $\hat{s}$ where we have collected those annoying side loops from 1 to n:

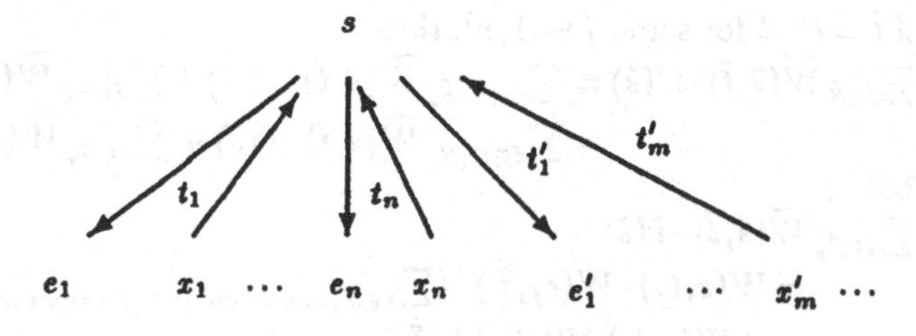

Fig. 4. An interface place and its side loops.

for each $j \in [1,n]$, t_j forms a side loop with s and no (nonnegative 1-conservative) S-invariant is known for Σ_{t_j} which simultaneously gives a weight 1 to e_j and x_j; the other children are such that either they do not occur in a side loop (so that there is an S-invariant giving them a weight 1) or there is an S-invariant covering them simultaneously with 1; there is also an S-component covering s.
If $n = 0$, from theorem 2, this will give us an S-invariant covering $\hat{s}$ with a weight 1 and the problem is solved.
If this is not the case, for each $j \in [1,n]$, there is an S-invariant ν_j^e giving a weight 1 to e_j and to some exit place x_j'' of Σ_{t_j}, and an S-invariant ν_j^x giving a weight 1 to x_j and to some entry place e_j'' of Σ_{t_j}; the S-invariant $\nu_j = \frac{1}{2}(\nu_j^e + \nu_j^x)$ will then cover simultaneously e_j, x_j, e_j'' and x_j'', but only with a weight $\frac{1}{2}$; this will lead from theorem 2 to an S-invariant $\tilde{\nu}$ of $\tilde{\Sigma}$ covering $\hat{s}$, but only with a weight 2^{-2n}.
Now we will use a grouping technique in order to reach our aim: let $\tilde{S}_g$ be the subset of $\tilde{S}$ with the same form as $\hat{s}$ but where $\forall j \in [1,n]$ e_j *may* be replaced by e_j'' and x_j *may* be replaced by x_j''; we may notice that $\tilde{\nu}$ gives the same weight 2^{-2n} to all the 2^{2n} places of $\tilde{S}_g$.
let us then define $\tilde{\nu}'$ by:

$\tilde{\nu}'(\hat{s}) = \frac{1}{2} = \tilde{\nu}'(\hat{s}')$, where $\hat{s}'$ has the same form as $\hat{s}$ but for any $j \in [1,n]$, e_j *is* replaced by e_j'' and x_j *is* replaced by x_j'';
$\tilde{\nu}'$ is 0 elsewhere in $\tilde{S}_g$ and is identical to $\tilde{\nu}$ otherwise.

We may now check that $\tilde{\nu}'$ is also a (nonnegative 1-conservative) S-invariant of $\tilde{\Sigma}$:

- if $\tilde{t} = t$,
  $$\begin{aligned}\textstyle\sum_{\tilde{s}\in\tilde{S}} \widetilde{W}(\tilde{s},\tilde{t}) \cdot \tilde{\nu}'(\tilde{s}) &= \textstyle\sum_{s\in S} \sum_{\tilde{s}\in S^\bullet} W(s,t) \cdot \tilde{\nu}'(\tilde{s}) \\ &= \textstyle\sum_{s\in S} W(s,t) \cdot \sum_{\tilde{s}\in S^\bullet} \tilde{\nu}'(\tilde{s}) = \sum_{s\in S} W(s,t) \cdot \sum_{\tilde{s}\in S^\bullet} \tilde{\nu}(\tilde{s}) \\ &= \textstyle\sum_{\tilde{s}\in\tilde{S}} \widetilde{W}(\tilde{s},\tilde{t}) \cdot \tilde{\nu}(\tilde{s}),\end{aligned}$$
  since $\tilde{\nu}'$ does not modify the global weight of $\tilde{S}_g$;
- if $\tilde{t} = t \cdot \hat{t}$ and t is none of the t_j's for $j \in [1,n]$,
  then again $\sum_{\tilde{s}\in\tilde{S}} \widetilde{W}(\tilde{s},\tilde{t}) \cdot \tilde{\nu}'(\tilde{s}) = \sum_{\tilde{s}\in\tilde{S}} \widetilde{W}(\tilde{s},\tilde{t}) \cdot \tilde{\nu}(\tilde{s})$ since either none of the places in $\tilde{S}_g$ will occur in the sum or they will all occur and may be grouped to form terms of the kind $f \cdot \sum_{\tilde{s}\in\tilde{S}_g} \tilde{\nu}'(\tilde{s}) = f \cdot \sum_{\tilde{s}\in\tilde{S}_g} \tilde{\nu}(\tilde{s})$, where f is some common factor, so that replacing $\tilde{\nu}'$ by $\tilde{\nu}$ does not change the weighted amount of absorbed tokens;
- if $\tilde{t} = t_j \cdot \hat{t}$ for some $j \in [1,n]$, then
  $$\begin{aligned}\textstyle\sum_{\tilde{s}\in\tilde{S}} \widetilde{W}(\tilde{s},\tilde{t}) \cdot \tilde{\nu}'(\tilde{s}) &= \textstyle\sum_{\tilde{s}\in\tilde{S}\setminus\tilde{S}_g} \widetilde{W}(\tilde{s},\tilde{t}) \cdot \tilde{\nu}'(\tilde{s}) + \sum_{\tilde{s}\in\tilde{S}_g} \widetilde{W}(\tilde{s},\tilde{t}) \cdot \tilde{\nu}'(\tilde{s}) \\ &= \textstyle\sum_{\tilde{s}\in\tilde{S}\setminus\tilde{S}_g} \widetilde{W}(\tilde{s},\tilde{t}) \cdot \tilde{\nu}(\tilde{s}) + \sum_{\tilde{s}\in\tilde{S}_g} \widetilde{W}(\tilde{s},\tilde{t}) \cdot \tilde{\nu}'(\tilde{s})\end{aligned}$$
  but
  $$\begin{aligned}&\textstyle\sum_{\tilde{s}\in\tilde{S}_g} \widetilde{W}(\tilde{s},\tilde{t}) \cdot \tilde{\nu}(\tilde{s}) \\ &\quad= W(s,t_j) \cdot W(e_j,\hat{t}\,) \cdot (\textstyle\sum_{\tilde{s}\in\tilde{S}_g \text{with a child } e_j \text{ in (2j-1)th position}} 2^{-2n}) \\ &\qquad + W(s,t_j) \cdot W(e_j'',\hat{t}\,) \cdot (\textstyle\sum_{\tilde{s}\in\tilde{S}_g \text{with a child } e_j'' \text{ in (2j-1)th position}} 2^{-2n}) \\ &\quad= W(s,t_j) \cdot W(e_j,\hat{t}\,) \cdot \tfrac{1}{2} + W(s,t_j) \cdot W(e_j'',\hat{t}\,) \cdot \tfrac{1}{2}\end{aligned}$$

$= \sum_{\tilde{s} \in \tilde{S}_g} \widetilde{W}(\tilde{s}, \tilde{t}) \cdot \tilde{\nu}'(\tilde{s});$

and $\forall \tilde{s} \notin \tilde{S} \setminus \tilde{S}_g : \tilde{\nu}'(\tilde{s}) = \tilde{\nu}(\tilde{s});$

- the weighted amount of produced tokens evaluates similarly.

Hence our claim. ■ 6

For instance, since Σ_* and $\Sigma_{\|}$ in Figure 2 are covered by 1-conservative S-components, we may deduce that the net $\Sigma_*[Y \leftarrow \Sigma_{\|}]$ shown in Figure 3 has the 1/2-property and each of its places may be covered by an S-invariant with a weight 1 or $\frac{1}{2}$; in particular the places 12 and 21 may be grouped by the technique described above, leading to the S-invariant $\tilde{\nu}' = \{1 \rightarrow 1, 11 \rightarrow 0, 12 \rightarrow \frac{1}{2}, 21 \rightarrow \frac{1}{2}, 22 \rightarrow 0, 2 \rightarrow 1\}$.

In order to exhibit the wide interest of this result, let us consider families of systems obtained through refinements from some set of initial nets.

Definition 7. *Family of Refined Nets*

If $\mathcal{S}$ is a set of nets (basically considered as non-refined), the family $\mathcal{R}(\mathcal{S})$ of nets generated through refinements from $\mathcal{S}$ is defined as the smallest family such that

- if $\Sigma \in \mathcal{S}$ then $\Sigma \in \mathcal{R}(\mathcal{S})$,
- if Σ and each Σ_i (for $i \in I$) belong to $\mathcal{R}(\mathcal{S})$, then for any family $\{X_i \mid i \in I\}$ of distinct variables: $\Sigma[X_i \leftarrow \Sigma_i \mid i \in I] \in \mathcal{R}(\mathcal{S})$. ■ 7

In this definition, we allowed both successive and embedded refinements, but it is rather easy to see that we could as well restrict ourselves to embedded refinements only:

Proposition 8. *Embedded Refinements*

If $\mathcal{S}$ is any set of nets, the family $\mathcal{R}(\mathcal{S})$ of nets generated through refinements from $\mathcal{S}$ is the smallest family such that

- *if $\Sigma \in \mathcal{S}$ then $\Sigma \in \mathcal{R}(\mathcal{S})$,*
- *if $\Sigma \in \mathcal{S}$ and $\forall i \in I : \Sigma_i \in \mathcal{R}(\mathcal{S})$, then for any family $\{X_i \mid i \in I\}$ of distinct variables: $\Sigma[X_i \leftarrow \Sigma_i \mid i \in I] \in \mathcal{R}(\mathcal{S})$.*

Proof: This results immediately from an iterative use of the expansion law for successive refinements

$$\Sigma[X_i \leftarrow \Sigma_i \mid i \in I][Y_j \leftarrow \Sigma'_j \mid j \in J] = \Sigma[X_i \leftarrow \Sigma_i[Y_j \leftarrow \Sigma'_j \mid j \in J], Y_k \leftarrow \Sigma'_k \mid i \in I, k \in K]$$

if $K = \{k \in J \mid Y_k \notin \{X_i \mid i \in I\}\}$

proved in [1], which is also valid at the net level up to isomorphism. ■ 8

This result may be interpreted as saying that each net in a refined family $\mathcal{R}(\mathcal{S})$ may be represented by a tree of the form shown in Figure 5, where each node belongs to $\mathcal{S}$ and each set of children means 'refine our father by replacing each variable labelling the arcs from it to us by the corresponding child'; we may notice that this form is not unique, however.

We may now state the main result of this section:

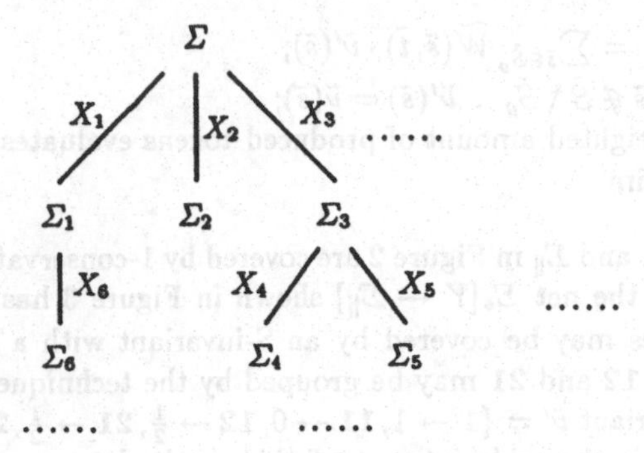

Fig. 5. The general embedded form of a net in a refined family

Theorem 9. ***2-Safeness of Refined Families***

> ***If each $\Sigma \in \mathcal{S}$ is S-covered by S-components, then each $\Sigma \in \mathcal{R}(\mathcal{S})$ has the 1/2-property; as a consequence it is at most 2-safe and exhibits the generalised emptiness property (from the natural initial marking).***

Proof: **This immediately results from propositions 6 and 8, by induction on the depth of the refining tree (see Figure 5), if we remark that the S-coveredness by S-components implies the 1/2-property; the consequences arise from what we said about the properties of S-invariants at the beginning of section 3.** ■ **9**

In particular, this is true for the family obtained from the basic and operative nets shown in Figure 2 used to model Box expressions.

In the same spirit, we may derive:

Theorem 10. ***Self-Concurrency Freeness of Refined Families***

> ***If each $\Sigma \in \mathcal{S}$ is T-covered by S-components, then so is each $\Sigma \in \mathcal{R}(\mathcal{S})$; as a consequence it is free of any self-concurrency (from the natural initial marking).***

Proof: **This immediately results from corollary 3(8) and proposition 8, by induction on the depth of the refining tree; the consequence arises from what we said about the properties of S-invariants at the beginning of section 3.** ■ **10**

5 Infinite nets, Box classes and other operators.

Let us now examine the relaxation of some of the restrictions used in the previous section.

First, one may want to drop the finiteness hypothesis, i.e., to allow infinite nets. This leads however to basic difficulties, already in the definition of S-invariants, as exhibited in Figure 6.

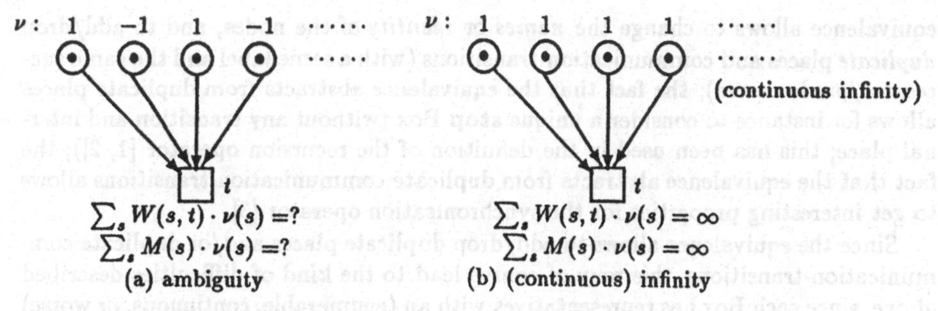

Fig. 6. Basic problems with infinite nets.

A simple solution is of course to add the restriction that all the evaluations of weighted markings or absorbed/produced tokens lead to finite sums, or to absolutely converging series; one may then speak about *finite* or *absolutely* converging S-invariants. But even this does not always allow to directly apply the construction used in theorem 2, as exhibited in Figure 7.

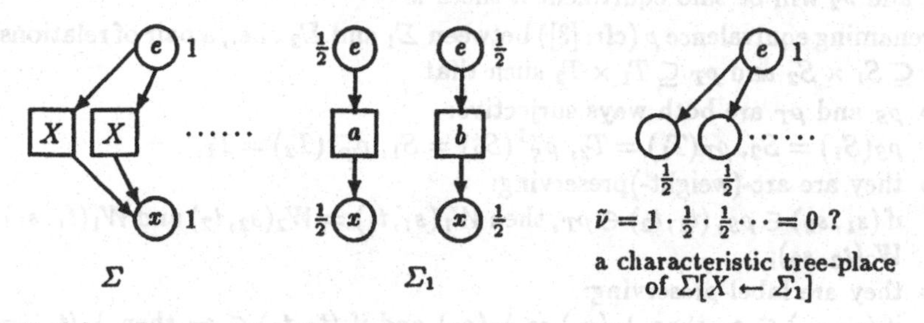

Fig. 7. S-invariant construction and cardinality explosion.

The problem here arises from the fact that, while none of the above problem arises in Σ nor in Σ_1, the refined net exhibits a continuous infinity of e-places and of x-places, each of them receives a weight corresponding to an infinite product usually considered as converging to 0 , and the sum of a continuous number of such zeroes has no reason to tend to 1.

As a consequence, the construction only works fine if for each $s \in S$, almost all $t \in s^\bullet \cap T^X$ are such that ν_t assigns a weight 1 to one place in ${}^\bullet\Sigma_t$ and 0 elsewhere (in ${}^\bullet\Sigma_t$) and almost all $t \in {}^\bullet s \cap T^X$ are such that ν_t assigns a weight 1 to one place in $\Sigma_t^\bullet$ and 0 elsewhere (in $\Sigma_t^\bullet$). In particular, it works for the construction of S-components, since then all the ν_t are 1-conservative and have exactly one designated input place and one designated output place with weight 1. The extension of the results of section 4 to the infinite case needs extensions of the grouping technique; this problem is solved and detailed in a further report [11].

Up to now, we only considered labelled nets, but our true aim is to define S-invariants and S-components for Boxes, i.e., for equivalence classes of nets [3]. The equivalence allows to change the ***names*** or ***identity*** of the nodes, and to add/drop ***duplicate*** places and communication-transitions (with a same label and the same successors/predecessors); the fact that the equivalence abstracts from duplicate places allows for instance to consider a unique stop Box (without any transition and internal place; this has been used in the definition of the recursion operator [1, 2]); the fact that the equivalence abstracts from duplicate communication-transitions allows to get interesting properties for the synchronisation operator [3].

Since the equivalence allows to add/drop duplicate places and/or duplicate communication-transitions, this may of course lead to the kind of difficulties described above, since each Box has representatives with an (enumerable, continuous, or worse) infinity of entry/exit places, some Boxes do not have finite representatives, and the situations described in Figures 6 and 7 are perfectly valid. However, we may define an equivalence relation between ***valid*** (finite or absolutely convergent) S-invariants of equivalent Boxes:

Definition 11. ***S-Invariant Classes.***

Let ν_1 be a valid S-invariant for a net Σ_1 and ν_2 be a valid S-invariant for a net Σ_2; ν_1 and ν_2 will be said equivalent if there is

- a renaming equivalence ρ (cfr. [3]) between Σ_1 and Σ_2 , i.e., a pair of relations $\rho_S \subseteq S_1 \times S_2$ and $\rho_T \subseteq T_1 \times T_2$ such that
  - ρ_S and ρ_T are both ways surjective: $\rho_S(S_1) = S_2$, $\rho_T(T_1) = T_2$, $\rho_S^{-1}(S_2) = S_1$, $\rho_T^{-1}(T_2) = T_1$;
  - they are arc-(weight-)preserving: if $(s_1, s_2) \in \rho_S$, $(t_1, t_2) \in \rho_T$, then $W_1(s_1, t_1) = W_2(s_2, t_2)$ and $W_1(t_1, s_1) = W_2(t_2, s_2)$;
  - they are label-preserving: if $(s_1, s_2) \in \rho_S$ then $\lambda_1(s_1) = \lambda_2(s_2)$ and if $(t_1, t_2) \in \rho_T$ then $\lambda_1(t_1) = \lambda_2(t_2)$;
  - ρ_T is bijective on hierachical transitions: if $\lambda_1(t_1) = X$ then $|\rho_T(t_1)| = 1$ and $|\rho_T^{-1}(\rho_T(t_1))| = 1$.
- and a function $f : \rho_S \rightarrow \Re$ such that $\forall s_1 \in S_1 : \nu_1(s_1) = \sum_{s_2 \in \rho_S(s_1)} f(s_1, s_2)$ and $\forall s_2 \in S_2 : \nu_1(s_2) = \sum_{s_1 \in \rho_S^{-1}(s_2)} f(s_1, s_2)$ (finite or absolutely convergent sums). ■ 11

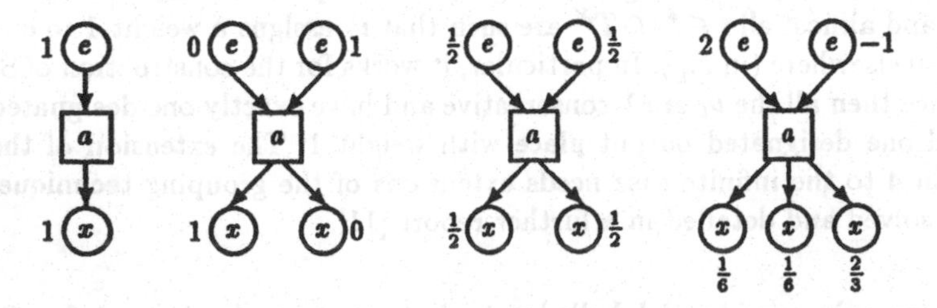

Fig. 8. Four equivalent S-invariants.

It may then be checked that, when it works, the construction used in theorem 2 when applied to equivalent S-invariants lead to equivalent synthesized invariants. However, it is no longer strictly possible to consider S-components as a special case of S-invariants (classes), since in the S-invariant class of an S-component we may find non-S-components, as illustrated in Figure 8. But it is immediate to restrict definition 11 to S-components, and so to get an equivalence relation on S-components, and thus S-component classes (in this case we may even restrict f to be a binary function $f : \rho_S \rightarrow \{0,1\}$ and the sums in the last part of definition 11 are always finite).

Finally, one may wonder if we may synthesize S-invariants and S-components for the result of other Box operators (see [3, 1, 2, 9]).

- Since sequentialisation, parallel composition, choice and iteration are special cases of refinements (see Figure 2), the problem may be considered as solved for them;
- relabelling modifies the label of some transitions, but not the structure of the net so that the S-invariants and S-components are not modified;
- restriction drops some transitions: again, old invariants are still valid (but new ones may be discovered);
- synchronisation adds new transitions corresponding to multisets of old ones, with the corresponding connectivity, so that again this does not modify the old S-invariants; but since the weights of the new transitions are (multi-) sums of the old ones, it may happen that an S-component leads to an S-invariant which is no longer an S-component (there are greater integer weights but they simply correspond to dead transitions, if we start from the natural initial marking);
- Recursion is a much more complex operator (detailed in [1, 2, 9]), but it may be observed that it also introduces tree-places and sequence-transitions, and their connectivity is driven by their structure, much like it is the case for refinement, so that the construction described in theorem 2 may be directly transported to this operator. However, even from finite nets, in the unguarded case, recursion may lead to infinite trees (in depth) so that the problems mentioned for the refinement of infinite nets will generally be more crucial here. But again, if we start from S-components all the problems vanish and the theory is rather satisfactory. The grouping technique used in section 4 may still be extended to this case but the developments are more delicate; this has been solved however and will be the subject of a further paper [11].

6 Conclusion

We have shown how the tree-structure of the interface places introduced in the construction of a refined net, immediately leads to the synthesis of S-invariants and S-components from similar characteristics of the composite nets. This may lead to interesting conclusions about the safeness, the self-concurrency freeness and the emptiness properties of the constructed nets, and sheds more light on the behaviour analysis of Petri Boxes. In particular it was shown that important families of refined

Boxes, like the one used to model Box expressions, essentially correspond to 1-safe or 2-safe systems without self-concurrency satisfying the generalised emptiness property.

The techniques used in this paper have been developed for finite refinements, but they may be extended to infinite refinements and recursions.

Acknowledgements

I want here to thank Eike Best who raised the problem of conducting an S-component or S-invariant analysis for constructed Petri Boxes, during the numerous discussions we had while elaborating the first reports on the definition of general refinement and recursion operators for the Petri Box Calculus.

The paper was written during a research stay at the L.R.I., while preparing a seminar on it; my thanks for this invitation go to Elisabeth Pelz.

The comments of four anonymous referees also helped in improving the presentation and (hopefully) the readability of the present paper.

References

1. E.Best, R.Devillers and J.Esparza: General Refinement and Recursion for the Box Calculus. Hildesheimer Informatik-Berichte 26/92 (1992).
2. E.Best, R.Devillers and J.Esparza: General Refinement and Recursion Operators for the Petri Box Calculus. Proceedings Stacs-93, P. Enjalbert et al. (eds.). Springer-Verlag Lecture Notes in Computer Science Vol. 665, pp.130-140 (1993).
3. E.Best, R.Devillers and J.Hall: The Box Calculus: a New Causal Algebra with Multi-label Communication. Advances in Petri Nets 1992, G. Rozenberg (ed.). Springer-Verlag Lecture Notes in Computer Science Vol. 609, pp.21-69 (1992).
4. E.Best, R.Devillers, A.Kiehn and L.Pomello: Concurrent Bisimulations in Petri Nets. Acta Informatica 28, pp.231-264 (1991).
5. E.Best, C.Fernández: Notations and Terminology on Petri Net Theory. GMD Arbeitspapiere 195 (1986).
6. E.Best and R.P.Hopkins: $B(PN)^2$ - A Basic Petri Net Programming Notation. To appear in the Proceedings of PARLE-93. Springer-Verlag Lecture Notes in Computer Science (June 1993).
7. R.Devillers: Maximality Preserving Bisimulation. TCS 102, pp.165-183 (1992).
8. R.Devillers: Maximality Preservation and the ST-idea for Action Refinements. Advances in Petri Nets 1992. Lecture Notes in Computer Science 609, pp.108-151 (1992).
9. R.Devillers: Tree Interfaces in the Petri Box Calculus. Draft Report LIT-265. Université Libre de Bruxelles (1992).
10. R.Devillers: General Refinement and Recursion for the Petri Box Calculus (Extended Abstract). Proceedings of the Workshop "What Good are Partial Orders ?", E.Best (ed.). Hildesheimer Informatik-Berichte 13/92, pp.57-76 (1992).
11. R.Devillers: S-invariant Analysis of Petri Boxes. Draft Report LIT-273. Université Libre de Bruxelles (December 1992).
12. R.Devillers: Towards a General Relabelling Operator for the Petri Box Calculus. Draft Report LIT-274. Université Libre de Bruxelles (January 1993).
13. R.P.Hopkins, J.Hall and O.Botti: A Basic-Net Algebra for Program Semantics and its Application to occam. Advances in Petri Nets 1992, G. Rozenberg (ed.). Springer-Verlag Lecture Notes in Computer Science Vol. 609, pp. 179-214 (1992).

14. R.Milner: *Communication and Concurrency.* Prentice Hall (1989).
15. W.Reisig: *Petri Nets, an Introduction.* EATCS Monographs on Theoretical Computer Science. Vol 4. Springer Verlag (1985).

Compositional Liveness Properties of EN-Systems*

D. Gomm[1] E. Kindler[1] B. Paech[2] R. Walter[1]

[1]Technische Universität München
Institut für Informatik
Arcisstr. 21
D-8000 München 2
Germany

[2]Ludwig-Maximilians-Universität München
Institut für Informatik
Theresienstr. 39
D-8000 München 2
Germany

Abstract

Modular design principles have gained great importance for the development of distributed systems. Compositional system properties are regarded as technical foundation for modular design methods. This paper introduces a compositional operator "changes to" for the expression of liveness properties, where the notion of composition is formalized by merging of places of Petri nets. "Changes to" is one operator of a temporal proof calculus which combines Petri nets with a UNITY-like temporal logic. The logic is interpreted on partial order semantics of Petri nets which allows the formalization of progress in distributed systems without a global fairness assumption.
In order to apply the operator "changes to" for proving a property of an example system, we introduce some additional operators and a part of the proof calculus.

Introduction

Petri net theory has successfully been applied to the modelling and analysis of distributed systems. S-Invariants, deadlocks, and traps are well known analysis methods for the verification of so called *safety properties*. *Liveness properties* such as "if a marking satisfies a formula φ, then eventually a marking will be reached which satisfies ψ" (φ *leads to* ψ) have as yet not received their due attention. Proving liveness properties involves different techniques [Lam77] which rarely exist in net theory.

As a first step towards a uniform treatment of safety and liveness properties we proposed in [DGK+92] a specification calculus for elementary net systems based on temporal logic. Standard analysis methods from net theory can be integrated into the proof calculus.

In this paper we lay emphasis on the compositional operator for liveness properties. We call a property *compositional* if it holds for a composed system, in case it holds for its components. We consider this notion of compositionality as a sound foundation

*supported by the Deutsche Forschungs Gemeinschaft SFB 342, TP A3: SEMAFOR

to develop a methodology for a modular design of distributed systems. In order to support the modelling of asynchronously communicating modules, subsystems are composed by merging of places. As an example we verify a module which interacts with an environment only specified by some system properties. This exemplifies the use of the new operator within the general context of the calculus.
System properties are formulated in a UNITY-like unnested temporal logic [CM88]. In contrast to other logics like [Pnu86], [CM88], and [Lam90] which interpret the operators on interleaving sequences, we consider partial orders as models for our logic. Partial order semantics permits for refinement of actions [Vog91], and for parallel composition of liveness properties [Rei91]. Moreover, there is no need for a global fairness assumption. Rather, we assume local *progress*. We argue about the local causes and effects of events in a partial order by interpreting propositional formulae on co-sets.

This paper is organized as follows: Section 1 introduces the basic formal notions used throughout the paper. In Sect. 2 we define the new operator *changes to* and prove its compositionality. Section 3 introduces those operators and rules from the calculus necessary to cope with the example verified in Sect. 4.

1 Basic Definitions

In this section we introduce the basic notions of *Elementary Net Systems (EN-systems)*, their *semantics*, and a *propositional logic* over their state space. Moreover, we use the *entailment relation* of propositional logic to interpret formulae on nonglobal states.

1.1 Elementary Net Systems

Our notion of EN-systems slightly differs from the usual definition of [Thi86]. Transitions with loops may fire if the corresponding places are marked.

Definition 1 (Net, Marking, EN-system)

1. A triple $N = (S,T;F)$ is a *net* iff S and T are disjoint sets and $F \subseteq (S \times T) \cup (T \times S)$ such that for every $t \in T$ the set $\{s \in S | (s,t) \in F \text{ or } (t,s) \in F\}$ is not empty. The elements of S and T are called *places* and *transitions*, respectively. F is the *flow relation*.
2. A subset $M \subseteq S$ is a *marking* of N.
3. An *EN-system* $\Sigma = (N, M_0)$ is a net with an initial marking M_0.

As usual, we graphically represent places by circles, transitions by boxes, and the flow relation by arcs. The initial marking M_0 of an EN-system is indicated by black dots (tokens) in the places which belong to M_0.

Notation 2 (Pre-, Postset) The set of elements $S \cup T$ of a net N will be denoted by N, too. The *pre-* and *postset* of an element $x \in N$ are defined by ${}^\bullet x = \{y \in N | (y,x) \in F\}$ and $x^\bullet = \{y \in N | (x,y) \in F\}$, respectively.

Definition 3 (Occurrence Rule) Let N be a net and M a marking of N. A transition $t \in T$ is *enabled at* M iff ${}^\bullet t \subseteq M$ and $M \cap (t^\bullet \setminus {}^\bullet t) = \emptyset$.

An enabled transition may *occur*, resulting in a new marking $M' = (M \setminus {}^\bullet t) \cup t^\bullet$. The occurrence of t at marking M is denoted by $M[t\rangle M'$.

Since our logic is based on states, the concept of composition is formalized as merging of places of two nets. For technical convenience we define the composition for EN-systems with the same set of places and the same initial marking, only.

Definition 4 (Composition of EN-systems)

Two EN-systems $\Sigma_1 = ((S_1, T_1; F_1), M_1)$ and $\Sigma_2 = ((S_2, T_2; F_2), M_2)$ are *composable* iff $S_1 = S_2$, $M_1 = M_2$, and $T_1 \cap T_2 = \emptyset$.

Then composition of Σ_1 and Σ_2 is defined by $\Sigma_1 \Box \Sigma_2 = ((S_1, T_1 \cup T_2; F_1 \cup F_2), M_1)$.

1.2 Runs of EN-systems

In the previous section we defined the occurrence rule as a semantics for EN-systems. But, as motivated in the introduction, our logic is based on partial order semantics. For that purpose we extend the notion of processes (cf. [Rei85]): With regard to compositionality (cf. Sect. 2) we formalize runs of noncontact-free systems, too.

Occurrence nets capture the two main aspects of runs: Runs are *acyclic* and *conflict free*.

Definition 5 (Occurrence Net)

A net $K = (B, E; F_K)$ is an *occurrence net* iff the transitive closure F_K^+ of F_K is irreflexive and for each $b \in B$ holds $|{}^\bullet b| \leq 1$ and $|b^\bullet| \leq 1$.

Notation 6 Let $K = (B, E; F_K)$ be an occurrence net.

1. F_K^+ induces a partial order on B, which will be denoted by $\prec$.
2. A subset $Q \subseteq B$ is a *co-set* of K iff for all $q, q' \in Q$ neither $q \prec q'$ nor $q' \prec q$.
3. A co-set $C \subseteq B$ is a *slice* of K iff there exists no co-set $Q \supset C$.
4. The maximal and minimal elements of K wrt. $\prec$ are denoted by $K^\circ = \{b \in B | \neg\exists b' \in B : b \prec b'\}$ and ${}^\circ K = \{b \in B | \neg\exists b' \in B : b' \prec b\}$, respectively. By definition K° and ${}^\circ K$ are co-sets, but in general need not to be slices.
5. The partial order $\prec$ of K defines a partial order $<$ on the set of all slices of K in a canonical way: $C < C'$ iff $C \neq C'$ and for all $b \in C$ there exists $b' \in C'$ such that $b \prec b'$ or $b = b'$.
6. The *immediate successor* relation on slices is defined by $C \lessdot C'$ iff $C < C'$ and there exists no slice C'' such that $C < C'' < C'$.
   Obviously, $C \lessdot C'$ implies that there is exactly one transition $e \in E$ between both slices.

A run of an EN-system Σ is an annotated occurrence net K such that every transition of K represents an occurrence of a transition of Σ. Usually, places of K are annotated by places of Σ in order to establish a correspondence between the pre- and postset of a transition of K and the pre- and postset of a transition of Σ. This notion of processes is adequate for contact free systems, only. In order to cope with contact situations, we do not only represent the existence of a token at place s by an inscription s, but we represent the absence of a token at place s by an inscription $\neg s$, too.

Definition 7 (Implicit Complementation)

Let $N = (S, T; F)$ be a net and M a marking of N.

1. The set of *literals* $L(S)$ of N is defined by $L(S) = S \cup \{\neg s | s \in S\}$, such that for every $s \in S$ there exists a unique $\neg s \in L(S)$.
2. For a set $X \subseteq S$ we define the set of corresponding *complements* $\overline{X} = \{\neg x | x \in X\}$.
3. For each transition $t \in T$, $\mathbf{act}(t) = {}^\bullet t \cup \overline{t^\bullet \setminus {}^\bullet t}$ is the *activation condition* and $\mathbf{eff}(t) = t^\bullet \cup \overline{{}^\bullet t \setminus t^\bullet}$ is the *effect* of t.
4. $\widehat{M} = M \cup \overline{S \setminus M}$ is the *complemented* marking of M. The occurrence rule for complemented markings is defined by $\widehat{M}[t\rangle\widehat{M'}$ iff $M[t\rangle M'$.

Proposition 8 Let $N = (S, T; F)$ be a net and M, M' markings of N. For each transition $t \in T$ holds: $\widehat{M}[t\rangle\widehat{M'}$ iff $\mathbf{act}(t) \subseteq \widehat{M}$ and $\widehat{M'} = (\widehat{M} \setminus \mathbf{act}(t)) \cup \mathbf{eff}(t)$.

A *run* of a system is defined as an occurrence net inscribed over the literals of the net. In order to guarantee that a run does not stop without being forced to, we require that runs are maximal. A run is *maximal* if there is no transition of the system which can be appended to the end of the run. This requirement is the usual liveness assumption of partial order semantics and is called *progress assumption*. A run of an EN-system is *initialized* iff its initial slice corresponds to the initial marking of the system.

Definition 9 (Run, Initialized Run)

1. Let $N = (S, T; F)$ be a net, $\Sigma = (N, M_0)$ an EN-system, $K = (B, E; F_K)$ an occurrence net, and $l : B \rightarrow L(S)$ an *inscription* of B over places of N. $R = (K, l)$ is a *run* of Σ iff
   (a) for every slice C of K and every place $s \in S$ there exists exactly one $b \in C$ such that $l(b) = s$ or $l(b) = \neg s$.
   (b) for every $e \in E$ there exists a transition $t \in T$ such that $l({}^\bullet e) = \mathbf{act}(t)$ and $l(e^\bullet) = \mathbf{eff}(t)$.
   (c) there exists no transition $t \in T$ such that $\mathbf{act}(t) \subseteq l(K^\circ)$. (progress assumption)
2. A run R of an EN-system $\Sigma = (N, M_0)$ is called *initialized* iff ${}^\circ K$ is a slice of K such that $l({}^\circ K) = \widehat{M_0}$.

Condition 1a of this definition is a requirement which must be satisfied by *every* slice of the run. For finite nets this will be implied by (b) if a *single* slice satisfies that requirement.

The interleaving semantics (occurrence rule) and the partial order semantics of EN-systems are closely related. This relation is illustrated in the following proposition.

Proposition 10 For every pair of slices $C < C'$ in a run R of a system Σ there exists a transition t and a unique pair of corresponding markings such that $M[t\rangle M'$ and $l(C) = \widehat{M}$, $l(C') = \widehat{M'}$, and $l(C \cap C') = \widehat{M} \setminus \mathbf{act}(t)$.

Vice versa, for each transition occurrence $M[t\rangle M'$ there exists a run with corresponding slices $C < C'$.

1.3 Propositional Logic

The temporal logic for EN-systems will be based on a propositional logic over states. The nontemporal part of this logic is introduced in this section. The definition of propositional formulae and their interpretation on markings is a straightforward adaptation of standard propositional logic.

Additionally, we define the interpretation of formulae on *nonglobal* states (co-sets) of runs. This is one of the essentially new features used in the definition of the temporal operators.

Definition 11 (Propositional Formulae)

Let Σ be an EN-system with places S. The *propositional formulae* $PF(\Sigma)$ over (states of) Σ are inductively defined by:

1. $S \subset PF(\Sigma)$
2. If $\varphi \in PF(\Sigma)$, then $\neg\varphi \in PF(\Sigma)$
3. If $\varphi, \psi \in PF(\Sigma)$, then $(\varphi \wedge \psi) \in PF(\Sigma)$

The set of propositional variables $\mathbf{var}(\varphi)$ occurring in a formula φ is recursively defined:

1. $\mathbf{var}(s) = \{s\}$ for $s \in S$
2. $\mathbf{var}(\neg\varphi) = \mathbf{var}(\varphi)$
3. $\mathbf{var}(\varphi \wedge \psi) = \mathbf{var}(\varphi) \cup \mathbf{var}(\psi)$

The other classical operators $\vee$, $\Rightarrow$, and $\Leftrightarrow$ can be expressed by $\wedge$ and $\neg$ in the usual way. In the following definition we show how formulae are interpreted over markings of a system.

Definition 12 (Validity of Formulae at Markings, Entailment)

Let Σ be an EN-system and M a marking of Σ.

1. The *validity* of a formula φ at marking M is recursively defined and denoted by $M \models \varphi$:

   (a) $M \models s$ iff $s \in M$
   (b) $M \models \neg\varphi$ iff $M \not\models \varphi$
   (c) $M \models (\varphi \wedge \psi)$ iff $M \models \varphi$ and $M \models \psi$

2. A set of formulae Φ is valid at M (denoted by $M \models \Phi$) iff each formula $\varphi \in \Phi$ is valid at M.

3. A set of formulae Φ *entails* a formula φ (denoted by $\Phi \Vdash \varphi$) iff for each marking with $M \models \Phi$ holds $M \models \varphi$.

We use a symbol different from $\models$ for the entailment $\Phi \Vdash \varphi$, because a marking M can be regarded as a set of formulae, too. Then $M \models \varphi$ and $M \Vdash \varphi$ have quite a different meaning[1], since in $M \models \varphi$, M is interpreted as a marking and in $M \Vdash \varphi$, M is interpreted as a set of formulae. However, there is a strong relationship: for every marking M holds $M \models \varphi$ iff $\widehat{M} \Vdash \varphi$.

By now, we have defined the validity of formulae at markings. The validity of formulae at co-sets is defined by means of the entailment relation.

Definition 13 (Validity of Formulae at Co-sets)

Let Σ be an EN-system, $R = (K, l)$ a run of Σ, and Q a co-set of R. A formula φ is valid at Q (again denoted by $Q \models \varphi$) iff $l(Q) \Vdash \varphi$.

Co-sets and slices satisfying a formula φ are called *φ-sets* and *φ-slices*, respectively.

There is an important difference between the validity relation on slices (global states) and the validity relation on co-sets (nonglobal states): For each slice C either $C \models \varphi$ or $C \models \neg\varphi$ holds. That does not hold for arbitrary co-sets because nonglobal states do not contain full information about the global state and therefore are not able to determine the truth value of an arbitrary formula. Formally, this difference is reflected by the entailment relation in the definition of the validity at co-sets.

Another vital consequence of the above definition is that substituting equivalent propositional formulae does not change the validity on co-sets.

Proposition 14

Let R be a run of an EN-system Σ, Q a co-set of R, and $\varphi \Leftrightarrow \varphi'$ a propositional tautology. Then Q is a φ-set iff Q is a φ'-set.

Now, we are prepared for defining some temporal operators. So we can conclude this section with a formal definition of the *leads to* and *invariant* operator; the former was already addressed in the introduction informally.

Definition 15 (Leads To, Invariant)

Let Σ be an EN-system and φ, ψ be two propositional formulae over Σ.

1. A run R of Σ satisfies the temporal formula $\varphi \rhd \psi$ (φ leads to ψ) iff for each φ-slice C of R there exists a ψ-slice C' of R such that $C \leq C'$.

   $\varphi \rhd \psi$ is valid in Σ iff every initialized run R of Σ satisfies that formula.

2. A run R of Σ satisfies the temporal formula **invariant** φ iff every slice C of R is a φ-slice.

   Σ satisfies **invariant** φ iff every initialized run R of Σ satisfies **invariant** φ.

The validity of a temporal formula τ on runs and systems will again be denoted by $R \models \tau$ and $\Sigma \models \tau$, respectively.

[1] E.g. $\emptyset \models \neg a$, but $\emptyset \not\Vdash \neg a$.

2 The compositional liveness operator cht_C

In this section we are heading for a compositional operator which implies $\rhd$. The following counterexample shows that $\rhd$ itself is not compositional: Figure 1 depicts two components which both satisfy the property $a \rhd c$. In the composed system, however, $a \rhd c$ is no longer valid, because the token can get lost by the occurrence of transition t. More generally, a liveness property cannot be compositional if the

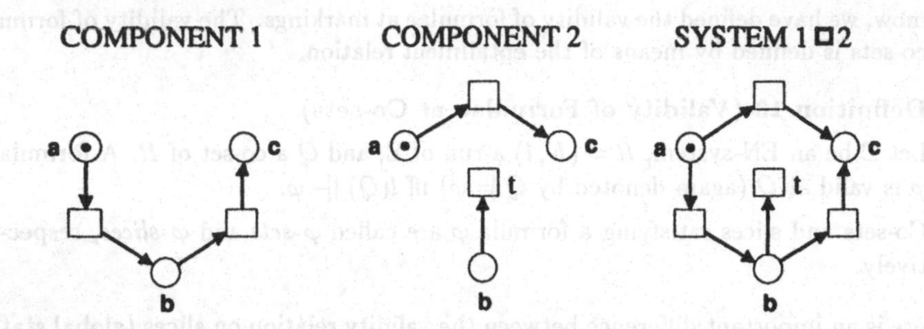

Figure 1: *a leads to c* is not a compositional property

state transformations are accomplished in several steps. Thus, we first introduce the operator *allows* (alo_C) which formalizes an adequate "one step" property. An additional liveness requirement, viz. the existence of an appropriate transition, yields the compositional liveness operator in demand.
Since this approach is similar to UNITY we compare the operators to the corresponding ones from UNITY and point out some consequences of these differences.

2.1 Defining alo_C

The "one step" property $\varphi\ \mathrm{alo}_C\ \psi$ holds if each transition dependent on φ establishes ψ when it occurs. To achieve compositionality of alo_C this is required for *all* runs (not only for the initialized ones) since nonreachable states of a component may become reachable when composed with another component.

Definition 16 (alo_C) Let Σ be an EN-system, R a run of Σ, and $\varphi, \psi \in PF(\Sigma)$.

$R \models \varphi\ \mathrm{alo}_C\ \psi$ iff for all slices $C \lessdot C'$ of R such that $C \models \varphi$ holds $C' \models \psi$ or $C \cap C' \models \varphi$

$\Sigma \models \varphi\ \mathrm{alo}_C\ \psi$ iff for every run R of Σ holds $R \models \varphi\ \mathrm{alo}_C\ \psi$

Figure 2 illustrates this definition: An occurrence of a transition at a φ-slice C either results in a ψ-slice C' or does not "touch" φ. The latter means that $C \cap C'$ is a φ-set.

The unless operator of UNITY permits a transition to touch φ. It can be expressed by the alo_C operator as follows: $\Sigma \models \varphi$ unless ψ iff $\Sigma \models \varphi \wedge \neg\psi\ \mathrm{alo}_C\ \varphi \vee \psi$. The inverse, however, is not true: alo_C cannot be expressed by unless, because UNITY does not distinguish *untouched* from *invariant*.

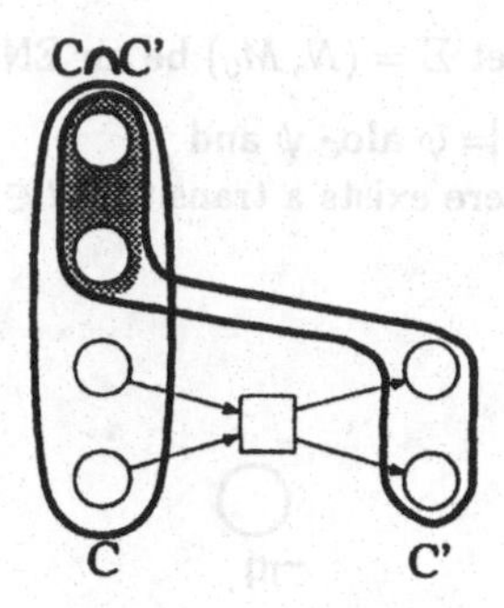

Figure 2: Two immediately successing slices $C < C'$

As a consequence of this definition, φ $\mathbf{alo}_C$ ψ does not hold for systems with an additional loop on φ. In the system of Fig. 3, a $\mathbf{alo}_C$ b does not hold, because a may be touched by t' without establishing b.

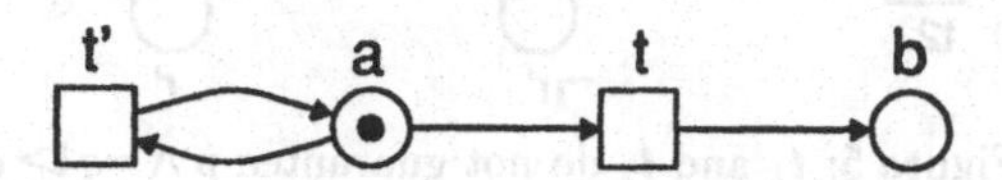

Figure 3: System with a loop on a

We will motivate briefly why this is sensible: $\mathbf{alo}_C$ should permit the incorporation of refinement techniques. In the system of Fig. 3, $a \rhd b$ should not hold since in the refined system depicted in Fig. 4, a stronger fairness requirement is necessary to achieve $a \rhd b$.

As mentioned before, we want that a $\mathbf{alo}_C$ b and the bare existence of a transition which can make the step from a to b yields a $\mathbf{cht}_C$ b. Such a transition exists in the system of Fig. 3. Thus, assuming a $\mathbf{alo}_C$ b would imply a $\mathbf{cht}_C$ b. Therefore, $\mathbf{cht}_C$ cannot imply $\rhd$, because $a \rhd b$ does not hold.

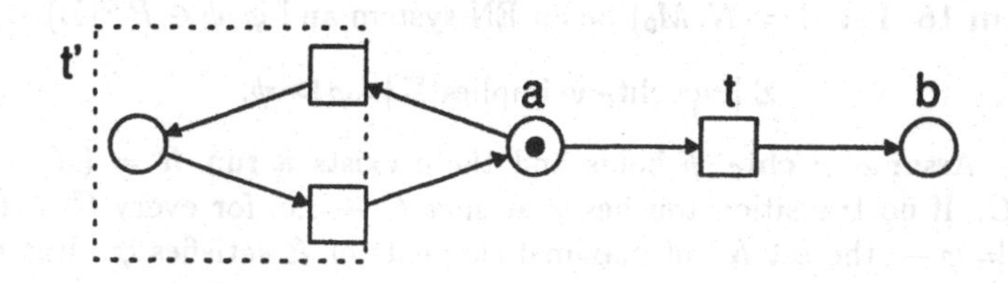

Figure 4: System with a refined loop on a

2.2 Defining $\mathbf{cht}_C$

$\mathbf{alo}_C$ only expresses safety properties: a real change from φ to ψ is not enforced. For the definition of the compositional liveness operator $\mathbf{cht}_C$ we additionally require that there exists a transition transforming φ into ψ.

Definition 17 (cht$_C$) Let $\Sigma = (N, M_0)$ be an EN-system and $\varphi, \psi \in PF(\Sigma)$.

$\Sigma \models \varphi$ **cht**$_C$ ψ iff $\Sigma \models \varphi$ **alo**$_C$ ψ and
there exists a transition $t \in N$ such that $\varphi \Vdash$ **act**(t)

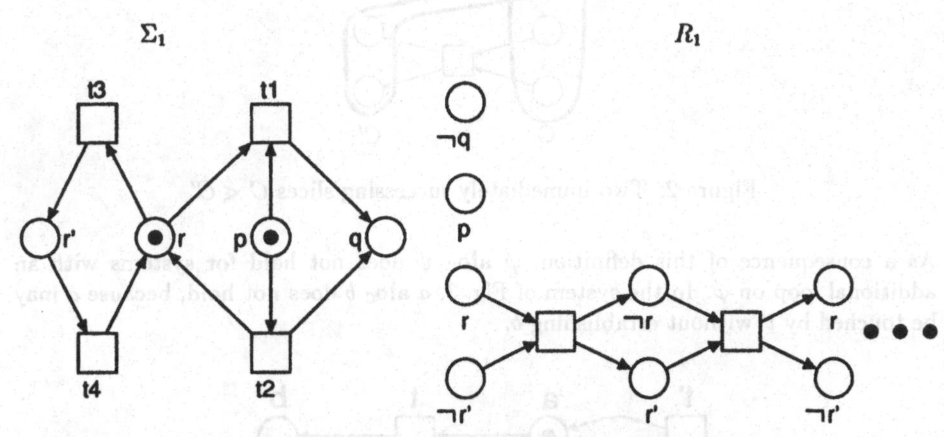

Figure 5: t_1 and t_2 do not guarantee $p \wedge \neg q \rhd q$

Definition 17 requires the existence of a transition that is activated in *every* φ-slice. The system Σ_1 which is depicted in Fig. 5 indicates that this strong requirement is necessary since **cht**$_C$ should imply $\rhd$. Σ_1 satisfies $p \wedge \neg q$ **alo**$_C$ q and for every $p \wedge \neg q$-slice either t_1 or t_2 is enabled which both could establish q. But $p \wedge \neg q \rhd q$ is not valid for Σ_1, because it does not hold in its run R_1.

> **cht**$_C$ is based on **alo**$_C$ in the same way **ensures** is based on **unless**. In particular, both UNITY operators are compositional. However, **ensures** guarantees liveness only because of a global fairness assumption incorporated into the semantics of UNITY programs whereas **cht**$_C$ is solely based on the progress assumption.

The existence of a single transition, however, is sufficient for **cht**$_C$ to imply $\rhd$.

Theorem 18 Let $\Sigma = (N, M_0)$ be an EN-system and $\varphi, \psi \in PF(\Sigma)$.

$$\Sigma \models \varphi \text{ cht}_C\ \psi \text{ implies } \Sigma \models \varphi \rhd \psi.$$

Proof: Assume φ **cht**$_C$ ψ holds and there exists a run $R = (K, l)$ with a φ-slice C. If no transition touches φ at slice C — i.e. for every $C' \gg C$ holds $C \cap C' \models \varphi$ — the set K° of maximal elements of R satisfies φ. But then, R is not a run since there exists a transition $t \in N$ such that $\varphi \Vdash$ **act**(t) and therefore **act**$(t) \subseteq l(K^\circ)$. □

2.3 Compositionality of cht$_C$

The compositionality of the **cht**$_C$ operator mainly relies on the compositionality of the **alo**$_C$ operator. The compositionality of the **alo**$_C$ operator can be easily shown since **alo**$_C$ properties can be proven for each transition of a system separately. This is formalized in Lemma 19 which expresses **alo**$_C$ in terms of the occurrence rule.

Lemma 19 Let $\Sigma = (N, M_0)$ be an EN-system.

$$\Sigma \models \varphi \ \mathbf{alo}_C\ \psi \text{ iff } \text{ for every transition } t \in N \text{ and all markings } M[t\rangle M' \text{ holds: } M \models \varphi \text{ implies } (M' \models \psi \text{ or } \widehat{M} \setminus \mathbf{act}(t) \Vdash \varphi).$$

Proof: For the proof we exploit the correspondence of the occurrence rule and the runs of Σ exhibited in Prop. 10. Both directions of the above equivalence are shown by contradiction.

"$\Rightarrow$" Suppose that there exists $M[t\rangle M'$ such that $M \models \varphi$, $M' \not\models \psi$, and $\widehat{M} \setminus \mathbf{act}(t) \not\Vdash \varphi$. By Prop. 10 there exists a run R and slices $C \lessdot C'$ such that $l(C) = \widehat{M}$, $l(C') = \widehat{M'}$, and $l(C \cap C') = \widehat{M} \setminus \mathbf{act}(t)$. This implies that $C \models \varphi$ but neither $C' \models \psi$ nor $C \cap C' \models \varphi$ holds.

"$\Leftarrow$" Suppose that there exist $C \lessdot C'$ such that $C \models \varphi$, but $C' \not\models \psi$ and $C \cap C' \not\models \varphi$. Then there exists a transition t such that $l(C)[t\rangle l(C')$ and $l(C \cap C') = l(C) \setminus \mathbf{act}(t)$.

□

By Lemma 19 it is straightforward to prove that $\mathbf{cht}_C$ is compositional.

Theorem 20 Let $\Sigma_i = (N_i, M_0), i = 1, 2$ be two composable EN systems. If $\Sigma_1 \models \varphi\ \mathbf{cht}_C\ \psi$ and $\Sigma_2 \models \varphi\ \mathbf{alo}_C\ \psi$, then $\Sigma_1 \Box \Sigma_2 \models \varphi\ \mathbf{cht}_C\ \psi$.

Proof: By the premise and Lemma 19 we know $\Sigma_1 \Box \Sigma_2 \models \varphi\ \mathbf{alo}_C\ \psi$, because $\mathbf{alo}_C$ properties can be verified for each transition separately. Since the transitions of $\Sigma_1 \Box \Sigma_2$ are given by the union of the transitions of Σ_1 and Σ_2 we can deduce from $\Sigma_1 \models \varphi\ \mathbf{cht}_C\ \psi$ that there exists a transition $t \in \Sigma_1 \Box \Sigma_2$ such that $\varphi \Vdash \mathbf{act}(t)$. Thus, $\Sigma_1 \Box \Sigma_2 \models \varphi\ \mathbf{cht}_C\ \psi$ holds. □

Theorem 20 does not require that the $\mathbf{cht}_C$ property holds in all components. Only one component must contain a transition which transforms φ into ψ. For the other components it suffices that φ remains untouched, which is expressed by the $\mathbf{alo}_C$ property.

3 Embedding $\mathbf{alo}_C$ and $\mathbf{cht}_C$ into a Calculus

The $\mathbf{alo}_C$ and $\mathbf{cht}_C$ properties serve as a basis for the derivation of **invariant** and $\rhd$ properties. In this section we introduce some proof rules of the derivation calculus. Particularly, we show those rules which are necessary for the example in Sect. 4.

3.1 Additional Temporal Operators

For a uniform presentation of the rules we introduce two additional temporal operators.

Definition 21

Let $\Sigma = (N, M_0)$ be an EN-system and $\varphi, \psi \in PF(\Sigma)$.

$\Sigma \models \varphi$ alo ψ iff for every initialized run R of Σ holds $R \models \varphi$ alo$_C$ ψ
$\Sigma \models$ invariant$_C$ φ iff every run R of Σ satisfies invariant φ

We introduce both, the compositional and the noncompositional version of alo and invariant, because they are needed for different purposes: Compositional properties are necessary for compositional proofs, but substitution of invariants into those properties yields noncompositional properties. Thus, both kinds of properties must be distinguished by different operators (cf. [San91], [Kin92]).
Moreover, both versions of invariant operators can be expressed in terms of the corresponding version of the alo operator.

Proposition 22 **Let $\Sigma = (N, M_0)$ be an EN-system and $\varphi, \psi \in PF(\Sigma)$.**

$\Sigma \models$ invariant φ iff $\Sigma \models \varphi$ alo φ and $M_0 \models \varphi$
$\Sigma \models$ invariant$_C$ φ iff $\Sigma \models \varphi$ alo$_C$ φ and $M_0 \models \varphi$

By Prop. 22, invariant$_C$ φ is a compositional property, too. The relation between the compositional and noncompositional properties is summarized in the following proposition: Compositional properties imply their noncompositional version.

Proposition 23

Let $\Sigma = (N, M_0)$ be an EN-system and $\varphi, \psi \in PF(\Sigma)$.

1. **$\Sigma \models \varphi$ alo$_C$ ψ implies $\Sigma \models \varphi$ alo ψ**
2. **$\Sigma \models$ invariant$_C$ φ implies $\Sigma \models$ invariant φ**

We introduce the following abbreviations for frequently used temporal formulae:

Notation 24

stable φ stands for φ alo φ, and stable$_C$ φ stands for φ alo$_C$ φ.

untouched φ stands for φ alo$_C$ *false*.

untouched $\varphi_1, \ldots, \varphi_n$ is an abbreviation for untouched $\varphi_1, \ldots,$ untouched φ_n.

3.2 Proof Calculus

By the use of a proof calculus, system properties can be derived in a purely syntactic way. We distinguish two different types of rules:

- ***Pickup rules* allow the "reading off" of simple properties from the syntactic structure of a system.**
- ***Derivation rules* allow the derivation of new properties from already known ones.**

In order to ease the derivation of properties we additionally introduce a third type of rules which translates net theoretic notions like S-invariants, deadlocks, and traps into logical formulae.
The proof rules will be presented in steps corresponding to the different types of rules.

3.2.1 Pickup Rules

In order to read off properties from a given EN-system we introduce the graphical notion of *patterns*. Each pattern has an associated formula which is true for the system if the pattern fits to the system in the following sense: The transitions of the pattern correspond to transitions of the system; the arcs which are crossed out in the pattern indicate which additional arcs are not allowed in the system.

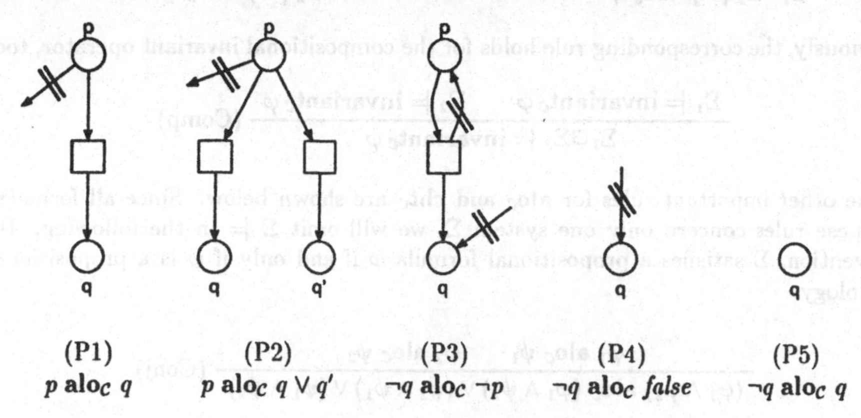

Figure 6: Pickup rules

For example, the pattern of pickup rule (P1) from Figure 6 requires that every transition in the system with p in its preset has q in its postset[2] (which is indicated by the crossed out arc). Every system with this structural property satisfies p **alo**$_C$ q, the formula associated with (P1).

The other pickup rules depicted in Figure 6 can be interpreted similarly. Note, that the pattern of (P5) only requires that q is a place of the corresponding system.

Remark 1 Pickup rules are supposed for a "graphical" derivation of properties of a system. But, checking the validity of the compositional operators can be done automatically: The validity of **alo**$_C$ and **cht**$_C$ formulae can be reduced to proving that a corresponding propositional is a tautology. This reduction is similar to the one for checking Hoare triples, but a bit more intricate.

Therefore, there exists a decision procedure for the validity of compositional formulae at a given system.

3.2.2 Derivation Rules

The derivation rules are introduced in order to derive new properties from already known ones. Therefore, derivation rules consist of a premise and a conclusion. The derivation rules are noted in the usual way: The premise is a set of formulae above a line, the conclusion is a formula below this line.

[2]Additional places in the pre- and postset of the transition are allowed.

Rules for compositional operators

In this notation the compositionality theorem for $\mathbf{alo}_C$ and $\mathbf{cht}_C$ can be reformulated as rules:

$$\frac{\Sigma_1 \models \varphi \ \mathbf{alo}_C \ \psi \qquad \Sigma_2 \models \varphi \ \mathbf{alo}_C \ \psi}{\Sigma_1 \Box \Sigma_2 \models \varphi \ \mathbf{alo}_C \ \psi} \qquad \frac{\Sigma_1 \models \varphi \ \mathbf{cht}_C \ \psi \qquad \Sigma_2 \models \varphi \ \mathbf{alo}_C \ \psi}{\Sigma_1 \Box \Sigma_2 \models \varphi \ \mathbf{cht}_C \ \psi} \ (\mathrm{Comp})$$

Obviously, the corresponding rule holds for the compositional invariant operator, too:

$$\frac{\Sigma_1 \models \mathbf{invariant}_C \, \varphi \qquad \Sigma_2 \models \mathbf{invariant}_C \, \varphi}{\Sigma_1 \Box \Sigma_2 \models \mathbf{invariant}_C \, \varphi} \ (\mathrm{Comp})$$

Some other important rules for $\mathbf{alo}_C$ and $\mathbf{cht}_C$ are shown below. Since all formulae in these rules concern only one system Σ, we will omit $\Sigma \models$ in the following. By convention, Σ satisfies a propositional formula φ if and only if φ is a propositional tautology.

$$\frac{\varphi_1 \ \mathbf{alo}_C \ \psi_1 \qquad \varphi_2 \ \mathbf{alo}_C \ \psi_2}{(\varphi_1 \wedge \varphi_2) \ \mathbf{alo}_C \ (\varphi_1 \wedge \psi_2) \vee (\varphi_2 \wedge \psi_1) \vee (\psi_1 \wedge \psi_2)} \ (\mathrm{Conj})$$

$$\frac{\varphi \ \mathbf{alo}_C \ \psi \qquad \psi \Rightarrow \psi'}{\varphi \ \mathbf{alo}_C \ \psi'} \ (\mathrm{Weak}) \qquad \frac{\varphi_1 \ \mathbf{alo}_C \ \psi_1 \qquad \varphi_2 \ \mathbf{alo}_C \ \psi_2}{(\varphi_1 \wedge \varphi_2) \ \mathbf{alo}_C \ (\psi_1 \vee \psi_2)} \ (\mathrm{SConj})$$

Though (*SConj*) is a "simplified" version of the conjunction rule (*Conj*) which can easily be derived from (*Conj*) by applying the weakening rule (*Weak*), we present it as a separate rule since it is frequently used.
A similar rule holds for the disjunction of $\mathbf{alo}_C$-properties:

$$\frac{\varphi_1 \ \mathbf{alo}_C \ \psi_1 \qquad \varphi_2 \ \mathbf{alo}_C \ \psi_2}{(\varphi_1 \vee \varphi_2) \ \mathbf{alo}_C \ (\psi_1 \vee \psi_2)} \ (\mathrm{Disj})$$

All of the above rules hold for the unless-operator of UNITY(cf. [CM88]), analogously. Since $\mathbf{alo}_C$ satisfies stronger requirements there are some rules which do not hold for unless. For example we present two kinds of 'synchronization rules':

$$\frac{\varphi \ \mathbf{alo}_C \ \psi \qquad \varphi \ \mathbf{alo}_C \ \psi'}{\varphi \ \mathbf{alo}_C \ \psi \wedge \psi'} \ (\mathrm{Sync})$$

(*Sync*) expresses that the transition from φ to ψ resp. ψ' must be established in the same step, if both $\mathbf{alo}_C$ properties hold. Another rule is introduced with respect to synchronization conditions such as illustrated in Fig. 7: A token will only be removed from p if there is a token on q. This can be expressed by $p \ \mathbf{alo}_C \ \neg q$[3]. Additionally, once q is unmarked, it remains unmarked forever which is formalized by $\neg q \ \mathbf{alo}_C \ \mathit{false}$. Thus, a token can never be removed from p if q is unmarked which

[3]In [DGK+92] we introduce a separate operator to characterize this situation.

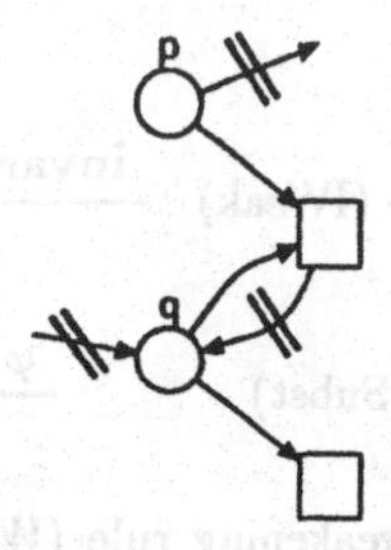

Figure 7: Motivation for the "lock rule"

means $p \wedge \neg q$ $\mathbf{alo}_C$ *false*. The idea of the above argumentation is captured in the following rule, where φ corresponds to the place p of the above example:

$$\frac{\varphi\ \mathbf{alo}_C\ \neg q \quad \neg q\ \mathbf{alo}_C\ false \quad q \notin \mathbf{var}(\varphi)}{\varphi \wedge \neg q\ \mathbf{alo}_C\ false}\ (\text{Lock})$$

Though, most of the above $\mathbf{alo}_C$-rules are correct for $\mathbf{cht}_C$, too[4], we do not present rules for $\mathbf{cht}_C$. In order to prove φ $\mathbf{cht}_C$ ψ for a system it is neccessary and sufficient to prove φ $\mathbf{alo}_C$ ψ (using the above rules) and the existence of a transition t such that $\varphi \Vdash \mathbf{act}(t)$. For proving $\varphi \Vdash \mathbf{act}(t)$ standard methods from propositional logic can be used.
Similarly, $\mathbf{invariant}_C$ φ can be proven by showing φ $\mathbf{alo}_C$ φ and $M_0 \models \varphi$. Nevertheless, we give a rule to derive compositional invariants from propositional tautologies:

$$\frac{\varphi}{\mathbf{invariant}_C\ \varphi}\ (\text{Taut})$$

This rule can be easily derived from $\mathbf{invariant}_C$ *true* and the implictly used rule of ***substitution of propositionally equivalent formulae*** within temporal formulae. This rule (TautEquiv) is justified since all temporal operators are defined by the validity of propositional formulae on co-sets. By Prop. 14 the validity on co-sets is invariant to substitution of equivalents.

Rules for noncompositional operators

At first we give those rules which allow to derive noncompositional properties from compositional ones. These rules are only a reformulation of Prop. 23 and Theorem 18.

$$\frac{\varphi\ \mathbf{alo}_C\ \psi}{\varphi\ \mathbf{alo}\ \psi} \qquad \frac{\mathbf{invariant}_C\ \varphi}{\mathbf{invariant}\ \varphi} \qquad \frac{\varphi\ \mathbf{cht}_C\ \psi}{\varphi \rhd \psi}\ (\text{Impl})$$

Of course, the composition rules do not hold for the noncompositional operators. Instead, we get some kind of substitution rules which do not hold for the corresponding

[4]Only (*Disj*) is not valid for $\mathbf{cht}_C$.

compositional operators.

$$\frac{\varphi \text{ alo } \psi \qquad \textbf{invariant}\, \psi \Rightarrow \psi'}{\varphi \text{ alo } \psi'} \text{ (IWeak)} \qquad \frac{\textbf{invariant}\, \varphi \qquad \textbf{invariant}\, \varphi \Rightarrow \psi}{\textbf{invariant}\, \psi} \text{ (Subst)}$$

$$\frac{\varphi \rhd \psi \qquad \textbf{invariant}\, \chi \Rightarrow \varphi}{\chi \rhd \psi} \text{ (Subst)} \qquad \frac{\varphi \rhd \psi \qquad \textbf{invariant}\, \psi \Rightarrow \chi}{\varphi \rhd \chi} \text{ (Subst)}$$

(*IWeak*) is very similar to the weakening rule (*Weak*) for the compositional $\textbf{alo}_C$. However, in (*Weak*) $\psi \Rightarrow \psi'$ must be a propositional tautology, whereas in (*IWeak*) it is sufficient that $\psi \Rightarrow \psi'$ is an invariant of the system.

The last rules formalize some addtitional properties of $\rhd$ which do not hold for $\textbf{cht}_C$: $\rhd$ is *reflexive*, *transitive* and *disjunctive*.

$$\frac{}{\varphi \rhd \varphi} \text{ (Refl)} \qquad \frac{\varphi \rhd \psi \qquad \psi \rhd \chi}{\varphi \rhd \chi} \text{ (Trans)} \qquad \frac{\varphi \rhd \psi \qquad \varphi' \rhd \psi'}{\varphi \vee \varphi' \rhd \psi \vee \psi'} \text{ (Disj)}$$

3.2.3 Net Rules

This section exemplifies how notions from net theory can be incorporated with benefit to prove system properties. Especially, when proving invariants of a system, structural properties of a net can be exploited. We only give two examples here:

A *deadlock* of a system Σ is a set of places $D \subseteq S$ such that $^\bullet D \subseteq D^\bullet$. Once a deadlock $D = \{s_1, \ldots, s_n\}$ is unmarked, it will never become marked again. This is exactly expressed by the formula **untouched** $\neg s_1 \wedge \ldots \wedge \neg s_n$.

Similarly, a *trap* $D = \{s_1, \ldots, s_n\}$ (a set of places such that $D^\bullet \subseteq {}^\bullet D$) is characterized by $\textbf{stable}_C\ s_1 \vee \ldots \vee s_n$.

Proposition 25

Let Σ be an EN-system and $D = \{s_1, \ldots, s_n\}$ a subset of its places.

1. D is a deadlock of Σ iff $\Sigma \models$ **untouched** $\neg s_1 \wedge \ldots \wedge \neg s_n$
2. D is a trap of Σ iff $\Sigma \models \textbf{stable}_C\ s_1 \vee \ldots \vee s_n$

4 Verification of "proper termination"

Let us consider the module Σ_M depicted in Fig. 8. Initially, the module is active and

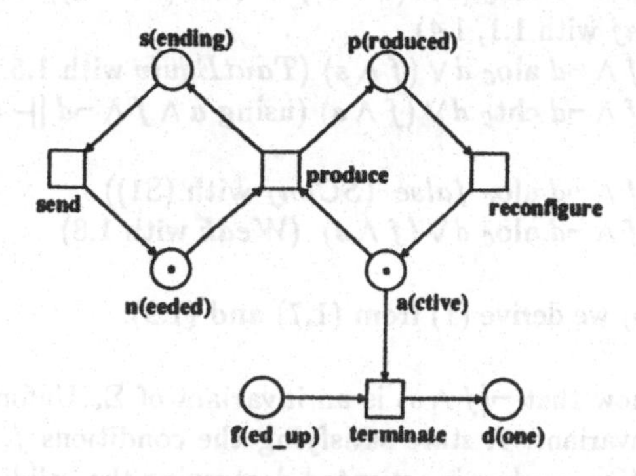

Figure 8: The module Σ_M

produces some output, if needed. After production it proceeds by sending the output to an environment and, concurrently, reconfigures itself to produce another output. We we want to verify proper termination of this module in an environment which once may decide that no more output is needed. Then it establishes the condition *fed up*. We want to prove that the composed system eventually terminates (*done* holds) if the local conditions *active* and *fed up* hold.
This property is intuitively valid, but not easy to prove in the context of an environment which is not modelled as a net. A formal specification of an appropriate environment module Σ_E is given below:

$$\Sigma_E \models \textbf{untouched } a, f, \neg n, \neg s, \neg p, \neg d \quad (S1)$$
$$\neg f \textbf{ alo}_C \neg n \quad (S2)$$

The environment must be initialized equally to Σ_M. (S1) determines the conditions that are not touched by the environment. (S2) specifies the expected behaviour on the interface: When the environment signals *fed up*, it invalidates the needed flag. Proposition 28 proves the property of termination ($a \wedge f \rhd d$) for any system Σ composed of Σ_M and an environment module Σ_E that satisfies the above specification. The proof is in two main steps. In Prop. 26 we show that the weaker compositional property $a \wedge f \wedge \neg d$ $\textbf{cht}_C$ $d \vee (f \wedge s)$ holds. Then we derive in Prop. 27 that $\neg(f \wedge s)$ is an invariant of the system. By this invariant $(f \wedge s)$ can be eliminated from the above $\textbf{cht}_C$ property and termination can easily be derived.

Proposition 26

$$\Sigma \models a \wedge f \wedge \neg d \textbf{ cht}_C\ d \vee (f \wedge s) \tag{1}$$

Proof: We verify (1) for the module and show that the environment preserves its safety part.

$\Sigma_M \models$ (1.1) f **alo**$_C$ d (by P1)
(1.2) a **alo**$_C$ $s \vee d$ (by P2)
(1.3) $\neg d$ **alo**$_C$ d (by P5)
(1.4) $a \wedge \neg d$ **alo**$_C$ $s \vee d$ (*SConj* with 1.2,1.3)
(1.5) $a \wedge f \wedge \neg d$ **alo**$_C$ $(d \wedge (s \vee d)) \vee (d \wedge (a \wedge \neg d)) \vee (f \wedge (s \vee d))$
(*Conj* with 1.1, 1.4)
(1.6) $a \wedge f \wedge \neg d$ **alo**$_C$ $d \vee (f \wedge s)$ (*TautEquiv* with 1.5)
(1.7) $a \wedge f \wedge \neg d$ **cht**$_C$ $d \vee (f \wedge s)$ (using $a \wedge f \wedge \neg d$ $\Vdash$ **act**(*terminate*))

$\Sigma_E \models$ (1.8) $a \wedge f \wedge \neg d$ **alo**$_C$ $false$ (*SConj* with (S1))
(1.9) $a \wedge f \wedge \neg d$ **alo**$_C$ $d \vee (f \wedge s)$ (*Weak* with 1.8)

Applying (*Comp*) we derive (1) from (1.7) and (1.9): □

Now, we want to show that $\neg(f \wedge s)$ is an invariant of Σ. Unfortunately, $\neg(f \wedge s)$ is no compositional invariant: A state satisfying the conditions $f, \neg s, n$ and a satisfies $\neg(f \wedge s)$, but transition *produce* is activated destroying the validity of $\neg(f \wedge s)$. So we first prove the stronger compositional invariant that *sending* and *needed* never hold simultaneously and, if *fed up* holds, then neither of them holds.

Proposition 27

$$\Sigma \models \textbf{invariant}_C\, \neg(s \wedge n) \wedge (f \Rightarrow (\neg s \wedge \neg n)) \quad (2)$$
$$\Sigma \models \textbf{invariant}\, \neg(f \wedge s) \quad (3)$$

Proof: Note that (2) is equivalent to $\Sigma \models \textbf{invariant}_C(\neg f \wedge \neg s) \vee (\neg f \wedge \neg n) \vee (\neg n \wedge \neg s)$. We have to prove (2) for both system components.

$\Sigma_M \models$ (2.1) $\neg f$ **alo**$_C$ $false$ (by P4)
(2.2) $\neg s$ **alo**$_C$ $\neg n$ (by P3)
(2.3) $\neg n$ **alo**$_C$ $\neg s$ (by P3)
(2.4) $\neg f \wedge \neg s$ **alo**$_C$ $\neg f \wedge \neg n$ (*Conj* with 2.1, 2.2)
(2.5) $\neg f \wedge \neg n$ **alo**$_C$ $\neg f \wedge \neg s$ (*Conj* with 2.1, 2.3)
(2.6) **stable**$_C$ $\neg n \wedge \neg s$ (deadlock rule with $D = \{n, s\}$)
(2.7) $\neg f \wedge \neg s$ **alo**$_C$ $(\neg f \wedge \neg s) \vee (\neg f \wedge \neg n) \vee (\neg n \wedge \neg s)$ (*Weak* with 2.4)
(2.8) $\neg f \wedge \neg n$ **alo**$_C$ $(\neg f \wedge \neg s) \vee (\neg f \wedge \neg n) \vee (\neg n \wedge \neg s)$ (*Weak* with 2.5
(2.9) $\neg n \wedge \neg s$ **alo**$_C$ $(\neg f \wedge \neg s) \vee (\neg f \wedge \neg n) \vee (\neg n \wedge \neg s)$ (*Weak* with 2.6)
(2.10) **stable**$_C(\neg f \wedge \neg s) \vee (\neg f \wedge \neg n) \vee (\neg n \wedge \neg s)$ (*Disj* with 2.7,2.8,2.9)
(2.11) **stable**$_C$ $\neg(s \wedge n) \wedge (f \Rightarrow (\neg s \wedge \neg n))$ (*TautEquiv* with 2.10)

$\Sigma_E \models$ (2.12) $\neg s$ **alo**$_C$ $false$ (by S1)
(2.13) $\neg n$ **alo**$_C$ $false$ (by S1)
(2.14) $\neg f \wedge \neg s$ **alo**$_C$ $\neg n \wedge \neg s$ (*Conj* with S2, 2.12)
(2.15) $\neg f \wedge \neg n$ **alo**$_C$ $false$ (*Lock* with S2, 2.13)
(2.16) $\neg n \wedge \neg s$ **alo**$_C$ $false$ (*SConj* with 2.12, 2.13)
(2.17) **stable**$_C$ $\neg(s \wedge n) \wedge (f \Rightarrow (\neg s \wedge \neg n))$
(same derivation as 2.7 to 2.11 based on 2.14 to 2.16)

Hence, we conclude with (*Comp*) that $\Sigma \models \textbf{stable}_C \neg(s \wedge n) \wedge (f \Rightarrow (\neg s \wedge \neg n))$.
Due to the initial marking also $\Sigma \models \textbf{invariant}_C \neg(s \wedge n) \wedge (f \Rightarrow (\neg s \wedge \neg n))$.
(3) is derived finally by application of (*Subst*) with (2) and the tautology $(\neg(s \wedge n) \wedge (f \Rightarrow (\neg s \wedge \neg n))) \Rightarrow \neg(f \wedge s)$, which is an invariant by (*Taut*).
□

Now we are ready to show, that any system $\Sigma = \Sigma_M \Box \Sigma_E$ eventually terminates, if the conditions *active* and *fed up* hold.

Proposition 28 $\qquad \Sigma \models a \wedge f \rhd d \qquad$ (4)

Proof:

$\Sigma \models$ (3.1) $a \wedge f \wedge \neg d \rhd d \vee (f \wedge s)$ (*Impl* with (1))
(3.2) $\textbf{invariant}_C$ $\neg(f \wedge s) \Rightarrow ((d \vee (f \wedge s)) \Rightarrow d)$ (*Taut*)
(3.3) **invariant** $d \vee (f \wedge s) \Rightarrow d$ (*Subst* with (3), 3.2)
(3.4) $a \wedge f \wedge \neg d \rhd d$ (*Subst* with 3.3, (1))
(3.5) $d \rhd d$ (*Refl*)
(3.6) **invariant** $(a \wedge f \wedge d) \Rightarrow d$ (*Taut*)
(3.7) $a \wedge f \wedge d \rhd d$ (*Subst* with 3.5 and 3.6)
(3.8) $a \wedge f \rhd d$ (*Disj* with 3.4 and 3.7)

□

An implementation of the environment need not achieve the switch to *fed up*. But, we have shown that any composed system consisting of Σ_M and an appropriate environment module satisfies proper termination if *active* and *fed up* eventually hold.

We used the calculus to demonstrate that the proof can be done without sacrificing rigour. In case we do not want to rely on informal arguments the existence of such a calculus is necessary. Moreover, it allows for the support of computer aided verification.

5 Conclusion

We have introduced the elementary operator $\textbf{cht}_C$ in order to derive liveness properties of distributed systems. This operator was influenced by the following constraints:

- compositionality and
- progress assumption as the only overall liveness requirement.

Similar to ensures of UNITY, $\textbf{cht}_C$ has a strong safety part ($\textbf{alo}_C$). It turns out that this safety part is the even more sophisticated aspect of $\textbf{cht}_C$.

Of course, we are not heading for a single liveness operator, but for a specification, proof, and design calculus for distributed systems. The full set of operators and proof rules have been introduced in [DGK+92]. Here, we only introduce those operators

(Sect. 3.1) which are necessary to prove an example system. Additionally, as shown by the example of deadlocks, classical net analysis methods can be integrated into this calculus.

Composition is a central issue of this paper. Nevertheless, the composition of EN-systems seems to be a rather technical operation. Still, we claim that it is a sound technical foundation for a methodology of modular design. To this end two additional features must be introduced: Composition of modules with different but not disjoint state space and a rely/guarantee-like specification technique for modules. This methodology as well as refinement techniques and an extension to high level Petri nets will be worked out in forthcoming papers.

On the one hand, the presented calculus can be seen as an adaptation of UNITY to partial order semantics. The structure of the temporal formulae (unnested formulae) as well as the way in which the elementary liveness operators are defined are similar. However, we avoid the problems with the substitution axiom (cf. [San91]) from the beginning by distinguishing compositional from noncompositional operators. Additionally, we will benefit from partial order semantics when refinement concepts and nonglobal fairness assumptions are considered.

On the other hand there are other temporal logics which are interpreted on partially ordered executions (e.g. [PW84, Pel87, Pen88, Rei88, MT89, KP90, Pae91b]). However, these logics are not concerned with compositionality aspects. Besides, the logics of [Pel87, Rei88, KP90] are based on global states, excluding arguments about nonglobal states. By contrast, the logics in [PW84, Pen88, MT89] are based on a relation between local states or actions and cannot express properties of global states like invariants. The logic of [Pae91a] includes an operator for decomposing runs into concurrent parts. Therefore, it can also distinguish between ***invariant*** and ***untouched***, global and nonglobal states. However, this is only possible, if the structure of the system is known.

References

[CM88] K. M. Chandy and J. Misra. *Parallel Program Design: A Foundation*. Addison-Wesley, 1988.

[DGK+92] J. Desel, D. Gomm, E. Kindler, B. Paech, and R. Walter. Bausteine eines kompositionalen Beweiskalküls für netzmodellierte Systeme. SFB-Bericht Nr. 342/16/92 A, Technische Universität München, July 92.

[Kin92] Ekkart Kindler. Invariants, compositionality, and substitution. SFB-Bericht Nr. 342/25/92 A, Technische Universität München, November 1992.

[KP90] Shmuel Katz and Doron Peled. Interleaving set temporal logic. *Theoretical Computer Science*, 75:263–287, 1990.

[Lam77] Leslie Lamport. Proving the correctness of multiprocess programs. *IEEE Transactions on Software Engineering*, SE-3(2):125–143, March 1977.

[Lam90] Leslie Lamport. A temporal logic of actions. Research Report SRC57, Digital Equipment Corporation, Systems Research Center, April 1990.

[MT89] M. Mukund and P.S. Thiagarajan. An axiomatization of event structures. *LNCS 405*, pages 143–160. Springer, 1989.

[Pae91a] Barbara Paech. Concurrency as a modality. SFB-Bericht Nr. 342/1/91 A, Technische Universität München, Institut für Informatik, January 1991.

[Pae91b] Barbara Paech. Extending temporal logic by explicit concurrency. In *MFCS, LNCS 520*, pages 377–386. Springer, 1991.

[Pel87] David Peleg. Concurrent dynamic logic. *Journal of the ACM*, 34(2):450–479, 1987.

[Pen88] W. Penczek. A temporal logic for event structures. *Fundamenta Informaticae*, 11:297–326, 1988.

[Pnu86] Amir Pnueli. Specification and development of reactive systems. In H.-J. Kugler, editor, *Information Processing*, pages 845–858. IFIP, Elsevier Science Publishers B.V. (North-Holland), 1986.

[PW84] Shlomit S. Pinter and Pierre Wolper. A temporal logic for reasoning about partially ordered computations. In *Proceedings Third Symposium on Principles of Distributed Computations*, pages 28–37. ACM, August 1984.

[Rei85] Wolfgang Reisig. *Petri Nets*, volume 4 of *EATCS Monographs on Theoretical Computer Science*. Springer-Verlag, 1985.

[Rei88] Wolfgang Reisig. Towards a temporal logic for causality and choice in distributed systems. In *REX workshop, LNCS 354*, pages 603–627. Springer, 1988.

[Rei91] Wolfgang Reisig. Parallel composition of liveness. SFB-Bericht Nr. 342/30/91 A, Technische Universität München, 1991.

[San91] Beverly A. Sanders. Eliminating the substitution axiom from UNITY logic. *Formal Aspects of Computing*, 3:189–205, 1991.

[Thi86] P.S. Thiagarajan. Elementary net systems. In W. Brauer, W. Reisig, and G. Rozenberg, editors, *Petri Nets: Central Models and Their Properties, LNCS 254*, pages 26–59. Springer-Verlag, September 1986.

[Vog91] Walter Vogler. Failure semantics based on interval semiwords is a congruence for refinement. *Distributed Computing*, 4:139–162, 1991.

Analysis of Place/Transition Nets with Timed Arcs and its Application to Batch Process Control

Hans-Michael Hanisch

Universität Dortmund, Fachbereich Chemietechnik, Lehrstuhl Anlagensteuerungstechnik
Postfach 500 500, D-44221 Dortmund
Germany

Abstract. ***The paper presents an analysis method for Place/Transition nets with timing of arcs directing from places to transitions. Based on this class of timed nets, the corresponding state graph, called dynamic graph, and a method to compute the state graph are defined.***
By means of the dynamic graph, the complete dynamic behaviour of the modeled system can be studied, objective functions can be formulated, and optimal control strategies can be computed. The concept is applied to two different problems of supervisory control of batch plants in the chemical industry.

Keywords

Timed Petri nets, system design, analysis, performance evaluation, batch process control

1 Introduction

Batch production systems usually consist of several process units (reaction, distillation, drying, filtering etc.) which perform sequences of technological operations and are coupled by flows of mass or energy and shared resources. The coupling pattern is time-dependent and many of the control actions are of a discrete nature such as opening/closing valves, starting/stopping pumps or agitators etc. Thus, discrete models must be applied to batch process systems.
Although the application of Petri nets to batch processes is still very sparse [1-6, 14-16], it has been shown by several applications which are summarized in [2] that Petri nets are an excellent tool to model and analyse the interactions between the process units in a batch plant and to describe the desired mode of operation of the plant as a basis for supervisory controller design.
The method presented here is based on a control and modeling hierarchy. Each single process in its single equipment is controlled by a local sequential controller and runs independently from other processes with the exception of interactions by

- flows of mass or energy from one process unit to another one and
- allocation of limited, shared resources (auxiliary devices, such as metering

and storage tanks, mains for steam and vacuum etc., which are temporarily used by the process units).

It is supposed that the local controllers ensure that the behaviour of the process units correspond to the technological requirements.

The *goal* of the modeling and analysis process is *not* to model the continuos chemical processes in the process units together with discrete control processes by means of hybrid models [8]. We suppose that the continuous processes are controlled by the single controllers for each process unit.

The *supervisory control* has to ensure that

- the flows of mass and energy between the single processes are performed according to the desired mode of operation,
- overloadings of limited resources are prevented, and
- an objective function (throughput, costs, or other) is optimized.

In order to achieve such a global behaviour, the supervisory control starts or delays sequences of operations in the process units and allocates shared resources to the process units.

Models of the desired behaviour require that a control strategy (i.e. the resolution of all conflicts, see Section 4) is pre-determined. In some cases, however, no initial knowledge about a feasible control strategy is available. Only a goal (throughput maximization, minimization of startup or shutdown times) is given which is to be achieved by the control. A concept to solve such problems must ensure that a suitable (or optimal) control strategy can be computed *as a result* of a modelling and analysis process. The modeling capability of causal, classical Petri nets and the corresponding analysis methods [12] are not sufficient since the goals which are to be achieved by the control explicitly include process times. Hence, timed Petri nets must be applied.

The most common concept of timed Petri nets is that of timed transitions [11], where each transition has a firing duration greater than zero. However, the models must be able to describe disturbances which can occur while a transition is firing and can cause states which are different from the states normally reached when the firing is finished. After a short period of using nets with timed places [1,14], the concept of timed arcs was chosen, which is described below. It has larger modeling capabilities than timed transitions or places (which are included as a special case of timed arcs) and is well suited for the modeling of batch processes as well as for other timed systems. By means of an appropriate software tool [13], models which are larger than the examples in Section 5 can be analysed.

2 Place/Transition Nets with Timed Arcs

A Place/Transition net with timed arcs is a tuple ATPN = (P,T,F,V,m_0,ZB,D_0) where (P,T,F,V,m_0) is a (causal) Place/Transition net [12].
For each $t \in T$ two |P|-dimensional vectors t^+ and t^- are defined which describe the change of the marking when t fires:

$$t^-(p) = \begin{cases} V(p,t), & \text{if } p \in Ft \\ 0, & \text{otherwise} \end{cases} \qquad t^+(p) = \begin{cases} V(t,p), & \text{if } p \in tF \\ 0, & \text{otherwise .} \end{cases} \tag{1}$$

Each arc (p,t) from a place to a transition is labeled with a time interval ZB(p,t):

$$ZB\colon (P \times T) \cap F \rightarrow \mathbf{N} \times (\mathbf{N} \cup \{\infty\}) \tag{2}$$

that describes the *permeability* of this arc. The permeability of an arc $ZB=[ZB_r,ZB_l]$ is a tuple of two integer values, where the first value $ZB_r(p,t)$ denotes the begin of the permeability and the second one $ZB_l(p,t)$ denotes the end of the permeability of the arc relative to the local clock of p.
The *state* z=(m,D) of a timed net consists of two components, the *marking* m of the net and *local clocks* D denoting the age of the marking of the places:

$$D\colon P \rightarrow \mathbf{N}. \tag{3}$$

It depends on the marking *and* the local clocks whether a transition is enabled or not. *A transition t is enabled* if all places $p \in Ft$ carry enough tokens according to the token weight of the arc and if all arcs directing to the transition are permeable. For each transition enabled at the marking m, we can compute the earliest firing time eft(t), which is the time the last pre(t)-arc becomes permeable:

$$eft(t) = \mathrm{Max}\,\{\, sgn(ZB_r(p,t) - D(p)) \mid p \in Ft \,\} \tag{4}$$

$$sgn(i) := \begin{cases} i, & \text{if } i>0 \\ 0, & \text{otherwise.} \end{cases} \tag{5}$$

A transition, however, can only fire if no pre(t)-arc becomes unpermeable before the last pre(t)-arc becomes permeable. The time lft(t) when the first pre(t)-arc of transition t becomes unpermeable can be computed as follows:

$$lft(t) = \mathrm{Min}\,\{\, ZB_l(p,t) - D(p) \mid p \in Ft \,\} . \tag{6}$$

Note that eft(t) and lft(t) depend on the local clocks of the places $p \in Ft$.

The set K of transitions enabled in state z is given by:

$$K = \{ t \mid t^- \leq m \wedge eft(t) \leq lft(t) \} . \quad (7)$$

We use the earliest and maximal firing rule.
The *step delay* $\Delta\tau$ denotes the earliest time a transition can fire:

$$\Delta\tau = \mathrm{Min} \{ eft(t) \mid t \in K \} .$$

The elements of E are the enabled transitions which fire exactly after $\Delta\tau$:

$$E = \{ t \mid eft(t) = \Delta\tau \wedge t \in K \} . \quad (9)$$

The set E must be partitioned into maximal subsets of transitions which can fire simultaneously after $\Delta\tau$ time units. Such subsets are called *maximal steps* u^*. For each maximal step u^* the vector Δu^* of dimension |P| denotes the change of marking by firing u^*:

$$\Delta u^* = \sum_{t \in u^*} t^+ - t^- . \quad (10)$$

The firing of a maximal step u^* after $\Delta\tau$ time units creates a new state z' (abbreviated: $z[(u^*,\Delta\tau)>z'$) by changing the marking *and* the local clocks of the timed net:

$$m' := m + \Delta u^* \quad (11)$$

$$D'(p) := \begin{cases} D(p) + \Delta\tau , & \text{if } m(p)>0 \wedge p \notin F u^* \cup u^* F \\ 0 , & \text{otherwise.} \end{cases} \quad (12)$$

Equation (12) describes the change of the clock positions. Each clock of a marked place which is not connected with an element of u^* is increased by $\Delta\tau$, since the marking of these places is not changed and the marking therefore becomes $\Delta\tau$ time units "older". The other clocks are put back to zero since the marking of the places is changed by firing u^*.
The concept includes the special cases:
1. no delay of permeability ($ZB_r(p,t)=0$)
2. no limitation of permeability ($ZB_l(p,t)=\infty$).

If all arcs have a time interval $ZB=[0..\infty]$, the net behaves like a causal Petri net under the maximal firing rule.

3 The State Graph for Place/Transition Nets with Timed Arcs

Similar to the definition of the reachability graph of classical, causal Petri nets [12], we can define the *state graph* for arc-timed Place/Transition nets as follows.
Let G be the set of all state transitions:

$$G \subseteq \wp(T) \times \mathbf{N}, \quad \wp(T)\text{...Power Set of } T. \tag{13}$$

The abbreviation $z[g>z'$ denotes that state z' can be reached from state z by state transition g, where g is a tuple of the maximal step u^* and the step delay $\Delta\tau$.
The set of all possible firing sequences of state transitions g is the set of words $W_{ATPN}(G)$. We call a state z' *reachable from state z* (abbreviated: $z[*>z'$) if there is a sequence $q = g_1, g_2, ..., g_n$ with

- q is a word
- q transforms state z into state z':

$$z[*>z' : \Leftrightarrow \exists q \; (q \in W_{ATPN}(G) \wedge z[q>z'). \tag{14}$$

Let $R_{ATPN}(z)$ be the set of all states in the arc-timed Petri net which are reachable from state z. We can then define the state graph (dynamic graph) of an arc-timed Petri net as $DG_{ATPN}=(R_{ATPN}(z_0),A_{ATPN})$ with

$$A_{ATPN} = \{ (z,g,z') \mid z,z' \in R_{ATPN}(z_0) \wedge g \in G \wedge z[g>z' \}. \tag{15}$$

Each element of A_{ATPN} describes an arc directing from node z to node z' and labeled with a maximal step $u^*(g)$ and the corresponding step delay $\Delta\tau(g)$.
In order to be able (at least potentially) to compute the dynamic graph for a given arc-timed Petri net, each of the two components of the state (the marking as well as the clock positions) must be bounded. Since the set of reachable markings (not states!) of a timed Petri net is always a subset of the reachable markings of the corresponding causal Petri net, any bounded causal Petri net with arbitrary time intervals of the arcs will have a finite number of markings.

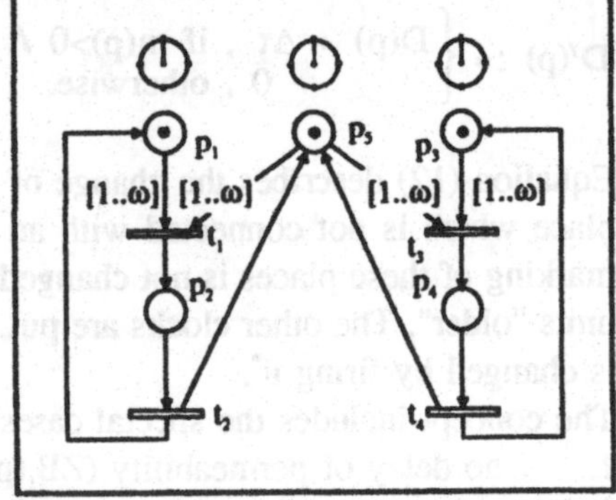

Fig. 1. Arc-timed Petri net with initial state

However, even for bounded Petri nets, the number of clock positions can be infinite, and some additional studies are necessary to prevent this. At first we give a small example of a net with an infinite number of clock positions and then describe some mechanisms to prevent infinite clock positions.
Fig. 1. shows a bounded (safe) Petri net with unbounded clock positions. The initial

state is given by the initial marking and by the clock positions (graphically represented by the clock symbols at places p_1, p_3 and p_5) which are all zero. Only the arcs (p_1,t_1), (p_5,t_1), (p_5,t_3), and (p_3,t_3) are labeled with the time interval $[1..\infty]$. Consequently, the arcs (p_2,t_2) and (p_4,t_4) have the time interval $[0..\infty]$.

Fig. 2. shows the corresponding dynamic graph with an infinite number of nodes caused by infinite clock positions. The nodes of the graph describe the state vectors $(m(p_1),D(p_1),...,m(p_5),D(p_5))$, where the marking durations for the marked places are given in parentheses. The maximal steps and - in parentheses - the corresponding step delays $\Delta\tau$ are given as arc labels.

It can be seen easily that firing of t_1 and t_2 increases the clock position of p_3 by one. For an arbitrary long firing sequence which only consists of t_1 and t_2, the clock position of p_3 will exceed any bound. The same goes for a firing sequence of t_3 and t_4 and the local clock of p_1. On the other hand, values greater than one of the local clocks $D(p_1)$, $D(p_3)$ and $D(p_5)$ will obviously never influence the permeability of the corresponding arcs (after one time unit the arcs are permeable and stay permeable forever). Consequently, the clocks of places p_1, p_3 and p_5 do not need to indicate durations greater than one.

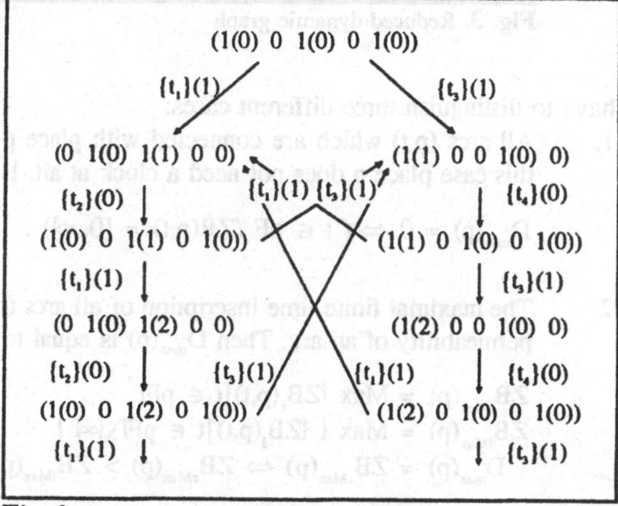

Fig. 2. Dynamic graph of the net of Fig. 1.

Therefore, we can modify the second part of the firing rule (12) as follows:

$$D'(p) := \begin{cases} D(p)+\Delta\tau\ , & \text{if } m(p)>0 \wedge p \notin F u \cup uF \wedge D(p)+\Delta\tau \leq D_{max}(p) \\ D_{max}(p)\ , & \text{if } m(p)>0 \wedge p \notin F u \cup uF \wedge D(p)+\Delta\tau > D_{max}(p) \\ 0\ , & \text{otherwise}\ . \end{cases} \quad (16)$$

where $D_{max}(p)$ is one for places p_1, p_3 and p_5 and zero for places p_2 and p_4 in the example of Fig. 1..

With the modified firing rule, the dynamic graph of Fig. 3. is computed which describes the same language as the dynamic graph in Fig. 2..

Based of the above idea, we can find for any place and arbitrary time inscriptions a finite bound $D_{max}(p)$ which denotes the maximal value of the local clock $D(p)$ that influences the permeability of the arcs (p,t), $t \in pF$ which are connected with p. We

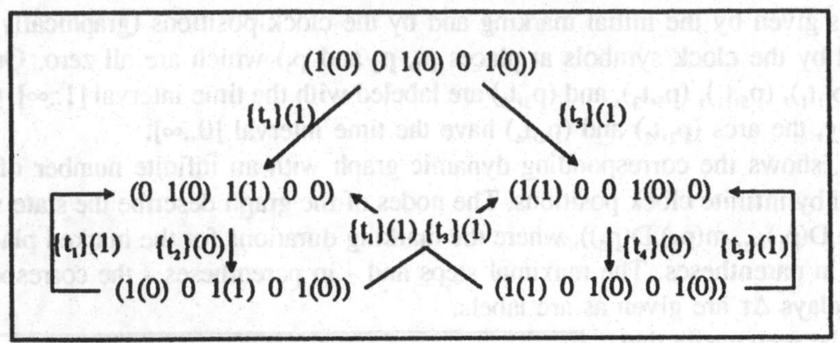

Fig. 3. Reduced dynamic graph

have to distinguish three different cases:

1. All arcs (p,t) which are connected with place p are labeled with [0..∞]. In this case place p does not need a clock at all. Hence, $D_{max}(p)$ is zero:

$$D_{max}(p) = 0 \Leftrightarrow \forall\, t \in pF\ (ZB(p,t) = [0..\infty])\ . \tag{17}$$

2. The maximal finite time inscription of all arcs (p,t) denotes the *begin* of the permeability of an arc. Then $D_{max}(p)$ is equal to this value:

$$\begin{aligned} ZB_{rMax}(p) &= \mathrm{Max}\ \{ZB_r(p,t)\,|\,t \in pF\} \\ ZB_{lMax}(p) &= \mathrm{Max}\ \{\ \{ZB_l(p,t)\,|\,t \in pF\} \setminus \{\infty\}\ \} \\ D_{max}(p) &= ZB_{rMax}(p) \Leftrightarrow ZB_{rMax}(p) > ZB_{lMax}(p)\ . \end{aligned} \tag{18}$$

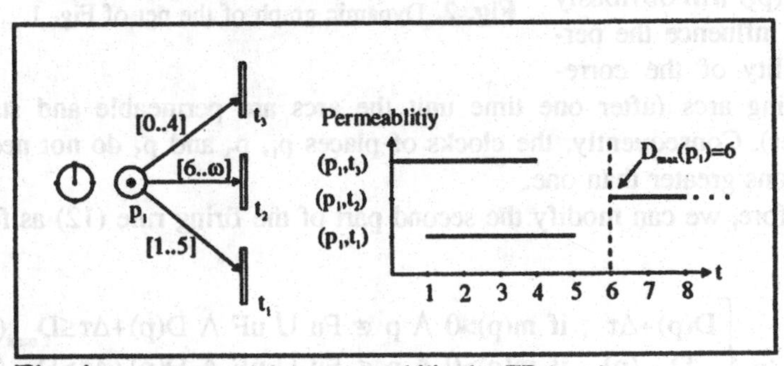

Fig. 4. Last change of the permeability by $ZB_r(p_1,t_2)$

Example

Fig. 4. shows such a case. The last change of the time-dependent permeability occurs when the arc (p_1,t_2) becomes permeable after 6 time units. Any value of $D(p_1)>6$ will not change the permeability of the arcs since the arcs (p_1,t_3) and $p_1,t_1)$ are already unpermeable and the arc (p_1,t_2) stays permeable forever. This situation can only be changed by a change of the marking of place p_1 which resets the clock to zero.

3. The maximal finite time inscription of all arcs (p,t) denotes the *end* of the permeability of an arc. Then $D_{max}(p)$ is equal to the maximal finite value plus one:

$$D_{max}(p) = ZB_{lMax}(p)+1 \Leftrightarrow ZB_{lMax}(p) \geq ZB_{rMax}(p) \; . \qquad (19)$$

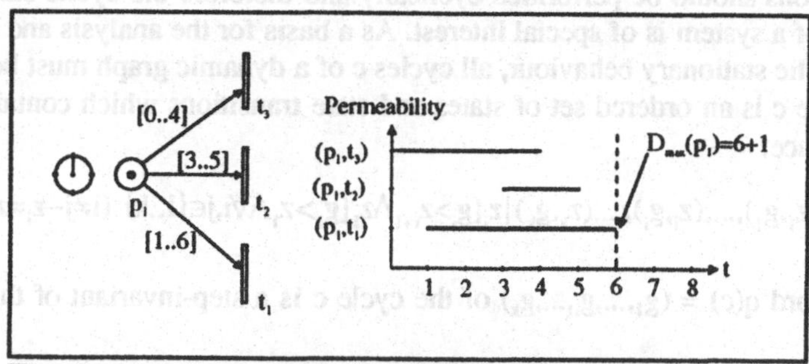

Fig. 5. Last change of the permeability by $ZB_l(p_1,t_1)$

Example

Fig. 5. shows a case where the maximal finite time inscription is given by $ZB_l(p_1,t_1)$. However, the arc is still permeable after 6 time units and becomes unpermeable one time unit later. After seven time units, all arcs are unpermeable. This situation can only be changed by a change of the marking of place p_1 which resets the clock to zero.

The method has some advantages over other timing concepts and analysis methods [9,12]:

1. A new state is generated only by firing of transitions. This reduces the number of states and state transitions compared with methods that generate a new state by any tick of a clock.
2. The concept does not require that all of the arcs $(P \times T) \cap F$ *must* be timed. We can also analyse models which include undelayed firing sequences. Such steps fire "at the same time" but in a causal order.
3. By the modification of the firing rule (16) we define an equivalence of all states with the same markings and clock positions greater than D_{max}. This additionally reduces the number of states and state transitions that must be computed and does not cause a loss of information. For a given initial state and a sequence of state transitions, we can always compute the real clock positions (even if they are greater than D_{max}) by applying equation (12) instead of (16) to compute the resulting clock positions and thereby "unfolding" the equivalent states. This can be performed without additional knowledge of the corresponding Petri net.

Any formal property which is known from the theory of Petri nets can be tested for a timed Petri net by means of the corresponding dynamic graph, but this is beyond the scope of this paper. In the following section, dynamic graphs will only be applied in the sense of control of the modeled system.

4 Performance Analysis and Optimization by Means of Dynamic Graphs

Obviously, any possible firing sequence of state transitions in the dynamic graph represents a discrete process. However, especially for production systems, the state transitions should be performed cyclically and therefore the cyclic stationary behaviour of a system is of special interest. As a basis for the analysis and evaluation of the cyclic stationary behaviour, all cycles c of a dynamic graph must be determined. A cycle c is an ordered set of states and state transitions which contains each state only once:

$$c = ((z_1,g_1),\dots,(z_i,g_i),\dots,(z_k,g_k) \mid z_i[g_i>z_{i+1} \wedge z_k[g_k>z_1 \wedge \forall i,j \in \{1..k\}\ (i \neq j \rightarrow z_i \neq z_j)). \tag{20}$$

The word $q(c) = (g_1,\dots,g_i,\dots,g_k)$ of the cycle c is a step-invariant of the Petri net:

$$\sum_{g \in q(c)} \Delta u^*(g) = \underline{0}\,. \tag{21}$$

The corresponding sum of all step delays $\Delta\tau$ is not zero but denotes the cycle time t_{Cycl} of cycle c:

$$t_{Cycl}(c) = \sum_{g \in q(c)} \Delta\tau(g)\,. \tag{22}$$

If a function B defines a value for the contribution of each transition $t \in T$ to an objective function (for instance the output of product):

$$B: T \rightarrow \mathbf{N} \tag{23}$$

the throughput of the system can be computed for each cycle:

$$d(c) = \frac{\sum_{t \in T} B(t)\ occ(t,c)}{t_{Cycl}(c)} \tag{24}$$

where occ(t,c) denotes how often t fires in cycle c.
The cyclic stationary behaviour of the system $\bar{q} = q(c)^\infty$ is the infinite iteration of the word q(c).
Of course, different cycles can be combined to another cyclic stationary behaviour (see Section 5.1).

Modeling of Control Decisions

The dynamic graph of a persistent Petri net model at most contains one cycle. Such a system cannot be influenced by an external control since all conflicts are resolved internally. Since the dynamic graph analysis is applied to compute control strategies, the models which describe the behaviour of the uncontrolled batch process and which are analysed must contain conflicts. Such conflicts represent decisions of a control (see Section 5). Let $T_s \subseteq T$ be the set of controlled transitions. To be controllable, each of this transitions must be in conflict with another controlled transition in at least one state:

$$\forall t \in T_s(\exists t' \in T_s, \exists z \in Z(t' \in cfl(t,z))) \tag{25}$$

where cfl(t,z) denotes the set of transitions which are in conflict with t in state z. A *control strategy* q_s is a word which only contains controlled transitions and therefore describes a sequence of conflict resolutions (control decisions).
If $T_s \subseteq T$ is the set of controllable transitions, the control strategy $q_s(q,T_s)$ for any process $q = (g_1,...,g_i,...,g_k)$ can be determined:

$$\begin{aligned} u^*(g_{is}) &:= u^*(g_i) \cap T_s \\ \Delta\tau(g_{is}) &= \Delta\tau(g_i). \end{aligned} \tag{26}$$

Modelling of Control Restrictions

Existing *restrictions* for the control strategy can be expressed by *facts*. Facts are transitions which never may be enabled. They have a special graphical symbol (see Section 5). Since facts must be dead, a control strategy must never reach states where facts are enabled. Hence, if the dynamic graph is computed to find out a suitable control strategy, no following states need to be computed when a state is reached where a fact is enabled. The results are smaller dynamic graphs and less effort for the computation of the cycles.

5 Application to Batch Process Control

The application of the above method is based on Petri net models of the batch production systems which must be developed by the chemical engineer, since the chemical engineer has the deepest knowledge of the process which is to be controlled. We will not discuss here how such models can be developed in a systematic way but we will concentrate on the benefit of dynamic graph analysis to determine suitable control strategies.

5.1 Control of a Multiproduct Plant

Fig. 6. shows the flowsheet of a part of a multiproduct batch plant. Product 1 is produced in reactor R1 and product 2 is produced in reactor R2. Both reactors use the metering tanks A and B which contain raw material A and B, respectively. The metering tank C for raw material C is only used by reactor R2.

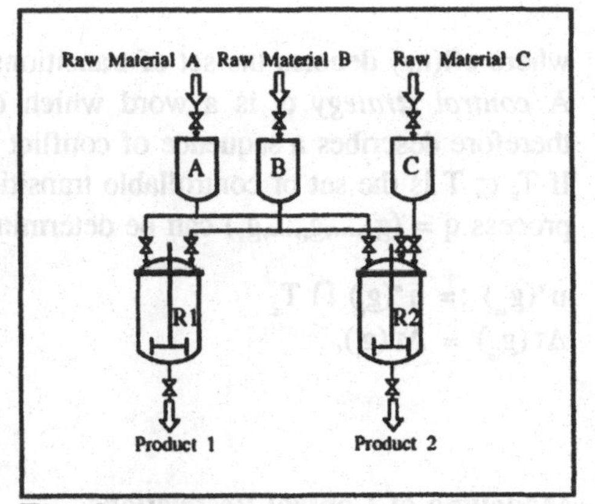

Fig. 6. Flowsheet

The production of product 1 is performed as follows:
At first, raw material A is transfered from metering tank A into reactor R1 (duration: one time unit). After this, raw material B from metering tank B is added (duration: one time unit). Then the reaction is started. When the reaction is completed, the product is discharged (duration: 4 time units), and the cycle starts anew.
For the production of product 2, at first raw material from metering tank C is transfered into reactor R2 (duration: one time unit), then raw material from metering tank B and then from metering tank A is added (duration: each one time unit). Then the reaction is performed, the product is discharged, and the cycle is started anew (duration: 4 time units).
The metering tanks A and B are shared resources and can be used either by reactor R1 or by reactor R2 but never by both reactors simultaneously.
Fig. 7. shows the Petri net model of the production system. The interpretation of the model is given in Table 1.. Places p_{13}, p_{23} and p_{25} describe that a production process is waiting for an available metering tank. These places must be included into the model to avoid deadlocks.
The initial state is given by the initial marking in Fig. 7.. Since all arcs connecting initially marked places with transitions have the time interval $[0..\infty]$, the local clocks of the marked places are always zero and the clocks need not to be graphically re-

presented.
The model contains structural conflicts between the transitions

- t_{11} and t_{25} and
- t_{13} and t_{23}.

These conflicts must be resolved by a control strategy which must be determined.
Fig. 8. shows a part of the dynamic graph of the net model of Fig. 7.. The complete dynamic graph consists of 47 nodes. The states are not given in detail since they are not necessary for the solution of the control problem. The dotted arcs and the inscriptions $q_{d1}(16)$ and $q_{d2}(6)$ represent sequences of states and state transitions without conflicts. The numbers in parentheses give the times of these two sequences. We can see that the dynamic graph has three cycles (see also Table 2.). These cycles describe elementary stationary control strategies.
Transition t_{15} represents the completion of one batch of product 1 whereas transition t_{27} models the completion of one batch of product 2.

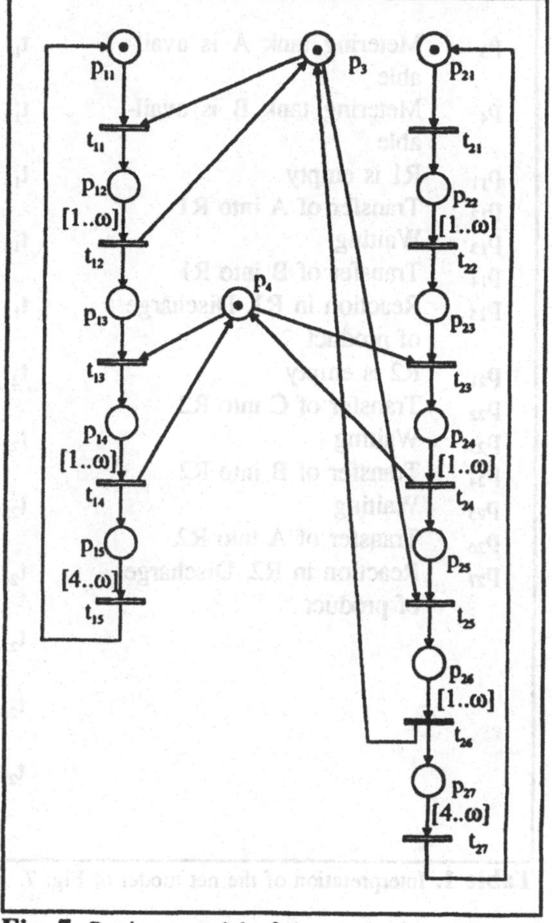

Fig. 7. Petri net model of the production system of Fig. 6.

Hence, $B(t_{15}) = B(t_{27}) = 1$, $B(t_j) = 0$ for all other transitions. We can compute the throughputs of product 1 and 2, respectively, for each cycle (see Table 2.). The corresponding firing sequences for each cycle can be seen from the dynamic graph (Fig. 8.):

$q(c_1) = (\{t_{11},t_{21}\}(0), \{t_{12},t_{22}\}(1), \{t_{23}\}(0), \{t_{24}\}(1), \{t_{13},t_{25}\}(0), \{t_{14},t_{26}\}(1), \{t_{15},t_{27}\}(4))$

$q(c_2) = (q_{d2}(6), \{t_{25}\}(0), \{t_{26}\}(1))$

$q(c_3) = (\{t_{11},t_{21}\}(0), \{t_{12},t_{22}\}(1), \{t_{13}\}(0), \{t_{14}\}(1), q_{d1}(16), q_{d2}(6), \{t_{11}\}(0), \{t_{12}\}(1), \{t_{13},t_{25}\}(0), \{t_{14},t_{26}\}(1), \{t_{15},t_{27}\}(4))$.

The strategies of cycles c_1 and c_2 resolve the conflicts between t_{11} and t_{25} and between t_{13} and t_{23} in favour of production process 2, cycle c_3 gives priority to production process 1. Consequently, the cycles c_1 and c_2 maximize the throughput of product 2 and the cycle c_3 maximizes the throughput of product 1 (see Table 2.). By combining the different cycles, any throughput of one product between the

p_3	Metering tank A is available	t_{11}	Start of transfer of A into R1
p_4	Metering tank B is available	t_{12}	End of transfer of A into R1
p_{11}	R1 is empty	t_{13}	Start of transfer of B into R1
p_{12}	Transfer of A into R1		
p_{13}	Waiting	t_{14}	End of transfer of B into R1
p_{14}	Transfer of B into R1		
p_{15}	Reaction in R1, Discharge of product	t_{15}	End of product discharge of R1
p_{21}	R2 is empty	t_{21}	Start of transfer of C into R2
p_{22}	Transfer of C into R2		
p_{23}	Waiting	t_{22}	End of transfer of C into R2
p_{24}	Transfer of B into R2		
p_{25}	Waiting	t_{23}	Start of transfer of B into R2
p_{26}	Transfer of A into R2		
p_{27}	Reaction in R2, Discharge of product	t_{24}	End of transfer of B into R2
		t_{25}	Start of transfer of A into R2
		t_{26}	End of transfer of A into R2
		t_{27}	End of product discharge of R2

Table 1. Interpretation of the net model of Fig. 7.

maximal and the minimal value from Table 2. can be realized. Finally, we require that a throughput of product 1 of 0,15 batches per time unit must be realized. A corresponding control strategy shall be determined.

We can see from Table 2. that cycle c_3 causes a production of 5 batches of product 1 in 30 time units and that cycles c_1 and c_2 cause a production of one batch in 7 time units. Let m be the number of realizations of cycle c_3 and n be the number of realizations of cycles c_1 or c_2. The ratio of m to n can be computed by means of the following equation:

$$0{,}15\frac{\text{Batches of Product 1}}{\text{Time Unit}} = \frac{(5m+n)\ \text{Batches of Product 1}}{(30m+7n)\ \text{Time Unit}}\ . \tag{27}$$

The solution is m:n=1:10. The total time for this strategy is 100 time units and the resulting throughput of product 2 is 0,14 batches per time unit.

The according firing sequence $\bar{q}$ begins with state z_0 and is given as follows: $\bar{q} = (q(c_3),q(c_1)^{10})^{\infty}$. The firing sequences $q(c_3)$ and $q(c_1)$ are given above.

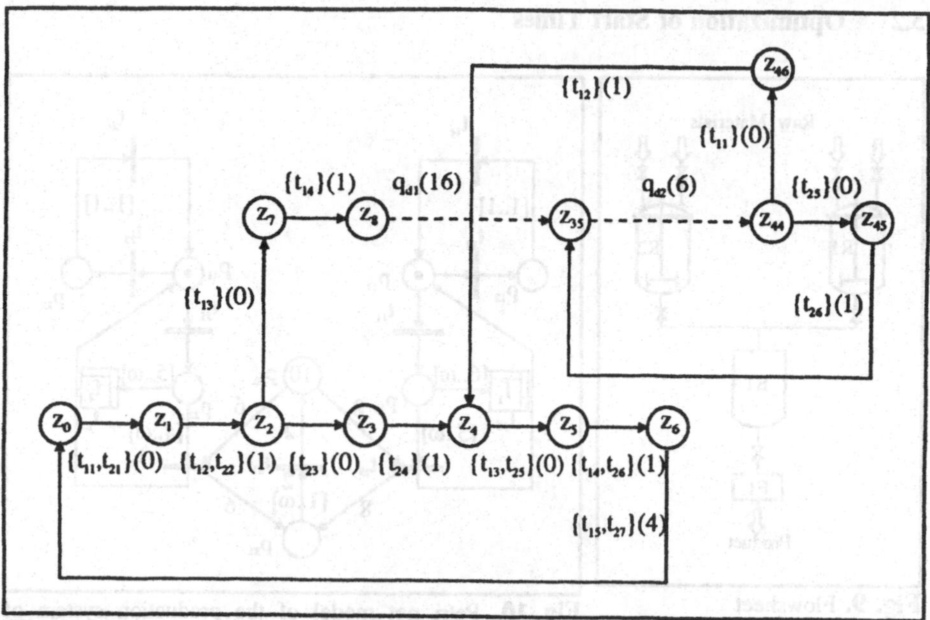

Fig. 8. Dynamic graph of the net model of Fig. 7.

No.	Nodes	t_{Cycl} [Time Units]	Occurrence of		Throughput [Batches/Time Unit]	
			t_{15}	t_{27}	Product 1	Product 2
1	$Z_0..Z_6$	7	1	1	0,143	0,143
2	$Z_{35}..Z_{45}$	7	1	1	0,143	0,143
3	$Z_0..Z_2$, $Z_7..Z_{44}$, Z_{46}, Z_4 .. Z_6	30	5	4	0,167	0,133

Table 2. Comparison of the cycles of the dynamic graph of Fig. 8.

The set of controllable transitions is $T_s = \{t_{11}, t_{13}, t_{23}, t_{25}\}$.
Thus, the corresponding control strategy is (see (26)):
$\bar{q}_s(q,T_s)$ = $(\{t_{11}\}(0), \{\}(1), \{t_{13}\}(0), \{\}(23), \{t_{11}\}(0), \{\}(1), \{t_{13},t_{25}\}(0), \{\}(5), (\{t_{11},t_{21}\}(0), \{\}(1), \{t_{23}\}(0), \{\}(1), \{t_{13},t_{25}\}(0), \{\}(5))^{10})^{\infty}$.
The empty steps reflect that these steps do not require control operations.

5.2 Optimization of Start Times

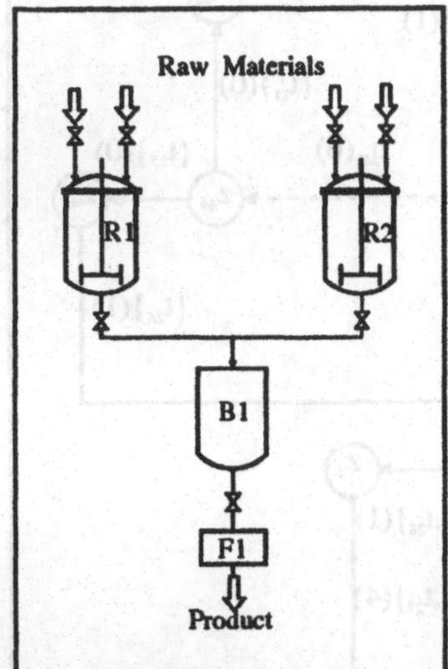

Fig. 9. Flowsheet

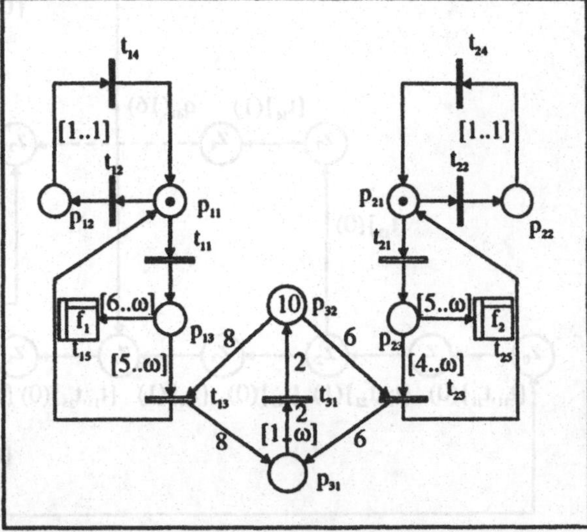

Fig. 10. Petri net model of the production system of Fig. 9.

Fig. 9. shows a plant for the production of a single product. It is produced in two batch reactors (R1 and R2) simultaneously. After completion of the reaction, the product is discharged from the reactors into a storage tank (B1) with a maximal storage capacity of 10 mass units. The product is separated continuously by means of a filter (F1) with a throughput of two mass units per time unit.

p_{11}	Reactor 1 is ready	t_{11}	Start of reaction process R1
p_{12}	Reactor 1 waits		
p_{13}	Reaction in reactor 1	t_{12}	Delay of reaction process R1
p_{21}	Reactor 2 is ready		
p_{22}	Reactor 2 waits	t_{13}	End of reaction process R1
p_{23}	Reaction in reactor 2	t_{21}	Delay of reaction process R2
p_{31}	Storage tank contains $m(p_{31})$ mass units of product	t_{22}	Waiting of reaction process R2
p_{32}	Storage tank has free storage capacity for $m(p_{32})$ mass units of product	t_{23}	End of reaction process R2
		t_{31}	Two mass units of product are separated

Table 3. Interpretation of the net model of Fig. 10.

The reaction processes in the two reactors have different durations (R1: 5 time units, R2: 4 time units) and produce different amounts of product (R1: 8 mass units,

R2: 6 mass units). The control has to ensure that the product in reactor R1 or R2, respectively, is discharged immediately after the completion of the reaction. Otherwise, the batch is unusable.
Fig. 10. shows the corresponding Petri net model. The interpretation is given in Table 3.. Control decisions consist of starting the reaction processes (t_{11}, t_{21}) or delaying the start by one time unit (t_{12}, t_{22}) and then decide anew on start or further delay. The technological restrictions described above are modeled by means of facts f_1 and f_2. A suitable control strategy must ensure that these facts are never enabled. The corresponding dynamic graph was computed by means of ATNA [13]. The result of an analysis of the dynamic graph is a word which describes a stationary behaviour of the system and which is permissible (both facts are never enabled) and optimal (t_{31} fires as often as possible). Fig. 11. shows such a behaviour. The corresponding word q(c) is:

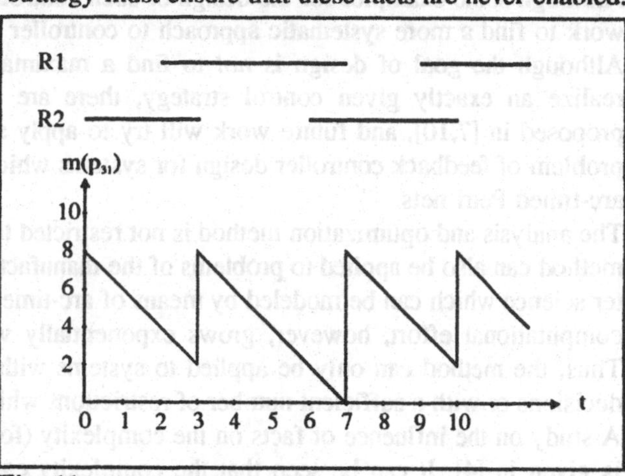

Fig. 11. Control strategy

$q(c) = ((\{t_{12}\}(0), \{t_{14},t_{31}\}(1))^2, \{t_{11}\}(0), \{t_{23},t_{31}\}(1), (\{t_{22}\}(0), \{t_{24},t_{31}\}(1))^3, \{t_{21}\}(0), \{t_{13},t_{31}\}(1)).$

The set of controlled Transitions is $T_s = \{t_{11}, t_{12}, t_{21}, t_{22}\}$.
The control strategy can be computed by means of (26):

$\bar{q}_s(q,T_s) = ((\{t_{12}\}(0), \{\}(1))^2, \{t_{11}\}(0), \{\}(1), (\{t_{22}\}(0), \{\}(1))^3, \{t_{21}\}(0), \{\}(1))^\infty.$

However, the initial state which is given in Fig. 10. is not reproduced by the stationary control strategy. Therefore, a control strategy for the startup phase must be found which ensures that the production is started as soon as possible. A suitable strategy (which was computed by means of ATNA) is to start with reaction process R2 since it takes less time than reaction process R1. This is realized by the word $q(p) = (\{t_{12},t_{21}\}(0), \{t_{14}\}(1), \{t_{12}\}(0), \{t_{14}\}(1), \{t_{11}\}(0), \{t_{23}\}(2), (\{t_{22}\}(0), \{t_{24},t_{31}\}(1))^2, \{t_{21}\}(0), \{t_{13},t_{31}\}(1))$. The duration of the startup strategy is 7 time units. The first product, however, is separated (t_{31} fires) after 5 time units.
The corresponding control strategy is not cyclic but time-optimal for the startup-phase:

$q_s(q,T_s) = (\{t_{12},t_{21}\}(0), \{\}(1), \{t_{12}\}(0), \{\}(1), \{t_{11}\}(0), \{\}(2), (\{t_{22}\}(0), \{\}(1))^2, \{t_{21}\}(0), \{\}(1)).$

One can simulate the token flow in the net of Fig. 10. to prove these results.

6. Conclusions

The result of the analysis of a batch production system by means of dynamic graphs is a suitable or optimal control strategy, given as a time-sequence of firings of controllable transitions. The strategy can be realized by a human operator or an automatic controller. In the second case, the controller must be designed based on the process model (the Petri net) and the control strategy which must be performed. Although some examples for the design of such controllers are given in [3], further work to find a more systematic approach to controller design is necessary.
Although the goal of design is *not* to find a maximal permissive control but to realize an exactly given control strategy, there are close relations to methods proposed in [7,10], and future work will try to apply some of these results to the problem of feedback controller design for systems which are modeled by means of arc-timed Petri nets.
The analysis and optimization method is not restricted to batch process control. The method can also be applied to problems of the manufacturing industry or of computer science which can be modeled by means of arc-timed Place/Transition nets. The computational effort, however, grows exponentially with the size of the model. Thus, the method can only be applied to systems with a small number of control decisions or with a sufficient number of restrictions which can be modeled by facts. A study on the influence of facts on the complexity (for the system of Section 5.2) is given in [4]. It can be seen that the complexity can be reduced drastically by facts, which represent technological knowledge of the process.
Another direction for further research is the problem of online control for disturbance rejection. The basic algorithms of ATNA [13] will be implemented on a process control system for a pilot batch plant to realize a decision support system for the operator. The basic idea is to fire the transitions in the model when the corresponding events in the real process occur. So the marking always represents the current state of the process. If a disturbance occurs, the behaviour of the system can be predicted and appropriate control operations can be performed to prevent dangerous states *before* the system reaches dangerous states. If the time horizon is not too tight, also control strategies for the minimization of losses caused by the disturbance can be computed.

Acknowledgement

The work is supported by the Deutsche Forschungsgemeinschaft under grant Ha 1886/1-1.
The author wishes to thank Prof. P. Starke for the cooperation in the development of ATNA.

References

[1] Hanisch, H.-M.: Dynamik von Koordinierungssteuerungen in diskontinuierlichen verfahrenstechnischen Systemen. at-Automatisierungstechnik 38 (1990), 399-405.

[2] Hanisch, H.-M.: Coordination control modelling in batch production systems by means of Petri nets. Computers and Chemical Engineering, 16 (1992), 1-10.

[3] Hanisch, H.-M.: Petri-Netze in der Verfahrenstechnik. Modellierung und Steuerung verfahrenstechnischer Systeme. R. Oldenbourg Verlag, 1992.

[4] Hanisch, H.-M.: Berechnung optimaler diskreter Koordinierungssteuerungen auf der Grundlage zeitbewerteter Petri-Netze. at-Automatisierungstechnik 40 (1992), 384-390.

[5] Helms, A.; Hanisch, H.-M.: Darstellung einer diskreten Steuerungsaufgabe für flexible Mehrproduktenanlagen mit Hilfe von Petri-Netzen. Chemische Technik 37 (1985), 236-239.

[6] Helms, A.; Hanisch, H.-M.; Stephan, K.: Steuerung von Chargenprozessen. Verlag Technik, Berlin, 1989.

[7] Holloway, L.E.; Krogh, B.H.: Synthesis of feedback control logic for a class of controlled Petri nets. IEEE Transactions on automatic control, Vol. 35, No. 5, May 1990, 514-523.

[8] Le Bail, J.; Alla, H.; David, R.: Hybrid Petri Nets. Proceedings of the European Control Conference, Hermes, Paris, 1991, 1472-1477.

[9] Quäck, L.: Aspekte der Modellierung und Realisierung der Steuerung technologischer Prozesse mit Petri-Netzen. at-Automatisierungstechnik 39 (1991), 116-120, 158-164.

[10] Ramadge, P.J.; Wonham, W.M.: The Control of Discrete Event Systems. Proceedings of the IEEE, Vol. 77, No.1, 1989, pp. 81-98.

[11] Ramchandani, C.: Analysis of asynchronuos concurrent systems by timed Petri nets. MIT, Project MAC, Technical Report 120, 1974.

[12] Starke, P.: Analyse von Petri-Netz-Modellen. B.G. Teubner, Stuttgart, 1990.

[13] Starke, P.: ATNA-Arc Timed Net Analyser. Petri Net Newsletter 37, Dezember 1990, 27-33.

[14] Thiemicke, K.; Hanisch, H.-M.: Prozeßanalyse einer diskontinuierlichen Anlage zur Kunstharzproduktion mit Petri-Netzen. msr, Berlin, 34 (1991), 416-419.

[15] Yamalidou, E.C.; Patsidou, E.P. and Kantor, J.: Modelling discrete-event systems for chemical process control - a survey of several new techniques. Computers and Chemical Engineering 14 (1990), 281-299.

[16] Yamalidou, E.C.; Kantor, J.: Modelling and optimal control of discrete-event chemical processes using Petri nets. Computers and Chemical Engineering 15 (1990), 503-519.

ON WELL-FORMED NETS AND OPTIMIZATIONS IN ENABLING TESTS(*)

Jean-Michel Ilié

MASI - CNRS UA 818
Université Pierre et Marie CURIE
4, place Jussieu - F75252 Paris, Cedex 05, France
Tel.: +33(1)44.27.31.92
Fax: +33(1)44.27.62.86

Omar Rojas

IUT - Feredico Rivero Palacio
Apartado postal 40347,
Caracas 1040,
Venezuela
Tel.: +58(2)69.18.81
Fax: +58(2)68.12.754

Abstract

Simulation techniques are frequently used to verify behavioural properties of complex systems. The tools which enhance these techniques have to cope with combinatorial explosion problems, even for small system specifications. Several optimizations have been proposed in order to perform exhaustive simulations. In well-formed nets, they may be complemented by optimized management of transition firings. We would like to propose a methodological approach to take into account dependency relations between tokens to avoid useless combinations. A simulation service has been implemented to enhance this work.

1. Introduction

Various and complementary validation techniques have been proposed to verify behavioural properties of complex systems. They include temporal logic model checkers, algebraic transformations of CSP and CCS, invariant methods about Petri nets and simulations. Nowadays, workshops highlight these techniques allowing designers to improve their productivities [1][7][14].

Simulation techniques are frequently used because they clearly bring out the behaviour of systems. Sequences of system computation steps (traces) give an immediate idea of certain behaviour while exhaustive simulation enables the verification of properties on the whole system state space. Simulations of discrete event systems have already been proposed from high level Petri net models [10][13] because they take parallelism and synchronization concepts into account. Moreover, coloration mechanisms allow one to distinguish instances of the same objet type by colouring tokens.

The simulation power is constrained by the combinatorial explosion of the state space of even small systems. To cope with these problems, several improvements have been proposed including reduced state space analysis in which reasoning about the complete state space is replaced by the analysis of reduced representations [11][12][15]. Recently, several efficient and optimized

(*) This work has been partially supported by the ECC EUREKA project IRENA (EU-389)

tools have been presented on a high level Petri net form, called well-formed nets (WN). In particular, symbolic simulation tools [3][5] take into account the ability of this model to exploit model symmetries by means of the concept of symbolic marking.

Colour management in high level Petri nets introduces new complexities with respect to the Petri net transition firing rules, because a transition may be enabled by several combinations of colours. However, we propose to enhance optimizations that make the enabling test more efficient. Trace or marking graph techniques may directly benefit from such optimizations. An efficient simulation service has been defined to enhance these enabling test optimizations.

This paper is organized as follows : section 2 recalls some useful basic WN concepts and states the needs of optimizations in the enabling test algorithms ; section 3 introduces definitions to manage the enabling test in an optimal way, then, a methodological approach is proposed ; section 4 proposes an optimized enabling test algorithm ; section 5 briefly introduces the simulation services ; section 6 demonstrates some experimental results; and the last section contains our concluding remarks.

2. The Enabling Test Complexity

The enabling test of a transition in a high level Petri net [10] [13] consists in taking into account the set of variables which are related to a transition and to its input places. First, each of these variables is instanciated with respect to their relative definition domains in order to apply the colour functions defined on input arcs. The input place marking must be large enough with respect to the marking returned by the colour functions. Predicates may reduce the definition domains of functions. As a formalism extension, an inhibitor arc may connect a place to a transition (instead of a classical arc) causing the firing of the transition to be blocked when the place marking is large enough [6].

The variable management leads to introduce a notion of "transition instances" which are defined by the cartesian product of the variable definition domains. Thus, any instance of the transition may be defined as a tuple of values. . The technique, that consists in testing the whole set of possible tuples, leads to a problem of complexity in space and time and it is as inconsistent as the variable domains are big.

We propose to optimize the enabling test by taking into account the markings of input places in order to determine as closely as possible the enabled instances. For this optimization, we need to define reverse functions from place to transition and to introduce a notion of "instance classes of a transition", for which only some components are defined. Indeed, each arc may be associated with only a subset of the variables which are bound to a transition and, therefore, the analysis of an input place may only determine values for such a subset. For example, the set of variables which are bound to the t transition in figure 2.1 are equal to <X,Y,U,V,T> and the (p1,t) arc is related to X and Y only. The marking analysis of p1 leads to the computation of tuples for which only the two first components are defined. Each of these tuples represents an instance class. The instance classes are completed by analysing the other input arcs. One has to take into account that a variable may appear several times in colour functions which causes dependencies between token values. In figure 2.1, X appears four times within the arcs adjacent with t. In fact, we may take advantage of these dependencies to reduce the combination problem.

The well-formed nets (WN) are original coloured Petri nets for which the colour management is well defined. The WN ables one to formalize the above optimizing approach, in particular, by defining reverse functions.

Unformally, colour domains in WN are tuples of basic colour domains called colour classes (i.e. finite sets of colours). Each colour class is possibly partitioned in static subclasses. Place and transition colour domains are defined as cartesian products of colour classes. Arc functions are expressed using the four following colour class functions : identity, successor, constant and diffusion (diffusion means that every colour of a colour class must be in a place). Several identity and successor functions may be defined by using different variables. Predicates may be used either bounding them to transitions or within colour function definitions.

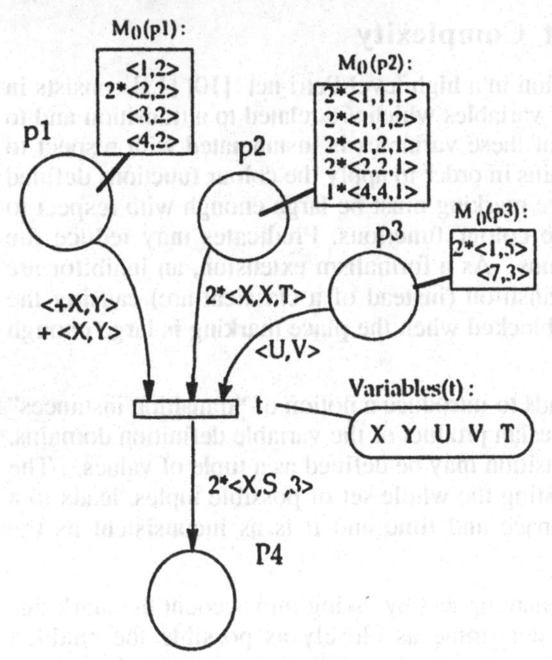

Figure 2.1 : A WN net and its colour functions.

<u>WN definitions :</u>

Colour classes :

C1 = C2 = 1..4,

C3 = 1..7.

Variables :

X: C1; Y : C2;

U, V, T : C3.

Place colour domains :

C(p1) = C1 x C2,

C(p2) = C1 x C1 x C3,

C(p3) = C3 x C3,

C(p4) = C1 x C1 x C3.

Colour functions :

W^-(p1,t)=<+X,Y>+<X,Y>,

W^-(p2,t) = 2*<X,X,T>,

W^-(p3,t) = <U,V>,

W^+(p4,t) = 2*<X,S,3>.

where X,Y,U,V,T are Identity functions, +X is a successor function, S represents a diffusion function.

The net of figure 2.1 is an example of WN with its M_0 initial marking. The colour definitions are to the right. It illustrates the management complexity of the WN caused by its high level expression power.

In WN, enabling test optimization techniques, which are based on an analysis of places, are available because identity and successor functions have simple inverse functions : respectively, identity and predecessor functions. Moreover, constant and diffusion functions may be viewed as existance tests of tokens because they do not induce specific colours for a related transition.

Practically, optimizations are not so easy to perform because colour functions are structured into linear combinations of tuples of colour class functions. The cardinality of each tuple is directly related to the colour domain of the connected place in order to define a colour class function for every component of the place tokens. For reasons of clarity, the function tuples will be called **elementary colour functions.**

In the following definitions, management of variables is enhanced due to their use in the present enabling test technique. The reader may consult [3] in order to have a complete definition of WN.

2.1. Definition of colour functions

Let t be a transition and C(t) its colour domain. Let s be a place, the colour domain of which is assumed to be C(s)=C1xC2...xCm. Let VAR = {X1, X2, ..., Xn} be a set of distinct variables such that every variable represents a colour class of C(s). (However, several variables of Var may represent the same colour class).

An elementary colour function , ef, from C(t) to Bag(C(s)) is defined as follows :

$ef(Var) = \langle bf_1(Y1), bf_2(Y2), ..., bf_m(Ym)\rangle$

where Var is a subset of VAR,

bf_i is a colour class function of Ci : Identity, Successor, Constant or Diffusion;

$Yi \in Var$ ($i = 1..m$), iff bf_i is an identity or a successor function, otherwise it is undefined. ♦

Thus, a **colour function** labelling an arc may be expressed as :

$F = \sum ef.coef \langle ef\rangle\ ef.pred$,

where ef.coef is a coefficient and ef.pred is a predicate function. ♦

Example :

The colour functions of the net of figure 2.1 are now expressed in a new way to take into account the above definitions.

$W^-(p1,t)=ef_1(X,Y)+ ef2(X,Y)$, with : $ef_1(X,Y)= \langle +X,Y\rangle$,

$W^-(p2,t) = 2*ef_3(X,T)$, $ef2(X,Y)= \langle X,Y\rangle$,

$W^-(p3,t) = ef_4(U,V)$ $ef_3(X,T) = \langle X,X,T\rangle$,

$W^+(p4,t) = 2*ef_5(X)$, $ef_4(U,V) = \langle U,V\rangle$,

$ef_5(X) = \langle X,S,3\rangle$.

The set of variables bound with t is **Variables(t)** $= \bigcup_{i=1..n} (Var_i)$, where $ef(Var_1)$, ... $ef(Var_n)$ are the elementary colour functions which are associated with t. The variables of this set are always considered in the same order during the analysis. For reasons of clarity, we assume in this paper that any variable of Variable(t) appears at least once on the input arcs of t. In general, some variables may be quoted on output arcs only but such variables do not act in enabling tests.

The computation of enabled instances has to take into account that several adjacent colour functions may take as input the same colour. This is due to the ability to use the same variable name for several functions. As a consequence, one must verify **dependencies between token sets.**

2.2. Token dependencies

Let t a transition, and let $bf_1(Y1)$ and $bf_2(Y2)$ two colour class functions associated with t.

> A token dependency is caused by (bf1,bf2) iff Y1=Y2. ♦
>
> We distinguish three levels of token dependency :
>
> (0) **Intra-token dependency** : a variable is used several times in a same elementary colour function,
>
> (1) **Intra-place token dependency** : a variable is used several times in a same colour function but in distinct elementary colour functions,
>
> (2) **Inter-place token dependency** : a variable is used in distinct colour functions.

Example

In the previous example of well-formed functions (see section 2.1), there are the following token dependencies:

- ef1(X,Y) and ef2(X,Y) causes an intra-place token dependency w.r.t. X and Y,
- ef3(X,T) causes an intra-token dependency w.r.t. X,
- W^-(p1,t) and W^-(p2,t) causes an inter-place token dependency w.r.t. X.

Table 2.1 presents the six variable instances which are enabled with respect to the current marking and the token dependencies. For each i instance, i•mk(p,ef) defines the colours of a "p" place which is associated with "i" by applying an ef function. Hence, ef•coef is the number of i•mk(p,ef) tokens which may be taken in p.

Instances (t)	Associated tokens			
	place "p1"		place "p2"	place "p3"
Variables(t) = {X,Y,U,V,T}	i1•mk(p1,ef1) (ef1•coef = 1)	i1•mk(p1,ef2) (ef2•coef = 1)	i1•mk(p2,ef3) (ef3•coef = 2)	i1•mk(p3,ef4) (ef4•coef = 1)
i1 =<1,2,1,5,1>	<2,2>	<1,2>	<1,1,1>	<1,5>
i2 =<1,2,7,3,1>	<2,2>	<1,2>	<1,1,1>	<7,3>
i3 =<1,2,1,5,2>	<2,2>	<1,2>	<1,1,2>	<1,5>
i4 =<1,2,7,3,2>	<2,2>	<1,2>	<1,1,2>	<7,3>
i5 =<2,2,1,5,1>	<3,2>	<2,2>	<2,2,1>	<1,5>
i6 =<2,2,7,3,1>	<3,2>	<2,2>	<2,2,1>	<7,3>

Table 2.1 : Variable instances and their associated marking sets

2.3. Remark on symbolic markings and tools

Symbolic markings have been proposed in WN, either for reachable marking graph techniques [3] or for trace techniques [5]. The main idea is to represent classes of tokens instead of tokens themselves by exploiting model symmetries. The goal is to consider the least number of tokens in order to restrain combinatorial explosion problems as much as possible. Moreover, trace techniques may profit of the symbolic representation because different concurrently enabled instances are related to the same symbolic marking. In this case, only one of these instances must be considered due to the fact that they lead to the same behaviour.

The representation of a symbolic marking is very close to the classical view due to the fact that each symbolic token is built as a tuple associated with a coefficient. The difference is due to the fact that each tuple's component refers to a subset of component values, namely dynamic subclasses, instead of being a single value. Each dynamic subclass refers itself to a static subclass of a class and to its cardinality. Enabling test must take the cardinality values into account in order to know the effective quantity of a token . For example, a symbolic token, $<Z_1,Z_2>$ where Z_1 and Z_2 are two dynamic subclasses of respective cardinalities 2 and 3, represents 6 tokens.

The four following stages are performed to fire a transition from a symbolic marking :
- enabling test of the transition,
- splitting of the dynamic subclasses in order to bring out the tokens which are associated with one chosen enabled instance,
- firing of the transition,
- grouping of dynamic subclasses in order to represent the new symbolic marking in a minimal form.

In the next section, enabling test optimization techniques are demonstrated. They may be applied starting from both marking and symbolic marking due to their similarities. For reasons of symplicity, classical markings are considered.

3. Optimizations techniques

Our method to search a convenient optimization technique is expressed in five stages :

(1) To minimize the number of analyses with regards to the place markings; indeed, such well-formed net elements highly influence the enabling test cost due to their classical representation by dynamical structures. (2) To deal with dependency relations between tokens as early as possible in order not to compute useless variable instances, and to partition the independent places in order to replace the enabling test of a transition by smaller enabling tests. (3) To propose heuristic techniques to schedule the places and the different kinds of WN elements which act in place analysis. (4) To stock and update with each treated transition the resulting set of its variable instances, and to

improve the internal data representation. (5) To transform the net structurally in order to take better advantage of the above techniques.

3.1. Optimizing of place marking analysis

The analysis of input places is not independent of others according to the inter-place token dependencies. So, any algorithm which deals with places independently, may perform a set of instance classes for each place and must perform useless combinations of them to achieve the transition instances. We have chosen to take into account the inter-place token dependency relation as early as possible. Therefore, each input place is analysed exhaustively before dealing with another. Thus it is possible to analyse a new place, eliminating any previously computed instance class which denies the inter-place token dependency relation.

Since in WN, a colour function is expressed as a linear combination of elementary functions, this technique must be refined in dealing with each elementary coloured function separately.

In each arc, a variable previously treated may appear again, requiring a verification of intra-place and inter-place token dependencies for this variable. Our choice is to deal with these token dependencies as early as possible again. In other words, the set of variables which is associated with an arc is dynamically split into two current subsets : treated and untreated variables. A treated variable is distinguished from the others by the fact that it has an assigned value in each computed "instance class".

A new * symbol is now introduced to express un-defined components in colours. Then, classes of instances and classes of tokens are defined formally as having * components.

3.1.1. Augmented colour classes

Let C be any WN class and let * be a symbol which does not belong to any WN class.

An augmented colour class for C is defined by C^* such as $C^*=C\cup\{*\}$. ♦

In the following, * denotes an undefined value either for a variable or for a colour class function.

3.1.2. Augmented colour domains

Let C1 , C2 ..., Cn be WN colour classes.

The augmented colour domain of C1xC2...xCn is defined as $C1^*xC2^*...xCn^*$ ♦

An instance class ic of t denotes an element of the augmented colour domain of C(t).

A token class tc of p denotes an element of the augmented colour domain of C(p).

According to the splitting of the transition's variables, into treated and untreated parts, two kinds of computations must be performed with each place : for a previously treated variable, one has to verify that the marking is sufficient in the respective place and, for an untreated variable, new instance classes may be computed from the current place. The correspondance between the current marking to be analysed and the current set of instance classes is made, on one hand, by using the following constructions : projections from transition to place, namely direct-projections, are used in case of treated variables to compute a token class that will be compared, using a © operator, with each token in current place. Thus, unsuited instance classes are removed, causing only the suited ones to be submitted to further combination operations. Projections from place to transition, namely retro-projections, are used to perform inverse colour class functions for each occurrence of a untreated variable and to apply the ⊗ operator in order to define a new instance class. Both © and ⊗ operators take into account the intra-token dependencies resulting from the fact that a variable may occur several times in a projection function.

We now define these new operators and projection functions.

3.1.3. Augmented class operators (⊗,©)

Let C be a colour class.

©: C*x C → {True, False} such that : x©y=True iff x=y or x=*.
⊗ : Cx C → C such that : x ⊗ x = x,
x ⊗ y is undefined iff x≠y.♦

3.1.4. Augmented colour domain operators

Let C1x C2...x Cn be a colour domain and x any coulour in it. Let $C1^*$x $C2^*$...x Cn^* be the augmented colour domain of C1x C2...x Cn, and let y be any colour of it.

Let x be any colour of a $C1^*$x $C2^*$...x Cn^* colour domain and let y be any colour of C1x C2...x Cn.

- x © y is true iff $x]_i$ © $y]_i$ is true for any i of 1..n.

($\alpha]_i$ stands for the ith component of any α compounded colour)

- x ⊗ y is defined by z iff $x]_i$ ⊗ $y]_i$ is defined, for any i of 1..n .
  In this case, $z]_i = x]_i \otimes y]_i$.♦

Moreover, for any set $\{f_1, \ldots, f_g\}$ of colour functions, $(f_1 \otimes \ldots \otimes f_g)(x)$ means $f_1(x) \otimes \ldots \otimes f_g(x)$.

3.1.5. Direct (>) and retro (<) projections on a variable

Let t be a transition such that : Variable(t) = <X1, X2, ... , Xn>.
Let ef(Var) be a WN elementary colour function associated with t such that : ef(Var)= $<bf_1(Y1), bf_2(Y2),\ldots,bf_m(Ym)>$ where Y1,...,Ym ∈ Variable(t).

Let us focus on any X variable of Var such that $j_1 \bullet\bullet j_k$ is its associated set of subscripts with respect to the ef components, such that : $Yj_g = X$ and $j_g \subseteq \{1\bullet\bullet m\}$ with $g = 1\bullet\bullet k$.

A direct-projection of ef with respect to a variable
X of Var is defined as follows :

$$(ef]X)^{>} = <f_1(Y1), f_2(Y2), \ldots, f_m(Ym)>$$
$$\text{with } f_j = bf_j \text{ iff } Yj = X \text{ , otherwise } f_j = * \quad \blacklozenge$$
$$j = 1\bullet\bullet m$$

A retro-projection of ef with respect to a variable
X of Var is defined as follows :

- $(ef]X)^{<} = <rf_1, rf2, \ldots, rf_n>$
  with $rf_i = (bf_{j_1})^{-1} \otimes \ldots \otimes (bf_{j_k})^{-1}$ iff $X = X_i$ otherwise $rf_j = *$
  $i = 1\bullet\bullet n$
- $(ef]X)^{<}$ is defined iff rf_i is defined. ♦

Example

Let us consider the p1 and p2 places in figure 2.1 in order to apply some projections on the X variable.

- $(ef_1]x)^{<}$ $(<2,2>) = (<\text{-}X,*>)\ (<2,2>) = <1,*,*,*,*>$ $(\Rightarrow X=1)$
- $(ef_3]x)^{<}$ $(<1,2,1>) = (<X,X,*>)\ (<1,2,1>)$
  $= <1 \otimes 2,*,*,*,*>$ is undefined.
- Finally, let us assumed that, at some step of the enabling test algorithm, we have an instance class represented by X=1, Y=2 . Then :

  $(ef_3]x)^{>}$ $(<1,2,*,*,*>) = (<X,X,*>)\ (<1,2,*,*,*>) = <1,1,*>$

One may note that two p2 tokens give a true value by using the © operator :
<1,1,*>©<1,1,1>=TRUE and <1,1,*>©<1,1,2> = TRUE
(The three other p2 tokens yield false values).

3.2. Input place partitions

The input places of a transition are partitioned, applying an operation of transitive closure from the inter-place token dependency relation.

Equivalence classes of places
Let (p1,t) and (p2,t) be two arcs of a transition $t \in T$.
p1 and p2 are in the same equivalence class iff the colour class functions associated with these arcs cause some inter-place token dependencies. ♦

As a primary consequence, computations from places which are not in the same equivalence class are independent due to the fact that they do not deal with the same variables. Thus, instance classes are performed from each equivalence class and the final result is their combination. As a second consequence, computations in the same equivalent class are dependent. Hence, a dynamical splitting of variables (see section 4.1) is made within each equivalent class.

As a particular case, two subsets of input places which are not associated with any variable are managed separately : on one hand, coloured places which are related to a transition with constant or diffusion functions only, and on the other hand, classical Petri net places. Classical Petri nets are accepted due to their ability to be understood with respect to a neutral colour class. Clearly, these subsets must be considered first since they lead to inexpensive token tests (with the data structure we adopted).

Moreover, Coloured inhibitor arcs have been proposed as a formalism extension [5][6]. Since they act in enabling tests, we take them into account in each partition. Classical arcs are analysed before inhibitor arcs due to the fact that inhibitions are related to absence of marking.

Example

In the well-formed net of figure 2.1, the inter-token dependency relation causes two input place partitions to be built :
PART1 = {P1,P2} is associated with the variables : X and Y,
PART2 = {P3} is associated with the variable : T.

Let us consider the PART1 partition in order to illustrate the construction of instance classes. The algorithm starts with P1 as it is the least marked place. us assume that ef1 is the first function to be analysed.

• P1, ef_1 :

As X and Y are untreated, the following retro-projections are applied :
$(ef_1]X)^< = $ <-X,*> and $(ef_1]Y)^< = $ <*,Y>

By applying these functions to every token of p1, the first set of instance classes is as follows :

Instance classes (PART1)	PART1•Var= {X,Y,T}	associated tokens (p1)
ic1	<0, 2, *>	<1,2>
ic2	<1, 2, *>	<2,2>
ic3	<2, 2, *>	<3,2>
ic4	<3, 2, *>	<4,2>

Table 3.1 : Instance classes and their associated tokens

• P1, ef_2 :

As X and Y are treated, the direct-projections on X and Y result in :

Instance classes (PART1)	PART1•Var= {X,Y,T}	associated tokens (p1)
ic2	<1, 2, * >	<1,2>+<2,2>
ic3	<2, 2, * >	<2,2>+<3,2>
ic4	<3, 2, * >	<3,2>+<4,2>

Table 3.2 : Instance classes and their associated tokens

Because of the intra-place token dependency management, ic1 has disappeared because there is no token in p1 which corresponds to the <0,*> token class.

Then, p2 is analysed :

• P2, ef_3 :

The following projections are applied : $(ef_3]X)^{>} = \langle X,X,* \rangle$

$(ef_3]T)^{<} = \langle *,*,T \rangle$

Instance classes (PART1)	PART1•Var = {X,Y,T}	associated tokens (p1)	associated tokens (p2)
ic2'	<1,2, 1>	<1,2>+<2,2>	2*<1,1,1>
ic2"	<1, 2, 2>	<1,2>+<2,2>	2*<1,1,2>
ic3'	<2, 2, 1>	<2,2>+<3,2>	2*<2,2,1>

Table 3.3 : Instance classes and their associated tokens

In table 3.3, ic2 has been replaced by ic2' and ic2" because there are two tokens in the p2 place which match with the <1,1,*> token class, obtained with X=1.

Moreover, there are not enough tokens in the p2 place corresponding to the f3.coef coefficient with X=4. Therefore, ic4 does not produce any new instance. As the value X=4 has been computed from the study of another arc, this removal is a direct consequence of the inter-place token dependency.

After considering the PART2 partition similarly to PART1, the combination of their instance class sets is performed. Then, the instances represented in table 2.1 are obtained.

3.3. Heuristic techniques

The fact that the algorithm deals with each place separately leads us to propose heuristic techniques to schedule the places and the different kinds of net elements which act in place analysis : elementary colour functions, variables, tokens and instance classes. The goal is to dynamically schedule the input places of each equivalence class in order to always choose the place which causes the smallest set of transition instance classes. Several features must be taken into account :

- **the number of tokens in places,**
- **the three-level dependencies and their respective numbers.**
- **the number of tokens that must be selected from a place through a colour function.**

The relevance of these features depends on the net structure. Of course, the combination operation is as expensive as the marking is large. However, the outcome of a marking size is tenuous in the face of high levels of dependencies and large numbers of selected tokens. Moreover, when there is no expression of token dependency, the result is that a place scheduling technique is quite useless.

Faced with this conflicting situation, we have chosen only to schedule the input places of each equivalence class such that the place having the smallest number of tokens is analysed first. This implementation may change by giving weight to places according to another feature.

3.4. Stock techniques

Since a transition may be tested several times within a simulation session, it is interesting to stock the resulting set of its instances with each treated transition. A boolean function is associated with each transition to indicate whether this set is correct (i.e. the input place marking is unchanged).

Finally, the internal data representation such as instance classes or marking may also be improved. These very technical features are not reported in this paper.

3.5. Structural transformations

WN structures may be transformed in order to make the WN analysis easier [4]. In particular, predicates on WN arcs may be avoided duplicating transitions and updating transition predicates. This transformation is interesting, despite of the transition number increasing, due to the fact that predicates on arcs leads heavy magement of instances. Indeed, colour functions must be taken into account, only for the instances which yield true values.

4. An enabling test algorithm

When the simulation kernel is requested for a net simulation, it generates and stores the data structures for what concerns the net topology and its management. Partitions of places are computed according to the sets of variables which are bound to transitions. A boolean variable, t.InstanceDone, is associated with each t transition in order to know whether the transition instances are performed.

This section presents the "EnablingTest" function and its main function, "ElemFctAnalysis".

4.1 Enabling test procedure

The "EnablingTest" function is called with a t transition and a PLACES subset of t input places according to the proposed place partition. This procedure performs a t.IC set of instance classes with respect to the different p places of PLACES, their marking, p.marking and to the type of arcs. Each instance class is associated with a set of token references in order to know immediately what marking is selected in each place of PLACES. Finally, a cartesian product is performed to obtain the transition instances. ELEM is the current set of untreated elementary functions for a specific arc; ef.var represents the variables which are associated with any elementary current variable, ef ; TREATED is a current set of variables boolean indicating whether a variable is currently

treated. $IC_{[p]}$ temporarily stocks the current t.IC set of instance classes before analysing an inhibitor arc.

```
Function EnablingTest (t, PLACES)                         [see remark]
If t.InstanceDone =False Then

   ForAll v of ef.var do TREATED[v] = FALSE
   Init_Instances_WithStar(t.IC)   (t.IC is initialized with a single
                                    instance class made of '*')
   While PLACES ≠ Ø and t.IC≠ Ø do     (loop on the places of PLACES)
      If ArcType(p,t)=Inhibitor Then IC=t.IC
      p = ChoicePlace (PLACES) ; PLACES \= {p} (choice of a p place)
      ELEM = ElemFn (p,t)
      While ELEM ≠ Ø do (loop on the elem. colour functions of (p,t))

         ef = ChoiceElem (ELEM ) ; ELEM \= {ef}     (choice of an ef)

         t.IC = ElemFctAnalysis (ef, t.IC, p.marking, TREATED)
      EndWhile   (loop on the elementary colour functions )
      If ArcType(p,t)=Inhibitor Then AnalysisInhibitor(IC,t.IC)
   EndWhile    (loop on the input places)
   AnalysisPredicate(t.pred,t.IC)
   t.InstanceDone=True
End If
End EnablingTest
```

Remark

Seven internal functions are called : (1) "<u>Init_Instances_WithStar</u>", initializes a set of instance classes with a single augmented token only constructed with the * symbol ; "<u>ElemFn</u>" gives the elementary colour functions associated with an arc ; (2,3) "<u>ChoicePlace</u>" and "<u>ChoiceElem</u>", allow one to choose an element, respectively, within a set of places and within a set of elementary colour functions ; (4) "<u>ElemFctAnalysis</u>" aims at computing a set of instance classes, t.IC, for every elementary colour function of a current place ; (5) "<u>ArcType</u>" boolean function yielding a different value for each type of arc : "Classical", "Inhibitor" ; (6) "<u>AnalysisInhibitor</u>" performes the new set of instances according to the instance set obtained by an inhibitor arc analysis ; (7) "<u>Predicate_Analysis</u>", boolean function which tests a predicate for every instance class of a transition. The wrong instance classes are removed.

4.2 Elementary colour function analysis

The "ElemFctAnalysis" function will now be present (see below). It loops over each instance class, ic, of t.IC and each marking, m, of p.Marking. For each elementary colour function, ef, a new set of instance classes, $IC_{[ef]}$, is built up to finally update t.IC.

For each current (ic,m) pair, only one analysis is performed, dealing with all the variables associated with the elementary colour function and with the token reference set of ic, ic.marking. In other words, a new instance class, ic',

is created from a copy of ic and the (ic',m) pair is tested in relation to the direct- and the retro-projection fonctions.

$IC_{[ic]}$ temporarily stocks the suitable instance class subset generated from an ic instance class of t.IC. With the $IC_{[ic]}$ set, it is possible to verify whether an instance class has yet been computed for another token. Such a verification is needed because several (ic',m) pair may lead to a same instance class. Thus, the computation of an ic' which is already a member of $IC_{[ic]}$, induces that a previous computation yields ic' such that a token reference is bound to it. In that case, it is necessary to test the existance of diffusion functions before adding another m reference to ic'.marking.

```
Function ElemFctAnalysis (ef, t•IC, p•marking, treated)     [see remark]
IC[ef] = Ø
ForAll ic of t.IC    do
 (loop on the instances of t)
 ic'= ic
 IC[ic] = Ø
 ForAll m of p.Marking do       (loop on the tokens of the p place)
   ForAll v of ef.variables while (ic'≠null) do  (loop on the variables of ef)
                    If TREATED[v]= FALSE
                         Then ic'= Direct_Projection_Analysis (v, m, ic', ef)
                         Else  ic'= Retro_Projection_Analysis (v, m, ic', ef)
   EndForAll

   If (ic' ≠ null)                    (there is an effective ic')
      And Constants_Analysis (m,ef) = TRUE
      And Verif_Token_Quantity (ic.marking_ref, ef.coef, m)
      Then
         If ic' ∉ IC[ic]
            then   (ic' is new and its marking references must be created)
                 Copy_Refered_Tokens(ic,ic') ;
                 Add_Selected_tokens (ic', ef.coef, p, m)

                 IC[ic] ∪= ic'
            Else    (the marking references of ic' must be updated)
                 If Diffusion_present(ef)
                    Then Add_Selected_tokens (ic', ef.coef, p, m)
         EndIf
   EndIf
   EndForAll          (loop on tokens)

   IC[ef] ∪= IC[ic]
 EndForAll       (loop on instance classes)
 For All ic of IC[ef] do
    If Not Diffusion_Analysis(ef,ic.marking) Then IC[ef] \= {ic}
 t.IC = IC[ef]
 ForAll  v of ef.variables  Do TREATED[v] = TRUE
End ElemFctAnalysis
```

Remark :

Nine functions procedures are called : (1) "Direct Projection Analysis", performs a direct-projection $(ef_{[V})^{>}$ from an ic instance class. If $ef_{[V}(ic)$ © m = true with respect to an m token then the final result is ic, else null; (2) "Retro Projection Analysis", performs a retro-projection $(ef_{[V})^{<}$ from an m token. If $(ef_{[V})^{<}(m)$ is defined, then the final result is ic ⊗ $(ef_{[V})^{<}(m)$, else null ; (3) "Constants Analysis", verifies whether the constant functions of an elementary colour function are verified with respect to a specific token ; (4) "Find_Instance" tests whether an ic instance class is in some set ; in this case, the final result is ic, otherwise null ; (5) "Verif TokenQuantity", verifies whether the quantity of a token is enough with respect to the coefficient of an elementary colour function and to the quantity taken by the previously studied elementary colour functions (if there are any) ; (6) "Add Selected tokens", adds a token reference in a token reference set which is associated with an instance class ; (7) "Copy Refered Tokens", copies the associated token references of an instance class to initialize another one ; (8) "Diffusion present", boolean function which returns true if there is some diffusion function in the current elementary colour function, otherwise false ; (9) "Diffusion Analysis", boolean function which returns true if ic.marking is enough according to the diffusion functions;

5. The Simulation Services

Several services have been defined in order to manage a well-formed net structure and to perform optimized simulations. Figure 5.1 illustrates such services.

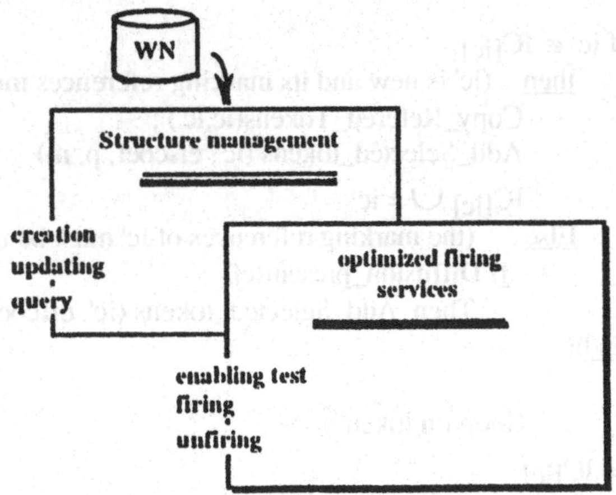

Figure 5.1 : The well-formed net simulation services

With creation primitives, an internal structure of a well-formed net may be set : classes, domains, places, transitions, arcs ... . Net information is assumed to be valid since a well-formed net compiler may be coupled with

application managing simulations. There are two ways to introduce a net, either from a primitive which reads a text modelling a well-formed net or directly from more explicit primitives like createPlace, createTransition and so on.

Updating primitives allow one to remove or add any net element as well as to update attributes like a place marking, an arc colour function or a transition priority value.

With query services, information about structural or computed data may be known. For instance, one may ask for the current marking or the set of computed transition's colours as well as for the name of a transition or for the components of a colour domain.

With firing services, a transition may be tested and fired. Classically, this transition must be enabled and have the highest priority value among the set of enabled transitions. An undeterministic choice is performed if several transitions are enabled with respect to a current marking. The study may be constrained to a subset of transitions. It is possible to perform only an enabling test . One may also suppress the marking modifications caused by a transition firing, in order to return to a previous state ("unfiring primitive").

The net of figure 5.2 illustrates the implementation of our simulation services in the AMI environment [1][8]. This environment is an Interactive Modeling CASE which handles multiple specification formalisms. It has been created to manage graphs based on any kind of graph formalism (WN for instance).

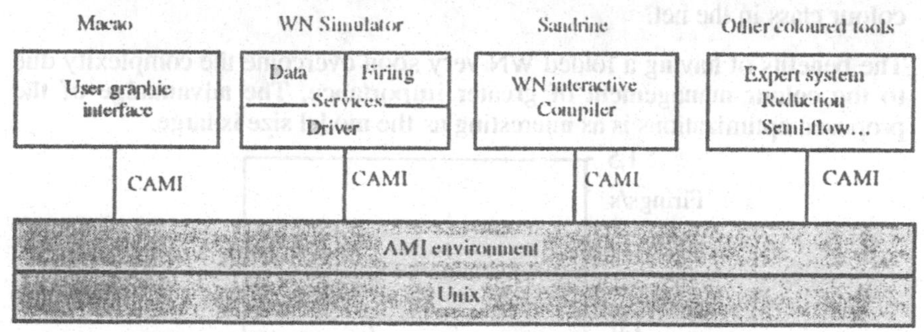

Figure 5.2 : The AMI environnement for simulations

External management in AMI makes the integration of new formalisms and new tools easier. This is particularly interesting because of the rapid evolution of the theory in formal specification areas. AMI also manages communication between the implemented tools, taking into account concurrency and cooperation problems. Every application communicates with others using a common language, CAMI, expressing queries and responses. The AMI-design enables one to profit from the whole services that are offered by the implemented application set. In particular, any customer that wishes to simulate a WN or a Petri net can use the graphic editor, MACAO. The simulation driver treats the requests made by the customer and uses a WN compiler in order to validate WN specifications [2]. Finally, it returns simulation results in terms of the MACAO interface. Tools can be exploited by several users or group users.

6. Some experimental results

In the context of this paper, we want to demonstrate the efficiency of considering dependency relations as earlier as possible. One way is to compare a WN to an equivalent unfolded Petri net. This last net is built up, according to some WN, with one transition for each <transition,instance> pair and one place for each <place, colour> pair. Intuitively, this unfolding makes the transittion enabling test easier, however, there are two major drawbacks : the obvious one is the size of the Petri net which increases with respect to the colour domain cardinalities, the second drawback is that every possible instance must be tested due to its structural representation.

The diagram of figure 6.1 summarizes experimental results obtained from models of the well-known distributed data base specification with multiple copies (see [8] for example). Since we do not have a tool which unfolds a WN net automatically, we have chosen to partially unfold the net in order to produce comparisons. The same engine is run for the WN and for the associated partially unfolded net in a SPARCstation SLC. It is run from the same initial marking and using the same transition firing sequence. In both cases, the list of all enabled instances is systematically updated and we measure the CPU-time needed to perform one given transition sequence. The experiment is performed several times increasing the cardinality of one given colour class in the net.

The benefits of having a folded WN very soon overcome the complexity due to the colour management of greater importance. The advantadge of the proposed optimizations is as interesting as the model size is large.

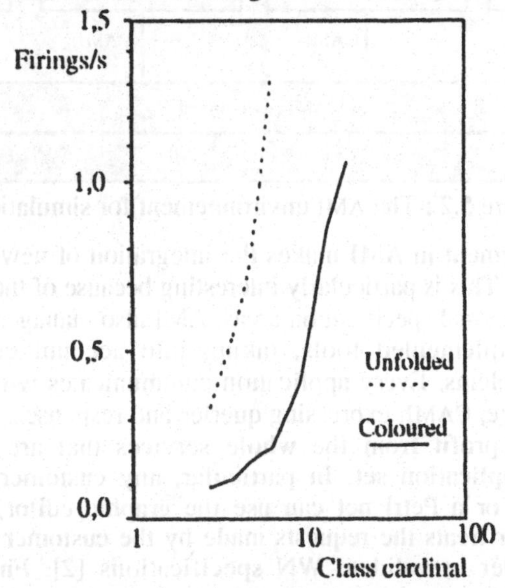

figure 6.1

7. Conclusion

We have presented a simulation service based on well-formed nets. In these models, colour functions are well-defined, which allows one to deal with the complexity due to colour management. In particular, optimized solutions were able to be demonstrated because of the existance of some reversable functions and of dependency relations between tokens.

Trace or marking graph tools (even symbolic) may directly benefit from such optimizations. Classical Petri nets are accepted due to their ability to be understood with respect to a neutral colour class. As a formalism extension, inhibitor arcs are integrated too.

Acknoledgements. The authors are grateful to the ones which have contributed to the integration of an efficient simulation environement in the AMI workshop : in particular, D. Poitrenaud for the implementation of a simulation driver, "Desir", X. Bonnaire for the WN compiler, "Sandrine" and J.L. Mounier for the graphic editor "Macao".

REFERENCES

1. Bernard J.M., Mounier J.L., Beldiceanu N., Haddad S.; "AMI : An extensible Petri Net Interactive Workshop"; 9th European Workshop on ATPN, Venice, Italy, pp- 101-117; June 1988.
2. Bonnaire X., Dutheillet C., Haddad S., "SANDRINE : An Analysis System for the declaration of AMI-Nets, Internal report of the Blaise Pascal Institute, MASI 92.20, June 1992.
3. Chiola G. , C. Dutheillet, G. Franceschinis, S. Haddad; "On Well-Formed Coloured Petri nets and their Symbolic Reachability Graph". Proceedings of the 11th International Conference on Application and Theory of Petri nets, Paris, June 1990.
4. Chiola G. ,Franceschinis G. : "A Structural Colour Simplification in Well-Formed Coloured Nets". In proc. of the 4th International Workshop on Petri Nets and Perfomance Models (PNPM 91), IEEE CS Press, Melbourne,Australia, December 1991, pp 144-153.
5. Chiola G. ,Franceschinis G., Gaeta R. : A symbolic simulation mechanism for Well-Formed Coloured Petri Nets. In proc. of the 25th simulation symposium, Orlando, Florida, 1992.
6. Christensen S.; "Coloured Petri Nets Extended with Place Capacities, Test Arcs and Inhibitor Arcs". In proc. of the 14th International Conference on Application and Theory of Petri nets, Chicago, June 1992.
7. Doldi L., Gauthier P. ; "VEDA2, Power for the designer" ; FORTE'92, Fifth International Conference on Formal Description Techniques ; Lannion ; France ; Oct. 1992.
8. Dutheillet C., Haddad S. ; "Regular Stochastic Petri Nets", In High-level Petri Nets. Theory and Application, K. Jensen and G. Rozenberg eds., Springer-Verlag, pp 470-493, 1991.

9. Fischer Y., Foughali K., "Integrating Tools in the AMI v.1.3 Environment", Internal report of the Blaise Pascal Institute, MASI 93.15, Mars 1993.

10. Genrich H.J. ; "Predicate / Transition Nets". In High-level Petri Nets. Theory and Application, K. Jensen and G. Rozenberg eds., Springer-Verlag, 1991, pp 3-43.

11. Godfroid P.; "Using Partial Orders to improve Automatic Verification Methods ; LNCS 531, Springer, 176-185; 1991.

12. Janicki R.; "Optimal Simulation, nets and reachability Graphs" ; Advances in Petri Nets ; LNCS 524, pp 205-226 ; 1991.

13. Jensen K., Rozenberg G. (Eds) ; "High Level Petri Nets, Theory and Appplication", Springer Verlag, 1991.

14. Nutt G.J., "A Simulation System Architecture for Graph Models", Advances in Petri Nets, LNCS 483 , pp 417-435, 1990.

15. Valmari A. ; "Stubborn Sets of Coloured Petri Nets". Proceedings of the 12th International Conference on Application and Theory of Petri nets, Denmark, June 1991.

Linear Time Algorithm to Find a Minimal Deadlock in a Strongly Connected Free-Choice Net

Peter Kemper

Informatik IV, Universität Dortmund
Postfach 50 05 00, 4600 Dortmund 50, Germany

Abstract. This paper presents an improved algorithm compared to the one given in [7], which finds a minimal deadlock containing a given place p in a strongly connected Free-Choice net (FC-net). Its worst case time complexity is linear in the size of the net. The interest in finding such deadlocks arises from recognising structurally live and bounded FC-nets (LBFC-nets), where finding structural deadlocks efficiently is crucial for the algorithm's time complexity. Employing this new algorithm within [7] LBFC-nets can be recognised in $O(|P|^2|T|)$, which is a reduction by one order of magnitude. Furthermore this marks a lower limit for the complexity of algorithms based on the rank theorem as long as the computation of a matrix rank requires $O(n^3)$.

1 Introduction

Petri Nets have been successfully used for modelling discrete concurrent systems. Within the different classes of Petri Nets the class of FC-nets is theoretically well analysed as far as liveness and boundedness are concerned. In this context the notion of 'deadlock' traditionally refers to a structural not a behavioral deadlock. Since [6] the idea of deadlocks and traps has always been a central point in characterising live FC-nets. Recent research papers reveal a certain progress towards analysis algorithms with polynomial time complexity of low order. In [1] K. Barkaoui and M. Minoux present an algorithm to check liveness for bounded Extended Free-Choice Nets (EFC-nets) and Non Self-Controlling Nets (NSF-nets). This algorithm searches for a strongly connected deadlock which is not a trap and leads to a time complexity of $O(|P|^2|F|^2)$[1].

In [7] an algorithm to recognise structurally life and structurally bounded FC-nets (LBFC-nets) is given. FC-nets are a subclass of EFC-nets, but EFC-nets can be easily transformed into FC-nets, see [3]. This approach has a lower time complexity of $O(|P||T||F|)$. The algorithm is based on the rank theorem given in [5] and decides state-machine-decomposibility of a given FC-net by computing a sufficient set of minimal deadlocks to cover the net. Finding a minimal deadlock for a given place p is crucial in this algorithm and dominates the algorithm's time complexity. The effort for finding a minimal deadlock in

[1] P = set of places, T = set of transitions and F = set of arcs

[7] is $O(|T|(|P|+|T|+|F|))$. The algorithm presented in this paper reduces this effort to $O(|P|+|T|+|F|)$, which is linear in the size of the net. Employing this new algorithm within [7] LBFC-nets can be recognised in $O(|P|^2|T|)$. All algorithms based on the rank theorem require the calculation of the rank of the incidence matrix. As long as the calculation of a matrix rank takes an $O(n^3)$-algorithm, a further reduction of $O(|P|^2|T|)$ is not possible for these algorithms. In this sense our new algorithm reaches an optimum up to constant factors for recognising LBFC-nets.

This article is structured as follows: Sect. 2 contains basic definitions. In Sect. 3 some previous algorithms and ideas are summarized and by a small example the main idea of the new algorithm is introduced, such that we can present it in Sect. 4 as an improvement of existing algorithms. Section 5 illustrates the behaviour of the new algorithm by an example. Termination is proved in Sect. 6, while Sect. 7 contains a proof of correctness. Section 8 concerns the algorithm's time complexity and finally Sect. 9 finishes the paper with some conclusions and future prospects.

2 Basic Definitions and Theorems

Definition 1 Net.
A net is a triple $N = (P, T, F)$ where

1. P and T are non-empty, finite sets and $|P| = n$, $|T| = m$,
2. $P \cap T = \emptyset$,
3. $F \subseteq (P \times T) \cup (T \times P)$.

The next definition summarizes some well known notions for nets.

Definition 2.
Let $N = (P, T, F)$ be a net.

1. Let $x \in P \cup T$. The **preset** $\bullet x$ and **postset** $x\bullet$ are given by $\bullet x := \{y \in P \cup T | (y, x) \in F\}$, $x\bullet := \{y \in P \cup T | (x, y) \in F\}$. The preset (postset) of a set of nodes is the union of presets (postsets) of its elements.
2. A **path** of a net N is a sequence $(x_1, \ldots, x_n)$ of nodes $x_i \in P \cup T, i \in \{1, \ldots, n\}$, such that $(x_i, x_{i+1}) \in F$ for all $i \in \{1, \ldots, n-1\}$.
3. N is a **Free-Choice Net** (FC-Net) iff $\forall p \in P : |p\bullet| > 1 \Rightarrow \bullet(p\bullet) = \{p\}$.
4. A net $N' = (P', T', F')$ is a **subnet** of N, $(N' \subseteq N)$, iff $P' \subseteq P, T' \subseteq T$ and $F' = F \cap ((P' \times T') \cup (T' \times P'))$.
5. $P' \subseteq P$ is a **deadlock** of N iff $P' \neq \emptyset$ and $\bullet P \subseteq P\bullet$. A deadlock is **minimal** iff it does not contain a deadlock as a proper subset. A deadlock P' is **strongly connected** iff $N' = (P', \bullet P', F')$ with $F' = F \cap ((P' \times \bullet P') \cup (\bullet P' \times P'))$ is strongly connected.

3 On the Computation of Minimal Deadlocks

This section summarizes ideas and algorithms given in [4] and [7] in order to describe Algorithm 10 in Sect. 4 as a further development. This background allows an easier understanding of the correctness of Algorithm 10.

The precise problem is to find a minimal deadlock in a strongly connected FC-net which contains a given place p. In [4] J. Esparza describes an algorithm which exploits the following characterization of minimal deadlocks.

Theorem 3 [4].
Let $N = (P, T, F)$ be a FC-Net and $D \subseteq P$ a deadlock in N.
D is minimal iff D is strongly connected and $\forall t \in \bullet D : |\bullet t \cap D| = 1$.

The central idea for searching a minimal deadlock is to find a handle for any place of the minimal deadlock (starting with p) that has an input transition not being an output transition.

Definition 4 Handle.
Let $N = (P, T, F)$ be a net with two non-empty sets $S, S' \subseteq P \cup T$, $S \cup S' = P \cup T$ and $S \cap S' = \emptyset$. A path $H = (x_0, x_1, \ldots, x_{n-1}, x_n)$ in N is a handle iff $x_0, x_n \in S$, $x_1, \ldots, x_{n-1} \in S'$, $(x_i, x_{i+1}) \in F$, $\forall i \in \{0, \ldots, n-1\}$ and furthermore $x_i \neq x_j$, $\forall i, j \in \{1, \ldots, n-1\}, i \neq j$.

Informally a handle starts at a node in S, follows a circlefree path in the complement set S' and terminates in S. Note that Def. 4 includes the case $x_0 = x_n$. Given two such sets S and S' a handle always exists in a strongly connected net for any fixed node x_n. The following algorithm computes a minimal deadlock by searching handles for all places that belong to the minimal deadlock.

Algorithm 5. get-minimal-deadlock(N,p), [4] and [7]

Input: N=(P,T,F) strongly connected FC-Net with p ∈ P.
Output: minimal deadlock D ⊆ P containing p
Initiate: P̂ = { p } ; T̂ = ∅ ;

Function: get-handle(S,S',N,p,t), see Algorithm 6
This function computes a handle $(x_0, x_1, \ldots, x_{n-2}, t, p)$ with $x_0, p \in S$ and $x_1, \ldots, x_{n-2}, t \in S'$. It returns the handle as a set $\{x_0, x_1, \ldots, x_{n-2}, t, p\}$.

```
Program:
begin
    while (∃ p' ∈ P̂ : ∃ t ∈ • p' : t ∉ T̂ )
    begin
        H := get-handle((P̂ ∪ T̂), (P ∪ T)-(P̂ ∪ T̂),N,p',t) ;
        P̂ := P̂ ∪ ( H ∩ P ) ; T̂ := T̂ ∪ ( H ∩ T ) ;
    end
    D := P̂;
end
```

The set $\hat{P}$ of places belonging to the minimal deadlock is initiated with $\{p\}$ and successively all places of the computed handles are inserted into $\hat{P}$. The algorithm terminates if all places in $\hat{P}$ have handles covering all of their input transitions.

The problem of finding a handle is solved in [7] by Depth-First-Search (DFS). The main idea is that the search stops if an element of S is reached, and that it does a backtrack step if it reaches an element of S' which has been reached before. Due to the fact that no node is checked twice, the search path (stored in a list) does not contain an element of S' twice. The strong connectedness ensures the existence of a handle and the successful termination of DFS. Note that in the following algorithm the search proceeds **backwards** on the net due to the fact that the last place of the handle is given, while its starting node in S has to be computed. Furthermore the input transition which the handle shall cover is fixed. The number num of a node denotes whether a node was visited before or not.

Algorithm 6. get-handle(S, S', N, p, t),[7]

```
Input: N=(P,T,F) strongly connnected FC-Net, with S,S' ⊆ P ∪ T
    S ∪ S' = P ∪ T, S ∩ S' = ∅, p ∈ S, t ∈ •p
Output: Handle H=(x0,...,xn-2,t,p)
Initiate: i := 1 ; list := empty ; num(p) := 1 ;
    num(x) := 0, ∀ x ∈ S' ;
Function: dfs(v)
begin
    i := i + 1 ; num(v) := i ; append(list,v) ;
    if ( ∃ w ∈ •v : w ∈ S )                  { element of S is reached }
    then append(list,w) ; return Yes ;       { search stopped }
    forall ( w ∈ •v with num(w) = 0 )        { for unvisited elements of S'}
    begin                                    { recursive search }
        if ( dfs(w) = Yes )
        then return Yes ;
    end
    delete(list,v) ;
    return No ;
end

Program:
begin
    append(list,p) ;
    dfs(t) ;
    Output list ;
end
```

Algorithm 5 employing Algorithm 6 leads to a time complexity of $O(|T|(|P| + |T| + |F|))$, because at most $|T|$ handles have to be computed for a deadlock and computing a handle requires $O(|P| + |T| + |F|)$; see [7] for details. Is there a possibility to reduce this time complexity? In Algorithm 6 those paths are regarded as unsuccessful which lead back to a node w in S', which was visited before (num(w) > 0). These "unsuccessful" paths can be suitable as handles later on. The following example shows the effect we are interested in.

Example 1. Figure 1 shows a strongly connected FC-net. Let p be the place a minimal deadlock shall be computed for. The numbering of nodes show the num-values given by Algorithm 6 which started on p with transition a. If we focus on the list content whenever a node is reached that was visited before or when p is reached, we get the following four list contents:

1. p a b c d e f
2. p a b c g h
3. p a b c g i j k
4. p a b c g l p

List content 4 gives the handle we are looking for. In Algorithm 5 three further handles have to be computed. Two starting at place g, one for input transition h the other one for input transition i. The third one starts in p for transition h and becomes irrelevant after covering h with a handle from g. The interesting point is, that these handles were computed within the first get-handle call but have been regarded as not interesting. Part of list content 3 hold the corresponding information for the required handle (g i j k b). But the information for the handle (g h f a) is spread over list contents 1 and 2. Furthermore the list content 1 gives an example of the irrelevant path (c d e f a) which starts at a transition.

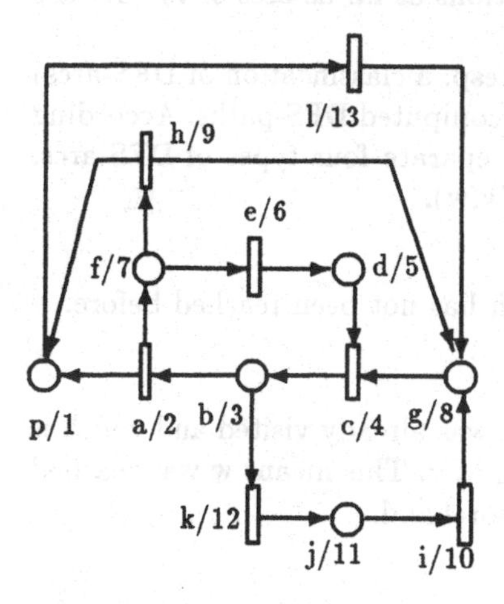

Fig. 1. Strongly connected FC-net after applying Algorithm 5 for place p, node description: <id/num-value>

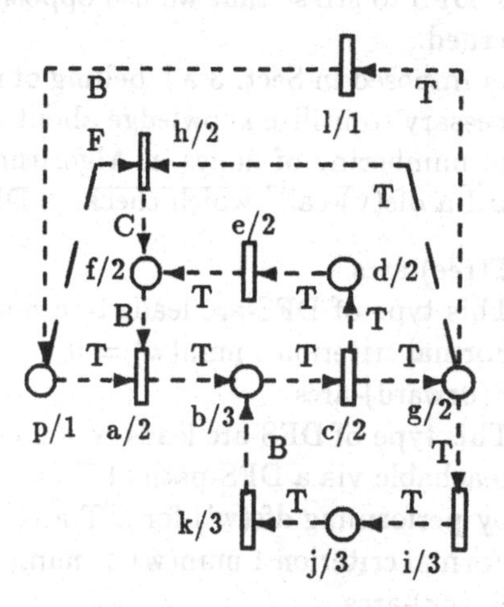

Fig. 2. Corresponding DFS-arcs after applying Algorithm 10 for place p, node description: <id/ll-value>

This small example shows that some of the knowledge we earn from former unsuccessful attempts is useful later on. A more sophisticated way of computing a minimal deadlock should exploit this knowledge. Previously unsuccessful paths must be labelled in such a way, that handles contained in these paths, are tractable without further searching. An algorithm which follows this idea naturally splits up into 3 phases:

1. Searching for a handle,
2. marking a handle as part of the deadlock and
3. evaluating old knowledge in order to mark directly those handles which consist of paths computed before.

This approach leads to an algorithm that computes a minimal deadlock within a single DFS-run. A detailed description of this algorithm follows within the next section.

4 Improved Algorithm to Find a Minimal Deadlock

This section presents an algorithm to compute a minimal deadlock for a given place of a strongly connected FC-net within a single DFS-run. Since the search for handles proceeds backwards on the net, we define DFS-arcs in order to avoid confusion on the direction of arcs.

Definition 7 DFS-arc. Let $N = (P, T, F)$ be a net.
f = (v,w) is a DFS-arc if (w,v) ∈ F.

By analogy with DFS-arc we use terms like DFS-successor, DFS-path etc. with prefix DFS to stress that we use opposite directions as far as arcs of the net are concerned.

As imposed in Sect. 3 a labelling of nodes (resp. a classification of DFS-arcs) is necessary to utilize knowledge about already computed DFS-paths. According to the numbering of nodes in Algorithm 6 we separate four types of DFS-arcs. Regard a dfs(v)-call[2] which checks a DFS-arc (v,w).

1. T(ree)-arcs
   This type of DFS-arc leads to a node which has not been reached before.
   Formal criterion : num(w) = 0.
2. F(orward)-arcs
   This type of DFS-arc leads to a node which was already visited and which is reachable via a DFS-path of T-arcs starting at v. This means w was reached by performing dfs(w') for a T-arc(v,w') beforehand.
   Formal criterion : num(w) > num(v).
3. B(ack)-arcs
   This type of DFS-arc leads back to a node w from which v was reached by T-arcs and whose dfs(w)-call has not terminated yet. It closes a circle.
   Formal criterion: num(w) < num(v) ∧ w ∈ list.

[2] Function dfs corresponds to function get-handle in the context of Algorithm 10.

4. C(ross)-arcs
   This type of DFS-arc leads to a node w whose dfs(w)-call terminated before v was found. This means that no DFS-path of T-arcs from w to v exists.
   Formal criterion: num(w) < num(v) $\wedge$ w $\notin$ list.

This classification of DFS-arcs is taken from [8], [10], where it is used for finding strongly connected components. Figure 2 shows the classification of DFS-arcs according to the net in Fig. 1, the ll-values in the node description refer to Def. 9. T-arcs form a tree on the net, while a B-arc closes a circle within this tree and C-arcs cross over subtrees, see also Fig. 5. An F-arc is a shortcut for a DFS-path of T-arcs. In FC-nets only one type of F-arcs can occur.

Lemma 8. *Apply Algorithm 6 on a strongly connected FC-net.*
If (v,w) is an F-arc $\Rightarrow$ v is a place and w is a transition.

Note that an F-arc is a DFS-arc and its corresponding net arc has the opposite direction.

Proof. Assume $v \in T, w \in P$
$\Rightarrow \exists x \in \bullet v : x \neq w$ and $\exists y \in w\bullet : y \neq v$ with (v, x) and (y, w) are T-arcs (reachability by T-arcs)
$\Rightarrow |w\bullet| > 1$ and $\{w, x\} \subseteq \bullet(w\bullet)$ contradicting FC-condition. □

The information which shall be stored at a node is its low-link value. This value for a node x holds the minimal num-value of any node reachable from x via a DFS-path on the net that consists of an arbitrary number of T- and/or C-arcs and finally one B-arc. Note that this definition differs from a similar definition in [10] with respect to C-arcs.

Definition 9 low-link value (ll-value).

$$ll(x) := min(\ \{\ num(x)\ \} \cup \{\ num(z) | (x,z) \text{ is B-arc }\} \cup \{\ ll(y) | (x,y) \text{ is T-arc or C-arc }\})$$

F-arcs cannot reduce the low-link value of a node any further, because any node reachable by an F-arc is also reachable via a DFS-path of T-arcs. This ensures that any num-value reachable by an F-arc is already considered.

Figure 2 corresponds to the net in Fig. 1 and shows DFS-arcs according to a search starting at place p for transition a. The labelling of a node gives its identifier and its ll-value. The corresponding num-values are the same ones as in Fig. 1. It displays the situation when handle (p a b c g l p) is found. Missing handles (g h f a) and (g i j k b) reveal constant ll-values 2, resp. 3. So marking handles following a DFS-path with constant ll-values is promising. This is not sufficient in general, because this DFS-path stops at a B-arc (v,w), with w not necessarily being a suitable termination node. Furthermore this DFS-path might get trapped in a circle. An appropriate strategy turns out to be "follow a DFS-successor with minimal ll-value" which the following algorithm demonstrates.

Algorithm 10.

```
Input: N=(P,T,F) is a strongly connected FC-Net with p ∈ P
Output: Minimal Deadlock D with p ∈ D
Initiate:
      i := 0 ; max-num := 1 ; list := empty ; D(p) := True ;
      ∀ x ∈ T ∪ P \{p} : num(x) := 0 ; ll(x) := 0 ; D(x) := False ;
Program:
begin
      get-handle(p) ;
      ∀ x ∈ P, D(x) = True : output x ;
end

Function: get-handle(v)
begin
      i := i + 1 ; num(v) := i ; ll(v) := i ; append( list, v) ;
      forall w ∈ • v
      begin
            if num(w) = 0                                        {T-arc}
            then get-handle(w) ;
                  ll(v) := min(ll(v),ll(w)) ;
                  if D(w) = True ∧ v ∈ T                         { handle was found }
                  then delete(list,v) ; return ;                 { stop search for transition }

            if num(w) > num(v)                                   {F-arc}
            then if num(v) ≤ max-num
                  then eval-old(w) ;                             { to complete handle }

            if 0 < num(w) < num(v) ∧ w ∈ list                    {B-arc}
            then ll(v) := min(ll(v), num(w)) ;
                  if D(w) = True                                 { handle found }
                  then mark-handle() ;
                        if v ∈ T                                 { stop search for transition }
                        then delete(list, v) ; return ;

            if 0 < num(w) < num(v) ∧ w ∉ list                    {C-arc}
            then ll(v) := min(ll(v),ll(w)) ;
                  if num(w) ≤ max-num
                  then eval-old(w) ;                             { to complete handle }
                        mark-handle() ;
                        if v ∈ T                                 { stop search for transition }
                        then delete(list, v) ; return ;
      end forall
      delete(list, v) ;
end function
```

```
Function: eval-old(v)
begin
    if ( D(v) = True or v ∈ list)
    then return ;                                              { stop recursion }
    D(v) := True ;
    if v ∈ T
    then choose one w ∈ • v with minimal ll(w) ;             { need not be unique }
         eval-old(w) ;
    else ∀ w ∈ • v : eval-old(w) ;
end function

Function: mark-handle()
begin
    forall ( x ∈ list with max-num < num(x) )
    begin
         D(x) := True ;                                        { mark element }
         if x ∈ P                                              { exploit knowledge }
         then ∀ y ∈ • x, num(y) > 0, y ∉ list : eval-old(y) ;
    end forall
    max-num := i ;                                             { update max-num }
end function
```

According to the three phases mentioned in Sect. 3 the algorithm splits into three functions:

get-handle to search for a handle with DFS and classify DFS-arcs
It checks fresh, yet unexplored nodes and converts them to old, explored ones. If it reaches a node, which terminates a handle (by a B-arc) or which can surely be completed to a handle (by C- or F-arc), function get-handle makes sure that this handle gets marked by function mark-handle. If necessary it completes the handle with function eval-old. Furthermore it tries to cover all DFS-successors of places, i.e. their input transitions, with handles (according to Def. of deadlocks). If it does not succeed, it leaves this to function eval-old. At a transition it tries to ensure that only one DFS-successor, i.e. only one input place, becomes element of the deadlock (according to Theorem 3).

mark-handle to mark elements of a handle
Along the new handle it invokes function eval-old for those DFS-successors of places, for which function get-handle did not succeed beforehand. This ensures that already explored input transitions of these places are covered with handles by function eval-old.

eval-old to mark handles tractable from previously unsuccessful attempts
Its duty is to follow handles on already explored parts of the net and to mark them. Under the assumption that its current node v belongs to a handle (which turns out to be true), it follows all DFS-successors for $v \in P$

(according to Def. of deadlocks) or only one DFS-successor for $v \in T$ (according to Theorem 3). It chooses the one DFS-successor in such a way that it cannot get trapped in a circle.

Attributes of a node v are:

num (integer) holds num-value,
ll (integer) stores ll-value and
D (boolean) denotes membership of minimal Deadlock D, resp. • D.

Whenever a node x is called marked, this means: x is marked D(x) = True. The variables used in Algorithm 10 have the following meaning:

i (integer) is a counter to give any reached node a unique number.
list stores nodes of the recursion stack of function get-handle.
max-num (integer) holds the maximal number of nodes reached when the last handle was found. It has two duties. On the one hand it splits list in such a way that all nodes in the first part (num(x) ≤ max-num) are marked with D(x)=True and all nodes in the second part (num(x) > max-num) remain D(x)=False until a handle is found. On the other hand it splits those nodes exploitable as old knowledge (num(x) ≤ max-num) from nodes found too recently to be exploited (num(x) > max-num). Note that node x with num(x)=max-num is not necessarily element of list.

5 Example Computation

This section illustrates the behaviour of Algorithm 10 by an example. The following lemma simplifies the naming of nodes.

Lemma 11. *Any node v can at most cause one get-handle(v)-call and obtains a unique number num(v) within this call.*

Proof. In both cases, the get-handle(p)-call in program and the get-handle(w)-call in get-handle(v), num(p)= 0, resp. num(w)=0 holds beforehand. But within any get-handle(v), num(v) is immediately set to the incremented counter i. This has the effect, that num(v) is unique. It is not set anywhere else and i is used exclusively for this purpose. A recursive call get-handle(v) with num(v) > 0 is not possible. □

Example 2. Figure 3 shows a strongly connected FC-net. Algorithm 10 computes a minimal deadlock for place 1. Nodes reached by the algorithm are labelled by their num-values, which are unique according to Lemma 11. Unlabelled nodes are not reached and have a num-value 0 which is omitted in Fig. 3. Figure 4 gives the corrsponding DFS-arcs. B-arcs and C-arcs are labelled B, resp. C. F-arcs do not occur. Unlabelled arcs denote T-arcs. Nodes in Fig. 4 are labelled <num-value/ll-value>. Note that these ll-values are the final ones. During computation temporarily higher ll-values occur, which are represented in the list contents

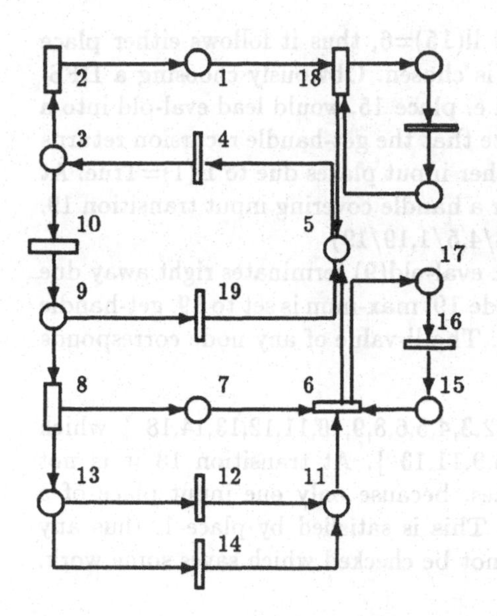

Fig. 3. Strongly connected FC-net

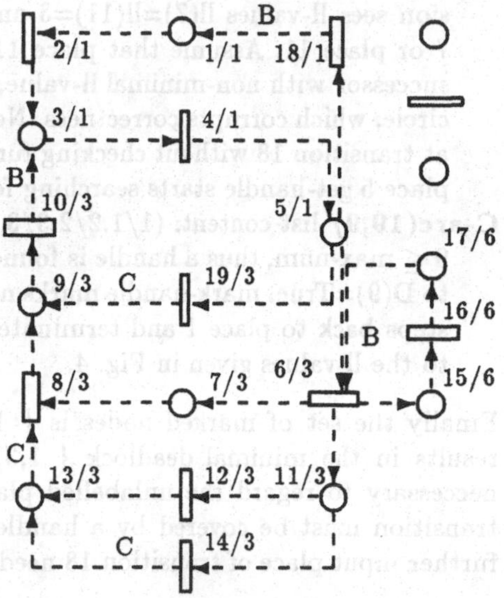

Fig. 4. Corresponding DFS-arcs after computation of a minimal deadlock containing place 1

below. Note that list keeps track of the get-handle-recursion. Let us highlight certain steps of the computation by giving the list content with <num-value/ll-value> and additional explanations:

start max-num = 1, list empty, marked nodes { 1 }

B-arc(10,3) list content: (1/1,2/2/,3/3,4/4,5/5,6/6,7/7,8/8,9/9,10/10)
The num-value 3 is catched from node 3 and by termination within recursion of get-handle it is inherited as ll-value to nodes 10, 9, 8, 7 and 6. get-handle steps back to transition 6.

C-arc(13,8) list content: (1/1,2/2,3/3,4/4,5/5,6/3,11/11,12/12,13/13)
The ll-value 3 from node 8 is inherited to nodes 13,12 and 11. get-handle steps back to place 11.

C-arc(14,13) list content: (1/1,2/2,3/3,4/4,5/5,6/3,11/3,14/14)
The ll-value 3 from node 13 is inherited to node 14. get-handle steps back to transition 6.

B-arc(17,6) list content: (1/1,2/2,3/3,4/4,5/5,6/3,15/15,16/16,17/17)
The num-value 6 from node 6 is inherited as ll-value to nodes 17,16 and 15. get-handle steps back to place 5.

B-arc(18,1) list content: (1/1,2/2,3/3,4/4,5/5,18/18)
The first handle is found. It must be found via a B-arc, because 1 is the only node that is marked and max-num = 1. The num-value 1 from node 1 is inherited as ll-value to nodes 18,5,4,3 and 2. The mark-handle-call marks nodes 2,3,4,5 and 18. max-num is set to 18. eval-old(6) is called, which either marks { 6,7,8,9,10 } or { 6,11,12,13,14,8,9,10 }. Note that at transition 6 the recur-

sion sees ll-values ll(7)=ll(11)=3 and ll(15)=6, thus it follows either place 7 or place 11. Assume that place 11 is chosen. Obviously choosing a DFS-successor with non-minimal ll-value, i.e. place 15, would lead eval-old into a circle, which corrupts correctness. Note that the get-handle recursion returns at transition 18 without checking further input places due to D(1)=True. At place 5 get-handle starts searching for a handle covering input transition 19.

C-arc(19,9) list content: (1/1,2/2,3/3,4/4,5/1,19/19)
$9 \leq$ max-num, thus a handle is found. eval-old(9) terminates right away due to D(9)=True. mark-handle marks node 19. max-num is set to 19. get-handle steps back to place 1 and terminates. The ll-value of any node corresponds to the ll-values given in Fig. 4.

Finally the set of marked nodes is { 1,2,3,4,5,6,8,9,10,11,12,13,14,18 } which results in the minimal deadlock { 1,3,5,9,11,13 }. At transition 18 it is not neccessary to regard the unlabelled places, because only one input place of a transition must be covered by a handle. This is satisfied by place 1, thus any further input place of transition 18 need not be checked which saves some work.

6 Termination of Algorithm

Prooving termination makes use of the two following observations:

Lemma 12. list *is an ordered list by increasing numbers* num.

Proof. A node v is appended to list only at the beginning of its get-handle(v) call. As num-values are unique by Lemma 11 and increasing, list is automatically sorted. □

Lemma 13. *Any node v can cause recursion in eval-old(v) at most once.*

Proof. D(v)=False is a precondition for recursion in eval-old(v). D(v):=True is set ahead of recursion. Reset D(v) does not happen, because v is set D(v):=False only by initialisation. □

Theorem 14. *Algorithm 10 terminates for any finite net N.*

Proof. 1. The recursion in eval-old terminates due to Lemma 13 and to the finiteness of the net.
2. mark-handle terminates because list is finite (ensured by the finiteness of the net, Lemma 11 and Lemma 12) and eval-old terminates.
3. get-handle terminates due to termination of eval-old, mark-handle and finite depth of recursion of get-handle ensured by Lemma 11.
4. program terminates due to termination of get-handle(p)-call.

□

7 Correctness of Algorithm

There are at least two ways to prove correctness of Algorithm 10. One way is to show that its output is a deadlock and furthermore a minimal deadlock. Another way is to show that Algorithm 10 realizes the idea of Algorithm 5, which exhaustively calculates handles to cover all input transitions of every place that belongs to the deadlock. In this section we follow the second way, i.e. we show first that eval-old marks a set of handles and as a second, main step that the "handles" found in get-handle actually are handles and no input transition of a deadlock place remains uncovered.

The following properties of Algorithm 10 turn out to be useful for proving its correctness. As mentioned in Sect. 4 **max-num** is used to separate **list** in a first part with marked nodes that already belong to a handle and a second part with nodes that might belong to a handle. These nodes are unmarked when appended to **list** and marked as soon as a handle is found.

Lemma 15. $x \in$ **list** $\wedge$ $num(x) \leq$ **max-num** $\Rightarrow D(x) = True$

Proof. by induction on updates of **max-num** (max-num := i)
Start: by initialisation **max-num** = 1 and **list** = empty, which fulfills the lemma. append(list,p) in get-handle(p) does not disturb the lemma due to D(p)=True by initialisation. As num(p) = 1, Lemma 11 and Lemma 12 ensure that for any further append(list, x) within get-handle(x), num(x) > 1 holds. Thus the lemma holds until the first update of **max-num**.
Step: update of **max-num** occurs in mark-handle
if x ∈ **list** ∧ num(x) ≤ **max-num** ⇒D(x) = True by assumption
if x ∈ **list** ∧ num(x) > **max-num** ⇒D(x):= True in mark-handle before the update.
Thus ∀ x ∈ **list** : D(x) = True at update of **max-num**. □

Lets regard the effect of Algorithm 10 from a node's point of view. Let v be a node different from p. At initialisation num(v)=0 and v belongs to an unexplored part of the net. Once reached by get-handle, it gets num(v) > 0, ll(v) = num(v), v ∈ **list** and num(v) > **max-num**. If a handle is found while v ∈ **list**, i.e. get-handle(v) has not terminated, then v gets marked in mark-handle and num(v) ≤ **max-num** holds after update of **max-num**. Otherwise, if no handle is found while v ∈ **list**, get-handle(v) terminates with D(v) = false. Once v is deleted from **list**, v ∉ **list** holds forever and v belongs to an old, explored part of the net. If v is still unmarked, it can be reached and marked once by eval-old, which just operates on explored parts of the net. During computation, the net separates in unexplored and explored parts. This is formalized in the following notation.

Definition 16 $S_{list}, S_{old}, S_{new}$.
$S_{list} := \{x | x \in \textsf{list} \vee D(x) = True)\}$, $S_{old} := \{x | x \notin S_{list} \wedge num(x) > 0\}$, and $S_{new} := \{x | x \notin S_{list} \wedge num(x) = 0\}$.

Obviously $S_{list} \dot\cup S_{old} \dot\cup S_{new} = P \cup T$ holds. The following lemma states that S_{list} forms a borderline between explored parts (S_{old}) and unexplored parts (S_{new}).

Lemma 17. *In computation of Algorithm 10 no DFS-arc (v,w) exists with $v \in S_{old}$ and $w \in S_{new}$.*

Proof. Assume the contrary: DFS-arc (v,w) with $v \in S_{old}$ and $w \in S_{new}$ exists. Due to num(v) > 0 and $v \notin$ list, get-handle(v) has occured and terminated to reach the assumed situtation.

Case $v \in P$ There is no way to hinder get-handle(x) for all $x \in \bullet v$ during computation of get-handle(v).
$\Rightarrow$num(w) > 0, $w \notin S_{new}$, contradicting the assumption.

Case $v \in T$ num(v) > 0, num(w) = 0 by assumption, i.e. no get-handle(w) occured. This could happen only in case of D(w') = True for some other w' $\in \bullet$ v. This is caused by finding a suitable B- or C-arc in recursion from get-handle(w') and calling mark-handle. During get-handle(w'), v $\in$ list and w' $\in$ list hold. Thus w' is marked by mark-handle. Hence, if num(v) > max-num, then v gets marked in mark-handle; otherwise Lemma 15 ensures that D(v) = True holds already. So $v \notin S_{old}$, contradicting the assumption.

□

Obviously get-handle operates on S_{list} and S_{new}, mark-handle on S_{list} and eval-old on S_{old}.

Strong connectedness has an impact on the algorithm's correctness. Figure 5 shows a typical situation within a computation. list contains nodes (a,b,c,d,e) with D(a)=D(b)=True and D(c)=D(d)=D(e)=False. Dotted lines to the unexplored part denote the possibility of further arcs, which were not followed yet. Triangles $\bar{b}, \bar{c}, \bar{d}, \bar{e}$ below nodes denote subtrees of nodes which were reached by T-arcs and whose get-handle-calls already terminated. Strong connectedness requires that these subtrees have B- or C-arcs as "exits". Note that C-arcs can only lead to subtrees previously explored. The B-arc from subtree $\bar{b}$ completes a handle, because it leads back to the starting node a. num(b) < max-num < num(c) holds, since get-handle(c) occurs after the update of max-num. The B-arc in subtree $\bar{c}$ does not complete a handle because it leads to c which is on list, but not yet element of a handle (D(c)=False). The C-arc (d',c') with d' $\in \bar{d}$ and c' $\in \bar{c}$ does not reach a subtree, which can lead to an element of a handle. This is recognised by max-num < num(c'). The F-arc does not reach an element, from which a handle can yet be found, because the B-arc of $\bar{e}$ does only lead to e which is not element of a handle (D(e)= False). If this B-arc would reach further back, say back to b, a handle would run through c,d,e and D(e)=True. By D(e)=False it is recognizable that from the DFS-successors of e so far no handle can be reached.

Lemma 18. *Regard a situation in Algorithm 10 where eval-old is called in get-handle or mark-handle. Let $S := S_{list}$, $S' := S_{old} \cup S_{new}$ and $v \in (\bullet d) \cap S_{old}$ for $d \in$ list. Then eval-old(v) marks a handle H covering v and handles for all places marked by eval-old(v).*

Proof. Note that S is defined ahead of the eval-old(v)-call, such that this set does not automatically get new elements when nodes get marked within eval-old(v).

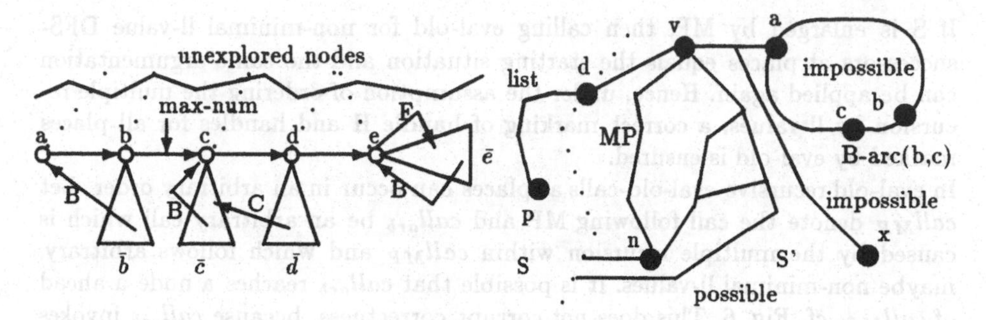

Fig. 5. Typical situation occuring in computation

Fig. 6. Possible and impossible behaviour of eval-old(v)-call

Assume that in eval-old the multiple recursion at places is ordered by ll-values of their DFS-successors in a way, that eval-old(w) for a DFS-successor with minimal ll-value is called first. This causes eval-old to follow first a DFS-path on the net directed by its minimal ll-values. This DFS-path is called minimal path MP. First we show that MP is a handle.

- No circles in MP
  The recursion starting at eval-old(v) cannot proceed over nodes of S. Thus Lemma 17 and v $\in S_{old}$ ensure that MP cannot contain a node x $\in S_{new}$. Furthermore x $\notin$ list holds for all nodes x in MP. Thus their get-handle(x)-call has terminated and their ll-values are not changed anymore.
  Obviously function get-handle reflects Def. 9, thus for any node x in MP, num(p) = ll(p) = 1 $\leq$ ll(x) $\leq$ num(x) holds.
  Regard a DFS-arc (y,z) within MP. The case z $\in$ S is uncritical, because it finishes MP in S. Let z $\in$ S' then. Def. 9 states ll(y)=ll(z) for a T- or C-arc and ll(y)=num(z) for a B-arc. For an F-arc (y,z) z is reachable via a DFS-path of T-arcs. Thus a T-arc (y,z') with ll(z') $\leq$ ll(z) exists. By (y,z) in MP and Def. 9 we get ll(y)=ll(z)=ll(z'). For a node z apart from node p (with num(p)=1) strong connectedness requires ll(z) < num(z) for final ll-values, because otherwise DFS-predecessors of z would not be reachable from z. Thus at a B-arc we get ll(y) > ll(z). B-arcs are the only arcs that close circles within the tree of T-arcs. Thus if MP contains a circle, this circle contains at least one B-arc. Constant ll-values at T-, C- and F-arcs in combination with decreasing ll-values at B-arcs contradict the existence of a circle-closing arc in MP, e.g. in Fig. 6 circle(a,b,c) requires ll(a) $\geq$ ll(b) > ll(c) $\geq$ ll(a).
- Termination in S
  Assume the contrary: an eval-old-call terminates in some x $\in$ S', cf. Fig. 6. Due to strong connectedness and Lemma 17, x must have at least one DFS-successor y with num(y) > 0. As x $\notin$ S, eval-old(x) would proceed in MP with the DFS-successor of minimal ll-value contradicting the assumption.

Altogether MP does not contain a circle and terminates in S. Hence, if d $\in$ list $\subseteq S$ is added to MP, handle H is MP regarded in opposite direction.

If S is enlarged by MP, then calling eval-old for non-minimal ll-value DFS-successors at places equals the starting situation and the same argumentation can be applied again. Hence, under the assumption of ordering the multiple recursion by ll-values, a correct marking of handle H and handles for all places marked by eval-old is ensured.
In eval-old recursive eval-old-calls at places can occur in an arbitrary order. Let $call_{MP}$ denote the call following MP and $call_{arb}$ be an arbitrary call which is caused by the multiple recursion within $call_{MP}$ and which follows arbitrary, maybe non-minimal ll-values. It is possible that $call_{arb}$ reaches a node n ahead of $call_{MP}$, cf. Fig. 6. This does not corrupt correctness, because $call_{arb}$ invokes eval-old(n) which is identical to the effect of $call_{MP}$ reaching n first. In fact $call_{arb}$ picks up the role of $call_{MP}$ and proceeds at node n, while $call_{MP}$ terminates when reaching n in second place. Thus finally the same set of nodes get marked. Therefore the order is not required for correctness and can be neglected. The multiple recursion at places reached by eval-old ensures, that all input transitions of these places are covered by handles and get marked. □

Theorem 19. *Algorithm 10 is correct, i.e. its output is a minimal deadlock.*

Proof. For correctness we have to show that

1. whatever node x is marked D(x) := True within mark-handle or eval-old, it actually belongs to a handle and
2. any input transition of a place d with D(d)=True is covered by a handle.

If these requirements are fulfilled, Algorithm 10 behaves just like Algorithm 5, which exhaustively calculates handles to cover all input transitions of any place which belongs to the deadlock.
at 1) Induction on finding handles in get-handle. Handles are found in 3 situations:

1. B-arc (v,w) with D(w)=True.
2. C-arc (v,w) with num(w) $\leq$ max-num.
3. F-arc (v,w) with num(v) $\leq$ max-num.

Induction assumption:
Any node, that gets marked in mark-handle or eval-old until the i-th handle is found in get-handle, belongs to a handle.
Induction start:
The first handle is found by a B-arc (v,p) due to the fact that p $\in$ list, D(p) = True and max-num = num(p) = 1. In this case list obviously contains the new handle. So marking all elements of list except p (max-num=1) is correct. According to Lemma 18 handles are marked by calling eval-old(y) for all nodes y $\in \bullet x$ forall nodes x $\in$ list with num(x) > max-num. These handles terminate in S_{list} which is correct because in this case $S_{list} = \{ x \mid D(x)=\text{True} \}$. Thus any node that gets marked belongs to a handle and this handle gets marked completely.
Induction step:

1. **Handle found with B-arc (v,w)**
   w ∈ list and marked correctly (by induction assumption), i.e. w is element of a handle. Lemma 15 ensures that elements of list with a num-value ≤ max-num are marked. Thus all elements of list with num > max-num form the new handle and marking these nodes is correct. As far as eval-old-calls within mark-handle are concerned, the same argumentation as for the induction start holds again. Thus all nodes marked D(x)=True belong to a handle.
2. **Handle found with C-arc (v,w)**
   num(w) ≤ max-num and w ∉ list imply, that w ∈ S_{old} and w was found before the last update of max-num.
   If D(w)=True, w belongs to a handle according to induction assumption and the unmarked elements of list form the new handle.
   If D(w)=False, strong connectedness together with num(w) ≤ max-num ensures that a minimal path, which starts at w, cannot lead into a circle (according to proof of Lemma 18). Finiteness of the net and Lemma 17 force this minimal path to end at a node x ∈ S_{list}. num(w) ≤ max-num and D(w)=False imply that w ∈ S_{old} when the last handle, which caused the update of max-num, was found. At that stage in the past of calculation, S_{list} formed a borderline and v ∈ S_{new}, i.e. v is not reachable from w without crossing S_{list}. This borderline was marked when the last handle was found. Thus at the current stage of calculation any termination node x ∈ S_{list} reachable by eval-old(w) is marked, more formally num(x) ≤ max-num and D(x)=True by Lemma 15. Thus eval-old marks the minimal DFS-path and all further required handles along this DFS-path according to Lemma 18 correctly. The unmarked elements of list and the minimal DFS-path form the new handle and the marking of unmarked elements of list by mark-handle is correct.
3. **Handle found with F-arc (v,w)**
   An F-arc (v,w) ensures existence of a DFS-path of T-arcs $(v, a_1, a_2, \ldots, a_n, w)$.
   num(v) ≤ max-num implies D(v)=True by Lemma 15.
   If D(w) = True, w is marked correctly by induction assumption, eval-old(w) terminates without recursion and no nodes have to be marked.
   If D(w) = False, the recursive structure of get-handle requires that get-handle(w) has terminated and all DFS-successors of w are fully explored. Thus ll(w) is fixed. The strong connectedness of the net ensures that v is reachable from w. According to Lemma 18 eval-old(w) terminates at nodes x ∈ S_{list}. This is correct if S_{list} = { x | D(x)=True }. The recursive structure of get-handle -delete(list,x) occurs at termination of get-handle(x)- together with Lemma 11 and Lemma 12 ensure that num(v) = max{num(x) | x ∈ list}. Thus by Lemma 15 and D(v)=True holds D(x)=True forall x ∈ list. Thus S_{list} = { x | D(x)=True }. So eval-old marks the new handle H covering w and furthermore it marks also other handles required by the places of H.

at 2) Assume the contrary, i.e. $x \in P$, $y \in T$ and $y \in \bullet x$, D(x)=True and D(y)=False. There are only two points in Algorithm 10, where D(x) can be set true:

1. D(x):= True in eval-old(x)
   If $y \notin$ list, then eval-old(x) with $x \in P$ implies eval-old(y). So D(y)=True.
   If $y \in$ list, then calling eval-old from finding a handle with B- or C-arcs ensures that mark-handle marks all elements of list with num-value > max-num. If num(y) $\leq$ max-num then Lemma 15 implies D(y) = True anyway.
   If eval-old is invoked from finding a handle with an F-arc (a,b), then $a \in$ list and num(a) $\leq$ max-num. Thus by Lemma 15 D(a)=True and as get-handle(a) is actually checked, a must be the maximal element of list. Thus num(y) $\leq$ num(a) and D(y)=True due to Lemma 15. In all cases D(y)=True contradicting the assumption.
2. D(x):=True in mark-handle
   If num(y) $\neq$ 0 then eval-old(y) occurs, because $x \in P$.
   If num(y) = 0 a get-handle(y) occurs due to $x \in P$. Thus $y \in$list for a while. Due to the strong connectedness of the net, get-handle(y) finds a handle and thus D(y) := True in mark-handle before update of max-num.

□

8 On the Algorithm's Time Complexity

The central idea in calculating complexity is to charge nodes and arcs for the effort they cause. This leads to constant costs per node and arc. Thus we get $O(|P| + |T| + |F|)$.

Lemma 11 ensures that there is at most one get-handle-call per node. All assignments and list handling are assigned to v. The list handling requires very restricted operations which are easily implemented in a way that they cause only constant effort. Any DFS-successor of v is reached via one arc and can cause:

1. a get-handle(w)-call
   The costs are handed over to DFS-successor w.
2. an eval-old(w)-call
   Calling eval-old(w) leads to constant costs for assignments and if-statements and costs for recursive calls of eval-old. The constant costs are assigned to the arc (v,w). It is never followed again, due to the fact that it is only followed once within get-handle when a handle is found and v gets D(v)=True. $v \in$ list or D(v) = True stops any recursive calls of eval-old in v. Lemma 13 states that any node can cause recursive eval-old-calls at most once, thus the total cost of a recursion is shared between those arcs which the recursion follows. As any arc can take part in a recursion once, its costs remain constant. Choosing a DFS-successor with minimal ll-value at a transition can be avoided by storing this DFS-successor when fixing the ll-value within get-handle.

3. a mark-handle-call
   Updates of **max-num** (max-num:=i) imply that **max-num** only increases. In mark-handle only those elements of **list** with num-value > **max-num** get marked. Thus these elements get never marked in mark-handle again. So these costs are constant per node and are assigned to the corresponding node. The costs for eval-old-calls within mark-handle are assigned according to eval-old-calls within get-handle.

Altogether any node and arc carry constant costs, which leads to $O(|P| + |T| + |F|)$.

9 Conclusions

In this paper we have shown that a particular DFS-algorithm allows to find a minimal deadlock in a strongly connected FC-net containing a given place p. The time complexity of the new algorithm is linear in the size of the net. This has positive consequences on the complexity of deciding structural liveness and boundedness of FC-nets. Embedding this new algorithm within the one given in [7] leads to the following calculation of time complexity:

check-str-connected	$O(\|P\| + \|T\| + \|F\|)$
get-minimal-deadlock (old)	$O(\|T\|(\|P\| + \|T\| + \|F\|))$
get-minimal-deadlock (new)	$O(\|P\| + \|T\| + \|F\|)$
check-s-component	$O(\|P\|\|T\|)$
matrix-rank & check-initial-marking	$O(\|P\|^2\|T\|)$

The functions get-minimal-deadlock and check-s-component are called at most $|P|$ times, all others are called at most once.

$$O([|P| + |T| + |F|] + |P|[(|P| + |T| + |F|) + |P||T|] + |P|^2|T|) =$$
$$O(|P||F| + |P|^2|T|) = \qquad (2|P||T| \geq |F| \text{ due to flow matrices})$$
$$O(|P|^2|T|)$$

The time complexity is reduced from $O(|P||T||F|)$ in [7] to $O(|P|^2|T|)$ here. This moves the focus point from finding minimal deadlocks to checking them for being S-components, calculating the rank of the incidence matrix and checking the initial marking for being live. Further improvements can easily be achieved by integrating check-s-component into get-minimal-deadlock and using other techniques for checking the initial marking, see [9] or [[5], proposition 4.3]. This leads to a reduction of constant factors but does not lower the worst case time complexity since the calculation of a matrix rank requires $O(n^3)$.

The algorithm is implemented and tested for recognising LBFC-nets within the QPN-Tool, a tool for qualitative and quantitative analysis of Queueing Petri Nets, cf. [2].

Acknowledgement

The author would like to thank F. Bause, H. Beilner, B. Müller-Clostermann and four anonymous referees for their valuable comments to improve this paper.

References

1. K. Barkaoui and M. Minoux. A polynomial-time graph algorithm to decide liveness of some basic classes of bounded petri nets. In K. Jensen, editor, *Application and Theory of Petri Nets 1992*, LNCS 616, pages 62–75, Berlin Heidelberg, 1992. Springer.
2. F. Bause and P. Kemper. *QPN-Tool Version 1.0 User's Guide*. Universität Dortmund, LS Informatik 4, 1991.
3. E. Best and M.W. Shields. Some equivalence results for free choice nets and simple nets and on the periodicity of live free choice nets. In *CAAP'83, Trees in Algebra and Programming, 8th Colloquium, L'Aquila*, LNCS 159, pages 141–154, Berlin Heidelberg, 1983. Springer.
4. J. Esparza. Minimal deadlocks in free choice nets. Hildesheimer Informatikberichte 1/89, Universität Hildesheim, Institut für Informatik, 1989.
5. J. Esparza. Synthesis rules for petri nets, and how they lead to new results. In J.C.M. Baeten and J.W. Klop, editors, *Concur '90*, LNCS 458, pages 182–198, Berlin Heidelberg, 1990. Springer.
6. M.H.T. Hack. Analysis of production schemata by petri nets. Technical Report TR-94, MIT, Boston, 1972 corrected 1974.
7. P. Kemper and F. Bause. An efficient polynomial-time algorithm to decide liveness and boundedness of free-choice nets. In K. Jensen, editor, *Application and Theory of Petri Nets 1992*, LNCS 616, pages 263–278, Berlin Heidelberg, 1992. Springer.
8. K. Mehlhorn. *Data Structures and Algorithms 2: Graph Algorithms and NP-Completeness*. EATCS Monographs on Theoretical Computer Science. Springer, Berlin Heidelberg, 1984.
9. M. Minoux and K. Barkaoui. Polynomial algorithm for finding deadlocks, traps and other substructures relevant to petri net analysis. Technical Report 212, Laboratoire MASI, Universite P. et M. Curie, Paris, 1988.
10. R. Tarjan. Depth-first search and linear graph algorithms. In *SIAM. Jour. Comput.*, volume 1, pages 146–160. 1972.

Exploiting T-invariant Analysis in Diagnostic Reasoning on a Petri Net Model*

Luigi Portinale

Dipartimento di Informatica - Universita' di Torino
C.so Svizzera 185 - 10149 Torino (Italy)
e-mail: portinal@di.unito.it

Abstract. This paper copes with the application of T-invariant analysis to diagnostic reasoning based on a Petri net model. In particular, it is formally shown how the notion of diagnostic solution can be related to that of Petri net T-invariant, partially transforming a problem traditionally solved by means of symbolic techniques into a linear algebraic one. The approach has been inspired by some previous works concerning the use of T-invariant analysis on Petri net models of logic programs and it takes its place among recent approaches aiming at integrating artificial intelligence and Petri net-based techniques. A diagnostic algorithm exploiting the idea is proposed and the role of the Petri net model in the resulting diagnostic architecture is discussed.

1 Introduction

Automatic diagnostic problem solving is an important area of computer science and especially of artificial intelligence where big efforts have been spent in the attempt to automate the reasoning processes leading to the identification and/or localization of faults in physical systems. In particular, one of the approaches that is gaining a lot of popularity is the *model-based approach* to diagnosis [15]. The starting point of the approach is to directly model the structure and function or the behavior of the device to be diagnosed. A lot of different formalizations have been proposed, both for modeling and representation purposes and in order to give a precise semantics to the diagnostic process itself. However, the attention focused essentially on logical formalizations and on declarative characterizations of the set of solutions for a diagnostic problem and only in recent time more attention has been paid to computational aspects [5,9].

In order to address more directly procedural aspects of physical system diagnosis, we recently proposed a novel approach to the problem in which the diagnostic process is defined within a framework based on a Petri net model of the causal behavior of the system to be diagnosed [4,25]. Indeed, the possibility of modeling causal relationships for describing the evolution of a device has been recognized as fundamental in order to guide the diagnostic system to explain a given set of symptoms [12].

The possibility of using Petri nets in artificial intelligence applications is recently receiving a lot of attention [18,29,30] and the adoption of net models as tools for

* This work has been partially supported by CNR under grant n. 92.01601.PF69 and MURST.

knowledge representation and reasoning has been investigated by many researchers [11,14,21]; however, the problem of exploiting Petri nets to perform diagnostic reasoning has not received much attention so far. In [4,25] we showed that a Petri net model can be used to capture what is usually intended to be the causal behavior of a physical system, with a view of diagnostic problem solving[2]. The basic goal of the mentioned approach was to redefine the logical notion of diagnostic problem in terms of reachability in the Petri net and to propose some simulation algorithms on the model, in order to exploit the parallelism of concurrent evolutions which are made explicit by the model itself[3].

In the present paper we propose an alternative approach to the problem of performing diagnostic reasoning on a Petri net model, exploiting the notion of T-invariant [17,20]. A previous attempt to perform diagnostic reasoning through T-invariant analysis has been presented in [22] where the attention was essentially focused on single-fault diagnosis using a set of production rules modeled as a Petri net. In the present paper we rely on a more general and formal framework where the problem of multiple-fault diagnosis is naturally addressed.

We formally show how the notion of diagnostic solution can be related to that of T-invariant, adapting an idea presented in [23,24] where T-invariant analysis is applied to the answer extraction problem in logic programming. In this way, a problem traditionally tackled using symbolic manipulation techniques can be partially reformulated in algebraic terms.

The remainder of the paper is organized as follows: in section 2 we discuss the diagnostic problem by presenting a general definition that will be used throughout the paper; in section 3 the use of Petri nets to model the behavior of the system to be diagnosed is proposed; in section 4 we show how the notion of T-invariant can be related to that of diagnostic solution and a diagnostic procedure, exploiting the idea, is outlined. Finally, we discuss how the approach can be implemented in a diagnostic architecture by also discussing a formalism that can be used as interface between the knowledge engineer (i.e. the builder of the diagnostic knowledge base) and the Petri net model used by the diagnostic algorithm.

2 The Diagnostic Problem

In the following, we will concentrate on models of the behavior of systems to be diagnosed, named *behavioral models*. Given a behavioral model of a system to be diagnosed and a set of observations about such a behavior, a *model-based diagnosis* informally consists in a set of assumptions about the presence of certain malfunctions that can explain the observations using the model. There are essentially two logical notions of explanations proposed in the literature: in *consistency-based diagnosis* [27] the set of assumptions must just be consistent with the observations (i.e. if a value has been observed for a parameter, a diagnosis must not predict any value different from the observed one for the same parameter), while in *abductive diagnosis* [7] the assumptions, in conjunction with the model, must entail the observations.

[2] Furthermore, in [26] it is shown how the model can be validated following some correctness criteria.

[3] An implementation of the algorithms on a parallel architecture is currently under study.

In [8] a general definition, having the two approaches mentioned above as particular cases, has been proposed. The definition can be sketched as follows: a ***diagnostic problem*** is a tuple $DP = (BM, \Phi, <\Psi^+, \Psi^->)$ where BM is the behavioral model of the system to be diagnosed, Φ is the set of possible faults of the modeled system, Ψ^+ is the set of parameter values that must be predicted (i.e. entailed) by a diagnostic hypothesis and Ψ^- is the set of parameter values for which consistency is required; if D is the set of observed data, Ψ^+ will be a suitable subset of D, while Ψ^- will contain all the parameter values that are known to be absent in the case under examination. In particular, if a given instance of a parameter has been observed, all the other instances of the same parameter will belong to Ψ^-[4]. From a logical point of view a ***diagnosis*** can be regarded as a set of assumptions $H \subseteq \Phi$ about the presence of some faults or malfunctions such that

$$\forall m \in \Psi^+ \; BM \cup H \vdash m$$

$$\forall n \in \Psi^- \; BM \cup H \not\vdash n$$

where $\vdash$ is the derivation symbol. This clearly means that elements in Ψ^+ (respectively in Ψ^-) must (respectively must not) be predicted by the assumptions H using the model. H represents a ***diagnostic solution*** or ***explanation***, for the problem.

This definition is able to capture the classical characterizations of diagnosis mentioned above (i.e. consistency-based and abductive diagnosis). In the following, we will show how this characterization of diagnostic problem solving can be interpreted in a Petri net framework.

3 Petri nets and Behavioral Models

When dealing with a diagnostic reasoning problem, there are essentially two basic aspects: the model that is used and consequently what the assumptions, in terms of which to explain the observations, really mean. As mentioned in the introduction, one of the most popular approaches to model the behavior of a system is certainly the use of ***causal models***. In such models the behavior of a system can be described by means of a set of states, representing either a normal or an abnormal condition of the system, and a set of cause-effect relationships among them. Notice that entities denoted as states actually represent partial states or component of states, i.e. conditions concerning part of the modeled system; in the following we will use the generic term state to indicate a partial state. For diagnostic purposes, it is useful to distinguish among at least three types of entities in the model[5]:

- ***initial states*** which have no causes in the model (in the case of a fault model they represent the initial perturbations leading to a given malfunction);

[4] For example, in a mechanical domain, if we observe, among the others things, than the value of the parameter ***oil_pressure*** is ***very_low*** and we do not consider the temporal aspect, we cannot accept as a diagnosis a set of assumptions which predict for the parameter ***oil_pressure*** another value.

[5] For the sake of simplicity, in this paper we do not consider the role of contextual information (see [8] for a discussion).

– ***internal states*** **corresponding to conditions which are not directly observable or measured;**

– ***manifestations*** **corresponding to observable or measurable conditions (i.e. to observable manifestations of states).**

In this view, performing a diagnosis means to explain a set of *manifestations* in terms of *initial states*, using the cause-effect relationships described in the model. The *initial states* are thus the entities in terms of which the solutions to a diagnostic problem have to be provided.

In the following, we will assume that the behavioral model of the system to be diagnosed is of the type described above and that the modeled entities may assume different values from a finite and predefinite set of admissible ones (i.e. different instantiations of the modeled entities are possible). As shown in [6] a model of this type can be described, from a logical point of view, by means of a set of *definite clauses* without recursion[6]. A definite clause is a formula of the type

$$A_1 \wedge \ldots \wedge A_n \rightarrow B$$

where $\wedge$ is the conjunction symbol and $\rightarrow$ the implication symbol. $\{A_1 \wedge \ldots \wedge A_n\}$ is a (possibly empty) set of atoms called the *body* and B is an atom called the *head* of the clause. A set of definite clause is called *definite logic program*. As part of a behavioral model, the clause above is interpreted as the relation *"A_1 and ... and A_n causes B"*. The set of symbols is partitioned, according to the classification introduced above, in *initial_state, internal_state* and *manifestation* symbols; an *initial_state* symbol cannot occur in the head of a clause (it cannot be caused by anything), while a *manifestation* symbol cannot occur in the body of a clause (it cannot cause anything). The absence of recursion means that we impose that the behavioral model is acyclic; from the technical point of view the set of definite clauses must form a *hierarchical definite logic program* [3]. Since the number of different values that each state can assume is finite, we can deal with propositional models, however we will assume, for notational convenience, to adopt a first-order formalism where each A_i or B will represent a particular instance of a state or manifestation. More precisely, such instances will be represented by ground predicates like $s(v)$ where s is either an *initial_state*, an *internal_state* or a *manifestation* symbol and v an admissible value. Moreover, we will define as *admissible_values(s)* the set of possible values s can assume and as *modeled_values(s)* $\subseteq$ *admissible_values(s)* the set of modeled ones[7]. We are interested in showing that, instead of directly using a logical model for performing diagnostic reasoning (and so adopting some form of symbolic computation), we can use a Petri net as behavioral model for the system to be diagnosed and then exploit some peculiar techniques, such as T-invariant analysis, to face the problem.

Since the behavior of the systems of interest can be described by means of sets of definite clauses, the net model we need can be defined by means of a translation procedure from a set of definite clauses. Let us briefly review some basic definitions concerning net models [1]:

---

[6] The logical description used in the present paper can also be applied to behavioral models which do not explicitly rely on causal knowledge (see [8]).

[7] For instance, the admissible values of the state s could be $\{normal, abnormal\}$, but if the model were a pure fault model, the only modeled value would be the second one.

Definition 1. A net is a triple N=(S,T,F) where

- $S \cap T = \emptyset$
- $S \cup T \neq \emptyset$
- $F \subseteq (S \times T) \cup (T \times S)$
- $dom(F) \cup cod(F) = S \cup T$

S is the set of places, T the set of transition and F is the flow relationship whose elements are called arcs. Given an element $x \in S \cup T$, we will use the classical notation ${}^\bullet x = \{y : yFx\}$ and $x^\bullet = \{y : xFy\}$.

Definition 2. A Place-Transition (P/T) system is a sixtuple $\Sigma = (S, T, F, K, W, M_0)$ where

- (S,T,F) is a net
- $K : S \rightarrow \mathcal{N}^+ \cup \{\infty\}$ is a capacity function
- $W : (S \times T) \cup (T \times S) \rightarrow \mathcal{N}$ is a weight function such that $W(f) \in \mathcal{N}^+$ if $f \in F$ and $W(f) = 0$ if $f \notin F$
- $M_0 : S \rightarrow \mathcal{N}$ is an initial marking function which satisfy: $\forall s \in S$: $M_0(s) \leq K(s)$

Definition 3. An ordinary Petri net (or simply Petri net) is a P/T system such that: $\forall s \in S\ K(s) = \infty$ and $\forall f \in F\ W(f) = 1$; it can be denoted as $\Sigma = (S, T, F, M_0)$ (we will use $\Sigma = (S, T, F, W, M_0)$ when W needs to be explicit).

Since we are interested in (ordinary) Petri nets, we use the following definition adapted from definition 8 of [1].

Definition 4. Let $\Sigma = (S, T, F, W, M_0)$ be a Petri net; a marking is a function $M : S \rightarrow \mathcal{N}$; a transition $t \in T$ is enabled at M iff $\forall s \in S\ M(s) \geq 1$; if t is enabled at M, then t may occur (fire) yielding a new marking M' such that for every place $s \in S$ we have $M'(s) = M(s) - W(s,t) + W(t,s)$ and we write $M[t > M'$; the reachability set from a marking M_0, indicated as $R[M_0 >$, is the smallest set of markings such that:
1) $M_0 \in R[M_0 >$; 2) if $M_1 \in R[M_0 >$ and $M_1[t > M_2$ for some $t \in T$, then $M_2 \in R[M_0 >$.

In [23], Murata and Zhang showed how to translate a set of definite clauses into a *Predicate-Transition (Pr/T) Net* [13]; that procedure can be adapted to our case with some minor changes concerning the different kind of problem to be solved (diagnostic vs query answering problem) and the final net model obtained (Petri net vs Pr/T net).

First of all, notice that we do not have in the behavioral model anything corresponding to a set of facts from which to prove a goal as in logic programming, however, determining the solution to a diagnostic problem involves to find a set of instances of initial states from which to derive a suitable set of observations (manifestations). This implies that we have to consider which are the possible consequences of the initial states; for this reason we can add to the behavioral model BM the set

$$IS = \{\rightarrow p(v),\ \forall p \in \textit{initial\_state} \text{ symbols and } v \in \textit{modeled\_values}(p)\}$$

where $\rightarrow p(v)$ is a clause representing the assertion of the fact $p(v)$ (a similar solution is adopted in [22]). We will call $P = BM \cup IS$ a *complete behavioral model*.
The translation procedure, adapted from [23], is the following:

Procedure Translation
INPUT: a complete behavioral model (hierarchical definite logic program) $P = BM \cup IS$
OUTPUT: a Petri net model $\Sigma_P = (S, T, F, \emptyset)$;
begin
K:= *number_of_clauses(P)*;
define S,T,F as sets;
S:=T:=F:= $\emptyset$;

for i:=1 **to** K **do**
read i*th* clause;
let $p_1 \wedge \ldots \wedge p_n \rightarrow p$ be the i*th* clause;
T:= T $\cup \{t_i\}$;
S:= S $\cup \{p_1, \ldots, p_n, p\}$;
F:= F $\cup \{(p_1, t_i), \ldots (p_n, t_i), (t_i, p)\}$;
od

end.

It can be observed that this procedure produce the unfolded version of the Pr/T net model produced by the Murata and Zhang procedure[8]; moreover, since the input is a subclass of the inputs allowed in that procedure, the equivalence result proved in [31] still holds.

Theorem 1. (adapted from theorem 4.2 of [31]). *Given a definite logic program P and its Petri net model Σ_P, we have that*

$$P \vdash p \Leftrightarrow \Sigma_P \vdash p \Leftrightarrow \exists M \in R[M_0 >: M(p) \neq 0$$

where M_0 is the empty initial marking and $R[M_0 >$ is the reachability set from M_0.

Example 1. Let us consider the problem of modeling the faulty behavior of a car; as an example we will consider an oversimplified model in which the following entities are modeled:

initial states
piston_state: admissible_values={*normal*, *worn*}
ground_clearance: admissible_values={*normal*, *low*}
oil_sump_state: admissible_values={*normal*, *worn*}
spark_plug_mileage: admissible_values={*normal*, *high*}
carbur_tuning: admissible_values={*regular*, *irreg*}

internal states
oil_consumption: admissible_values={*normal*, *high*}

[8] Notice that the Pr/T net model produced by the Murata and Zhang procedure can always be unfolded into an ordinary Petri net, because they assume function free programs with a finite set of variables and constants.

***oil_sump*: admissible_values={*normal*, *holed*}**
***oil_lack*: admissible_values={*normal*, *intense*}**
***engine_temp*: admissible_values={*normal*, *high*}**
***incr_cool_temp*: admissible_values={*normal*, *high*}**
***cool_leakage*: admissible_values={*absent*, *high*}**
***spark_ign*: admissible_values={*normal*, *irreg*}**
***mixt*: admissible_values={*regular*, *irreg*}**
***mixt_ign*: admissible_values={*normal*, *irreg*}**

manifestations
***exhaust_smoke*: admissible_values={*normal*, *black*}**
***hole_in_oil_sump*: admissible_values={*no*, *yes*}**
***oil_light*: admissible_values={*off*, *on*}**
***temp_indic*: admissible_values={*normal*, *red*}**
***smoke_from_eng*: admissible_values={*no*, *yes*}**
***acc_resp*: admissible_values={*normal*, *irreg*}**

Let us consider a pure fault model: this means that only the faulty behavior of the system is modeled and that there is no modeling of normal or correct evolutions. The clauses forming the behavioral model *BM* are the following:

piston_state(worn) → oil_consumption(high)
ground_clearance(low) ∧ oil_sump_state(worn) → oil_sump(holed)
spark_plug_mileage(high) → spark_ign(irreg)
spark_ign(irreg) → mixt_ign(irreg)
oil_consumption(high) → exhaust_smoke(black)
oil_consumption(high) → oil_lack(intense)
oil_sump(holed) → oil_lack(intense)
oil_sump(holed) → hole_in_oil_sump(yes)
oil_lack(intense) → oil_light(on)
oil_lack(intense) → engine_temp(high)
engine_temp(high) → incr_cool_temp(high)
incr_cool_temp(high) → temp_indic(red)
incr_cool_temp(high) → cool_leakage(high)
cool_leakage(high) → smoke_from_eng(yes)
carbur_tuning(irreg) ∧ engine_temp(high) → mixt(irreg)
mixt(irreg) → mixt_ign(irreg)
mixt_ign(irreg) → acc_resp(irreg)

To obtain the complete behavioral model we have to add the following set of clauses:

→ piston_state(worn)
→ oil_sump_state(worn)
→ ground_clearance(low)
→ spark_plug_mileage(high)

The corresponding Petri net model is show in figure 1. The net model has the following characteristics:

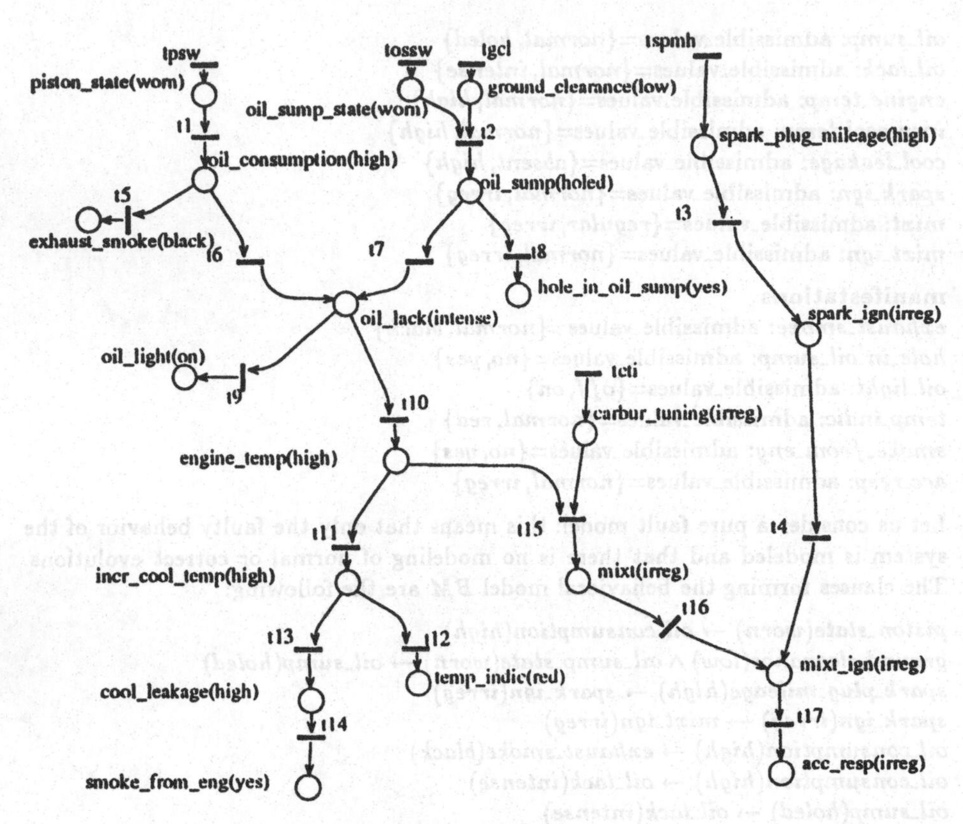

Fig. 1. Petri net model of the partial faulty behavior of a car

- the output of a *source* transition (i.e. $t : {}^{\bullet}t = \emptyset$) is a place representing an initial state instance (e.g. *piston_state*(*worn*));
- manifestation instances are represented by places having no transition in the output (e.g. *oil_light*(*on*)); the vice versa is not true in general, because we may have state instances without observable manifestations and without causal consequences;
- alternative causes to a given state or manifestation instance are represented by means of more than one transition in the input of a place (e.g. *oil_lack*(*intense*) can be caused by either *oil_consumption*(*high*) or *oil_sump*(*holed*));
- the net is acyclic, because of the requirement of the absence of recursion in the set of clauses; this means that cyclic processes cannot be modeled[9].

As we will see in the next section, the classification of the observed manifestations on such a model plays the fundamental role in the diagnostic process in general and, in particular, in the T-invariant analysis.

[9] This is a common assumption when modeling the causal behavior of a system without taking into account temporal aspects.

4 Relating T-invariants and Diagnostic Solutions

Let us briefly review some definitions and properties that will be useful in the following.

The incidence matrix for a Petri net $\Sigma = (S, T, F, W, M_0)$ with n transitions and m places, $A=[a_{ij}]$ is an $n \times m$ matrix of integers where $a_{ij} = W(i,j) - W(j,i)$ $(i \in T, j \in S)$. An n-vector of integers, X, is a *T-invariant* iff $A^T X = 0$. The entry $X(i)$ corresponds to transition i. The *support* $\| X \|$ of a T-invariant $X \geq 0$ is the subset of transitions corresponding to nonzero entries of X. A T-invariant X is *minimal* iff there is no other T-invariant X' such that $\forall t \in T \; X'(t) \leq X(t)$. A support is *minimal* iff no proper nonempty subset of the support is also a support. Not every minimal T-invariant has a support which is minimal. A minimal T-invariant corresponding to a minimal support is called *minimal support T-invariant*; it is well known that any T-invariant can be obtained as a linear combination of minimal support T-invariants [20].

A sequence of transition σ is called a *firing sequence*; the *firing count vector* $\bar{\sigma}$ of σ is the *Parikh mapping* of the sequence (i.e. the mapping associating with each transition its number of occurrences in σ). A firing sequence σ is said to be *executable* from a marking M_0 iff $\sigma = < t_1 \ldots t_n >$ and t_i is firable at M_{i-1} yielding M_i $(1 \leq i \leq n)$. A transition t is *potentially firable* from a marking M_0 iff there exists a firing sequence σ executable from M_0 containing t.

Characterizing a model-based diagnostic problem involves to distinguish between observations that must be predicted using the model and other parameters that are known to be absent and so that must not be predicted in the case under examination. In section 2, we indicated this two different sets of parameters respectively as Ψ^+ and Ψ^-. Since the goal of the paper is to characterize the notion of diagnostic solution in terms of T-invariants of a net model, we must introduce these parameters in the suitable way.

Definition 5. Given a complete behavioral model $P = BM \cup IS$, its Petri net model $\Sigma_P = (S, T, F, \emptyset)$ and a diagnostic problem $DP = (M, IS, < \Psi^+, \Psi^- >)$, a Diagnostic Petri net $(DiPN)$ corresponding to DP is a Petri net
$\Sigma_D = (S, \{T_{is}, T'', T^+, T^-\}, F', \emptyset)$ where: $T_{is}=\{t \in T : ^\bullet t = \emptyset\}$, $T'' = T - T_{is}$, $T^+=\{t_m : t_m^\bullet = \emptyset \wedge ^\bullet t_m = \{m\} \wedge m \in \Psi^+\}$, $T^-=\{t_m : t_m^\bullet = \emptyset \wedge ^\bullet t_m = \{m\} \wedge m \in \Psi^-\}$, $F' = F \cup \{(m, t_m) : (m \in \Psi^+ \wedge t_m \in T^+) \vee (m \in \Psi^- \wedge t_m \in T^-)\}$.

Example 2. Consider the net model in figure 1 and suppose to observe the manifestations *oil_light(on)*, *hole_in_oil_sump(no)*, *temp_indic(red)*, *acc_resp(normal)*. Notice that, since the net is a fault model (i.e. it represents just the faulty behavior of the system), there is no modeling of normal situations like *hole_in_oil_sump(no)* and *acc_resp(normal)*, however, this information is useful in order to determine what kind of parameters are absent in the case under examination. If we suppose that all observed abnormal values of the manifestations must be predicted by a diagnosis, then we have the following subdivision:

- $\Psi^+ = \{$*oil_light(on)*, *temp_indic(red)*$\}$
- $\Psi^- = \{$*hole_in_oil_sump(yes)*, *acc_resp(irreg)*$\}$

We model the observations by adding to the model the set $\{t_{olo}, t_{tir}, t_{hosy}, t_{ari}\}$ (corresponding respectively to *oil_light(on)*, *temp_indic(red)*, *hole_in_oil_sump(yes)* and *acc_resp(irreg)*) such that $T^{+}=\{t_{olo}, t_{tir}\}$ and $T^{-}=\{t_{hosy}, t_{ari}\}$. In figure 2 we

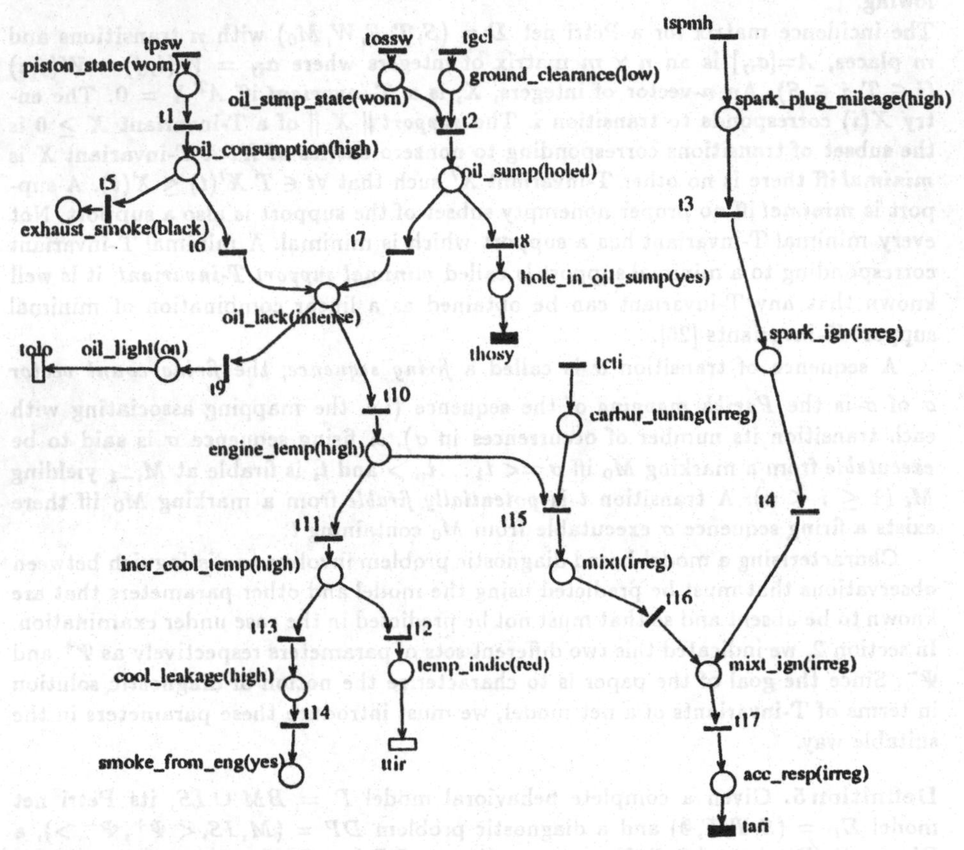

Fig. 2. A Diagnostic Petri Net Model

show the *DiPN* corresponding to this case; transitions in the set T^{+} are empty thick bars and transitions in T^{-} are full thick bars.

Since a diagnosis consists of a set of initial state instances predicting the manifestations in Ψ^{+} and not predicting those in Ψ^{-}, by considering the net model we have to characterize the firing sequences starting from source transitions (whose firing represents the instantiations of an initial state) and ending up with sink transitions (whose firing may represent the prediction of manifestation instances). The notion of T-invariant can clearly help in this task.

In [23,24,31], a *goal transition* is a transition corresponding to a goal clause (i.e. a clause with empty head) and so it is a sink transition. The following theorem has been proved in [24].

Theorem 2. *Let $\Sigma = (P, T, F, W, M_0)$ be a Petri net such that $\forall t \in T\ |t^\bullet| \leq 1$ ($|A|$=cardinality of set A). Let t_g be a goal transition in T; t_g is potentially firable from $M_0 = \emptyset$ iff Σ has a T-invariant X such that $X \geq 0$, $X(t_g) \neq 0$.*

Remark 1. Since every transtion of a *DiPN* $\Sigma_D = (S, \{T_{is}, T'', T^+, T^-\}, F, \emptyset)$ satisfies the condition of theorem 2, we can apply it to every transition $t_m \in (T^+ \cup T^-)$ and so Σ_D has a T-invariant X such that $X \geq 0$, $X(t_m) \neq 0$ iff t_m is potentially firable from the empty marking.

Let us now consider the following theorem which has been proved in [31].

Theorem 3. *Given a definite logic program P with a goal G, its net model Σ_P and a goal transition t_g, we have that $P \vdash G$ iff t_g is potentially firable from the empty marking in Σ_P.*

Remark 2. If we restrict P to be a complete behavioral model $P = BM \cup IS$ (hierarchical definite logic program) and we consider $\Sigma_D = (S, \{T_{is}, T'', T^+, T^-\}, F, \emptyset)$ to be the *DiPN* corresponding to the diagnostic problem $DP = (BM, IS, < \Psi^+, \Psi^- >)$, then the previous theorem can be restated as follows:
$\forall m \in (\Psi^+ \cup \Psi^-)\ P \vdash m$ iff $t_m \in (T^+ \cup T^-)$ (where $m^\bullet = \{t_m\} \wedge {}^\bullet t_m = \{m\}$) is potentially firable from the empty marking in Σ_D.

We can now prove the following important theorem.

Theorem 4. *Given a complete behavioral model $P = BM \cup IS$, a diagnostic problem $DP = (BM, IS, < \Psi^+, \Psi^- >)$ and a DiPN $\Sigma_D = (S, \{T_{is}, T'', T^+, T^-\}, F, \emptyset)$ corresponding to DP, there exists a T-invariant X of Σ_D such that $X(t_1) \ldots X(t_k) \neq 0$ and $t_1 \ldots t_k \in (T^+ \cup T^-)$ iff $P \vdash m_1 \ldots P \vdash m_k$ where $m_i^\bullet = \{t_i\} \wedge {}^\bullet t_i = \{m_i\}$ and $m_i \in (\Psi^+ \cup \Psi^-)$ $(1 \leq i \leq k)$.*

Proof. By Remark 1, there exists a T-invariant X such that $X(t_1) \ldots X(t_k) \neq 0$ with $t_1 \ldots t_k \in (T^+ \cup T^-)$ iff $t_1 \ldots t_k$ are potentially firable from the empty marking iff (by Remark 2) $P \vdash m_1 \ldots P \vdash m_k$ where $m_i^\bullet = \{t_i\} \wedge {}^\bullet t_i = \{m_i\}$ and $m_i \in (\Psi^+ \cup \Psi^-)$ $(1 \leq i \leq k)$. □

This theorem allow us to conclude that the supports of the T-invariants of the net characterize the evolutions (in logical terms the derivations) from initial states to parameters in the sets Ψ^+ and Ψ^-. In fact, the subset of T_{is} present in a support represents a set of initial state instances whose conjunction is a *sufficient condition* for the derivation of manifestation instances represented by transitions $t' \in (T^+ \cup T^-)$ that are in the support itself.

Definition 6. A set of transitions τ_1 covers a set of transitions τ_2 iff every transition in τ_2 occurs in τ_1.

It should be clear that if a subset τ of T_{is} represents a sufficient condition for the derivation of a manifestation instance m (i.e. its elements are present in a support where also t_m, corresponding to m, is present), then every subset of T_{is} that covers τ also represent a sufficient condition for the derivation of m.

Since the whole set of T-invariants of a net can be characterized by the set of *minimal support T-invariants*, we will be interested in computing these ones and in particular the minimal supports corresponding to them. One of the most popular method for this task is certainly the Martinez-Silva algorithm [19] able to compute all the minimal support invariants of a P/T net; this method has been applied to the computation of invariants in the *GreatSPN* package [2] which is the tool we used to analyse the examples in this paper.

Example 3. The minimal supports of the T-invariants for the net in figure 2 are the followings:

$\sigma_1=\{t_{ossw}, t_{gcl}, t_2, t_7, t_{10}, t_{11}, t_{12}, t_{tir}\}$

$\sigma_2=\{t_{ossw}, t_{gcl}, t_2, t_7, t_9, t_{olo}\}$

$\sigma_3=\{t_{psw}, t_1, t_6, t_{10}, t_{11}, t_{12}, t_{tir}\}$

$\sigma_4=\{t_{psw}, t_1, t_6, t_9, t_{olo}\}$

$\sigma_5=\{t_{ossw}, t_{gcl}, t_2, t_8, t_{hosy}\}$

$\sigma_6=\{t_{psw}, t_{cti}, t_1, t_6, t_{10}, t_{15}, t_{16}, t_{17}, t_{ari}\}$

$\sigma_7=\{t_{ossw}, t_{gcl}, t_{cti}, t_2, t_7, t_{10}, t_{15}, t_{16}, t_{17}, t_{ari}\}$

$\sigma_8=\{t_{spmh}, t_3, t_4, t_{17}, t_{ari}\}$

There are two alternative evolutions predicting *temp_indic(red)* (i.e. σ_1 and σ_3 containing t_{tir}) and two alternative evolutions predicting *oil_light(on)* (i.e. σ_2 and σ_4 containing t_{olo}); however, σ_1 and σ_2 contains transitions t_{ossw} and t_{gcl} which are also contained in σ_5 that predicts *hole_in_oil_sump(yes)* $\in \Psi^-$ (it contains transition t_{hosy}). This means that the evolutions starting from the conjunction of initial states *ground_clearance(low)* and *oil_sump_state(worn)* (or from every conjunction corresponding to a subset of T_{is} that covers $\{t_{ossw}, t_{gcl}\}$) cannot be considered diagnostic explanations and so supports σ_1, σ_2, σ_5 and σ_7 must be discarded. For a similar reason (i.e. prediction of *acc_resp(irreg)* represented by transition t_{ari}), also σ_6 and σ_8 must be discarded. Supports σ_3 and σ_4 contains transition t_{psw} and no other source transition; this transition, alone, does not appear in any support containing a transition of the set T^-, so $\{piston\_state(worn)\}$ is the only possible diagnosis for this diagnostic problem. Moreover, the set $\sigma_{3,4} = \sigma_3 \cup \sigma_4$ can be taken as a track of the derivation relating the obtained diagnosis to the observed manifestations. The general diagnostic algorithm based on the T-invariant analysis can then be informally sketched as follows:

Procedure Diagnosis

INPUT: a Diagnostic Petri net $\Sigma_D = (S, \{T_{is}, T', T^+, T^-\}, F, \emptyset)$
OUTPUT: a set of diagnostic solutions H in terms of transitions in T_{is}

begin
compute the minimal supports of the T-invariants for Σ_D;
let L be the list of such minimal supports;

for each $t \in T^-$ do
    for each support $\sigma : t \in \sigma$ do
        $DT := \{t' \in T_{is} : t'$ occurs in $\sigma\}$;
        delete from L all supports where
        $\tau \subseteq T_{is}$ covering DT occurs
        od
    od

$X := \emptyset$;

for each $t \in T^+$ do
    $H' := \emptyset$
    for each $\sigma \in L$: $t \in \sigma$ do
        $IT := \{t' \in T_{is} : t' \in \sigma\}$;
        $H' := H' \cup \{IT\}$
        od
    $X := X \cup \{H'\}$
    od

combine elements in X to produce H
end.

In order to consider in more detail how the set H is built, suppose to change the observations in the problem of example 2 so that *hole_in_oil_sump(yes)* $\in \Psi^+$ and that Ψ^- ={*acc_resp(irreg)*}; this means that we have to consider all the minimal supports from σ_1 to σ_5 of the net in figure 2 (the others are discarded because they contains t_{ari}). Looking at example 3 we have that the sets of source transitions from which t_{olo} and t_{tir} can fire are $\{t_{psw}\}$ and $\{t_{gcl}, t_{ossw}\}$ (so *oil_light(on)* and *temp_indic(red)* can be explained either by the single fault {*piston_state(worn)*} or by the double fault {*ground_clearance(low)*, *oil_sump_state(worn)*}); transition t_{hosy}, on the other hand, can only fire from the set $\{t_{gcl}, t_{ossw}\}$ (*hole_in_oil_sump(yes)* is only explained by the double fault above).

By combining them we get two possible diagnosis expressed in terms of source transitions: $\{t_{psw}, t_{gcl}, t_{ossw}\}$ and $\{t_{gcl}, t_{ossw}\}$. Notice that the first one is not minimal with respect to the assumed faults and so, as many approaches to model-based diagnosis suggest, it can be considered as redundant (the problem of characterizing in a suitable way the set of diagnoses is very important, but it is outside the scope of the present paper; a detailed discussion on the topic can be found in [10]).

5 Discussion

In this paper we presented a novel approach to the problem of performing model-based diagnosis, relying on a Petri net model of the behavior of the system to be diagnosed and exploiting T-invariant analysis. We showed how a very general characterization of the diagnostic problem can be captured by this approach, partially transforming a problem classically tackled by means of symbolic computation methods into a linear algebraic one.

It is worth noting that we do not aim at using the Petri net model as a direct tool of diagnostic knowledge representation, but rather as a modeling tool that can be derived from some other forms of knowledge representation such as a logical model or other higher level formalisms. For example, one of these formalisms is represented by the *causal networks* used in the CHECK system, a diagnostic architecture developed at the Dipartimento di Informatica of the University of Torino [6,28]. A causal network is a useful graphical formalism used to represent causal knowledge; a CHECK causal network is composed by a set of nodes connected by arcs representing different kinds of relationships. Let us consider the causal network reported in figure 3 representing part of a more detailed model used for the detection of car faults and corresponding to the Petri net model shown in figure 1. Different types of nodes are defined in the network[10].

- STATE nodes (elliptic boxes) represent non-observable internal states of the modeled system.
- FINDING nodes (rhomboidal boxes) represent observable parameters (manifestations) in the modeled system.
- INITIAL_CAUSE nodes (double-lined elliptic boxes) represent the initial perturbations (initial states) that may lead the system to a faulty behavior.

Each one of these nodes is characterized by a set of admissible values so that we can identify different instances for each node; however, for simplicity of description, we will consider that the network in figure 3 is a pure fault model where the name of the nodes denotes an implicit *abnormal* value (such a value is however explicitly given in figure 1).

Two main types of arcs are defined in the network: CAUSAL (continuous lines in figure 3) and HAM (Has As a Manifestation) arcs (dashed lines). Notice that more than one CAUSAL arc can enter a state: such multiple causes can be ANDed or ORed as indicated in figure 3. No cyclic processes can be modeled since CAUSAL relationships are assumed to form no loops. Each HAM arc connects a STATE node S to a FINDING node m and represents the fact that the finding m is an observable manifestation of the internal state S. Because of the discussion in section 3, it should be clear that a causal network can be described, in logical terms, by a set of definite clauses (see also [6]), so it is also possible to obtain a Petri net model corresponding to it.

[10] Other types of nodes can be defined in the formalism that is actually more structured than as presented here (see [6,28]); we will describe here only the types of nodes which are relevant for our discussion.

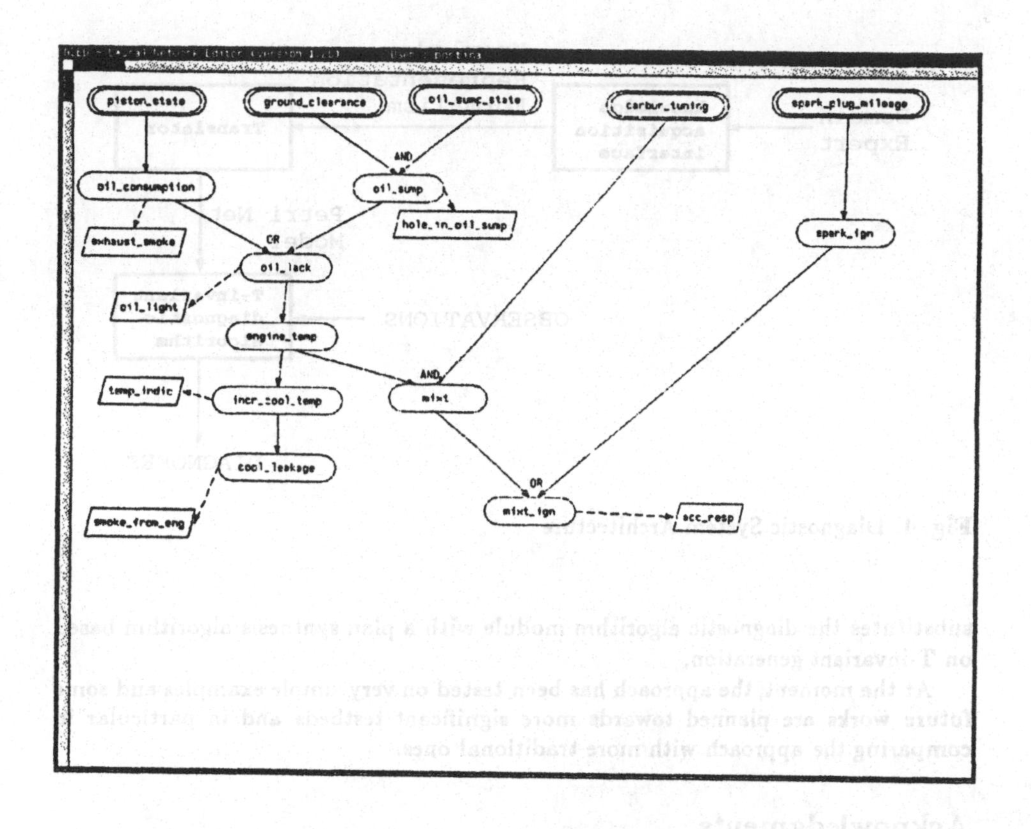

Fig. 3. Causal Network

A different possibility could be that of adopting a High-Level Net formalism such as *Predicate-Transition Nets* or *Coloured Petri Nets* [16] as knowledge representation tool; in this case it could be possible to choose between performing T-invariant analysis directly on it or to unfold the net into an ordinary Petri net (this is possible because each state or manifestation can assume a finite number of different instantiations). For instance, an approach very similar to that described here is the use of *Pr/T nets* to model robot plans applied in the ROPES planning tool [30]; ROPES uses T-invariant analysis on the model in order to generate plans, that is sequences of actions to be performed in order to obtain a given goal.

In conclusion, a diagnostic architecture based on the approach described in this paper can be represented as shown in figure 4. In fact, the Petri net model can be viewed as a model on which to implement the diagnostic algorithm, but hidden to the user by more suitable knowledge representation formalisms. Notice that, as the ROPES tool suggests, this kind of architecture can be applied to tasks different than diagnosis, for which T-invariant analysis is relevant; for instance ROPES simply

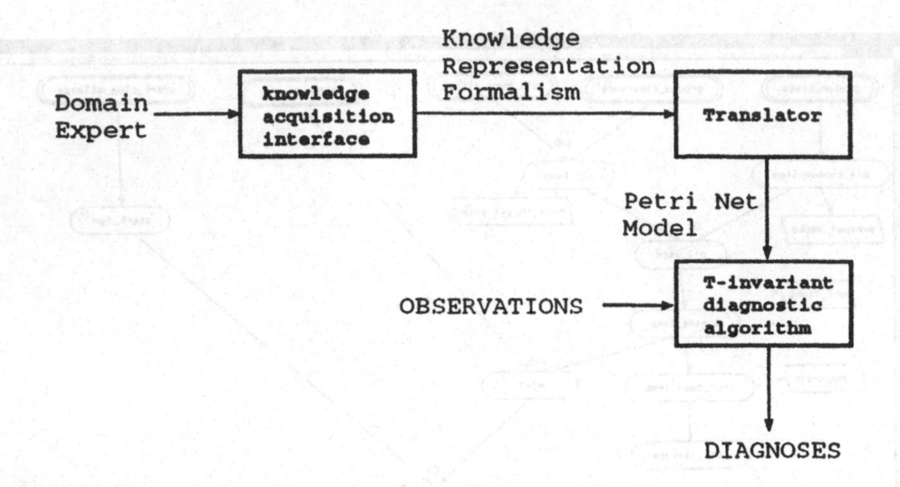

Fig. 4. Diagnostic System Architecture

substitutes the diagnostic algorithm module with a plan synthesis algorithm based on T-invariant generation.

At the moment, the approach has been tested on very simple examples and some future works are planned towards more significant testbeds and in particular in comparing the approach with more traditional ones.

Acknowledgments

I am very grateful to G. Chiola and his working group for having made me available the *GreatSPN* software tool. Special thanks to G. Franceschinis for helpful comments about the paper and to D. Zhang and T. Murata for having pointed out to me some of their interesting papers I was not acquainted with.

References

1. L. Bernardinello and F. De Cindio. A survey of basic net models and modular net classes. In G. Rozemberg, editor, *Advanced in Petri Nets 1992, LNCS 609*, pages 304–351. Springer Verlag, 1992.
2. G. Chiola. GreatSPN 1.5 software architecture. In G. Balbo and G. Serazzi, editors, *Proc. 5th Int. Workshop on Modeling Techniques and Tools for Computer Performance Evaluation*, Torino, 1991. North Holland.
3. K. Clark. Negation as failure. In H. Gallaire and J. Minker, editors, *Logic and Data Bases*, pages 293–322. Plenum Press, 1978.
4. L. Console and L. Portinale. Model based diagnosis of system malfunction with Petri nets. In S.G. Tzafestas and J.C. Gentina, editors, *Robotics and Flexible Manufacturing Systems*, pages 417–426. Elsevier Science, 1992. A preliminary version appeared in *Proc. 13th IMACS World Congress on Computation and Applied Mathematics*, Dublin, 1991.

5. L. Console, L. Portinale, and D. Theseider Dupré. Focusing abductive diagnosis. In *Proc. 11th Int. Conf. on Expert Systems and Their Applications (Conf. on 2nd Generation Expert Systems)*, pages 231–242, Avignon, 1991. Also in *AI Communications* 4(2/3):88-97, 1991.
6. L. Console, L. Portinale, D. Theseider Dupré, and P. Torasso. Combining heuristic and causal reasoning in diagnostic problem solving. In J.M. David, J.P. Krivine, and R. Simmons, editors, *Second Generation Expert Systems*. Springer Verlag. forthcoming.
7. L. Console, D. Theseider Dupré, and P. Torasso. A theory of diagnosis for incomplete causal models. In *Proc. 11th IJCAI*, pages 1311–1317, Detroit, 1989.
8. L. Console and P. Torasso. A spectrum of logical definitions of model-based diagnosis. *Computational Intelligence*, 7(3):133–141, 1991.
9. J. de Kleer. Focusing on probable diagnoses. In *Proc. AAAI 91*, pages 842–848, Anaheim, CA, 1991.
10. J. de Kleer, A. Mackworth, and R. Reiter. Characterizing diagnoses and systems. *Artificial Intelligence*, 56(2–3):197–222, 1992.
11. Y. Deng and S.K. Chang. A G-net model for knowledge representation and reasoning. *IEEE Trans. on Knowledge and Data Engineering*, KDE 2(3):295–310, 1990.
12. R. Milne (ed.). Special issue on causal and diagnostic reasoning. *IEEE Trans. on Systems, Man and Cybernetics*, 17(3), 1987.
13. H.J. Genrich and K. Lautenbach. System modeling with high level petri nets. *Theoretical Computer Science*, 13:109–136, 1981.
14. A. Giordana and L. Saitta. Modeling production rules by means of Predicate Transition Networks. *Information Sciences*, 35:1–41, 1985.
15. W. Hamscher, L. Console, and J. de Kleer. *Readings in Model-Based Diagnosis*. Morgan Kaufmann, 1992.
16. K. Jensen. Coloured Petri Nets and the invariant method. *Theoretical Computer Science*, 14:317–336, 1981.
17. K. Lautenbach. Linear algebraic techniques for Place/Transition nets. In W. Brauer, W. Reisig, and G. Rozenberg, editors, *Petri Nets: Central Models and Their Properties*, pages 142–167. Springer Verlag, 1987. LNCS 254.
18. J. Martinez, P.R. Muro, M. Silva, S.F. Smith, and J.L. Villaroel. Merging artificial intelligence techniques and Petri nets for real-time scheduling and control of production systems. In *Proc. 12th IMACS World Congress on Scientific Computation*, pages 528–531, Paris, 1988.
19. J. Martinez and M. Silva. A simple and fast algorithm to obtain all invariants of a generalized Petri net. In W. Reisig C. Girault, editor, *Informatik-Fachberichte, Applications and Theory of Petri Nets*, pages 301–310. Springer Verlag, 1982.
20. G. Memmi and G. Roucairol. Linear algebra in net theory. In *Lecture Notes in Computer Science*, volume 84, pages 213–223. Springer Verlag, 1980.
21. T. Murata, V.S. Subrahmanian, and T. Wakayama. A Petri net model for reasoning in the presence of inconsistency. *IEEE Transactions on Knowledge and Data Engineering*, KDE 3(3):281–292, 1991.
22. T. Murata and J. Yim. Petri-net deduction methods for propositional-logic rule-based systems. Technical Report UIC-EECS-89-15, University of Illinois at Chicago, 1989.
23. T. Murata and D. Zhang. A Predicate-Transition Net model for parallel interpretation of logic programs. *IEEE Transactions on Software Engineering*, SE 14(4):481–497, 1988.
24. G. Peterka and T. Murata. Proof procedure and answer extraction in Petri net model of logic programs. *IEEE Trans. on Software Eng.*, SE 15(2):209–217, 1989.

25. L. Portinale. Behavioral Petri Net: a model for diagnostic knowledge representation and reasoning. Technical report, Dip. Informatica, Universita' di Torino, 1992. (submitted for publication).
26. L. Portinale. Verification of causal models using Petri nets. *International Journal of Intelligent Systems*, 7(8):715–742, 1992.
27. R. Reiter. A theory of diagnosis from first principles. *Artificial Intelligence*, 32(1):57–96, 1987.
28. P. Torasso and L. Console. *Diagnostic Problem Solving: Combining Heuristic, Approximate and Causal Reasoning.* Van Nostrand Reinhold, 1989.
29. R. Valette and M. Courvoisier. Petri nets and artificial intelligence. In *Proc. Int. Workshop on Emerging Technologies for Factory Automation*, North Queensland, Australia, 1992.
30. D. Zhang. Planning with Pr/T nets. In *Proc. IEEE Int. Conference on Robotics and Automation*, pages 769–775, Sacramento, CA, 1991.
31. D. Zhang and T. Murata. Fixpoint semantics for Petri net model of definite clause logic programs. Technical Report UIC-EECS-87-2, University of Illinois at Chicago, 1987. also to appear in *Advances in the Theory of Computation and Computational Mathematics*, Ablex Publ.

Marking Optimization of Stochastic Timed Event Graphs

Nathalie SAUER and Xiaolan XIE*

SAGEP Project, INRIA
Technopôle Metz 2000, 4 Rue Marconi, 57070 METZ, FRANCE

Abstract. This paper addresses the performance evaluation and marking optimization of stochastic timed event graphs. The transition firing times are generated by random variables with general distribution. The marking optimization problem consists in obtaining a given cycle time while minimizing a p-invariant criterion (or S-invariant). Some important properties have been established. In particular, the average cycle time is shown to be non-increasing with respect to the initial marking while the p-invariant criterion is non-decreasing. We further prove that the criterion value of the optimal solutions is non-increasing in transition firing times in stochastic ordering's sense. Lower bounds and upper bounds of the average cycle time are proposed. Based on these bounds, we show that the p-invariant criterion reaches its minimum when the firing times become deterministic. A sufficient condition under which an optimal solution for the deterministic case remains optimal for the stochastic case is given. These new bounds are also used to provide simple proof of the reachability conditions of a given cycle time. Thanks to the tightness of the bounds for the normal distribution case, two algorithms which leads to near-optimal solutions have been proposed to solve the stochastic marking optimization problem with normal distributed firing times.

1. Introduction

Petri nets have been proven to be a powerful tool for modelling complex dynamic systems and for evaluating their performance. It provides an unified language for system modelling, property checking and performance evaluation of dynamic systems with synchronization, concurrency and common resources. Successful application areas include communication systems, computer systems, manufacturing systems, etc. Excellent survey can be found in [10, 15].

This paper is part of our work which intends to provide tools for optimizing the marking of the Petri net model of a real-life system. To better understand the problem, let us consider a manufacturing system. The marking of its Petri net model corresponds to the resources (transportation resources, machines and tools for example) and the work-in-process. Roughly speaking, the marking optimization is equivalent to the optimal utilization of the manufacturing resources.

In this paper we limit ourselves to stochastic timed event graphs in which the firing times generated by random variables. An event graph, also called marked graph, is a Petri net in which each place has exactly one input transition and one output transition. A strongly connected event graph has some important properties,

* Please address all correspondence to the second author

specifically: (i) the number of tokens in any elementary circuit remains constant whatever the transition firings, and (ii) the system is live iff each elementary circuit contains at least one token (see for instance [4, 5, 7]).

In the deterministic case, it has been proven [4, 13] that: (i) the cycle time of an elementary circuit is given by the ratio of the sum of the firing times of the transitions of the circuit by the number of tokens in the circuit; (ii) the cycle time of a strongly connected event graph is equal to the greatest cycle time among the ones of all the elementary circuits. Furthermore, a cycle time being given, algorithms have been proposed in [8, 9] to find an initial marking which leads to an average cycle time less than the given cycle time while minimizing a linear criterion function of the initial marking.

In the stochastic case, it is no more possible to take advantage of the elementary circuits to evaluate the behaviour of the event graph and to reach a given performance. Previous work mainly focused on ergodicity conditions and performance bounds. Ergodicity conditions have been obtained for timed event graphs [1], for stochastic timed Petri nets [6] and for max-plus algebra models of stochastic discrete event systems [14].

For a strongly connected stochastic timed event graph, it has been proven that an average cycle time exists under some fairly weak conditions (see Section 2). Both upper bounds and lower bounds have been proposed (see [2, 3, 11, 12, 17]). In particular, tight upper bounds and lower bounds have been obtained by using stochastic comparison properties (see [2]) and superposition properties (see [17]).

The marking optimization problem for strongly connected stochastic timed event graphs, which consists in obtaining a given cycle time while minimizing a linear criterion depending on the initial marking, has been addressed in [11, 12]. It was proven that any cycle time greater than the maximal mean transition firing time can be reached provided that enough tokens are available. Heuristic algorithms for solving the stochastic marking optimization problem have been proposed.

The purpose of this paper is to present new properties of the stochastic marking optimization problem and to propose near-optimal solutions.

To this end, we give in Section 2 the formal definition of the marking optimization problem of stochastic timed event graphs which consists in obtaining a given cycle time while minimizing a p-invariant criterion. We notice that a p-invariant is also called P-semi flow or S-invariant.

Some important properties are proposed in Section 3. In particular, we show that mutually reachable markings have the same average cycle time and the same p-invariant criterion value. The average cycle time is shown to be non-increasing with respect to the initial marking while the p-invariant criterion is non-decreasing. We further prove that the criterion value of the optimal solutions is non-increasing in transition firing times in stochastic ordering's sense.

Section 3 also presents performance bounds of stochastic timed event graph and establishes properties of the optimal solutions. Based on these bounds, we show that the p-invariant criterion reaches its minimum when the firing times become deterministic. A sufficient condition under which an optimal solution for the deterministic case remains optimal for the stochastic case is given. These new bounds are then used to provide simple proof of the reachability conditions proposed in [11, 12]

Thanks to the tightness of the bounds for the normal distribution case, two algorithms which leads to near-optimal solutions have been proposed in Section 4 to solve the stochastic marking optimization problem.

2. Notations, Assumptions and Problem Setting

Let us consider a strongly connected event graph $N = (\mathcal{P}, T, F)$. $\mathcal{P}$ is the set of places, T is the set of transitions, and $F \subseteq (\mathcal{P} \times T) \cup (T \times \mathcal{P})$ is the set of directed arcs connecting places to transitions and transitions to places. We denote by $\mathcal{M}_0$ the initial marking of N.

We assume that no transition can be fired by more than one token at any time (i.e. recycled transitions). This implies that there is a self loop place with one token related to each transition, i.e. $(t, t) \in \mathcal{P}$ and $\mathcal{M}_0((t,t)) = 1$, $\forall\, t \in T$ where (t, s) indicates the place connecting transition t to transition s. We further assume that, when a transition fires, the related tokens remain in the input places until the firing process ends. They then disappear and one new token appears in each output place of the transition.

As a result, the set of places $\mathcal{P}$ can be written as $\mathcal{P} = P \cup Pt$ where Pt denotes the set of self loop places and P the other places, i.e. $Pt = \{(t, t), \forall t \in T\}$ and $P \cap Pt = \emptyset$. Furthermore, since there is always exactly one token in each place belonging to Pt, only the marking of the places belonging to P will be considered in the following.

Since N is an event graph, each place has exactly one input transition and one output transition. Without loss of generality, we assume that there exists at most one place between any two transitions. The following notations will be used :

$^{\bullet}t$ (resp. $t^{\bullet}$) : set of input (resp. output) places of transition t
$^{\bullet}p$ (resp. $p^{\bullet}$) : unique input (resp. output) transition of place p
in(t) : set of transitions which immediately precede transition t, i.e.

$$in(t) = \{s \in T / \exists\, p \in P,\ ^{\bullet}p = s \text{ and } p^{\bullet} = t\}$$

out(t) : set of transitions which immediately follow transition t, i.e.

$$out(t) = \{s \in T / \exists\, p \in P,\ ^{\bullet}p = t \text{ and } p^{\bullet} = s\}$$

(t, s) : place connecting transition t to transition s
Γ: set of elementary circuits of N
M_0: initial marking of the places belonging to P
$M_0(\gamma)$: total number of tokens contained initially in $\gamma \in \Gamma$
$R(M_0)$: the set of markings reachable from M_0

The following notations related to the transition firing times will be used:

$X_t(k)$: non-negative random variable which generates the time required for the k-th firing of transition t
$S_t(k)$: instant of the k-th firing initiation of transition t

By convention, $X_t(k) = 0, \forall k \le 0$ and $S_t(k) = 0, \forall k \le 0$. As shown in [4], the transition firing initiation instants can be determined by the following recursive equations :

$$S_t(k) = \max_{\tau \in in(t)} \left\{ S_\tau\left(k - \mathcal{M}_0((\tau,t))\right) + X_\tau\left(k - \mathcal{M}_0((\tau,t))\right) \right\} \quad (1)$$

We assume that for each transition t, the sequence of its firing times $\{X_t(k)\}_{k=1}^{\infty}$ is a sequence of independent identically distributed (i.i.d.) integrable random variables and that $\{X_t(k)\}_{k=1}^{\infty}$ for all $t \in T$ are mutually independent sequences.

It was proven in [1] that, under the foregoing assumptions, there exists a positive constant $\pi(M_0)$ such that:

$$\lim_{k \to \infty} S_t(k)/k = \lim_{k \to \infty} E\left[S_t(k)\right]/k = \pi(M_0), \quad \text{a.s. } \forall t \in T \quad (2)$$

$\pi(M_0)$ is the average cycle time of the event graph.

Since $\{X_t(k)\}_{k=1}^{\infty}$ are sequences of i.i.d. random variables, the index k is often omitted and we use X_t to denote the firing time of transition t whenever k is not necessary. We further assume that the first and second moments of X_t exist and denote by m_t its mean value and by σ_t its standard deviation, i.e. $m_t = E[X_t]$ and $\sigma_t^2 = E[(X_t - m_t)^2]$.

Since any stochastic timed event graph is completely characterized by its net structure, its initial marking and the set of transition firing time sequences, it can be denoted by the triplet $(N, M_0, \{X_t(k)\})$.

Finally, the marking optimization problem can be defined as follows. Given a positive real value C, the marking optimization problem consists in finding an initial marking $M_0 \in \mathbb{N}^{|P|}$ in order to

minimize $f(M_0) = U^T . M_0$ (3)

subject to the following constraint :

$\pi(M_0) \le C$ (4)

where $U = (u_1, u_2, ..., u_{|P|})^T \in (\mathbb{R}^+)^{|P|}$ is a p-invariant (also called P-semi flow or S-invariant).

In the marking optimization problem, the value 1/C can be considered as the required throughput rate or productivity value. The vector U can be considered as the unit cost of the resources (or tokens). The marking optimization problem consists in reaching the required performance while minimizing the total resource cost $U^T . M_0$.

The use of the p-invariant criterion should not be surprising. As a matter of fact, the number of tokens in each elementary circuit of the event graph indicates the number of resources of the related type. Hence, two initial markings with identical circuit counts correspond to the use of the same set of manufacturing resources and give the same total resource cost. This implies that the cost function is a p-invariant criterion.

3. Properties

This section is devoted to the properties of the optimal solutions of the stochastic marking optimization problem. We first present some basic properties. Tight upper bounds and lower bounds of the average cycle time are then presented. These bounds allow to establish properties of the optimal solutions. Finally, we establish the reachability conditions of a given cycle time C.

3.1 Some Basic Properties

Property 1 ([5]).
An initial marking M_0 is a live marking iff each elementary circuit contains at least one token, i.e. $M_0(\gamma) \geq 1, \forall \gamma \in \Gamma$.

Property 2.
Assume that M and M_0 are two markings with identical circuit counts, i.e. $M(\gamma) = M_0(\gamma), \forall \gamma \in \Gamma$. In this case, it holds that :

(a) M_0 and M are mutually reachable;
(b) $\pi(M) = \pi(M_0)$;
(c) $f(M) = f(M_0)$.

Proof: The claim (c) is obvious since $f(M) = U^T . M$, $f(M_0) = U^T . M_0$ and U is a p-invariant. Claim (a) is Theorem 11 in [5] and claim (b) can be shown by using Property 3 in [17].

Q.E.D.

Property 3.
Let M_1 and M_2 be two initial markings such that $M_1(\gamma) \leq M_2(\gamma), \forall \gamma \in \Gamma$. Then,

(a) $\exists M \in R(M_1)$, $M \leq M_2$, i.e. $M(p) \leq M_2(p), \forall p \in P$.
(b) $\pi(M_1) \geq \pi(M_2)$;
(c) $f(M_1) \leq f(M_2)$.

Proof: The claim (a) is Theorem 10 in [5]. From Property 2, we have :

$$\pi(M_1) = \pi(M) \text{ and } f(M_1) = f(M).$$

The monotonicity property with respect to the marking in [2] implies that $\pi(M_2) \leq \pi(M) = \pi(M_1)$. Since U is a vector of non-negative numbers, claim (a) implies that $f(M_2) = U^T . M_2 \geq U^T . M = f(M) = f(M_1)$.

Q.E.D.

In the following, we need some stochastic ordering relations introduced by Stoyan [16]. In particular, the convex ordering relation and the strong ordering relation will be used. Let $\leq_{icx}$ denote the convex ordering relation and $\leq_{st}$ the strong ordering. Two random variables X and Y are said to satisfy the strong ordering relation (resp. the convex ordering relation), i.e. $X \leq_{st} Y$ (resp. $X \leq_{icx} Y$) if the following inequality

$$E[f(X)] \leq E[f(Y)]$$

holds for all monotone nondecreasing (resp. convex monotone nondecreasing) function f: $\mathbb{R} \to \mathbb{R}$ provided that the expectations exist.

Property 4.

Consider two stochastic timed event graphs with the same net structure STEG1 = (N, •, $\{X_t(k)\}$) and STEG2 = (N, •, $\{Y_t(k)\}$) and let M_1 (resp. M_2) be an optimal solution for STEG1 (resp. STEG2). If

$$X_t \leq_{icx} Y_t, \qquad \forall t \in T,$$

then

$$f(M_1) \leq f(M_2).$$

Since the strong ordering $\leq_{st}$ implies $\leq_{icx}$ ordering, the property still holds if the $\leq_{icx}$ ordering is replaced by the $\leq_{st}$ ordering.

Proof: Let $\pi^1(M_0)$ (resp. $\pi^2(M_0)$) be the average cycle time of STEG1 (resp. STEG2) when the initial marking is M_0. From the definition of M_1 and M_2, we have :

$$\pi^1(M_1) \leq C \text{ and } \pi^2(M_2) \leq C.$$

Let us prove the property by contradiction and assume that $f(M_1) > f(M_2)$. From the monotonicity property of the average cycle time with respect to the transition firing times in [2],

$$\pi^1(M_2) \leq \pi^2(M_2) \leq C$$

which implies that M_2 is also a solution for STEG1 and that M_1 cannot be the optimal solution for STEG1. This is in contradiction with our assumption.

Q.E.D.

Property 2 claims that if M_0 is an optimal solution to the marking optimization problem, then any marking reachable from M_0 is also an optimal solution. Property 3 claims that the average cycle time is non-increasing with regard to the initial marking while the p-invariant criterion is non-decreasing. Property 4 claims that the criterion value of the optimal solutions is non-decreasing in transition firing times in stochastic ordering's sense.

3.2 Performance Bounds and Properties of the Optimal Solutions

This subsection is devoted to the performance bounds of stochastic timed event graphs and properties of the optimal solutions of the marking optimization problem. We first present a lower bound of the average cycle time and show that the p-invariant criterion reaches its minimum when the firing times become deterministic. Subsection 3.2.2. presents upper bounds obtained by using superposition properties in [17] and establishes a sufficient condition under which an optimal solution for the deterministic case remains optimal. Subsection 3.2.3. provides an upper bound for a special case in which the marking is the multiple of another live marking. Finally, we address the normal distribution case.

3.2.1 A Lower Bound

Given the initial marking and the net structure, it was proven in [2, 3] that the minimal average cycle time is obtained in the deterministic case as indicated in the following property.

Property 5 ([2, 3]).

$$\pi(M_0) \geq \pi^D(M_0)$$

where $\pi^D(M_0)$ denotes the average cycle time of the deterministic timed event graph $(N, M_0, \{Y_t(k)\})$ with $Y_t(k) = m_t$, i.e.

$$\pi^D(M_0) = \operatorname*{Max}_{\gamma \in \Gamma} \frac{\sum_{t \in \gamma} m_t}{M_0(\gamma)} \tag{5}$$

From this property, the following corollary can be shown.

Corollary 1.

$$f(M^S) \geq f(M^D)$$

where M^S (resp. M^D) denotes an optimal solution of the marking optimization problem of the timed event graph $(N, M_0, \{X_t(k)\})$ (resp. $(N, M_0, \{m_t\})$).

Remark that this corollary can also be derived from Property 4 by considering the relation $X_t \leq_{icx} E[X_t] = m_t$.

3.2.2 Upper Bounds

The superposition properties in [17] are particularly useful in deriving upper bounds of the average cycle time. In particular, the main superposition property claims that the average cycle time is sub-additive when the transition firing times are generated by the superposition of two sets of sequences of random variables.

Main superposition property ([17]).

Assuming that $X_t \le X_t^1 + X_t^2$, $\forall t \in T$ where X_t^1 and X_t^2 are non-negative random variables, let $\pi^1(M_0)$ (resp. $\pi^2(M_0)$) be the average cycle time of the stochastic timed event graph (N, M_0, $\{X_t^1(k)\}$) (resp. (N, M_0, $\{X_t^2(k)\}$)). Then it holds that :

$$\pi(M_0) \le \pi^1(M_0) + \pi^2(M_0)$$

Using this property, the following upper bound can be easily derived.

Property 6.

$$\pi(M_0) \le \pi^D(M_0) + \operatorname*{Inf}_{Z \in \mathcal{E}} \left\{ \sum_{t \in T} E\left[(X_t - z_t)^+ \right] \right\} \tag{6}$$

where

$$\mathcal{E} = \left\{ Z / z_t \ge m_t, \forall t \in T \text{ and } \sum_{t \in \gamma} z_t \le \pi^D(M_0) . M_0(\gamma), \forall \gamma \in \Gamma \right\}$$

This bound shows that the firing time randomness of transitions belonging to non-critical elementary circuits has little effect on the average cycle time of the whole system. The effect of their randomness can almost be completely canceled by taking large values for z_t.

Proof: Consider two other timed event graphs (N, M_0, $\{z_t\}$) and (N, M_0, $\{Y_t(k)\}$) with $Y_t(k) = (X_t(k) - z_t)^+$ and let $\pi^1(M_0, Z)$ (resp. $\pi^2(M_0, Z)$) be the average cycle time of (N, M_0, $\{z_t\}$) (resp. (N, M_0, $\{Y_t(k)\}$)). Since $X_t(k) \le z_t + Y_t(k)$ for all t and k, the main superposition property implies that:

$$\pi(M_0) \le \operatorname*{Inf}_{z_t \in \mathcal{E}} \left\{ \pi^1(M_0, Z) + \pi^2(M_0, Z) \right\}$$

$$\le \operatorname*{Inf}_{z_t \in \mathcal{E}} \left\{ \operatorname*{Max}_{\gamma \in \Gamma} \frac{\sum_{t \in \gamma} z_t}{M_0(\gamma)} + \pi^2(M_0, Z) \right\}$$

By using the upper bound in [3], we obtain :

$$\pi(M_0) \le \operatorname*{Inf}_{z_t \in \mathcal{E}} \left\{ \operatorname*{Max}_{\gamma \in \Gamma} \frac{\sum_{t \in \gamma} z_t}{M_0(\gamma)} + \sum_{t \in T} E\left[(X_t - z_t)^+ \right] \right\}$$

By considering the fact that for all $Z \in \mathcal{E}$,

$$\operatorname*{Max}_{\gamma\in\Gamma}\frac{\sum_{t\in\gamma}z_t}{M_0(\gamma)}\le\pi^D(M_0)$$

we obtain the following upper bound :

$$\pi(M_0)\le\pi^D(M_0)+\operatorname*{Inf}_{Z\in\Sigma}\left\{\sum_{t\in T}E\left[(X_t-z_t)^+\right]\right\}$$

Q.E.D.

From Property 6, we can established the following property which provides an upper bound by using the two first moments of the related random variables.

Property 7.

$$\pi(M_0)\le\pi^D(M_0)+\sum_{t\in T}\sigma_t \tag{7}$$

Proof: By taking $z_t = m_t$, $\forall t\in T$, Property 6 implies that:

$$\begin{aligned}\pi(M_0)&\le\pi^D(M_0)+\sum_{t\in T}E\left[(X_t-E[X_t])^+\right]\\&\le\pi^D(M_0)+\sum_{t\in T}E\left[|X_t-E[X_t]|\right]\\&\le\pi^D(M_0)+\sum_{t\in T}\sqrt{Var(X_t)}\\&\le\pi^D(M_0)+\sum_{t\in T}\sigma_t\end{aligned}$$

Q.E.D.

From these bounds, the following corollary can be easily shown.

Corollary 2.

Let M^S and M^D be the markings defined in corollary 1. If the following relation holds

$$\operatorname*{Inf}_{Z\in\Sigma}\left\{\sum_{t\in T}E\left[(X_t-z_t)^+\right]\right\}\le C-\pi^D\left(M^D\right), \tag{8}$$

then

$$M^S = M^D$$

Recall that C is the average cycle time to be reached. This corollary provides a sufficient condition under which an optimal solution for the deterministic timing case remains optimal.

3.2.3 A Special Case

In this subsection, we consider a special case in which the initial marking is a multiple of another live marking. By using superposition property, the following tighter upper bound can be derived.

Property 8.

$$\pi(n.M_0) \le \pi^D(M_0) + \operatorname*{Inf}_{Z\in\mathcal{Z}} \left\{ \sum_{t\in T} E\left[\left(\frac{1}{n}\sum_{k=1}^{n} X_t(k) - z_t\right)^+\right] \right\} \tag{9}$$

where $\mathcal{Z}$ is the set of vectors defined in Property 6.

Recall that M_0 denotes the initial marking of the places which are not self-loop places, i.e. for all $p \in P$ (see Section 2). The marking $n . M_0$ assigns one token to each self-loop place and the number of tokens in any place p which is not a self-loop place is equal to $n.M_0(p)$.

Proof: The proof is based on a constrained operating mode in which each transition is forced to fire without interruption n times whenever it starts to fire. Each of these n consecutive firings is called a batch of n firings. In this operating mode, the tokens in the net can be available and unavailable. At the beginning, all the tokens are available.

A transition t can start a batch of n firings if available tokens are present at each of its input places. In this case, it fires n times consecutively. Before the end of this batch, the tokens appeared in the output places of t are frozen, i.e. unavailable. These tokens become available when the batch is completed. At this moment, n new available tokens appear in each of the output places of transition t.

Let $V_t(k)$ be the instant at which transition t starts the firing of its k-th batch in the constrained operating mode. At time $V_t(k + 1)$ for all $k \ge 0$, the transition t starts its (nk + 1)-th firing initiation in the constrained operating mode. Since $S_t(nk +1)$ is the instant of the (nk + 1)-th firing initiation of transition t in the earliest operating mode, we have

$$V_t(k+1) \ge S_t(nk+1), \forall t \in T, \forall k \ge 0 \tag{10}$$

Consider also a new stochastic timed event graph $(N, M_0, \{Y_t(k)\})$ with

$$Y_t(k) = \sum_{i=1}^{n} X_t((k-1)n+i)$$

Of course, $Y_t(k)$ is equal to the time required to complete the k-th batch by transition t in the constrained operating mode. Furthermore, the number of tokens in each place in the new event graph is exactly the same as the number of batches in each place of the original event graph. Hence, the k-th firing initiation instant of transition t denoted as $W_t(k)$ in this new stochastic timed event graph coincides with the instant at which transition t starts the firing of its k-th batch in the constrained operating mode, i.e.

$$W_t(k) = V_t(k),\ \forall t \in T,\ \forall k \tag{11}$$

From the definition of the new stochastic timed event graph, its average cycle time exists and let $\pi^n(M_0)$ be this average cycle time. Thus,

$$\pi^n(M_0) = \lim_{k\to\infty} \frac{W_t(k)}{k},\ \forall t \in T \tag{12}$$

From relations (2, 10, 11, 12), we have :

$$\pi^n(M_0) \geq n\pi(n \,.\, M_0) \tag{13}$$

Since $E[Y_t(k)] = n.m_t$, $\forall t \in T$, Property 6 with X_t replaced by Y_t implies that :

$$\pi^n(M_0) \leq \operatorname*{Inf}_{Z\in\mathcal{Z}} \left\{ n\pi^D(M_0) + \sum_{t\in T} E\left[(Y_t - nz_t)^+\right] \right\}$$

Combining with relation (13),

$$\pi(n \,.\, M_0) \leq \operatorname*{Inf}_{Z\in\mathcal{Z}} \left\{ \pi^D(M_0) + \sum_{t\in T} E\left[\left(\frac{Y_t}{n} - z_t\right)^+\right] \right\}$$

$$= \pi^D(M_0) + \operatorname*{Inf}_{Z\in\mathcal{Z}} \left\{ \sum_{t\in T} E\left[\left(\frac{1}{n}\sum_{k=1}^{n} X_t(k) - z_t\right)^+\right] \right\}$$

Q.E.D.

From this property, we can prove the following corollary which shows that increasing the number of tokens can reduce the variance of the transition firing times.

Corollary 3.

$$\pi(n \,.\, M_0) \leq \pi^D(M_0) + \sum_{t\in T} \frac{\sigma_t}{\sqrt{n}} \tag{14}$$

Proof: Property 8 with $z_t = m_t$ implies that

$$\pi(n \,.\, M_0) \leq \pi^D(M_0) + \sum_{t\in T} E\left[\left(\frac{1}{n}\sum_{k=1}^{n} X_t(k) - m_t\right)^+\right]$$

$$\leq \pi^D(M_0) + \sum_{t\in T} E\left[\left|\frac{1}{n}\sum_{k=1}^{n} X_t(k) - m_t\right|\right]$$

$$\leq \pi^D(M_0) + \sum_{t\in T} \sqrt{Var\left(\frac{1}{n}\sum_{k=1}^{n} X_t(k)\right)}$$

$$\leq \pi^D(M_0) + \sum_{t\in T} \frac{\sigma_t}{\sqrt{n}}$$

Q.E.D.

3.2.4 Normal Distribution Case

As the random variables with normal distribution may not be positive, we assume that the transition firing times are generated as follows :

$$X_t = \left(X_t^*\right)^+ \tag{15}$$

where X_t^* is a random variable with normal distribution with parameters (m_t, σ_t). We assume that $m_t/\sigma_t >> 1$ which ensures that X_t and X_t^* have almost the same probability distribution.

In the following, we denote by $\Phi(x)$ the standard normal distribution function and by $\phi(x)$ its density, i.e.

$$\phi(x) = \frac{1}{\sqrt{2\pi}} \exp\left(-\frac{x^2}{2}\right), \quad \Phi(x) = \int_{-\infty}^{x} \phi(s)ds$$

Property 9 ([17]).

$$\pi(M_0) \leq \underset{z_t \geq 0}{Min}\left\{\underset{\gamma\in\Gamma}{Max} \frac{\sum_{t\in\gamma} z_t}{M_0(\gamma)} + \sum_{t\in T} \sigma_t f\left(\frac{z_t - m_t}{\sigma_t}\right)\right\} \tag{16}$$

where $f(x) = \phi(x) - x\,(1 - \Phi(x))$.

The proof of this property can be found in [17]. Let us consider the case with $z_t = m_t + \alpha\,\sigma_t$. The upper bound becomes :

$$\pi(M_0) \leq \underset{\gamma\in\Gamma}{Max} \frac{\sum_{t\in\gamma}(m_t + \alpha\sigma_t)}{M_0(\gamma)} + \sum_{t\in T} \sigma_t f(\alpha) \tag{17}$$

Let us observe the values of function f(.) given in table 1. We notice that the value of f(.) converges quickly to zero.

α	0	1	2	3	4
f(•)	0.398	0.083	0.008	0.001	0.000

Table 1. The values of function f(•)

Taking $\alpha = 3$, we obtain a very tight bound :

$$\pi(M_0) \le \underset{\gamma\in\Gamma}{\mathrm{Max}} \frac{\sum_{t\in\gamma}(m_t + 3\sigma_t)}{M_0(\gamma)} + 0.001 \sum_{t\in T} \sigma_t \tag{18}$$

As the second term of this upper bound is small enough in many real-life systems, this new upper bound further confirms that the firing time randomness of transitions belonging to elementary circuit with small average cycle time has little impact on the cycle time of the whole system.

We observe that, since the value of $f(\alpha)$ is really small for $\alpha \ge 3$, the upper bound cannot be improved by making α greater than 3 in most real-life systems.

By using similar arguments than in the proof of Property 8, the following upper bound can be established.

Property 10.

$$\pi(nM_0) \le \underset{\gamma\in\Gamma}{\mathrm{Max}} \frac{\sum_{t\in\gamma}\left(m_t + \alpha\frac{\sigma_t}{\sqrt{n}}\right)}{M_0(\gamma)} + \sum_{t\in T} \frac{\sigma_t}{\sqrt{n}} f(\alpha) \tag{19}$$

3.3 Cycle Time Reachability

This subsection is devoted to the reachability of a given cycle time C. In the following, we prove that any cycle time C which is strictly greater than the maximal average transition firing time can be reached, while any C which is strictly smaller than the maximal transition firing time cannot be reached. Necessary and sufficient condition of the reachability of a given cycle time C equal to the maximal average transition firing time is also given.

Let C* be the maximal average transition firing time, i.e.

$$C^* = \underset{t\in T}{\mathrm{Max}}\{m_t\} \tag{20}$$

Property 11.

$\pi(M_0) \ge C^*, \forall M_0$

Proof: From Property 5,

$\pi(M_0) \geq \pi^D(M_0)$

Since each transition is connected to a self-loop place and each self-loop place contains one token, we obtain :

$\pi(M_0) \geq \pi^D(M_0) \geq C^*$

Q.E.D.

Property 12.
Let $M_0 = n\ 1_{|P|}$ where $1_{|P|}$ is a vector with all its components equal to 1. Then,

$$\pi(M_0) \leq C^* + \sum_{t \in T} \frac{\sigma_t}{\sqrt{n}} \tag{21}$$

Proof: From Property 8,

$$\pi(M_0) \leq \pi^D(1_{|P|}) + \sum_{t \in T} \frac{\sigma_t}{\sqrt{n}}$$

Since the number of tokens in each elementary circuit γ is equal to the number of transitions in it for marking $1_{|P|}$, we have:

$$\pi^D(1_{|P|}) = \underset{\gamma \in \Gamma}{\mathrm{Max}} \left\{ \frac{\mu(\gamma)}{L(\gamma)} \right\} \leq \underset{\gamma \in \Gamma}{\mathrm{Max}} \left\{ \frac{L(\gamma).C^*}{L(\gamma)} \right\} = C^*$$

where $L(\gamma)$ is the number of transitions in γ and $\mu(\gamma)$ is the sum of mean firing times of the transitions in γ. This implies that:

$$\pi(M_0) \leq C^* + \sum_{t \in T} \frac{\sigma_t}{\sqrt{n}}$$

Q.E.D.

Recall that similar property has also been shown in [11] by means of a so called N-POM constrained operating mode.

Property 13.
If the transition firing times are generated by random variables of truncated normal distribution defined in equation (15), it holds that for the marking $M_0 = n\ 1_{|P|}$

$$\pi(M_0) \leq \underset{t \in T}{\mathrm{Max}} \left\{ m_t + \alpha \frac{\sigma_t}{\sqrt{n}} \right\} + \sum_{t \in T} \frac{\sigma_t}{\sqrt{n}} f(\alpha) \tag{22}$$

The proof of Property 13 is similar to the one of Property 12 and we omit it.

The following property presents the necessary and sufficient condition for the reachability of C* given in [12].

Property 14 ([12]).
C* is reachable iff there exists t* ∈ T such that

$$P\left[X_{t^*} = \underset{t\in T}{\text{Max}}\, X_t\right] = 1 \tag{23}$$

Furthermore,

(a) If this condition holds, $\pi(n\, 1_{|P|}) = C^*, \forall n \geq 1$.

(b) If it does not hold, $\pi(M_0) > C^*, \forall M_0 \in IN^{|P|}$.

4. Solving the Normal Distribution Case

In this section, we assume that the transition firing times are generated by truncated normal distributed random variables defined as in relation (15).

In this case, we have the following two upper bounds :

$$\pi(M_0) \leq \underset{\gamma\in\Gamma}{\text{Max}} \frac{\sum_{t\in\gamma}(m_t + 3\sigma_t)}{M_0(\gamma)} + 0.001 \sum_{t\in T} \sigma_t \tag{24}$$

$$\pi(nM_0) \leq \underset{\gamma\in\Gamma}{\text{Max}} \frac{\sum_{t\in\gamma}\left(m_t + \alpha\frac{\sigma_t}{\sqrt{n}}\right)}{M_0(\gamma)} + \sum_{t\in T} \frac{\sigma_t}{\sqrt{n}} f(\alpha) \tag{25}$$

The tightness of these upper bounds suggests their application in solving the stochastic marking optimization problem.

Let

$$\rho(n) = C^*(n) + \varepsilon(n) \tag{26}$$

where

$$C^*(n) = \underset{t\in T}{\text{Max}}\left\{m_t + 3\frac{\sigma_t}{\sqrt{n}}\right\} \tag{27}$$

and :

$$\varepsilon(n) = 0.001 \sum_{t\in T} \frac{\sigma_t}{\sqrt{n}} \tag{28}$$

In the following, we first present a near-optimal solution for the case of $C \geq \rho(1)$ and then extend to the general case. Finally, a numerical example is provided.

4.1 Case : $C \geq \rho(1)$

First remark that from relation (26),

$$C - \varepsilon(1) \geq C^*(1) \tag{29}$$

Consider a deterministic timed event graph $DTEG_\alpha = (N, \bullet, \{m_t + \alpha\, \sigma_t\})$. Consider the marking optimization problem for obtaining a cycle time smaller than $C - \varepsilon(1)$ in

$DTEG_\alpha$. Relation (29) implies that the optimal solution of this problem exists for $\alpha \leq 3$ and let D_α be this optimal solution.

The following property claims that the marking D_3 is also a solution to the stochastic marking optimization problem.

Property 15.
$\pi(D_3) \leq C$

Proof: From relation (24),

$$\pi(D_3) \leq \underset{\gamma \in \Gamma}{\mathrm{Max}} \frac{\sum_{t \in \gamma}(m_t + 3\sigma_t)}{D_3(\gamma)} + 0.001 \sum_{t \in T} \sigma_t \tag{30}$$

Remark that the first term of RHS term is the average cycle time of $DTEG_3$. Thus, from the definition of D_3, we have

$$\underset{\gamma \in \Gamma}{\mathrm{Max}} \frac{\sum_{t \in \gamma}(m_t + 3\sigma_t)}{D_3(\gamma)} \leq C - \varepsilon(1) \tag{31}$$

Combining relations (28, 30, 31), we obtain :

$\pi(D_3) \leq C$

Q.E.D.

From property 4, the following property can be easily shown :

Property 16.
$f(D_0) \leq f(D_1) \leq f(D_2) \leq f(D_3)$

Unfortunately, the markings D_0, D_1 and D_2 may not be the solution of the stochastic marking optimization problem.

In the following, we propose a near-optimal solution of the stochastic optimization problem. To this purpose, we compute the three markings D_0, D_1, D_2 and D_3 and choose the best among the markings which are also solution of the stochastic optimization problem.

Algorithm 1.

Step 1. Let $\alpha = 0$

Step 2. Compute the optimal solution D_α for $DTEG_\alpha$ by using algorithms proposed in [8, 9]

Step 3. Compute the average cycle time $\pi(D_\alpha)$ by simulation

Step 4. If $\pi(D_\alpha) \leq C$, then stop and keep D_α as the optimal solution; otherwise, let $\alpha := \alpha+1$ and return to step 2.

4.2 General Case

Without loss of generality, let us assume that $\rho(n-1) \geq C \geq \rho(n)$ with $n \geq 2$. Consider a deterministic timed event graph DTEGn,α = $(N, \bullet, \{m_t + \alpha.\sigma_t/\sqrt{n}\})$. Consider the marking optimization problem for obtaining a cycle time smaller than C - $\varepsilon(n)$.

Clearly, the optimal solution of this problem exists for $\alpha \leq 3$ and let Dn,α be this optimal solution.

The following property claims that the marking $n.D_{n,3}$ is also a solution to the stochastic marking optimization problem.

Property 17.

$$\pi(n.D_{n,3}) \leq C$$

Proof: From relation (25),

$$\pi(nD_{n,3}) \leq \underset{\gamma\in\Gamma}{\text{Max}} \frac{\sum_{t\in\gamma}\left(m_t + 3\frac{\sigma_t}{\sqrt{n}}\right)}{D_{n,3}(\gamma)} + 0.001 \sum_{t\in T} \frac{\sigma_t}{\sqrt{n}} \tag{32}$$

Remark that the first term of RHS term is the average cycle time of $DTEG_{n,3}$. Thus, from the definition of $D_{n,3}$, we have

$$\underset{\gamma\in\Gamma}{\text{Max}} \frac{\sum_{t\in\gamma}\left(m_t + 3\frac{\sigma_t}{\sqrt{n}}\right)}{D_{n,3}(\gamma)} \leq C - \varepsilon(n) \tag{33}$$

Combining relations (28, 32, 33), we obtain :

$$\pi(n.D_{n,3}) \leq C$$

Q.E.D.

Similarly to the Property 16, we have

Property 18.

$$f(D_{n,0}) \leq f(D_{n,1}) \leq f(D_{n,2}) \leq f(D_{n,3})$$

In the following we propose a heuristic algorithm which provides a near optimal solution. To this end, we first compute the minimum multiple n* such that the marking ($n^* D_{n,3}$) is a solution of the stochastic marking optimization problem. We then consider the markings ($n^* D_{n,\alpha}$) for $\alpha \leq 3$ and choose the best among those which are solution of the stochastic marking optimization problem.

Algorithm 2.

Step 1. For $\alpha = 0$ to 3, compute the optimal solution $D_{n,\alpha}$ for $DTEG_{n,\alpha}$ by using algorithms proposed in [8, 9]

Step 2. Determine by simulation n* defined as follows :

$$n^* = \min \{ i / \pi(i\, D_{n,3}) \le C \}$$

Step 3. Let $\alpha = 0$

Step 4. Compute the average cycle time $\pi(n^*\, D_{n,\alpha})$ by simulation

Step 5. If $\pi(n^*\, D_{n,\alpha}) \le C$, then stop and keep $(n^*\, D_{n,\alpha})$ as the optimal solution; otherwise, let $\alpha := \alpha+1$ and return to step 4.

Finally, the computation of the minimum multiple n* can be simplified by the following property which is a corollary of the monotonicity property with respect to the initial marking in [2].

Property 19.

$$\pi(kM_0) \ge \pi((k+1)M_0),\ \forall k \ge 1 \text{ and } \forall M_0$$

4.3 A Numerical Example

Consider the event graph presented in figure 1. The transition firing times are generated by truncated random variables with normal distribution and the parameters of the random variables are given by :

t1 (5, 0.5), t2 (2, 0.1), t3 (8, 0.2)
t4 (5, 0.4), t5 (3, 0.1), t6 (3, 0.2)

where the numbers in the parenthesis are mean values and standard deviations.

The stochastic marking optimization problem consists in reaching a cycle time smaller than C = 8.7 while minimizing the total number of tokens, i.e. $U = 1_{|P|}$.

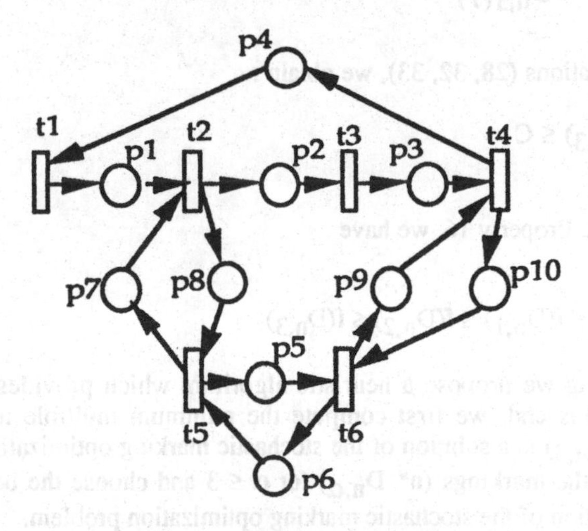

Fig. 1. A numerical example

In this example,

$$C^*(1) = 8.6, \varepsilon(1) = 0.0015 \text{ and } \rho(1) = C^*(1) + \varepsilon(1) = 8.6015 \leq C$$

Hence, algorithm 1 can be used to solve the marking optimization problem. The optimal solution obtained by the algorithm is as follows :

$$M^S = D_0 = (1, 0, 1, 1, 0, 1, 0, 1, 0, 1) \text{ and } \pi(M^S) = 8.034$$

which means that the optimal solution for the deterministic timing case remains optimal.

5. Conclusion

In this paper, we have investigated the marking optimization problem of stochastic timed event graphs which consists in obtaining a given cycle time while minimizing a p-invariant criterion.

Some important properties were established. In particular, it was shown that the average cycle time is non-increasing with respect to the initial marking while the p-invariant criterion is non-decreasing. We further proved that the criterion value of the optimal solution is non-increasing in transition firing times in stochastic ordering's sense.

Lower bounds and upper bounds of the average cycle time were proposed. Based on these bounds, we show that the p-invariant criterion reaches its minimum when the transition firing times become deterministic. A sufficient condition under which an optimal solution for the deterministic case remains optimal for the stochastic case is given. These new bounds have also been used to provide simple proof of the reachability conditions proposed in [11, 12].

Thanks to the tightness of the bounds for the normal distribution case, two algorithms which leads to near-optimal solutions have been proposed to solve the stochastic marking optimization problem.

Further research directions are multiple. First we believe that the near-optimal solution proposed in this paper can be further improved. For example, it seems possible to improve this solution by iteratively removing tokens from places with high waiting times. The second research direction consists in solving the stochastic marking optimization with general distribution. Of course, the algorithms presented in this paper can be adapted to solve the general case. Some numerical experiences indeed provide interesting results. However, no theoretic support have been found by now. The third research direction includes the extensions to other Petri net models.

Acknowledgement

The authors would like to thank the anonymous referees for the constructive comments in improving the presentation of this paper.

References

1. F. BACCELLI, "Ergodic Theory of Stochastic Petri Networks", *Annals of Probability*, Vol. 20, No. 1, pp. 350-374, 1992.
2. F. BACCELLI and Z. LIU, "Comparison Properties of Stochastic Decision Free Petri Nets", to appear in *IEEE Transactions on Automatic Control*, 1992.
3. J. CAMPOS, G. CHIOLA and M. SILVA, "Properties and Performance Bounds for Closed Free Choice Synchronized Monoclass Queueing Networks", *IEEE Transactions on Automatic Control*, Vol. 36, No. 12, pp. 1368-1382, December 1991.
4. P. CHRETIENNE, "Les réseaux de Petri temporisés", Université Paris VI, Paris, France, Thése d'Etat, 1983.
5. F. COMMONER, A. HOLT, S. EVEN and A. PNUELI, "Marked Directed Graphs", *Journal of Computer and System Science*, Vol. 5, No. 5, pp. 511-523, 1971.
6. P.J. HAAS and G.S. SHEDLER, "Stochastic Petri Nets: Modeling Power and Limit Theorem", *Probability in Engineering and Informational Sciences*, Vol. 5, pp. 477-498, 1991.
7. H.P. HILLION and J.M. PROTH, "Performance Evaluation of Job-Shop Systems Using Timed Event-Graphs", *IEEE Transactions on Automatic Control*, Vol. 34, No. 1, pp. 3-9, January 1989.
8. S. LAFTIT, J.M. PROTH and X.L. XIE, "Marking Optimization in Timed Event Graphs", Proc. of 12th Int. Conf. on Applications and Theory of Petri Nets, Denmark, June 1991.
9. S. LAFTIT, J.M. PROTH and X.L. XIE, "Optimization of Invariant Criteria for Event Graphs", *IEEE Transactions on Automatic Control*, Vol. 37, No. 5, pp. 547-555, May 1992.
10. T. MURATA, "Petri Nets: Properties, Analysis and Applications," Proceedings of the IEEE, vol. 77, N° 4, pp. 541-580, April 1989.
11. J.M. PROTH, N. SAUER and X.L. XIE, "Stochastic Timed Event Graphs: Bounds, Cycle Time Reachability and Marking Optimization", to appear in the Proc. of the 12th IFAC World Congress, 1993.
12. J.M. PROTH and X.L. XIE, "Performance Evaluation and Optimization of Stochastic Timed Event Graphs", submitted for publication in *IEEE Transactions on Automatic Control*, June 1992.
13. C.V. RAMAMOORTHY and G.S. HO, "Performance Evaluation of Asynchronous Concurrent Systems using Petri Nets", *IEEE Trans. Software Eng.*, Vol. SE-6, No. 5, pp. 440-449, 1980.
14. J.A.C. RESING, R.E. de VRIES, G. HOOGHIEMSTRA, M.S. KEANE and G.J. OLSDER, "Asymptotic Behaviour of Random Discrete Event Systems", *Stochastic Processes and their Applications*, Vol. 36, pp. 195-216, 1990.
15. M. SILVA, "Petri Nets and Flexible Manufacturing," in Advances in Petri Nets 1989, G. Rozenberg (ed.), Lecture Notes of Computer Science, Springer Verlag, pp. 374-417, 1989.
16. D. STOYAN, "Comparison Methods for Queues and Other Stochastic Models", English translation (D.j. Daley editor), J. Wiley and Sons, New York, 1984.
17. X.L. XIE, "Superposition Properties and Performance Bounds of Stochastic Timed Event Graphs", Submitted to IEEE Trans. on Automatic Control, 1992

A Client-Server Protocol for the Composition of Petri Nets

C. Sibertin-Blanc
Université Toulouse 1
Place A. France, 31042 Toulouse Cedex, France
E-mail: sibertin@cix.cict.fr

Abstract. Modelling the behavior of a system as a set of cooperating nets requires to define a high-level communication protocol which takes into account the very nature of their interactions. This paper proposes to adapt the client-server protocol promoted by the object-oriented approach to Petri nets, and to compose Petri nets according to this protocol.This protocol relies upon four basic rules which assert the honesty and discretion of clients and servers. A class of nets respecting these rules, called client-server nets, is defined, as is the composition of these nets according to a Use function. The possibility to compose client-server nets while preserving the nets' language and liveness is studied. This possibility comes down to very simple relationships between the main characteristics of client-server nets: the demand and the confidence degree as a client, and the supply and the reliability degree as a server. These relationships are preserved by the composition of nets, so the client-server protocol allows for the incremental design of systems and favors the reuse of nets.

1 Introduction

The wish to model and analyze large systems by means of Petri nets has shown the need for a modular approach within this formalism [1]. The main advantages which are expected from this approach are:

- a better command of the complexity of systems, thanks to a rigourous structure of the model and the possibility to consider different parts of the model independently of each other;
- a greater ease to adapt, correct, analyze or reuse a model, thanks to the localization of these tasks at the level of components.

But such benefits are really gained only if the components fulfil two dual properties [2]: from an external point of view, the components are loosely coupled, so that they are as independent as possible and have only a few relationships; from an internal point of view, the components have a high functional cohesion, so that the role and the contribution of each component to the whole are clearly defined.
Applying these general principles to Petri nets has some implications on the communication between nets (how are they connected) and on the design of nets (what is the behavior and the structure of a net). Nets may be connected by merging of transitions, by merging of places, and also by arcs (for instance [3], [4]). Transitions merging has the advantage to ease the system's analysis, because many properties are preserved when nets are composed in this way ([5], [6], [7], [8] among many others). But in this case the nets are very tightly coupled; for instance, it is not possible to consider a net as modelling an autonomous process having its own token game player, since it is not possible to locally decide whether a transition is enabled or not. Moreover, the functional unity of the nets is very difficult to define because their internal behavior is mixed with their interactions with other nets and it is not possible to distinguish which net initiates a communication. Composition of nets by places merging corresponds to communication by variable sharing. This way of proceeding does not have the inconveniences of the previous one, but the management of the shared place requires an additional synchronization between the nets. Finally, connecting nets by arcs ensures the smallest coupling and it corresponds to communication by "message sending" (arc from a transition to a place) and "message taking" (arc from a place to a transition). This kind of communication may also be implemented by fusionned places, provided that they are a one-way channel, that is no net may both put and take tokens in

these places. This is the solution adopted for our client-server protocol.
But a modular approach improves the structure of models and favors the design of reusable nets only if it is based upon a high level protocol which defines the communication between nets in terms of their functional abilities and not solely in terms of their graph structure. If not, how nets communicate cannot be dissociated from the purpose of their communication, and the designer is not provided with an abstract view of the nets' communication.
In agreement with the object-oriented approach, the functional unity of nets may be obtained by designing them as components which put some services at the disposal of other nets, and apply to other nets for the services they offer. Then nets communicate according to the client-server protocol which consists of the four following steps:

- the client requests a service;
- the server accepts the service request;
- the server processes the service request and provides a result;
- the client retrieve the result.

This very intuitive protocol is an adaptation to Petri nets of the Ada's rendez-vous and of the inter-objects communication mechanism of many object-oriented languages. Its suitability to the design of well-structured systems has been widely demonstrated and will be not discussed in this paper. The aim of this paper is to define this protocol and to show that it is a basis for a good design calculus.

The nets we consider are *clients* - nets able to send service requests and to retrieve service results -, or *servers* - nets able to accept requests, to process them and to provide results -, or nets which are both client and server. Given a client and a server, a Use function may be defined which maps the services requested by the client onto the services offered by the server, and allow to link them into a *composed net*.
The client-server protocol relies upon four basic rules:
Client honesty: a client seeks a service result only if it has previously requested that service.
Client discretion: a client requests a service only if it is able to retrieve the result.
Server honesty: a server accepts a service request only if it is sure to be able to process it.
Server discretion: a server issues a service result only if it has been requested for that service.
In section II, clients and servers are defined in such a way that they fulfil these four rules. This definition may be considered as one among all the conceivable implementations of this protocol, since the results shown further rely not on the specific structure of clients and servers but on these basic rules. Honesty and discretion ensure the independence of the different services both in the client and in the server, and they allow to dissociate the enabling of a service from its processing.

In order to be allowed to incrementally design and analyze a system by adding new components, the composition of nets must be a stable operation, that is the composition of two client-server nets must still be a client-server.
The composition must also preserve general properties of the components so that the properties of the whole system may be deduced from the properties of its components. According to the respective roles of clients and servers, we are mainly interested, when linking of a client C with a server S results in a compound net CPN, by preservation of the properties of C in CPN. The properties whose preservation is investigated in this paper are the place invariants, the enabled sequences of transitions and the liveness. It turns out that the preservation of these properties is gained by a (very simple) relationship between two characteristics of clients and two characteristics of servers, which are the following:

- the *demand* of a client is the set of sequences of service requests it may issue,
- the *supply* of a server is the set of sequences of service requests it may accept,
- the *confidence degree* of a client measures how tightly it controls the processing of its requests by the server,
- the *reliability degree* of a server measures to what extent it is actually able to process the sequences of service requests which belongs to its supply.

Finally, these characteristics must be preserved by the composition of nets, so that they do not have to be checked again when a new component is added. It is shown that composing nets whose characteristics match does not affect these characteristics: if CPN is the result of the composition of a client-server C with a client-server S whose characteristics match, then CPN keeps the server characteristics of C and the client characteristics of S.

This paper considers only static links, i.e. the server and the offered service involved by a service request of a client are predefined. The Cooperative Objects formalism [9] adopts this client-server protocol to compose high-level Petri nets where clients and servers are dynamically created and have communication dynamically set up. This formalism has been used to model flexible manufacturing systems [10], organizational information systems [11] and graphical user interface [12].

The remainder of the paper is organized as follows.
Section II is devoted to the definition of clients, servers and of their composition. Section III concerns place invariants: they are preserved in any case. Section IV studies the preservation of the set of the enabled sequences of transitions of a client; it is shown that it is equivalent to the compatibility of the server with the client -that is the client's demand is included into the server's supply-, and that composing compatible nets does not alter their compatibility with other nets. Section V addresses the transparency of servers for clients, a relationship which implies the preservation of the client's liveness. It is shown that it comes down to the compatibility and the agreement between the confidence and reliability degrees, and again that these degrees are not decreased when transparent nets are composed.

2 Clients, Servers and their composition

We will first define some notations which will be used throughout this paper.
Let $N = <P, T, F>$ be a Petri net, where P is the set of places, T the set of transitions, $T \cap P = \emptyset$, and $F \subset T \times P \cup T \times P$ is the flow relation.
The set of successors of a node $x \in P \cup T$ is $x^\bullet = \{y \in P \cup T; (x, y) \in F\}$, and the set of its predecessors is $^\bullet x = \{y \in P \cup T; (y, x) \in F\}$.
If X is a set of nodes, $^\bullet X = \cup_{x \in X} {}^\bullet x$ and $X^\bullet = \cup_{x \in X} x^\bullet$.
When it is useful to precise to which net belongs the node, we note $\Gamma^N(x) = x^\bullet$.
Let $M \in \mathbb{N}^P$ be a marking of N. If $P' \subset P$, then $M|P'$ stands for the restriction of M to P'.
Let $\sigma \in T^*$ and $T' \subset T$; then $\sigma \mid T' \in T'^*$ is the substring of σ obtained by removing the occurrences of elements of $T \setminus T'$.
If $\sigma, \sigma' \in T^*$, we note $\sigma' \leq \sigma$ if σ' is a prefix of σ, i.e. if there exists $\sigma'' \in T^*$ such that $\sigma'.\sigma'' = \sigma$.
If $t \in T$, then $\underline{\sigma}(t) \in \mathbb{N}$ is the number of occurrences of t in σ,
$\underline{\sigma}(T') = \Sigma_{t \in T'} \underline{\sigma}(t)$, and $\underline{\sigma} = (\underline{\sigma}(t))_{t \in T}$.
$L(N, M) = \{\sigma \in T^* ; M[\sigma>\}$, and $L(N, M)|T' = \{\sigma|T' ; \sigma \in L(N, M)\}$.
If $S = (N, M, X)$, where M is a marking of the Petri net N, L (S) denotes L (N, M).
We note $M \dashrightarrow M'$ if there exists $\sigma \in T^*$ such that $M[\sigma > M'$.
$I \in \mathbb{N}^P$ is a place invariant of N if $I.C = 0$, where C is the incidence matrix of N.
Let $N^i = <P^i, T^i, F^i>$, $i = 1, 2$ be two nets, and M^i a marking of N^i.
The intersection of N^1 and N^2 is the net $N = N^1 \cap N^2 = <P^1 \cap P^2, T^1 \cap T^2, F^1 \cap F^2>$, and they are disjoint if $P^1 \cap P^2 = \emptyset$ and $T^1 \cap T^2 = \emptyset$.
Their union is the net $N = N^1 \cup N^2 = <P^1 \cup P^2, T^1 \cup T^2, F^1 \cup F^2>$.
$M = M^1 \cup M^2$ is defined if $\forall p \in P^1 \cap P^2$, $M^1(p) = M^2(p)$.

2.1 Definition of Client

A client is a net including transitions which are associated to services supplied by other nets, the occurrence of which causes a call to the service it is associated to. Since the processing of a service cannot be assumed to be instantaneous, a call transition is broken

down into a request-transition which sends the service request to the server and a get-transition which retreive the results issued by the service (cf. Figure 1).
To be composable with servers, a client has to be provided with an interface made of places to hold the service requests and the service results. Thus each request-transition receives an extra output place, each get-transition receives an extra input place as shown in Figure 2, and the resulting net is called the client completion. A client completion will be composed with a server by merging these interface places with appropriate places of the server.

Definition 2.1 ***client***

$C = (N, M, Serv)$ is a client iff:

1) $N = (P, T, F)$ is a Petri net;
2) $Serv \subset T \times T$ is a bijective relation between two disjoint subsets of T, where

   $T_r = \{t_r \in T \,;\, \exists\, t_g \in T, (t_r, t_g) \in Serv\}$ is the set of *request-transitions*,

   $T_g = \{t_g \in T \,;\, \exists\, t_r \in T, (t_r, t_g) \in Serv\}$ is the set of *get-transitions*, and $T_r \cap T_g = \emptyset$;

   We note $T_{rug} = T_r \cup T_g$, and for $s = (t_1, t_2) \in Serv$, $t_1 = t_r^s$ and $t_2 = t_g^s$.
3) For any service $s = (t_r, t_g)$, there exists a waiting place $p_W \in P$ such that

   $p_W \in t_r^{\bullet}$, ${}^{\bullet}p_W = \{t_r\}$, ${}^{\bullet}t_g = \{p_W\}$ and $p_W^{\bullet} = \{t_g\}$;
4) M is a marking of N such that $M(p_W) = 0$ for any waiting place p_W.

The waiting place of a service corresponds to the state where the client is waiting for the server to complete a service request. According to 3), it is connected to the request- and get-transitions of the service, as shown in Figure 1. Figure 3 shows a client featuring two services requests.

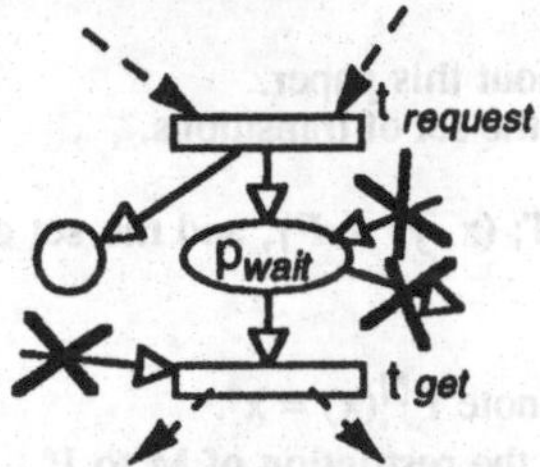

Figure 1: connections of the request- and get-transitions with the waiting place

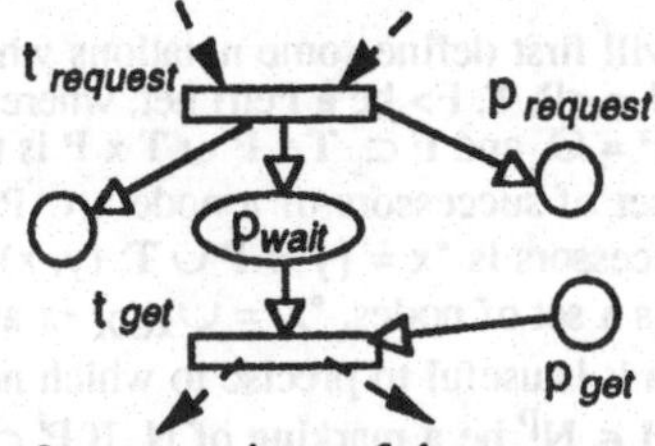

Figure 2: connections of the request- and get-places

Definition 2.2 ***client completion***

Let $C = (N, M, Serv)$ be a client, where $N = (P, T, F)$.
Its completion is the marked net $\underline{C} = (\underline{N}, \underline{M})$, where:

1) $\underline{N} = (\underline{P}, T, \underline{F})$;
2) $\underline{P} = P \cup P_r \cup P_g$, where $P_r = \{p_r^s \,;\, s \in Serv\}$ is the set of *request-places*,

   $P_g = \{p_g^s \,;\, s \in Serv\}$ is the set of *get-places*, and P, P_r and P_g are disjoint;
3) $\underline{F} = F \cup \{(s_r, p_r^s) \,;\, s = (s_r, s_g) \in Serv\} \cup \{(p_g^s, s_g) \,;\, s = (s_r, s_g) \in Serv\}$;
4) $\underline{M}$ is the marking of $\underline{N}$ such that $\underline{M}(p) = M(p)$ for $p \in P$, and $\underline{M}(p) = 0$ for $p \in P_r \cup P_g$.

Provided with these definitions, we are able to give a precise meaning to the concepts of client discretion and honesty. An honest client never performs a get-transition for a service if it has not previously required this service, and a discreet client never requests a service if it is not sure to be able to get back the result of this request.

Definition 2.3 ***client honesty and discretion***

Let $C = (N, M, Serv)$ be a client.
C is said to be honest iff $\forall\, \sigma \in L(C), \forall\, s \in Serv, \underline{\sigma}(s_g) \leq \underline{\sigma}(s_r)$.

C is said to be **discreet** iff:

1) $\forall$ s $\in$ Serv, $\forall \sigma.t_r^s \in L(C)$, $\exists \alpha \in T^*$ such that $\sigma.t_r^s.\alpha.t_g^s \in L(C)$;

2) $\forall$ s $\in$ Serv, $\forall \sigma^1.t_r^s.\sigma^2.t_g^s \in L(C)$, $\exists \alpha^1 \in (T \setminus T_{rug})^*$, $\exists \alpha^2 \in T^*$ such that $\sigma^2 = \alpha^1.\alpha^2$ and $\sigma^1.t_r^s.\alpha^1.t_g^s.\alpha^2 \in L(C)$.

In this definition, condition 1) is not sufficient to ensure the discretion of the client. Indeed, if α includes a get-transition t_g^o, it may happen that t_g^o will never occur if the requested server does not issue the expected result; in this case C is blocked at transition t_g^o and will not be able to perform transition t_g^s. According to condition 2), the interactions (i. e. request- or get-transitions) which take place after a service resquest may always be delayed after the get of that service; thus, getting back the result of a service only depends on the achievement of that service and cannot be disrupted by some trouble in the achievement of another service.
The client discretion and honesty are ensured by the intuitive structural pattern shown in Figure 1, and a weaker condition to warrant the discretion and honesty of the client would not extend the expressive power of the client-server protocol.

We will now state the two fundamental properties resulting from discretion and honesty. The ability for a client to perform a get-transition on a service depends only on the number of request-and get-transitions which have previously occured for this service. A transition sequence is normalized if each request-transition is just followed by the get-transition on that service, that is there are never two service requests pending at the same time; any transition sequence may be normalized without change in the order of its request-transitions.

Proposition 2.1 ***normalized transition sequences***

Let C = (N, M, Serv) be a client, and $\sigma \in L(C)$.

a) $\forall$ s $\in$ Serv $[\sigma.t_g^s \in L(C) \Longleftrightarrow \underline{\sigma}(t_g^s) < \underline{\sigma}(t_r^s)]$.

b) There exists $\sigma' \in L(C)$ such that

1) $\underline{\sigma'} = \underline{\sigma}$, 2) $\sigma' | T_r = \sigma | T_r$,

3) $\forall$ s $\in$ Serv, $\forall \sigma^1, \alpha \in T^*$ $[\sigma^1.t_r^s.\alpha.t_g^s < \sigma' \Longrightarrow \alpha$ is the empty string]

and σ' is called the **normalized** sequence associated to σ.

The proof is immediate. If only discretion and honesty are assumed instead of 3) in definition 2.1, $\alpha \in (T \setminus T_{rug})^*$ instead of being empty (cf. proposition 2.4 below).

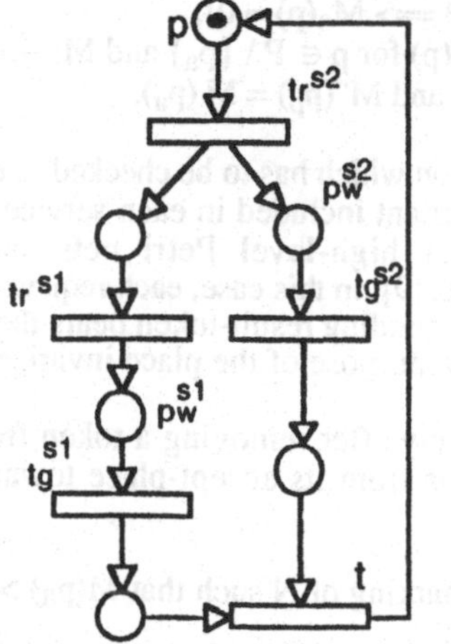

Figure 3: a client requesting two services

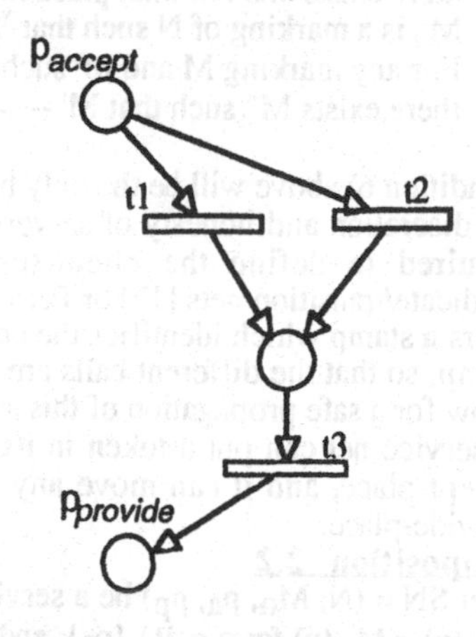

Figure 4: a simple service net which may process service requests in two ways

2.2 Definition of Server

A server is a net offering services to clients: it is able to accept requests, to process them, and to provide results. The interface of servers matches with the clients' one: for each service offered, the completion of a server includes one accept-place for receiving the service's requests, and one provide-place for delivering the results of the service's processing. Thus, a server and a client will be composed by merging both

- the request-places of the client completion with accept-places of the server completion, and
- the get-places of the client completion with provide-places of the server completion.

A server must provide its services in a discreet and honest way. But contrarily to clients, it is not desirable to warrant these properties by a fully structural definition of servers. Indeed, processing a service may require complex computations which cannot be constrained to be a single processing-place. Moreover, the structure of a server is deeply transformed when it uses a sub-server for achieving a service it offers, whereas the structure of a client is not altered when it is used by a super-client. Thus the definition of servers will explicitly require discretion and honesty as behavioral conditions.

In fact, we will always consider servers build from elementary servers whose definition is much more structural. ***Elementary servers*** are servers which are not composed with sub-servers, and their structure is based upon the two functions that any server must fulfil:

- to achieve the work expected when a service is called;
- to coordinate these services' processing.

According to this view, an elementary server is the gathering of a set of ***service nets*** (one service net for each service provided by the server, in charge of processing the requests for that service) with a ***coordinator net*** in charge of deciding in which cases a request for a service may be accepted.

What is required from a service net is to provide a result if and only if it has accepted a request. Thus, a service net is always a refinement of a net similar to the one shown in Figure 4, where $T_a = \{t1, t2\}$ and $T_p = \{t3\}$. .

Definition 2.4 ***Service Nets***

$SN = (N, M_o, p_a, p_p)$ is a service net iff:

1) $N = (P, T, F)$ is a Petri net;
2) $p_a \in P$ is the ***accept-place*** of SN, $^{\bullet}p_a = \emptyset$, and $T_a = p_a^{\bullet}$ is called the set of ***accept-transitions*** of SN;
3) $p_p \in P$ is the ***provide-place*** of SN, $p_p^{\bullet} = \emptyset$ and $T_p = {}^{\bullet}p_p$ is called the set of ***provide-transitions*** of SN;
4) there exists one minimal place invariant I such that $I(p_a) = I(p_p) = 1$;
5) M_o is a marking of N such that $\forall p \in P\ [I(p) > 0 \Longrightarrow M_o(p) = 0]$;
6) For any marking M and M' such that $M(p) = M_o(p)$ for $p \in P \setminus \{p_a\}$ and M ---> M', there exists M" such that M' ---> M", $M''(p_a) = 0$ and $M''(p_p) = M(p_a)$.

Condition 6) above will be the only behavioral condition which has to be checked to ensure the discretion and honesty of servers. The place invariant included in each service net is required to define the client-server protocol for high-level Petri nets such as predicate/transition nets [13] or Petri Nets with Objects [9]: in this case, each request-token bears a stamp which identifies the call, and the corresponding result-token bears the same stamp, so that the different calls are not confused; the purpose of the place invariant is to allow for a safe propagation of this stamp.

A service net can put a token in its provide-place only after removing a token from its accept-place, and it can move any number of tokens from its accept-place towards its provide-place.

Proposition 2.2

Let $SN = (N, M_o, p_a, p_p)$ be a service net, M be a marking of N such that $M(p_a) > 0$ and $M(p) = M_o(p)$ for $p \in P \setminus \{p_a\}$, and $\sigma \in L(N, M)$.

1) $\underline{\sigma}(T_p) \leq \underline{\sigma}(T_a)$;
2) $\exists \sigma' \in T^* [\sigma.\sigma' \in L(N, M)$ and $\underline{\sigma.\sigma'}(T_p) = \underline{\sigma}(T_a)]$.

Proof

1) Since $I.M_0 = 0$, we have $I.M = M(p_a)$. Let M' be such that $M[\sigma> M'$.
Thus $\underline{\sigma}(T_a) = M(p_a) - M'(p_a) = I.M - M'(p_a) = I.M' - M'(p_a) \geq M'(p_p) = \underline{\sigma}(T_p)$.
2) results from point 6) of definition 2.4, since $M''(p_p) = M(p_a)$ is equivalent to $\underline{\sigma.\sigma'}(T_p) = \underline{\sigma.\sigma'}(T_a)$, and we may consider M such that $M(p_a) = \underline{\sigma}(T_a) = \underline{\sigma.\sigma'}(T_a)$. ♦

A server is built from a set of disjoint service nets and from a net for coordinating the acceptation of service requests. They are connected with two kinds of arcs:

- arcs from a place of the coordinator toward an accept-transition of a service net, and
- arcs from a transition of a service net toward a place of the coordinator.

The resulting net is in fact the completion of a server, and the server itself is obtained by removing the accept- and provide-places.

Definition 2.5 ***Elementary server and their completion***

S = (N, M, Serv) is an *elementary server* and $\underline{S} = (\underline{N}, \underline{M})$ is its *completion* iff there exists a set of service nets $SN^s = (N^s, M_0^s, p_a^s, p_p^s)$, $s \in Serv$, and a coordinator net Coord such that:

1) the N^ss and Coord are pairwise disjoint nets;
2) If $\underline{N} = (\underline{P}, \underline{T}, \underline{F})$, $N^s = (P^s, T^s, F^s)$ and Coord = (P^{co}, T^{co}, F^{co}), we have

$$\underline{P} = \cup_{s\in Serv} P^s \cup P^{co},\ \underline{T} = \cup_{s\in Serv} T^s \cup T^{co},\ \underline{F} \supseteq \cup_{s\in Serv} F^s \cup F^{co},$$

and $\forall (x, y) \in \underline{F} \setminus (\cup_{s\in Serv} F^s \cup F^{co})$, $\exists s \in Serv$

$[x \in P^{co}$ and $y \in T_a^s]$ or $[x \in T^s$ and $y \in P^{co}]$;

we note $\mathbf{P_a} = \{p_a^s ; s \in Serv\}$, $\mathbf{P_p} = \{p_p^s ; s \in Serv\}$, $\mathbf{T_a} = \cup_{s\in Serv} T_a^s = P_a^\bullet$,

$\mathbf{T_p} = \cup_{s\in Serv} T_p^s = {}^\bullet P_p$, and $\mathbf{T_{aup}} = T_a \cup T_p$;

3) $\forall s \in Serv, \forall p \in P^s\ [\underline{M}(p) = M_0^s(p)]$;
4) If N = (P, T, F), we have

$$P = \underline{P} \setminus (P_a \cup P_p),\ T = \underline{T},\ F = \underline{F} | (PxT \cup TxP) \text{ and } M = \underline{M} | P.$$

Figure 5 shows an elementary server offering three services. s^1 is processed by a single transition, whereas s^2 is processed by two transition sequences which are alternately performed. s^3 may be accepted from two different states of the server.

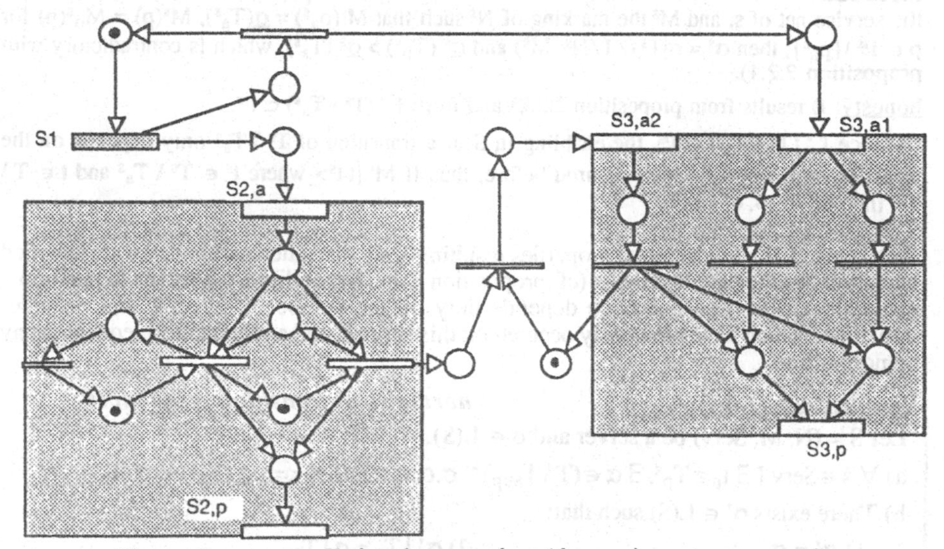

Figure 5: a server offering three services (the service nets are greyed)

When an elementary server is composed with another server becoming its sub-server, the resulting net is still a net offering services, but most often it no longer satisfies the structural condition of elementary servers. Thus we need a more functional definition of servers than definition 2.5, as a net offering services in a discreet and honest way. Discretion means that it never provides a result for a service which has not been previously requested; honesty means that it accepts a service request only if it is sure to be able to provide a result, without needing to receive other service requests. Of course elementary servers are servers.

Definition 2.6 ***server and server completion***

$S = (N, M, Serv)$ is a *server* and $\underline{S} = (\underline{N}, \underline{M})$ is its *completion* iff:

1) $N = (P, T, F)$ and $\underline{N} = (\underline{P}, T, \underline{F})$ are Petri nets;
2) Serv is a bijective relation between two disjoint subsets of $\underline{P}$, that is there exists $\mathbf{P_a} = \{p_a{}^s ; s \in Serv\}$, $\mathbf{P_p} = \{p_p{}^s ; s \in Serv\}$, and $P_a \cap P_p = \varnothing$;
3) $\forall p \in P_a, {}^\bullet p = \varnothing$, $\forall p \in P_p, p^\bullet = \varnothing$, and we define: $\mathbf{T_a{}^s} = p_a{}^{s\bullet}$, $\mathbf{T_p{}^s} = {}^\bullet p_p{}^s$, $\mathbf{T_a} = \cup_{s \in Serv} T_a{}^s = P_a{}^\bullet$, $\mathbf{T_p} = \cup_{s \in Serv} T_p{}^s = {}^\bullet P_p$, and $\mathbf{T_{aup}} = T_a \cup T_p$;
4) $\forall p \in P_a \cup P_p, \underline{M}(p) = 0$;
5) $P = \underline{P} \setminus (P_a \cup P_p)$, $F = \underline{F} \mid (PxT \cup TxP)$, and $M = \underline{M} \mid P$;
6) for each service s, there exists a minimal place invariant I such that $I(p_a{}^s) = I(p_p{}^s) = 1$;
7) S is *discreet*: $\forall \sigma \in L(S), \forall s \in Serv, \underline{\sigma}(T_p{}^s) \leq \underline{\sigma}(T_a{}^s)$;
8) S is *honest*:
   i) $\forall s \in Serv, \forall t_a \in T_a{}^s, \forall \sigma.t_a \in L(S), \exists \alpha \in T^*, \exists t_p \in T_p{}^s\ [\sigma.t_a.\alpha.t_p \in L(S)]$;
   ii) $\forall s \in Serv, \forall (t_a, t_p) \in T_a{}^s x\, T_p{}^s, \forall \sigma^1.t_a.\sigma^2.t_p \in L(S), \exists \alpha^1 \in (T \setminus T_{aup})^*$, $\exists \alpha^2 \in T^*$ such that $\sigma^2 = \alpha^1.\alpha^2$ and $\sigma^1.t_a.\alpha^1.t_p.\alpha^2 \in L(S)$.

Proposition 2.3

Any elementary server is a server

Proof

It is clear that any elementary server fulfils conditions 1) to 6) of definition 2.6.

discretion: let's consider $\sigma \in L(S)$ and $s \in Serv$ such that $\underline{\sigma}(T_p{}^s) > \underline{\sigma}(T_a{}^s)$, $(N^s, M_o{}^s, p_a{}^s, p_p{}^s)$ the service net of s, and M^s the marking of N^s such that $M^s(p_a{}^s) = \underline{\sigma}(T_a{}^s)$, $M^s(p) = M_o{}^s(p)$ for $p \in P^s \setminus \{p_a{}^s\}$; then $\sigma^s = \sigma|T^s \in L(N^s, M^s)$ and $\underline{\sigma}^s(T_p{}^s) > \underline{\sigma}^s(T_a{}^s)$, which is contradictory with proposition 2.2.1).

honesty: i) results from proposition 2.2.2) and from $\Gamma^-(T^s \setminus T_a{}^s) \subset P^s$.

ii) since $\Gamma^-(T^s \setminus T_a{}^s) \subset P^s$, the enabling in S of a transition of $T^s \setminus T_a{}^s$ only depends on the transitions of $T^s \setminus T_a{}^s$ which occured before; thus if $M'\ [t.t^s>$ where $t^s \in T^s \setminus T_a{}^s$ and $t \in T \setminus T^s$, then $M'\ [t^s.t>$. ♦

As expected, the fundamental properties resulting from discretion and honesty are satisfied by servers as they are by clients (cf. proposition 2.1). The ability for a server to perform a provide-transition on a service depends only on the number of accept- and provide-transitions which have previously occured on this service, and any transition sequence may be normalized.

Proposition 2.4 ***normalized transition sequences***

Let $S = (N, M, Serv)$ be a server and $\sigma \in L(S)$.

a) $\forall s \in Serv\ [\exists t_p \in T_p{}^s, \exists \alpha \in (T \setminus T_{aup})^*\ \ \sigma.\alpha.t_p \in L(S)] \Leftrightarrow \underline{\sigma}(T_p{}^s) < \underline{\sigma}(T_a{}^s)$.

b) There exists $\sigma' \in L(S)$ such that:

1) $\underline{\sigma'} = \underline{\sigma}$, 2) $\sigma' \mid T_a = \sigma \mid T_a$,

3) $\forall s \in Serv, \forall t_a \in T_a{}^s, \forall t_p \in T_p{}^s, \forall \sigma^1, \alpha \in T^* [\sigma^1.t_a.\alpha.t_p < \sigma' \Rightarrow \alpha \in (T \setminus T_{aup})^*]$
and σ' is called the **normalized sequence associated to σ.**

Proof

a) If $\sigma.\alpha.t_p \in L(S)$ and $\alpha \in (T \setminus T_a)^*$, $\underline{\sigma}(t_p) + 1 = \underline{\sigma.\alpha.t_p}(t_p) \le \underline{\sigma.\alpha.t_p}(t_a) = \underline{\sigma}(t_a)$ results from discretion.

If $\underline{\sigma}(t_p) < \underline{\sigma}(t_a)$, there exists $\alpha \in T^*$ such that $\sigma.\alpha.t_p \in L(S)$ from honesty.i), and we may suppose $\alpha \in (T \setminus T_a)^*$ from honesty.ii).

b) is immediate from honesty.ii). ♦

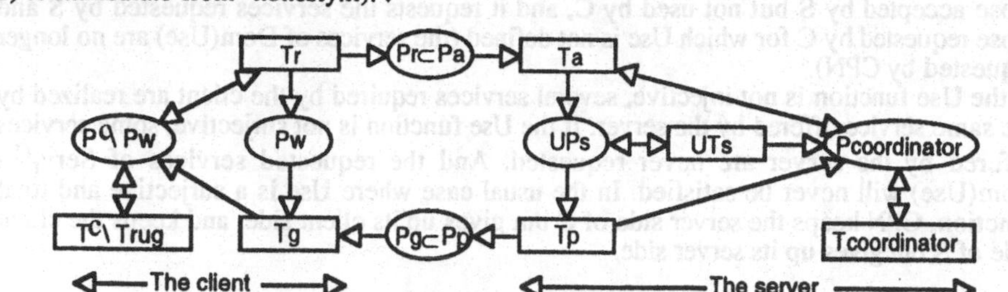

Figure 6: connections between places and transitions of clients and servers

2.3 Composition of clients and servers

Composing a client with a server requires to specify, for each service requested by the client, by which service offered by the server it is taken in charge. The composed net results from merging the request- with the accept-places and the provide- with get-places of the client and server completions according to a Use function mapping the requested services into the offered ones, as shown in Figure 6.

In order to be able to incrementally compose a number of nets, we have to define nets which are both client and server; we also must know if the composition of a client and a server is still a client or a server, and if the order of the composition has an incidence upon the resulting net.

Definition 2.7 ***client-server***

CS = (N, M, $Serv_a$, $Serv_r$) is a *client-server* net if (N, M, $Serv_a$) is a server according to definition 2.6, (N, M, $Serv_r$) is a client according to definition 2.1, and $T_a \cap T_g = \emptyset$.

Of course, each property which holds for clients also holds for a client-server considered as a client, and the same for servers' properties. By the way, a client [resp. a server] may by defined as a client-server such that $Serv_a = \emptyset$ [resp. $Serv_r = \emptyset$].

Definition 2.8 ***composition of client-servers***

Let C = (N^C, M^C, $Serv_a{}^C$, $Serv_r{}^C$) be a client-server and $\underline{C}$ = ($\underline{N}^C$, $\underline{M}^C$) its client-completion, S = (N^S, M^S, $Serv_a{}^S$, $Serv_r{}^S$) be a client-server disjoint[1] of C and $\underline{S}$ = ($\underline{N}^S$, $\underline{M}^S$) its server-completion, and a (partial) function Use : $Serv_r{}^C$ ----> $Serv_a{}^S$.
The completion of the composition of C and S by Use is the structure
$\underline{CP}$ (C, S, Use) = ($\underline{N}$, $\underline{M}$, $Serv_a{}^{CP}$, $Serv_r{}^{CP}$), where:

1) $Serv_a{}^{CP} = Serv_a{}^C \cup (Serv_a{}^S \setminus Use(Serv_r{}^C))$,
2) $Serv_r{}^{CP} = (Serv_r{}^C \setminus Dom(Use)) \cup Serv_r{}^S$,
3) $\underline{N} = \underline{N}^C \cup \underline{N}^S$ and, for any service $s \in Dom(Use)$, the place $p_r{}^s \in P_r{}^C$ is merged with the place $p_a{}^{Use(s)} \in P_a{}^S$, and the place $p_g{}^s \in P_g{}^C$ is merged with $p_p{}^{Use(s)} \in P_p{}^S$.

[1] The disjointness of C and S is not a necessary condition, to the extent that each sub-net corresponding to a service call or a service processing of one of the client-servers is not overloaded in the other one, and a client-server may be composed with itself.

4) $\underline{M} = \underline{M}^C \cup \underline{M}^S$.
The composition of C and S by Use is CP (C, S, Use) = (N, M, $Serv_a^{CP}$, $Serv_r^{CP}$), where N results from $\underline{N}$ by eliminating the interface places P_a^C and P_r^S, and where $M = M^C \cup M^S$.

Figure 8 shows a very simple example where the server models a stack (the accept- and provide-transitions of services are gathered) and the client is a stack user.
The client-server CPN = CP (C, S, Use) accepts the services accepted by C together with those accepted by S but not used by C, and it requests the services requested by S and those requested by C for which Use is not defined (the services of Dom(Use) are no longer requested by CPN).
If the Use function is not injective, several services required by the client are realized by the same service offered by the server. If the Use function is not surjective, some services offered by the server are never requested. And the requested services of $Serv_r^C \setminus Dom(Use)$ will never be satisfied. In the usual case where Use is a surjective and total function, CPN keeps the server side of C but gives up its client side, and keeps the client side of S but gives up its server side.

If $Serv_r^S$ is not empty, S is a client; CPN satisfies definition 2.1 and it is a client as well. If $Serv_a^C$ is not empty, C is a server; CPN offers services and it satisfies structural conditions 1) to 6) of the server's definition 2.6; but it may not be an elementary server even if C and S are: the services nets of services accepted by C may extend to the services nets of S so that they are no longer disjoint, and CPN has no coordinator net[2].

Theorem 2.5

Let CPN be the composition of two client-servers by a Use function.
1) CPN is a client;
2) CPN is a server except for the honesty condition.

The proof consists mainly in relating definitions 2.1, 2.6 and 2.8; the minimal place invariant of each offered service is ensured by proposition 3.1.3, and the server discretion results from lemma 4.ii). On the other hand, CPN is honest only if C and S are suitably composed, and one of the main results of this paper is theorem 5.6 which gives the conditions ensuring that the composition of a server with a sub-server is still a server.

But a much simpler way to build a client-server from two different ones is to union them.

Definition 2.9 ***union of client-servers***

Let $CS^1 = (N^1, M^1, Serv_a^1, Serv_r^1)$ and $CS^2 = (N^2, M^2, Serv_a^2, Serv_r^2)$ be two disjoint client-servers. Their union is the client-server

$$CS^1 \cup CS^2 = (N^1 \cup N^2, M^1 \cup M^2, Serv_a^1 \cup Serv_a^2, Serv_r^1 \cup Serv_r^2).$$

The proof that the union of client-servers is a client-server is straightforward.

If we put aside the condition of server honesty, the composition of client-servers appears to be an operator

$$CP : \mathit{CS} \times \mathit{CS} \times \mathit{USE} \dashrightarrow \mathit{CS}$$

where ***CS*** is the set of client-server nets and ***USE*** the set of Use functions.
The model of a system is made up of a number of client-servers which are gathered either by union or by composition. Thus two crucial questions for incremental design arise:
1) what are the distributivity rules between union and composition?
2) does the order upon which client-servers are composed have an effect on the resulting net?

[2] This is the reason for introducing the distinction between servers and elementary servers.

As expected, a number of client-server nets may be unioned and composed in any order. If you consider a set $\{CS^i = (N^i, M^i, Serv_a^i, Serv_r^i)\ ;\ i \in I\}$ of client-servers and a function Use : $\cup_{i\in I} Serv_r^i$ ---> $\cup_{i\in I} Serv_a^i$, then the result of the composition of the CS^i does not depend on the order in which they are composed but only on the Use function. Indeed, given a net, a set of requested services $Serv_r$ and a set of accepted services $Serv_a$, there are several ways to break down this net into a set of client-servers such that $\cup_{i\in I} Serv_r^i = Serv_r$ and $\cup_{i\in I} Serv_a^i = Serv_a$ (because a client-server may be non-connected, and not disjoint client-servers may be considered). If these decompositions have the same Use function, the composition of their elements leads to the same given net.

Proposition 2.6 ***composition is distributive wrt union***

1) For i = 1... 2, let $C^i = (N^{ci}, M^{ci}, Serv_a^{ci}, Serv_r^{ci})$ and $S^i = (N^{si}, M^{si}, Serv_a^{si}, Serv_r^{si})$ be pairwise disjoint client-servers, and $Use^i : Serv_r^{ci}$ ---> $Serv_a^{si}$. Then
$CP(C^1 \cup C^2, S^1 \cup S^2, Use^1 \cup Use^2) = CP(C^1, S^1, Use^1) \cup CP(C^1, S^2, Use^2)$
$= CP(CP(C^j, S^j, Use^j) \cup C^i, S^i, Use^i)$
where $i \neq j$;
2) Let $C = (N^c, M^c, Serv_a^c, Serv_r^c)$, $CS = (N^{cs}, M^{cs}, Serv_a^{cs}, Serv_r^{cs})$ and $S = (N^s, M^s, Serv_a^s, Serv_r^s)$ be disjoint client-servers, and $Use^c : Serv_r^c$ ---> $Serv_a^{cs}$, $Use^{cs} : Serv_r^{cs}$ ---> $Serv_a^s$. Then
$CP(C, CP(CS, S, Use^{cs}), Use^c) = CP(CP(C, CS, Use^c), S, Use^{cs})$
$= CP(C \cup CS, CS \cup S, Use^c \cup Use^{cs})$
$= CP(C, CS, Use^c) \cup CP(CS, S, Use^{cs})$.

Practically, there are three basic ways to compose nets which are illustrated in Figure 7:

- A client uses several servers: a new server is added to the client;
- Several clients use the same server: a new client is added to the server;
- Hierarchic use: a client C uses a server S, and a client is added to C or a server to S.

We will see later that the first and third ways are safe, because adding a new server to a client and hierarchic use do not modify the properties of the previously composed nets. This does not hold when a new client it added to a server, so that a server will have to be composed with the union of its clients, and not individually with one after the other.

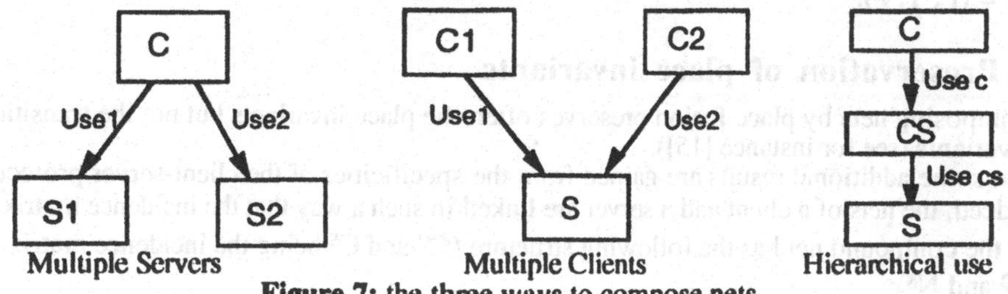

Figure 7: the three ways to compose nets

2.4 Two restrictive hypotheses

Since the clients of a server must be considered as a single one, we will suppose from now on that the Use function relating a client and a server is a surjective function. We will also suppose that Use is a total function, i.e. Dom (Use) = $Serv_r^C$, since to consider services which may be accepted by no server makes no sense and propositions 4.3 and 5.4 show that the services of a client may be partitionned without damage. This hypothesis is not required for the results given in the remainder of this paper, but we keep it in order to avoid examining cases which are of no practical interest.

According to the definition of the sets of requested and accepted services of client-servers and to the idea that nets communicate by arcs, the Use function may be defined as:

$$\text{Use} : T_r \cup T_g \;\text{------>}\; P_a \cup P_p$$
$$t \;\text{:------>}\; p_a^{\text{Use}(s)} \text{ if } t = t_r^s, s \in \text{Serv}_r^C$$
$$p_p^{\text{Use}(s)} \text{ if } t = t_g^s, s \in \text{Serv}_r^C$$

In order to study how much a server fulfils the demand of a client, we will need to compare the sequences of request- and get-transitions enabled in the client with the sequences of accept- and provide-transitions enabled in the server. The simplest way is to define a labelling

$$l : T_a \cup T_p \;\text{------>}\; P_a \cup P_p$$
$$t \;\text{:------>}\; p \in P_a \text{ such that } t \in p\bullet \text{ if } t \in T_a$$
$$p \in P_p \text{ such that } t \in \bullet p \text{ if } t \in T_p$$

allowing to compare $\text{Use}(\sigma^C)$ with $l(\sigma^S)$ if $\sigma^C \in (T_r \cup T_g)^*$ and $\sigma^S \in (T_a \cup T_p)^*$. In order to avoid this labelling, we will suppose $|p^\bullet| = 1$ for any $p \in P_a$ and $|{}^\bullet p| = 1$ for any $p \in P_p$ in the following. Now Use may be defined as:

$$\text{Use} : T_r \cup T_g \;\text{------>}\; T_a \cup T_p$$
$$t \;\text{:------>}\; (p_a^{\text{Use}(s)})^\bullet \text{ if } t = t_r^s, s \in \text{Serv}_r^C$$
$${}^\bullet(p_p^{\text{Use}(s)}) \text{ if } t = t_g^s, s \in \text{Serv}_r^C$$

and $\text{Use}(\sigma^c)$ may be directly compared with σ^s. This hypothesis brings a significative simplification in the statement of the results, but does not alter the structure of the proofs.

In the following, we will use the following conventions:
$C = (N^C, M^C, \text{Serv}_a^C, \text{Serv}^C)$ is a *client*-server such that $\text{Serv}^C \neq \emptyset$, $N^C = (P^C, T^C, F^C)$,
$\underline{C} = (\underline{N}^C, \underline{M}^C)$ is its client-completion and $\underline{N}^C = (P^C \cup P_r^C \cup P_g^C, T^C, \underline{F}^C)$,
$S = (N^S, M^S, \text{Serv}^s, \text{Serv}_r^S)$ is a client-*server* such that $\text{Serv}^S \neq \emptyset$, $N^S = (P^S, T^S, F^S)$,
$\underline{S} = (\underline{N}^S, \underline{M}^S)$ is its server-completion and $\underline{N}^S = (P^S \cup P_a^S \cup P_p^S, T^S, \underline{F}^S)$,
$\text{Use} : \text{Serv}^C \text{ ---> } \text{Serv}^S$ is a function defining how C uses S,
$CPN = CP(C, S, \text{Use}) = (N, M, \text{Serv}_a^C, \text{Serv}_r^S)$ is the composition of C and S, and
$N = (P, T, F)$.

3 Preservation of place invariants

Composing nets by place fusion preserves often the place invariants but not the transition invariants (see for instance [15]).
But some additional results are gained from the specificities of the client-server protocol. Indeed, the nets of a client and a server are linked in such a way that the incidence matrix $\mathbb{C}$ of the compound net has the following structure ($\mathbb{C}^C$ and $\mathbb{C}^S$ being the incidence matrix of N^C and N^S):

$$\mathbb{C} = \begin{array}{r|c|c|} & T^C & T^S \\ P^C & \mathbb{C}^C & O \\ P_r^C = P_a^S & 0\;1 \ldots 0\;0\;1 & -1\;0 \ldots -1\;0\;0\;1 \\ P_g^C = P_p^S & -1\;0 \ldots -1\;0\;0 & 0\;1 \ldots 0\;0\;1\;0 \\ P^S & O & \mathbb{C}^S \end{array}$$

A net is called conservative if each place belongs to the support of a place invariant.

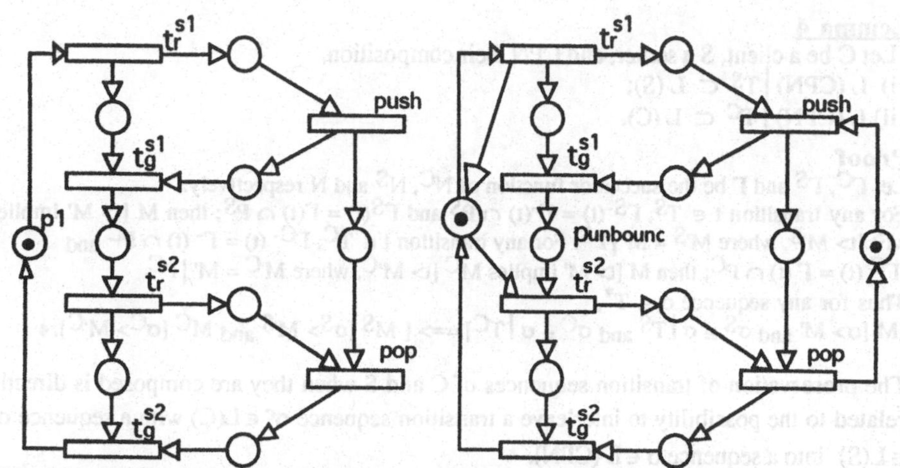

Figure 8: N is bounded whereas S is not; the Demand of C is strictly contained in the Supply of S

Figure 9: N is bounded whereas C is not; S is not compatible with C

Proposition 3.1 ***conservativity is preserved***

Let C be a client, S a server and CPN their composition.

1) Let I^C be a place invariant of N^C, and $I \in \mathbb{N}^P$ such that $I(p) = I^C(p)$ if $p \in P^C$, $I(p) = 0$ if $p \in P \setminus P^C$. Then I is an invariant of N.
2) In the same way, any place invariant of N^S gives rise to an invariant of N.
3) Let $s \in Serv^C$ be such that $p_w{}^s$ belongs to the support of a place invariant of N^C; then $p_r{}^s$ and $p_g{}^s$ belong to the support of a place invariant of N.
4) If N^C and N^S are conservative, then N is conservative.
5) If N is conservative, then $\underline{N}^C$ and $\underline{N}^S$ are conservative.

1) and 2) are obvious from the structure of the incidence matrix of N shown above. The proof of 3) - 5) is in [14] and results from the existence of a place invariant for each service of S.

4 Compatibility and preservation of enabled transition sequences

After they have been composed, a server S and a client C are mutually dependent: an accept-transition of S may occur only if a request-transition has previously occurred in C for that service, and conversely for get-transitions of C with regard to provide-transitions of S. But we are more interested in preserving the client's behavior than in preserving the server's behavior, since servers are designed to be at the client's disposal and not the opposite.

In this section we will study the preservation of the language (that is the set of enabled transition sequences) of a client as well as the preservation of the set of its reachable markings. To this end, we will introduce a main characteristic of a client - its *Demand* which is the set of request-transitions sequences it may fire - , and a main characteristic of a server - its *Supply* which is the set of accept-transitions sequences it may fire. It will be shown that the language of a client is preserved if and only if its Demand matches with the Supply of its server, and that composition of client-servers does not alter their Supply and may only reduce their Demand.

The following lemma states that the behavior of the composition of a client and a server is always a restriction of their respective behavior [15]. The preservation of transition sequences is nothing else than the converse of this lemma.

Lemma 4

Let C be a client, S a server, and CPN their composition.
i) $L(CPN) \mid T^S \subset L(S)$;
ii) $L(CPN) \mid T^C \subset L(C)$.

Proof

Let Γ^C, Γ^S and Γ be the successor function of N^C, N^S and N respectively.
For any transition $t \in T^S$, $\Gamma^{S-}(t) = \Gamma^-(t) \cap P^S$ and $\Gamma^S(t) = \Gamma(t) \cap P^S$; then $M\ [t\!\!>\ M'$ implies $M^S\ [t\!\!>\ M'^S$, where $M'^S = M' \mid P^S$. For any transition $t \in T^C$, $\Gamma^{C-}(t) = \Gamma^-(t) \cap P^C$ and $\Gamma^C(t) = \Gamma(t) \cap P^C$; then $M\ [t\!\!>\ M'$ implies $M^C\ [t\!\!>\ M'^C$, where $M'^C = M' \mid P^C$.
Thus for any sequence $\sigma \in T^*$,
$[M\ [\sigma\!\!>\ M'$ and $\sigma^S = \sigma \mid T^S$ and $\sigma^C = \sigma \mid T^C] \Longrightarrow [M^S\ [\sigma^S\!\!>\ M'^S$ and $M^C\ [\sigma^C\!\!>\ M'^C]$.♦

The preservation of transition sequences of C and S when they are composed is directly related to the possibility to interleave a transition sequence $\sigma^c \in L(C)$ with a sequence $\sigma^s \in L(S)$ into a sequence $\sigma \in L(CPN)$.

Definition 4.1 ***interleaving***

Let C be a client, S a server, CPN = CP (C, S, Use), $\sigma^C \in L(C)$ and $\sigma^S \in L(S)$.
σ^C and σ^S may be interleaved iff there exists $\sigma \in L(CPN)$ such that $\sigma^C = \sigma \mid T^C$ and $\sigma^S = \sigma \mid T^S$.
Then σ is an interleaving of σ^C and σ^S, and we note $\sigma = \sigma^C * \sigma^S$.

Interleaving lemma

Let C be a client, S a server, CPN = CP (C, S, Use), $\sigma^C \in L(C)$ and $\sigma^S \in L(S)$.
The following properties are equivalent:

1) σ^C and σ^S may be interleaved.
2) a) $\forall s \in Serv^S$, $\underline{\sigma}^S(t_a^s) \leq \underline{\sigma}^C(Use^{-1}(t_a^s))$;
b) $\forall \sigma_1^C \leq \sigma^C, \exists \sigma_1^S \leq \sigma^S$ such that
$\forall s \in Serv^S$, $\underline{\sigma}_1^S(t_a^s) \leq \underline{\sigma}_1^C(Use^{-1}(t_a^s))$ and $\underline{\sigma}_1^C(Use^{-1}(t_p^s)) \leq \underline{\sigma}_1^S(t_p^s)$.
3) a) $\forall s \in Serv^S$, $\underline{\sigma}^C(Use^{-1}(t_p^s)) \leq \underline{\sigma}^S(t_p^s)$;
b) $\forall \sigma_1^S \leq \sigma^S, \exists \sigma_1^C \leq \sigma^C$ such that
$\forall s \in Serv^S$, $\underline{\sigma}_1^S(t_a^s) \leq \underline{\sigma}_1^C(Use^{-1}(t_a^s))$ and $\underline{\sigma}_1^C(Use^{-1}(t_p^s)) \leq \underline{\sigma}_1^S(t_p^s)$.

Proof

1) ⇒ 2) & 3)
If σ^C and σ^S may be interleaved, 2.a) and 3.a) hold; moreover, any prefix sequence of σ^C may be interleaved with a prefix sequence of σ^S for which 2.a) and 3.a) also hold, so that 2.b) holds for σ^C and σ^S. Symmetrically, any prefix sequence of σ^S may be interleaved with a prefix sequence of σ^C and 3.b) holds.
2) ⇒ 1) The proof is by recurrence on $\underline{\sigma}^C(T_g) + \underline{\sigma}^S(T_a)$.
• If σ^C and σ^S satisfy 2) and $\sigma^S = \sigma_0^S.t_a^{s1}$ for some service $s1 \in Serv^S$, then σ^C and σ_0^S also satisfy 2) and they may be interleaved. Indeed $\forall s \in Serv^S\ [\underline{\sigma}_0^S(t_a^s) \leq \underline{\sigma}^S(t_a^s) \leq \underline{\sigma}^C(Use^{-1}(t_a^s))]$ and a) holds. Let's consider $\sigma_1^C \leq \sigma^C$ and $\sigma_1^S \leq \sigma^S$ such that $\forall s \in Serv^S\ [\underline{\sigma}_1^S(t_a^s) \leq \underline{\sigma}_1^C(Use^{-1}(t_a^s))$ and $\underline{\sigma}_1^C(Use^{-1}(t_p^s)) \leq \underline{\sigma}_1^S(t_p^s)]$; if $\sigma_1^S = \sigma^S = \sigma_0^S.t_a^{s1}$, then $\forall s \in Serv^S$ $[\underline{\sigma}_0^S(t_a^s) \leq \underline{\sigma}_1^S(t_a^s) \leq \underline{\sigma}_1^C(Use^{-1}(t_a^s))$ and $\underline{\sigma}_1^C(Use^{-1}(t_p^s)) \leq \underline{\sigma}_0^S(t_p^s) = \underline{\sigma}_1^S(t_p^s)]$ and b) holds. Thus σ^C and σ_0^S may be interleaved into $\sigma_0 = \sigma^C * \sigma_0^S$ such that $M\ [\sigma_0\!\!>\ M'$; but $\underline{\sigma}_0^S(t_a^{s1}) < \underline{\sigma}^C(Use^{-1}(t_a^s))$, so that $M'(p_a^{s1}) > 1$, $M'\ [t_a^{s1}\!\!>$, and $\sigma_0.t_a^{s1}$ is an interleaving of σ^C and

σ^S.

• If σ^C and σ^S satisfy 2) and $\sigma^C = \sigma_o{}^C.t_g$ where $t_g \in Use^{-1}(t_p{}^{s1})$ for some service s1 $\in Serv^S$, then σ^S and $\sigma_o{}^C$ also satisfy 2) and they may be interleaved into σ_o. Since $\underline{\sigma}_o{}^C(Use^{-1}(t_p{}^{s1})) < \underline{\sigma}^S(t_p{}^{s1})$, M $[\sigma_o>$ M' $[t_g>$ and $\sigma_o.t_g$ is an interleaving of σ^C and σ^S.

3) $\Rightarrow$ 1) The proof is symmetric to the one of 2) $\Rightarrow$ 1). ♦

Now we are able to state the main result of this section. It asserts that the transition sequences enabled in C are still enabled in CPN if and only if S can accept the service sequences that C may request, and in the same order. Thus the client-server protocol preserves the enabled transition sequences as if the client and the server were composed by transitions merging, whereas they are composed by places merging.

Definition 4.2 ***Demand, Supply and compatibility***

Let C be a client, S a server and a function Use : $Serv^C$ ---> $Serv^S$.

1) Dem (C) = L(C) $|$ T_r is called the *Demand* of C.
2) Sup (S) = L(S) $|$ T_a is called the *Supply* of S.
3) S is said to be *compatible* with C for Use iff Use (Dem (C)) $\subset$ Sup (S).

For instance, the supply of the server of figure 5 is

$$\{\sigma \in \{s^1, s^2, s^3\}^* ; \forall \sigma' \leq \sigma, \underline{\sigma}'(s^2) \leq \underline{\sigma}'(s^1) \text{ and } \underline{\sigma}'(s^3) \leq \underline{\sigma}'(s^1) + \underline{\sigma}'(s^2)/2\}.$$

As another example, the server and the client of Figure 8 are compatible since Use (Dem (C)) = {push.pop}* $\subset$ Sup(S) = $\{\sigma \in \{push, pop\}^*; \forall \sigma' \leq \sigma, \underline{\sigma}'(pop) \leq \underline{\sigma}'(push)\}$; but the ones of Figure 9 are not since the demand of C is $\{s^1, s^2\}^*$ whereas the supply of S is that of a bounded stack.

Theorem 4.1 ***preservation of enabled transition sequences***

Let C be a client, S a server, and CPN = CP (C, S, Use).

1) L (CPN) $|$ T^C = L (C) $\Longleftrightarrow$ S is compatible with C for Use;
2) L (CPN) $|$ T^S = L (S) $\Leftarrow$ Sup (S) $\subset$ Use (Dem(C)).

Proof

1) $\Rightarrow$ We have to show that Use (L(C) $|$ T_r) $\subset$ L (S) $|$ T_a. Let $\sigma^C \in$ L(C); we can suppose that σ^C is normalized since this does not modify $\sigma^C | T_r$. From the hypothesis, there exists $\sigma \in$ L(CPN) such that $\sigma^C = \sigma | T^C$. In σ, a request-transition must be accepted as soon as it is issued, since the get-transition, which needs the result of the provide-transition, must occur before the next request. Thus Use $(\sigma^C | T_r) = \sigma | T_a = \sigma^s | T_a$, where $\sigma^s = \sigma | T^s \in$ L(S).

1) $\Leftarrow$ L(CPN) $|$ $T^C \subset$ L(C) from lemma 4.ii), and we only have to show the converse inclusion.

Let $\sigma^C \in$ L(C). From the hypothesis there exists $\sigma^S \in$ L(S) such that Use $(\sigma^C | T_r) = \sigma^S | T_a$, and we can suppose that σ^S is normalized since this does not change $\sigma^S | T_a$. σ^C and σ^S satisfy 2) of the interleaving lemma. Indeed $\forall$ s $\in Serv^S$, $\underline{\sigma}^S(t_a{}^s) = \underline{Use(\sigma^C)}(t_a{}^s) = \underline{\sigma}^C(Use^{-1}(t_a{}^s))$ and a) holds; for any $\sigma_1{}^C \leq \sigma^C$, let's consider the longest $\sigma_1{}^S \leq \sigma^S$ such that Use $(\sigma_1{}^C | T_r) = \sigma_1{}^S | T_a$ and s $\in Serv^S$; then $\underline{\sigma}_1{}^S(t_a{}^s) = \underline{Use(\sigma_1{}^C)}(t_a{}^s) = \underline{\sigma}_1{}^C(Use^{-1}(t_a{}^s))$ and $\underline{\sigma}_1{}^C(Use^{-1}(t_p{}^s)) \leq \underline{\sigma}_1{}^C(Use^{-1}(t_a{}^s)) = \underline{\sigma}_1{}^S(t_a{}^s) = \underline{\sigma}_1{}^S(t_p{}^s)$ and b) holds. Since σ^C may be interleaved with a $\sigma^S \in$ L(S), then $\sigma^C \in$ L(CPN) $|$ T^C.

2) $\Leftarrow$ The proof is very similar to the one of 1) $\Leftarrow$.

2) $\Rightarrow$ does not hold because the communication between nets is asynchronous; thus S may accept requests in an order which is different from the order in which they are issued by C. ♦

Lemma 4 and theorem 4.1 could also be formulated in terms of reachable markings, and they have the following consequences.

Corollary 4.2 ***preservation of boundedness***

Let C be a client, S a server, and CPN = CP (C, S, Use).

1) If (N^C, M^C) and (N^S, M^S) are bounded, then (N, M) is bounded.
2) If S is compatible with C and (N, M) is bounded, then (N^C, M^C) is bounded.
3) If Sup (S) $\subset$ Use (Dem(C)) and (N, M) is bounded, then (N^S, M^S) is bounded.

The nets of Figure 8 and 9 illustrate 2) and 3) above. In both cases N is (structurally) bounded; but in Figure 8 S is not bounded, C is bounded and Use (Dem (C)) is strictly contained in Sup (S); and in Figure 9, C is not bounded, S is bounded and S is not compatible with C.

Let's now examine how compatibility is preserved when several nets are composed together. Except when another client is added to a server, compatibility is not reconsidered when a new net is composed with a set of already composed nets and compatibility is a local property in the following sense: to decide if a client C is compatible with a server S, neither the eventual clients of C nor the eventual servers of S have to be taken into account if they are compatible with C or S respectively.

A client compatible with each of its servers is compatible with their union.

Proposition 4.3 ***multiple servers***

Let C = (N^C, M^C, $Serv^1 \cup Serv^2$) be a client, $Serv^1 \cap Serv^2 = \emptyset$, and S^1 and S^2 two disjoint servers.

If S^i is compatible with C^i = (N^C, M^C, $Serv^i$) for Use^i, for i = 1, 2,

then $S^1 \cup S^2$ is compatible with C for $Use^1 \cup Use^2$.

Proof: results from $L(S^1) \cup L(S^2) \subset L(S^1 \cup S^2)$. ♦

Composing a client-server with a compatible server does not alter its Supply (and thus does not alter the set of clients it is compatible with), and composing a client-server with a client reduces its Demand (and thus does not reduce the set of servers compatible with it).

Proposition 4.4 ***hierarchic use***

Let C and S be two client-servers, and CPN = CP (C, S, Use).

1) If C is compatible with S for Use, then Sup (CPN) = Sup (CS);
2) Dem (CPN) $\subset$ Dem (CS).

Proof: 1) results from theorem 4.1, and 2) results from lemma 4.i). ♦

5 Transparency and preservation of liveness

Compatibility is a necessary condition for a server to be able to fulfil the Demand of a client.

But it is not a sufficient condition for definitively preventing a client to get into trouble when using a server, since CP (C, S, Use) may reach a deadlock even if (N^C, M^C) and (N^S, M^S) are live. This happens when S reaches a state where it can no longer accept a service request whereas it could in a previous state. This may be due to a "silent move" which disables an accept-transition as in the net of Figure 10: occurrence of transition t disables the service s2. This may also be due to some kind of indeterminism, if a service request may be processed in differents ways which act differently upon the marking of the coordinator net.

The perfect server is the one that its client may ignore because it will process any service request in any situation; from the clients point of view, this server behaves as if it had no coordinator net and was made up only of transitions linking the request- to the get-places of the service calls. Such a server is said to be *transparent* to the client. We will show that transparency comes down to compatibility and to the comparison of the second main characteristic of a client - its degree of confidence -, with the second main characteristic of a server - its degree of reliability. Moreover, transparency is preserved by net composition in the same way than compatibility, and it is the property thanks to which the honesty of servers is preserved; the proofs are given in [14].

S is transparent to C for a given Use function if it may always accept a new request from C, whatever are the previous requests and its actual state.

Definition 5.1 ***transparency***

Let C be a client, S a server, CPN = CP (C, S, Use).
S is said to be transparent to C for Use iff
$\forall t \in T^C, \forall \sigma \in L(CPN)\ [\sigma^c = \sigma | T^C \text{ and } \sigma^c.t \in L(C) \Rightarrow \exists \alpha \in T^{S*}\ [\sigma.\alpha.t \in L(CPN)]]$.

If S is transparent to C for Use, C and CPN satisfy the various sequential equivalence notions with regard to T^C [16]: they are failure equivalent [8], behavior equivalent [6], CPN is deadlock-free if C is, and the same transitions of C are live in C and in CPN.

Theorem 5.1 ***transparency preserves the client's liveness***

Let S be transparent to C for Use, and $t \in T^c$. Then t is live in CPN iff it is in C.

Proof

$t \in T$ is live in (N, M) iff $\forall M'\ [M \dashrightarrow M' \Rightarrow \exists \sigma \in T^*\ [M'[\sigma.t\rangle]]$.
According to lemma 4.ii), any transition of T^C live in CPN is live in C.
Conversely, let $\sigma \in L$ (CPN), and $t \in T^C$ be live in C. Then there exists $t^1....t^r \in T^{C*}$ such that $\sigma^c.t^1....t^r.t \in L(C)$ where $\sigma^c = \sigma | T^C$. If S is transparent to C, there exist $\alpha^1, \cdots, \alpha^r, \alpha \in T^{S*}$ such that $\sigma.\alpha^1.t^1....\alpha^r.t^r.\alpha.t \in L$ (CPN); thus t is live in CPN. ♦

In Figure 9, S is not transparent to C, althought CPN is live. Transparency is stronger than liveness preservation since definition 5.1 requires $\alpha \in T^{S*}$ instead of $\alpha \in T^{CPN*}$.

Compatibility characterizes the cooperation between a client and a server from a functional point of view. Transparency requires to also examine, from an operational point of view, how they cooperate. On the client side, the confidence degree assesses how nearly it controls the execution by its servers of the services it requests. On the server side, the reliability degree measures to what extent it is able to actually process any transition sequence of its Supply even when its client does not control much of this processing.

Definition 5.2 ***confidence***

Let C be a client and $k = \mathrm{Sup}\ \{\underline{\sigma}(T_r) - \underline{\sigma}(T_g)\ ;\ \sigma \in L(C)\} \in \mathbb{N} \cup \{\omega\}$.
k is called the confidence degree of C, and C is said to be k-confident.

A client is k-confident if it never reaches a marking where the waiting places hold more than k tokens, that is more than k get-transitions are enabled. For instance, the client of Figure 8 is 1-confident and the one of Figure 9 is ω-confident, whereas the client of Figure 3 is 2*n-confident where $M_0(p) = n$.

A 0-confident client does not request any service; thus it is compatible and transparent to any server! 1-confidence means that C never requests a service before getting back the result of the previous request; thus it does not take advantage of the asynchronous communication with the server which allows to have several service requests concurrently in progress.

The confidence degree of a client is $k = \mathrm{Sup}\ \{\Sigma_{p \in P_w} M'(p)\ ;\ M^C \dashrightarrow M'\}$. It may be computed from the covering graph, and it is ω if and only if one of the waiting places is unbounded. A bound may also be computed when each waiting place belongs to a conservative component of N^C. The confidence degree of a client may always be decreased to an integer k without altering its sequential behavior, by adding a place holding k tokens at the initial marking and with arcs to each request-transition and from each get-transition.

<u>Definition 5.3</u> ***reliability***

Let S be a server and k a strictly positive integer.
$\forall\, \sigma^1, \sigma^2 \in T^S$, we define $\Delta(\sigma^1, \sigma^2) = \mathrm{Max}_{\sigma 1' \leq \sigma 1}\,(\mathrm{Min}_{\sigma 2' \leq \sigma 2}\,(\Sigma_{t \in T_a} \mid (\underline{\sigma}^{1'}(t) - \underline{\sigma}^{2'}(t) \mid))$
S is *k-reliable* iff $\forall\, \sigma^1, \sigma^2 \in L(S)$
$[\Delta(\sigma^1, \sigma^2) < k \Rightarrow \exists\, \alpha^1, \alpha^2\, [(\underline{\sigma}^1.\underline{\alpha}^1(t))_{t \in T_a} = (\underline{\sigma}^2.\underline{\alpha}^2(t))_{t \in T_a}$ and $\sigma^1.\alpha^1, \sigma^2.\alpha^2 \in L(S)]]$.
The reliability degree of S is the greatest integer such that S is k-reliable, 0 if S is not reliable for any k, or ω if S is k-reliable for any k.

We have $\Delta(\sigma^1, \sigma^2) = 0 \Leftrightarrow \sigma^1 | T_a = \sigma^2 | T_a$, and $\Delta(\sigma^1, \sigma^2) < k \Longleftrightarrow \forall\, \sigma^{1'} \leq \sigma^1, \exists\, \sigma^{2'} \leq \sigma^2$ $[\Sigma_{t \in T_a} \mid (\underline{\sigma}^{1'}(t) - \underline{\sigma}^{2'}(t) \mid < k]$; for instance, $\Delta(t^1.t^2.t^3.t^4.t^5, t^2.t^3.t^1.t^4) = 1$ and $\Delta(t^1.t^2.t^3.t^4.t^5, t^2.t^4.t^5.t^3) = 2$.
A server is k-reliable if for any sequence of accept-transitions $\sigma^1 | T_a$ of its Supply, a transition sequence σ^2 which includes some of these accept-transitions - in an order which may be different but never "forgets" more than k of them - may always be extended to recover the missed accept-transitions. The servers of Figures 5, 8 and 9 are ω-reliable, whereas the one of Figure 10 is not 1-reliable.
1-reliability is equivalent to the "behaviour condition" of [6] which may be stated in the more intuitive following way: $\forall\, \sigma^1, \sigma^2 \in L(S)\, [\sigma^1 | T_a = \sigma^2 | T_a \Longrightarrow$

$$\{t \in T_a\, ; \exists\, \alpha \in (T \setminus T_a)^*\ \sigma^1.\alpha.t \in L(S)\} = \{t \in T_a\, ; \exists\, \alpha \in (T \setminus T_a)^*\ \sigma^2.\alpha.t \in L(S)\}].$$

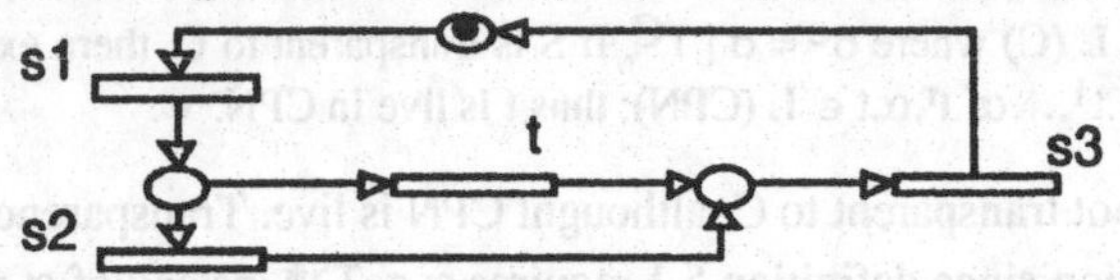

Figure 10: a not 1-reliable server (t is a silent move)

<u>Theorem 5.2</u> ***characterization of transparency***

Let C be a client, S a server, and Use : $\mathrm{Serv}^C \dashrightarrow \mathrm{Serv}^S$. S is transparent to C for Use if
i) S is compatible with C for Use, i.e. Use (Dem(C)) $\subset$ Sup(S), and
ii) S is more reliable than C is confident,
i.e. $k_c \leq k_r$, where k_c is the confidence degree of C and k_r is the reliability degree of S.

This result reduces transparency to comparisons between the two main characteristics of the client - its Demand and its confidence degree -, with the two main characteristics of the server - its Supply and its reliability degree. Thanks to it the client-server protocol favors reusability: once the characteristics of a client and of a server are computed, it is quite easy to verify if they are compatible and/or transparent for a given Use function.

The following theorem shows that the sufficient conditions of transparency are also necessary if we do not want to worry about occasional relationships between C, S and Use. Transparency implies compatibility, and a server which is not k-reliable will not be transparent to some k-confident clients with which it is compatible. Thus any server must be built in such a way that it is reliable for some k. Then its composition with a client requires only to check their compatibility and eventually to decrease the confidence of the client.

<u>Theorem 5.3</u> ***necessity of compatibility and reliability***

1) Let C be a client such that Dem (C) $\neq \emptyset$, S a server, and Use : $\mathrm{Serv}^C \dashrightarrow \mathrm{Serv}^S$;
If S is transparent for C and Use, then S is compatible with C for Use.
2) Let S be a server and k a positive integer such that
for any client C and any Use function,
[S is compatible with C and C is k-confident => S is transparent to C];

Then S is k-reliable.

We have now to examine how confidence and reliability (and thus transparency) are preserved by composition of nets. The results are the same than for compatibility (cf. Propositions 4.3 and 4.4). If each server of a client is transparent to the client, their union is also transparent to the client.

Proposition 5.4 ***multiple servers***

Let $C = (N^C, M^C, Serv^1 \cup Serv^2)$ be a client, $Serv^1 \cap Serv^2 = \emptyset$, and S^1 and S^2 two disjoint servers.
If S^i is transparent to $C^i = (N^C, M^C, Serv^i)$ for Use^i, with $i = 1, 2$,
then $S^1 \cup S^2$ is transparent to C for $Use^1 \cup Use^2$.

Composing a client-server with a transparent server does not alter its reliability, and composing a client-server with a client may only reduce its confidence degree.

Proposition 5.5 ***hierarchic use***

Let C and S be two client-servers, and CPN = CP (C, S, Use).
1) If S is transparent to C for Use, CPN has the same reliability degree than C;
2) Let k_{CPN} and k_S be the confidence degree of CPN and S respectively; then $k_{CPN} \leq k_S$.

Last but not least, transparency preserves the server's honesty (cf. point 8 of definition 2.6) and it is the property thanks to which a compound net is a server. Thus composition is a stable operation in the set of client-server nets as we claimed in section II.3, provided that transparency is respected.

Theorem 5.6 ***server's honesty is preserved***

Let C be a client, and S a server transparent to C for Use.
Then CPN = CP (C, S, Use) is a honest server.

6 Conclusion

The aim of this paper was to introduce into the Petri nets formalism the client-server protocol which has proved to be suitable to structure large systems. To this end, we have set basic rules to axiomize this protocol (the honesty and discretion of clients and servers), given a definition of client-server satisfying these rules, and investigated some of their consequences.
The nets we consider have a client face - they are able to send service requests and to retrieve service results -, and/or a server face - they are able to accept requests and to provide results.
The server's structure dissociates the processing of service requests (the service nets) from controlling the acceptability of these requests (the coordinator net). Thus services are defined at a high-level and they may be achieved differently according to the server's state.
A client is composed with a server by means of a Use function mapping the services requested by a client onto the services offered by a server. Links between clients and servers are thus defined in an abstract manner and, for instance, several client's services may request the same server's service.
Communication is asynchronous; it is realized by means of places which are one-way channel, so that the enabling of a transition in a net may be decided without regard to what happens in other nets; thus the nets are loosely coupled and a designer may consider each net as modelling an autonomous process having its own token game player.
Four main characteristics of client-server nets have been identified: the Demand and the confidence degree of a client, and the Supply and the reliability degree of a server. These are quite easy to compute, except for the reliability degree, and they are preserved by the composition of nets. Although the equivalence point of view is not addressed in this paper,

two servers having the same supply and the same reliability degree are equivalent with regards to many features, as are two clients having the same demand and confidence degree.
Finally, it is shown that for CPN = CP(C, S, Use), the enabled transition sequences of C are still enabled in CPN if and only if Use(Demand(C)) $\subseteq$ Supply(S), and the liveness of C is preserved in CPN if S is more reliable than C is confident.
A consequence of these results is that the tuning of the synchronization degree between nets, through the client's confidence degree, is essential for the preservation of dynamic properties such as liveness. They also make clear that the price to pay to design nets as autonomous components which may be reused in different contexts is the reliability of the server.

References

[1] P. Huber, K. Jensen, R. M. Shapiro. Hierarchies in coloured Petri nets. APN 1990, LNCS 483.

[2] B. Meyer. Object-Oriented Software Construction; Prentice Hall, 1988

[3] B. Baumgarten. On internal and external characterization of PT-net building bloc behaviour ; Advances in Petri Nets 88, LNCS 340

[4] A. Maggiolo-Schetteni, J. Winkowski. A compositional semantics for timed Petri Nets; Fundamenta Informaticae XIII, IOS Press, 1990

[5] R. Valette. Analysis of Petri nets by stepwise refinements; Journal of Computer and System Science 18, 3; 1979

[6] C. André. Behaviour of a place-transition net on a subset of transition. Informatik-Fachberichte 52, Springer 1982

[7] F. De Cindio, G. De Michelis, L. Pomello, C. Simone. Superposed automata nets ; Informatik-Fachberichte 52, Springer 1982

[8] W. Vogler. Asynchronous communication of Petri nets and the refinement of transitions; Report TUM I9112, Inst. Informatik, Techn. Univ. München, 1991

[9] R. Bastide, C. Sibertin-Blanc. Object-oriented design of parallel systems. 2nd International Workshop on Software Engineering and its Applications; Toulouse (France), Dec. 1989

[10] R. Bastide, C. Sibertin-Blanc. Modelling flexible manufacturing systems by means of CoOperative Objects. Computer Applications in Production and Engineering CAPE 91, IFIP (G. Doumeingts, J. Browne, M. Tomjanovich Editors, North-Holland); Bordeaux (F), Sept. 1991

[11] C. Sibertin-Blanc. Cooperative Objects for the conceptual modelling of organizational information systems. The Object-Oriented Approach in Information Systems, IFIP TC8 Conf.; Quebec, 28-31 Oct. 1991

[12] R. Bastide, P. Palanque. Modelisation de l'interface d'un logiciel de groupe par objets cooperatifs; 3ème journées sur l'ingénierie des IHM, dec 91, Dourdan, France

[13] H. J. Genrich. Predicate/transition Nets; in Petri Nets : Applications and relationships to other models of concurrency (W. Brauer, W. Reisig, G. Rosenberg editor), LNCS 254, Springer

[14] C. Sibertin-Blanc. Analysis of Petri nets communicating through a client-server protocol. Technical report of University Toulouse 1, 1992

[15] Y. Souissi, G. Memmi. Composition of nets via a communication medium. 10th International Conference on Applications and Theory of Petri Nets, Bonn, June 1989

[16] R. Van Glabbeek, U. Goltz. Equivalence notions for concurrent systems and refinement of actions; MFCS 89, LNCS 379, 1989

Analysis of Dynamic Load Balancing Strategies Using a Combination of Stochastic Petri Nets and Queueing Networks

C.R.M.Sundaram
Department of Computational Science
University os Saskatchewan, Saskatoon, Canada
Y.Narahari
Department of Computer Science and Automation
Indian Institute of Science, Bangalore, India

Abstract

This paper is concerned with the analytical evaluation of two well known dynamic load balancing strategies, namely, *shortest queue routing* (SQR) and *shortest expected delay routing* (SEDR). We overcome the limitations of existing analysis methodologies, using a well known hybrid performance model that combines *generalized stochastic Petri nets* and *product form queueing networks*. Our methodology is applicable to both open queueing network and closed queueing network models of load balancing in distributed computing systems. The results show that for homogeneous distributed systems, SQR outperforms all other policies. For heterogeneous systems, SEDR surprisingly performs worse than SQR at low levels of imbalance in loads. However, with increase in imbalance in load, SEDR expectedly performs better than SQR.

1 Introduction

The problem of balancing the load over the entire system so that its overall performance is optimized is an important research problem [20] in Distributed Computing Systems(DCSs). Routing an arriving job to the processor with the shortest queue length(SQR, shortest queue routing) [17] in the case of homogeneous systems and routing to the processor with the shortest expected delay (SEDR, shortest expected delay routing) [4] in the case of heterogeneous systems have been found to be close to optimal heuristic load balancing strategies.

Analysis of the performance of SQR and SEDR policies through simulation will be computationally expensive. It has been reported by Nelson and Philips in [17] that, when the number of parallel queues is 32, it took several days of computing on an IBM 3090 main frame to get a mere 5 samples. So, in this paper we make an analytical evaluation of SQR and SEDR policies. Analytical evaluation through Markov chains by setting up exact global balance equations will be practically infeasible as the number of equations grow exponentially with increase in the number of parallel queues. Exact analysis through queueing networks is not possible as the individual queues are no longer independent of one another, and the system loses the product form property. Because of the difficulties in making an exact analysis of SQR and SEDR policies, approximate analysis has been attempted by several researchers [5], [8], [10], [11], [12], [13], [14], [15], [17], [22], [23]. In most of the above papers and also in this paper, we model a DCS as a set of parallel servers, each with its own queue. This is the model that is most commonly used in the literature [4], [6], [9], [22].

Kingman [11] and Flatto [14] studied the SQR problem via transform methods and found expressions for the mean number in the system and the occupancy probabilities. The mean number of jobs is expressed as an infinite sum and the sum simiplifies only under a heavy traffic assumption. Halfin [13] discusses the two queue version of this problem with homogeneous exponential servers and employing linear programming techniques obtains bounds for the probability distribution of the number of customers in the system and its expected value in equilibirium. Blanc [5] considers a numerical method based on power series expansions and recursions to calculate state probabilities and moments of queue length distributions. Conolly [9] discusses a queueing model fed by a Poisson arrival stream of jobs, and served by two identical expenential servers, each with a queue, with the customers choosing the shortest queue at the instant of their arrival. He proposes a method to calculate state probabilties by means of inversion of tridiagonal matrices. He also shows that there is hardly any difference between the distribution of the sum of the lengths of two parallel queues and the queue length distribution for the M/M/2 queue. Gubner [12] shows that expectation of the difference of the queue lengths for any two queues is uniformly bounded for all loads. Based on the evidences given in [9] and [12], Nelson [17] derives an approximate closed form expression for the mean response time of a multiple queueing system consisting of identical exponential servers fed by a Poission stream of jobs. Yuan [22] proposes an approximate numerical solution method to determine equilibirium state probabilities of a queueing system consisting of two heterogeneous exponential servers, wherein, an arriving job to the system is routed to the queue with the shortest expected delay.

In spite of a wealth of results available on approximate analysis, we find that these approximations have the following drawbacks:

1. Different techniques are employed to compute different average performance measures such as mean response time [17], and moments of individual queue length distributions [5]. There is no unifying technique to compute all the desirable average performance measures.

2. Most of the approximations [10],[15],[23] proposed so far lead to accurate results only under light traffic or heavy traffic assumptions.

3. Several approximations such as in [5],[10],[13],[15] have been developed for the 2-server case but they are not generalizable to more than 2-servers.
4. Several approximations such as in [13],[17],[23] assume that the arrival stream of jobs constitute a Poisson stream. However, if SQR or SEDR is employed only in a part of the system, the arrivals to this subsystem in general need not be Poisson.
5. Most of the approximations [13],[17],[22] assume that service times of the individual servers are exponentially distributed.

In this paper, we present an efficient, accurate, and general analytical methodology for the evaluation of the performance of SQR and SEDR policies in DCSs. The methodology is a hybrid technique based on a combination of Product Form Queueing Networks (PFQNs) and Generalized Stochastic Petri Nets (GSPNs) [1] proposed by Balbo, Bruell, and Ghanta [3]. Such an integrated technique is necessary because: (a) PFQNs cannot capture these dynamic routings. (b) Though GSPNs can elegantly model these policies, the resulting models are intractable due to state space explosion. Since GSPNs are used to capture dynamic policies, modelling using hybrid technique overcomes drawbacks 1, 2, and 3 mentioned above. Drawback 4 can be overcome by explicitly modelling that part of the subsystem which sends jobs to the subsystem employing SQR or SEDR policies. Drawback 5 can be overcome in a limited sense by allowing Erlang, hypo-exponential, and hyper-exponential distributions. GSPNs are used as a model for only that part of the system employing SQR (or SEDR) policies and hence the method is computationally efficient.

To investigate the efficacy of this approach, we have considered, in this paper, closed central server systems with SQR (or SEDR) routings. Central server models with state dependent routing functions have been considered in the literature [2],[16],[21]. In these papers, queueing networks with state-dependent routing functions based on rational functions of queue lengths (SQR and SEDR do not belong to this category) have been shown to be product form. Even though we have considered closed queueing models, our results are applicable to open queueing models of load balancing in which the maximum number of jobs that can be present in the submodels employing these dynamic routings is finite.

The rest of the paper is organized as follows. In Section 2, we give a brief introduction to the Integrated PFQN-GSPN Models, and then present the integrated PFQN-GSPN methodology for modelling DCSs with SQR and SEDR routings. In Section 3, we present detailed numerical results and make a comparison of SQR and SEDR policies with some simple state-dependent routing functions. The major results of the numerical experimentation are as follows:

- SQR performs the best among all load balancing policies, for homogeneous DCSs
- In heterogeneous DCSs, SQR surprisingly performs better than SEDR for very low levels of load imbalances, but is expectedly outperformed by SEDR in other conditions.

The results obtained have been found to be remarkably accurate. The models were analyzed using software packages developed by us for PFQN analysis and GSPN analysis.

2 Integrated Models for SQR and SEDR

The integrated queueing network - Petri net (IQP) modelling originated by Balbo, Bruell, and Ghanta [3] is based on the concept of flow-equivalence. The basic idea is to isolate one or more subsystems of a given system and compute the flow-equivalents for these subsystems. The overall model (also called the high level model) of the given system will then have these subsystems represented by the flow-equivalents. The flow-equivalents of these subsystems (also called the submodels) are computed by evaluating the throughput of these subsystems in isolation for all possible populations. The throughputs thus computed together with associated populations constitute the flow-equivalent. The IQP modelling can be of two types : (i) IQP model with GSPN as the high level model and (ii) IQP model with PFQN as the high level model. If the high level model is a PFQN model, then some of its stations will be the flow-equivalent representations (derived using GSPNs) of the submodels that incorporate non-product form features. If the high level model is a GSPN model, then some of its timed transitions will be the flow-equivalent representations (derived using PFQNs) of the submodels that are product form solvable.

Consider a DCS containing a subsystem in which SQR (or SEDR) is employed. Assume that the rest of the DCS contains only product form components. The basic idea behind modelling the above system using IQP models is to compute the flow-equivalent of the SQR (or SEDR) subsystem using a GSPN model. The DCS can then be evaluated as a PFQN, with the SQR (or SEDR) subsystem represented as a state-dependent node derived using the GSPN model of the subsystem. Since it is always possible to obtain an exact representation of the flow-equivalent server, the only source of error in this technique could be due to the interaction between the subsystem and the rest of the DCS.

To illustrate the IQP modelling technique for SQR and SEDR, we consider a central server model of a DCS as shown in Figure 1. It is to be noted that by considering the central server model, we are only simplifying the presentation, and not undermining the general applicability of our results. Also, we stress that our results are applicable not only to closed queueing models but also to those open queueing models, in which the subnetworks with non-product form components have limited populations, such as in [4]. A typical state of the central server network in Figure 1 is given by $\underline{n} = (n_0, n_1, n_2, ..., n_m)$ where n_i is the number of jobs in the ith node, i = 0,1,2,..., m. In the state $\underline{n}$, the jobs after getting executed from node 0 are routed to the ith node with probability $q_i(\underline{n})$, i =1,2,..., m in such a way that the load across the subsystem is balanced. In the following, we mathematically describe the heuristic policies SQR and SEDR. We also describe a few other simple heuristic policies for load balancing. We describe totally three policies A1, A2, and A3 for homogeneous systems, and four policies,

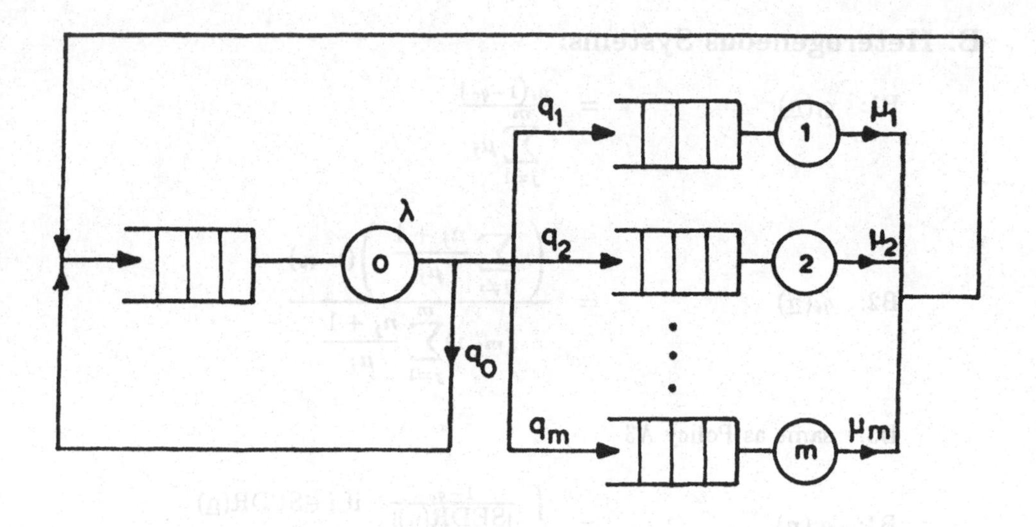

Figure 1: Closed central server model with (m+1) nodes.

B1, B2, B3, and B4, for heterogeneous systems. In the following expressions, μ_i denotes the service rate of the ith node, S denotes the state space of the central server network, $\underline{n}$ is any state in S, and the index i takes the values, 1,2,..., m.

A. Homogeneous Systems:

A1: $q_i(\underline{n}) \quad = \quad \frac{1-q_0}{m}$

A2: $q_i(\underline{n}) \quad = \quad \dfrac{(1-q_0)\sum_{j\neq i}^{m} n_j}{(m-1)\sum_{j=1}^{m} n_j}$

A3: $q_i(\underline{n}) \quad = \quad \begin{cases} \frac{1-q_0}{|\mathrm{SQR}(\underline{n})|} & \text{if } i \in \mathrm{SQR}(\underline{n}) \\ 0 & \text{otherwise} \end{cases}$

where $\mathrm{SQR}(\underline{n}) = \{\ k\colon n_k = \min(n_1, n_2, \ldots, n_m)\}$

B. Heterogeneous Systems:

B1: $$q_i(\underline{n}) = \frac{\mu_i(1-q_0)}{\sum_{j=1}^{m}\mu_j}.$$

B2: $$q_i(\underline{n}) = \frac{\left(\sum_{j\neq i}\frac{n_j+1}{\mu_j}\right)(1-q_0)}{(m-1)\sum_{j=1}^{m}\frac{n_j+1}{\mu_j}}.$$

B3: same as Policy A3

B4: $$q_i(\underline{n}) = \begin{cases} \frac{1-q_0}{|\mathrm{SEDR}(\underline{n})|} & \text{if } i \in \mathrm{SEDR}(\underline{n}) \\ 0 & \text{otherwise} \end{cases}$$

where $\mathrm{SEDR}(\underline{n}) = \{k: \frac{n_k+1}{\mu_k} = \min(\frac{n_1+1}{\mu_1}, \frac{n_2+1}{\mu_2}, \ldots, \frac{n_m+1}{\mu_m})\}$

The intuition behind policy A1 is that routing with equal probabilities to one of the nodes 1,2,...,m will balance the loads across the subsystem , since all the server speeds are identical. This can be analyzed by product form queueing networks. In policy A2, it is to be observed that, in a given state $\underline{n}$, the expressions computed are in such a way that, if $n_k \leq n_l$ then $q_k(\underline{n}) \geq q_l(\underline{n})$. So, the shortest queue will have the maximum probability and the longest queue will have the minimum probability, and the queues having queue lengths in between will have probabilities in such a way that the queue having a queue length close to that of the shortest queue will be assigned greater probability than the one having queue length close to that of the longest queue. The central server model employing this policy is non-product form and can be analyzed by GSPNs. The intuiton behind policy B1 is that allocating a job to a faster server with greater probability than to a slower server, would minimize the overall expected response time. The central server model employing this policy is product form solvable. Policy B2 is a generalization of policy A2 presented for the homogeneous systems, and again the intuition is that allocating an arriving job with greater probability to the queue having less expected delay would minimize the overall average response time of a job. Central server model employing this policy is not product form solvable, but can be analyzed by GSPNs.

Figure 2 shows the GSPN model of the subsystem with dynamic routing for a closed central server model with 5 nodes ($m = 4$). The reader is referred to [1] for an excellent review of GSPNs. Table 1 gives the interpretation of the places and transitions of the GSPN model. The table also specifies the dynamic random switch in the GSPN model, which models the dynamic routing in the network.

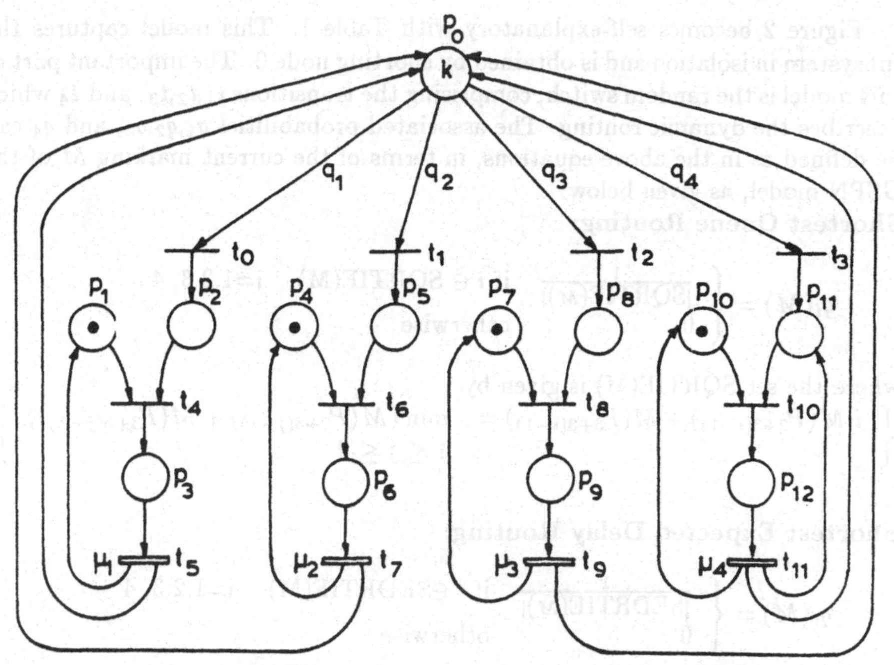

Figure 2: GSPN model for computing flow-equivalents.

Places:

P_0 : jobs just entered into the subsystem.
$P_{1+3(i-1)}$,i=1,2,3,4 : availability of node i.
$P_{2+3(i-1)}$,i=1,2,3,4 : jobs waiting for node i.
$P_{3+3(i-1)}$,i=1,2,3,4 : jobs getting processed in node i.

Immediate Transitions:

$t_{0+(i-1)}$,i=1,2,3,4: job is routed to node i after a burst from node 0.
$t_{4+2(i-1)}$,i=1,2,3,4 jobs start processing in node i

Exponential Transitions:

$t_{5+2(i-1)}$,i=1,2,3,4: processing of jobs in node i.

Dynamic Random Switch:

t_0-q_1; t_1-q_2; t_2-q_3; t_3-q_4.

These probabilities are computed using the equations in Section 2.

Initial Marking:

P_0-k; P_1-1; P_4-1; P_7-1; P_{10}-1;

Table 1: Legend for GSPN Model in Figure 2.

Figure 2 becomes self-explanatory with Table 1. This model captures the subsystem in isolation and is obtained by shorting node 0. The important part of this model is the random switch, comprising the transitions t_1,t_2,t_3, and t_4 which describes the dynamic routing. The associated probabilities q_1,q_2,q_3, and q_4 can be defined as in the above equations, in terms of the current marking M of the GSPN model, as given below:

Shortest Queue Routing:

$$q_i(M) = \begin{cases} \frac{1}{|\mathrm{SQRTIE}(M)|} & \text{if } i \in \mathrm{SQRTIE(M)}, \quad i=1,2,3,4 \\ 0 & \text{otherwise} \end{cases}$$

where the set SQRTIE(M) is given by

$$\left\{ i{:}M(P_{2+3(i-1)}) + M(P_{3+3(i-1)}) = \min_{1 \leq j \leq 4} \left(M(P_{2+3(j-1)}) + M(P_{3+3(j-1)})\right) \right\}$$

Shortest Expected Delay Routing:

$$q_i(M) = \begin{cases} \frac{1}{|\mathrm{SEDRTIE}(M)|} & \text{if } i \in \mathrm{SEDRTIE(M)}, \quad i=1,2,3,4 \\ 0 & \text{otherwise} \end{cases}$$

where the set SEDRTIE(M) is given by

$$\left\{ i{:}\frac{M(P_{2+3(i-1)})+M(P_{3+3(i-1)})+1}{\mu_i} = \min_{1 \leq j \leq 4} \left(\frac{M(P_{2+3(j-1)})+M(P_{3+3(j-1)})+1}{\mu_j}\right) \right\}$$

To compute the flow-equivalent of the subsystem using the GSPN model, one has to evaluate the GSPN model for each possible population. If the population of the closed central server network is N, then the possible populations in the subsystem are 0,1,2,..., N. For each population k, the initial marking of the GSPN model can be set by making the number of tokens in place P_0 as k. The sum of the throughput rates of the transitions t_9,t_{10},t_{11}, and t_{12} will then give the throughput rate of the subsystem. The throughput rates thus computed, together with the associated populations constitute the flow-equivalent of the subsystem.

Figure 3 gives the high level PFQN model of the central server network, with the flow-equivalent of the subsystem constituting just a single node of the PFQN. The PFQN model of Figure 3 can now be solved using standard computational algorithms for PFQNs [7],[18]. It would now be very efficient to study, for example, the effect of variation of the scheduling discipline, service distribution, and the service rate of node 0 on the performance of the overall DCS, since the experimentation is required on a 2-node PFQN. Effectively, we have reduced the evaluation of an overall GSPN model into an evaluation of the GSPN submodel whose state space a cardinality much smaller than that of the overall GSPN model, followed by a 2-node PFQN evaluation.

To see the computational advantage obtained, we show in Table 2, a comparison of the state space size between the GSPN model of Figure 2 (which models

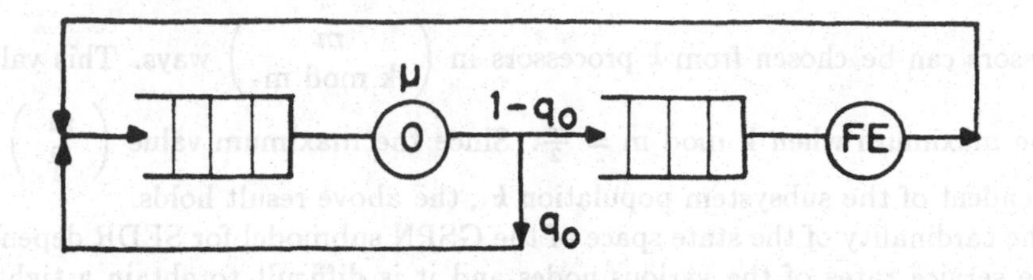

Figure 3: PFQN high level model.

only the subsystem) and the exact GSPN model for the entire system(i.e., the model that captures all the 5 nodes in a GSPN). The significant reduction in the size of the state space is to be noted.

Population	Size of state space of exact GSPN for SQR	Size of state space of subsystem GSPN for SQR	Size of state space of exact GSPN for SEDR	Size of state space of subsystem GSPN for SEDR
4	16	1	12	1
5	48	4	30	2
6	72	6	36	1
7	80	4	48	1
8	81	1	72	1
9	189	4	96	1
10	243	6	120	1
11	255	4	160	1

Table 2: Comparison of state space sizes for detailed and subsystem models.

Generalizing from the example of 5 nodes(m=4), we get the following result about the cardinality of the state spaces of the GSPN submodel.

Result: For any population, the cardinality of the state space of the GSPN submodel for SQR is less than or equal to $\binom{m}{\frac{m}{2}}$.

Proof: Let the subsystem contain k jobs. Let S be the state space of the GSPN submodel. Take any state $(k_1, k_2, ..., k_m) \in S$. Since the routing is based on the SQR policy, $|k_i - k_j| \leq 1 \; \forall \; 1 \leq i, j \leq m$. If not, jobs can be exchanged between them making the load more balanced. So, each processor must get $\lceil \frac{k}{m} \rceil + x$ jobs where x can be either zero or 1. Since the total number of jobs is equal to k, x will be equal to 1 for exactly $k - \lceil \frac{k}{m} \rceil = k \bmod m$ processors. These k mod m

processors can be chosen from k processors in $\binom{m}{k \bmod m}$ ways. This value will be maximum when k mod m = $\frac{m}{2}$. Since the maximum value $\binom{m}{\frac{m}{2}}$ is independent of the subsystem population k , the above result holds.

The cardinality of the state space of the GSPN submodel for SEDR depends on the service rates of the various nodes and it is difficult to obtain a tighter bound. However, in all the numerical experiments we carried out, we get much less number of states than for SQR.

3 Numerical Results

For the purpose of numerical experiments, we consider the central server model of Figure 1 with $m = 4$. We first present the results for the homogeneous case ($\mu_1 = \mu_2 = \mu_3 = \mu_4$) and then for the heterogeneous case. Note that SQR is same as SEDR for the homogeneous case. We shall study the variation of response time as a function of a parameter that we shall call as the *imbalance in load*, denoted by ρ. ρ is the ratio of the processing capacity of the subsystem ($\mu_1+\mu_2+\mu_3+\mu_4$) to that of node 0 (λ). ρ was chosen as a parameter of interest since the performance of the central server model is crucially affected by this ratio. We present below the important results [19].

3.1 Results for Homogeneous DCS

Here we compare three routing policies A1, A2, and A3. Policy A1 assigns equal probailities to all the queues. Policy A2 assigns greater probabilities to shorter queues, whereas policy A3 is the SQR policy. Figure 4 shows the percentage difference in response time between policies A1 and A3, while Figure 5 shows the difference in response time between policies A2 and A3. Two populations N=4 and N=6 are considered.

As the figures show, the difference in performance is appreciable only for small values of ρ.This is due to the following reason: When ρ is low, the processing capacity of the subsystem is much less than that of node 0. In that case, queues are formed in each of the service centers, and the policies that try to utilize this information perform appreciably better than the policies that either do not use this information or use in a limited way. However, when ρ is high, the processing capacity of the subsystem is much higher than that of node 0. In this case, no queues are formed in the service centers of the subsystem, and the nodes are idle, virtually all the time. So, the policies that use queue length information do not produce significant difference in performance from the ones that do not use this information.

As the figures show, for a given ρ, the percentage difference in response time decreases as J increases. This is due to the following reason: The shortest queue routing tries to balance the load across the subsystem and tries to obtain maximum throughput for a given ρ. Due to this, node 0 is utilized to the

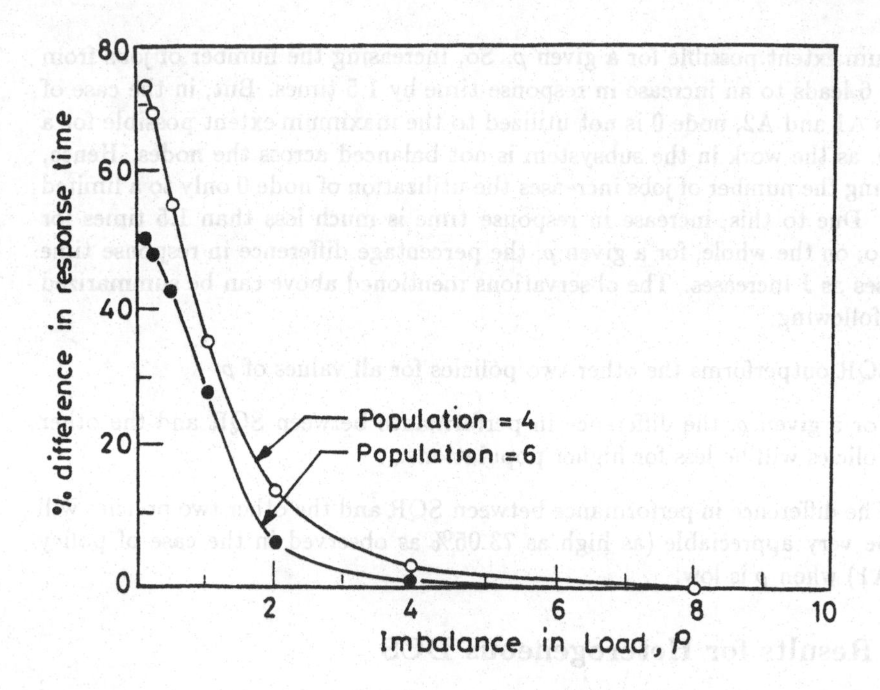

Figure 4: Percentage difference in average response times for Policies A1 and A3.

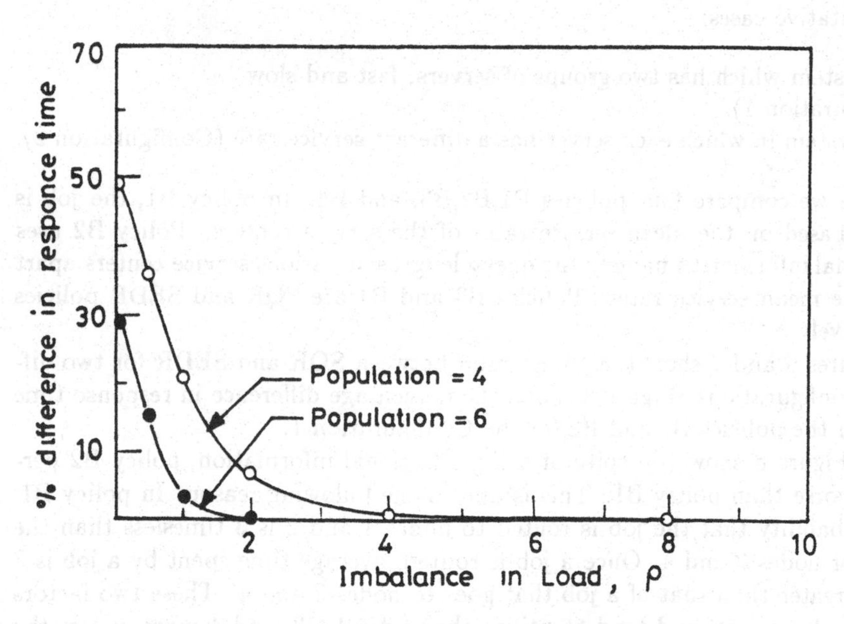

Figure 5: Percentage difference in average response times for Policies A2 and A3.

maximum extent possible for a given ρ. So, increasing the number of jobs from J=4 to 6 leads to an increase in response time by 1.5 times. But, in the case of policies A1 and A2, node 0 is not utilized to the maximum extent possible for a given ρ, as the work in the subsystem is not balanced across the nodes. Hence, increasing the number of jobs increases the utilization of node 0 only to a limited extent. Due to this, increase in response time is much less than 1.5 times for J=6. So, on the whole, for a given ρ, the percentage difference in response time decreases as J increases. The observations mentioned above can be summarized to the following:

1. SQR outperforms the other two policies for all values of ρ.
2. For a given ρ, the difference in performance between SQR and the other policies will be less for higher populations.
3. The difference in performance between SQR and the other two policies will be very appreciable (as high as 73.05% as observed in the case of policy A1) when ρ is low.

3.2 Results for Heterogeneous DCS

For the purpose of modelling heterogeneous distributed systems, we assumed that service centers in the subsystem have different speeds. Since we cannot exhaust all the possible combinations of service speeds, we consider the following representative cases:

(i) A system which has two groups of servers, fast and slow
(Configuration 1).
(ii) A system in which each server has a different service rate (Configuration 2).

Here we compare four policies B1,B2,B3, and B4. In policy B1, the job is routed based on the mean service rates of the service centers. Policy B2 uses additional information namely the queue lengths at various service centers apart from the mean service rates. Policies B3 and B4 are SQR and SEDR policies respectively.

Figures 6 and 7 show the comparison between SQR and SEDR for two different configurations. Figure 8 shows the percentage difference in response time between the policies B1 and B2 for the Configuration 1.

As Figure 8 shows, in spite of using additional information, policy B2 performs worse than policy B1. This is due to the following reason: In policy B1, the probability that the job is routed to nodes 1 and 2 is 3 timesless than the same for nodes 3 and 4. Once a job is routed, average time spent by a job is 3 times greater than that of a job that goes to nodes 3 and 4. These two factors are complementary and tend to utilize the nodes 1,2,3, and 4 more or less the same in policy B1. However, in policy B2, utilizations of nodes 1 and 2 are more than those of nodes 3 and 4. This is because, in all states, the ratio of routing probabilities to nodes 1 and 2 to those of nodes 3 and 4 is not as high as 3 as in

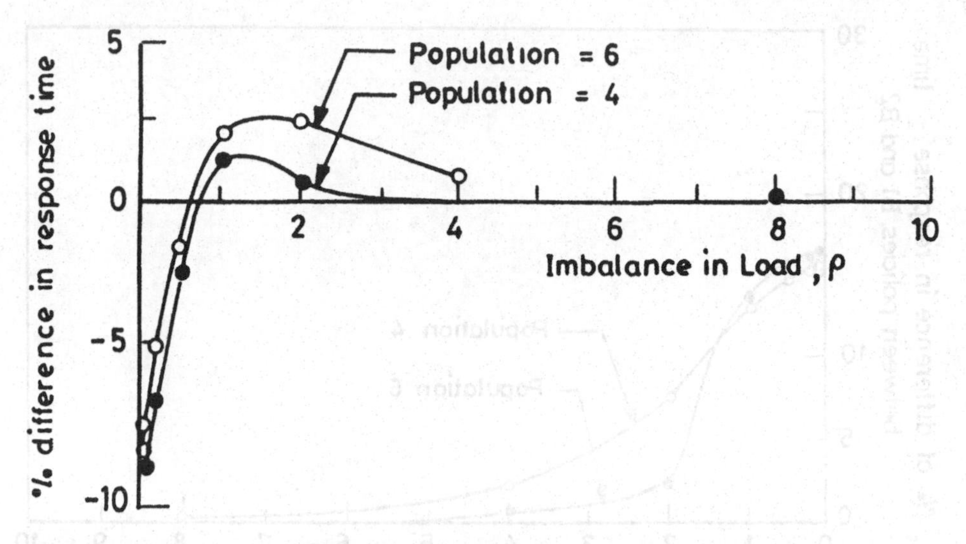

Figure 6: Percentage difference in average response time for policies SQR and SEDR, with $\mu_1 = \mu_2; \mu_3 = \mu_4 = 3\mu_1$.

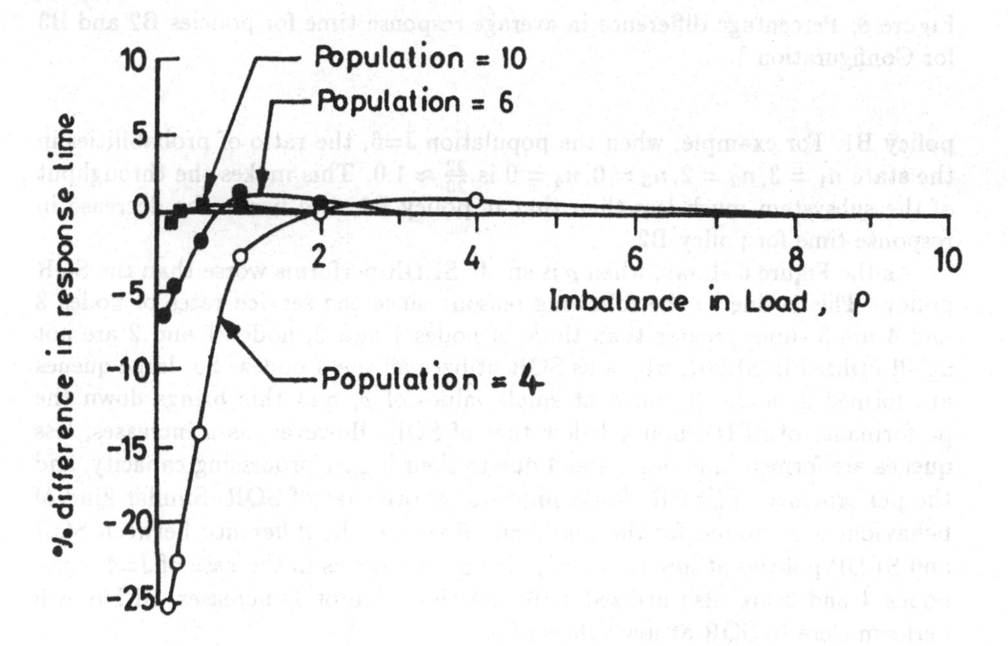

Figure 7: Percentage difference in average response time for policies SQR and SEDR, with $\mu_1 = 2\mu_2; \mu_3 = 3\mu_1; \mu_{4=4\mu_1}$

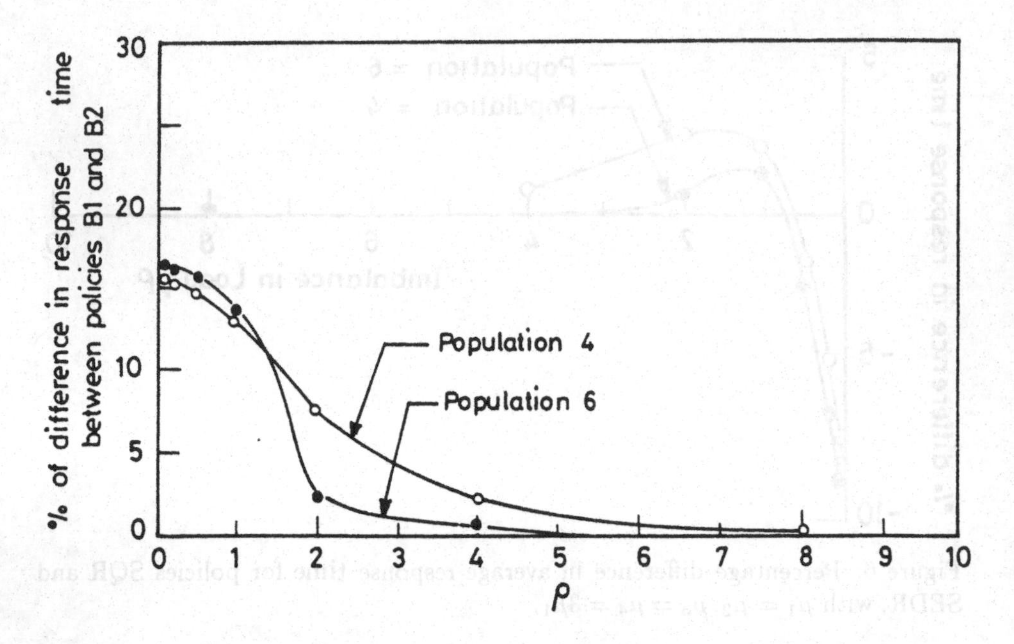

Figure 8: Percentage difference in average response time for policies B2 and B3 for Configuration 1.

policy B1. For example, when the population J=6, the ratio of probabilities in the state $n_1 = 3, n_2 = 2, n_3 = 0, n_4 = 0$ is $\frac{22}{23} \approx 1.0$. This makes the throughput of the subsystem much less than that in policy B1, and hence the decrease in response time for policy B2.

As the Figure 6 shows, when ρ is small, SEDR performs worse than the SQR policy. This is due to the following reason: since the service rates of nodes 3 and 4 are 3 times greater than those of nodes 1 and 2, nodes 1 and 2 are not at all utilized in SEDR, where as SQR utilizes all the 4 nodes. So, large queues are formed in nodes 3 and 4 at small values of ρ, and this brings down the performance of SEDR policy below that of SQR. However, as ρ increases, less queues are formed in nodes 3 and 4 due to their higher processing capacity, and the performance of SEDR slowly improves above that of SQR. Similar kind of behaviour is exhibited for the case J=6. However, the difference between SQR and SEDR policies at low values of ρ, is not as large as in the case of J=4, since nodes 1 and 2 are also utilized J=6. As the value of J increases, SEDR will perform close to SQR at low values of ρ.

For configuration 2, we assumed that the service rate of node 2 is twice that of node 1, the service rate of node 3 is thrice that of node 1, and the service rate of node 4 is 4 times that of node 1. In this case also, SEDR performs worse than SQR at low values of ρ. However, the differences in performance are not as high as configuration 1, since the nodes in the system get more or less equally utilized, whereas in the configuration 1, nodes 1 and 2 are not at all utilized.The observations mentioned above can be summarized to the following:

1. Policy B2 performs worse than policy B1.
2. SQR performs even better than SEDR at very low values of ρ for the configurations having service centers with widely varying service rates.
3. SEDR outperforms all other policies for higher values of ρ.

3.3 Accuracy of IQP Models

It is reasonable to expect that the accuracy of the IQP models will depend only on the interaction between the PFQN portion and the GSPN portion of the model. In particular, in the case of central server models, the parameters q_0 and ρ will influence the accuracy of the IQP models. Extensive numerical investigation has shown that the estimates produced by IQP modelling are remarkably accurate for the central server model. Even for very low values of q_0 and ρ, for which the error was the highest, the errors turned out to be very small and were within 3%. Table 3 shows for different populations, the errors in average response time values obtained using an exact GSPN model for the entire system and using the IQP modelling technique.

population of network	%error in Mean response time SQR Policy ($q_0 = 0.1; \rho = 0.2$)	%error in mean response time SEDR Policy ($\lambda = 5, \mu_1 = 1, \mu_2 = 2, \mu_3 = 3, \mu_4 = 4$)
4	2.700	2.717
5	1.244	0.975
6	0.431	0.316
7	0.143	0.105
8	0.146	0.034
9	0.015	0.011
10	0.0048	0.004
11	0.00155	0.0012

Table 3: Percentage error in mean response time values.

4 Summary

In this paper, we have applied an efficient analytical methodology for the evaluation of two heuristic dynamic load balancing policies namely shortest queue routing and shortest expected delay routing in distributed computing systems. To investigate the efficacy of this approach, we developed integrated analytical models for the closed central server systems employing these heuristic policies. We compared the performance of these two policies with simple policies which can be modelled exactly using product form queueing networks at different levels of ρ, the imbalance in load and developed results on their relative performance.

For the central server models considered in the paper, extensive numerical results have shown that the hybrid technique employed yields remarkably accurate results. More general models have to be investigated, however. The technique can also be applied to open queueing network models in which the non-product form subnetwork has bounded population.

References

[1] M.Ajmone Marsan, G.Balbo and G.Conte, A Class of Generalized Stochastic Petri Nets for the Performance Evaluation of Multiprocessor Systems, **ACM Transactions of Computer Systems**, Vol.2, No.2, May 1984, pp.93-122.

[2] I.F.Akyildiz, Central Server Models with Multiple Job Classes, State Dependent Routing, and Rejection Blocking, **IEEE Transactions on Software Engineering**, Vol.15, No.10, October 1989, pp.1305-1312.

[3] G.Balbo, S.C.Bruell and S.Ghanta, Combining Queueing Networks and Generalized Stochastic Petri Nets for the Solution of Complex Models of Computer Systems, **IEEE Transactions On Computers**, Vol.17, No.10, October 1988, pp.1251-1268.

[4] S.A.Banawan and J.Zahorjan, Load Sharing in Heterogeneous Queueing Systems, **Proceedings of the IEEE INFOCOM'89**, 1989, pp.731-739.

[5] J.P.C.Blanc, A Note on Waiting Times in Systems with Queues in Parallel, **Journal of Applied Probability**, Vol.24, 1987, pp.540-546.

[6] F.Bonomi and A.Kumar, Adaptive Optimal Load Balancing in a Heterogeneous Multiserver System with a Central Job Scheduler, **IEEE Transactions on Computers**, Vol.39, No.10, October 1990, pp.1232-1250.

[7] J.P.Buzen, Computational Algorithms for Closed Queueing Networks with Exponential Servers, **Communications of the ACM**, Vol.16, No.9, September 1973, pp.527-531.

[8] L.Flatto and H.P.Mckean, Two Queues in Parallel, **Communications on Pure and Applied Mathematics**, Vol.30, 1977, pp.255-263.

[9] B.W.Conolly, An Autostrada Queueing Problem, **Journal of Applied Probability**, Vol.21, 1984, pp.394-403.

[10] D.L.Eager, D.Edward, E.D.Lazowska, and John Zahorjan, Adaptive Load Sharing in Homogeneous Distributed Systems, **IEEE Transactions on Software Engineering**, Vol.12, No.5, May 1986, pp.662-675.

[11] G.Foschini and J.Salz, A Basic Dynamic Routing Problem and Diffusion, **IEEE Transactions on Communications**, Vol.26, No.3, March 1978.

[12] J.A.Gubner, B.Gopinath, and S.R.S.Varadhan,Bounding Functions of a Markov Process and the Shortest Queue Problem, **Technical Report**, Department of Computer Science, University of Maryland, U.S.A., 1988.

[13] S.Halfin, The Shortest Queue Problem, **Journal of Applied Probability**, Vol.22, 1985, pp.865-878.

[14] J.F.C.Kingman, Two Similar Queues in Parallel, **Biometrika**, Vol.48, 1961, pp.306-310.

[15] G.Knessl, B.J.Mattowsky, Z.Schuss and C.Tier, Two Parallel M/G/1 Queues where Arrivals Join the System with the Smaller Buffer Content, **IEEE Transactions on Communications**, Vol.35, No.11, November 1987, pp.1153-1158.

[16] A.E.Krzesinski, Multiclass Queueing Networks with State Dependent Routing, **Performance Evaluation**, Vol.7, No.2, June 1987, pp.125-145.

[17] R.D.Nelson and T.K.Philips, An Approximation to the Response Time for Shortest Queue Routing, **Performance Evaluation Review**, Vol.17, No.1, May 1989, pp.181-189.

[18] M.Reiser and S.S.Lavenberg, Mean Value Analysis of Closed Mutichain Queueing Networks, **Journal of ACM**, Vol.27, No.2, April 1980, pp.313-322.

[19] .R.Meenakshi Sundaram, **Integrated Analytical Models for Parallel and Distributed Systems**, M.S.Thesis, Department of Computer Sciene and Automation, Indian Institute of Science, Bangalore, October 1990.

[20] A.N.Tantawi and D.Towsley, Optimal Static Load Balancing in Distributed Computer Systems, **Journal of the ACM** , Vol.32, No.2, April 1985, pp.445-465.

[21] D.Towsley, Queueing Models with State Dependent Routing, **Journal of the ACM**, Vol.27, No.2, April 1980, pp.323-337.

[22] Yuan Chow and Walter H.Kohler, Models for Dynamic Load Balancing in a Heterogeneous Multiprocessor System, **IEEE Transactions on Computers**, Vol.28, No.5, May 1979, pp.354-361

[23] T.Yung, Wang and R.J.T.Morris, Load Sharing in Distributed Systems, **IEEE Transactions on Computers**, Vol.34, No.3, March 1985, pp.204-217.

Liveness and Home States in Equal Conflict Systems *

Enrique Teruel Manuel Silva

Departamento de Ingeniería Eléctrica e Informática
Centro Politécnico Superior de Ingenieros de la Universidad de Zaragoza
María de Luna 3, E-50015 Zaragoza (Spain)
Tel: + 34 76 517274 Fax: + 34 76 512932
e-mail: eteruel@cc.unizar.es

Abstract

Structure theory is a branch of net theory devoted to investigate the relationship between the structure and the behaviour of net models. Many of its powerful results have been derived for some *subclasses* of *ordinary* (i.e. all weights one) Place/Transition net systems. Nevertheless, weights may be very convenient to properly model systems with *bulk* services and arrivals. They can be implemented by means of ordinary subnets, but then the original structure of the net becomes too complex, even in the simplest cases, to be amenable of some interesting structural analysis techniques.

We present here some results regarding the structure theory of a subclass of (weighted) Place/Transition systems that naturally generalizes the ordinary subclass of Free Choice systems: *Equal Conflict systems.* We concentrate on a couple of important results concerning the relationship between the structure (in particular, represented by the state equation) and the behaviour. Actually, we show that liveness of bounded Equal Conflict systems can be decided verifying that there is no solution to a linear system of inequalities in the integer domain, and that live and bounded Equal Conflict systems have home states.

Keywords: Place/Transition Net Systems, Structure Theory, Equal Conflict Systems.

Topics: Analysis and synthesis, structure and behaviour of nets; System design and verification using nets.

*This work has been partially supported by the projects P IT-6/91 of the Aragonese CONAI (DGA), CICYT TIC-91-0354 of the Spanish Plan Nacional de Investigación, and Esprit W.G. 6067 (CALIBAN).

1 Introduction

One of the indigeneous techniques for the analysis of Petri net models is the so called ***structural analysis.*** This analysis method focuses on the relationship between the net structure and its behaviour. Net structure can be studied using ***graph theory***, or ***linear algebra***-based arguments using the ***incidence matrices.*** In general, structural techniques present a good trade-off between computational efficiency and decision power.

From the early seventies, much effort has concentrated on studying limited ***subclasses*** of Petri nets to obtain a better understanding of more general systems. One of the most studied of these subclasses, presenting an outstanding trade-off between generality and "nice" theory are ***Free Choice*** systems, introduced and studied in [Hack 72], and further investigated by many researchers (see, for instance, [TV 84, BV 84, Best 87, BCDE 90, ES 91]). The study of subclasses usually considers ***ordinary*** Place/Transition systems, i.e all arcs weighted one. Nevertheless, ***weights*** are convenient to properly model systems with ***bulk*** services and arrivals. They can be implemented by means of ordinary subnets, but then the original structure of the net becomes too complex, even in the simplest cases, to be amenable of some interesting structural analysis techniques. An attempt to generalize the structure theory of a well-studied (ordinary) subclass, namely Marked Graphs, to the weighted case is [TCCS 92].

We mainly consider structural analysis techniques based on linear algebra, using the ***state equation*** of the net system. Let $(\mathcal{N} = (P, T, W), M_0)$ be a net system, and σ a sequence of transitions firable from M_0:

$$M_0 \xrightarrow{\sigma} M \Rightarrow (M, \vec{\sigma}) \in \mathbb{N}^{|P|+|T|} \text{ is a solution to } M = M_0 + C \cdot \vec{\sigma}$$

where $\vec{\sigma}$ is the Parikh (or firing count) vector of sequence σ, and C is the incidence matrix of the net $\mathcal{N}$. In order to study a property, it could be expressed as a (set of) linear condition(s) and then it would be checked whether the solutions to the state equation fulfill it. To do so, Integer Programming techniques can be used [NW 88]. Unfortunately, in general, the state equation has integer solutions, $(M, \vec{\sigma})$, not reachable on the net system. They are called ***spurious*** solutions. In general, due to the existence of spurious solutions, only ***necessary*** or ***sufficient*** conditions are obtained when this relaxation is used to study behavioural properties like boundedness or deadlock-freeness. Spurious solutions can be removed using different approaches [CS 91], but, in general, it isn't still known whether all of them are removed.

We present here some results regarding the structure theory of a subclass of (weighted) Place/Transition systems that naturally generalizes the ordinary subclass of Free Choice systems: ***Equal Conflict systems.*** We concentrate on a couple of important results concerning the relationship between the structure (namely the state equation) and the behaviour. Actually, we show that liveness of bounded Equal Conflict systems can be decided verifying that there is no solution to a linear system of inequalities in the integer domain, and that live and bounded Equal Conflict systems have home states, as live and bounded Free Choice systems do [Vogler 89]. Both results are theoretically and practically interesting. On the practical side, they provide an algebraic technique to check liveness, and a strong property of the behaviour which is helpful for the analysis of these systems, respectively. We find them theoretically interesting because they provide a deeper understanding of the basic causes of some of the outstanding results about Free Choice (in fact, only the equality of conflicts is used in the proofs).

The rest of the paper is organized as follows. In section 2, we recall the basic concepts of Petri nets. Equal Conflict systems are defined in section 3. A couple of technical results about the structure of the potential reachability graph are developed in section 4, and they are exploited in sections 5 and 6.

2 Basic Concepts and Notation

In this section we recall the basic concepts and notation that are going to be used. For the sake of readability, whenever a net or system is defined it "inherits" the definition of all the characteristic sets, functions, parameters,... with names conveniently marked to identify whose is which. More comprehensive presentations of Petri net concepts are, for instance, [Silva 85, Murata 89].

Place/Transition net and related concepts. A *Place/Transition net* (P/T-$\mathcal{N}$) is a triple $\mathcal{N} = (P, T, W)$ where P and T are disjoint sets of ***places*** and ***transitions***, with $|P| = n, |T| = m$, and $W : (P\times T) \cup (T\times P) \mapsto \mathbb{N}$ defines the ***weighted flow relation***. (Sometimes we write that $(u, v) \in W$ meaning that $W(u, v) > 0$.) Since a P/T-$\mathcal{N}$ can be seen, and drawn, as a bipartite weighted directed graph, several graph concepts, like paths, circuits, strong connectedness, etc., can be extended to nets. In particular, let $v \in P \cup T$; its ***preset*** and ***postset*** are given, respectively, by:

$$^\bullet v \stackrel{\text{def}}{=} \{u \in P \cup T \mid W(u, v) > 0\} \quad v^\bullet \stackrel{\text{def}}{=} \{u \in P \cup T \mid W(v, u) > 0\}$$

The preset (postset) of a set of nodes is the union of presets (postsets) of its elements. A place (transition) v such that $|v^\bullet| > 1$ [$|^\bullet v| > 1$] is a ***choice*** ***[attribution] (fork [join])***.

The weighted flow relation can alternatively be defined ($\forall p \in P, \forall t \in T$) by: $Pre(p, t) \stackrel{\text{def}}{=} W(p, t)$, $Post(p, t) \stackrel{\text{def}}{=} W(t, p)$. These functions can be represented by matrices[1]. If $\mathcal{N}$ is *pure* (i.e. $\forall p \in P, \forall t \in T: W(p,t) \cdot W(t,p) = 0$; we assume without loss of generality that all nets are pure), then the weighted flow relation is represented by the *incidence matrix* $C \stackrel{\text{def}}{=} Post - Pre$. *Ordinary* nets are those where $W: (P \times T) \cup (T \times P) \mapsto \{0,1\}$.

Place/Transition system and related concepts. A function $M: P \mapsto \mathbb{N}$ is called *marking*, and can be represented by a vector. A *Place/Transition system* (P/T-Σ) is a pair $(\mathcal{N}, M_0)$ where $\mathcal{N}$ is a P/T-$\mathcal{N}$ and M_0 is the *initial* marking. A transition t is *enabled* at M iff $M \geq Pre[P,t]$. Being enabled, t may *occur* (or *fire*) yielding a new marking $M' = M + C[P,t]$, and this is denoted by $M \xrightarrow{t} M'$ and $M' = M \xrightarrow{t}$.

An *occurrence (or firing) sequence* from M is a sequence $\sigma = t_1 \cdots t_k \in T^*$ such that $M \xrightarrow{t_1} M_1 \ldots M_{k-1} \xrightarrow{t_k} M'$, denoted by $M \xrightarrow{\sigma} M'$ and $M' = M \xrightarrow{\sigma}$. The set of all the occurrence sequences from M_0 (the *language*) is denoted by $L(\mathcal{N}, M_0)$.

The *reachability set* is $R(\mathcal{N}, M_0) \stackrel{\text{def}}{=} \{M_0 \xrightarrow{\sigma} \mid \sigma \in L(\mathcal{N}, M_0)\}$. The *reachability graph* is a labelled directed graph $RG(\mathcal{N}, M_0) = (V, E, l)$ with $l: E \mapsto T$ given by:

- $V = R(\mathcal{N}, M_0)$
- $((M, M') \in E \wedge l(M, M') = t) \Leftrightarrow M \xrightarrow{t} M'$

The set $succ(U)$ of *successors* of a subset of nodes $U \subseteq R(\mathcal{N}, M_0)$, is the minimal set such that $U^\bullet \subseteq succ(U)$ and $succ(U)^\bullet \subseteq succ(U)$. The *subgraph* of (V, E, l) induced by U is defined by $G(U) = (U, (U \times U) \cap E, l|_{(U \times U) \cap E})$ [2] Finally, the *set of labels* of $G(U)$ is $l(U) = \{t \in T \mid \exists e \in (U \times U) \cap E : l(e) = t\}$. We say that U is a *final* iff $(U = \{v\} \wedge v^\bullet = \emptyset)$ or $(G(U)$ is strongly connected $\wedge\, U = succ(U))$. Finals of the first kind wil be called *deadlocks* and those of the second *active finals* or, when $l(U) = T$, *live finals*. Observe that there are systems without any final, and that (active) finals could be infinite.

$(\mathcal{N}, M_0)$ is *bounded* iff $\exists max_{p \in P, M \in R(\mathcal{N}, M_0)}\{M[p]\} \in \mathbb{N}$. This property is required if the system has to be implemented. Its structural version, *structural boundedness*, tells us about its fulfilment for *any* initial marking. If $(\mathcal{N}, M_0)$ is bounded, then $R(\mathcal{N}, M_0)$ is finite, and $RG(\mathcal{N}, M_0)$ *has* (finite) finals.

---

[1] Places and transitions are assumed to be arbitrarily, but fixedly, ordered. Therefore the rows and columns can be indexed by the places and transitions respectively. $A[\Pi, \Theta]$ denotes the submatrix of A corresponding to rows of places in $\Pi \subseteq P$ and columns of transitions in $\Theta \subseteq T$.

[2] $\varphi|_\Sigma$ denotes the restriction of φ to subset Σ.

$(\mathcal{N}, M_0)$ is *live* iff every transition can ultimately occur from any reachable marking. $(\mathcal{N}, M_0)$ is live iff $\forall v \in R(\mathcal{N}, M_0) : l(succ(\{v\})) = T$. A necessary condition for liveness is that every final of $RG(\mathcal{N}, M_0)$ is live. If $(\mathcal{N}, M_0)$ is bounded then it is also sufficient. $(\mathcal{N}, M_0)$ is *deadlock-free* iff at least one transition is enabled at every reachable marking. $(\mathcal{N}, M_0)$ is deadlock-free iff $RG(\mathcal{N}, M_0)$ has no finals being deadlocks. These two properties (the second being a necessary condition for the first) are usually part of the specification of systems. Their *structural* versions tell us about the *existence* of a suitable initial marking such that they hold. We say that $(\mathcal{N}, M_0)$ is *well-behaved* when it is live and bounded. Strong connectedness of the net is a necessary condition for well-behavedness[Murata 89]. As our main interest are well-behaved systems, we will usually deal with strongly connected nets.

M is a *home state* in $(\mathcal{N}, M_0)$ iff it is reachable from every reachable marking. $(\mathcal{N}, M_0)$ is *reversible* iff M_0 is a home state. M is a home state iff there is a unique final in $RG(\mathcal{N}, M_0)$, and this is active and contains M.

The state equation. (See also [CS 91].) Let $(\mathcal{N}, M_0)$ be a P/T-Σ with incidence matrix C. The *state equation* is an algebraic equation that gives a *necessary* condition for a marking to be reachable: a vector $M \in \mathbb{N}^n$ such that $\exists \vec{\sigma} \in \mathbb{N}^m : M = M_0 + C \cdot \vec{\sigma}$ is said to be *potentially reachable*, denoted by $M \in PR(\mathcal{N}, M_0) \supseteq R(\mathcal{N}, M_0)$. The *potential reachability graph* is defined like the reachability graph, but the set of nodes is now $PR(\mathcal{N}, M_0)$ instead of $R(\mathcal{N}, M_0)$. The definitions of successors, subgraphs, and finals are extended to deal with potential reachability graphs. Observe that $RG(\mathcal{N}, M_0)$ is a subgraph of $PRG(\mathcal{N}, M_0)$, defined by the subset of nodes $R(\mathcal{N}, M_0) \subseteq PR(\mathcal{N}, M_0)$.

3 Definition of Equal Conflict Net Systems

We call *syntactical* to subclasses of nets (or systems) that can be defined by imposing certain restrictions on the net, and perhaps the initial marking. Well known examples of syntactical subclasses are Marked Graphs, or Free Choice systems, for instance. Equal Conflict systems are another syntactical subclass of P/T-Σ. As the name suggests, their definition has much to do with conflicts, so let us start by formally defining several *conflict relations* on the set of transitions of a net.

Definition 3.1 (Structural conflicts.)
Let $\mathcal{N}$ be a P/T-$\mathcal{N}$.

1. *Two transitions, t and t', are in* Structural Conflict Relation *(denoted by $(t,t') \in$ SCR) iff $t = t'$ or ${}^\bullet t \cap {}^\bullet t' \neq \emptyset$. For instance, every transition that is output of a choice place is in* SCR *with each other. This relation is not transitive.*
2. *Two transitions, t and t', are in* Coupled Conflict Relation *(denoted by $(t,t') \in$ CCR) iff there exists $t_0, \ldots, t_k \in T$ such that $t = t_0$, $t' = t_k$ and for $1 \leq i \leq k : (t_{i-1}, t_i) \in$ SCR. This is an* equivalence relation *on the set of transitions, and each equivalence class is a* Coupled Conflict Set *(CCS). A transition that is not in structural conflict with any other is itself a* CCS. *(Such* CCS *will be called* trivial.*)*
3. *Two transitions, t and t', are in* Equal Conflict Relation *(denoted by $(t,t') \in$ ECR) iff $t = t'$ or $Pre[P,t] = Pre[P,t'] \neq 0$. This is also an* equivalence relation *on the set of transitions, and each equivalence class is an* Equal Conflict Set *(ECS). Note that whenever two transitions are in* ECR *they are also in* CCR, *but not the other way round.*

The interpretation of these definitions follows. A conflict is a situation in which several actions are enabled but not all of them can occur simultaneously. Structural Conflicts are the "topological construct" making possible the existence of conflicts. Coupled Conflict Sets capture the concept of the set of transitions that might be in conflict, so in order to decide firing one we need not only know that it is enabled but also what happens with the others. An Equal Conflict Set is a particular kind of Coupled Conflict Set such that whenever any transition belonging to it is enabled all of them are.

Equal Conflict systems are defined as those whose net is such that every structural conflict is equal.

Definition 3.2 (Equal Conflict systems.)
A P/T-Σ, $(\mathcal{N}, M_0)$, *is an* Equal Conflict system *(EC-Σ) iff $\forall t, t' \in T$*

$$(t,t') \in \text{ECR} \Leftrightarrow (t,t') \in \text{CCR}$$

Let us compare this definition with that for (Extended) Free Choice systems:

Definition 3.3 (Extended Free Choice systems.)
An ordinary P/T-Σ, $(\mathcal{N}, M_0)$, *is an* Extended Free Choice system *iff* $\forall t, t' \in T$

$${}^\bullet t \cap {}^\bullet t' \neq \emptyset \Rightarrow {}^\bullet t = {}^\bullet t'$$

Ordinary Equal Conflict systems trivially coincide with Extended Free Choice systems. Nevertheless, if we drop the word "ordinary" from definition 3.3 we get a larger class of weighted P/T-Σ than EC-Σ. (We would call this other class *topological* Extended Free Choice systems or *topological* EC-Σ meaning that they have the same graph, appart from the weights.)

The reason why we say that EC-Σ naturally generalize Free Choice is because they capture a major feature of these, namely that whenever a transition in conflict with others is enabled, *all* of them are, so any one of them is *freely chosen* to occur, what implies that *non-determinism and synchronization are neatly separated.* That all conflicts are equal is indeed a central fact in the properties we will deal with in the next sections.

4 Reachability and the State Equation

This section contains a proof of the *directedness* property of the reachability graph of a live EC-Σ, which generalizes a similar result already known for Free Choice systems [BV 84, Vogler 89]. An important feature of our proof is that it is done at the level of *potential* reachability, that is, we proof not only that the reachability but also the potential reachability graph is directed. The directedness result, as in [BV 84, Vogler 89], is the main fact leading to existence of home states, as will be shown in section 6. One of the advantadges of proving it at the potential reachability level is that it also implies the impossibility of a spurious solution being non-live, which is central in section 5.

We will derive the pursued results from the following rather technical theorem, which says that, in a live EC-Σ, every potentially reachable marking has successors that are reachable.

Theorem 4.1 *Let* $(\mathcal{N}, M_0)$ *be a live* EC-Σ.
If $M \in PR(\mathcal{N}, M_0)$ *then* $R(\mathcal{N}, M_0) \cap R(\mathcal{N}, M) \neq \emptyset$.

Proof Take any $M \in PR(\mathcal{N}, M_0)$, that is, there exists $\vec{\sigma} \in \mathbb{N}^m$ such that $M = M_0 + C \cdot \vec{\sigma}$. It must be shown that there exists a common successor of M_0 and M. Let $\vec{\sigma} = \vec{\sigma}' + \vec{\sigma}''$ be such that there exists a sequence σ', eventually empty, with firing count vector $\vec{\sigma}'$ such that $M_0 \xrightarrow{\sigma'} M_0'$, and none of transitions in the support[3] of $\vec{\sigma}''$ is firable at M_0'. (It is clear that such decomposition of $\vec{\sigma}$ does exist. It can be constructed starting form σ' empty and $\vec{\sigma}'' = \vec{\sigma}$, and then sucessively firing transitions in the successive $\|\vec{\sigma}''\|$ until no one is firable.) Therefore we can write $M = M_0' + C \cdot \vec{\sigma}''$.
Since M_0' is reachable from M_0, $(\mathcal{N}, M_0')$ is of course live. If some transition $t \notin \|\vec{\sigma}''\|$ is enabled at M_0' (and since M_0' is live, there is at least one), it is also enabled at

[3]The set of the non-zero entries of a vector, $\mathcal{V}$, denoted by $\|\mathcal{V}\|$.

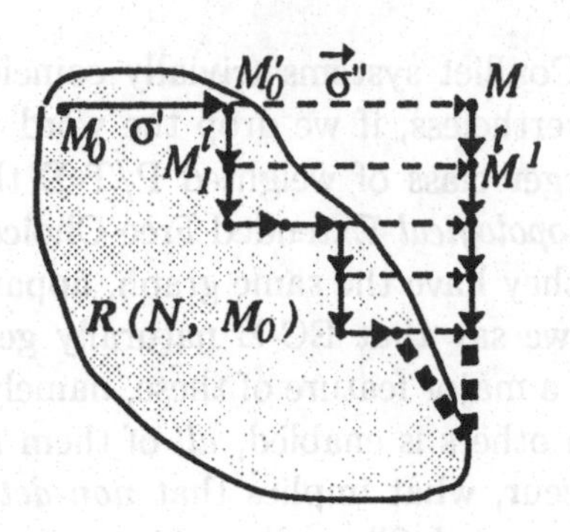

Figure 1: Illustrating the proof of theorem 4.1

M because, due to the EC assumption, in $\|\vec{\sigma}''\|$ there is no transition which takes tokens from input places of t (otherwise such transition in $\|\vec{\sigma}''\|$ being in ECR with t would be enabled too). Let $M_0' \xrightarrow{t} M_1$ and $M \xrightarrow{t} M^1$. We can rename M_1 as M_0 and M^1 as M and apply again the same reasoning. That is, we can go on firing transitions from the successive descendants of M_0, and

- if a given transition does not belong to the corresponding $\|\vec{\sigma}''\|$, then it is also firable from the corresponding successor of M (we fire it advancing in parallel)
- if it does belong to the corresponding $\|\vec{\sigma}''\|$, then the new successor of M_0 is "closer" to the corresponding successor of M with respect to $\sum_{t \in T} \vec{\sigma}''[t]$.

Since M_0 is live, every transition in $\|\vec{\sigma}''\|$ can be ultimately fired. This process advances from M_0 and M either "in parallel" by firing a transition not belonging to $\|\vec{\sigma}''\|$ or "reducing" $\|\vec{\sigma}''\|$, so finally a common successor is found. (Figure 1 illustrates this proof: The arrows represent firing sequences; The horizontal dashed arrows pictorially represent successive vectors $\vec{\sigma}''$.)

If every potentially reachable marking has reachable successors, obviously no potentially reachable marking can be a deadlock. In fact, from every potentially reachable marking every transition can ultimately be fired, at least after getting into the reachable subgraph.

Corollary 4.2 (Absence of killing spurious solutions.)
If $(\mathcal{N}, M_0)$ *is a live* EC-Σ *and* $M \in PR(\mathcal{N}, M_0)$, *then* $(\mathcal{N}, M)$ *is live.*

Theorem 4.1 can be generalized: *every* two potentially reachable markings have common successors, which is the usual statement of the directedness property mentioned above.

Theorem 4.3 (Directedness of the potential reachability graph.)
Let $(\mathcal{N}, M_0)$ be a live EC-Σ.

$$\forall M_1, M_2 \in PR(\mathcal{N}, M_0) : R(\mathcal{N}, M_1) \cap R(\mathcal{N}, M_2) \neq \emptyset$$

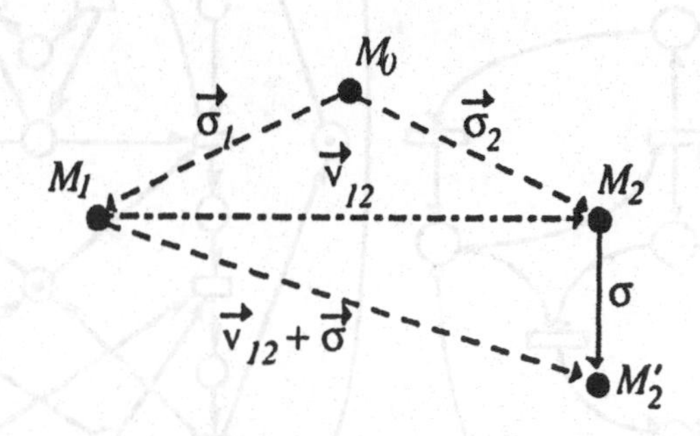

Figure 2: Illustrating the proof of theorem 4.3

Proof Take any $M_1, M_2 \in PR(\mathcal{N}, M_0)$. It must be shown that there exists a common successor of M_1 and M_2. Let $\vec{\sigma}_1, \vec{\sigma}_2 \geq 0$ be vectors such that $M_i = M_0 + C \cdot \vec{\sigma}_i, i = 1, 2$. Let $\vec{v}_{12} = \vec{\sigma}_2 - \vec{\sigma}_1$. Clearly $M_2 = M_1 + C \cdot \vec{v}_{12}$.
The vector $\vec{v}_{12}$ may contain negative coordinates. Since $(\mathcal{N}, M_2)$ is live (by corollary 4.2, because $(\mathcal{N}, M_0)$ is live), there is a successor $M_2' = M_2 \xrightarrow{\sigma}$ satisfying $M_2' = M_1 + C \cdot (\vec{\sigma} + \vec{v}_{12}), \vec{\sigma} + \vec{v}_{12} \geq 0$ (to find such σ it is enough to fire any sequence having at least $-\vec{v}_{12}[t]$ occurrences of t for each negative coordinate t in $\vec{v}_{12}$). It follows that $M_2' \in PR(\mathcal{N}, M_1)$. As $(\mathcal{N}, M_1)$ is live (by corollary 4.2) there exists (by theorem 4.1) a common successor of M_1 and M_2', so a common successor of M_1 and M_2. (Figure 2 illustrates this proof. We have used different dashes for $\vec{v}_{12}$ in order to distinguish it from potentially firable firing count vectors.)

Figure 3 contains several examples showing that the Equal Conflict assumption is needed. The empty marking is isolated in the potential reachability graph of the top-left system. The system below (taken form [CS 91]) shows an ordinary 1-bounded and live system with potentially reachable markings (for instance, two tokens in the leftmost place) without reachable successors, although the reachability graph is directed. The rightmost system (taken from [BV 84]) shows an ordinary 1-bounded and live system where even the reachability graph is not directed (it has two *livelocks*, it doesn't have home states).

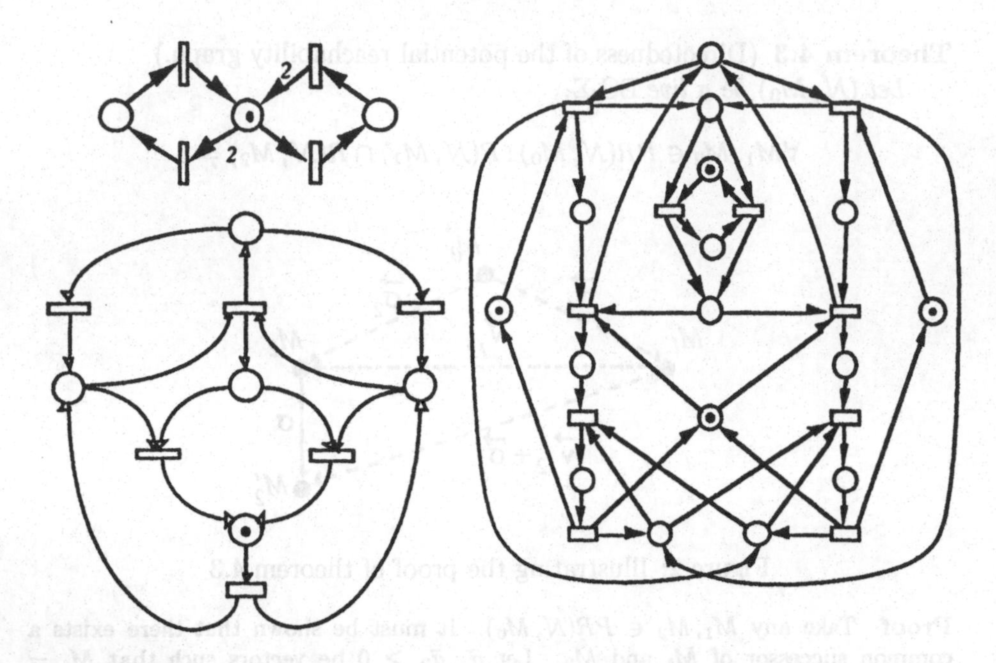

Figure 3: Some live systems whose potential reachability graph does not enjoy the directedness property.

5 Liveness and Deadlock-Freeness

As a first consequence of the results in section 4, liveness of a bounded EC-Σ can be decided proving that a system of linear inequalities hasn't got a solution in the integer domain. To do so, we show first the equivalence of liveness and deadlock-freeness in bounded EC-Σ, which is based on identical arguments to those for Free Choice [Best 87]. Afterwards, we give a sufficient condition for deadlock-freeness, which is an adaptation of the work presented in [Colom 89, TCS 92]. Thanks to corollary 4.2, this technique gives not only a sufficient but a necessary and sufficient condition for deadlock-freeness. Finally, the simplicity of EC-$\mathcal{N}$ ensures that this condition can be written as the check for solution of a single linear system of inequalities in the integer domain.

Theorem 5.1 *Let* $(\mathcal{N}, M_0)$ *be a bounded strongly connected* EC-Σ.

$$(\mathcal{N}, M_0) \text{ is live} \Leftrightarrow (\mathcal{N}, M_0) \text{ is deadlock-free}$$

Proof Only "$\Leftarrow$" must be proven. Assume $(\mathcal{N}, M_0)$ is not live. Let t be dead at some reachable marking. By EC, every t' in CCR with t is also dead. Every input place of a dead CCS will retain all the tokens that arrive to it, so boundedness implies that every input transition to these places will die eventually. By strong connectedness, every transition will finally die, hence $(\mathcal{N}, M_0)$ is not deadlock-free.

A general technique to check properties in an algebraic manner (see, for instance, [Silva 85, Colom 89]) consists in relaxing the set of reachable markings by the set of solutions of the state equation. The property to be checked is expressed as some linear inequalities/equations on the marking and firing count variables and then it is checked whether there is a solution to the state equation satisfying the property or not. In particular, the following is a sufficient condition for deadlock-freeness based on the state equation of the net system, and containing $|T|$ "complex conditions", each one concerning the non-firability of a transition of the net.

Proposition 5.2 *Let* $(\mathcal{N}, M_0)$ *be a* P/T-Σ. *If*

$$\nexists\,(M,\vec{\sigma}) \in \mathbb{N}^{n+m} : (M = M_0 + C\cdot\vec{\sigma}) \wedge \left(\bigwedge_{t\in T}\left(\bigvee_{p\in {}^\bullet t} M[p] < W(p,t)\right)\right) \quad (1)$$

then $(\mathcal{N}, M_0)$ *is deadlock-free.*

Proof A system $(\mathcal{N}, M_0)$ is deadlock-free iff

$$\nexists\, M \in \mathbb{N}^n : (M \in R(\mathcal{N}, M_0)) \wedge (\forall t \in T : \exists p \in {}^\bullet t : M[p] < W(p,t)) \quad (2)$$

Since $R(\mathcal{N}, M_0) \subseteq PR(\mathcal{N}, M_0)$, condition (2) is implied by the following one:

$$\nexists\, M \in \mathbb{N}^n : (M \in PR(\mathcal{N}, M_0)) \wedge (\forall t \in T : \exists p \in {}^\bullet t : M[p] < W(p,t)) \quad (3)$$

It is immediate to see that condition (3) is equivalent to condition (1).

The above condition is, in general, only sufficient for deadlock-freeness, since it is based on the set of *potentially* reachable markings. Nevertheless, using the absence of killing spurious solutions (corollary 4.2) and the equivalence between liveness and deadlock-freeness, it is shown next that condition (1) provides a characterization of liveness in bounded strongly connected EC-Σ.

Theorem 5.3 *A bounded strongly connected* EC-Σ, $(\mathcal{N}, M_0)$, *is live iff*

$$\nexists\,(M,\vec{\sigma}) \in \mathbb{N}^{n+m} : (M = M_0 + C\cdot\vec{\sigma}) \wedge \left(\bigwedge_{t\in T}\left(\bigvee_{p\in {}^\bullet t} M[p] < W(p,t)\right)\right) \quad (1)$$

Proof Only necessity needs to be proven. Assume that $(\mathcal{N}, M_0)$ is live and M_d is (the marking part of) a solution of the above condition. It follows that it is a deadlock (i.e. a final consisting of a single node without successors) in the potential reachability graph. Since $M_d \in PR(\mathcal{N}, M_0)$ and $(\mathcal{N}, M_d)$ is not live, $(\mathcal{N}, M_0)$ is not live either, according to (the contrapositive of) corollary 4.2, against the hypothesis.

Up to now we have obtained a characterization of liveness in bounded strongly connected EC-Σ in terms of the set of solutions of the state equation. Let us concentrate now in the expression of condition (1). Since the different linear equations/inequalities are not all linked conjunctively, condition (1) is not a linear system, but a set of linear systems. In fact, we could write it as a (logical) sum of products applying the distributive property to the actual product of sums, leading to:

$$\begin{aligned}&\nexists\ (M, \vec{\sigma}) \in \mathbb{N}^{n+m}:\\ &\bigvee_{\alpha : T \mapsto P} ((M = M_0 + C \cdot \vec{\sigma}) \wedge (\textstyle\bigwedge_{t \in T} M[\alpha(t)] < W(\alpha(t), t)))\end{aligned} \quad (4)$$

where $\alpha : T \mapsto P$ is a mapping that assigns to each transition one of its input places. It follows that evaluating condition (4) involves checking for solution a large number of linear systems, namely $\prod_{t \in T} |{}^\bullet t|$ of them. There are several ways to alleviate this problem, from conventional logical simplifications to Petri net based specific rules [TCS 92]. Fortunately, in many interesting particular cases, including that of EC-Σ, condition (1) can be written in terms of a single linear system.

In order to prove that, for EC-Σ, condition (1) results in a single linear system, let us introduce a sufficient condition for the "linearity" of the non-firability condition of a transition. We shall first recall that the ***structural bound*** of place p in $(\mathcal{N}, M_0)$ is defined as

$$SB(p) \stackrel{\text{def}}{=} \max\{M[p] \mid M \in PR(\mathcal{N}, M_0)\}$$

From now on we assume the net is structurally bounded, so structural bounds are well defined for every place.

Theorem 5.4 *Let $(\mathcal{N}, M_0)$ be a P/T-Σ, and let $t \in T$ be a transition such that ${}^\bullet t = \pi \cup p' \wedge \forall p \in \pi : SB(p) \leq W(p, t)$. The non-firability condition for transition t can be written as an integer linear inequality as follows:*

$$SB(p') \sum_{p \in \pi} M[p] + M[p'] < SB(p') \sum_{p \in \pi} W(p, t) + W(p', t) \quad (5)$$

Proof Let us rewrite first the original non-firability condition for transition t:

$$\left(\bigvee_{p \in \pi} M[p] < W(p, t) \right) \vee (M[p'] < W(p', t)) \quad (6)$$

It must be shown that, in the integer domain, equation (6) $\Leftrightarrow$ equation (5).

$\Rightarrow$ We distinguish two cases for t disabled:

Case 1: Some $p'' \in \pi$ is such that $M[p''] < W(p'', t)$, so $M[p''] \leq W(p'', t) - 1$. Using also that $\forall p \in P: M[p] \leq SB(p)$ and that $\forall p \in \pi: SB(p) \leq W(p, t)$ we get:

$$\begin{aligned}
& SB(p') \sum_{p \in \pi} M[p] + M[p'] \\
= \; & SB(p') \sum_{p \in \pi \setminus \{p''\}} M[p] + SB(p') M[p''] + M[p'] \\
\leq \; & SB(p') \sum_{p \in \pi \setminus \{p''\}} W(p, t) + SB(p')(W(p'', t) - 1) + M[p'] \\
= \; & SB(p') \sum_{p \in \pi \setminus \{p''\}} W(p, t) + SB(p') W(p'', t) - (SB(p') - M[p']) \\
\leq \; & SB(p') \sum_{p \in \pi} W(p, t) < SB(p') \sum_{p \in \pi} W(p, t) + W(p', t)
\end{aligned}$$

Case 2: $M[p'] < W(p', t)$. Using also that $M[p] \leq SB[p]$ and that $\forall p \in \pi: SB(p) \leq W(p, t)$ the result follows immediately.

$\Leftarrow$ Let us rewrite first equation (5):

$$SB(p') \sum_{p \in \pi} (M[p] - W(p, t)) < W(p', t) - M[p'] \tag{7}$$

Assume contrary: equation (7) holds and t is enabled, so, in particular, $M[p'] \geq W(p', t)$. Therefore $\sum_{p \in \pi} (M[p] - W(p, t)) < 0$, and, since $\forall p \in \pi: M[p] - W(p, t) \leq 0$, it follows that $\exists p \in \pi: M[p] - W(p, t) < 0$, hence t is not enabled, against the hypothesis.

If it was the case that every transition in the net fulfilled the condition of the above theorem, then every non-firability condition could be written linearly, and therefore condition (1) would be a linear system. What we are going to show next is that any (structurally bounded) EC-Σ can be transformed, preserving deadlock-freeness, into another EC-Σ enjoying such property. Our aim here is to show that it is *possible* to achieve this, disregarding efficiency considerations. (There are indeed more sensible ways, and more rules, to rewrite the original condition reducing the number of linear systems [TCS 92] without adding so many auxiliary variables — places and transitions — as we are going to add here for the sake of simplicity. The interested reader can also find examples of the application of these techniques in the cited work.)

We first perform the classical separation of the synchronization and choice in the "general" (non trivial and having several input places) Equal Conflict

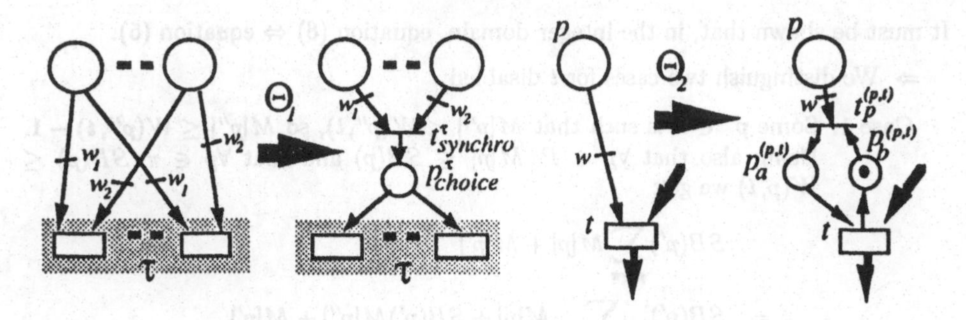

Figure 4: Illustration of transformations Θ_1 and Θ_2.

Sets, identical to the transformation from an Extended Free Choice to a Free Choice, as shown in figure 4, left. This transformation can be more precisely defined as follows:

Definition 5.1 (Transformation Θ_1.)

Let $(\mathcal{N}, M_0)$ be a P/T-Σ. The P/T-Σ $(\mathcal{N}, M_0)$ transformed by Θ_1, denoted $(\mathcal{N}^{\Theta_1}, M_0^{\Theta_1})$ is obtained as follows:

$(\mathcal{N}^{\Theta_1}, M_0^{\Theta_1}) := (\mathcal{N}, M_0)$
forall $\tau \subseteq T$ ECS of $(\mathcal{N}, M_0)$ **such that** $|\tau| > 1$ **and** $|{}^{\bullet}\tau| > 1$ **do**
  Add to P^{Θ_1} **one place:** p^{τ}_{choice} $(M_0^{\Theta_1}[p^{\tau}_{choice}] := 0)$
  Add to T^{Θ_1} **one transition:** $t^{\tau}_{synchro}$
  Add to W^{Θ_1} **the following arcs:**
    $\forall p \in {}^{\bullet}\tau : W^{\Theta_1}(p, t^{\tau}_{synchro}) := W^{\Theta_1}(p, t)$, **where** $t \in \tau$
    $W^{\Theta_1}(t^{\tau}_{synchro}, p^{\tau}_{choice}) := 1$
    $\forall t \in \tau : W^{\Theta_1}(p^{\tau}_{choice}, t) := 1$
  Remove from W^{Θ_1} **the following arcs:**
  $\forall p \in {}^{\bullet}\tau \setminus \{p^{\tau}_{choice}\}, \forall t \in \tau : (p, t)$
od

After applying transformation Θ_1 to an EC-Σ, obviously every transition in a non-trivial ECS will have just one input place, so ***every transition having more than one input place will be the only output transition of all its input places***, or, in other terms, ***every place being input of a join transition will be a non-choice (i.e. persistent) place.*** Still it is possible, though, that some join transition has more than one input place having a structural bound bigger than the weight of the arc, violating the condition of theorem 5.4. But since every input place of a join transition has been made non-choice, we can "sequentialize" such synchronizations without affecting the language, by means of the following transformation, illustrated in figure 4, right.

Definition 5.2 (Transformation Θ_2.)

Let $(\mathcal{N}, M_0)$ be a P/T-Σ. *The* P/T-Σ $(\mathcal{N}, M_0)$ *transformed by Θ_2, denoted $(\mathcal{N}^{\Theta_2}, M_0^{\Theta_2})$ is obtained as follows:*

$(\mathcal{N}^{\Theta_2}, M_0^{\Theta_2}) := (\mathcal{N}, M_0)$
forall arc $(p,t) \in W$ **such that** $|{}^\bullet t| > 1$ **and** $p^\bullet = \{t\}$ **do**
  Add to P^{Θ_2} two places: $p_a^{(p,t)}$ and $p_b^{(p,t)}$ $(M_0^{\Theta_2}[p_a^{(p,t)}] := 0, M_0^{\Theta_2}[p_b^{(p,t)}] := 1)$
  Add to T^{Θ_2} one transition: $t_p^{(p,t)}$
  Add to W^{Θ_2} the following arcs:
    $W^{\Theta_2}(p, t_p^{(p,t)}) := W^{\Theta_2}(p,t)$
    $W^{\Theta_2}(t_p^{(p,t)}, p_a^{(p,t)}) := W^{\Theta_2}(p_a^{(p,t)}, t) := W^{\Theta_2}(t, p_b^{(p,t)}) := W^{\Theta_2}(p_b^{(p,t)}, t_p^{(p,t)}) := 1$
  Remove from W^{Θ_2} the arc (p,t)
od

It is clear that, being $x \in \{1,2\}$, every occurrence sequence in $(\mathcal{N}, M_0)$ is firable in $(\mathcal{N}^{\Theta_x}, M_0^{\Theta_x})$, and viceversa, disregarding the occurrences of the added (silent) transitions. In particular, after the definitions, the equivalence of deadlock-freeness in the original and the transformed net follows immediately.

Property 5.5 *Let $(\mathcal{N}, M_0)$ be a* P/T-Σ, *and let $(\mathcal{N}^{\Theta_x}, M_0^{\Theta_x})$ be $(\mathcal{N}, M_0)$ transformed by Θ_x, where $x \in \{1,2\}$.*
$(\mathcal{N}, M_0)$ is deadlock-free iff $(\mathcal{N}^{\Theta_x}, M_0^{\Theta_x})$ is.

Finally, the announced result can be stated as follows:

Theorem 5.6 *Let $(\mathcal{N}, M_0)$ be a structurally bounded* EC-Σ. *Liveness of $(\mathcal{N}, M_0)$ can be decided verifying that there is no solution to a single linear system of inequalities in the integer domain.*

Proof Let $(\mathcal{N}', M_0')$ be $(\mathcal{N}, M_0)$ transformed by Θ_1 and Θ_2 (in this order). Both transformations preserve the Equal Conflict property. It is also clear that the structural bound of "$p_a^{(p,t)}$" and "$p_b^{(p,t)}$" places (added by transformation Θ_2) is one, equal to the weight of their output arcs. Therefore, *every transition* in the transformed EC-Σ fulfills the condition of theorem 5.4. (In particular, every input place to the original join transitions and the added "$t^T_{synchro}$" is now structurally bounded by one, since all of them are places "$p_a^{(p,t)}$", and the added "$t_p^{(p,t)}$" transitions have only one input place with a structural bound eventually bigger than the corresponding weight, because the other one is a "$p_b^{(p,t)}$" place.) It follows that analysing deadlock-freeness in the transformed EC-Σ involves checking that there isn't any solution of a single linear system. Since this is equivalent to analysing deadlock-freeness of the original system, according to corollary 5.5, and this is equivalent to analysing its liveness, according to theorem 5.1, we are done.

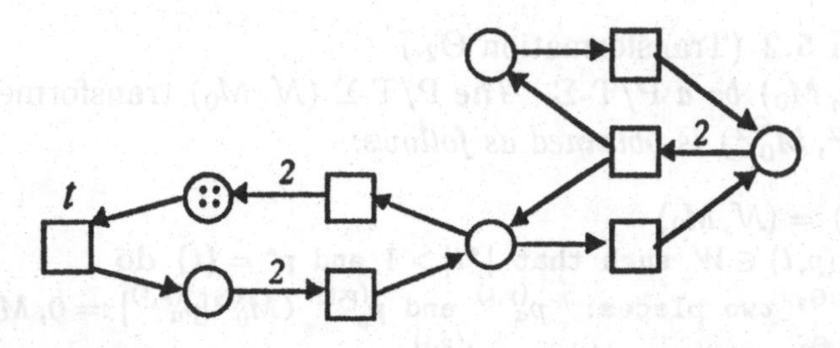

Figure 5: A non reversible live and bounded EC-Σ.

6 Equal Conflict Systems Also Have Home States

The existence of home states in live and bounded EC-Σ basically means that these kind of systems always have states that can be reached after whichever possible evolution. This result generalizes the results appearing in [BV 84] (live and 1-bounded Free Choice systems have home states) and in [Vogler 89] (live and bounded Free Choice systems have home states). The generalization of the result from 1-bounded to just bounded was foreseen by the authors in [BV 84], *"however it is not entirely obvious how the proof might be generalised, because the usual way of transforming bounded systems into 1-bounded does not preserve the Free Choice property"*. An analogous observation applies to our new generalization, because the transformations from weighted to ordinary systems do not preserve the Free Choice property either. In fact we base our proof also in directedness, as in both cited works, but this time we refer to the *potential* reachability graph instead of the reachability graph, what has proved useful for other different purposes in the previous section.

As pointed out in [BV 84], the existence of home states is a strong property, sometimes appearing in the specification of reactive systems: usually they require states to return to from any point, in order to be able to resume their activity. It also facilitates the analysis, because the reachability graph is known to be rather particular: it has only one final, that is active (a strongly connected subgraph). Moreover, many performance analyses or partial simulations are meaningless if there isn't a home space, because it could be the case that the system behaves in completely different ways depending on the initial course of events.

Theorem 6.1 (Well-behaved EC-Σ have home states)
Let $(\mathcal{N}, M_0)$ be a live and bounded EC-Σ. *There exists a nonempty subset of reachable markings $\emptyset \neq HS \subseteq R(\mathcal{N}, M_0)$ such that*

$$\forall M_h \in HS, \forall M \in R(\mathcal{N}, M_0) : M_h \in R(\mathcal{N}, M)$$

(HS is called the home space.*)*

Proof Boundedness ensures the existence of finals in the reachability graph. Theorem 4.3 ensures that there is only one (which is, indeed, a live final). Every marking in this final can be reached from every reachable marking, hence it is the home space.

In [BV 84] the authors conjectured that a firing sequence containing every transition at least once would necessarily produce a home state in a live and bounded Free Choice, which was proven in [BCDE 90]. This is not true for EC-Σ, as the example in figure 5 shows (t must occur at least twice in order to reach the home space). We conjecture that an occurence sequence σ with vector $\vec{\sigma}$ bigger than or equal to *every* minimal T-semiflow of the (consistent) net, leads necessarily to the home space.

7 Conclusion and Future Work

We have reported here some results regarding the structure theory of a subclass of (weighted) Place/Transition systems that generalizes the ordinary subclass of Extended Free Choice systems: *Equal Conflict systems.* The basic technical results concern the directedness of the potential reachability graph of live EC-Σ. From this fact it has been shown that liveness of bounded EC-Σ can be decided checking the existence of solutions to a linear system inequalities in the integer domain. Secondly, the existence of a home space in well-behaved EC-Σ has been proved, generalizing with rather different arguments the result in [BV 84, Vogler 89]. These results are both theoretically and practically interesting. On the practical side, they provide an algebraic technique to check liveness, and a strong property which is helpful for the analysis of these systems, respectively. We find them theoretically interesting because they provide a deeper understanding of the basic causes of some of the outstanding results about Free Choice-like systems.

We conjecture that many other important results of the structure theory of Free Choice systems have their counterpart for EC-Σ (for instance the beautiful decomposition theory, naturally after adapting the concept of *components*), and that their generalization will not only be an extension but will also shed new light on the understanding of the relationships between the structure and the behaviour of concurrent systems. Some of our ongoing research is devoted to this topic.

Acknowledgement

We wish to thank J.M. Colom and P. Chrzastowski-Wachtel for fruitful discussions, and also the anonymous referees for their comments.

References

[BCDE 90] E. Best, L. Cherkasova, J. Desel, J. Esparza. Characterisation of Home States in Free Choice Systems. *Technical Report 9/90, Institut für Informatik, Universität Hildesheim.*

[Best 87] E. Best. Structure Theory of Petri Nets: the Free Choice Hiatus. In *APN 86, LNCS 254:168-205.* Springer-Verlag.

[BV 84] E. Best, K. Voss. Free Choice Systems Have Home States. *Acta Informatica 21:89-100.*

[Colom 89] J.M. Colom. *Análisis Estructural de Redes de Petri. Programación Lineal y Geometría Convexa.* PhD Thesis, Dpto. Ing. Eléctrica e Informática, Universidad de Zaragoza.

[CS 91] J.M. Colom, M. Silva. Improving the Linearly Based Characterization of P/T Nets. In *APN 90, LNCS 483:113-145.* Springer-Verlag.

[ES 91] J. Esparza, M. Silva. On the Analysis and Synthesis of Free Choice Systems. In *APN 90, LNCS 483:243-286.* Springer-Verlag.

[Hack 72] M.H.T. Hack. *Analysis of Production Schemata by Petri Nets.* M.S.Thesis, Project MAC TR-94, MIT, Cambridge, Massachusetts. (Corrections in Computation Structures Note 17, 1974.)

[Murata 89] T. Murata. Petri Nets: Properties, Analysis and Applications. In *Procs. IEEE 77,4:541-580.*

[NW 88] G.L. Nemhauser, L.A. Wolsey. *Integer and Combinatorial Optimization.* Wiley.

[Silva 85] M. Silva. *Las Redes de Petri: en la Automática y la Informatica.* AC.

[TCCS 92] E. Teruel, P. Chrzastowski-Wachtel, J.M. Colom, M. Silva. On Weighted T-systems. In *Applications and Theory of Petri Nets 1992, LNCS 616:348-367.* Springer-Verlag.

[TCS 92] E. Teruel, J.M. Colom, M. Silva. Linear Analysis of Deadlock-Freeness of Petri Net Models. *Technical Report GISI-RR-92-29, Dpto. Ing. Eléctrica e Informática, Universidad de Zaragoza.* (To appear in *Procs. European Control Conference, ECC '93.*)

[TV 84] P.S. Thiagarajan, K. Voss. A Fresh Look at Free Choice Nets. *Information and Control 61,2:85-113.*

[Vogler 89] W. Vogler. Live and Bounded Free Choice Nets have Home States. *Petri Net Newsletter 32:18-21.*

Bridging the Gap Between Place- and Floyd-Invariants with Applications to Preemptive Scheduling

Rüdiger Valk

Fachbereich Informatik, Universität Hamburg
Vogt-Kölln-Str. 30, D-2000 Hamburg 54
e-mail: valk@informatik.uni-hamburg.de

Abstract:
The notion of linear place-invariants for coloured nets is extended to sums of non-linear functions. The extension applies to such places where all tokens are removed by the occurrence of an output transition. It is shown how this covers the case of variable assignments and invariants in traditional programs. The result helps in understanding the relation of place-invariants of coloured nets in comparison with traditional Floyd-invariants of programs. In the second part the property of token clearing is introduced to the occurrence rule, showing that the results of the first part are still valid. Such types of nets are important for the modelling of fault tolerant applications.

Keywords :
analysis and synthesis, structure and behavior of Petri nets, higher-level net models, coloured Petri nets, program verification, place-invariants, Floyd-invariants, selfmodifying coloured Petri nets

1. Introduction

Invariants have been well studied for ALGOL-like programs [Floyd 67, Hoare 69, Gries 81] and for Petri nets (place-invariants) [Jensen 81]. In formal verification techniques such invariants are used to prove properties that hold invariant over an infinity of possible system states.

On the other hand rather less is known on the relation between these two types of invariants. Floyd-invariants allow formulation of very general properties, but there are no general procedures to compute them. (For a given ALGOL-like program partial correctness is undecidable.) Place-invariants of Petri nets are computable (even in polynomial time for ordinary place/transition nets) but are required to have a linear form.

```
Program ODD-SUM
{m≥1}
input m : INT
var s,i,odd : INT ;
(s,i,odd ) := (0,1,1)
while [i < m+1] do
      begin
      {s = (i-1)^2 ∧ 2*i - 1 = odd}
      (s,i,odd) := (s+odd,i+1,odd+2)
      end;
output (m,s) ;
{s = m^2}
```

step	m	s	i	odd
0	4	0	1	1
1		1	2	3
2		4	3	5
3		9	4	7
4		16	5	9

Fig. 1.1: Program ODD-SUM with assertions and a trace

In this paper we will start to bridge the gap between these two formalisms. To begin with, consider the program of fig. 1.1. It computes the square of a given integer m by successive addition of odd numbers. Traditionally this property is proved by the given pre- and post-conditions and the invariant assertion at the beginning of the while loop. There is also shown a trace including the first four steps.

In section 2 we will formulate this program as a coloured Petri net. Since the given invariant assertion is not linear, it does not fit into the classical place-invariance calculus of coloured nets. But we will prove a general and sufficient condition for computing such invariants from the incidence matrix. The problem has been studied in [Vautherin 85] for a restricted class of nets.

We will give a sufficient local condition for such places, for which the invariant equation contains a non-linear term. This allows us to formulate two theorems that cover (traditional) linear place-invariants and their non-linear extensions and to clarify the different nature of place-invariants of nets on the one hand and of non-linear invariants of programs on the other hand. Moreover by this approach we are able to consider an extended class of coloured nets that is very useful for the modelling of systems requiring the removal of all objects under certain circumstances, e.g. fault tolerant modelling or transitions with preemptive priority. We will show in section 4 that the classical place-invariance calculus as well as the extensions of section 3 still hold for this class of selfmodifying coloured nets.

2. Coloured Nets

In this section the technical definition of a coloured net is given. First we recall the notion of a multi-set and of a marking.

Definition 2.1 : A *multi-set* m, over a non-empty set S, is a function $m: S \to \mathbb{N}$, sometimes denoted as a formal sum $\sum_{s\in S} m(s)`s$. S_{MS} is the set of all multi-sets over S. Extending set union to S_{MS} we define the operation of sum (+) and difference (-). Since in this paper this can lead to confusion, addition and difference in the sets $\mathbb{Z}$ of integers and in the subset of non-negative integers $\mathbb{N}$ will be denoted by $+$ and $-$, respectively, and Σ is used for general summation.

If m, m_1 and m_2 are multisets over S, then we define :

$m_1 + m_2 := \sum_{s\in S} (m_1(s) + m_2(s))`s$

$m_1 \le m_2 :\Leftrightarrow \forall s\in S : m_1(s) \le m_2(s)$,

and if $m_2 \le m_1$ then $m_1 - m_2 := \sum_{s\in S} (m_1(s) - m_2(s))`s$.

$|m| := \sum_{s\in S} m(s)$ is the *size* of m and $\emptyset$ denotes the *empty multi-set* (when $|m| = 0$).

If $S \subseteq \mathbb{Z}$ is a set of integers, then $\#m := \sum_{s\in S} m(s)*s$ is the *sum* of S, in particular $\#\emptyset = 0$.

Definition 2.2 Let P be a finite set of *places*. To each place $p\in P$ we associate a *colour set* C(p). A *marking* M is a mapping defined on P such that $M(p) \in C(p)_{MS}$. If P $= \{p_1, p_2, ..., p_n\}$ is a totally ordered set, then M can be written as a vector :

$$\begin{pmatrix} M(p_1) \\ M(p_2) \\ \vdots \\ M(p_n) \end{pmatrix} \in C(p_1)_{MS} \times C(p_2)_{MS} \times ... \times C(p_n)_{MS} =: \mathcal{C}_P = \prod_{p\in P} C(p)_{MS}$$

$\mathcal{C}_P$ denotes the set of all possible markings.

To define coloured Petri nets we follow the formalism of [Jensen 87] in the form of a "CP-matrix".

Definition 2.3 A *coloured Petri net* (CPN) $N = (P,T,C,\Sigma,I_-,I_+,M_o)$ is given by

- a finite set P of *places*,
- a finite set T of *transitions*, disjoint with P : $P \cap T = \emptyset$
- a set Σ of *colour sets*,
- a *colour function* $C : P \cup T \rightarrow \Sigma$, where

  C(p) is called the *colour set* of p and

  C(t) is said to be the *colour set* (or *occurrence modes*) of t
- I_+ and I_- are the *positive* and *negative incidence functions* on $P \times T$:

  $\forall\, (p,t) \in P{\times}T : I_-(p,t),\ I_+(p,t) \in [C(t)_{MS} \rightarrow C(p)_{MS}]_L$

  (As usual, for sets of multisets A and B $[A \rightarrow B]_L$ denotes the set of linear functions, cf. [Jensen 87])
- M_o is a marking on P, called the *initial marking*.

Definition 2.4 For $P = \{p_1,p_2,...,p_n\}$, $t \in T$ and $b \in C(t)$ we will use the following notions of vectors $I_+(-,t)$ and $I_+(-,t)(b)$:

$$I_+(-,t) = \begin{pmatrix} I_+(p_1,t) \\ I_+(p_2,t) \\ \vdots \\ I_+(p_n,t) \end{pmatrix} \quad \text{and} \quad I_+(-,t)\,(b) = \begin{pmatrix} I_+(p_1,t)(b) \\ I_+(p_2,t)(b) \\ \vdots \\ I_+(p_n,t)(b) \end{pmatrix}$$

$I_+(-,t)(b)$ is extended to hold for multisets $b \in C(t)_{MS}$. The corresponding definitions for $I_-(-,t)$ and $I_-(-,t)(b)$ are omitted.

Example 2.1. Fig. 2.1 gives the incidence functions of the CPN *ODD_SUM* in the form of a common matrix. The values of I_- are placed in left upper corners of the entries and are represented in negative form. The corresponding lower right corners give the values of I_+. The colours of places and transitions are also given. The entries in the incidence matrix refer to the corresponding projections, e.g.: $I_+(p1,t2) = s{+}odd = pr_{C(p2)}(C(t2)) + pr_{C(podd)}(C(t2))$, where $pr_{C(p2)}(C(t2))$ is the projection on the first component of C(t2), i.e. on C(p2). To avoid confusion, the function $s{+}odd : C(t2) \rightarrow C(p1)$ is explicitly defined by $(s{+}odd)(s,i,odd) := s + odd$.

The initial marking is given in the first column :

$M_o = (M_o(p1), M_o(p2), M_o(pi), M_o(pm), M_o(podd), M_o(p3)) =$

$(1`0, \emptyset, 1`1, 1`4, 1`1, \emptyset)$

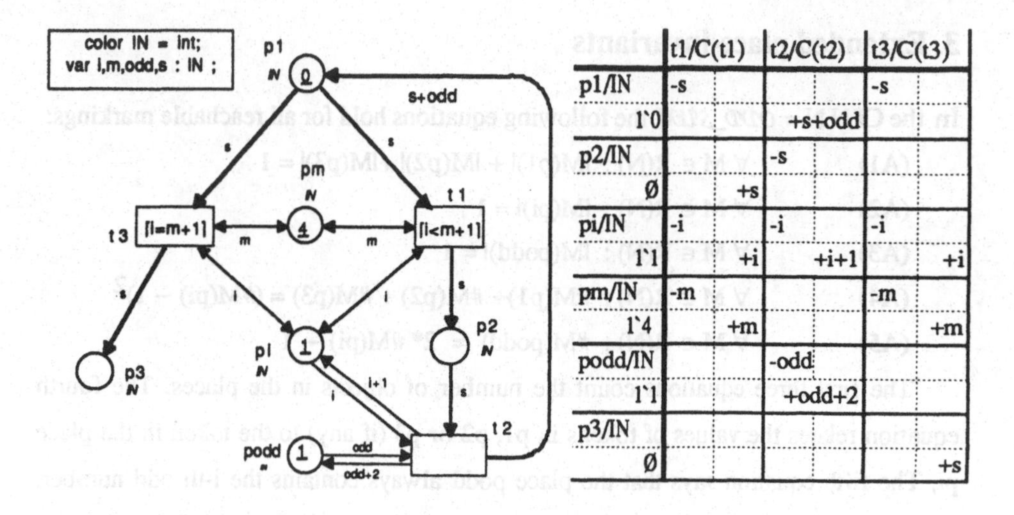

	t1/C(t1)	t2/C(t2)	t3/C(t3)
p1/IN	-s		-s
1`0		+s+odd	
p2/IN		-s	
Ø	+s		
pi/IN	-i	-i	-i
1`1	+i	+i+1	+i
pm/IN	-m		-m
1`4	+m		+m
podd/IN		-odd	
1`1		+odd+2	
p3/IN			
Ø			+s

$$C(t1) = \{ (s,m,i) \in C(p1)\times C(pm)\times C(pi) \mid i < m + 1 \}$$
$$C(t2) = \{ (s,i,odd) \in C(p2)\times C(pi)\times C(podd) \mid \text{true} \}$$
$$C(t3) = \{ (s,m,i) \in C(p1)\times C(pm)\times C(pi) \mid i = m + 1 \}$$

Fig. 2.1: Coloured net *ODD_SUM* with incidence functions I_+ and I_- and colours.

Fig. 2.1 also shows the same CPN in graphical form. For the generation of this form see [Jensen 87].) The conditions in the transitions of this graph are represented as restricted domains in the matrix representation. Arrows with two heads stand for two arcs in opposite orientation, but with the same inscription (like "side condition"). Readers familiar with coloured Petri nets will recognize that it has been constructed in a straight forward way from the program of the introduction. End of **Example 2.1**

Using I_+ and I_- as matrices, the occurrence rule is defined as follows.

Definition 2.5 • A transition t is *enabled* at a marking M with binding (colour) $b \in C(t)$ if $M \geq I_- * X$ and $X = (t,b)$ ($I_- * (t,b)$ is the vector $I_-(-,t)(b)$). We write M [X> in this case.

- If t is enabled in M, then the follower marking relation M[t,b>M' or M[X>M' is defined by the *follower marking* $M' := (M - I_- * X) + I_+ * X$ ($I_+ * X = I_+ * (t,b)$ is defined in the same way as $I_- * (t,b)$)
- As usual, the relation M[X>M' is extended to words u over $Z = \{(t,b) \mid t \in T, b \in C(t)\}$ and the reachability set is $R(N) := \{ M \mid \exists u \in Z^* : M_0[u>M \} \subseteq \mathcal{C}_P$.

3. Extended place-invariants

In the CPN N = *ODD_SUM* the following equations hold for all reachable markings:

(A1) $\forall M \in R(N) : |M(p1)| + |M(p2)| + |M(p3)| = 1$

(A2) $\forall M \in R(N) : |M(pi)| = 1$

(A3) $\forall M \in R(N) : |M(podd)| = 1$

(A4) $\forall M \in R(N) : \#M(p1) + \#M(p2) + \#M(p3) = (\#M(pi) - 1)^2$

(A5) $\forall M \in R(N) : \#M(podd) = 2 * \#M(pi) - 1$

The first three equations count the number of objects in the places. The fourth equation relates the values of tokens in p1, p2 or p3 (if any) to the token in the place pi. The fifth equation says that the place podd always contains the i-th odd number. Invariant equations (A4) and (A5) correspond to the loop invariant of the program ODD-SUM.

As usual these invariants are used to prove the partial correctness of the net: in a terminating marking M_E there is an object q in p3 (and by (A1) only one) and from the guard [i=m+1] of t3 we have $M_E(pm) = M_E(pi) - 1$. By equation (A1) p1 and p2 are empty, hence by (A4) : $q = \#M_E(p3) = (\#M_E(pi) - 1)^2 = (M_E(pm))^2 = m^2$ i.e. the net computes the square of the "input" m.

Ordinary place-invariants can be deduced directly from the incidence functions. This is a very important property since they can be checked by inspection of the static description of the net. Hence it is not necessary to compute the reachability set, which is very complex in most cases. This property strongly depends on the linearity of ordinary place-invariants and on an invariance of weighted flow for each transition.

Different to this situation equation (A4) is not linear as in numerous similar cases. Therefore the technique mentioned does not apply in general. However, case studies have shown that for many coloured nets modelling classical sequential programs, the invariance of weighted flow is fulfilled.

To prove the invariant equation (A4) by induction, the fifth equation (A5) is also used, as will be shown at the end of this section. Opposite to this fact, classical place invariants can be independently verified. In the following, we will give a number of sufficient conditions for also including non-linear equations into the invariance calculus of coloured Petri nets.

Definition 3.1 A pair $(p,t)\in P\times T$ of a CPN is a *clearing arc* if $\forall\, M\in R(N)\ \forall\, b\in C(t) : M[(t,b)> \Rightarrow I_-(p,t)(b) = M(p)$. A place $p\in P$ is a *clearing place* if for all $t\in T$ and $c\in C(t)$ either $I_-(p,t)(b) = I_+(p,t)(b) = \emptyset$ or (p,t) is a clearing arc. If p does not have this property, it is called *non-clearing.*

By this definition an occurring transition removes all tokens from input clearing places. Obviously, this definition is not "static" in the sense mentioned above, but can be easily checked by formal computation in many examples, e.g. in the net *ODD_SUM* place pi obviously has this property.

Equations (A1) to (A5) have the form $\sum_{p\in P} W_p(M(p)) = c$ where $W_p(M(p))$ has values in an additive group $\mathbb{A}$ and $c\in \mathbb{A}$ is a constant.

Definition 3.2 Let $(\mathbb{A},+)$ be an (additively written) commutative group. For a place p a function $W_p : C(p)_{MS} \rightarrow \mathbb{A}$ is linear if $W_p(\emptyset) = 0$ and
$W_p(x+y) = W_p(x) + W_p(y)$ for all $x,y \in C(p)_{MS}$.
For a marking M the *weight function* $W_P : \mathcal{C}_P \rightarrow \mathbb{A}$ is defined by $W_P(M) := \sum_{p\in P} W_p(M(p))$. "$W_P(M) = c$" for a constant $c \in \mathbb{A}$ is called a *weighted equation.* The weighted equation is called an *invariant equation* of a CPN N if $W_P(M)=c$ for all reachable markings $M \in R(N)$.

Remark 3.1: $|M|$ and $\#M$, as introduced in definition 2.1 , are examples of linear functions. Note that the definition of a clearing arc (definition 3.1) also covers the case where $I_-(p,t)(b) = M(p) = \emptyset$.

Theorem 3.3 Let W_P be a weight function for a CPN $N = (P,T,C,\Sigma,I_-,I_+,M_o)$, such that W_p is linear on all non-clearing places $p\in P$.
If $W_P(M_o) = c$ and $W_P[I_+(-,t)(b)] = W_P[I_-(-,t)(b)]$ for all $t \in T$ and $b \in C(t)$, then $W_P(M) = c$ is an invariant equation of N. The theorem remains true if "=c" is replaced by "$\leq$ c" or "$\geq$ c".

Proof (by induction on the reachability set R(N)) :

a) $W_P(M_o) = c$ (or $\leq c$, $\geq c$, resp.) by assumption.

b) Let $W_P(M) = c$ (or $\leq c$, $\geq c$, resp.) and $M[(t,b)>M'$ be an occurrence of $X = (t,b)$ such that *not* $I_-(p,t)(b) = I_+(p,t)(b) = \emptyset$. We have to prove $W_P(M') = c$.
In fact, $W_P(M') = \sum_{p\in P} W_p[M'(p)] = \sum_{p\in P} W_p[\ M(p) - I_-(p,t)(b) + I_+(p,t)(b)\ ]$ (I)

b1) If p is a non-clearing place, then W_p is linear :

$W_p[\, M(p) - I_-(p,t)(b) + I_+(p,t)(b)\,] = W_p[M(p)] - W_p\,[I_-(p,t)(b)] + W_p\,[I_+(p,t)(b)\,]$

b2) If p is a clearing place, then since M[(t,b)> we have :

$$M(p) - I_-(p,t)(b) = \emptyset \qquad (*)$$

and $\quad W_p[M(p)] \;-\; W_p[I_-(p,t)(b)] = 0 \qquad (**)$

By calculating

$W_p[\, M(p) - I_-(p,t)(b) + I_+(p,t)(b)\,] =$ (by $(*)$)

$W_p\,[I_+(p,t)(b)\,] =$ (by $(**)$)

$W_p[M(p)] \;-\; W_p\,[I_-(p,t)(b)] + W_p\,[I_+(p,t)(b)]$ we obtain the same result as in case b1). Now we are able to continue with line (I) :

$W_P(M') = \sum_{p\in P} W_p[M'(p)] = \sum_{p\in P} W_p[\, M(p) - I_-(p,t)(b) + I_+(p,t)(b)\,] =$ (by b1 and b2)

$\sum_{p\in P} (W_p[\, M(p)\,] - W_p[\,I_-(p,t)(b)] + W_p[I_+(p,t)(b)\,]\,) =$ (by commutativity in **A**)

$\sum_{p\in P} W_p[\, M(p)\,] - \sum_{p\in P} W_p[\,I_-(p,t)(b)] + \sum_{p\in P} W_p[I_+(p,t)(b)\,] \;=$ (by definition)

$W_P(M) - W_P(I_-(\text{-},t)(b)) + W_P(I_+(\text{-},t)(b)) = W_P(M) = c$ (or $\le c, \ge c$, resp.)

(by assumption of $W_P[I_+(\text{-},t)(b)] = W_P[I_-(\text{-},t)(b)]$ and by induction) . ❐

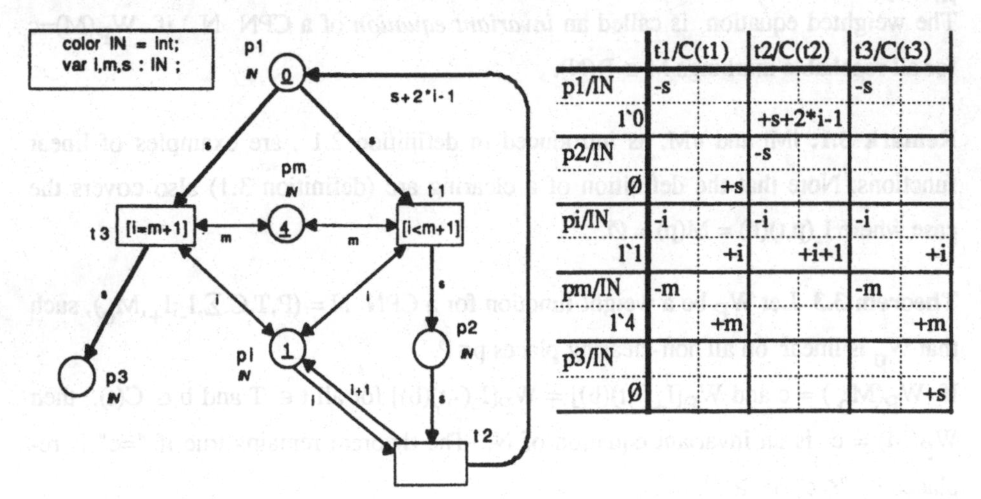

	t1/C(t1)	t2/C(t2)	t3/C(t3)
p1/IN	-s		-s
1`0		+s+2*i-1	
p2/IN		-s	
Ø	+s		
pi/IN	-i	-i	-i
1`1	+i	+i+1	+i
pm/IN	-m		-m
1`4	+m		+m
p3/IN			
Ø			+s

Fig. 3.1: Coloured net *ODD_SUM*-1 with incidence functions I_+ and I_- and colours.

Example 3.1: The theorem does *not* allow the proof of the most interesting invariant (A4). This is however the case for the following, slightly modified CPN *ODD_SUM-1*, where the place podd is omitted and the function $I_+(p1,t2)$ is replaced by s+2*i-1. Since (A4) has the form $\forall M \in R(N) : \#M(p1) + \#M(p2) + \#M(p3) - (\#M(pi) - 1)^2 = 0$, the functions W_p are defined as follows : $W_{p1}(a) = W_{p2}(a) = W_{p3}(a) = \#a$, $W_{pi}(a) = -(\#a - 1)^2$, $W_{pm}(a) = 0$.

Condition $W_P[I_+(-,t)(b)] = W_P[I_-(-,t)(b)]$ of the theorem is satisfied for t=t2 and $b = (s,i) \in \mathbb{N}^2$ (we have to look to the nonempty entries of I_- and I_+ only) :

left-hand side : $W_{p1}[I_+(p1,t2)(s,i)] + W_{pi}[I_+(pi,t2)(s,i)] =$
$W_{p1}[s + 2*i - 1] + W_{pi}[i + 1] = s + 2*i - 1 - ((i+1) - 1)^2 = s + 2*i - 1 - i^2$.

right-hand side : $W_{p2}[I_-(p2,t2)(s,i)] + W_{pi}[I_-(pi,t2)(s,i)] = W_{p2}[s] + W_{pi}[i] = s - (i - 1)^2 = s - i^2 + 2*i - 1$. The analogous calculations for t1 and t3 are simple, since pi is only trivially involved. **End of Example 3.1.**

Example 3.2: Let us return to the example 2.1 with the CPN *ODD_SUM*. The analogous calculation as in the preceding example yields (observe that $W_{podd}(a)=0$):

left-hand side : $W_P[I_+(-,t2)(b)] = W_{p1}[I_+(p1,t2)(s,i)] + W_{pi}[I_+(pi,t2)(s,i)] = W_{p1}[s + odd] + W_{pi}[i + 1] = s + odd - ((i + 1) - 1)^2 = s + odd - i^2$.

right-hand side : $W_P[I_-(-,t2)(b)] = W_{p2}[I_-(p2,t2)(s,i)] + W_{pi}[I_-(pi,t2)(s,i)] = W_{p2}[s] + W_{pi}[i] = s - (i - 1)^2 = s - i^2 + 2*i - 1$.

Hence for b = (s,i,odd) the difference $W_P[I_+(-,t2)(b)] - W_P[I_-(-,t2)(b)] = odd - 2*i + 1$ is not zero, and the condition $W_P[I_+(-,t2)(b)] = W_P[I_-(-,t2)(b)]$ of the theorem is *not* satisfied. However, in this example we have the invariant equation

(A5) $\forall M \in R(N) : \#M(podd) - 2* \#M(pi) + 1 = 0$

In all reachable markings by (A2) and (A3) there is exactly one token in pi and podd, having values i and odd, respectively. By (A5) they satisfy odd - 2*i + 1 = 0, hence also (A4) is an invariant equation. On the other hand, (A5) can be verified by application of theorem 3.3. This is done by defining $W_{pi}(a) := 1 - 2*\#a$, $W_{podd}(a) := \#a$ and $W_p(a) = 0$ for $p \notin \{pi,podd\}$ and the calculation (e.g. for t = t2 and b = (s,i,odd)): $W_P[I_+(-,t2)(b)] - W_P[I_-(-,t2)(b)] = W_{pi}[i + 1] + W_{podd}[odd+2] - (W_{pi}[i] + W_{podd}[odd]) = 1 - 2*(i+1) + odd + 2 - ((1 - 2*i) + odd) = 0$. **End of Example 3.2**

This example is of particular interest, since different from the case of ordinary and linear place-invariants, some equations cannot be verified by the incidence function independently from other equations. Note that (A4) can be transformed by substituting #M(pi) by its value from (A5). Then the problem would not occur. Such transformations, however, require special insight, and may lead to clumsy equations.

When formalizing the method we will use a property that is frequently satisfied by coloured Petri-nets arising from practical applications, i.e. the colour sets C(t) of transitions are subsets of products of place colours (c.f. example 2.1).

Theorem 3.4 Let $N = (P,T,C,\Sigma,I_-,I_+,M_0)$ be a CPN, where all transition colours C(t) are subsets of Cartesian products of some place colours :

$C(t) \subseteq C(p_{i_1}) \times ... \times C(p_{i_m})$, and let

(C1) $W_P{}^1(M) = c_1$

(C2) $W_P{}^2(M) = c_2$

.....

(Cr) $W_P{}^r(M) = c_r$

be a set of weight functions (r≥2) such that for each equation (Ci) (1≤i≤r) the following holds :

- $W_P{}^i$ is linear on all non-clearing places,
- $W_P{}^i(M) = c_i$ for the initial marking $M = M_0$, and
- either $\forall t \in T\ \forall b \in C(t) : W_P{}^i[I_+(-,t)(b)] - W_P{}^i[I_-(-,t)(b)] = 0$

  or $[\forall t \in T\ \forall b \in C(t) : W_P{}^i[I_+(-,t)(b)] - W_P{}^i[I_-(-,t)(b)] = q(b) \neq 0$

  and $b = (x_1, ... ,x_v)$

  and all x_i are inscriptions of clearing arcs

  and $\forall M \in R(N)$: $q(b)=W_P{}^j(M)=0$ for some j with $1 \leq j < i$]

Then (C1) ... (Cr) are invariant equations of N.

Technical remark : $q(b) = W_P{}^j(M) = 0$ is supposed to be the zero mapping, and not a constant.

Proof : Theorem 3.4 follows by successive verification of (C1), (C2), ...

If $\forall t \in T\ \forall b \in C(t) : W_P{}^i[I_+(-,t)(b)] - W_P{}^i[I_-(-,t)(b)] = 0$ then (Ci) holds by theorem 3.3. If $\forall t \in T\ \forall b \in C(t) : W_P{}^i[I_+(-,t)(b)] - W_P{}^i[I_-(-,t)(b)] = q(b) \neq 0$, then $q(b) = W_P{}^j(M) = 0$ for some j with $1 \leq j < i$ and theorem 3.3. can also be applied. This is justified by the condition that all x_i are inscriptions of clearing arcs. $b = (x_1, ... ,x_v)$ is evaluated in such a way that x_u has the value of the token in the input place p_u of the corresponding arc. Then $q(b) = q(x_1, ... ,x_v) = W_{p_u}{}^j(M) = 0$. (The evaluation may be a formal one, as in the following example). □

Example 3.3: Theorem 3.4 is illustrated by the equations of example 3.2 : take (A5) as (C1) and (A4) as (C2). For t = t2 and b = (s,i,odd) ∈ C(t2) we have q(b) = odd-2*i+1. By (C1) = (A5) we have #M(podd) - 2* #M(pi) + 1 = 0 and odd-2*i+1 = 0.

End of Example 3.3

Remark 3.2: The net *ODD_SUM* can be transformed in such a way that concurrent behavior is possible. By changing the initial marking of place p1 (see fig. 2.1) to 3'0 three objects are created, each computing part of the sum to be calculated. Finally we might have the end marking (M(p1),M(p2),M(pm),M(pi),M(podd),M(p3)) = (Ø,Ø,1`4,1`5,1`7,1`1+1`8+1`7). This could be interpreted as three processes, each of which is computing a part of $4^2 = 16$ by sharing the places pm, pi and podd. Invariant equations (A4) and (A5) are still valid, since the operator # collects the sum of all objects in a place. The end marking above stems from a trace where the first token received the value of the first odd number (1), the second the sum of the second (3) and the third (5) and the third token the value of the fourth odd number (7). (Obviously, the loop exit control has to be changed for this extension.)

4. Clearing places by occurrence rule

The property of clearing places is frequently desirable in practical system modelling by coloured Petri nets. Often a situation appears, where all messages of a channel have to be removed, but where the actual number of messages is unknown. Similar situations occur if an indefinite number of resources needs an update. Furthermore, a similar case appears in modelling fault tolerant applications. In such cases an ordinary set of tasks or processes is stopped by a super-task with maximal priority. Then by some preemptive scheduling rule all tokens representing ordinary tasks are substituted by the priorized process.

In fig. 4.1 a node of a queuing network (called elementary waiting system) contains a waiting pool where tasks a_i are waiting to be processed by the functional unit

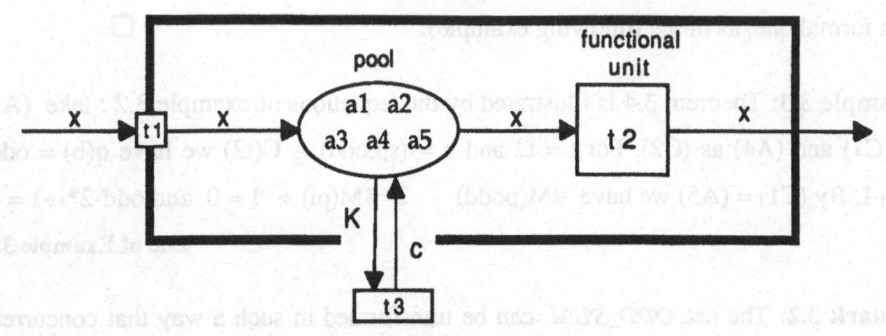

Fig. 4.1 : Elementary waiting system with preemptive scheduling

t2. In the case of system failure the priorized super-task c enters the pool and removes all tasks by a preemptive scheduling rule. Then by the arc K *all* the tasks in the pool have to be removed.

Fig. 4.2 shows a set of tasks or processes a_1 ... ,a_{100} in a processor pipeline. In case of a system failure all processes in the pools p1 to p4 are set back to their initial position by the recovery transition t5. This is done by arc expressions 'pi', which are evaluated to the multi-sets of the places pi in the actual marking. The arc from t5 to po adds the sum of all these multi-sets to the initial place p0.

The same system could be modelled as an ordinary coloured net by considering as tokens the multi-sets of tasks in the pools pi (see [Valk 93]). Such a modelling, however, is felt to be rather artificial, since individual objects (tasks) are not represented by individual tokens.

Arcs like (p_1,t_5) in fig. 4.2 are clearing arcs as introduced in definition 3.1. But

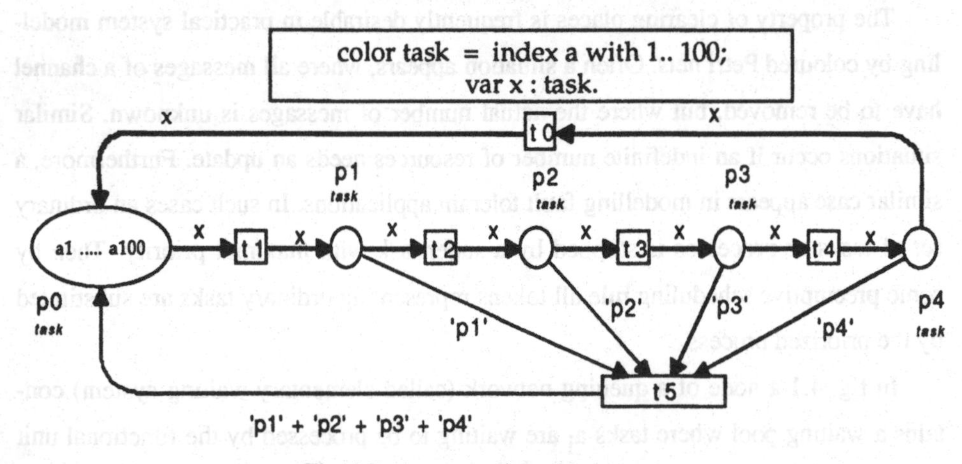

Fig. 4.2 : A process line with recovery

here this property is no longer a dynamical changing one, but is permanently holding by extension of the occurrence rule. This is an instance of selfmodifying coloured nets, which have been studied for the subclass of place/transition nets [Valk 78, Valk 83].

In a selfmodifying net the value of arc inscriptions are allowed to depend on the token content of some specified places. Hence, the structure of arcs between places and transitions is no longer time independent, but may change with the behavior of the net. This explains the name "selfmodifying net". Wellknown extensions like inhibitor arcs, priority transitions and reset arcs are subclasses of selfmodifying nets.

Although in this paper only the feature of clearing places will be used, it is easier in the definition to use selfmodifying coloured nets in general. Selfmodifying coloured nets (smCPN) differ from ordinary coloured nets (CPN, def. 2.3) only in the definition of the incidence functions which may depend on the actual marking. Later on, in definition 4.3, we will introduce a notation for expressing this dependence by arc inscriptions in the graphical representation of selfmodifying coloured nets.

Definition 4.1 A *selfmodifying coloured Petri net* (smCPN)

$N = (P,T,C,\Sigma,I_{M-},I_{M+},M_o)$ is given by

- a finite set P of *places*,
- a finite set T of *transitions*, disjoint with P : $P \cap T = \emptyset$
- a set Σ of *colour sets*,
- a *colour function* $C : P \cup T \rightarrow \Sigma$, where

  C(p) is called the *colour set* of p and

  C(t) is said to be the *colour set* (or *occurrence modes*) of t
- for every marking $M \in \mathcal{C}_P$ I_{M-} and I_{M+} are the *positive* and *negative incidence functions* on $P \times T$:

  $\forall (p,t) \in P \times T : I_{M-}(p,t),\ I_{M+}(p,t) \in [C(t)_{MS} \rightarrow C(p)_{MS}]_L$
- M_o is a marking on P, called *initial marking*.

Definition 4.2 • A transition t is *enabled* at a marking M with binding (colour) $b \in C(t)$ if $M \geq I_{M-} * X$ and $X = (t,b)$.

(Again, $I_{M-}* (t,b)$ is the vector $I_{M-}(-,t)(b)$.) We write M [X> in this case.

- If t is enabled in M, then the follower marking relation M[t,b>M' or M[X>M' is defined by the *follower marking* $M' := (M - I_{M-} * X) + I_{M+} * X$ ($I_{M+} * X = I_{M+} * (t,b)$ is defined in the same way as $I_{M-} * (t,b)$)
- As usual, the relation M[X>M' is extended to words u over $Z = \{(t,b) | t \in T, b \in C(t)\}$ and the reachability set is $R(N) := \{ M \mid \exists u \in Z^* : M_0[u>M \} \subseteq \mathcal{C}_P$.

While the definition of the "CP-matrix" of smCPN requires only few changes, for the definitions of concrete marking-dependent functions some notations are useful.

Definition 4.3 • For a multi-set $m : S \rightarrow \mathbb{N}$ and $s \in S$ let $m[s] := m(s)`s$ be the sub-multi-set of ocurrences of s.

- Let p be a place with colour set $C(p) = (X_1)_{MS} \times (X_2)_{MS} \times ... \times (X_n)_{MS}$ and M a marking $M \in \mathcal{C}_P$. Then for $x \in X_i$ we define
$M[p(x|i)] := \{ (m_1, m_2, ..., m_n) \in (X_1)_{MS} \times (X_2)_{MS} \times ... \times (X_n)_{MS}) \mid$
$\exists (q_1, ..., q_n) \in M(p) : q_j = m_j$ for $j \neq i$ and $m_i = q_i[x] \neq \emptyset \}$.
(M((p(x | i)) is the set of all tuples in M(p) having a (positive) multiple of x as i-th component.)
- For n =1 we write $M[p(x)] := M[p(x|1)]$.

As an example, let $X_1 := \{a_1, a_2\}$, $X_2 := \{f_1, f_2\}$ and $M(p) := \{(2`a_1, 3`f_2), (1`a_2, 2`f_1), (1`a_3, 2`f_1 + 1`f_2)\}$. Then $M[p(f_2|2)] = \{(2`a_1, 3`f_2), (1`a_3, 1`f_2)\}$

Definition 4.4 When using definition 4.3 as arc inscriptions of smCPN in graphical form or as entries of the incidence functions, the explicit reference to the actual marking M is omitted: i.e. instead of M[p(x|i)] we use p(x|i), since there is no confusion. In a similar way, we will use p() instead of M(p).
(This definition replaces the notion of K in fig. 4.1 and of 'pi' in fig. 4.2. The brackets distinguish p() from the name of the place $p \in P$.

Example 4.1: Fig. 4.4 shows an instance of the problem of readers and writers as a smCPN $\mathcal{RpW}$ in graphical form.

There are a set of 5 tasks $A = \{a_1, a_2, ..., a_5\}$ and two files $F = \{f_1, f_2\}$. Tasks entering the critical region as writers (transition t3) have priority. They start immediately with reading and writing a file $y = f_i$. By preemptive scheduling all tasks reading the

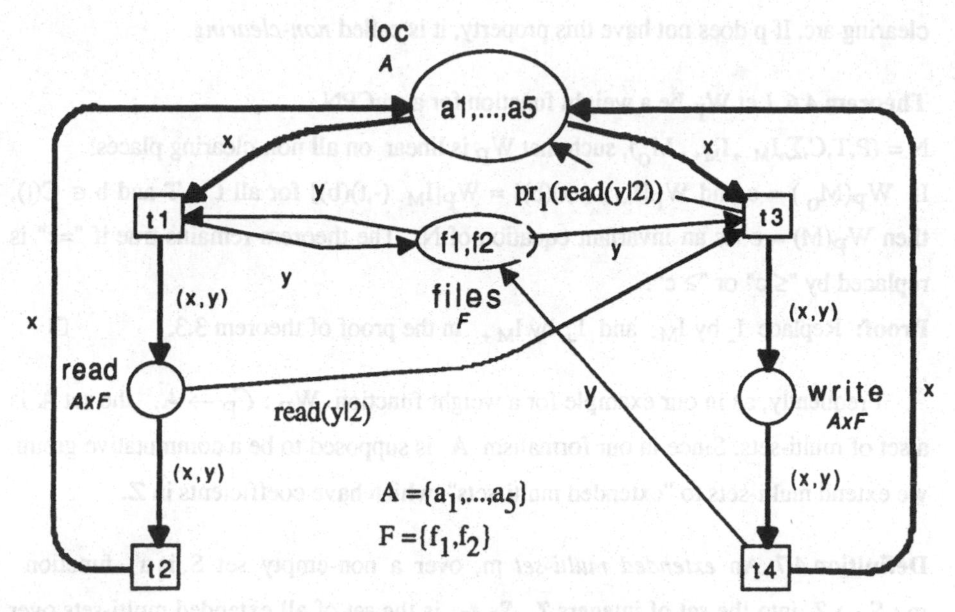

Fig. 4.3: $\mathcal{RpW}$: readers and preemptive writers

same file in the place "read" are interrupted and set back to their initial state in loc. Later they can try to repeat their action of reading. For instance, $I_{M+}(loc,t_3) = pr_1(read(y|2)) = \{ a_i | (a_i,f_j) \in M(read) \wedge f_j=y \}$ is the set of all tasks $a_j \in A$ that are reading the file y in "read". (Recall that by definition 4.4 $pr_1(read(y|2))$ stands for $pr_1(M[read(y|2)])$.

The smCPN $\mathcal{RpW}$ satisfies the following place-invariant equations (B1), (B2) and the inequality (B3):

(B1) $\quad \forall M \in R(N) : M(loc) + pr_1(M(read)) + pr_1(M(write)) = A$

(B2) $\quad \forall M \in R(N) : M(files) + pr_2(M(write)) = F$

(B3) $\quad \forall M \in R(N) : pr_2(M(read)) + 5\, pr_2(M(write)) \leq 5F$

It is easy to see that the following property can be verified by (B3): no task is reading a file f in the place "read" that is also used in the place "write" (in fact, 'f' in $pr_2(M(write))$ implies $pr_2(M[read(f|2)]) = \emptyset$).

To verify (B1) - (B3) we reformulate the theorem from section 3.

Definition 4.5 A pair $(p,t) \in P \times T$ of a smCPN is a *clearing arc* if $\forall M \in R(N)\ \forall b \in C(t) : M[(t,b)> \Rightarrow I_{M-}(p,t)(b) = M(p)$. A place $p \in P$ is a *clearing place* if for all $t \in T$ and $b \in C(t)$ either $I_{M-}(p,t)(b) = I_{M+}(p,t)(b) = \emptyset$ or (p,t) is a clearing arc. If p does not have this property, it is called *non-clearing*.

Theorem 4.6 Let W_P be a weight function for a smCPN $N = (P,T,C,\Sigma,I_{M-},I_{M+},M_o)$, such that W_P is linear on all non-clearing places. If $W_P(M_o) = c$ and $W_P[I_{M+}(-,t)(b)] = W_P[I_{M-}(-,t)(b)]$ for all $t \in T$ and $b \in C(t)$, then $W_P(M) = c$ is an invariant equation of N. The theorem remains true if "=c" is replaced by "$\leq c$" or "$\geq c$".

Proof: Replace I_- by I_{M-} and I_+ by I_{M+} in the proof of theorem 3.3. ❒

Frequently, as in our example for a weight function $W_P : \mathcal{C}_P \rightarrow \mathbb{A}$, the set $\mathbb{A}$ is a set of multi-sets. Since in our formalism $\mathbb{A}$ is supposed to be a commutative group, we extend multi-sets to "extended multi-sets", which have coefficients in $\mathbb{Z}$.

Definition 4.7 An *extended multi-set* m, over a non-empty set S, is a function $m : S \rightarrow \mathbb{Z}$ into the set of integers $\mathbb{Z}$. S_{MS} is the set of all extended multi-sets over S. The operation + and the relation $\leq$ are defined as in definition 2.1. Moreover $m_1 - m_2 := \sum_{s \in S} (m_1(s) - m_2(s))`s$ is defined without the precondition $m_2 \leq m_1$.

Example 4.1 (cont.): Applying the preceding theorem to the invariant equation

(B1) $\forall M \in R(N): M(loc) + pr_1(M(read)) + pr_1(M(write)) = A$

of the smCPN of example 4.1, the weight function $W_P : \mathcal{C}_P \rightarrow \mathbb{A}$ is selected as follows : $\mathbb{A}$ is the set of all extended multisets A_{MS} over A.

$W_{loc}(m) := m$, $W_{read}(m) = W_{write}(m) := pr_1(m)$ and $W_{files}(m) := \emptyset$.

Then $W_P(M_o) = A$ is satisfied and the equation $W_P[I_{M+}(-,t)(b)] = W_P[I_{M-}(-,t)(b)]$ is calculated for (e.g.) $t = t_3$ and $b = (a,f)$ as follows. (The incidence matrices are not given here. They are constructed in the same way as in section 2.)

left-hand side : $W_{loc}[I_{M+}(loc,t_3)(a,f)] + W_{read}[I_{M+}(read,t_3)(a,f)] + W_{write}[I_{M+}(write,t_3)(a,f)] + W_{files}[I_{M+}(files,t_3)(a,f)] = pr_1(read(fl2)) + \emptyset + pr_1((a,f)) + \emptyset = pr_1(read(fl2)) + 1`a$.

right-hand side : $W_{loc}[I_{M-}(loc,t_3)(a,f)] + W_{read}[I_{M-}(read,t_3)(a,f)] + W_{write}[I_{M-}(write,t_3)(a,f)] + W_{files}[I_{M-}(files,t_3)(a,f)] = 1`a + pr_1(read(fl2)) + \emptyset + \emptyset$.

In this example we observe that by symbolic evaluation the equation

$$W_P[I_{M+}(-,t)(b)] = W_P[I_{M-}(-,t)(b)]$$

is verified without explicitly referring to all reachable markings M. Repeating the same procedure for the inequality (B3) in

$$W_P[I_{M+}(-,t_3)(b)] = W_P[I_{M-}(-,t_3)(b)]$$

we obtain the *left-hand side* 5`f and the *right-hand side* $pr_2(read(fl2))$, which is different. The reason for this defect is some lack of information in the left-hand side of (B3). As is known from the verification of Floyd-invariants, such difficulties are eliminated by the introduction of "auxiliary variables". These auxiliary variables are introduced for the verification process and do not affect the working of the program. Hence, they can be removed afterwards without influencing the validity of the correctness proof.

In the remainder of this section we will extend the net $\mathcal{RpW}$ to nets $\mathcal{RpW}_1$ and $\mathcal{RpW}_2$ in such a way that theorem 4.6 can be applied to verify an equation (B3b) (also to be constructed). Then we will prove that (B3b) holds for $\mathcal{RpW}_2$ if and only if (B3) holds for $\mathcal{RpW}$. Then all of the equations (B1), (B2) and (B3) will be verified by theorem 4.6.

To start with, we introduce a place "co_read", which contains for every file f that is not in the place "write" but is read by $0 \le n \le 5$ tasks (in the place "read"), 5 - n instances of f. The auxiliary place allows to transform (B3) into an equality :

(B3a) $\quad \forall M \in R(N): \; pr_2(M(read)) + M(co\_read) + 5\, pr_2(M(write)) = 5F$

The corresponding smCPN $\mathcal{RpW}_1$ is shown in fig. 4.4. "co_read" behaves like an ordinary complementary place, when "write" is empty :

$$\forall M \in R(N): \quad M(write) = \emptyset \quad \Rightarrow \quad pr_2(M(read)) + M(co\_read) = 5F$$

With an occurrence of t, however, all files y are removed from "co_read". Since (B3a) implies (B3) it is sufficient to prove (B3a).

To use theorem 4.6, we introduce a place "deferred", where all instances of files f in "read" are collected while a task is writing on f. In the smCPN $\mathcal{RpW}_2$ of fig 4.5 all instances of y in the places "read" and "co_read" are moved to "deferred", when t3 is clearing them. The smCPN $\mathcal{RpW}_2$ satisfies the following invariant equation :

(B3b) $\quad \forall M \in R(N): \; pr_2(M(read)) + M(co\_read) + M(deferred) = 5F$

This is easily proved with theorem 4.6 by the corresponding incidence functions. (For the definition of deferred() see Def. 4.4.)

Our goal now is to deduce (B3a) for $\mathcal{RpW}_1$. To do this we use the fact that in

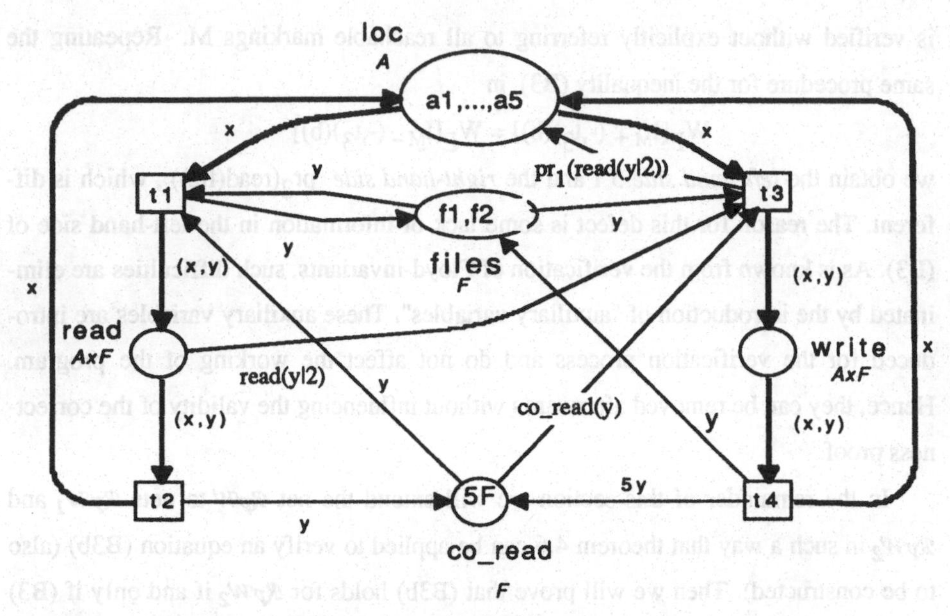

Fig. 4.4: $\mathcal{R}p\mathcal{W}_1$: readers and preemptive writers with place co_read

$\mathcal{R}p\mathcal{W}_2$ an occurrence of t3 in the colour (a,f) is clearing the places "read" and "co_read" with respect to "f", hence after the occurrence of t3:

(C1) $\quad M(read(f|2)) = \emptyset \quad$ and $\quad M(co\_read(f)) = \emptyset$

This is holding up to the next occurrence of t4, which is the only possibility to move token f. Therefore from (B3b) we conclude $M[deferred(f)] = 5`f$ for such an interval, where we also have $pr_2(M[write(f)]) = f$, hence for each $f \in F$:

(C2) $\quad M[deferred(f)] = 5`f \quad \Leftrightarrow \quad pr_2(M[write(f)]) = f \quad$ or

(C3) $\quad M[deferred(f)] = 5'pr_2(M(write))$

Since (C3) is true in $\mathcal{R}p\mathcal{W}_2$ we can omit the place "deferred" and replace the term M(deferred) in (B3b) by $5`pr_2(M(write))$. By this operation we have reconstructed the smCPN $\mathcal{R}p\mathcal{W}_1$ and the invariant equation (B3a). As mentioned above this implies (B3) for $\mathcal{R}p\mathcal{W}$, which ends the proof of (B3).

5. Conclusion and Ongoing Research

The subject of this paper has its origin in a student's question. We believe the work allows a better understanding of how place-invariants of coloured nets relate to

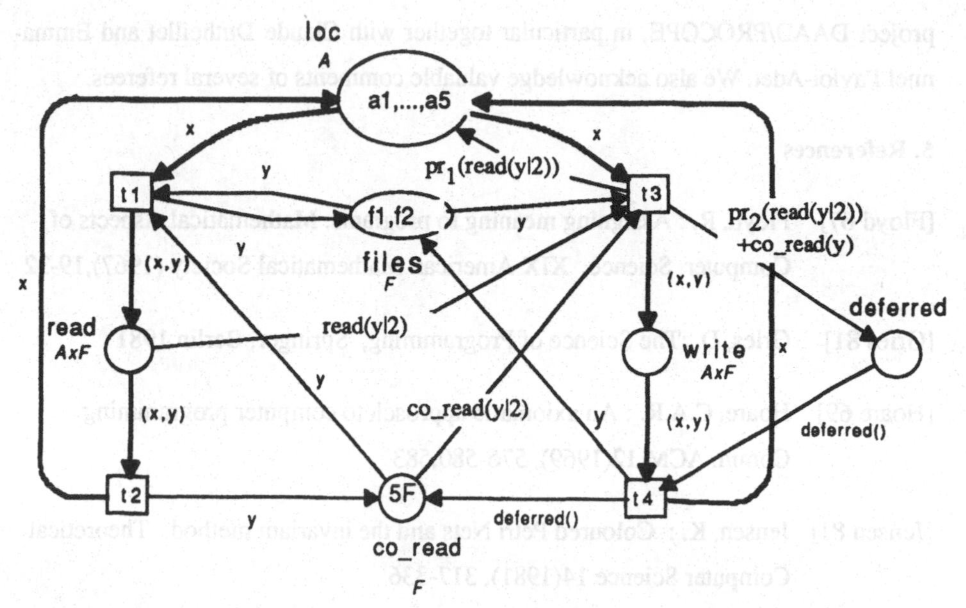

Fig. 4.5: $\mathcal{RpW}_2$: readers and preemptive writers with places "co_read" and "deferred"

Floyd-invariants of ALGOL-like programs. Since the extended form of invariant equations, introduced in this paper, allows to relate internal values of tokens, we have done a step towards an invariance calculus for coloured nets with complex objects as tokens.

The extension of coloured nets, investigated in section 4, is very useful in many system modelling applications, such as preemptive scheduling and systems reset, but preserves the important property of place-invariant verification from the incidence matrices.

The main theorem of this paper (theorem 3.3) has been extended to hold for a broader class of equations. The extended theorem [Valk 93] proves weight functions, where non-linear parts may depend on more than one place. Hence also equations can be verified expressing non-linear relations between distinct variables. By this extension other types of applications are included, for instance a coloured net modelling a distributed algorithm for computing the greatest common divisor of a set of natural numbers.

Acknowledgements

Part of this work has been done with Lab. MASI, Univ. Paris 6, in the common project DAAD/PROCOPE, in particular together with Claude Dutheillet and Emmanuel Paviot-Adet. We also acknowledge valuable comments of several referees.

5. References

[Floyd 67] Floyd, R. : Assigning meaning to programs. Mathematical Aspects of Computer Science, XIX American Mathematical Society (1967),19-32

[Gries 81] Gries, D. :The Science of Programming, Springer, Berlin 1981

[Hoare 69] Hoare, C.A.R. : An axiomatic approach to computer programming, Comm. ACM 12(1969), 576-580,583

[Jensen 81] Jensen, K. : Coloured Petri Nets and the invariant method. Theoretical Computer Science 14(1981), 317-336

[Jensen 87] Jensen, K. : Coloured Petri Nets, in Petri Nets: Central Models and Their Properties, in Brauer, W. et al. (Eds), Lecture Notes in Computer Science No 254, Springer, Berlin 1987, 248-299

[Valk 78] Valk, R. :Selfmodifying Nets, a Natural Extension of Petri Nets, Lecture Notes in Computer Science No 62, Springer, Berlin 1978, 464-476

[Valk 83] Valk, R. : Facts in Place/Transition Nets with Unrestricted Capacities, Annales Univ. Scie. Budapestensis R. Eötvös, Sect.Comput. Tom. IV, 1983, 97-105

[Valk 93] Valk, R. : Extending S-invariants for Coloured and Selfmodifying Nets, Techn. Report,Univ. Hamburg, Dep. of Computer Science, 1993

[Vautherin 85] Vautherin, J. : Non-linear invariants for safe coloured nets and appli cation to the proof of parallel programs, Proc. 6th European Workshop on Applications and Theory of Petri Nets, Espoo, Finland, 1985

Interval Timed Coloured Petri Nets and their Analysis

W.M.P. van der Aalst

Eindhoven University of Technology
Dept. of Mathematics and Computing Science

Abstract. Practical experiences show that only *timed* and *coloured* Petri nets are capable of modelling large and complex real-time systems. This is the reason we present the *Interval Timed Coloured Petri Net* (ITCPN) model. An interval timed coloured Petri net is a coloured Petri net extended with time; time is in tokens and transitions determine a delay for each produced token. This delay is specified by an upper and lower bound, i.e. an interval. The ITCPN model allows the modelling of the dynamic behaviour of large and complex systems, without losing the possibility of formal analysis. In addition to the existing analysis techniques for coloured Petri nets, we propose a new analysis method to analyse the temporal behaviour of the net. This method constructs a reduced reachability graph and exploits the fact that delays are described by an interval.

1 Introduction

Petri nets have been widely used for the modelling and analysis of concurrent systems (Reisig [25]). There are several factors which contribute to their success: the graphical nature, the ability to model parallel and distributed processes in a natural manner, the simplicity of the model and the firm mathematical foundation. Nevertheless, the basic Petri net model is not suitable for the modelling of many systems encountered in logistics, production, communication, flexible manufacturing and information processing. Petri nets describing real systems tend to be complex and extremely large. Sometimes it is even impossible to model the behaviour of the system accurately. To solve these problems many authors propose extensions of the basic Petri net model.

Several authors have extended the basic Petri net model with *coloured* or *typed tokens* ([10], [15], [16], [12], [13]). In these models tokens have a value, often referred to as 'colour'. There are several reasons for such an extension. One of these reasons is the fact that (uncoloured) Petri nets tend to become too large to handle. Another reason is the fact that tokens often represent objects or resources in the modelled system. As such, these objects may have attributes, which are not easily represented by a simple Petri net token. These 'coloured' Petri nets allow the modeller to make much more succinct and manageable descriptions, therefore they are called *high-level nets*.

Other authors have proposed a Petri net model with explicit quantitative time (e.g. [28], [24], [20], [19], [12], [26]). We call these models *timed Petri net* models.

In our opinion, only timed *and* coloured Petri nets are suitable for the modelling of large and complex real-time systems. Although there seems to be a consensus of opinion on this matter, only a few timed coloured Petri net models have been proposed in literature (e.g. Van Hee et al. [12, 11], Morasca [22]). Moreover, even fewer methods have been developed for the analysis of the temporal behaviour of these nets. This is one of the reasons we propose the *Interval Timed Coloured Petri Net* (ITCPN) model and an analysis method, called *MTSRT*, based on this model.
The ITCPN model uses a rather new timing mechanism where time is associated with tokens. This timing concept has been adopted from Van Hee et al. [12]. In the ITCPN model we attach a *timestamp* to every token. This timestamp indicates the time a token becomes available. Associating time with tokens seems to be the natural choice for high-level Petri nets, since the colour is also associated with tokens. The *enabling time* of a transition is the maximum timestamp of the tokens to be consumed. Transitions are *eager* to fire (i.e. they fire as soon as possible), therefore the transition with the smallest enabling time will fire first. Firing is an atomic action, thereby producing tokens with a timestamp of at least the firing time. The difference between the firing time and the timestamp of such a produced token is called the *firing delay*. The (firing) delay of a produced token is specified by an *upper* and *lower bound*, i.e. an *interval*.

Instead of using 'interval timing', we could have used a Petri net model with fixed delays or stochastic delays.
Petri nets with fixed (deterministic) delays have been proposed in [28], [24], [26] and [12]. They allow for simple analysis methods but are not very expressive, because in a real system the durations of most activities are variable.
One way to model this variability, is to assume certain delay distributions, i.e. to use a timed Petri net model with delays described by probability distributions. These nets are called *stochastic Petri nets* ([9], [19], [18]). Analysis of stochastic Petri nets is possible (in theory), since the reachability graph can be regarded, under certain conditions, as a Markov chain or a semi-Markov process. However, these conditions are severe: all firing delays have to be sampled from an exponential distribution or the topology of the net has to be of a special form (Ajmone Marsan et al. [18]). Since there are no general applicable analysis methods, several authors resorted to using simulation to study the behaviour of the net (see section 4).
To avoid these problems, we propose delays described by an *interval* specifying an upper and lower bound for the duration of the corresponding activity. On the one hand, interval delays allow for the modelling of variable delays, on the other hand, it is not necessary to determine some artificial delay distribution (as opposed to stochastic delays). Instead, we have to specify bounds. These bounds can be used to verify time constraints. This is very important when modelling time-critical systems, i.e. *real-time* systems with 'hard' deadlines. These hard (real-time) deadlines have to be met for a safe operation of the system. An acceptable behaviour of the system depends not only on the logical correctness of the results, but also on the time at which the results are produced. Examples of such systems are: real-time computer systems, process controllers, communication systems, flexible manufacturing systems and just-in-time manufacturing systems.

To our knowledge, only one other model has been presented in literature which also uses delays specified by an interval. This model was presented by Merlin in [20, 21]. In this model the enabling time of a transition is specified by a minimal and a maximal time. Another difference with our model is the fact that Merlin's model is not a high-level Petri net model because of the absence of typed (coloured) tokens. Compared to our model, Merlin's model has a rather complex formal semantics, which was presented in [7] by Berthomieu and Diaz. This is caused by a redundant state space (marking and enabled transitions are represented separately) and the fact that they use a relative time scale and allow for multiple enabledness of transitions. An additional advantage of our approach is the fact that our semantics closely correspond to the intuitive interpretation of the dynamical behaviour of a timed Petri net.

The main purpose of this paper is to present a high-level Petri net model extended with interval timing which allows for new methods of analysis. In section 2 we introduce the ITCPN model. The formal definition and semantics are given in section 3. Section 4 deals with the analysis of interval timed coloured Petri nets. In this section, we introduce a new and powerful analysis method.

2 Interval Timed Coloured Petri Nets

We use an example to introduce the notion of interval timed coloured Petri nets. Figure 1 shows an ITCPN composed of four **places** ($p_{in}, p_{busy}, p_{free}$ and p_{out}) and two **transitions** (t_1 and t_2). At any moment, a place contains zero or more **tokens**, drawn as black dots. In the ITCPN model, a token has three attributes: a position, a value and a timestamp, i.e. we can use the tuple $\langle\langle p, v\rangle, x\rangle$ to denote a token in place p with value v and timestamp x. The value of a token is often referred to as the **token colour**. Each place has a **colour set** attached to it which specifies the set of allowed values, i.e. each token residing in place p must have a colour (value) which is a member of the colour set of p.
The ITCPN shown in figure 1 represents a jobshop, jobs arrive via place p_{in} and leave the system via place p_{out}. The jobshop is composed of a number of machines. Each machine is represented by a token which is either in place p_{free} or in place p_{busy}. There are three colour sets $\mathcal{M} = \{M1, M2, ..\}$ and $\mathcal{J} = \{J1, J2, ..\}$ and $\mathcal{M} \times \mathcal{J}$. Colour set $\mathcal{J}$ (job types) is attached to place p_{in} and place p_{out}, colour set $\mathcal{M}$ (machine types) is attached to place p_{free}. Colour set $\mathcal{M} \times \mathcal{J}$ is attached to place p_{busy}.
Places and transitions are interconnected by **arcs**. Each arc connects a place and a transition in precisely one direction. Transition t_1 has two input places (p_{in} and p_{free}) and one output place (p_{busy}). Transition t_2 has one input place (p_{busy}) and two output places (p_{free} and p_{out}).
Places are passive components, while transitions are the active components. Transitions cause state changes. A transition is called **enabled** if there are 'enough' tokens on each of its input places. In other words, a transition is enabled if all input places contain (at least) the specified number of tokens (further details will be given later). An enabled transition may **occur** (**fire**) at time x if all the tokens to be consumed have a timestamp not later than time x. The **enabling time** of a transition is the

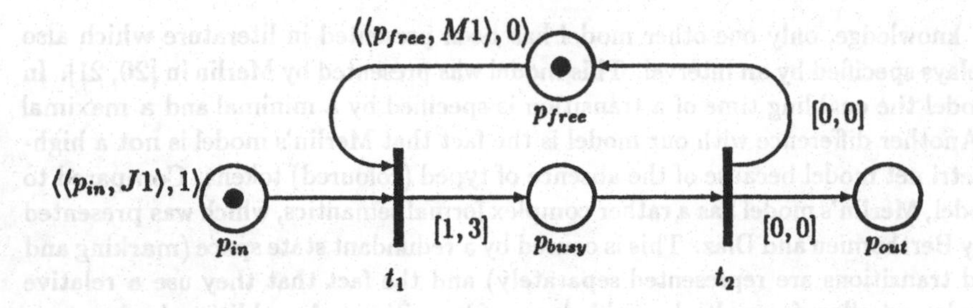

Figure 1: An interval timed coloured Petri net

maximum timestamp of the tokens to be consumed. Because transitions are eager to fire, a transition with the smallest enabling time will fire first.
Firing a transition means consuming tokens from the input places and producing tokens on the output places. If, at any time, more than one transition is enabled, then any of the several enabled transitions may be 'the next' to fire. This leads to a non-deterministic choice if several transitions have the same enabling time.
Firing is an atomic action, thereby producing tokens with a timestamp of at least the firing time. The difference between the firing time and the timestamp of such a produced token is called the firing delay. This delay is specified by an interval, i.e. only delays between a given upper bound and a given lower bound are allowed. In other words, the delay of a token is 'sampled' from the corresponding delay interval. Note that the term 'sampled' may be confusing, because the modeller does not specify a probability distribution, merely an upper and lower bound.
Moreover, it is possible that the modeller specifies a delay interval which is too wide, because of a lack of detailed information. In this case, the actual delays (in the real system) only range over a part of the delay interval.
The number of tokens produced by the firing of a transition may depend upon the values of the consumed tokens. Moreover, the values and delays of the produced tokens may also depend upon the values of the consumed tokens. The relation between the *multi-set* of consumed tokens and the *multi-set* of produced tokens is described by the transition function. Function $F(t_1)$ specifies transition t_1 in the net shown in figure 1:

$$dom(F(t_1)) = \{\langle p_{in}, j\rangle + \langle p_{free}, m\rangle \mid j \in \mathcal{J} \text{ and } m \in \mathcal{M}\}$$

For $j \in \mathcal{J}$ and $m \in \mathcal{M}$, we have:[1]

$$F(t_1)(\langle p_{in}, j\rangle + \langle p_{free}, m\rangle) = \langle\langle p_{busy}, \langle m, j\rangle\rangle, [1,3]\rangle$$

The domain of $F(t_1)$ describes the condition on which transition t_1 is enabled, i.e. t_1 is enabled if there is (at least) one token in place p_{in} and one token in p_{free}. This means that transition t_1 may occur if there is a job waiting and one of the machines is free. Note that, in this case, the enabling of a transition does not depend upon the values of the tokens consumed. The enabling time of transition t_1 depends upon the

[1] Note that $\langle p_{in}, j\rangle + \langle p_{free}, m\rangle$ and $\langle\langle p_{busy}, \langle m, j\rangle\rangle, [1,3]\rangle$ are multi-sets, see section 3.1.

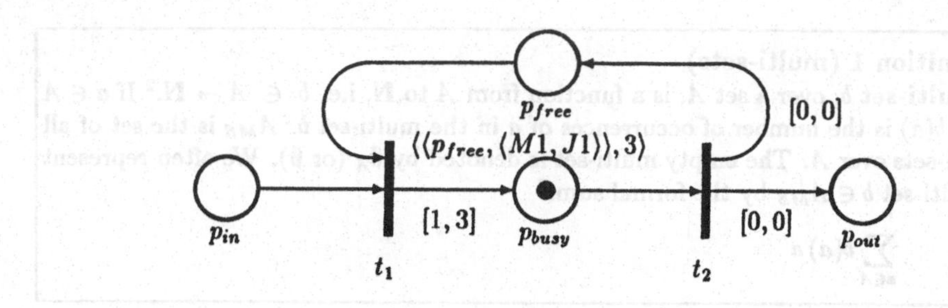

Figure 2: Transition t_1 has fired

timestamps of the tokens to be consumed. If t_1 occurs, it consumes one token from place p_{in} and one token from p_{free} and it produces one token for place p_{busy}. The colour of the produced token is a pair $\langle m, j\rangle$, where m represents the machine and j represents the job. The delay of this token is an arbitrary value between 1 and 3, e.g. 2, 2.55 or 4/3. The situation shown in figure 2 is the result of firing t_1 in the state shown in figure 1. In this case the delay of the token produced for p_{busy} was equal to 2.
Transition t_2 is specified as follows:

$$dom(F(t_2)) = \{\langle p_{busy}, \langle m, j\rangle\rangle \mid j \in \mathcal{J} \text{ and } m \in \mathcal{M}\}$$

For $j \in \mathcal{J}$ and $m \in \mathcal{M}$, we have:

$$F(t_2)(\langle p_{busy}, \langle m, j\rangle\rangle) = \langle\langle p_{free}, m\rangle, [0,0]\rangle + \langle\langle p_{out}, j\rangle, [0,0]\rangle$$

Transition t_2 represents the completion of a job. If t_2 occurs, it consumes one token from place p_{busy} and it produces two tokens (one for p_{free} and one for p_{out}) both with a delay equal to zero. If t_2 occurs in the state shown in figure 2, then the resulting state contains two tokens: $\langle\langle p_{out}, J1\rangle, 3\rangle$ and $\langle\langle p_{free}, M1\rangle, 3\rangle$.

3 Formal Definition

In this section we define interval timed coloured Petri nets in mathematical terms, such as functions, multi-sets and relations.

3.1 Multi-sets

A *multi-set*, like a set, is a collection of elements over the same subset of some universe. However, unlike a set, a multi-set allows multiple occurrences of the same element. Another word for multi-set is *bag*. Bag theory is a natural extension of set theory (Jensen [16]).

Definition 1 (multi-sets)
A multi-set b, over a set A, is a function from A to $\mathbb{N}$, i.e. $b \in A \rightarrow \mathbb{N}$.[2] If $a \in A$ then $b(a)$ is the number of occurrences of a in the multi-set b. A_{MS} is the set of all multi-sets over A. The empty multi-set is denoted by $\emptyset_A$ (or $\emptyset$). We often represent a multi-set $b \in A_{MS}$ by the formal sum:[3]

$$\sum_{a \in A} b(a)\, a$$

Consider for example the set $A = \{a, b, c, ..\}$, the multi-sets $3a$, $a + b + c + d$, $1a + 2b + 3c + 4d$ and $\emptyset_A$ are members of A_{MS}.

Definition 2
We now introduce some operations on multi-sets. Most of the set operators can be extended to multi-sets in a rather straightforward way. Suppose A a set, $b_1, b_2 \in A_{MS}$ and $q \in A$:

$q \in b_1$ iff $b_1(q) \geq 1$	(membership)
$b_1 \leq b_2$ iff $\forall_{a \in A}\, b_1(a) \leq b_2(a)$	(inclusion)
$b_1 = b_2$ iff $b_1 \leq b_2$ and $b_2 \leq b_1$	(equality)
$b_1 + b_2 = \sum_{a \in A} (b_1(a) + b_2(a))\, a$	(summation)
$b_1 - b_2 = \sum_{a \in A} ((b_1(a) - b_2(a)) \max 0)\, a$	(subtraction)
$\# b_1 = \sum_{a \in A} b_1(a)$	(cardinality of a finite multi-set)

See Jensen [16, 17] for more details.

3.2 Definition of Interval Timed Coloured Petri Nets

The ITCPN model presented in this paper is analogous to the model described in [3]. However, in this paper we give a definition which is closer to the definition of Coloured Petri Nets (CPN), see Jensen [15, 16, 17].
Nearly all timed Petri net models use a continuous time domain, so do we.

Definition 3
TS is the time set, $TS = \{x \in \mathbb{R} \mid x \geq 0\}$, i.e. the set of all non-negative reals.
$INT = \{[y, z] \in TS \times TS \mid y \leq z\}$, represents the set of all closed intervals.
If $x \in TS$ and $[y, z] \in INT$, then $x \in [y, z]$ iff $y \leq x \leq z$.

We define an interval timed coloured Petri nets as follows:

[2] $\mathbb{N} = \{0, 1, 2, ..\}$
[3] This notation has been adopted from Jensen [16].

Definition 4 (ITCPN)
An Interval Timed Coloured Petri Net is a five tuple ITCPN = (Σ, P, T, C, F) satisfying the following requirements:

(i) Σ is a finite set of types, called colour sets.

(ii) P is a finite set of places.

(iii) T is a finite set of transitions.

(iv) C is a colour function. It is defined from P into Σ, i.e. $C \in P \to \Sigma$.

(v) $CT = \{\langle p, v\rangle \mid p \in P \wedge v \in C(p)\}$ is the set of all possible coloured tokens.

(vi) F is the transition function. It is defined from T into functions. If $t \in T$, then:[4]

$$F(t) \in CT_{MS} \not\rightarrow (CT \times INT)_{MS}$$

(i) Σ is a set of types. Each type is a set of colours which may be attached to one of the places.
(ii) and (iii) The places and transitions are described by two disjoint sets, i.e. $P \cap T = \emptyset$.
(iv) Each place $p \in P$ has a set of allowed colours attached to it and this means that a token residing in p must have a value v which is an element of this set, i.e. $v \in C(p)$.
(v) CT is the set of all coloured tokens, i.e. all pairs $\langle p, v\rangle$ where p is the position of the token and v is the value of the token.
(vi) The transition function specifies each transition in the ITCPN. For a transition t, $F(t)$ specifies the relation between the multi-set of consumed tokens and the multi-set of produced tokens. The domain of $F(t)$ describes the condition on which transition t is enabled. Note that the produced tokens have a delay specified by an interval. In this paper, we require that both the multi-set of consumed tokens and the multi-set of produced tokens contain finitely many elements.
Apart from the interval timing and a transition function instead of incidence functions, this definition resembles the definition of a CP-matrix (see Jensen [15, 17]).

3.3 Dynamic Behaviour of Interval Timed Coloured Petri Nets

The five tuple (Σ, P, T, C, F) specifies the static structure of an ITCPN. In the remainder of this section we define the behaviour of an interval timed coloured Petri net, i.e. the semantics of the ITCPN model.

[4] $A \not\rightarrow B$ denotes the set of all partial functions from A to B.

Definition 5
A state is defined as a multi-set of coloured tokens each bearing a timestamp. S is the state space, i.e. the set of all possible states:

$$S = (CT \times TS)_{MS}$$

The marking of an ITCPN in state $s \in S$ is the 'untimed' token distribution: $M(s) \in CT_{MS}$ and

$$M(s) = \sum_{\langle\langle p,v\rangle,x\rangle \in CT \times TS} s(\langle\langle p,v\rangle,x\rangle)\, \langle p,v\rangle$$

A state of the ITCPN is a multi-set of coloured tokens bearing a timestamp, i.e. a multi-set of tuples $\langle\langle p,v\rangle,x\rangle$ ($p \in P$, $v \in C(p)$ and $x \in TS$). The state shown in figure 1 is $\langle\langle p_{in}, J1\rangle, 1\rangle + \langle\langle p_{free}, M1\rangle, 0\rangle$, that is a state with one token in p_{in} with value $J1$ and one token in p_{free} with value $M1$. The token in p_{in} bears timestamp 1, the token in p_{free} bears timestamp 0.

Definition 6
An event is a triple $\langle t, b_{in}, b_{out}\rangle$, which represents the possible firing of transition t while removing the tokens specified by the multi-set b_{in} and adding the tokens specified by the multi-set b_{out}. E is the event set:

$$E = T \times (CT \times TS)_{MS} \times (CT \times TS)_{MS}$$

An event $e = \langle t, b_{in}, b_{out}\rangle$ represents the firing of t while consuming the tokens specified by b_{in} and producing the tokens specified by b_{out}. If $\langle\langle p,v\rangle,x\rangle \in b_{in}$, then e consumes a token from p with value v and timestamp x. If $\langle\langle p',v'\rangle,x'\rangle \in b_{out}$, then e produces a token for p' with value v' and *delay* x'. Note that x' is relative to the firing time and x' is a member of one of the delay intervals specified by $F(t)(b_{in})$. To select arbitrary members of these delay intervals, we need the specialization concept.

Definition 7 (Specialization)
To relate multi-sets of tokens bearing timestamps with multi-sets of tokens bearing (time) intervals, we define the specialization relation, $\lhd \subseteq (CT \times TS)_{MS} \times (CT \times INT)_{MS}$. For $b \in (CT \times TS)_{MS}$ and $\bar{b} \in (CT \times INT)_{MS}$, $b \lhd \bar{b}$ if and only if each token in b corresponds to exactly one token in $\bar{b}$, such that they are in the same place, have the same value and the timestamp of the token in b is in the (time) interval of the token in $\bar{b}$.
More formally: $b \lhd \bar{b}$ if and only if ($b = \emptyset$ and $\bar{b} = \emptyset$) or

$$\exists_{\langle\langle p,v\rangle,x\rangle \in b}\ \exists_{\langle\langle p,v\rangle,[y,z]\rangle \in \bar{b}}\ (x \in [y,z]) \text{ and } (b - \langle\langle p,v\rangle,x\rangle) \lhd (\bar{b} - \langle\langle p,v\rangle,[y,z]\rangle)$$

Consider for example:

$\emptyset \lhd \emptyset$
$\langle\langle p_{in}, J1\rangle, 1\rangle \lhd \langle\langle p_{in}, J1\rangle, [0.5, 1.5]\rangle$
$\langle\langle p_{in}, J1\rangle, 1\rangle + 2\,\langle\langle p_{free}, M1\rangle, 0\rangle \lhd \langle\langle p_{in}, J1\rangle, [0.5, 1.5]\rangle + 2\,\langle\langle p_{free}, M1\rangle, [0, 1]\rangle$

Note that $\langle\langle p_{in}, J1\rangle, 1\rangle$ is not a specialization of $\langle\langle p_{in}, J2\rangle, [0.5, 1.5]\rangle$, because the

values of the two tokens differ.
If $b \lhd \bar{b}$, then there exists a bijection between the tokens in b and the tokens in $\bar{b}$ such that each token in b corresponds to exactly one token in $\bar{b}$ which is in the same place, has the same value and a 'matching' time-interval.

Definition 8
An event $\langle t, b_{in}, b_{out} \rangle \in E$ is **enabled** in state $s \in S$ iff:

(i) $b_{in} \leq s$

(ii) $M(b_{in}) \in dom(F(t))$

(iii) $b_{out} \lhd F(t)(M(b_{in}))$

An event is enabled iff:
(i) The tokens to be consumed are present in the current state.
(ii) A transition is enabled if there are 'enough' tokens on each of its input places, this is specified by the domain of $F(t)$. Note that the enabling may depend upon the values of the tokens to be consumed, but not on their timestamps.
(iii) The number and values of the tokens to be produced are determined by the multi-set $F(t)(M(b_{in}))$. This multi-set also specifies upper and lower bounds for the delays of these tokens.

Definition 9
The **enabling time** of an event $\langle t, b_{in}, b_{out} \rangle \in E$ is the maximum of all the timestamps of the tokens consumed, i.e.

$$ET(\langle t, b_{in}, b_{out} \rangle) = \max_{\langle (p,v),x \rangle \in b_{in}} x$$

An enabled event is **time enabled** iff no other enabled events have a smaller enabling time.

If an event is time enabled, it may occur. In fact, a transition fires as soon as possible (transitions are 'eager'). Although the time domain is continuous ($TS = \{x \in \mathbb{R} \mid x \geq 0\}$), time progresses discontinuously.

Definition 10
The **model time** of a state $s \in S$ is the minimum of all enabling times, i.e.

$$MT(s) = \min\{ET(e) \mid e \in E \text{ and } e \text{ is enabled in state } s\}$$

The model time only changes if something happens. Note that an enabled event e is time enabled in state s iff $ET(e) = MT(s)$.
If $\langle t, b_{in}, b_{out} \rangle \in E$ an enabled event, then a timestamp in b_{out} represents the delay of the token instead of an absolute timestamp. Therefore we need a function to 'scale' timestamps.

Definition 11
Function $SC \in ((CT \times TS)_{MS} \times TS) \rightarrow (CT \times TS)_{MS}$ scales the timestamps in a multi-set of timed coloured tokens. For $b \in (CT \times TS)_{MS}$ and $y \in TS$, we have:

$$SC(b,y) = \sum_{\langle\langle p,v\rangle,x\rangle \in CT \times TS} b(\langle\langle p,v\rangle,x\rangle)\,\langle\langle p,v\rangle,x+y\rangle$$

A time enabled event e in state s may occur at time $ET(e) = MT(s)$.

Definition 12
When an enabled event $\langle t, b_{in}, b_{out}\rangle$ is time enabled in state s_1, it may occur, i.e. transition t fires while removing the tokens specified by b_{in} and adding the tokens specified by b_{out}.
If $\langle t, b_{in}, b_{out}\rangle$ occurs in state s_1, then the net changes into the state s_2, defined by:

$$s_2 = (s_1 - b_{in}) + SC(b_{out}, ET(\langle t, b_{in}, b_{out}\rangle))$$

State s_2 is said to be directly reachable from s_1 by the occurrence of event $e = \langle t, b_{in}, b_{out}\rangle$, this is also denoted by:

$$s_1 \xrightarrow{e} s_2$$

Moreover, $s_1 \longrightarrow s_2$ means that there exists a time enabled event e such that $s_1 \xrightarrow{e} s_2$. Transitions fire as soon as possible, i.e. if an event occurs, then it occurs at its enabling time. Note that the state shown in figure 2 is directly reachable from the state shown in figure 1.

Definition 13
A firing sequence is a sequence of states and events:

$$s_1 \xrightarrow{e_1} s_2 \xrightarrow{e_2} s_3 \xrightarrow{e_3} s_4 \xrightarrow{e_4} ..$$

State s_n is reachable from s_1 iff there exists a firing sequence of finite length starting in s_1 and ending in s_n:

$$s_1 \xrightarrow{e_1} s_2 \xrightarrow{e_2} s_3 \xrightarrow{e_3} .. \xrightarrow{e_{n-1}} s_n$$

An important property of the ITCPN model is the monotonicity of time, i.e. time can only move forward.

Theorem 1 (Monotonicity)
If $s_1, s_2 \in S$ and $s_1 \longrightarrow s_2$, then $MT(s_1) \leq MT(s_2)$.

Proof
Let $s_1, s_2 \in S$ and $e = \langle t, b_{in}, b_{out}\rangle \in E$ such that $s_1 \xrightarrow{e} s_2$.
$ET(e) = MT(s_1)$ and $s_2 = (s_1 - b_{in}) + SC(b_{out}, ET(e))$.
Deleting tokens (b_{in}) may disable events, but it will never enable new ones.
Adding tokens ($SC(b_{out}, ET(e))$) may enable new events. However, in this case these events have an enabling time of at least $ET(e)$.
Since $MT(s_2)$ is the minimum of all enabling times, we have $MT(s_2) \geq MT(s_1)$. □

In [3], we prove a number of other properties of the ITCPN model (e.g. progressiveness, etc.). In that monograph we also show how to model other timed Petri nets (e.g. Merlin's timed Petri nets) in terms of our model.

4 An Analysis Method

In this section we present an approach to verify certain properties and to calculate bounds for all sorts of performance measures. This approach is based on an analysis method, called *Modified Transition System Reduction Technique* (MTSRT), which generates a *reduced reachability graph*. The MTSRT method can be applied to arbitrary ITCPNs.
Existing techniques which can be used to analyse the dynamic behaviour of timed and coloured Petri nets, may be subdivided into three classes: simulation, reachability analysis and Markovian analysis.
***Simulation* is a technique to analyse a system by conducting controlled experiments. Because simulation does not require difficult mathematical techniques, it is easy to understand for people with a non-technical background. Simulation is also a very powerful analysis technique, since it does not set additional restraints. However, sometimes simulation is expensive in terms of the computer time necessary to obtain reliable results. Another drawback is the fact that (in general) it is not possible to use simulation to *prove* that the system has the desired set of properties.**
Recent developments in computer technology stimulate the use of simulation for the analysis of timed coloured Petri nets. The increased processing power allows for the simulation of large nets. Modern graphical screens are fast and have a high resolution. Therefore, it is possible to visualize a simulation graphically (i.e. animation).
***Reachability analysis* is a technique which constructs a reachability graph, sometimes referred to as reachability tree or occurrence graph (cf. Jensen [15, 17]). Such a reachability graph contains a node for each possible state and an arc for each possible state change. Reachability analysis is a very powerful method in the sense that it can be used to prove all kinds of properties. Another advantage is the fact that it does not set additional restraints. Obviously, the reachability graph needed to prove these properties may, even for small nets, become very large (and often infinite). If we want to inspect the reachability graph by means of a computer, we have to solve this problem. This is the reason several authors developed reduction techniques (Hubner et al. [14] and Valmari [27]). Unfortunately, it is not known how to apply these techniques to timed coloured Petri nets.**
For timed coloured Petri nets with certain types of stochastic delays it is possible to translate the net into a *continuous time Markov chain*. This Markov chain can be used to calculate performance measures like the average number of tokens in a place and the average firing rate of a transition.
If all the delays are sampled from a negative exponential probability distribution, then it is easy to translate the timed coloured Petri net into a continuous time Markov chain. Several authors attempted to increase the modelling power by allowing other kinds of delays, for example mixed deterministic and negative exponential distributed delays, and phase-distributed delays (see Ajmone Marsan et al. [18]). Nearly all stochastic Petri net models (and related analysis techniques) do not allow for coloured tokens, because the increased modelling power is offset by computational difficulties.

This is the reason stochastic high-level Petri nets are often used in a simulation context only.

Besides the aforementioned techniques to analyse the behaviour of timed coloured Petri nets, there are several analysis techniques for Petri nets without 'colour' or explicit 'time'.

An interesting way to analyse a coloured Petri net is to calculate (or verify) *place and transition invariants*. Place and transition invariants can be used to prove properties of the modelled system. Intuitively, a place invariant assigns a weight to each token such that the weighted sum of all tokens in the net remains constant during the execution of any firing sequence. By calculating these place invariants we find a set of equations which characterizes all reachable states. Transition invariants are the duals of place invariants and the main objective of calculating transition invariants is to find firing sequences with no 'effects'.
Note that we can calculate these invariants for *timed* coloured Petri nets (e.g. an ITCPN). However, in this case, we do not really use the timing information. Therefore, in general, these invariants do not characterize the dynamic behaviour of the system. On the other hand, they can be used to verify properties which are time independent. For more information about the calculation of invariants in a coloured Petri net, see Jensen [15, 17].

In our ITCPN model, a delay is described by an interval rather than a fixed value or some delay distribution. On the one hand, interval delays allow for the modelling of variable delays, on the other hand, it is not necessary to determine some artificial delay distribution (as opposed to stochastic delays). Instead, we have to specify bounds. These bounds are used to specify and to verify time constraints. This is very important when modelling time-critical systems, i.e. *real-time* systems with 'hard' deadlines. These deadlines have to be met for a safe operation of the system. An acceptable behaviour of the system depends not only on the logical correctness of the results, but also on the time at which the results are produced. Therefore, we are interested in techniques to verify these deadlines and to calculate upper and lower bounds for all sorts of performance criteria.

4.1 Reachability Graphs

In section 3.3, we defined the behaviour of an ITCPN. The definitions of that section can be used to construct the so-called **reachability graph**. The basic idea of a reachability graph is to organize all reachable markings in a graph, where each node represents a state and each arc represents an event transforming one state into another state.
Consider for example the reachability graph shown in figure 3. Suppose that s_1 is the initial state of the ITCPN we want to consider. This state is connected to a number of states $s_{11}, s_{12}, s_{13}, ..$ reachable from s_1 by the firing of some transition, i.e. $s_1 \longrightarrow s_{1i}$. These states are called the 'successors' (or children) of the s_1. Repeating this process produces the graphical representation of the reachability graph, see figure 3. Such a reachability graph contains all relevant information about the dynamic behaviour of

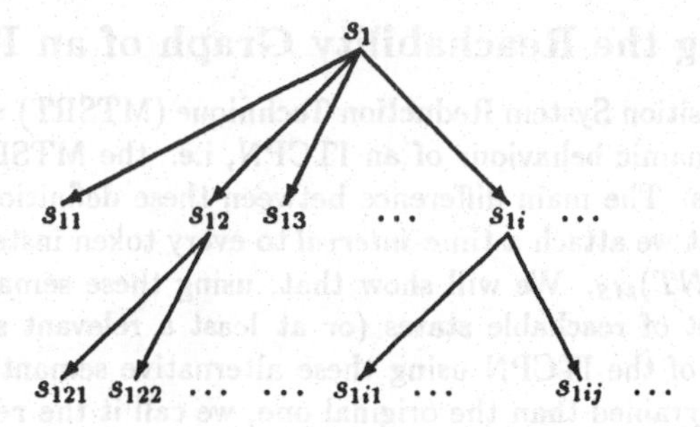

Figure 3: A reachability graph

the system. If we are able to generate this graph, we can answer 'any' kind of question about the behaviour of the system.

Obviously such a reachability graph may, even for small nets, become very large (often infinite). Many authors have presented analysis techniques for the efficient calculation of a reachability graph of an *untimed coloured* Petri net (e.g. [27], [16], [14], [8]). In this section we focus on the reachability graph of an ITCPN.

In general the number of reachable states of an ITCPN (given an initial state) is infinite. This is mainly caused by the fact that we use interval timing. Consider an enabled transition. In general, there is an infinite number of allowed firing delays, all resulting in a different state. Consider for example the ITCPN shown in figure 4. If transition t occurs, then it consumes one token from p_1 and it produces one token for p_2 and one token for p_3. The delay intervals are given in the figure. Suppose the initial state is such that there is one token in p_1 with timestamp 0. The number of successors of this state is infinite, because all states with one token in p_2 having a timestamp $x \in [1,2]$ and one token in p_3 having a timestamp $y \in [3,4]$ are reachable. It may seem unreasonable that this simple example corresponds to a reachability graph with an infinite number of states. This is the reason we developed the Modified Transition System Reduction Technique described in this section. This technique generates the reachability graph and uses, for computational reasons only, alternative definitions for the dynamic behaviour of an ITCPN.

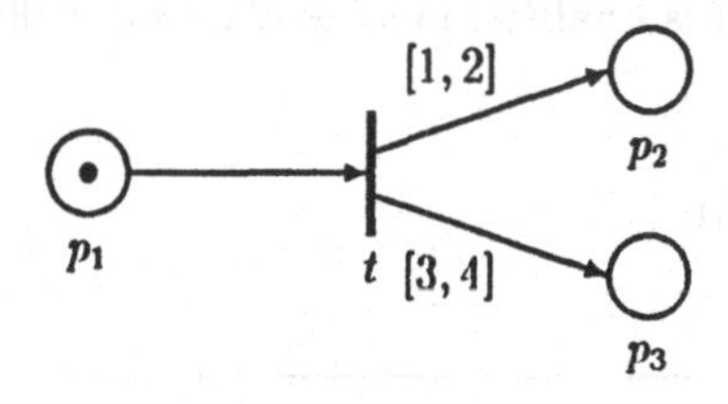

Figure 4: An ITCPN

4.2 Reducing the Reachability Graph of an ITCPN

The Modified Transition System Reduction Technique (MTSRT) uses alternative definitions for the dynamic behaviour of an ITCPN, i.e. the MTSRT method uses alternative semantics. The main difference between these definitions and the original ones is the fact that we attach a time-*interval* to every token instead of a timestamp, i.e. $\overline{S} = (CT \times INT)_{MS}$. We will show that, using these semantics, it is possible to calculate the set of reachable states (or at least a relevant subset). Since, the reachability graph of the ITCPN using these alternative semantics is much smaller and more coarsely grained than the original one, we call it the **reduced reachability graph.** Every state in the reduced reachability graph corresponds to a (infinite) number of states in the original reachability graph. One may think of these states as *equivalence* or *state classes.* One state class $\overline{s} \in \overline{S}$ corresponds to the set of all states being a specialization of $\overline{s}$, i.e. $\{s \in S \mid s \lhd \overline{s}\}$. Informally speaking, state classes are defined as the union of 'similar' states having the same token distribution (marking) but different timestamps (within certain bounds).

In the remainder of this section we redefine the behaviour of an interval timed coloured Petri net, i.e. we give alternative semantics of the ITCPN model.
In section 4.3, we will show how these two semantics relate to each other. We will see that the alternative definitions given in this section can be used to answer questions about the behaviour of the ITCPN (as specified in section 3.3).

Definition 14
A state class is defined as a multi-set of coloured tokens each bearing a time-interval. $\overline{S}$ is the state space:[5]

$$\overline{S} = (CT \times INT)_{MS}$$

An event is a triple $\langle t, b_{in}, b_{out}\rangle$, which represents the possible firing of transition t while removing the tokens specified by the multi-set b_{in} and adding the tokens specified by the multi-set b_{out}. $\overline{E}$ is the event set:

$$\overline{E} = T \times (CT \times INT)_{MS} \times (CT \times INT)_{MS}$$

Note that tokens bear time-*intervals* instead of timestamps.

Definition 15
An event $\langle t, b_{in}, b_{out}\rangle \in \overline{E}$ is enabled in state class $s \in \overline{S}$ iff:

(i) $b_{in} \leq s$

(ii) $M(b_{in}) \in dom(F(t))$

(iii) $b_{out} = F(t)(M(b_{in}))$

The point of time a token becomes available is specified by an interval, therefore it is

[5]Symbols superscripted by a horizontal line are associated with the alternative semantics, this to avoid confusion.

impossible to specify *the* enabling time of an event. However, it is possible to give an upper and lower bound for the enabling time of an event $e \in \overline{E}$.

Definition 16
The minimum and maximum enabling time of an event $\langle t, b_{in}, b_{out}\rangle \in \overline{E}$ are defined as follows:

$$ET_{min}(\langle t, b_{in}, b_{out}\rangle) = \max_{\langle\langle p,v\rangle,[y,z]\rangle \in b_{in}} y$$
$$ET_{max}(\langle t, b_{in}, b_{out}\rangle) = \max_{\langle\langle p,v\rangle,[y,z]\rangle \in b_{in}} z$$

An enabled event e is time enabled iff the minimum enabling time of e is smaller than the maximum enabling time of any other enabled event.

If $\langle t, b_{in}, b_{out}\rangle \in \overline{E}$ an enabled event, then a time-interval in b_{out} represents the delay interval of the token instead of an absolute time-interval. Therefore we need a function to 'scale' time-intervals.

Definition 17
Function $\overline{SC} \in ((CT \times INT)_{MS} \times INT) \to (CT \times INT)_{MS}$ scales the time-intervals in a multi-set of timed coloured tokens. For $b \in (CT \times INT)_{MS}$ and $[y, z] \in INT$, we have:

$$\overline{SC}(b, [y, z]) = \sum_{\langle\langle p,v\rangle,[y',z']\rangle \in CT \times INT} b(\langle\langle p, v\rangle, [y', z']\rangle) \langle\langle p, v\rangle, [x' + x, y' + y]\rangle$$

A time enabled event e in state class s may occur at a time between $ET_{min}(e)$ and $MT_{max}(s) = \min\{ET_{max}(e) \mid e \in \overline{E} \text{ and } e \text{ is enabled in } s\}$.

Definition 18
When an enabled event $\langle t, b_{in}, b_{out}\rangle$ is time enabled in state class $s_1 \in \overline{S}$, it may occur, i.e. transition t fires while removing the tokens specified by b_{in} and adding the tokens specified by b_{out}.
If $\langle t, b_{in}, b_{out}\rangle$ occurs in state class s_1, then the net changes into the state class $s_2 \in \overline{S}$, defined by:

$$s_2 = (s_1 - b_{in}) + \overline{SC}(b_{out}, [ET_{min}(\langle t, b_{in}, b_{out}\rangle), MT_{max}(s_1)])$$

State class s_2 is said to be directly reachable from s_1 by the occurrence of event $e = \langle t, b_{in}, b_{out}\rangle$, this is also denoted by:

$$s_1 \overset{e}{\Longrightarrow} s_2$$

Note that we use double arrows to denote possible state (class) changes given the alternative semantics of an ITCPN. If we use the definitions given in this section for the generation of the reachability graph, then we obtain the **reduced reachability graph**. Comparing these definitions with the definitions given in section 3.3 shows that all differences stem from the fact that the alternative semantics associate a time-interval (instead of a timestamp) with each token. As a result of these intervals, the

enabling time of an event and the model time of a state class are both characterized by an upper and lower bound, etc.
To illustrate these alternative semantics we will use the example shown in figure 1. The initial state corresponds to the state class $\langle\langle p_{in}, J1\rangle, [1,1]\rangle + \langle\langle p_{free}, M1\rangle, [0,0]\rangle$. Note that there is only one state which is a specialization of this class. The minimum enabling time (ET_{min}) of the event which corresponds to the firing of t_1 is equal to 1. The maximum model time (MT_{max}) is also 1. The only state class directly reachable from $\langle\langle p_{in}, J1\rangle, [1,1]\rangle + \langle\langle p_{free}, M1\rangle, [0,0]\rangle$ is the state class $\langle\langle p_{busy}, \langle M1, J1\rangle, [2,4]\rangle$. For this state class, the minimum enabling time (ET_{min}) of the event which corresponds to the firing of t_2 is equal to 2. The maximum model time (MT_{max}) is 4. Transition t_2 will fire between 2 and 4. Consequently, the only state class directly reachable from $\langle\langle p_{busy}, \langle M1, J1\rangle, [2,4]\rangle$ is $\langle\langle p_{out}, J1\rangle, [2,4]\rangle + \langle\langle p_{free}, M1\rangle, [2,4]\rangle$.

4.3 Soundness

The alternative definitions of section 4.2 have been given for computational reasons. However, calculating the reduced reachability graph only makes sense if the reduced reachability graph can be used to deduce properties of the reachability graph which represents the behaviour of the ITCPN. Therefore, we investigate the relation between the two reachability graphs. Examples indicate that such a relation exists.
It is easy to see that the two reachability graphs are not equivalent. Moreover, there is no sensible morphism between them (see [3]). However, state $s \in S$ and state class $\overline{s} \in \overline{S}$ seem to be 'related' if s is a specialization of $\overline{s}$ (i.e. $s \lhd \overline{s}$). Recall that $s \lhd \overline{s}$ if and only if each token in s corresponds to exactly one token in $\overline{s}$, such that they are in the same place, have the same value and the timestamp of the token in s is in the (time) interval of the token in $\overline{s}$ (see definition 7).

Theorem 2 (Soundness)
Let (Σ, P, T, C, F) be an ITCPN and $s_1, s_2 \in S$ such that $s_1 \longrightarrow s_2$. If $\overline{s}_1 \in \overline{S}$ and $s_1 \lhd \overline{s}_1$, then there exists a state class $\overline{s}_2 \in \overline{S}$ such that $\overline{s}_1 \Longrightarrow \overline{s}_2$ and $s_2 \lhd \overline{s}_2$.

Proof [6]
Let $s_1, s_2 \in S$, $\overline{s}_1 \in \overline{S}$ and $e \in E$ such that $s_1 \stackrel{e}{\longrightarrow} s_2$ and $s_1 \lhd \overline{s}_1$. Now we have to prove that there exists an $\overline{e}$ such that $\overline{s}_1 \stackrel{\overline{e}}{\Longrightarrow} \overline{s}_2$ and $s_2 \lhd \overline{s}_2$ (see figure 5).
Suppose $e = \langle t, b_{in}, b_{out}\rangle$ and let $\overline{e} = \langle t, \overline{b}_{in}, \overline{b}_{out}\rangle$ where $\overline{b}_{out} = F(t)(M(b_{in}))$ and $\overline{b}_{in}$ such that $\overline{b}_{in} \leq \overline{s}_1$ and $b_{in} \lhd \overline{b}_{in}$ (this is possible because $s_1 \lhd \overline{s}_1$ and $b_{in} \leq s_1$).

Remains to prove that: (1) $\overline{e}$ is enabled, (2) $\overline{e}$ is time enabled and (3) $s_2 \lhd \overline{s}_2$.
(1) Event $\overline{e}$ is enabled, because $\overline{b}_{in} \leq \overline{s}_1$, $M(\overline{b}_{in}) \in dom(F(t))$ and $\overline{b}_{out} = F(t)(M(\overline{b}_{in}))$.[7]
(2) $ET_{min}(\overline{e}) \leq ET(e) = MT(s_1) \leq MT_{max}(\overline{s}_1)$, i.e. $\overline{e}$ is time enabled.
(3) Since $s_1 \lhd \overline{s}_1$, $b_{in} \lhd \overline{b}_{in}$ and $b_{out} \lhd \overline{b}_{out}$, we have that:
$SC(b_{out}, ET(\langle t, b_{in}, b_{out}\rangle)) \lhd \overline{SC}(\overline{b}_{out}, [ET_{min}(\langle t, \overline{b}_{in}, \overline{b}_{out}\rangle), MT_{max}(\overline{s}_1)])$ and
$s_2 = (s_1 - b_{in}) + SC(b_{out}, ET(\langle t, b_{in}, b_{out}\rangle))$ is a specialization of
$\overline{s}_2 = (\overline{s}_1 - \overline{b}_{in}) + \overline{SC}(\overline{b}_{out}, [ET_{min}(\langle t, \overline{b}_{in}, \overline{b}_{out}\rangle), MT_{max}(\overline{s}_1)])$, i.e. $s_2 \lhd \overline{s}_2$. □

---
[6] A more formal proof is given in [3].
[7] $M(\overline{b}_{in}) = M(b_{in})$, because $b_{in} \lhd \overline{b}_{in}$.

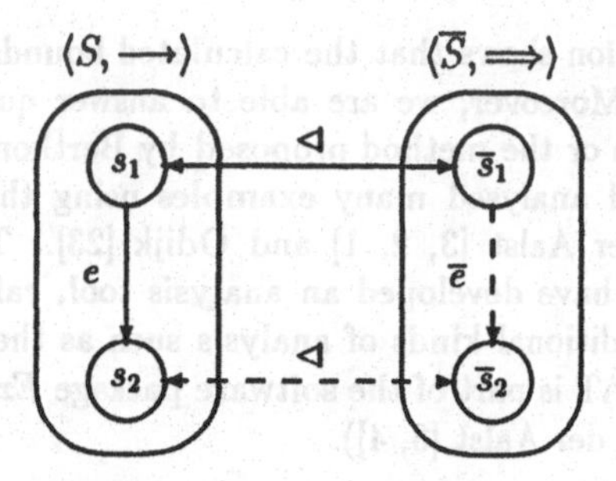

Figure 5: The soundness property

This theorem tells us that if an event occurs which changes state s_1 into s_2, then there is a corresponding event which changes any state class $\overline{s}_1$ that 'covers' s_1 into a state class which 'covers' s_2 (see figure 5). We say that the alternative semantics are 'sound'.

An implication of theorem 2 is that if $s_1 \lhd \overline{s}_1$ and there exists a firing sequence $s_1 \xrightarrow{e_1} s_2 \xrightarrow{e_2} s_3 \xrightarrow{e_3} .. \xrightarrow{e_{n-1}} s_n$, then there exists a firing sequence $\overline{s}_1 \xrightarrow{\overline{e}_1} \overline{s}_2 \xrightarrow{\overline{e}_2} \overline{s}_3 \xrightarrow{\overline{e}_3} .. \xrightarrow{\overline{e}_{n-1}} \overline{s}_n$ such that $s_n \lhd \overline{s}_n$. For any state reachable from s_1 there is a related state class reachable from $\overline{s}_1$.

If we compare the reachability graph and the reduced reachability of an arbitrary ITCPN, then we see that all the state changes possible in the reachability graph are also possible in the reduced reachability. Note that the opposite does not hold, because of *dependencies* between tokens are not taken into account. Consider for example the net shown in figure 4. Suppose there is one token in $p1$ with a time interval $[0, 1]$ and the other places are empty. In this case t fires between time 0 ($ET_{min}(e)$) and time 1 ($MT_{max}(s)$). The next state of the ITCPN using the alternative definitions given in section 4.2, will be the state with one token in $p2$ (with interval $[1, 3]$) and one token in $p3$ (with interval $[3, 5]$). This suggests that it is possible to have a token in $p2$ with timestamp 1 and a token in $p3$ with timestamp 5. However, this is not possible, because these timestamps are related (i.e. they where produced at the same time). We say that the alternative semantics are not 'complete'.

Despite the non-completeness, the soundness property allows us to answer various questions. We can *prove* that a system has a desired set of properties by proving it for the modified transition system. For example, we can often use the reduced reachability graph to prove boundedness, absence of traps and siphons (deadlocks), etc. The reduced reachability graph may also be used to analyse the *performance* of the system modelled by an ITCPN. With performance we mean characteristics, such as: response times, occupation rates, transfer rates, throughput times, failure rates, etc. The MTSRT method can be used to calculate bounds for performance measures. Although these bounds are sound (i.e. safe) they do not have to be as tight as possible, because of possible dependencies between tokens (non-completeness).

However, experimentation shows that the calculated bounds are often of great value and far from trivial. Moreover, we are able to answer questions which cannot be answered by simulation or the method proposed by Berthomieu et al. [7].
We have modelled and analysed many examples using the approach presented in this paper, see Van der Aalst [3, 2, 1] and Odijk [23]. To facilitate the analysis of real-life systems we have developed an analysis tool, called *IAT* ([3]). This tool also supports more traditional kinds of analysis such as the generation of place and transition invariants. IAT is part of the software package *ExSpect* (see ASPT [6], Van Hee et al. [12] and Van der Aalst [3, 4]).

5 Conclusion

In this paper, we presented a coloured Petri net model extended with time. This ITCPN model uses a new timing mechanism where time is associated with tokens and transitions determine a delay specified by an interval. The formal semantics of the ITCPN model have been defined in section 3. The fact that time is in tokens results in transparent semantics and a compact state representation. Specifying each delay by an interval rather than a deterministic value or stochastic variable is promising, since it is possible to model uncertainty without having to bother about the delay distribution.

From the analysis point of view, the ITCPN model is also interesting, since interval timing allows for new analysis methods. In this paper, the MTSRT method has been described. This is a powerful analysis method, since it can be applied to arbitrary nets and answers a large variety of questions. This method constructs a reduced reachability graph. In such a graph a node corresponds to a set of (similar) states, instead of a single state.

A lot of applications have been modelled and analysed using the approach described in this paper and the software package ExSpect which supports the MTSRT method. Experimentation shows that, in general, the results obtained using the MTSRT method are quite meaningful. A direction of further research is to incorporate other reduction techniques for coloured Petri nets into our approach (e.g. [27], [16], [14], [8]).

References

[1] W.M.P. VAN DER AALST, *Interval Timed Petri Nets and their analysis.* Computing Science Notes 91/09, Eindhoven University of Technology, Eindhoven, 1991.

[2] ——, *Modelling and Analysis of Complex Logistic Systems*, in Proceedings of the IFIP WG 5.7 Working Conference on Integration in Production Management Systems, Eindhoven, the Netherlands, 1992, pp. 203–218.

[3] ——, *Timed coloured Petri nets and their application to logistics*, PhD thesis, Eindhoven University of Technology, Eindhoven, 1992.

[4] W.M.P. VAN DER AALST AND A.W. WALTMANS, *Modelling logistic systems with EXSPECT*, in Dynamic Modelling of Information Systems, H.G. Sol and K.M. van Hee, eds., Elsevier Science Publishers, Amsterdam, 1991, pp. 269–288.

[5] C. ANDRE, *Synchronized Elementary Net Systems*, in Advances in Petri Nets 1989, G. Rozenberg, ed., vol. 424 of Lecture Notes in Computer Science, Springer-Verlag, New York, 1990, pp. 51–76.

[6] ASPT, *ExSpect 4.0 User Manual*, Eindhoven University of Technology, 1993.

[7] B. BERTHOMIEU AND M. DIAZ, *Modelling and verification of time dependent systems using Time Petri Nets*, IEEE Transactions on Software Engineering, 17 (1991), pp. 259–273.

[8] G. CHIOLA, C. DUTHEILLET, G. FRANCESCHINIS, AND S. HADDAD, *On well-formed coloured nets and their symbolic reachability graph*, in Proceedings of the 11th International Conference on Applications and Theory of Petri Nets, Paris, June 1990, pp. 387–411.

[9] G. FLORIN AND S. NATKIN, *Evaluation based upon Stochastic Petri Nets of the Maximum Throughput of a Full Duplex Protocol*, in Application and theory of Petri nets: selected papers from the first and the second European workshop, C. Girault and W. Reisig, eds., vol. 52 of Informatik Fachberichte, Berlin, 1982, Springer-Verlag, New York, pp. 280–288.

[10] H.J. GENRICH AND K. LAUTENBACH, *System modelling with high level Petri nets*, Theoretical Computer Science, 13 (1981), pp. 109–136.

[11] K.M. VAN HEE, *System Engineering: a Formal Approach* (to appear), 1993.

[12] K.M. VAN HEE, L.J. SOMERS, AND M. VOORHOEVE, *Executable specifications for distributed information systems*, in Proceedings of the IFIP TC 8 / WG 8.1 Working Conference on Information System Concepts: An In-depth Analysis, E.D. Falkenberg and P. Lindgreen, eds., Namur, Belgium, 1989, Elsevier Science Publishers, Amsterdam, pp. 139–156.

[13] K.M. VAN HEE AND P.A.C. VERKOULEN, *Integration of a Data Model and Petri Nets*, in Proceedings of the 12th International Conference on Applications and Theory of Petri Nets, Aarhus, June 1991, pp. 410–431.

[14] P. HUBNER, A.M. JENSEN, L.O. JEPSEN, AND K. JENSEN, *Reachability trees for high level Petri nets*, Theoretical Computer Science, 45 (1986), pp. 261–292.

[15] K. JENSEN, *Coloured Petri Nets*, in Advances in Petri Nets 1986 Part I: Petri Nets, central models and their properties, W. Brauer, W. Reisig, and G. Rozenberg, eds., vol. 254 of Lecture Notes in Computer Science, Springer-Verlag, New York, 1987, pp. 248–299.

[16] ——, *Coloured Petri Nets: A High Level Language for System Design and Analysis*, in Advances in Petri Nets 1990, G. Rozenberg, ed., vol. 483 of Lecture Notes in Computer Science, Springer-Verlag, New York, 1990, pp. 342–416.

[17] ——, *Coloured Petri Nets. Basic concepts, analysis methods and practical use.*, to appear in EATCS monographs on Theoretical Computer Science, Springer-Verlag, New York, 1992.

[18] M. Ajmone Marsan, G. Balbo, A. Bobbio, G. Chiola, G. Conte, and A. Cumani, *On Petri Nets with Stochastic Timing*, in Proceedings of the International Workshop on Timed Petri Nets, Torino, 1985, IEEE Computer Society Press, pp. 80–87.

[19] M. Ajmone Marsan, G. Balbo, and G. Conte, *A Class of Generalised Stochastic Petri Nets for the Performance Evaluation of Multiprocessor Systems*, ACM Transactions on Computer Systems, 2 (1984), pp. 93–122.

[20] P. Merlin, *A Study of the Recoverability of Computer Systems*, PhD thesis, University of California, Irvine, California, 1974.

[21] P. Merlin and D.J. Faber, *Recoverability of communication protocols*, IEEE Transactions on Communication, 24 (1976), pp. 1036–1043.

[22] S. Morasca, M. Pezzè, and M. Trubian, *Timed High-Level Nets*, The Journal of Real-Time Systems, 3 (1991), pp. 165–189.

[23] M. Odijk, *ITPN analysis of ExSpect specifications with respect to production logistics*, Master's thesis, Eindhoven University of Technology, Eindhoven, 1991.

[24] C. Ramchandani, *Performance Evaluation of Asynchronous Concurrent Systems by Timed Petri Nets*, PhD thesis, Massachusetts Institute of Technology, Cambridge, 1973.

[25] W. Reisig, *Petri nets: an introduction*, Prentice-Hall, Englewood Cliffs, 1985.

[26] J. Sifakis, *Use of Petri Nets for performance evaluation*, in Proceedings of the Third International Symposium IFIP W.G. 7.3., Measuring, modelling and evaluating computer systems (Bonn-Bad Godesberg, 1977), H. Beilner and E. Gelenbe, eds., Elsevier Science Publishers, Amsterdam, 1977, pp. 75–93.

[27] A. Valmari, *Stubborn sets for reduced state space generation*, in Proceedings of the 10th International Conference on Applications and Theory of Petri Nets, Bonn, June 1989.

[28] W.M. Zuberek, *Timed Petri Nets and Preliminary Performance Evaluation*, in Proceedings of the 7th annual Symposium on Computer Architecture, vol. 8(3) of Quarterly Publication of ACM Special Interest Group on Computer Architecture, 1980, pp. 62–82.

Integration of Specification for Modeling and Specification for System Design

Chang-Yu Wang[1] and Kishor S. Trivedi[2] *

[1] Department of Computer Science, Duke University, Durham, NC 27708
E-mail : cw@cs.duke.edu
[2] Department of Electrical Eng., Duke University, Durham, NC 27708
E-mail : kst@egr.duke.edu

Abstract. This paper presents a procedure of transforming an Estelle specification into Stochastic Reward Net(SRN) formalism. Estelle is an ISO standard formal specification language which can help avoid ambiguity, incompleteness and inconsistency in system development. SRN is a well-developed modeling technique that is used to carry out performance and reliability analysis. Integration of specification for system design like Estelle and specification for modeling like SRN can minimize the effort and the cost in designing a functionally correct as well as reliable and performance satisfied system. The objective of transforming Estelle into SRN is to have a system designer specify a system using Estelle and then the specification is automatically transformed into SRN to carry out the performance and reliability analysis.

1 Introduction

Today's computing systems are large and complex, therefore, informal and intuitive specifications are too vague and too imprecise to capture the complete semantics of the system's requirements. A formal specification language is a mathematically based language describing system properties to provide a systematic approach to avoid ambiguity, incompleteness and inconsistency in system development. Today, computing systems are also required to have high performance and high reliability. Separating the functional analysis from the performance and reliability analysis may increase the cost and the time in system development. Formal specifications provide a means to design a functionally correct system, but are weak at incorporating performance and reliability requirements. Modeling techniques like Markov chains and stochastic Petri nets are good at evaluating the performance and reliability of a system, but may be too abstract from the view of functional behavior. Therefore, integration of the modeling techniques and the formal specifications may help in designing a functionally correct as well as reliable and performance satisfied computing system.

Estelle [4, 17, 22] is a formal specification language developed by ISO (International Organization for Standardization) for specifying OSI (Open System

* The work was supported in part by the Naval Surface Warfare Center under contract N60921-92-C-0161.

Interconnection) protocols and services, however, it may be used to specify not only communication protocols but also any distributed system. Estelle is based on a finite state machine model described in a PASCAL-like language. Components of communication protocols are specified by a module whose functional behavior is described as a finite state machine. Estelle has become an international standard within ISO [17].

Stochastic reward nets (SRN) [8] provide a modeling paradigm derived by extension of generalized stochastic Petri nets [2]. They have been successfully applied for reliability analysis or performance analysis of computer systems [16, 20], local area networks [15], and concurrent and fault-tolerant software [9]. Not only performance and reliability measures but also combined performance and reliability measures can be derived from SRNs by generating an underlying Markov reward model [8]. A basis for the logical analysis of SRNs is also available in the Petri net theory [21].

In this paper, we present a procedure of transforming Estelle specifications into SRNs. The idea is to combine the merits of a well-defined specification language like Estelle and those of a powerful modeling technique like SRN. Therefore, a designer can specify a system using Estelle, and then the Estelle specification is automatically transformed into SRN so as to facilitate the automation of the performance and reliability analysis. We claim that complete transformation from Estelle specifications into SRNs can be achieved assuming that deterministic and uniformly distributed firing times are possible in SRNs. Possible methods of extending SRNs to have deterministic and uniformly distributed firing times are simulation [5], phase type expansions, and Markov Regenerative Stochastic Petri nets(MRSPN)[7].

In Section 2 and 3, we introduce the Estelle specification language and the SRN formalism respectively. Transformation from Estelle into SRN is presented in Section 4. In Section 5, we discuss how to model a FIFO queue in SRNs and how to incorporate deterministic and uniformly distributed firing times. Section 6 gives an example and concluding remarks are given in Section 7.

2 Estelle

Estelle [4, 22] is a formal specification language for distributed systems; especially for communication protocols and services. It is developed by ISO (International Organization of Standardization) for specifying OSI (Open System Interconnection) protocols and services. Estelle is based on a finite state machine model described in a PASCAL-like language. A distributed system specified in Estelle is considered to be a collection of communicating components which are called *tasks* or *module instances* in this paper. Tasks are organized hierarchically in a tree structure. Details of Estelle syntax and semantics are discussed in the following subsections.

2.1 Communication

Communication between module instances can be done via shared variables or message exchange. Variables are shared only between a task and its parent task, and are declared by the child task. Simultaneous access to shared variables are not allowed because the parent always has higher priority than the child.

Message exchange is the main method of communication in Estelle. The access points where a module instance can send and receive messages are called ***interaction points***. Two bound interaction points are ***connected*** if they belong to two sibling module instances. If an interaction point of a child task is bound to an interaction point of its parent task, they are said to be ***attached***.

There are some rules about how connections and attachments can be made between interaction points: first, an interaction point cannot connect to itself. Second, an interaction point may be connected to at most one other interaction point. Third, an interaction point of a task may be attached to at most one interaction point of its parent task and to at most one interaction point of its child task. Fourth, an interaction point cannot be attached to its parent task and be connected to its sibling simultaneously.

Messages arriving at an interaction point are deposited in an unbounded FIFO queue. For two connected interaction points, a message sent from one point directly arrives at the queue of the other interaction point. For two attached interaction points, the message received at the parent side is passed down to the queue of the attached child point. Each interaction point has its own queue (***individual queue***), or shares a queue (***common queue***) with other interaction points of the same module instance. The types of messages which can be transmitted between two interaction points are fixed. Messages may also have parameters. For example, an acknowledgment message may be declared as $ACK(ak\_no : integer)$, where ak_no denotes the sequence number of successfully received data packets.

2.2 Module

Instances of modules are the communicating components in Estelle. A module has a header and a body. The external visibility of a module, defined in the header, is characterized by the external interaction points, the hierarchical position in the system and the module's attribute. Each module has one of the following attributes: systemprocess, systemactivity, process, activity. The way modules are attributed will affect the sequence of transition firings. We will discuss the details in Section 2.3. The internal behavior of a module, defined in the body, is expressed as a finite state machine which is specified by giving the state space, the initial state and the transitions.

A transition consists of two parts: the transition conditions and the transition actions. If all the conditions of a transition are satisfied, the transition can ***fire***. Upon firing the transition, the actions are taken and the internal state of the module instance can be changed. Types of conditions and actions that can be specified in Estelle are listed below:

A. Condition

- From clause (*from S*): This clause is satisfied if the current state of the task is S (S is a single state) or belongs to S (S is a set of states).
- When clause (*when p.m*): This clause is satisfied when the first message in the queue of interaction point p is of type m.
- Provided clause (*provided* B): This clause is satisfied when the Boolean expression B evaluates to true.
- Priority clause (*priority n*): This clause assigns a non-negative integer n as the priority level of the transition.
- Delay clause (*delay* (E_1, E_2)): if the *from* clause, *when* clause and *provided* clause of a transition are all are satisfied, the transition is *enabled*. The *delay* clause "delay (E_1, E_2)" states that firing of this transition is to be delayed at least E_1 time units but no longer than E_2 time units. After the transition is enabled for E_2 time units, it must fire. During the time the transition is waiting, should any of its *from* clause or *provided* clause become false, the transition is *disabled* and the timer is turned off. A *delay* clause and a *when* clause cannot appear simultaneously in the condition part of a transition.

B. Action:

- Release child tasks: Upon firing a transition, a parent task can destroy its child task and all the descendent tasks of that child task.
- Disconnect or detach interaction points.
- Output messages : "output *p.m*" means output type m message via interaction point p. As each interaction point is connected to only one other point, there is no confusion about which interaction point the message goes to.
- PASCAL statements for updating values of variables.

2.3 Attributes

We now discuss the attributes of modules mentioned in Section 2.2. There are four types of attributes: systemactivity, systemprocess, activity and process. Tasks attributed as "systemprocess" or "systemactivity" are called *system tasks*. A subsystem in Estelle is a subtree of tasks rooted in a system task. A task is *active* if it includes at least one transition in its specification; otherwise, it is *inactive*. Because an inactive task does not have any transitions, the structure of its child tasks and the communication channels between them cannot be changed (Because there are no transitions, no actions can be taken; namely, child tasks cannot be released and no interaction points can be disconnected or detached). Therefore, the internal structure of an *inactive task* is static rather than dynamic.

Here are some rules about how tasks are attributed: first, every active task must be attributed. Second, system tasks cannot be nested within an attributed task. Third, tasks attributed "process" or "activity" must be nested within a system task. Fourth, child tasks of a task attributed "process" or "systemprocess" can be attributed as either "process" or "activity". Fifth, Children tasks of a task attributed "activity" or "systemactivity" must be attributed as "activity".

Since the transition firing sequence is decided by the attributes of tasks, every active task must be attributed in order to decide the firing sequence (Rule 1). Rule 2 implies that the relationship between system tasks is static. Rule 1, Rule 2 and Rule 3 together implies that if there are active tasks in the system, at least one of them is a system task; otherwise, all the tasks are inactive in the system.

2.4 Parallelism and Non-Determinism

If a set of transitions are *ready-to-fire*, there are two strategies in Estelle to decide the sequence of firing: non-determinism and parallelism.

- Non-determinism : Only one transition can fire at a time.
- Synchronous parallelism : A set of *firable* transitions can fire in parallel. If firing a transition triggers some other transitions to fire, these new *firable* transitions have to wait till all the old transitions have completed. Therefore, transition firings are "synchronized".
- Asynchronous parallelism : A set *ready-to-fire* transitions can fire simultaneously, but a new triggered transition can start as soon as it becomes *firable*.

Asynchronous parallelism is permitted only between transitions from different system tasks. Transitions from the same system task have to follow the following firing rules: first, inside a task, only one transition (decided by priorities) can fire at a time. Second, when both the parent task and child task have firable transitions, the parent has higher priority than the child. Third, among sibling tasks, tasks attributed as processes may run in synchronous parallelism, while tasks attributed as activities may run in non-determinism.

Note that it is meaningless to have synchronized transitions without specifying the firing times. Estelle itself does not specify firing times explicitly. The delay clause in transitions specifies the amount of time that a transition has to wait to be executed (enabling time), but the time to execute this transition is unknown. The execution time (firing time) of a transition is considered implementation dependent. In this paper, we will adopt the assumption in [11, 13] that firing a transition is viewed as an atomic activity; an executing transition cannot be interrupted or suspended by other transitions which are completed earlier.

3 Stochastic Reward Nets

Stochastic reward nets (SRN) [8] are an extension of Petri nets [21]. Stochastic Petri nets (SPN) [19] extend the Petri net model by assigning exponentially distributed firing times to its transitions. In Generalized Stochastic Petri nets (GSPN) [2], transitions are allowed to be either *timed* (exponentially distributed firing time) or *immediate* (zero firing time). GSPN also introduces *inhibitor arc* connecting a place to a transition. A transition with an inhibitor arc cannot fire if the input place of the inhibitor arc contains more tokens than the multiplicity of that arc.

Stochastic reward nets (SRN) are based on GSPN where each transition may have an enabling function (or a guard) such that a transition is enabled in a marking only if its enabling function evaluates to true. Marking dependent enabling function and arc multiplicities are allowed in SRN. In order to flush a place when a transition is fired, we can make this place an input place of the transition and assign the multiplicity of the corresponding input arc as the current number of tokens in that place. Then upon firing the transition, all the tokens from that place will be removed. Timed and immediate transitions can also have priorities in SRN.

To represent a SRN as a graph, places are represented by circles and tokens are black dots inside places. Immediate transitions are drawn as thin bars and timed transitions as white boxes. Inhibitor arcs have small hollow circles instead of arrows at their terminating ends. The current number of tokens in a place p is denoted as $\sharp p$.

4 Transformation from Estelle Specifications into SRNs

Transformation of Estelle specifications into SRNs is done in a bottom-up fashion. First, each module instance is treated as an independent unit and is transformed into a subnet of SRN. Second, pre-constructed subnets are connected together according to their communication relationship defined in the Estelle specification. Third, new places and transitions are added to control the firing sequence; either using parallelism or using non-determinism according to the attributes of tasks.

In order to be able to express the delay timers of Estelle in SRN, we assume that transitions with deterministic or uniformly distributed firing times are allowed in SRN (represented as filled boxes in this paper). SRN formalism is currently being extended to include these features[6, 7].

4.1 Module instance

To transform an Estelle task into a SRN, we start by converting the transitions. Transitions in Estelle can be divided into two sets, those with delay clauses and those without delay clauses.

A. Transitions without delay clauses:

Transitions without delay clauses may have from clauses, provided clauses, when clauses, and priority clauses in their condition part. For an Estelle transition with specification shown in the left part of Figure 1a, we can transform this transition into a SRN with places $S1$, $S2$, $in_p.m$ and $out_q.n$, and one transition t_i with enabling function B and priority K (center part of Figure 1a). A token in places $S1$ ($S2$) denotes the current state is $S1$ ($S2$). Tokens in places $in_p.m$ ($out_q.n$) denote the incoming (outgoing) messages of type m (n) at interaction point p (q). Therefore, transition t_i can fire if and only if there is a token in place $S1$, a token in place $in_p.m$, and condition B evaluates to true. When t_i fires, one token each is removed from places $S1$ and $in_p.m$ while a token is deposited in both the places $S2$ and $out_q.n$.

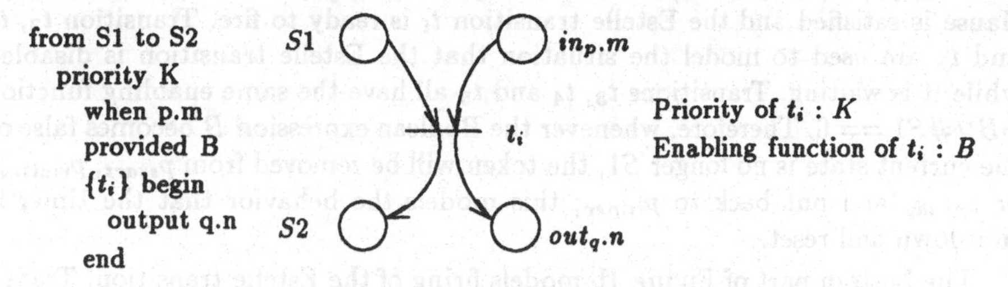

a. SRN of Estelle transition without delay clauses

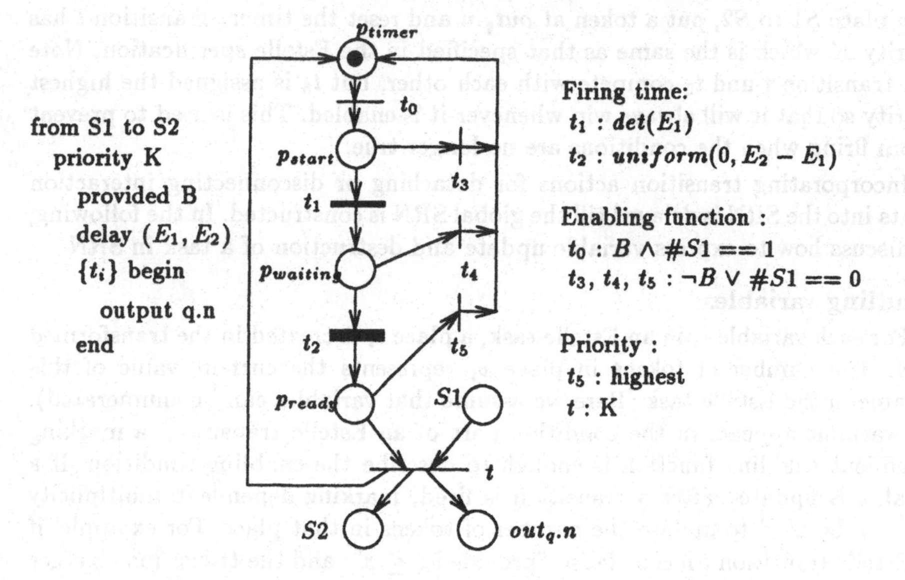

b. SRN of Estelle transition with delay clauses

Fig. 1. SRNs of Estelle transitions

B. Transitions with delay clauses:

If an Estelle transition has a from clause, a priority clause, a provided clause and a delay clause in its condition part (see the left part in Figure 1b), this transition can be transformed into a SRN shown in the center of Figure 1b.

The upper part of the SRN is used to model the timer of the delay clause "delay(E_1, E_2)". A token in place p_{timer} denotes that the timer has not been triggered yet. Immediate transition t_0 is associated with enabling function ($B \wedge \#S1 == 1$) so that t_0 is enabled whenever B is true and there is a token at place $S1$. When t_0 fires, a token is moved from place p_{timer} to p_{start}. Transition t_1 has deterministic firing time E_1. Transition t_2 has uniformly distributed firing time $uniform(0, E_2 - E_1)$. A token in place p_{ready} indicates that the delay clause is satisfied and the Estelle transition t_i is ready to fire. Transition t_3, t_4 and t_5 are used to model the situation that the Estelle transition is disabled while it is waiting. Transitions t_3, t_4 and t_5 all have the same enabling function $\neg B \vee \#S1 == 0$. Therefore, whenever the Boolean expression B becomes false or the current state is no longer $S1$, the token will be removed from p_{start}, $p_{waiting}$ or p_{ready} and put back to p_{timer}; this models the behavior that the timer is shutdown and reset.

The bottom part of Figure 1b models firing of the Estelle transition. Transition t fires if and only if there is a token in p_{ready} (the delay clause is satisfied and the other conditions are still true), and upon firing, it will move the token from place $S1$ to $S2$, put a token at $out_q.n$ and reset the timer. Transition t has priority K which is the same as that specified in the Estelle specification. Note that transition t and t_5 compete with each other, but t_5 is assigned the highest priority so that it will always win whenever it is enabled. This is used to prevent t from firing when the conditions are no longer true.

Incorporating transition actions for detaching or disconnecting interaction points into the SRN is delayed till the global SRN is constructed. In the following, we discuss how to express variable update and destruction of a task in SRN.

Handling variables:

For each variable v_i in an Estelle task, a place v_i is created in the transformed SRN. The number of tokens in place v_i represents the current value of this variable in the Estelle task (Here we assume that variables can be enumerated). If a variable appears in the condition part of an Estelle transition, a marking dependent enabling function is enough to describe the enabling condition. If a variable is updated after a transition is fired, marking dependent multiplicity arcs can be used to update the number of tokens in that place. For example, if an Estelle transition t has a clause "provided $v \leq x$", and the transition changes the value of v into y after it is fired. This can be modeled in SRN by setting the enabling function of t to $\#v \leq x$, and assign the multiplicity of the input arc to the current marking of place v and the multiplicity of the output arc to the new value y.

To handle arrays of variables, we can use arrays of places. The SRN tool SPNP [10] provides some convenient functions to manage arrays of places, transitions and arcs.

Destroying a task:

Destroying a task can be done by removing the token in the places representing control states of the task. For each task, the total number of tokens in all the control places should be at most one. After the token is removed, no transition in that task can fire since every transition needs a token from the place denoting the current control state to be enabled. Thus, the task is *deactivated* in the SRN which expresses that the task is *destroyed* in the Estelle specification.

The initial marking of the SRN is specified by putting one token at the place which denotes the initial state of the Estelle module instance, one token at the place p_{timer} for each transition with delay clause, and as many tokens as the initial value of a variable at each of the variable places.

4.2 Communication structure

After all the tasks are transformed into SRNs, the next step is to combine them together based on their communication relationship. Consider the situation that two interaction points are connected or attached :

A. Connect :

Assume that an interaction point x of task A is connected to an interaction point y of task B, let $M = \{m_1, m_2, \ldots\}$ be the set of message types which can be transmitted at x, and $N = \{n_1, n_2, \ldots\}$ be the set of message types transmitted at y. For each pair of places $(out_x.m_i, in_y.m_i)$, add a new immediate transition c_{xym_i} such that $out_x.m_i$ is the input place of c_{xym_i} and $in_y.m_i$ is the output place of c_{xym_i} (Figure 2a). Also do the similar thing for $(out_y.n_j, in_x.n_j)$. Now, whenever there is token in place $out_x.m_i$, the token will immediately go to $in_y.m_i$. This models the behavior that a type m_i message is transmitted from interaction point x to interaction point y.

B. Attach :

If an interaction point x of a task A is attached to an interaction point y of its child task B, and $M = \{m_1, m_2, \ldots\}$ is the set of message types which can be received at x, for each pair $(in_x.m_i, in_y.m_i)$, add a new immediate transition a_{xym_i} such that the input place of a_{xym_i} is $in_x.m_i$ and the output place of a_{xym_i} is $in_y.m_i$ (Figure 2b). Therefore, whenever there is a token in place $in_x.m_i$, the token will go to $in_y.m_i$.

C. Disconnect or detach:

If firing a transition t causes an interaction point p_i to be disconnected, a new place dc_i is added as an output place of transition t and inhibitor arcs are connected between the place dc_i and all the transitions (c_{ijm} and c_{jim}) representing the linking between interaction point p_i and other interaction points (see Figure 2c). Therefore, a token in place dc_i will disallow any communications between interaction point p_i and others. Detaching an interaction point can be done in a similar way (Figure 2d).

Note that the arriving order of messages with different types at an interaction point is ignored in this section, how to model the FIFO queue at an interaction point will be discussed in Section 5.1. If a common queue is shared between interaction points $\{p_1, p_2, \ldots, p_i, \ldots\}$, for each message type m_j, merge the places $in_{p_i}.m_j$ of all interaction points p_i. Therefore, the tokens are shared among these places, just like the messages are shared among the interaction points.

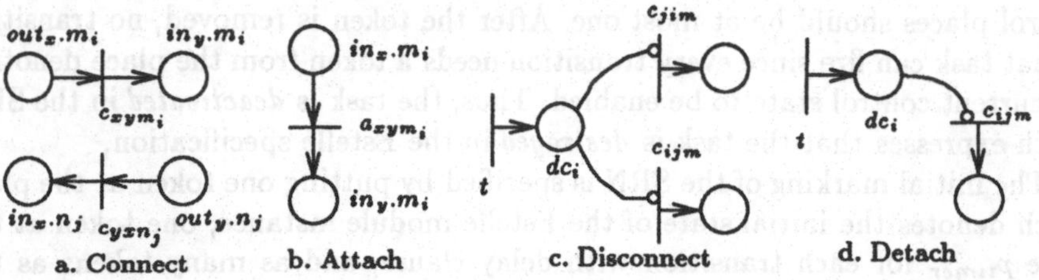

Fig. 2. Connect SRNs based on the communication relationship

An example

An example in Budkowski and Dembinski's paper [4] is used to demonstrate the procedure of transforming Estelle specifications into SRNs discussed earlier. The Estelle specification has two tasks Y and Z. Task Y has interaction point $p1$ which is connected to the interaction point $p2$ at task Z. The internal behavior of task Y and Z is illustrated as state diagrams shown in Figure 3a and 3b. Each state of an Estelle task is denoted by a circle, and a transition is denoted by a thin bar. The initial state of task Y is set to $Y1$, and that of task Z is set to $Z1$.

Figure 3c gives the transformed SRN. The subnet for task Y has four places $Y1$, $Y2$, $out_{p1}.a$, $in_{p1}.b$, and two transitions $t1$ and $t2$. The subnet for task Z has six places $Z1$, $Z2$, $Z3$, $Z4$, $in_{p2}.a$, $out_{p2}.b$, and four transitions $s1$, $s2$, $s3$ and $s4$. Communications between interaction points $p1$ and $p2$ are expressed by transitions $c_{p_1p_2a}$ and $c_{p_2p_1b}$.

4.3 Parallelism and non-determinism

To give an intuitive understanding, we use the previous example to show the reader how to incorporate non-determinism and synchronous parallelism between tasks in Estelle into the SRN. For asynchronous parallelism, we do not have to do anything on the pre-constructed SRNs because the SRNs themselves imply asynchronous parallelism.

A. Non-determinism:

If both task Y and Z are attributed as activities, these two tasks may run in non-determinism; only one transition can fire at a time. A new place p_{perm} is added in the transformed SRN (Figure 4a). For each transition, an input arc and an output arc are added between this transition and the place P_{perm}. Therefore, transitions must compete with each other to grab the token at P_{perm} to fire, and after a transition is fired, it will return the token and someone else can fire. This ensures that only one transition can fire at a time.

B. Synchronous parallelism:

If both tasks Y and Z are attributed as processes, they may run in synchronous parallelism. The new SRN is shown in Figure 4b. In order to make

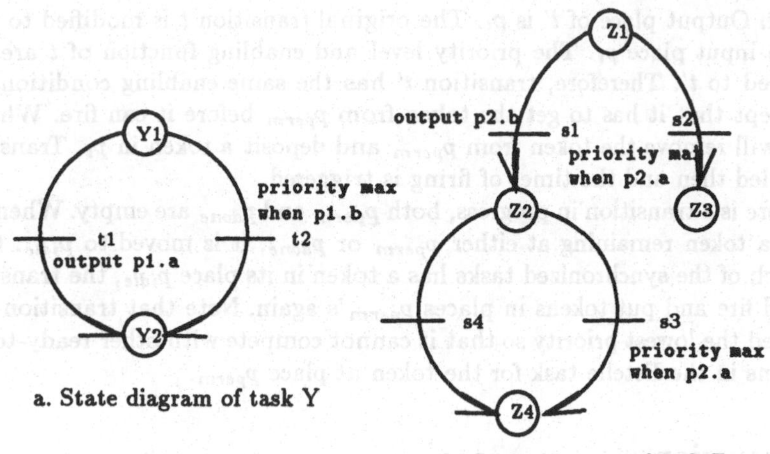

a. State diagram of task Y

b. State diagram of task Z

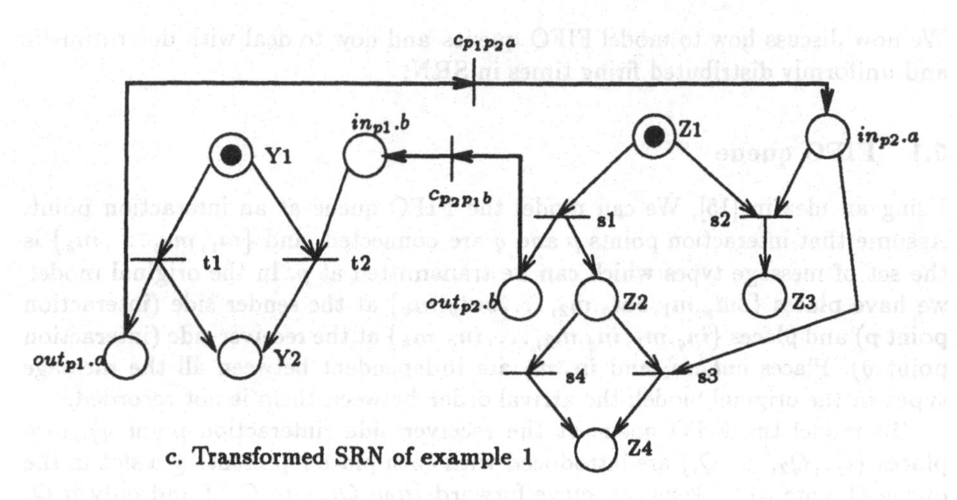

c. Transformed SRN of example 1

Fig. 3. Example 1

the synchronization between tasks meaningful, we assume that transition $t1$ and $t2$ are timed transitions (represented as a filled rectangle). For each task, three places, p_{perm}, p_{done} and p_{idle}, are added. A token in p_{perm} implies that the task is permitted to fire its next firable transition. A token at p_{done} says that the task has finished executing the current transition. A token at p_{idle} states that the task is idle; either the current transition is done and it is waiting for other tasks, or it does not have any firable transition now.

For each transition t, an immediate transition t' and a place p_t are added. Input places of transition t' are the same as those of t plus the place p_{perm} of this task. Output place of t' is p_t. The original transition t is modified to have only one input place p_t. The priority level and enabling function of t are also transferred to t'. Therefore, transition t' has the same enabling condition as t does except that it has to get the token from p_{perm} before it can fire. When t' fires, it will remove the token from p_{perm} and deposit a token in p_t. Transition t is enabled then and the timer of firing is triggered.

If there is a transition in progress, both p_{perm} and p_{done} are empty. Whenever there is a token remaining at either p_{perm} or p_{done}, it is moved to p_{idle}. Only when each of the synchronized tasks has a token in its place p_{idle}, the transition t_{sync} will fire and put tokens in places p_{perm}'s again. Note that transition t_{idle} is assigned the lowest priority so that it cannot compete with other ready-to-fire transitions in the Estelle task for the token at place p_{perm}.

5 Discussion

We now discuss how to model FIFO queues and how to deal with deterministic and uniformly distributed firing times in SRN.

5.1 FIFO queue

Using an idea in [15], We can model the FIFO queue at an interaction point. Assume that interaction points p and q are connected, and $\{m_1, m_2, \ldots, m_k\}$ is the set of message types which can be transmitted at p. In the original model, we have places $\{out_p.m_1, out_p.m_2, \ldots, out_p.m_k\}$ at the sender side (interaction point p) and places $\{in_q.m_1, in_q.m_2, \ldots, in_q.m_k\}$ at the receiver side (interaction point q). Places $out_p.m_i$ and $in_q.m_i$ are independent between all the message types in the original model; the arrival order between them is not recorded.

To model the FIFO queue at the receiver side (interaction point q), new places $\{Q_1, Q_2, \ldots, Q_l\}$ are introduced with each place representing a slot in the queue (Figure 5). Tokens can move forward from Q_{i-1} to Q_i if and only if Q_i is not occupied. The type of a message is identified by the number of tokens in places Q_j's; i tokens are in place Q_j if the message is of type m_i. Multiplicities of input and output arcs between places Q_i and Q_{i+1} are set to the current marking of Q_i to keep track of the message types in the queue. Here we assume that the FIFO queue is bounded with length l. This assumption contradicts the assumption of unbounded queue in Estelle, but in real systems, all queues must have limited capacities. Also note that figure 5a implies that messages arriving while the queue is full are buffered and will lose their arriving order while waiting in the place $out_p.m_i$. Modeling that messages are dropped when the queue is full can also be done by adding immediate transitions with enabling functions detecting whether the queue is full or not.

At the sender side, transition ts_i is used to deposit a type i message in the queue, and a number of tokens equal to i is deposited in place Q_1. At the receiver side, transition tr_i is used to receive a type m_i message and deposit a token at place $in_q.m_i$. By placing an arc of multiplicity i and an inhibitor arc of

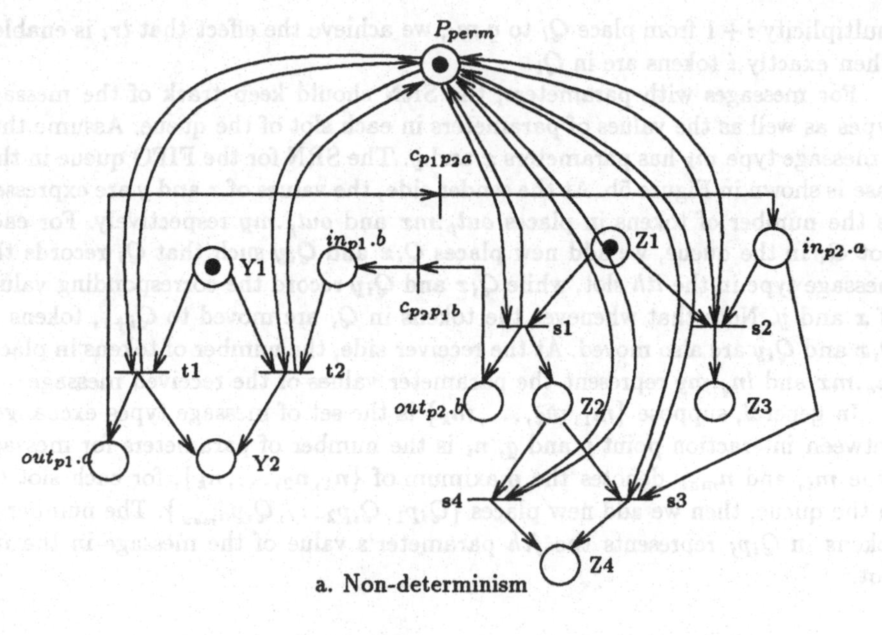

a. Non-determinism

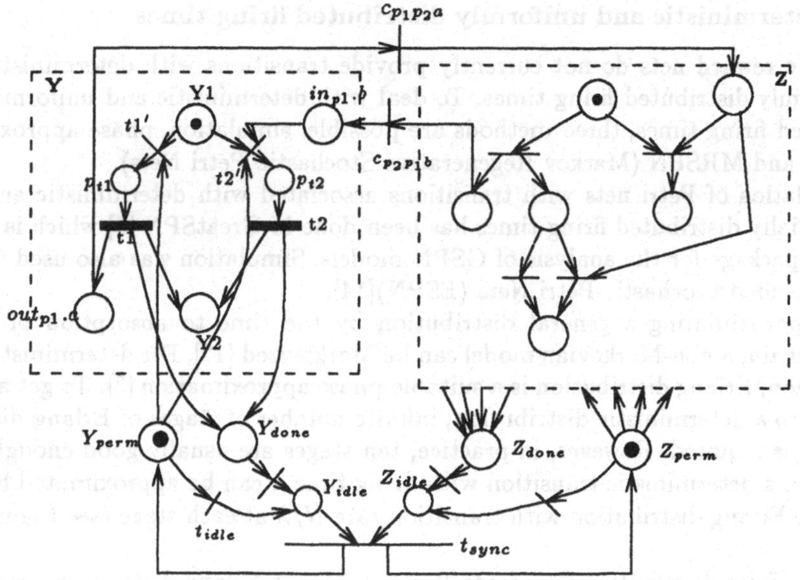

b. Synchronous parallelism

Fig. 4. Non-determinism and synchronous parallelism

multiplicity $i+1$ from place Q_l to $q.m_i$, we achieve the effect that tr_i is enabled when exactly i tokens are in Q_l.

For messages with parameters, the SRN should keep track of the message types as well as the values of parameters in each slot of the queue. Assume that a message type m_k has parameters x and y. The SRN for the FIFO queue in this case is shown in Figure 5b. At the sender side, the values of x and y are expressed as the number of tokens in places $out_p.mx$ and $out_p.my$ respectively. For each slot Q_i in the queue, we add new places Q_ix and Q_iy such that Q_i records the message type in the ith slot, while Q_ix and Q_iy record the corresponding values of x and y. Note that whenever the tokens in Q_i are moved to Q_{i+1}, tokens in Q_ix and Q_iy are also moved. At the receiver side, the number of tokens in places $in_q.mx$ and $in_q.my$ represent the parameter values of the received message.

In general, suppose $\{m_1, m_2, \ldots, m_k\}$ is the set of message types exchanged between interaction point p and q, n_i is the number of parameters for message type m_i, and n_{max} denotes the maximum of $\{n_1, n_2, \ldots, n_k\}$, for each slot Q_i in the queue, then we add new places $\{Q_ip_1, Q_ip_2 \ldots, Q_ip_{n_{max}}\}$. The number of tokens in Q_ip_j represents the jth parameter's value of the message in the ith slot.

5.2 Deterministic and uniformly distributed firing times

Stochastic reward nets do not currently provide transitions with deterministic or uniformly distributed firing times. To deal with deterministic and uniformly distributed firing times, three methods are possible: simulation, phase approximations, and MRSPN (Markov Regenerative Stochastic Petri Nets).

Simulation of Petri nets with transitions associated with deterministic and exponentially distributed firing times has been done in GreatSPN [5] which is a software package for the analysis of GSPN models. Simulation was also used to solve Extended Stochastic Petri Nets (ESPN)[14].

By approximating a general distribution by the time to absorption of a Markov chain, a non-Markovian model can be Markovized [12]. For deterministic distribution, Erlang distribution is a suitable phase approximation [3]. To get an exact fit to a deterministic distribution, infinite number of stages of Erlang distribution is required. However, in practice, ten stages are usually good enough. Therefore, a deterministic transition with firing time λ can be approximated by a N-stage Erlang distribution with transition rate N/λ at each stage (see Figure 6a).

For uniform distribution ($unifomr(0, b)$), a 3-stage Erlang distribution provides a reasonably good approximation [18]. Using the method of moments, we find that by concatenating three exponentially distributed transitions each with rate $6/b$, we obtain the desired results. (see Figure 6b).

Marsan and Chiola introduced deterministic and stochastic Petri nets (DSPN) [1] as an extension to GSPNs. Timed transitions in DSPN are allowed to have not only exponentially distributed firing times but also deterministic firing times. Steady state solution of DSPN is obtained numerically under the restriction that at most one deterministic transition is enabled at a time. Note that the DSPN

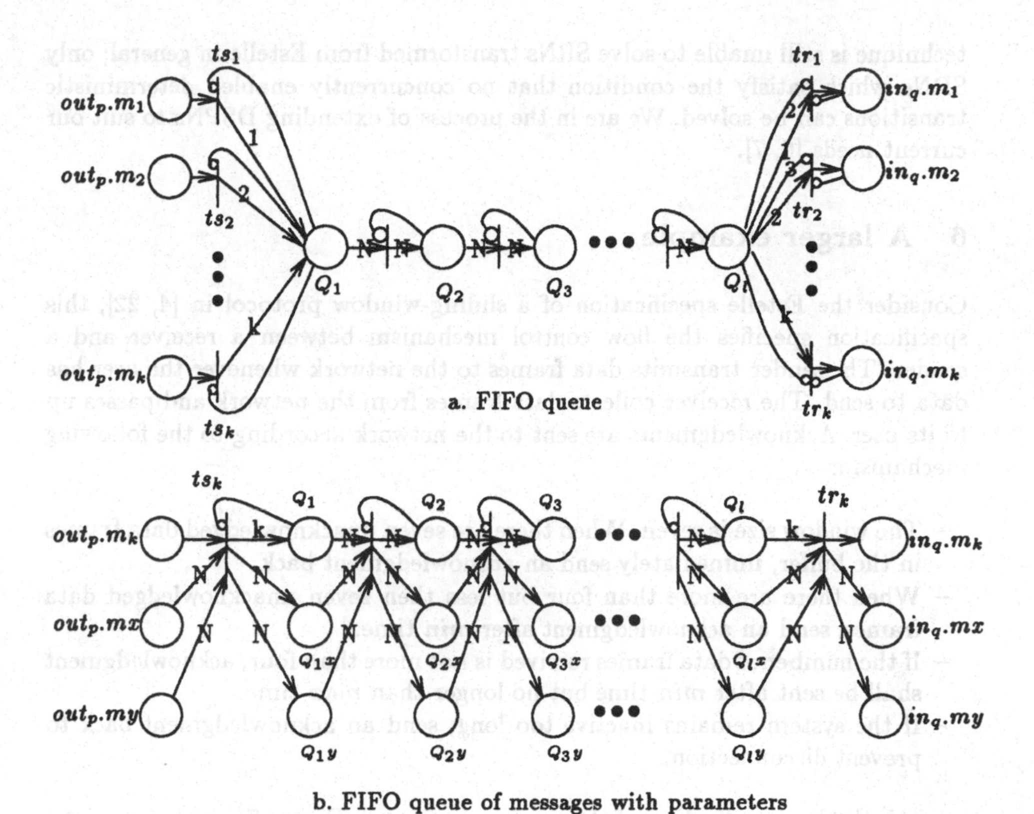

a. FIFO queue

b. FIFO queue of messages with parameters

Fig. 5. FIFO Queue

$det(\lambda)$ $t_1, N/\lambda$ $t_N, N/\lambda$

a. Phase approximation for deterministic transition

$uniform(0,b)$ $t_1, 6/b$ $t_2, 6/b$ $t_3, 6/b$

b. Phase approximation for transitions with uniformly distributed firing times

Fig. 6. Phase approximation for deterministic transition

technique is still unable to solve SRNs transformed from Estelle in general; only SRNs which satisfy the condition that no concurrently enabled deterministic transitions can be solved. We are in the process of extending DSPNs to suit our current needs [6, 7].

6 A larger example

Consider the Estelle specification of a sliding-window protocol in [4, 22], this specification specifies the flow control mechanism between a receiver and a sender. The sender transmits data frames to the network whenever the user has data to send. The receiver collects data frames from the network and passes up to its user. Acknowledgments are sent to the network according to the following mechanism:

- The window size is seven. When there are seven unacknowledged data frames in the buffer, immediately send an acknowledgment back.
- When there are more than four but less then seven unacknowledged data frames, send an acknowledgment after *min* time.
- If the number of data frames received is not more than four, acknowledgment shall be sent after *min* time but no longer than *max* time.
- If the system remains inactive too long, send an acknowledgment back to prevent disconnection.

The Estelle specification of the receiver is given in Figure 7a, and that of the sender is given in Figure 7b. For the receiver, there are two interaction points (N and U), two states ($IDLE$ and AK_SENT), and six transitions ($t1$ through $t6$) in the specification. Communication between the network and the receiver is through interaction point N, and communication between the user and the receiver is through interaction point U. The message type $DATA\_INDICATION$ indicates that a data frame has arrived, and the message type $SEND\_AK$ denotes the acknowledgment. Transition $t1$ fires when a data frame arrives. Transition $t2$ through $t5$ specify the action of sending acknowledgments. The variable *ak_no* is used to record the number of unacknowledged data frames. Transition $t6$ resets *ak_no* when an acknowledgment is sent. For the sender, there are one state $IDLE$, two interaction points (US and NS), and two transitions ($s1$ and $s2$). Interaction point US is used to communicated with the user, while NS is used to communicated with the network. Transition $s1$ is fired when the user has a data frame to send, and transition $s2$ is fired when an acknowledgment is received from the network. The variable *sent_no* is used to count the number of data frames which are already sent to the network but haven't been acknowledged. The transformed SRN is shown in Figure 8. Note that the place $SENT\_NO$ is duplicated to make the figure easier to understand, and the FIFO queue between interaction points N and NS is not shown in detail.

To solve the SRN, we can first approximate the uniformly distributed firing time of transition t_2^2 with a 3-stage Erlang distribution and then apply the DSPN solution technique. The DSPN condition is satisfied because deterministic transitions t_2^1, t_3^1, and t_5^1 cannot be simultaneously enabled; this can be proved by showing the existence of place invariant ($\#Pstart_2 + \#Pstart_3 + \#Pstart_5 == 1$). To use this transformed SRN for performance analysis, a timed transition with exponentially distributed time is added to model the arrival process of data frames (labeled as *DATA_ARRIVAL* in Figure 8). Transition t_{send} and t_{recv} are associated with exponentially distributed firing times to model the propagation delays in the network. At the receiver side, the average number of unacknowledged data frames can be obtained by computing the average number of tokens in place *AK_NO*, while at the sender side, the average number of unknowledged data frames is obtained from the average number of tokens in *SENT_NO*. If we assign the rates of transitions t_{send} and t_{recv} as the ratio of propagation delay to the transmission time of a data frame, the throughput of transition t_{send} is the utilization of this network protocol.

7 Conclusion

In this paper, we presented a procedure of transforming Estelle specifications into SRNs. Two examples are used to demonstrate the procedure. Limitations of SRNs for fully representing Estelle specifications are the inability to model unbounded FIFO queues, to express variable values which cannot be enumerated, and incorporating deterministic and uniformly distributed firing times. To deal with FIFO queues, we exclude the unboundedness property from Estelle. This assumption is reasonable because no real system can have unlimited capacities. For deterministic and uniformly distributed firing times, either simulation or phase approximation and DSPN can be used as solution methods.

Future work involves extending the current version of SRNs to have deterministic and uniformly distributed firing times and automating the transformation from Estelle to SRN.

References

1. M. Ajmone Marsan and G. Chiola. On Petri nets with deterministic and exponentially distributed firing times. In *Lecture Notes in Computer Science*, volume 266, pages 132–145. Springer-Verlag, 1987.
2. M. Ajmone Marsan, G. Conte, and G. Balbo. A class of generalized stochastic Petri nets for the performance evaluation of multiprocessor systems. *ACM Transactions on Computer systems*, 2:93–122, 1984.
3. D. Aldous and L. Shepp. The least variable phase type distribution is Erlang. *Stochastic Models*, 3(3):467–473, 1987.
4. S. Budkowski and P. Dembinski. An introduction of Estelle: A specification language for distributed systems. *Computer Networks and ISDN Systems*, 14:3–23, 1987.
5. G. Chiola. A grpahical Petri net tool for performance analysis. In S. Fdida and G. Pujolle, editors, *Modeling Techniques and Performance Evaluation*, pages 323–333, 1987.

6. H. Choi, V. G. Kulkarni, and K. S. Trivedi. Markov Regenerative Stochastic Petri Nets. Technical Report DUKE-CCSR-93-001, Center for Computer Systems Research, Duke University, 1993.
7. H. Choi, V. G. Kulkarni, and K. S. Trivedi. Transient analysis of deterministic and stochastic Petri nets. In *Proceedings of The 14th International Conference on Application and Theory of Petri Nets*, Chicago, U.S.A., Jun. 21-25 1993.
8. G. Ciardo, A. Blakemore, P.F. Chimento, JR. , J.K. Muppala, and K.S. Trivedi. Automated generation and analysis of Markov reward models using Stochastic Reward Nets. In C. Meyer and R.J. Plemmons, editors, *Linear Algebra, Markov Chains, and Queueing Models*, Heidelberg, 1992. Springer-Verlag.
9. G. Ciardo, J. Muppala, and K.S. Trivedi. Analyzing concurrent and fault-tolerant software using stochastic reward nets. *Journal of Parallel and Distributed Computing*, pages 255–269, 1992.
10. G. Ciardo, J.K. Muppala, and K.S. Trivedi. *Manual for the SPNP Package*. Department of Electrical Engineering, Duke University, Sept. 1991.
11. J. P. Courtiat, A. Pedroza, and J.M. Ayache. A simulation environment for protocol specifications described in Estelle. In M. Diaz, editor, *IFIP WG 6.1 6th International Workshop on Protocol Specification, Testing and Verification.* North-Holland, 1986.
12. A. Cumani. ESP — a package for the evaluation of stochastic Petri nets with phase-type distributed transition times. In *Proceedings of the International Workshop on Timed Petri Nets*, Torino Italy, July 1985.
13. P. Dembinski and S. Budkowski. Simulating Estelle specifications with time parameters. In H. Rudin and C. H. West, editors, *IFIP WG 6.1 7th International Workshop on Protocol Specification, Testing and Verification.* North-Holland, 1987.
14. J. Bechta Dugan, K.S. Trivedi, V. Nicola, and R. Geist. Extended stochastic petri nets: Applications and analysis. In E. Gelenbe, editor, *Performance'84*, pages 507–519. North-Holland, 1985.
15. O. C. Ibe, H. Choi, and K. S. Trivedi. Performance evaluation of client-server systems. *IEEE Transactions on Parallel and Distributed Systems*, To Appear.
16. O.C. Ibe, K.S. Trivedi, A. Sathaye, and R.C. Howe. Stochastic Petri net modeling of VAXcluster system availability. In *Proceedings of the International Conference on Petri Nets and Performance Models*, Kyoto, Japan, Dec. 1989.
17. ISO. *Information Processing -Open System Interconnection-Estelle: A formal description technique based on an extended state transition model.* ISO International Standard 9074, 1988.
18. M. Malhotra, 1992. Personal communication, Computer Science Dept., Duke University.
19. M.K. Molly. Performance analysis using stochastic Petri nets. *IEEE Transactions on Computers*, C-31:913–917, 1982.
20. J.K. Muppala, S. Woolet, and K.S. Trivedi. Real-time systems performance in the presence of failures. *IEEE computer*, May 1991.
21. J.L. Peterson. *Petri Net Theory and the Modeling of Systems.* Prentice-Hall, 1981.
22. Richard J. Linn, Jr. The features and facilities of Estelle. In M. Diaz, editor, *IFIP WG 6.1 5th International Workshop on Protocol Specification, Testing and Verification*, pages 271–296. North-Holland, 1985.

```
body RECEIVER_BODY for RECEIVER
type
  time_period = integer;
const
  high = 0;
  medium = 1;
  low = 2;
state
  IDLE, AK_SENT;
var
  ak_no:integer;
  min,max,inactive_period:time_period;
initialize
  to IDLE
    begin
      min := 1;
      max := 20;
      inactive_period := 60;
      ak_no := 0;
    end;
trans
  from IDLE
    to IDLE
      priority medium
        when N.DATA_INDICATION;
          {t1} begin
               output U.DATA_INDICATION;
               ak_no := ak_no + 1
               end;
    to AK_SENT
      provided (ak_no>0) and (ak_no<=4)
        priority low
          delay(min,max)
            {t2} begin
                 output N.SEND_AK(ak_no)
                 end;
      provided(ak_no>4) and (ak_no<7)
        priority high
          delay(min)
            {t3} begin
                 output N.SEND_AK(ak_no)
                 end;
      provided ak_no = 7
        priority high
          {t4} begin
               output N.SEND_AK(ak_no)
               end;
      provided ak_no = 0
        priority low
          delay(inactive_period)
            {t5} begin
                 output N.SEND_AK(ak_no)
                 end;
  from AK_SENT
    to IDLE
      {t6} begin
           ak_no = 0;
           end;
```

a. Estelle specification for the receiver

```
body SENDER_BODY for SENDER
const
  high = 0;
  medium = 1;
  low = 2;
state
  IDLE;
var
  sent_no:integer;
 initialize
   to IDLE
     begin
       sent_no := 0;
     end;
trans
  from IDLE
    to IDLE
      when US.DATA_INDICATION;
      provided sent_no < 7;
      priority medium
        {s1} begin
             output NS.DATA_INDICATION;
             sent_no := sent_no+1;
             end;
      when NS.SEND_AK(ak_no);
      priority high
        {s2} begin
             sent_no := sent_no-ak_no
             end;
```

b. Estelle specification for the sender

Fig. 7. Estelle specification for the receiver

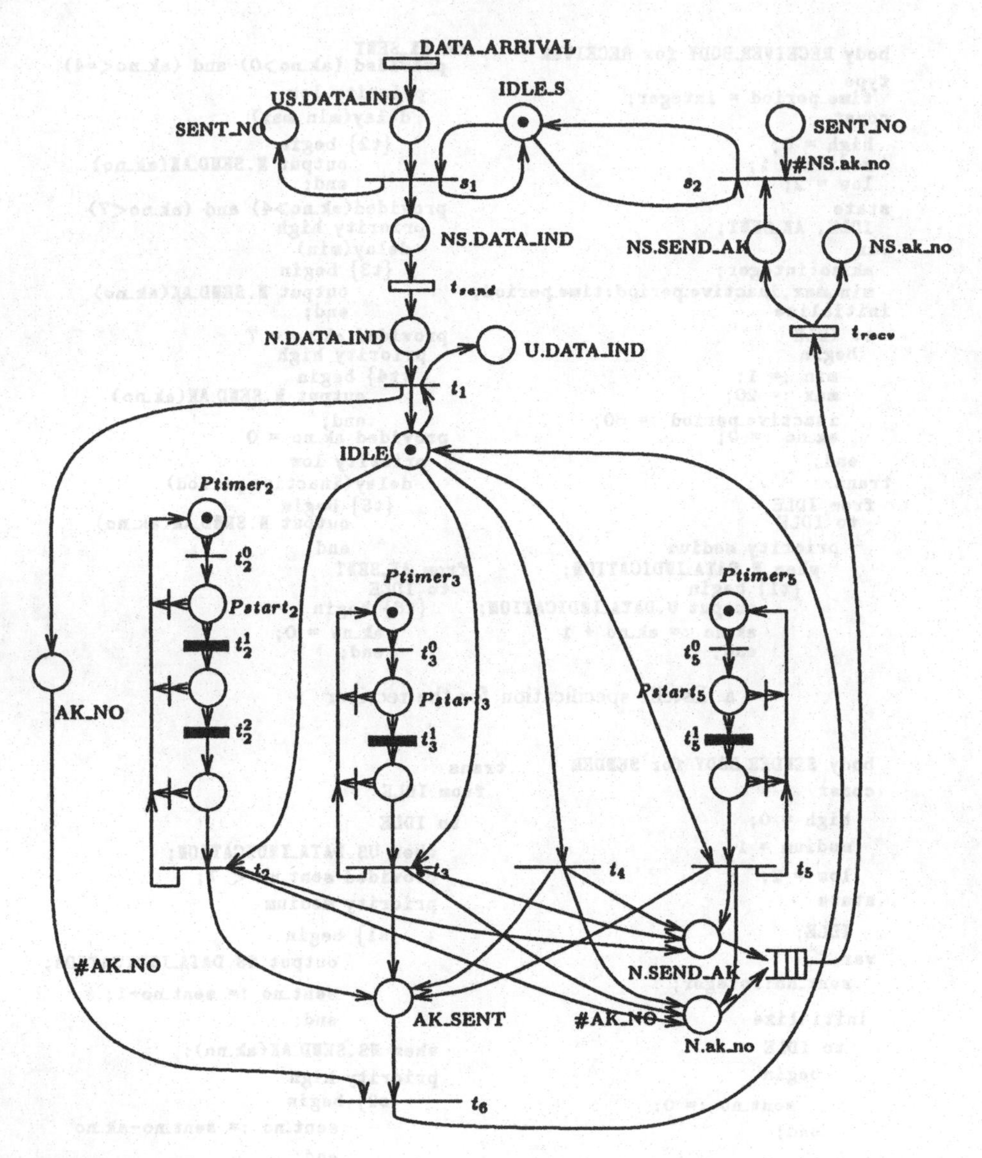

Enabling functions:

$t_2^0 : (\#IDLE == 1) \wedge (0 < \#AK\_NO \leq 4)$

$t_3^0 : (\#IDLE == 1) \wedge (4 < \#AK\_NO < 7)$

$t_4 : (\#IDLE == 1) \wedge (\#AK\_NO == 7)$

$t_5^0 : (\#IDLE == 1) \wedge (\#AK\_NO == 0)$

$s_1 : \#SENT\_NO < 7$

Priority:

t_1 : medium

t_2 : low

t_3 : high

t_4 : high

t_5 : low

s_1 : medium

s_2 : high

Firing times :

$t_2^1 : det(min)$

$t_2^2 : uniform(0, max - min)$

$t_3^1 : det(min)$

$t_5^1 : det(inactive\_period)$

Fig. 8. Transformed SRN

New Priority-Lists for Scheduling in Timed Petri Nets

Toshimasa Watanabe and Masahiro Yamauchi

Department of Circuits and Systems, Faculty of Engineering, Hiroshima University
4-1, Kagamiyama 1-chome, Higashi-Hiroshima, 724 Japan
E-mail:watanabe@huis.hiroshima-u.ac.jp

Abstract. The subject of the paper is to propose two new priority-lists for scheduling in timed Petri nets. Both of the proposed priority-lists are constructed by taking feasibility into consideration, and our experimental evaluation on 10200 test problems shows their superiority over those by the Sifakis bounds that have been widely used.

1. Introduction

Two new priority-lists for priority-list scheduling in timed Petri nets are proposed. It is experimentally evaluated that scheduling algorithms using these priority-lists produce better solutions than the four algorithms proposed in [26,27]. The total number of test problems we have tried so far is 10200 (6600 under infinite server semantics and 3600 under single-server semantics).

Scheduling theory is one of research fields that have been well investigated from both practical and theoretical viewpoints. The results are summarized in [4,16,17] for classical results; see [6,7,8] for bounding on approximate solutions and complexity results; [10,11] for scheduling in parallel processing. Although timed Petri nets are useful models in scheduling theory, related research results are much less than those using task graphs: see [5,9,23,24] for scheduling in marked graphs; [12,13,14,18,21,25] for minimum cycle time problems; [1,19] for periodic scheduling in timed Petri nets; [26,27] for priority-list scheduling in timed Petri nets, [28,29,30,31] for minimizing initial markings of ordinary or timed Petri nets.

Timed or ordinary Petri nets have two extreme possibilities in interpreting transition firing: *infinite-server semantics* and *single-server semantics.* The first semantics allows any transition to fire concurrently with itself, while this is not the case with the second one. Various processors and their total numbers are explicitly represented as places (called *processor pools*) and tokens residing within them, respectively, providing flexible models for scheduling problems. We often consider cyclic scheduling: scheduling for a repeatedly executing job that consists of some tasks, each of which is supposed to be executed prescribed number of times. If this job is repeated k times spending τ time units then the average τ/k is called *average completion time* (for this job). It is very likely that average completion time in cyclic scheduling can be reduced if cyclic structure of Petri net models is fully utilized. These explain some advantages of timed Petri nets over task graphs that have been used in classical scheduling problems.

We consider priority-list scheduling in timed Petri nets, that is, scheduling is done by choosing a transition of top priority from a priority-list. Priority-lists are constructed based on bottlenecks (each being a certain set of transitions defined later), which are counterparts of critical paths commonly used in classical scheduling. These lists are fixed as predetermined or can be changed dynamically. [26,27] proposed four algorithms *SPLA*, *DPLA*, *FM_SPLA* and *FM_DPLA*. *SPLA* and *FM_SPLA* (*DPLA* and *FM_DPLA*, respectively) are based on fixed (dynamic) priority-lists. It is reported in [26,27,32] that experimental results for more than 25290 total test problems show superiority of *FM_DPLA* among them. (Infinite-server semantics is handled in [26,27], while single-server semantics in [32].)

The bottlenecks used in constructing priority-lists in these algorithms are extracted by means of the well-known *Sifakis bounds* [25], which are lower bounds on completion time and which have been widely used in performance evaluation of Petri net models. Since no feasibility of scheduling (that is, firability of sequences of transitions in timed Petri nets) is considered in the computation of the Sifakis bound, actual minimum completion time tends to be much greater than this bound, possibly preventing scheduling algorithms, utilizing priority-lists constructed by means of this bound, from producing good approximate solutions. This is observed from experimental results on *FM_DPLA* and the other three algorithms: it is very often that, because of firability checking, transitions of middle priority are selected instead of those of high priority.

The subject of the paper is to propose two new ways of constructing priority-lists and to show experimentally that priority-list scheduling algorithms based on these new priority-lists are superior to the four algorithms of [26,27] under both infinite-server semantics and single-server semantics.

The first one is modification of markings used in the computation of the Sifakis bound. An initial marking M_0 has been used in computing this bound, while our computation replaces it by a marking M_a, called an *active marking*, which consists of only tokens having possibility to be used in subsequent firing of transitions. The bound obtained by this modification is no less than the Sifakis bound, improving a lower bound on completion time. This modification incorporates firability checking to some extent. It is experimentally observed that *FM_DPLAM* has better performance than *FM_DPLA*, where *FM_DPLAM* denotes *FM_DPLA* using priority-lists constructed from bottlenecks based on this modified bound. The running time is slightly greater than *FM_DPLA*.

The second one is completely new. For each transition t_f that can fire on a current marking of a Petri net, it finds a depth-first-search tree by starting from t_f and by searching edges in their direction. This tree intends to trace how tokens produced by firing t_f once are used by other transitions. Each of places p and transitions t of the tree has a weight supply(p) or rate(t), which is computed during the search. A weight, supply(p), of a place p intends to represent as a ratio how many tokens, among those that can be brought into p, are produced by firing t_f once. Another weight, rate(t), of a transition t intends to denote as a ratio how many tokens, among those deleted from input places of t, are produced by firing t_f once. The total sum over all rate(t) is denoted as effect(t_f). The value effect (t_f) is expected to show, as the sum of such ratios, to what extent firing t_f once hepls other transitions become firable. A new priority-list is constructed based on values effect(t_f) of all transitions t_f that can fire on a current marking: transitions t_f with larger values of effect(t_f) get higher priority. This priority on transitions considers their firability as the most significant measure rather than the time required by their subsequent firing. Let *YW_PLA* denote a priority-list scheduling algorithm based on this new priority-lists. Experimental evaluation on 10200 test problems shows that *YW_PLA* has better performance than *FM_DPLAM* and the four algorithms of [26,27]. The running time of *YW_PLA* is less than or greater than those of *FM_DPLA* and/or *FM_DPLAM*, depending upon input test problems.

The early version of the subject was reported in [33].

2. Basic Definitions

We assume that the reader is familiar with graph algorithms and Petri net theory (see [6,20,22], for example). A *digraph* is denoted by G=(V,A), where V and A are the sets of *vertices* and *directed edges* (often called *arcs*), respectively. We denote a directed edge e from u to v by e=(u,v). Let $^*u=\{w|(w,u)\in A\}$, $u^*=\{v|(u,v)\in A\}$ for $u\in V$. If $|^*u|=0$ ($|u^*|=0$) then u is called a *source* (a *sink*) of G. The graph obtained from G by replacing each directed edge with an undirected one is called the *underlying graph* of G. G is *weakly connected* if the underlying graph of G has an undirected path between any pair of vertices.

A *Petri net* is a simple bipartite digraph PN=(P,T,E,α,β), where P is the set of *places*, T is that of *transitions*

$$P\cap T=\phi,\ E=E_{in}\cup E_{out},\ E_{in}\subseteq K(T,P)=\{(u,v)|u\in T, v\in P\},$$

$$E_{out}\subseteq K(P,T)=\{(u,v)|u\in P, v\in T\},$$

$\alpha:E_{out}\rightarrow Z^+$ (nonnegative integers) and $\beta:E_{in}\rightarrow Z^+$ are weight functions. If $\alpha(e)=\beta(e')=1$ for any $e,e'\in E$, or if weight functions are independent of discussion then PN is denoted simply as PN=(P,T,E). We always consider PN to be a simple directed digraph unless otherwise stated. PN is a *marked graph* if $(\forall p\in P)|^*p|,|^*p|\le 1$. PN is a *state machine* if $(\forall t\in T)\ |^*t|,|t^*|\le 1$. Let $C=C^+-C^-=[c_{ij}^+]-[c_{ij}^-]$ denote a $|P|\times|T|$ matrix, called the *place-transition incidence matrix* of PN, which is defined by

$$c_{ij}^+=\begin{cases}\beta(t_j,p_i) & \text{if } (t_j,p_i)\in E_{in},\\ 0 & \text{otherwise,}\end{cases}\qquad c_{ij}^-=\begin{cases}\alpha(p_i,t_j) & \text{if } (p_i,t_j)\in E_{out},\\ 0 & \text{otherwise.}\end{cases}$$

A *marking* M of PN is a function $M:P\rightarrow Z^+$. We denote $|M|=\Sigma_{p\in P}M(p)$. A marking initially given is called an *initial marking*. A transition t is *firable* on a marking M consecutively k times (k≥1) if $(\forall p\in {}^*t)\ M(p)\ge k\cdot\alpha(p,t)$. *Firing* such a transition t on M consecutively k'(≤k) times is to define a marking M' such that, for $\forall p\in P$, $M'(p)=M(p)+k'\cdot\beta(t,p)$ if $p\in t^*-{}^*t$, $M'(p)=M(p)-k'\cdot\alpha(p,t)$ if $p\in {}^*t-t^*$, $M'(p)=M(p)-k'\cdot\alpha(p,t)+k'\cdot\beta(t,p)$ if $p\in t^*\cap{}^*t$ and $M'(p)=M(p)$ otherwise. We denote $M'=M[t>$ if k'=1. *Single-server semantics* is to restrict k' as k'=1 even if k≥2; *infinite-server semantics* is to allow k' to take any value with 1≤k'≤k. In this paper we use the term "Petri nets" under infinite-server semantics unless otherwise stated. Let $\delta=t_{i1}\ldots t_{is}$ be a sequence of transitions, called a *firing sequence*, and $\bar{\delta}(t)$ be the total number of occurrences of t in δ. $\bar{\delta}=[\bar{\delta}(t_1)\ldots\bar{\delta}(t_n)]^{tr}$ (transposition of a matrix or a vector) is called the *firing count vector* of δ. For a marking M, δ is *legal* on M if t_{ij} is firable on M_{j-1}, where $M_0=M$ and $M_j=M_{j-1}[t_{ij}>$, j=1, . . . ,s. M_s is denoted as M[δ>. We say that M[δ> is *reachable* from M. For a |T|-dimensional vector $X=[x_1\ldots x_n]^{tr}$ with n=|T|, δ is *legal* on M *with respect to* X if δ is legal on M and $\bar{\delta}=X$. We denote $|X|=\Sigma_{t\in T}X(t)$. For any subset $T'\subseteq T$, the subnet $PN_{T'}=(P',T',E')$ (generated by T') is defined by $P'=\{p\in P|{}^*p\cap T'\neq\emptyset$ or $p^*\cap T'\neq\emptyset\}$ and $E'=\{(p',t),(t,p'')|\ t\in T',\ p',p''\in P'\}$. For a subset $S\subseteq P\cup T$, let PN-S denote the Petri net obtained by deleting all element of S from PN, where deleting $v\in S$ means deletion of v as well as all edges incident upon v. Let **0** (**1**, respectively) denote a vector with every component equal to 0 (1). A |T|-dimensional vector X with every component being a nonnegative integer is called a *T-invariant* of PN if $X\neq\mathbf{0}$ and $C\cdot X=\mathbf{0}$. A |P|-dimensional vector Y with every component being a nonnegative integer is called a *P-invariant* of PN if $Y\neq\mathbf{0}$ and $Y^{tr}\cdot C=\mathbf{0}$. Any linear combination X' of some T-invariants of PN is also a T-invariant if all elements of X' are nonnegative. A T-invariant X is called *elementary* if no linear combination of other T-invariants of PN is equal to it. Similarly elementary P-invariant is defined.

3. Timed Petri Nets and Scheduling

3.1. Timed Petri Nets

A *timed Petri net* is a Petri net $PN=(P,T,E,\alpha,\beta)$ with a *delay function* $D:T\to Z^+$. D(t) is called the *delay* of $t\in T$. It is often denoted as $PN=(P,T,E,\alpha,\beta,D)$ in the following. In this paper we assume that any transition t has $D(t)>0$. When we consider a timed Petri net PN, time instant or time interval that is an integer is always associated with markings and firing of transitions of PN. We consider only markings at integer time instant. (A Petri net without time is sometimes called an *ordinary* Petri net in order to distinguish it from a timed one.) An initial marking means a marking at time instant $0\in Z^+$. A marking M at time instant $\lambda\in Z^+$ is often denoted as $M^{<\lambda>}$. Firability of transitions is the same as those of ordinary ones. The difference exists in firing of a transition and the resulting marking. Any transition can begin its firing only at integer time instant. If a transition t fires (that is, begins its firing) on a marking $M^{<\lambda>}$ then, at time instant $\lambda+\varepsilon$ for a very small rational number $0<\varepsilon<1/2$, M is changed to another marking M' such that, for $\forall p\in P$, $M'(p)=M(p)-\alpha(p,t)$ if $p\in{}^*t$, and $M'(p)=M(p)$ otherwise. The transition t ends its firing at time instant $\lambda+D(t)-\varepsilon$, and, at time instant $\lambda+D(t)$, we get a marking M" such that, for $\forall p\in P$, $M''(p)=M'(p)+\beta(t,p)$ if $p\in t^*$, and $M''(p)=M'(p)$ otherwise. We say that t ends its firing at time instant $\lambda+D(t)$. For notational simplicity, we handle M' and M" as if they were $M^{<\lambda>}$ and $M^{<\lambda+D(t)>}$, respectively, unless otherwise stated: introducing ε clearly shows that M' follows $M^{<\lambda>}$ and M" precedes $M^{<\lambda+D(t)>}$.

We formally define the relation between $M^{<\lambda>}$ and $M^{<\lambda+\omega>}$ for a positive integer $\omega\in Z^+$. Suppose that X' (X", respectively) is a |T|-dimensional vector such that, for each $t\in T$, X'(t) (X"(t)) denotes the total occurrence of firing of $t\in T$ that begins at time instant τ', $\lambda\leq\tau'<\lambda+\omega$ (that ends at time instant τ'', $\lambda<\tau''\leq\lambda+\omega$). We define two m-dimensional vectors

$$B(t,\alpha)=[X'(t)\cdot b_1(t,\alpha),\ldots,X'(t)\cdot b_m(t,\alpha)]^{tr},\ B(t,\beta)=[X''(t)\cdot b_1(t,\beta),\ldots,X''(t)\cdot b_m(t,\beta)]^{tr}$$

for each $t\in T$ and $P=\{p_1,\ldots,p_m\}$ (m=|P|), where

$$b_s(t,\alpha)=\begin{cases}-\alpha(p_s,t) & \text{if } p_s\in{}^*t,\\ 0 & \text{otherwise,}\end{cases}\qquad b_s(t,\beta)=\begin{cases}\beta(t,p_s) & \text{if } p_s\in t^*\\ 0 & \text{otherwise,}\end{cases}$$

s=1, ..., m. Then $M^{<\lambda+\omega>}$ is defined by

$$M^{<\lambda+\omega>}=M^{<\lambda>}+\Sigma_{t\in T}(B(t,\alpha)+B(t,\beta)).$$

We say that $M^{<\lambda+\omega>}$ is reachable from $M^{<\lambda>}$. Let δ be a firing sequence each of whose elements, t, fires simultaneously X'(t) times at time instant λ, and $D'=\max\{D(t)|t$ appears in $\delta\}$. We denote as $M^{<\lambda+D'>}=M^{<\lambda>}[\delta>$ if no other transition begins its firing at τ', $\lambda\leq\tau'<\lambda+D'$ and no other transition ends its firing at τ'', $\lambda<\tau''\leq\lambda+D'$.

In this paper we assume that a timed Petri net $PN=(P,T,E,\alpha,\beta,D)$ has a specified set $PL=\{h_1,...,h_r\}\subseteq P$, $r\geq1$, such that $\cup_{1\leq i\leq r}T(h_i)=T$, where $T(h_i)=\{t\in T|h_i\in {}^*t\cap t^*\}$ and $\alpha(h_i,t)=\beta(t,h_i)=1$ for $\forall t\in T(h_i)$, $i=1,...,r$. Each $h_i\in PL$ is called a *processor pool* of type i. PN'=PN-PL is called the *underlying Petri net* of PN.

3.2. Scheduling in Timed Petri Nets

We define scheduling in a timed Petri net PN with a set PL of processor pools. Suppose that nonnegative integers $q_1,...,q_r$ $(r\geq1)$ are given, and let M_0 be any initial marking of PN satisfying that $M_0(h_i)=q_i$, $i=1,...,r$. This means that there are r types 1,...,r of processors (represented by processors pools $h_1,...,h_r$) and that total q_i processors of type i are available initially for i=1,...,r. In the following, unless otherwise stated, we assume that any initial marking of PN is as above. All processors of type i have the same capability and are numbered $1,...,q_i$ for each i, $1\leq i\leq r$. Let $q_{max}=\max\{q_1,...,q_r\}$. For a given |T|-dimensional vector X, let

$$\Psi(T,X)=\{(t,1),..., (t,X(t))|t\in T\},$$

where X(t) denotes the element of X corresponding to $t\in T$.

We assume that timed Petri nets satisfy the following conditions (C-1) - (C-4) unless otherwise stated.

(C-1) *no wait*: any transition has to fire as soon as it becomes firable.

(C-2) *nonpreemptive*: once a transition t starts firing at some time instant $j\in Z^+$ then it keeps firing through j+D(t) and cannot be interrupted during this interval.

(C-3) Only time instant that is an integer is considered.

(C-4) Transitions can fire only at some time instant.

Suppose that we are given a timed Petri net $PN=(P,T,E,\alpha,\beta,D)$, an initial marking M_0 and a |T|-dimensional vector X. Let

$$Z_r=\{1,...,r\},\ \Delta=\{1,...,q_{max}\},\text{ and } L(t)=\{h_i\in PL|h_i\in {}^*t\cap t^*\}\text{ for each } t\in T.$$

For any subset

$$S=\{\{i_1,j_1\},...,\{i_{r'},j_{r'}\}\}\subseteq 2^{Z_r\times\Delta}\text{ with } r'\leq r,\ 1\leq i_1\leq...\leq i_{r'}\leq r,\ 1\leq j_k\leq q_{i_k}\ (k=1,...,r'),$$

let

$$Z(S)=\{i_1,...,i_{r'}\}\text{ and }\Delta(S)=\{j_1,...,j_{r'}\}.$$

A *scheduling* (under infinite-server semantics) is a function

$$\sigma:\Psi(T,X)\rightarrow Z^+\times 2^{Z_r\times\Delta}$$

that satisfies the following conditions (1)-(5), where $\sigma((t,x))$ is written as $\sigma(t,x)$ for notational simplicity, and the first or second element of $\sigma(t,x)$ is denoted as $\sigma(t,x)_1\in Z^+$ (time instant) or $\sigma(t,x)_2\in 2^{Z_r\times\Delta}$ (a set of processors of some types), respectively:

(1) if M is a marking at time instant $\sigma(t,x)_1$ then t is firable on M for any $(t,x)\in\Psi(T,X)$;

(2) $\sigma(t,x)_1$ is time instant when firing of t is supposed to begin for each $(t,x)\in\Psi(T,X)$;

(3) if $\sigma(t,x)_2=\{\{i_1,j_1\},...,\{i_{r'},j_{r'}\}\}$ for some $r'\leq r$ then (i) and (ii) hold:

(i) $\{h_{i_1},...,h_{i_{r'}}\}$=PL(t);

(ii) the x-th firing of t is associated with processing by the j_k-th processor of type i_k, which is selected from available ones, for each k, k=1,...,r';

(4) if there is any pair $\sigma(t_1,x_1)_2$ and $\sigma(t_2,x_2)_2$ such that $\{i_k,j_k\}\in\sigma(t_1,x_1)_2\cap\sigma(t_2,x_2)_2$ and $t_1\neq t_2$ then either $\sigma(t_1,x_1)_1\geq\sigma(t_2,x_2)_1+D(t_2)$ or $\sigma(t_1,x_1)_1+D(t_1)\leq\sigma(t_2,x_2)_1$ holds;

(5) $|\theta(j,\lambda)|\leq q_j$ for any type j, $1\leq j\leq r$, and any time instant $\lambda\in Z^+$, where $\theta(j,\lambda)=\{(t,x)\in\Psi(T,X)|j\in Z(\sigma(t,x)_2),\ \sigma(t,x)_1\leq\lambda\leq\sigma(t,x)_1+D(t)\}$.

Note that the following (6) is to be added in handling single-server semantics:

(Under single-server semantics)

(6) if there is any pair $\sigma(t,x_1)_2$ and $\sigma(t,x_2)_2$ with $x_1<x_2$ then $\sigma(t,x_1)_1+D(t)\leq\sigma(t,x_2)_1$.

If r'=1 then $\{\{i_1,j_1\}\}$ is denoted as $\{i_1,j_1\}$ to avoid extra brackets. Let

$$\tau(\sigma)=\max\{\sigma(t,x)_1+D(t)|(t,x)\in\Psi(T,X)\}$$

be called *completion time* of σ. We say that σ is called a *scheduling of completion time* $\tau(\sigma)$ with respect to M_0, X and PN. If we obtain a marking $M^{<\tau(\sigma)>}$ equal to a given initial marking $M_0(=M^{<0>})$ then σ is called a *cyclic scheduling of period* $\tau(\sigma)$ with respect to M_0, X and PN. The *scheduling problem of timed Petri nets* PLS($r;q_1,...,q_r$) is defined by:

Instance: A timed Petri net PN=(P,T,E,α,β,D) with a set PL=$\{h_1,...,h_r\}\subseteq$P(r≥1) of processor pools, r nonnegative integers $q_1,...,q_r$, a firing count vector X and an initial marking M_0.

Question: Find a scheduling σ of minimum completion time $\tau(\sigma)$ with respect to M_0, X and PN.

Example 1. We show an example of PLS($r;q_1,...,q_r$) with r=2, $q_1=q_2=1$. Suppose that there are two jobs J_1, J_2 with J_1(J_2 respectively) consisting of two (three) tasks, denoted as $J_1=\{t_1,t_2\}$, $J_2=\{t_3,t_4,t_5\}$, and that we have two types of processors, one processor for each type, denoted as Type$_1=\{h_1\}$, Type$_2=\{h_2\}$. The following constraints are imposed on the tasks:

(i) t_1 is to be processed in a unit time by h_1 after t_2 is finished.

(ii) t_2 is to be processed in a unit time by h_1 after both t_1 and t_3 are finished.

(iii) t_3 is to be processed in a unit time by h_1 and h_2 after t_5 is finished.

(iv) t_4 is to be processed in a unit time by h_2 after t_3 is finished.

(v) t_5 is to be processed in a unit time by h_2 after both t_2 and t_4 are finished.

A timed Petri net PN=(P,T,E,α,β,D), representing this situation schematically, is constructed as follows (see Fig. 1):

$$P=\{p_i|i=1,...,5\}\cup PL,\ PL=\{h_1,h_2\},\ T=\{t_i|i=1,...,5\},\ E=E'\cup E_{PL}$$

$$E'=\{(t_1,p_1),(p_1,t_2),(t_2,p_2),(p_2,t_1),(t_3,p_1),(p_2,t_5),(t_3,p_3),(p_3,t_4),(t_4,p_4),(p_4,t_5),(t_5,p_5),(p_5,t_3)\},$$

$$E_{PL}=\{(h_1,t_i),(t_i,h_1)|i=1,2,3\}\cup\{(h_2,t_i),(t_i,h_2)|i=3,4,5\},$$

$$\alpha(u,v)=\begin{cases}2 & \text{if } (u,v)=(p_1,t_2),\\ 1 & \text{otherwise,}\end{cases}\qquad \beta(u,v)=\begin{cases}2 & \text{if } (u,v)=(t_2,p_2),\\ 1 & \text{otherwise.}\end{cases}$$

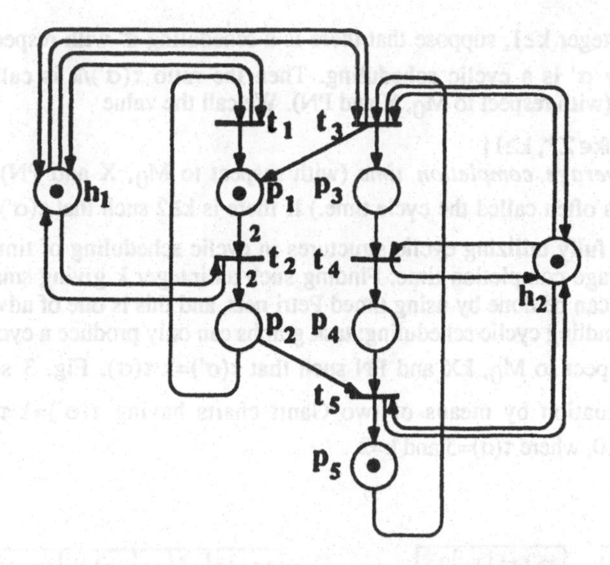

Fig. 1. An example of a (timed) Petri net. This also represents a set of tasks $\{t_1,t_2,t_3,t_4,t_5\}$ of Example 1.

Let $X=[1,1,1,1,1]^{tr}$ be the T-invariant of PN given in Example 1, $D(t)=1$ for $\forall t\in T$, and $\Psi(T,X)=\{(t_i,1)|i=1,2,3,4,5\}$. For the initial marking $M=[1,0,0,0,1,1,1]^{tr}$ with the i-th element denoting $M(p_i)$, the following σ is a scheduling with $\tau(\sigma)=3$ (see Fig. 2):

$\sigma(t_1,1)=(2,\{1,1\})$, $\sigma(t_2,1)=(1,\{1,1\})$, $\sigma(t_3,1)=(0,\{\{1,1\},\{2,1\}\})$, $\sigma(t_4,1)=(1,\{2,1\})$, $\sigma(t_5,1)=(2,\{2,1\})$.

Starting from $M^{<0>}(=M)$, we have $M^{<1>}=M^{<0>}[t_3>=[2,0,1,0,0,1,1]^{tr}$, $M^{<2>}=M^{<1>}[t_2t_4>=[0,2,0,1,0,1,1]^{tr}$, $M^{<3>}=M^{<2>}[t_1t_5>=M^{<0>}$, assuring that X is a T-invariant of PN. In this case $\tau(\sigma)=3$ is the minimum completion time, and σ is a solution to PLS(2;1,1). ♦

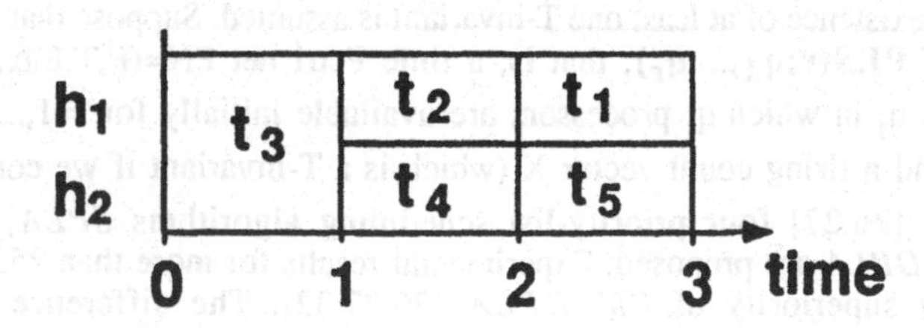

Fig. 2. A Gantt chart for the scheduling σ of Example 1.

3.3. Average Completion Time

Let σ be a cyclic scheduling with respect to M_0, X and PN, where X is a T-invariant of

PN. For some integer $k \geq 1$, suppose that there is a scheduling σ' with respect to M_0, kX and PN. Clearly σ' is a cyclic scheduling. Then the ratio $\tau(\sigma')/k$ is called *average completion time* (with respect to M_0, X and PN). We call the value

$$\min\{\tau(\sigma')/k \mid k \in Z^+, k \geq 1\}$$

the *minimum average completion time* (with respect to M_0, X and PN). (The value $lim_{k\to\infty}\tau(\sigma')/k$ is often called the cycle time.) If there is $k \geq 2$ such that $\tau(\sigma')/k < \tau(\sigma)$ then this implies that fully utilizing cyclic structures in cyclic scheduling of timed Petri nets may reduce average completion time. Finding such an integer k giving smaller average completion time can be done by using timed Petri nets, and this is one of advantages over task graphs in handling cyclic scheduling: task graphs can only produce a cyclic schduling σ' with with respect to M_0, kX and PN such that $\tau(\sigma')=k\cdot\tau(\sigma)$. Fig. 3 schematically explains this situation by means of two Gantt charts having $\tau(\sigma')=k\cdot\tau(\sigma)=10$ and $\tau(\sigma')=9<k\cdot\tau(\sigma)=10$, where $\tau(\sigma)=5$ and k=2.

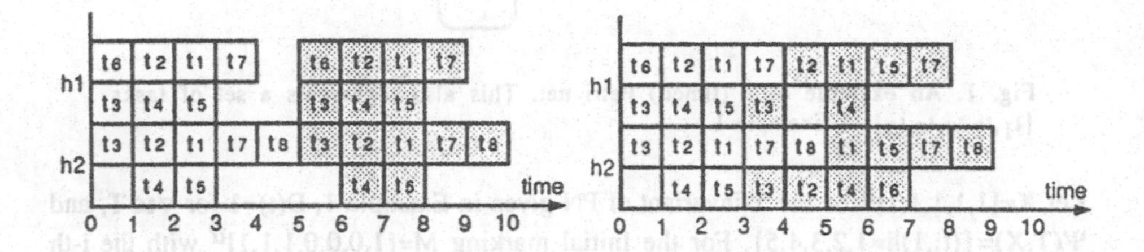

Fig. 3. Two Gantt charts schematically explaining the situation with $\tau(\sigma')=k\cdot\tau(\sigma)$ or $\tau(\sigma')<k\cdot\tau(\sigma)$. (1) $\tau(\sigma')=k\cdot\tau(\sigma)=10$; (2) $\tau(\sigma')=9<k\cdot\tau(\sigma)=10$, where $\tau(\sigma)=5$ and k=2.

4. Priority-List Scheduling

This and the following sections, except the subsection 5.4, consider only timed Petri nets having at least one P-invariant unless otherwise stated, and if cyclic scheduling is considered then existence of at least one T-invariant is assumed. Suppose that we are given any instance of $\mathbf{PLS}(r;q_1,...,q_r)$, that is, a time Petri net PN=(P,T,E,α,β,D) with r processor pools h_i in which q_i processors are available initially for i=1,...,r, an initial marking M_0, and a firing count vector X (which is a T-invariant if we consider cyclic scheduling). In [26,27] four priority-list scheduling algorithms *SPLA*, *FM_SPLA*, *DPLA* and *FM_DPLA* are proposed. Experimental results for more than 25290 total test problems show superiority of *FM_DPLA* [26,27,32]. The difference of the four algorithms exists in construction of priority-lists, and they are combined as procedure *PLS*(ξ) in [26,27], where we set ξ=1 if *SPLA*, ξ=2 if *DPLA*, ξ=3 if *FM_SPLA* and ξ=4 if *FM_DPLA*.

In this paper, priority-list scheduling algorithms are represented as a general scheme *GS*(θ) as follows, where priority-lists are fixed as predetermined if θ=1, while they are changed dynamically if θ=2. Note that θ=2 for each of the three algorithms *FM_DPLA*, *FM_DPLAM* and *YW_PLS* to be considered in the following. The priority-list L to be used in *GS*(θ) consists of transitions that are sorted in nonincreasing order of priority (the first element has the highest priority). The formal description of *GS*(θ) is as follows.

procedure $GS(\theta)$;

begin

1. $\tau \leftarrow 0$; $M^{<\tau>} \leftarrow M_0$;
2. for each $t \in T$ do $X'(t) \leftarrow 0$;
3. for i=1 to r do $j_i \leftarrow 1$; /* the smallest index of available processors of type i */
4. Construct a priority-list L.
5. Choose from L a transition t of highest priority among those having $X'(t)<X(t)$ and being firable on $M^{<\tau>}$;

   /* Note that $X'(t)<X(t)$ means $X(t)>0$ */

   /* t is firable if and only if $j_{i_k} \le q_{i_k}$, $k=1,\dots,r(t)$, where $L(t)=\{h_{i_1},\dots,h_{i_{r(t)}}\}$ */
6. **if** there is a transition t with $X'(t)<X(t)$ and a smallest time instant τ such that t is firable on $M^{<\tau>}$ **then**

   begin
7. $\sigma(t,X'(t)) \leftarrow (\tau,\{\{i_1 j_{i_1}\},\dots,\{i_{r(t)} j_{i_{r(t)}}\}\})$ for $PL(t)=\{h_{i_1},\dots,h_{i_{r(t)}}\}$;
8. $X'(t) \leftarrow X'(t)+1$;
9. for k=1 to r(t) do $j_{i_k} \leftarrow j_{i_k}+1$;
10. for each $p \in {}^*t$ do $M^{<\tau>}(p) \leftarrow M^{<\tau>}(p)-\alpha(p,t)$;
11. **if** θ is even **then goto** Step 4 **else goto** Step 5

    end

    else /* find the nearest time instant $\tau'(>\tau)$ at which a transition ends its firing */
12. **if** there is $t' \in T$ with $X'(t')<X(t')$ **then**

    begin
13. $\tau' \leftarrow \min\{\sigma(t,X'(t))_1+D(t) \mid t \in T,\ \sigma(t,X'(t))_1 < \tau < \sigma(t,X'(t))_1+D(t)\}$;
14. for each $t \in T$ with $\sigma(t,X'(t))_1+D(t)=\tau'$ do

    begin
15. for each $p \in P$ do
16. **if** $p \in t^*$ **then** $M^{<\tau'>}(p) \leftarrow M^{<\tau>}(p)+\beta(t,p)$
17. **else** $M^{<\tau'>}(p) \leftarrow M^{<\tau>}(p)$;
18. for each $h_{i_k} \in PL(t)=\{h_{i_1},\dots,h_{i_{r(t)}}\}$ do $j_{i_k} \leftarrow j_{i_k}-1$

    end;
19. $\tau \leftarrow \tau'$;
20. **goto** Step 5

    end

    else
21. **halt**

end;

Clearly $GS(\theta)$ produces a feasible scheduling σ if line 21 is not executed. Let $O(PL(\theta))$ denote time complexity for constructing L used in $GS(\theta)$ for $\theta=1,2$, and let $\chi=\Sigma_{t\in T} X(t)\cdot D(t)$ and $Q=\Sigma_{1\le i\le r} q_i$. Time complexity of $GS(\theta)$ is as follows:

$O(PL(1)+|P||T||X|+\chi Q)$ if $\theta=1$; $O((PL(2)+|P||T||X|)|X|+\chi Q)$ if $\theta=2$.

Remark 1. It should be noted that, with the above notations of time complexities using $|X|$, χ and Q, $GS(\theta)$ appears to be a polynomial-time algorithm. However this is not the case. Since each X(t) takes size proportional to $log_2 X(t)$ bits in the input, $|X|$ has size proportional to $\Sigma_{t\in T} log_2 X(t)$ (=Δ) bits. We have

$$|X|/\Delta \geq |X|/(|T|log_2\overline{X}) \geq |X|/(|T|log_2|X|),$$

and the last term is not bounded by any polynomial function of $|T|/log_2|X|$. Similarly for χ and Q. Nevertheless we use such representation as above for notational simplicity. ♦

5. Bottlenecks and Priority-Lists

We define S-bottlenecks, bottlenecks by the Sifakis bounds [25] on completion time of timed Petri nets, and explain some ways of constructing priority-lists based on S-bottlenecks. They are used in the four algorithms *SPLA*, *DPLA*, *FM_SPLA* and *FM_DPLA* [26, 27,32]. Two new ways of constructing priority-lists are also given.

5.1 S-Bottlenecks

Given an initial marking M_0, a P-invariant Y and a |T|-dimensional vector X of PN, let

$$\omega(X,Y)=(Y^{tr}\cdot C^{-}\cdot\overline{D}\cdot X)/Y^{tr}\cdot M_0$$

be called the *Sifakis bound* [25] with respect to X and Y, where $\overline{D}$ is a $|T|\times|T|$ diagonal matrix with elements d_{ij} such that

$$d_{ij}=\begin{cases} D(t_i) & \text{if } i=j,\\ 0 & \text{otherwise,}\end{cases}$$

for $i,j=1,\ldots,|T|$, where $T=\{t_1,\ldots,t_{|T|}\}$. Let

$$\omega(X,PN)=\max\{\omega(X,Y')|Y' \text{ is an elementary P-invariant of PN}\}.$$

We call $\omega(X,PN)$ the Sifakis bound of PN (with respect to X). Let Y be a P-invariant with

$$\omega(X,Y)=\omega(X,PN),\ P_Y=\{p\in P'|Y(p)>0\} \text{ and } T_Y=\bigcup_{p\in P_Y}({}^*p\cup p^*),$$

where P' is the place set of the underlying Petri net PN' of PN. The set T_Y is called an *S-bottleneck* of PN (or the S-bottleneck with respect to X and Y). The following proposition shows that $\omega(X,PN)$ is a lower bound on the period of any cyclic scheduling of timed Petri nets.

Proposition 1 [25]. Suppose that PN is a timed Petri net having a T-invariant X and at least one P-invariant, and let σ be a cyclic scheduling of period $\tau(\sigma)$ with respect to M, X and PN. Then $\tau(\sigma)\geq\omega(X,PN)$. ♦

The value $\omega(X,PN)$ can be computed by means of a linear programming as shown in the following proposition.

Proposition 2 [2,3]. The Sifakis bound $\omega(X,PN)$ of Proposition 1 can be computed from an optimum solution Y given by solving the following linear programming problem: maximize $\omega(X,Y)=Y^{tr}\cdot C^{-}\cdot\overline{D}\cdot X$ subject to $Y^{tr}\cdot C=0$, $Y\geq 0$ and $Y^{tr}\cdot M_0=1$. ♦

Example 2. PN of Fig. 1 has two elementary T-invariants

$$X_1=[2,1,0,0,0]^{tr},\ X_2=[0,1,2,2,2]^{tr},$$

where the j-th element of X_i denotes $X_i(t_j)$. It also has four elementary P-invariants

$$Y_1=[1,1,0,0,1,0,0]^{tr},\ Y_2=[0,0,1,1,1,0,0]^{tr}, Y_3=[0,0,0,0,0,1,0]^{tr},\ Y_4=[0,0,0,0,0,0,1]^{tr},$$

where the j-th element of Y_i denotes $Y_i(p_j)$. Let

$$X=[1,1,1,1,1],$$

which is a T-invariant of PN. Then we have

$\omega(X,Y_1)=5/2=2$, $\omega(X,Y_i)=3/1=3(i=2,3,4)$, $\omega(X,PN)=3$,

$P_{Y_1}=\{p_1,p_2,p_5\}$, $T_{Y_1}=\{t_1,t_2,t_3,t_5\}$; $P_{Y_2}=\{p_3,p_4,p_5\}$, $T_{Y_2}=\{t_3,t_4,t_5\}$,

$P_{Y_3}=\{h_1\}$, $T_{Y_3}=\{t_1,t_2,t_3\}$; $P_{Y_4}=\{h_2\}$, $T_{Y_4}=\{t_3,t_4,t_5\}$. ♦

5.2. Priority Lists by S-Botlenecks

We explain three measures MRi, i=1, 2, 3, used in determining priority among transitions in [26, 27]. We fix PN, M_0 and X in this section. The first measure is

MR1: if $D(t)>D(t')$ then t has priority over t'.

Let Y be a P-invariant with $\omega(X,PN)=\omega(X,Y)$ and T_Y be the bottleneck with respect to X and Y. The second measure MR2 is defined as follows:

MR2: if $t\in T_Y$ and $t'\notin T_Y$ then t has priority over t'.

The third measure MR3 is a little complicated, and is based on the costs given by the following procedure *AC*. We provide a $|T|\times\zeta$ matrix μ in which any element $\mu(i,j)=0$ initially, where ζ is the total number of elementary P-invariants of PN. Note that generally ζ is not polynomially bounded by |P|, |T| or |E|.

procedure *AC*;

1. Find all elementary P-invariants $Y_1,\ldots,Y_\zeta$ of PN by using the Fourier-Motzkin method (see [15] for example);
2. For each Y_j, repeat Steps 3 and 4;
3. Compute $\omega_j=\omega(X,Y_j)$;
4. For each $t_i\in T_{Y_j}$, $\mu(i,j)\leftarrow\omega_j$;
5. For each $t_i\in T$, sort $\mu(i,1),\ldots,\mu(i,\zeta)$ in nonincreasing order, and reindex them as $\mu(i,1)\geq\ldots\geq\mu(i,\zeta)$;
6. Let $\mu(i)$ denote the ζ-dimensional vector $[\mu(i,1),\ldots,\mu(i,\zeta)]^{tr}$, i=1,...,|T|;
   Sort $\mu(1),\ldots,\mu(|T|)$ in lexicographically nonincreasing order, and reindex them as $\mu(1)\geq,\ldots,\geq\mu(|T|)$;

The third measure MR3 is defined by using $\mu(1),\ldots,\mu(|T|)$:

MR3: if $\mu(i)>\mu(j)$ then t_i has priority over t_j.

Example 3. PN of Fig. 1 has four elementary P-invariants Y_j, j=1,...,4, as given in Example 2. Let $X=[1,1,1,1,1]^{tr}$, which is a T-invariant of PN. For each t_i, i=1,...,5, $\mu(i,j)=\omega(X,Y_j)$ (below left) and the lexicographically sorted result (below right) are given as follows, where the priority is denoted by the right most numbers in parentheses:

	Y_1	Y_2	Y_3	Y_4
t_1	5/2	0	3	0
t_2	5/2	0	3	0
t_3	5/2	3	3	3
t_4	0	3	0	3
t_5	5/2	3	0	3

→

t_1	3	5/2	0	0	(4)
t_2	3	5/2	0	0	(5)
t_3	3	3	3	5/2	(1)
t_4	3	3	0	0	(3)
t_5	3	3	5/2	0	(2)

♦

We define the following priority PR1, PR2 among MRi, i=1,2,3.

PR1: MR2>MR1; PR2: MR3>MR1,

where MR2>MR1, for example, means that if t and t' has the same priority concerning MR2 then the one of higher priority concerning MR1 is chosen; if t and t' has the same priority concerning MR1 then we assign priority such that the one with smaller index has higher priority, among all such transitions. Hence transitions of T are totally ordered. For each PRi, i=1,2, we construct a list $PL(i)=\{t_1,...,t_{|T|}\}$ of transitions of PN, where t_j has higher priority over t_{j+1} and is chosen before t_{j+1} according to PRi, for j=1,...,|T|-1. The list PL(i) is called the *priority list of type* i (with respect to X and PRi), i=1,2, and these lists have been used in the four approximation algorithms *SPLA*, *FM_SPLA*, *DPLA* and *FM_DPLA* proposed in [26,27,32].

5.3. Modification of S-Bottlenecks and Priority-Lists

We explain modification of the S-bottleneck computation. Computation of the Sifakis bound is modified so that it may reflect firability of transitions. After this modified bound is obtained, we find bottlenecks based on it and priority-lists are constructed as in 5.2. *FM_DPLA* using these modified priority-lists is denoted as *FM_DPLAM*. Only modification of the Sifakis bound is described in the following.

Let M_0, Y and X be an initial marking, a P-invariant and a firing count vector of PN, respectively. Let M be any marking reachable from M_0, and suppose that M is a marking at time instant τ_0. Let X_{done}, X_{fire} and X_{rest} be |T|-dimensional vectors whose elements are defined for t∈T as follows: $X_{done}(t)$ means that t has fired $X_{done}(t)$ times by time instant τ_0, $X_{fire}(t)$ denotes that t is being fired $X_{fire}(t)$ times simultaneously at τ_0 and firing of t is not yet finished, and

$$X_{rest}(t) = X(t) - (X_{done}(t)+X_{fire}(t)).$$

We have

$$Y^{tr}\cdot M_0 = Y^{tr}\cdot(M+C^+\cdot X_{fire}) \text{ and } \omega(X_{rest},Y) = (Y^{tr}\cdot C^-\cdot\overline{D}\cdot X_{rest}) / Y^{tr}\cdot(M+C^+\cdot X_{fire}).$$

This second formula equivalently computing $\omega(X_{rest},Y)$ implies that there are two kinds of tokens, *active tokens* and *dead ones*: an active one has possibility to be used by subsequent firing of some transitions, and a dead one has no such possibility. It seems that incorporating dead tokens in computing ω may make the value much less than actual minimum completion time. Define a marking M_a, called an *active marking*, as follows:

$$M_s = M + C^+\cdot X_{fire},\quad M_d = C^-\cdot X_{rest},$$

and, for each p∈P,

$$M_a(p) = \begin{cases} M_s(p) & \text{if } M_s(p) < M_d(p), \\ M_d(p) & \text{otherwise.} \end{cases}$$

Define a new bound $\omega'(X_{rest},Y)$ by

$$\omega'(X_{rest},Y) = (Y^{tr}\cdot C^-\cdot\overline{D}.X_{rest}) / Y^{tr}\cdot M_a.$$

We have

$$Y^{tr}\cdot M_0 \geq Y^{tr}\cdot M_a \text{ and } \omega'(X_{rest},Y)\geq\omega(X_{rest},Y).$$

M_a is computed by the following procedure, which runs in $O(|P|+|T|+|E|)$ time, and actual computation time for $\omega'(X_{rest},Y)$ is almost the same as that for $\omega(X_{rest},Y)$.

```
procedure ACTIVE_MARKING(M);
    begin
1.  for each p∈P do
```

begin

2. $M_a(p) \leftarrow 0$; $M_s(p) \leftarrow M(p)$; $M_d(p) \leftarrow 0$;
3. **for each** $t \in {}^*p$ **with** $X_{fire}(t)>0$ **do** $M_s(p) \leftarrow M_s(p)+X_{fire}(t) \cdot \beta(t,p)$;
4. **for each** $t \in p^*$ **with** $X_{rest}(t)>0$ **do** $M_d(p) \leftarrow M_d(p)+X_{rest}(t) \cdot \alpha(p,t)$;
5. **if**$(M_s(p)<M_d(p))$ **then** $M_a(p) \leftarrow M_s(p)$ **else** $M_a(p) \leftarrow M_d(p)$

end

end;

5.4. A New Priority-List

We propose a new method for determing priority on transitions, by taking firability into consideration. As mentioned in Section 1, this method is a depth-first-search during which weights are assigned to places and transitions. These weights intend to represent how firing a transition once on a current marking helps subsequent firing of other transitions.

It is not assumed that PN has a T-invariant or a P-invariant in this subsection: the discussion is independent of their existence. The outline determining priority on transitions based on values effect(t) of transitions is stated in the following, and the formal description computing values effect(t) will be given later as procedure *COMP_EFFECT*(M).

Let M be a marking at time instant τ_0 such that it is reachable from an initial marking M_0. Let X_{done}, X_{fire} and X_{rest} be defined as in 5.3. Let $t_f \in T$ be any transition having $X_{rest}(t_f)>0$ and being firable on M. We compute a value effect(t_f) by the following Steps 1-5, and then priority on transitions are determined in Step 6.

Step 1. Define a marking M_v by

$$M_v = M - M' + C^+ \cdot X_{fire},$$

where M' is another marking defined, for each $p \in P$, by

$$M'(p) = \begin{cases} \alpha(p,t_f) & \text{if } p \in {}^*t_f, \\ 0 & \text{otherwise.} \end{cases}$$

M'(p) represents tokens necessary for firing t_f once. $M-M'+C^+ \cdot X_{fire}$ is a marking at some time instant τ, with $\tau>\tau_0$, such that any firing that began at time instant $\tau' \leq \tau_0$ is finished by time instant τ, under the assumption that no other transitions begin their firing at any time instant τ'' with $\tau_0 \leq \tau'' \leq \tau$.

Step 2. Execute a depth-first-search starting at t_f and tracing unvisited edges in their direction. Each edge is originally marked "UNVISITED", and will be marked "VISITED" once it is visited by the search: every edge is visited at most once. At each place p visited from $t \in {}^*p$ during the search, we compute a value max(p) by executing

$$\max(p) \leftarrow \max\{\max(p), \beta(t,p)\}.$$

This means that we will obtain

$$\max(p) = \max\{\beta(t,p) \mid t \in {}^*p \text{ and } (t,p) \text{ is marked "VISITED"}\} \text{ for each } p \in P$$

after the completion of the search. Subsequent search from p to $t' \in p^*$ is stopped if

$$\max(p)+M_v(p)<\alpha(p,t') \text{ or } M_v(p) \geq \alpha(p,t').$$

This is because tokens produced by firing t_f once and by subsequent firing of other transitions visited so far cannot expand further through p or because t' can fire even if no such tokens are brought to p (firing of t' is independent of that of t_f), respectively.

Step 3. After the completion of the search we compute a value supply(p) for each $p \in P$ as follows:

$$supply(p)=\beta'(p)/\beta_{sum}(p),$$

where

$$\beta(p)=\Sigma_{t\in visit(p)}\beta(t,p),\ \ visit(p)=\{t\in {}^*p|(t,p) \text{ is marked "VISITED"}\},$$

$$\beta_{sum}(p)=\Sigma_{t\in {}^*p}\beta(t,p).$$

The value supply(p) intends to represent as a ratio how many tokens, among those that can be brought into p, are produced by firing t_f once.

Step 4. We compute a value rate(t) for each $t\in T$ as follows. Define rate(t) by

$$rate(t)=\begin{cases}\alpha'(t)\cdot d(t)/\alpha_{sum}(t) & \text{if } t\in T',\\ 0 & \text{otherwise,}\end{cases}$$

where

$$\alpha'(t)=\Sigma_{p\in {}^*t}\alpha(p,t)\cdot supply(p),\ \ \alpha_{sum}(t)=\Sigma_{p\in {}^*t}\alpha(p,t),\ T'=\{t\in T|X_{rest}(t)>0\}.$$

The value rate(t) intends to represent as a ratio how many tokens, among those deleted from input places of t, are produced by firing t_f once.

Step 5. Define a value effect(t_f) of t_f by

$$effect(t_f)=\Sigma_{t\in T}rate(t).$$

Note that we have computed effect(t_f) for every t_f having $X_{rest}(t_f)>0$ and being firable on M and that effect(t)=0 if t is not firable on M or if $X_{rest}(t)=0$. We are expecting the value effect(t_f) to show, as the sum of rate(t), to what extent firing t_f once helps other transitions become firable.

Step 6. Define priority on transitions of T by means of effect(t): those t with larger value of effect(t) get higher priority.

This completes the outline of constructing a new priority-list to be proposed. The formal description of procedure *COMP_EFFECT*(M) computing effect(t), $t\in T$, is given in the following, where procedure *SEARCH* is outlined in Step 2.

```
procedure SEARCH(t,Mv,state,max);
    begin
1.  for each p∈t* do
        begin
2.      if (state(t,p)=UNVISITED) then
            begin
3.          state(t,p)←VISITED;
4.          if (max(p)<β(t,p)) then max(p)←β(t,p);
5.          for each t'∈*p do
                begin
6.              if (state(p,t')=UNVISITED)∧(Mv(p)<α(p,t')≤max(p)+Mv(p))
                                                      ∧(Xrest(t')>0 ) then
                    begin
7.                  state(p,t')←VISITED;
8.                  SEARCH(t',Mv,state,max)
                    end
                end
            end
        end
    end;
```

```
procedure COMP_EFFECT(M);
    begin
1.  for each t∈T do effect(t)←0;
2.  F←{t∈T|X_rest(t) > 0, (∀p∈*t)M(p) ≥ α(p,t)};
3.  for each t_f∈F do
        begin
4.      for each p∈P do
            begin max(p)←0; supply(p)←0; β_sum(p)←0; β'(p)←0 end;
5.      for each t∈T do
            begin
6.          rate(t)←0; α_sum(t)←0; α'(t)←0;
7.          for each p∈*t do state(p,t)←UNVISITED;
8.          for each p∈t* do state(t,p)←UNVISITED
            end
9.      for each p∈P do if p∈*t_f then M'(p)←α(p,t) else M'(p)←0;
10.     M_v←M - M' + C+·X_fire;
11.     SEARCH(t_f,M_v,state,max); /* computing max(p) for ∀p∈P */
12.     for each p∈P do
            begin  /* computation of supply(p) */
13.         for each t∈*p with X_rest(t)>0 do
                begin
14.             β_sum(p)←β_sum(p)+β(t,p);
15.             if state(t,p)=VISITED then β'(p)←β'(p)+β(t,p)
                end;
16.         supply(p)←β'(p)/β_sum(p)
            end;
17.     for each t∈T with X_rest(t)>0 do
            begin  /* computation of rate(t) */
18.         for each p∈*t do
                begin α_sum(t)←α_sum(t)+α(p,t); α'(t)←α'(t)+α(p,t)·supply(p) end;
19.         if t≠t_f then rate(t)←α'(t)·d(t)/α_sum(t) else rate(t)←1
            end;
20.     for each t∈T do effect(t_f)←effect(t_f)+rate(t)
        end
    end;
```

Clearly procedure *COMP_EFFECT*(M) runs in $O(|T|(|P|+|T|+|E|))$ time, and a new priority-list can also be constructed in $O(|T|(|P|+|T|+|E|))$ time, which is $O(|T||E|)$.

6. Experimental Evaluation

We experimentally evaluate *FM_DPLAM* and *YW_PLA* by comparing with those results by the other four algorithms shown in [27,32]. All six algorithms are implemented on a workstation, DATA GENERAL AV300 (CPU: 88100; 16.7MHz), by means of C programming codes. Test nets are generated manually or randomly by the authors (see [26,27,30,31,32] for the details that are omitted here). D, X, |P|, |T|, |E| are as follows:

Table 1. A part of our experimental results for the case with r=2, RANDOM and k=5. The six columns show the results given by the corresponding algorithms, where *SPLA*, *DPLA*, *FM_SPLA*, *FM_DPLA*, *FM_DPLAM* and *YW_PLA* are denoted as S, D, FS, FD, FM and YW, respectively. The column "No." denotes the data identification: for example, sm3-eq means data #3 which is a state machine with equal delays (EQUAL) on all transitons, and "ra" does RANDOM (delays randomly generated), while "mg" and "gn" denote marked graphs and general Petri nets, respectively. The column "CT" gives average completion time $\tau(\sigma')/k$ (in the number of time units), where σ' is scheduling with respect to kX. The column "CPU" denotes CPU time in 1/60 second. The column "S-b" is the Sifakis bound, and "ratio" is the ratio CT(*)/(S-b) for each algorithm *. The notation "-" shows that datum is not available.

DATA				S			D			FS			FD			FM			YW			
No.	|P|	|T|	|E|	CT	CPU	ratio	CT	CPU	ratio	CT	CPU	ratio	CT	CPU	ratio	CT	CPU	ratio	CT	CPU	ratio	S-b
sm1-ra	15	16	72	46.4	12	1.51	47.2	48	1.54	42.8	19	1.40	42.8	75	1.40	42.8	82	1.40	39.2	60	1.28	30.7
sm7-ra	35	40	176	157.2	71	1.06	157.2	504	1.06	149.8	51	1.01	149.8	764	1.01	148.8	867	1.01	151.2	1441	1.02	148.0
sm8-ra	33	40	174	81.2	78	1.22	101.2	491	1.52	81.6	49	1.23	78.4	774	1.18	77.2	815	1.16	73.0	1113	1.10	66.5
sm14-ra	45	54	248	115.4	133	1.07	117.2	1174	1.09	116.0	71	1.08	115.8	1913	1.08	113.4	2054	1.05	109.0	5229	1.01	107.5
sm17-ra	43	54	250	75.2	111	1.11	75.2	1122	1.11	76.0	68	1.12	76.0	1679	1.12	75.6	1867	1.11	70.0	4561	1.03	68.0
sm20-ra	59	70	314	161.0	247	1.05	161.0	2477	1.05	-	-	-	-	-	-	160.0	4665	1.05	153.0	16722	1.00	153.0
sm24-ra	54	70	308	127.0	230	1.36	127.0	2359	1.36	-	-	-	-	-	-	-	-	-	95.4	10613	1.02	93.7
sm25-ra	76	85	382	177.8	439	1.04	179.6	4570	1.05	191.6	136	1.12	189.0	7435	1.11	182.2	8011	1.07	171.2	29138	1.00	171.0
sm28-ra	71	85	378	124.4	394	1.11	124.4	4350	1.11	138.8	128	1.24	138.8	7011	1.24	124.2	6991	1.11	113.8	20150	1.02	111.7
sm32-ra	82	94	418	228.0	533	1.16	228.0	5945	1.16	214.6	156	1.09	211.4	10278	1.08	217.8	10607	1.11	199.2	41378	1.02	196.0
sm47-ra	28	38	176	-	-	-	133.2	400	1.16	118.6	42	1.03	118.6	618	1.03	120.8	716	1.05	115.6	1334	1.01	115.0
sm52-ra	42	54	252	105.2	141	1.04	105.2	1135	1.04	107.2	68	1.06	107.2	1883	1.06	112.4	2028	1.11	101.8	6285	1.00	101.5
sm55-ra	59	69	316	-	-	-	-	-	-	-	-	-	-	-	-	269.0	4778	1.02	-	-	-	263.0
sm57-ra	56	69	308	149.4	271	1.12	149.4	2348	1.12	143.0	95	1.08	142.8	3640	1.07	144.6	3909	1.09	135.0	10597	1.02	133.0
sm62-ra	66	78	348	291.6	381	1.02	291.6	3444	1.02	308.8	118	1.08	308.8	5587	1.08	300.4	6943	1.05	295.0	26394	1.03	287.0
sm67-ra	84	94	420	183.8	573	1.26	183.8	5964	1.26	176.0	155	1.21	180.2	10282	1.23	166.0	10283	1.14	150.8	30247	1.03	146.0
sm70-ra	78	94	422	361.0	503	1.01	361.0	5816	1.01	361.4	154	1.01	362.4	10234	1.02	357.0	12317	1.00	359.4	39996	1.01	357.0
sm75-ra	11	14	62	48.4	9	1.15	48.4	32	1.15	44.2	14	1.05	44.2	50	1.05	44.2	55	1.05	43.6	53	1.04	42.0
sm78-ra	11	14	60	28.8	8	1.17	28.8	29	1.17	25.4	13	1.03	26.0	47	1.05	33.4	49	1.35	25.4	40	1.03	24.7
sm93-ra	44	54	254	196.0	130	1.00	196.0	1150	1.00	201.8	72	1.03	201.6	2080	1.03	202.0	2441	1.03	196.0	9409	1.00	196.0
mg4-ra	18	14	70	561.0	8	1.02	561.0	35	1.02	561.0	21	1.02	561.0	66	1.02	561.0	78	1.02	561.0	48	1.02	550.0
mg9-ra	42	38	176	1662.4	95	1.19	1599.2	530	1.15	1695.2	59	1.22	1747.4	832	1.26	1584.4	1073	1.14	1487.6	1702	1.07	1392.0
mg10-ra	42	36	186	795.2	95	1.26	757.8	510	1.20	735.6	59	1.16	720.2	738	1.14	708.0	934	1.12	711.4	884	1.12	633.0
mg31-ra	87	81	382	1866.0	527	1.11	1849.0	4438	1.10	1983.0	156	1.18	1917.8	8859	1.14	1898.2	10500	1.13	1830.0	20607	1.09	1678.0
mg40-ra	96	86	404	1245.0	671	1.08	1245.6	5585	1.08	1304.2	174	1.13	1332.2	10835	1.15	1274.4	13142	1.10	1164.2	38934	1.01	1154.7
mg41-ra	96	84	404	3589.8	587	1.06	3589.8	5207	1.06	3641.4	168	1.07	3632.8	10190	1.07	3590.6	12478	1.06	3400.2	33940	1.00	3388.0
mg42-ra	96	82	388	1646.4	621	1.09	1612.2	5034	1.07	1591.6	168	1.05	1623.2	9826	1.08	1637.2	11430	1.08	1554.6	27171	1.03	1509.0
mg58-ra	58	52	250	2429.8	169	1.03	2426.0	1286	1.03	2391.2	86	1.01	2366.0	2322	1.00	2394.0	2774	1.01	2381.6	5891	1.01	2366.0
mg59-ra	56	50	230	2029.4	187	1.01	2029.4	1176	1.01	2086.6	84	1.03	2107.4	1999	1.05	2027.8	2523	1.01	2029.4	4391	1.01	2012.0
mg72-ra	80	76	346	2241.8	468	1.12	2192.8	3899	1.10	2223.4	139	1.11	2225.6	7082	1.11	2207.4	8359	1.10	2120.4	15669	1.06	2001.0
mg75-ra	80	70	334	1247.4	399	1.03	1250.6	3150	1.03	1241.2	134	1.03	1272.0	6021	1.05	1237.4	7133	1.02	1219.0	20060	1.01	1209.5
mg76-ra	80	68	334	1541.4	376	1.06	1557.0	2907	1.07	1624.6	130	1.12	1642.2	5678	1.13	1666.2	6860	1.15	1559.8	13284	1.07	1454.0
mg80-ra	96	90	416	1872.4	696	1.04	1859.6	6168	1.03	2050.0	178	1.14	2045.4	11946	1.13	1993.0	14390	1.10	1805.2	44729	1.00	1804.5
mg83-ra	98	84	404	1389.0	685	1.24	1253.2	5566	1.12	1307.2	170	1.17	1295.2	10162	1.16	1341.4	12086	1.20	1213.4	21136	1.09	1116.3
mg86-ra	16	13	68	600.0	7	1.00	600.0	27	1.00	600.0	18	1.00	600.0	56	1.00	600.0	65	1.00	600.0	93	1.00	600.0
mg96-ra	32	24	118	540.2	24	1.01	540.2	158	1.01	573.2	42	1.07	573.2	364	1.07	573.2	421	1.07	540.2	223	1.01	536.0
mg98-ra	32	22	116	721.0	20	1.00	721.0	129	1.00	721.0	42	1.00	721.0	514	1.00	721.0	568	1.00	721.0	201	1.00	721.0
gn2-ra	18	15	72	438.0	16	1.49	438.0	54	1.49	454.8	23	1.55	404.4	98	1.38	371.4	114	1.27	513.6	77	1.75	293.0
gn9-ra	42	38	176	1662.4	114	1.19	1599.2	636	1.15	1695.2	63	1.22	1747.4	988	1.26	1584.4	1286	1.14	1487.6	2029	1.07	1392.0
gn10-ra	43	37	172	771.0	114	1.27	706.2	638	1.17	755.8	63	1.25	738.2	986	1.22	720.2	1263	1.19	668.4	1201	1.10	606.0
gn11-ra	43	36	166	605.2	126	1.16	605.2	592	1.16	605.6	63	1.16	605.0	917	1.16	576.8	1179	1.11	591.2	784	1.13	521.0
gn13-ra	40	36	168	-	-	-	-	-	-	1367.0	58	1.51	1367.0	829	1.51	-	-	-	1067.4	1121	1.18	907.0
gn16-ra	54	54	240	2126.8	242	1.08	2039.4	1635	1.03	2069.0	90	1.05	2158.8	2666	1.09	2015.4	3290	1.02	2062.4	6146	1.04	1975.0
gn17-ra	54	55	250	1111.6	294	1.18	1053.2	1781	1.12	1116.2	93	1.19	1262.2	2849	1.34	1175.0	3523	1.25	1047.0	4999	1.11	940.5
gn18-ra	53	55	250	976.4	286	1.19	876.8	1997	1.07	965.2	93	1.18	959.6	2852	1.17	921.2	3443	1.12	848.6	6060	1.03	820.3
gn21-ra	50	54	244	-	-	-	-	-	-	1625.2	86	1.06	1625.4	2565	1.06	1762.2	3066	1.15	1600.2	5027	1.04	1538.0
gn23-ra	70	70	322	2657.2	407	1.00	2657.2	3351	1.00	2680.2	129	1.01	2780.2	5963	1.05	2674.4	7532	1.01	2659.0	21696	1.00	2650.0
gn24-ra	70	70	320	2444.0	438	1.03	2424.2	3412	1.02	2466.0	128	1.04	2464.4	5941	1.04	2446.2	7259	1.03	2426.2	19658	1.02	2369.0
gn28-ra	67	70	316	-	-	-	-	-	-	-	-	-	-	-	-	-	-	-	2550.6	20199	1.00	2543.0
gn30-ra	86	85	386	1075.2	819	1.09	1071.2	6232	1.09	1130.6	174	1.15	1137.8	11711	1.16	1067.4	14044	1.08	1043.0	36179	1.06	985.0
gn31-ra	86	83	390	2082.0	743	1.07	2050.8	5728	1.07	2081.4	168	1.08	2084.0	10472	1.09	2032.4	13058	1.06	2087.0	29124	1.09	1920.0
gn34-ra	89	79	372	1553.0	713	1.07	1553.0	5300	1.07	1579.2	170	1.08	1616.8	10802	1.11	1527.8	12792	1.05	1466.6	33025	1.01	1456.0
gn35-ra	88	79	370	-	-	-	-	-	-	3049.0	164	1.00	3189.4	10160	1.05	3065.0	12281	1.01	3066.4	28662	1.01	3049.0
gn37-ra	95	94	428	3209.4	784	1.08	3048.2	7785	1.03	3013.8	199	1.02	3075.8	15647	1.04	3254.6	18594	1.10	3157.8	47814	1.07	2962.0

$D(t)=D(t')$ for $\forall t,t' \in T$ (EQUAL) or $D(t)$ is created randomly for $\forall t \in T$ (RANDOM),
$X=1$, $24 \le |P| \le 191$; $14 \le |T| \le 94$; $82 \le |E| \le 516$; $1 \le D(t) \le 100$.

The total number of test problems we have tried so far is 10200 (6600 problems under infinite-server semantics; 3600 ones under single-server semantics).

Under infinite-server semantics, 6600 problems are the combinatins of the following: 100 state machines(sm), 100 marked graphs(mg), and 100 general Petri nets(gn) as underlying Petri nets; two kinds of delays EQUAL and RANDOM; two combinations of processor pools and the number of processors ($1 \le r \le 4$; each q_i, $1 \le i \le r$, is rondomly chosen from $\{1,2,3\}$); three values of k, $k \in \{1,5,10\}$. (600 out of 7200 problems are not tried yet.)

Under single-server semantics, 3600 problems are as follows: 150 state machines, 200 marked graphs, and 50 general Petri nets as underlying Petri nets; two kinds of delays EQUAL and RANDOM; two combinations of processor pools and number of processors ($1 \le r \le 4$; each q_i, $1 \le i \le r$, is rondomly chosen from $\{1,2,3\}$); three values of k, $k \in \{1,5,10\}$. Tests nets created for infinite-server semantics are transformed into those (also under infinite-server semantics) that behave exactly as Petri nets under single-server semantics. The transformation is very simple: for each transition t, add a new place p_t having exactly one token in it and two edges (t,p_t), (p_t,t). (Only 3600 out of 9600 problems are tried.)

Table 1 shows a part of our experimental results for the case with r=2, RANDOM and k=5. Let CT(YW_PLS) and CPU(YW_PLS) denote average completion time and CPU time by *YW_PLS*, and similarly for others. Concerning average completion time,

CT(YW_PLS)≤CT(FM_DPLAM)≤CT(FM_DPLA)

in general. On the other hand, concerning CPU time, we have

CPU(FM_DPLA)≤CPU(FM_DPLAM),

while CPU(YW_PLS) is better than the others or worse than one or all of them, depending upon test problems. It is noted that there are many test problems, each having an integer k such that $\tau(\sigma')/k < \tau(\sigma)$, where σ or σ' is a scheduling with respect to X or with respect to kX, respectively. For example, the average (over all six algorithms) of the ratio $(\tau(\sigma')/k)/\tau(\sigma)$ over 6600 test problems under infinite-server semantics is 0.970 if k=5 and 0.964 if k=10. That is, utilizing cyclic structures in scheduling of timed Petri nets may lead to shorter average completion time. This assures one of advantages of timed Petri nets over task graphs in handling cyclic scheduling.

Tables 2 and 3 show some statistical data for each of the six algorithms. In Table 2, the column "ratio" denotes the ratio CT(*)/(S-b) and each integer appearing in each row denotes the total number of test problems for which feasible schedulings are found by the corresponding algorithm *. Table 3 shows other statistical data. The row "average" shows the average of the ratio CT(*)/(S-b) , "success" does the total number of test problems for which feasible schedulings are found by the corresponding algorithm *, "only_one" does that of those for which only the algorithm * has succeeded to produce feasible schedulings but any of others has failed, and "inferior" does that of those for which the algorithm * has failed to get feasible schedulings but at least one of the others has succeeded. These tables experimentally assure superiority of our new priority-lists and shows that *YW_PLA* has best performance among the six algorithms.

It should be mentioned that *YW_PLA* produces optimum solutions to examples (see Appendix of [27]) each of whose worst approximation by *FM_DPLA* or *FM_DPLAM* cannot be bounded by a constant.

7. Concluding Remarks

The followings are left for future research:

Table 2. The total number of test problems for which feasible schedulings are found by each of the six algorithms: (1) infinite-server semantics; (2) single-server semantics. The column "ratio" denotes the ratio CT(*)/(S-b), where ">2.00" shows the case with the ratio greater than 2.00. Each integer appearing in each row denotes the total number of test problems for which feasible schedulings are found by the corresponding algorithm *.

(1)

ratio	S	D	FS	FD	FM	YW
1.00	633	614	858	1235	1136	1030
1.05	1300	1505	1600	1495	1399	1932
1.10	789	807	706	563	645	624
1.15	538	514	415	358	400	340
1.20	320	259	201	175	207	187
1.25	235	178	128	115	135	120
1.30	127	113	82	73	86	56
1.35	85	54	47	41	66	33
1.40	43	40	34	31	29	20
1.45	41	34	27	31	25	15
1.50	18	21	17	14	14	6
1.55	37	34	33	38	36	26
1.60	10	5	9	5	7	2
1.65	9	8	17	14	13	7
1.70	11	12	13	12	11	8
1.75	8	8	13	6	7	4
1.80	2	2	3	2	1	2
1.85	1	0	3	1	1	1
1.90	5	5	4	3	3	0
1.95	2	1	0	0	1	0
2.00	3	2	1	1	1	0
>2.00	3	4	5	5	5	3

(2)

ratio	S	D	FS	FD	FM	YW
1.00	517	465	530	605	658	607
1.05	802	862	937	1130	821	1130
1.10	449	460	397	266	378	341
1.15	230	240	203	177	194	146
1.20	133	130	103	64	102	93
1.25	112	93	74	53	85	45
1.30	46	55	54	36	44	19
1.35	42	32	21	16	25	14
1.40	28	27	22	12	18	9
1.45	13	13	20	15	14	4
1.50	7	7	11	7	13	3
1.55	22	19	23	19	22	17
1.60	4	2	4	6	4	2
1.65	4	3	7	7	3	1
1.70	8	8	6	6	6	3
1.75	3	2	4	1	4	0
1.80	1	0	1	1	0	1
1.85	0	0	4	2	3	1
1.90	4	4	3	3	3	0
1.95	1	1	0	0	0	0
2.00	1	1	0	0	1	0
>2.00	2	2	4	4	3	2

Table 3. Other statistical data: (1) infinite-server semantics; (2) single-server semantics. The row "average" shows the average of the ratio CT(*)/(S-b), "success" does the total number of test problems for which feasible schedulings are found by the corresponding algorithm *, "only_one" does that of those for which only the algorithm * has succeeded to produce feasible schedulings but any of others has failed, and "inferior" does that of those for which the algorithm * has failed to get feasible schedulings but at least one of the others has succeeded.

(1)

	S	D	FS	FD	FM	YW
average	1.101	1.090	1.079	1.068	1.065	1.057
success	4220	4220	4216	4218	4228	4416
only_one	2	1	5	5	21	206
inferior	353	353	351	349	339	151

(2)

	S	D	FS	FD	FM	YW
average	1.085	1.081	1.078	1.059	1.073	1.050
success	2429	2426	2428	2430	2401	2438
only_one	4	0	4	4	8	63
inferior	271	187	177	158	195	130

(1) incorporating the number of time units required by subsequent firing of transitions as the second measure in constructing priority-lists to be used in *YW_PLA*;
(2) providing more experimental results for $r \geq 3$ or $q_i \geq 4$;
(3) theoretical evaluation of approximate solutions;
(4) estimating integers k with $\tau(\sigma')/k < \tau(\sigma)$, where σ or σ' is a scheduling with respect to X or with respect to kX, respectively.

Acknowledgements

The research of T. Watanabe was partly supported by the Telecommunications Advancement Foundation (TAF), Tokyo, Japan; and is partly supported by The Okawa Institute of Information and Telecommunication, Tokyo, Japan.

References

[1] H.Arai, A.Fujimori and T.Hisamura, Applications of Timed Petri Net to Scheduling Problems of Repetitive Processes and Their Restoring Strategies in Emergency Stops, Trans. Society of Instrument and Control Engineers, Vol.22, No.9(1986), pp.955-961. (in Japanese)
[2] J.Campos, G.Chiola and M.Silva, Ergodicity and Throughput Bounds of Petri Nets with Unique Consistent Firing Count Vector, IEEE Trans. Software Engineering, Vol.SE-17, No.2(1991), pp117-125.
[3] J.Campos, G.Chiola, J.M.Colom and M.Silva, Tight Polynomial Bounds for Steady-State Performance of Marked Graphs, Proc. 3rd International Workshop on Petri Nets and Performance Models (December 1989), pp.200-209.
[4] E. G. Coffman, Computer and Job-Shop Scheduling Theory, John Wiley & Sons, N. Y., 1976.
[5] F. Commoner, A. W. Holt, S. Even and A. Pnueli, Marked Directed Graphs, J. Computer and System Sciences, Vol.5, No.5(1971), pp.511-523.
[6] M. R. Garey and D. S. Johnson, Computers and Intractability: A Guide to the Theory of NP-completeness, Freeman, San Francisco, CA, 1978.
[7] R. L. Graham, Bounds on Multiprocessing Timing Anomalies, SIAM J. Appl. Math., Vol.17, No.2(1969), pp.416-429.
[8] R. L. Graham, E. L. Lawler, J. K. Lenstra and A. H. G. Rinnooy Kan, Optimization and Approximation in Deterministic Sequencing and Scheduling: A Survey, Annals of Discrete Mathematics Vol.5(1979), pp.287-326.
[9] H. P. Hillion and J. M. Proth, Performance Evaluation of Job-Shop Systems Using Timed Event-Graphs, IEEE Trans. Automatic Control, Vol.34, No.1(1989), pp.3-9.
[10] K. Iwano and S.Yeh, An Efficient Algorithm for Optimal Loop Parallelization, Research Report RT 0043, IBM Research, Tokyo Research Laboratory, Chiyoda-Ku, Tokyo, Japan (March, 1990).
[11] H. Kasahara and S. Narita, A Practical Optimal/Approximate Algorithm for Multi-Processor Scheduling Problem, Trans. IEICE of Japan, Vol.J67-D, No.7(1984), pp.792-799. (in Japanese)
[12] J. Magott, Performance Evaluation of Concurrent Systems Using Petri Nets, Information Processing Letters, 18(1984), pp.7-13.
[13] J. Magott, Performance Evaluation of Systems of Cyclic Sequential Processes with Mutual Exclusion Using Petri Nets, Information Processing Letters Vol.21(1985), pp. 229-232.

[14] J. Magott, New NP-Complete Problems in Performance Evaluation of Concurrent Systems Using Petri Nets, IEEE Trans. Software Engineering, Vol.SE-13, No.5(1987), pp.578-581.
[15] J.Martinez and M.Silva, A Simple and Fast Algorithm to Obtain All Invariants of a Generalized Petri Net, Fachberichte Informatik, Vol.52, (C.Girault and W.Reisig (Eds)), Springer-Verlag, Berlin, pp.301-310 (1982).
[16] R. R. Muntz and E. G. Coffman, Jr., Optimal Preemptive Scheduling on Two-Processor Systems, IEEE Trans. on Computers, Vol. C-18, No.11(1969), pp.1014-1020.
[17] I. Nabeshima, Sukejuringu-Riron (Theory of Scheduling), Morikita Shuppan Pub. Co. Ltd., 1974. (in Japanese)
[18] K. Onaga, Scheduling of Extended Marked Graphs, Trans. IEICE, Vol.J69-A, No.2(1986), pp.241-251. (in Japanese)
[19] K. Onaga, T. Watanabe and M. Silva, On Periodic Schedules for Deterministically Timed Petri Nets Systems, Proc. 4th International Workshop on Petri Nets and Performance Models (PNPM91) (Dec., 1991), pp.210-215. (See also: K.Onaga, M.Silva and T.Watanabe, Qualitative Analysis of Periodic Schedules for Deterministically Timed Petri Nets Systems, Trans. IEICE, Vol.E76-A, No.4(1993), to appear.)
[20] J. L. Peterson: Petri Net Theory and the Modeling of Systems, Prentice-Hall, Englewood Cliffs, N.J., 1981.
[21] C. V. Ramamoorthy and G. S. Ho, Performance Evaluation of Asynchronous Concurrent Systems Using Petri Nets, IEEE Trans. Software Eng., Vol. 6, No.5(1983), pp.440-449
[22] W. Reisig: Petri Nets/An introduction, Springer-Verlag, Berlin, 1982.
[23] R.Reiter, Scheduling Parallel Computations, J. Assoc. Computing Math., 15, pp.590-599 (1968).
[24] R.Reiter, On Assembly-Line Balancing Problems, Operations Research, Vol.17, No.4(1969), pp.685-700.
[25] J. Sifakis, Modeling and Performance Evaluation of Computer Systems, Proc. Third International Symposium on Measuring, Modeling and Evaluating Computer Systems, (H.Beilner and E.Gelenbe (Eds.)), North-Holland, Amstedam, pp.75-93(1977).
[26] T.Tanida, T.Watanabe, K.Masuoka and K.Onaga, Scheduling in a Timed Petri Net Model of a Repeatedly Executing Set of Tasks — Priority-List Scheduling —, Tech. Rep. IEICE of Japan, COMP92-94, pp.41-48 (March 1992).
[27] T.Tanida, T.Watanabe, M.Yamauchi and K.Onaga, Priority-List Scheduling in Timed Petri Nets, Trans. IEICE of Japan, Vol. E 75, No.10(1992), pp. 1394-1406.
[28] T. Watanabe, Y. Mizobata and K. Onaga, Minimum Initial Marking Problems of Petri Nets, Trans. IEICE of Japan, Vol.E72, No.12(1989), pp.1390-1399.
[29] T. Watanabe, Y. Mizobata and K. Onaga, Time Complexity of Legal Firing Sequence and Related Problems of Petri Nets, Trans. IEICE of Japan, E72, No.12(1989), pp.1400-1409.
[30] T.Watanabe, T.Tanida, M.Yamauchi and K.Onaga, The Minimum Initial Marking Problem for Scheduling in Timed Petri Nets, Trans. IEICE of Japan, Vol. E75, No.10(1992), pp. 1407-1421.
[31] M.Yamauchi, T. Tanida, T. Watanabe and K.Onaga, Scheduling in Timed Petri Net Model of a Repeatedly Executing Set of Tasks — Minimum Initial Marking Problems —, Tech. Rep. IEICE of Japan, COMP91-93, pp.29-40 (March 1992).
[32] M.Yamauchi and T. Watanabe, Experimental Evaluation of Priority-List Schedulings in Timed Petri Nets, Tech. Rep. IEICE of Japan, CASP92-46, pp.33-40(September 1992).
[33]M.Yamauchi and T.Watanabe, Constructing Priority-Lists for Scheduling of Timed Petri Nets, Tech. Rep. IEICE of Japan, COMP92-62, pp.11-20 (November 1992).

A Unified Approach for Reasoning about Conflict-Free Petri Nets *

Hsu-Chun Yen
Bow-Yaw Wang
Ming-Sheng Yang
Dept. of Electrical Engineering
National Taiwan University
Taipei, Taiwan, R. O. C.

Abstract. The aim of this paper is to develop a unified approach for deriving complexity results for problems concerning conflict-free Petri nets. To do so, we first define a class of formulas for paths in Petri nets. We then show that answering the satisfiability problem for conflict-free Petri nets is tantamount to solving a system of linear inequalities (which is known to be in P). Since a wide spectrum of Petri net problems (including various fairness-related problems) can be reduced to the satisfiability problem in a straightforward manner, our approach offers an umbrella under which many Petri net problems for conflict-free Petri nets can be shown to be solvable in polynomial time. As a side-product, our analysis provides evidence as to why detecting unboundedness for conflict-free Petri nets is easier (provided P $\neq$ NP) than for normal and sinkless Petri nets (which are two classes that properly contain that of conflict-free Petri nets).

1 Introduction

Despite the broad attention they have received during the past three decades, Petri nets remain one of the least understood computational models. Analyzing Petri nets, in many cases, is a costly endeavor. In fact, the decidability of the reachability problem, which is one of the most important questions concerning Petri nets, was left unsolved for many years until the work of [23] (see also [20]) which provided an answer in the affirmative. (Even so, the precise complexity of the reachability problem has not yet been established.) Other notable examples along this line of research include the EXPSPACE-completeness result of the boundedness and covering problems [21, 26], and the undecidability result of the containment and equivalence problems [2, 9]. Such a high degree of complexity is believed to be the key obstacle that limits the applicability of Petri nets to real-world problems [10]. To circumvent such a difficulty, one attempt has to do with lowering our expectations by focusing on various restricted subclasses of Petri nets instead. This line of research includes, e.g., [3, 8, 11, 22], and more recently, [4, 5, 6, 12, 13, 14, 15, 17, 33].

* This work was supported in part by the National Science Council of the Republic of China under Grants NSC-81-0408-E-002-01 and NSC-82-0408-E-002-025.

Of all of the Petri net subclasses for which completeness results have been shown concerning all four of the problems mentioned above, the class for which the decision procedures are most efficient is that of conflict-free Petri nets. In particular, the boundedness problem is complete for P (polynomial time) [15] (see also [1]), the reachability problem is NP-complete [12], and the containment and equivalence problems are Π_2^P-complete [12], where Π_2^P is the set of all languages whose complements are in the second level of the polynomial-time hierarchy [30]. The pairwise concurrency problem for 1-bounded conflict-free Petri nets is in P [33] (see also [5]). Free choice Petri nets feature another subclass for which several interesting complexity results have been obtained. For example, the liveness problem is co-NP complete for free choice Petri nets [18]; however, if we restrict ourselves to bounded free choice Petri nets, liveness can be determined in polynomial time [6]. For reversible live and bounded free choice Petri nets, the reachability problem is solvable in polynomial time [4].

Due to the fact that conflict-free Petri nets comprise such a simple class, their modeling power is somewhat limited [17]. In [32], two generalizations of conflict-free Petri nets, namely normal Petri nets and sinkless Petri nets, have been defined and studied. The containment relationship among the above three classes of Petri nets is {conflict-free} $\subsetneq$ {normal} $\subsetneq$ {sinkless}. Subsequent study [17] has shown that with the exception of the boundedness problem, the complexities of the reachability, containment and equivalence problems for normal and sinkless Petri nets are identical to those for conflict-free Petri nets. (The boundedness problem for these two classes of Petri nets is co-NP-complete.)

In light of the above, the following questions arise naturally:

1. What makes the boundedness problem 'easier' to solve (provided $P \neq NP$) for conflict-free Petri nets than for normal and sinkless Petri nets?
2. Within the class of conflict-free Petri nets, why boundedness is easier to detect than reachability?
3. Can the technique used for solving the boundedness problem for conflict-free Petri nets be generalized to a wider class of problems? (Due to practical considerations, problems solvable in P are more likely to find real-world applications.)

To answer the above questions, we will first define a class of formulas for paths in Petri nets, each of which is of the form

$$\exists \mu_1 \leq \mu_2 \leq \ldots \leq \mu_m \exists \sigma_0, \sigma_1, \ldots, \sigma_{m-1}$$
$$(\mu_0 \overset{\sigma_0}{\longmapsto} \mu_1 \overset{\sigma_1}{\longmapsto} \mu_2 \ldots \overset{\sigma_{m-1}}{\longmapsto} \mu_m) \wedge F(\sigma_0, \sigma_1, \ldots, \sigma_{m-1})$$

meaning that marking μ_i can be reached from μ_{i-1} $(1 \leq i \leq m)$ through the firing of transition sequence σ_{i-1} and predicate $F(\sigma_0, \ldots, \sigma_{m-1})$ holds. What makes our formulas useful is that it is powerful enough to express many Petri net properties. We then show that the satisfiability problem, i.e., the problem of, given a Petri net and a formula, determining whether there exists a path in the Petri net satisfying the given formula, is complete for P for conflict-free

Petri nets, whereas it is complete for NP for normal and sinkless Petri nets. Our approach can be thought of as an umbrella under which some previously known results can be explained, and more importantly, many new results can be derived immediately. Basically, the technique used for deriving the upper bounds for conflict-free and for normal (sinkless) Petri nets are identical; both rely on reducing to the integer linear programming problem. In this way, determining the existence of a path satisfying certain path formula is tantamount to solving an integer linear programming problem. (The reader is referred to [17] for the origins of this technique.) Unfortunately, NP turns out to be the best upper bound that can be concluded using such an approach, for integer linear programming problem is well-known to be NP-hard (in fact, NP-complete). With the help of a novel idea used in a recent article [5], and several unique properties possessed by conflict-free Petri nets (which will be derived in this paper), the integer constraint of linear programming can be eliminated for conflict-free Petri nets. As a result, a polynomial time algorithm follows. (Linear programming is known to be solvable in P [19].) As a side-product, our analysis also reveals the reason why reachability is harder to analyze than boundedness for conflict-free Petri nets. As far as we know, the class of conflict-free Petri nets, so far, is the only one in which the boundedness and the reachability problems are known to exhibit different completeness results (provided P $\neq$ NP). We feel that our analysis nicely explores the disparity between the characteristics of boundedness and reachability (for conflict-free Petri nets). Hopefully, this will allow us to have a better understanding of the intricate nature of Petri nets.

2 Definitions

Let Z (N) denote the set of (nonnegative) integers, and Z^k (N^k) the set of vectors of k (nonnegative) integers. For a k-dimensional vector v, let v(i), $1 \leq i \leq k$, denote the ith component of v. For a k $\times$ m matrix A, let $a_{i,j}$, $1 \leq i \leq k$, $1 \leq j \leq m$, denote the element in the ith row and the jth column of A, and let a_j denote the jth column of A. For a given value of k, let **0** denote the vector of k zeros (i.e., $\mathbf{0}(i) = 0$ for i = 1,...,k).

A *Petri net* is a 4-tuple (P,T,φ,μ_0), where P is a finite set of *places*, T is a finite set of *transitions*, φ is a *flow function* $\varphi : (P \times T) \cup (T \times P) \rightarrow \{0,1\}$, and μ_0 is the *initial marking* $\mu_0 : P \rightarrow N$. A *marking* is a mapping $\mu : P \rightarrow N$. A transition t $\in$ T is *enabled* at a marking μ iff for every p $\in$ P, $\varphi(p,t) \leq \mu(p)$. A transition t may *fire* at a marking μ if t is enabled at μ. We then write $\mu \stackrel{t}{\longmapsto} \mu'$, where $\mu'(p) = \mu(p) - \varphi(p,t) + \varphi(t,p)$ for all p $\in$ P. A sequence of transitions $\sigma = t_1 \dots t_n$ is a *firing sequence* from μ_0 (or a firing sequence of (P,T,φ, μ_0)) iff $\mu_0 \stackrel{t_1}{\longmapsto} \mu_1 \stackrel{t_2}{\longmapsto} \cdots \stackrel{t_n}{\longmapsto} \mu_n$ for some sequence of markings $\mu_1,\dots,\mu_n$. (We also write '$\mu_0 \stackrel{\sigma}{\longmapsto} \mu_n$'.) We write '$\mu_0 \stackrel{\sigma}{\longmapsto}$' to denote that σ is enabled and can be fired from μ_0, i.e., $\mu_0 \stackrel{\sigma}{\longmapsto}$ iff there exists a marking μ such that $\mu_0 \stackrel{\sigma}{\longmapsto} \mu$. For $\sigma, \sigma' \in T^*$, $\sigma' = t_1 \dots t_n$, let $\sigma \dot{-} \sigma'$ be inductively defined as follows. Let σ_0 be σ. If $t_i \in \sigma_{i-1}$, let σ_i be σ_{i-1} with the leftmost occurrence of t_i deleted; otherwise, let $\sigma_i = \sigma_{i-1}$. Finally, let $\sigma \dot{-} \sigma' = \sigma_n$.

Given a sequence of transitions σ, we define $\#_\sigma$ to be a mapping $\#_\sigma : \mathrm{T} \to \mathrm{N}$ such that $\#_\sigma(t)$ = the number of occurrences of t in σ. For convenience, we sometimes treat $\#_\sigma$ as an r-dimensional vector assuming that an ordering on T is established ($|T| = r$). Let $\mu \xmapsto{\sigma} \mu'$. The *value* of σ, denoted by $\Delta(\sigma)$, is defined to be $\mu' - \mu$ ($\in Z^k$, where k is the number of places in the Petri net). We let $S(\sigma)$ denote the set of transitions occurring in σ, i.e., $S(\sigma) = \{t | t \in T, \#_\sigma(t) > 0\}$.

Let $\mathcal{P} = (\mathrm{P}, \mathrm{T}, \varphi, \mu_0)$ be a Petri net. The ***reachability set*** of $\mathcal{P}$ is the set $\mathrm{R}(\mathcal{P}) = \{\mu \mid \mu_0 \xmapsto{\sigma} \mu$ for some $\sigma\}$. Given a place s, we let $s^\bullet = \{t | \varphi(s,t) = 1, t \in \mathrm{T}\}$ and $^\bullet s = \{t | \varphi(t,s) = 1, t \in \mathrm{T}\}$. A place s and a transition t are on a ***self-loop*** iff $t \in s^\bullet$ and $t \in {}^\bullet s$, i.e., s is both an input and output place of t. By establishing an ordering on the elements of P and T (i.e., $\mathrm{P} = \{\mathrm{p}_1, \ldots, \mathrm{p}_k\}$ and $\mathrm{T} = \{\mathrm{t}_1, \ldots, \mathrm{t}_m\}$), we define the k $\times$ m ***addition matrix*** A of $(\mathrm{P}, \mathrm{T}, \varphi, \mu_0)$ so that $a_{i,j} = \varphi(\mathrm{t}_j, \mathrm{p}_i) - \varphi(\mathrm{p}_i, \mathrm{t}_j)$. Thus, if we view a marking μ as a k-dimensional column vector in which the ith component is $\mu(\mathrm{p}_i)$, each column a_j of A is then a k-dimensional vector such that if $\mu \xmapsto{t_j} \mu'$, then $\mu' = \mu + a_j$. Also, if $\mu_0 \xmapsto{\sigma} \mu$, then $\mu_0 + \mathrm{A} \cdot \#_\sigma = \mu$ (note that the converse does not necessarily hold). (See [24, 25, 27] for more about Petri nets and their related problems.)

A *circuit* of a Petri net is simply a closed path in the Petri net graph. Given a Petri net $\mathcal{P}$, let $c = a_1 a_2 \cdots a_n a_1$ be a circuit and let m be a marking. Let $P_c = \{a_1, a_3, \cdots, a_{n-1}\}$ denote the set of places in c and let $T_c = a_2 a_4 \cdots a_n$ denote the sequence of transitions in c. (Without loss of generality, we assume that a_1 is a place.) We then define the token count of circuit c in marking m to be $m(c) = \sum_{p \in P_c} m(p)$. A circuit c is said to be ***token-free*** in m iff $m(c) = 0$. We say c is ***minimal*** iff P_c does not properly include the set of places in any other circuit. A circuit c is said to have a ***sink*** iff for some $\mu \in \mathrm{R}(\mathcal{P})$ and some σ and μ' such that $\mu \xmapsto{\sigma} \mu'$, $\mu(\mathrm{c}) > 0$, but $\mu'(\mathrm{c}) = 0$. c is said to be ***sinkless*** iff it does not have a sink. (See [32] for more details.)

In this paper, we will mainly focus on the following subclasses of Petri nets:

- *Conflict-free Petri net:*
  A Petri net $\mathcal{P}$ is said to be ***conflict-free*** iff for every place s, either
  1. $|s^\bullet| \leq 1$, <u>or</u>
  2. $\forall t \in s^\bullet$, t and s are on a self-loop.

  In words, a Petri net is conflict-free if every place which is an input of more than one transition is on a self-loop with each such transition ([18, 22]). In a conflict-free Petri net, once a transition becomes enabled, the only way to disable the transition is to fire the transition itself. (That is, $\forall t, t' \in \mathrm{T}$, $t \neq t'$, $\mu \xmapsto{t} \mu'$ and $\mu \xmapsto{t'}$ implies $\mu' \xmapsto{t'}$.)
- *Normal Petri net:*
  A Petri net is ***normal*** [32] iff for every minimal circuit c and transition t_j, $\sum_{p_i \in P_c} a_{i,j} \geq 0$. A Petri net is normal iff for every minimal circuit c and transition t, if one of t's input places is in c, then one of t's output places must be in c as well. Intuitively, a Petri net is normal iff no transition can

decrease the token count of a minimal circuit by firing at any marking.

– *Sinkless Petri net:*

A Petri net $\mathcal{P}$ is said to be ***sinkless*** [32] iff each minimal circuit of $\mathcal{P}$ is sinkless.

In what follows, we will define a class of Petri net ***path formulas***, for which the ***satisfiability problem*** will be shown to be P (respectively, NP) complete for conflict-free (respectively, normal and sinkless) Petri nets. Given transition sequences $\sigma_0, \sigma_1, ..., \sigma_k$, a ***predicate*** on $\sigma_0, \sigma_1, ..., \sigma_k$, denoted by $F(\sigma_0, \sigma_1, ..., \sigma_k) = \bigvee_{1 \le j \le r} \bigwedge_{1 \le i \le s} T_i^j$ (i.e., in disjunctive normal form) where each T_i^j is one of the following ***atomic predicates***.

1. $\Delta(\sigma_i)(p_j) > 0$, where $k \ge i \ge 1$ and p_j is a place.
2. $\#_{\sigma_0}(t_j) \ge (>) c$, where t_j is a transition and $c \ge 0$ is a fixed integer.
3. $\#_{\sigma_i}(t_j) \ge (\le, =, >) c$, where $k \ge i \ge 1$, t_j is a transition and $c \ge 0$ is a fixed integer.
4. $\#_{\sigma_i}(t_j) \ge \#_{\sigma_i}(t_k) + d$, where $k \ge i \ge 1$, t_j and t_k are transitions and $d \ge 0$ is a fixed integer.

(The reason why atomic predicates of the form $\#_{\sigma_0}(t_j) \le (=) c$ are not allowed in our path formulas will be explained in detail in the proof of Lemma 6.)

In this paper, we deal with formulas f of the following form (with respect to Petri net (P, T, φ, μ_0)):

$$\exists \mu_1 \le \mu_2 \le \cdots \le \mu_m \exists \sigma_0, \sigma_1, ..., \sigma_{m-1}$$
$$((\mu_0 \stackrel{\sigma_0}{\longmapsto} \mu_1 \stackrel{\sigma_1}{\longmapsto} \cdots \mu_{m-1} \stackrel{\sigma_{m-1}}{\longmapsto} \mu_m) \wedge F(\sigma_0, \sigma_1, ..., \sigma_{m-1}))$$

Given a Petri net $\mathcal{P}$ and a formula f, we use $\mathcal{P} \models f$ to denote that f is true in $\mathcal{P}$. The ***satisfiability problem*** is the problem of determining, given a Petri net $\mathcal{P}$ and a formula f, whether $\mathcal{P} \models f$. If $\sigma : \mu_0 \stackrel{\sigma_0}{\longmapsto} \mu_1 \stackrel{\sigma_1}{\longmapsto} \cdots \mu_{m-1} \stackrel{\sigma_{m-1}}{\longmapsto} \mu_m$ ($\mu_1 \le \mu_2 \le \cdots \le \mu_m$) is a path such that $F(\sigma_0, \sigma_1, ..., \sigma_{m-1})$ is true in $\mathcal{P}$, we say σ satisfies F. For the sake of simplicity, '$\mathcal{P} \models F_1 \vee F_2 \vee \cdots \vee F_r$?' will be considered as r questions '$\mathcal{P} \models F_1$?', '$\mathcal{P} \models F_2$?', ..., or '$\mathcal{P} \models F_r$?', where each F_j is the conjunction of some atomic predicates. Clearly, if each individual question is solvable in polynomial time, so is the disjunction of all r questions. Finally, it is important to point out that throughout this paper, each integer will be represented by its unary representation. (Recall that in our definition, the flow function φ of a Petri net is from (P × T) ∪ (T × P) to the set {0,1}; hence, at most one token can be added to or removed from a place by a transition. The unary representation is therefore justified under this circumstance.)

3 Previous results concerning conflict-free, normal, and sinkless Petri nets

To set the stage for presenting our polynomial time algorithm, we begin with a brief review of the strategy employed in [17], which deals with normal and sinkless Petri nets. We then show how a similar strategy, with a much improved efficiency, can be developed for reasoning about conflict-free Petri nets, taking advantage of the conflict-freedom property.

Let $\mathcal{P} = (P,T,\varphi,\mu_0)$ be a normal (sinkless) Petri net, and consider a sequence of Petri nets $\mathcal{P}_1,...,\mathcal{P}_n$, where each $\mathcal{P}_h = (P,T_h,\varphi_h,\mu_{h-1})$ such that $T_0 = \emptyset$, and for $1 \leq h \leq n$,

- $T_h = T_{h-1} \cup \{t_{j_h}\}$ for some $t_{j_h} \notin T_{h-1}$ enabled at μ_{h-1}, for $1 \leq h \leq n$;
- φ_h is the restriction of φ to $(P \times T_h) \cup (T_h \times P)$, for $0 \leq h \leq n$; and
- $\mu_{h-1} \overset{\sigma_{h-1}}{\longmapsto} \mu_h$ for some $\sigma_{h-1} \in T_h{}^*$, $1 \leq h \leq n$.

Notice that the set T_h, $0 \leq h \leq n$, grows as h increases, and the newly added transition t_{j_h} is enabled at μ_{h-1} but is not in T_{h-1}. An important property regarding the above sequence of Petri nets is that there is no token-free circuit in $\mathcal{P}_h$ for any marking in $R(\mathcal{P}_h)$ (see [17]). As a result, the reachability set of $\mathcal{P}_h$ can be expressed as the solution set of an integer linear programming instance, say L_h. (For more about this, see [17].) The initial marking of $\mathcal{P}_h$ is given by a solution to L_{h-1}. Using this idea, determining whether a path is executable can be equated with solving a system of linear inequalities over the integers. To give the reader a better feel for how this is done, consider the following path σ on which μ_{h-1}, $1 \leq h \leq n$, marks the first time at which transition t_{j_h} fires, where $t_{j_1},...,t_{j_h},...,t_{j_n}$ are those (distinct) transitions fired in σ (in the given order). Given a Petri net $\mathcal{P}$ and a sequence of transition $\tau = t_{j_1},...,t_{j_h},...,t_{j_n}$, we define the ***characteristic system of inequalities*** for $\mathcal{P}$ and τ as $L(\mathcal{P},\tau) = L_0 \cup ... \cup L_n$, where $L_0 = \{x_0 = \mu_0\}$, $L_h = \{x_{h-1}(i) \geq \varphi(p_i,t_{j_h}),\ x_h = x_{h-1} + A_h \cdot y_h \mid 1 \leq i \leq k\}$, and A_h is the $k \times h$ matrix whose columns are $a_{j_1},...,a_{j_h}$ for $1 \leq h \leq n$. The variables in L are the components of the k-dimensional column vectors $x_0,...,x_n$ and the h-dimensional column vectors y_h, $1 \leq h \leq n$. It was shown in [17] that for any marking μ, $\mu \in R(\mathcal{P})$ iff there is a sequence of distinct transitions $\tau = t_{j_1}...t_{j_n}$ such that $L(\mathcal{P},\tau)$ has a nonnegative integer solution in which $x_n = \mu$. Figure 1 gives a pictorial description of the above strategy.

Based on the above strategy, the reachability problem for normal and sinkless Petri nets can be shown to be solvable in NP in the following manner. Given a Petri net $\mathcal{P} = (P,T,\varphi,\mu_0)$ and a marking μ, first guess a sequence τ of n distinct transitions $t_{j_1}...t_{j_n}$ from T; then construct $L(\mathcal{P},\tau) \cup \{x_n = \mu\}$ (which can be done in polynomial time). Since integer linear programming is in NP [7], the NP upper bound for the reachability problem follows immediately. Boundedness can be determined similarly ([17]).

In light of the above, it is intuitively easy to observe that <u>nondeterminism</u> in the NP algorithm comes from

(N1) guessing a sequence of distinct transitions (i.e., τ), and
(N2) solving the integer version of linear programming. (It is important to note that without the integer requirement, linear programming is in P [19].)

Since the reachability and boundedness problems for normal and sinkless Petri nets can be shown to be NP-hard [17], the use of such 'nondeterminism' seems to

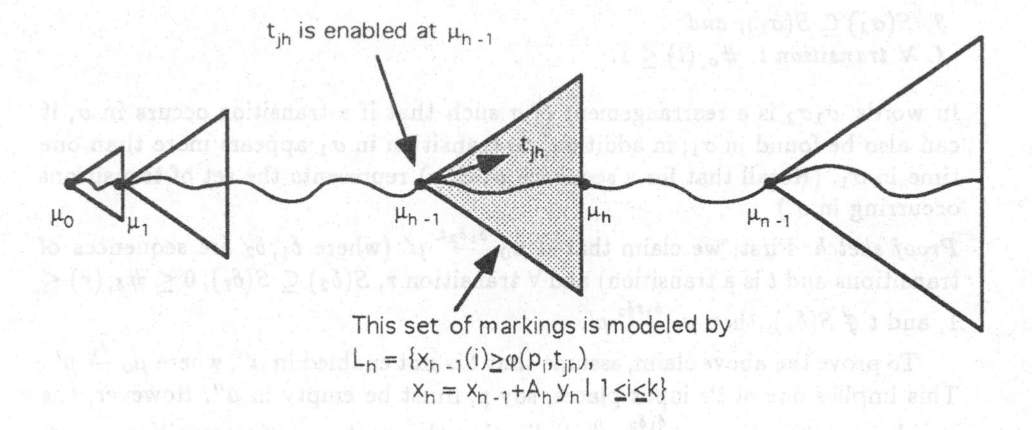

t_{jh} is enabled at μ_{h-1}
t_{jh}
μ_0
μ_1
μ_{h-1}
μ_h
μ_{n-1}
This set of markings is modeled by
$L_h = \{x_{h-1}(i) \geq \varphi(p_i, t_{jh}),$
$x_h = x_{h-1} + A_h y_h \mid 1 \leq i \leq k\}$

Fig. 1. Modeling a path by a system of linear inequalities.

be inevitable for those two classes of Petri nets. In view of the above, a natural question arises: For a more restricted class of Petri nets (such as conflict-free), can the above two 'hurdles' involving nondeterministic guesses be overcome, thus yielding a polynomial time algorithm? The rest of this section will be devoted to showing how such a goal can be accomplished. To a large extent, our work is built upon the work of [5, 15, 17, 22, 33], in which some nice properties regarding conflict-free Petri nets were discovered. We will explain how and to what extent the work of [5, 15, 17, 22, 33] affects our research as the discussion evolves.

Before deriving our main result, we require a few known results for conflict-free, normal, and sinkless Petri nets.

Lemma 1. *(from [15]) Given a conflict-free Petri net $\mathcal{P}=(P, T, \varphi, \mu_0)$, we can construct in polynomial time a sequence τ in which no transition in $\mathcal{P}$ is used more than once, such that if some transition t is not used in τ, then there is no path in which t is used.*

The following lemma will play an important role in the derivation of our main result. The detailed proof can be found in [33]. For the sake of completeness, a proof sketch will be given below.

Lemma 2. *(from [33]) Let $\mu_0 \xmapsto{\sigma} \mu$ be a path in a conflict-free Petri net $\mathcal{P}=(P, T, \varphi, \mu_0)$. Then there exist σ_1 and σ_2 such that*

1. $\#_\sigma = \#_{\sigma_1 \sigma_2}$,
2. $\mu_0 \xmapsto{\sigma_1 \sigma_2} \mu$,

3. $S(\sigma_2) \subseteq S(\sigma_1)$*, and*
4. $\forall$ *transition* t, $\#_{\sigma_1}(t) \leq 1$.

In words, $\sigma_1\sigma_2$ is a rearrangement of σ such that if a transition occurs in σ, it can also be found in σ_1; in addition, no transition in σ_1 appears more than one time in σ_1. (Recall that for a sequence σ, $S(\sigma)$ represents the set of transitions occurring in σ.)

Proof sketch: First, we claim that if $\mu_0 \overset{\delta_1\delta_2 t}{\rightarrow} \mu'$ (where δ_1, δ_2 are sequences of transitions and t is a transition) and $\forall$ transition r, $S(\delta_2) \subseteq S(\delta_1)$, $0 \leq \#_{\delta_1}(r) \leq 1$, and $t \notin S(\delta_1)$, then $\mu_0 \overset{\delta_1 t \delta_2}{\rightarrow} \mu'$.

To prove the above claim, assume that t is not enabled in μ'', where $\mu_0 \overset{\delta_1}{\rightarrow} \mu''$. This implies one of t's input places, say p, must be empty in μ''. However, t is enabled in μ''', where $\mu_0 \overset{\delta_1\delta_2}{\rightarrow} \mu'''$, indicating the existence of a transition, say t' in δ_2, which deposits a token to p. Since $S(\delta_2) \subseteq S(\delta_1)$, t' must be in δ_1. Since the Petri net is conflict-free and $t \notin S(\delta_1)$, $\mu''(p) \neq 0$ – a contradiction. So t must be enabled in μ''. Since the Petri net is conflict-free and $t \notin S(\delta_2)$ (because $S(\delta_2) \subseteq S(\delta_1)$ and $t \notin S(\delta_1)$), the firing of t in μ'' will not affect the enableness of the subsequent transition sequence δ_2. This completes the proof of the claim. □

Lemma 3. *(from [22]) Let* σ *and* τ *be two fireable sequences (at a marking* μ*) in a conflict-free Petri net. Then* $\sigma \cdot (\tau \dot{-} \sigma)$ *is fireable at* μ *as well.*

Lemma 4. *(from [17]) Let* $\mathcal{P} = (P, T, \varphi, \mu_0)$ *be a sinkless PN, and let* $\mathcal{P}' = (P, T', \varphi', \mu)$ *be such that* $\mu_0 \overset{\sigma}{\longmapsto} \mu$ *in* $\mathcal{P}$ *for some* σ, $T' \subseteq T$ *such that each* $t \in T'$ *is enabled at some point in the firing of* σ *from* μ_0*, and* φ' *is the restriction of* φ *to* $(P \times T') \cup (T' \times P)$*. Then* $\mathcal{P}'$ *has no token-free circuits in every reachable marking.*

Lemma 5. *(from [32]) Consider a Petri net* $\mathcal{P} = (P, T, \varphi, \mu_0)$ *which has no token-free circuits in every reachable marking. If* $\mu_0 + A \cdot x \geq \mathbf{0}$ *for some* $x \in N^m$*, where* m *is the number of transitions in* T*, then there exists a sequence* σ*,* $\#_\sigma = x$ *and* $\mu_0 \overset{\sigma}{\longmapsto}$.

4 Complexity analysis

In this section, we will show that the satisfiability problem for the path formulas defined in Section 2 is complete for P for conflict-free Petri nets, whereas it is complete for NP for normal and sinkless Petri nets. Basically, the technique used for deriving the upper bounds relies on reducing the satisfiability problem to integer linear programming. In this way, determining the existence of a path satisfying certain path formula is tantamount to solving an integer linear programming problem, which is well-known to be NP-hard (in fact, NP-complete). With the help of a novel idea used in [5], and several unique properties possessed by conflict-free Petri nets (which will be derived in this section), a speed-up is made possible by eliminating the integer constraint of linear programming for conflict-free Petri nets. As a result, a polynomial time algorithm follows. (Linear programming is known to be solvable in P [19].) Before presenting our main result, a few lemmas are in order.

As stated in the previous section, the firing sequence τ, which characterizes the firing order of transitions, plays a crucial role in the NP algorithm of the boundedness (or reachability) problem for normal and sinkless Petri nets. The following lemma suggests that for conflict-free Petri nets, τ can be constructed deterministically in polynomial time. Such a result allows us to get around the first hurdle (i.e., (N1)) stated earlier.

Lemma 6. *Given a conflict-free Petri net $\mathcal{P} = (N, \mu_0)$, if formula*

$$\exists \mu_1 \le \mu_2 \le \cdots \le \mu_m \exists \sigma_0, \sigma_1, ..., \sigma_{m-1}$$
$$(\mu_0 \xmapsto{\sigma_0} \mu_1 \xmapsto{\sigma_1} \mu_2 \xmapsto{\sigma_2} \cdots \xmapsto{\sigma_{m-1}} \mu_m) \wedge F(\sigma_0, \sigma_1, ..., \sigma_{m-1})$$

is satisfiable in $\mathcal{P}$, then $\exists$ a path $\mu_0 \xmapsto{\tau} \mu'_0 \xmapsto{\sigma'_0} \mu_1 \xmapsto{\sigma_1} \mu_2 \xmapsto{\sigma_2} \cdots \xmapsto{\sigma_{m-1}} \mu_m$ satisfying $F(\tau\sigma'_0, \sigma_1, ..., \sigma_{m-1})$ such that

1. *τ is the sequence of transitions guaranteed by Lemma 1,*
2. *$S(\tau)$=$S(\tau\sigma'_0\sigma_1 \cdots \sigma_{m-1})$, and*
3. *no token-free circuit is reachable from μ'_0 in Petri net (N', μ'_0), where N' is the restriction of Petri net N to $S(\tau)$.*

Proof: Suppose $\mu_0 \xmapsto{\sigma''_0} \mu'_1 \xmapsto{\sigma_1} \mu'_2 \xmapsto{\sigma_2} \cdots \xmapsto{\sigma_{m-1}} \mu'_m$ satisfies the given formula. Since $\mu'_m \ge \mu'_1$, $\mu_0 \xmapsto{\sigma} \mu''_1 \xmapsto{\sigma_1} \mu''_2 \xmapsto{\sigma_2} \cdots \xmapsto{\sigma_{m-1}} \mu''_m$ (where $\sigma = \sigma''_0\sigma_1 \cdots \sigma_{m-1}$ and $\mu''_1 = \mu'_m$) is clearly a legal sequence, for some markings $\mu''_2, \mu''_3, \cdots, \mu''_m$. It remains to show that σ can be rearranged and extended into $\tau\sigma'_0$ stated in the statement of the lemma.

Using Lemma 2, σ can be rearranged into $\delta\sigma'_0$, where $S(\delta) = S(\sigma)$ and each transition in δ occurs exactly once. It is then easy to see that $\mu_0 \xmapsto{\delta}$, $\mu_0 \xmapsto{\tau}$, and $S(\delta) \subseteq S(\tau)$ imply $\mu_0 \xmapsto{\delta\cdot(\tau \dot{-} \delta)}$ (Lemma 3). Furthermore, $S(\tau \dot{-} \delta) \cap S(\sigma) = \emptyset$. This, together with $\mathcal{P}$ being conflict-free, guarantees $\delta(\tau \dot{-} \delta)\sigma'_0\sigma_1 \cdots \sigma_{m-1}$ to be a legal firing sequence. (That is, the insertion of $\tau \dot{-} \delta$ does not affect the fireability of the remaining sequence.) Given the fact that atomic predicates in σ_0 are of the form $\#_{\sigma_0}(t) \ge (>)c$, $\mu_0 \xmapsto{\tau} \mu'_0 \xmapsto{\sigma'_0} \mu_1 \xmapsto{\sigma_1} \mu_2 \xmapsto{\sigma_2} \cdots \xmapsto{\sigma_{m-1}} \mu_m$ constitutes a path satisfying the given formula, for some $\mu'_0, \mu_1, \cdots, \mu_m$, (Note that τ and $\delta(\tau \dot{-} \delta)$ contain exactly the same set of transitions.)

(3) follows immediately from (2) and Lemma 4.

It is worth pointing out that τ (which might contain transitions not used in the original σ_0) is part of the new 'prefix'; as a result, predicates of the form $\#_{\sigma_0}(t) \le (=)c$ may no longer be valid. This explains why such predicates are excluded in our path formulas. □

Throughout the rest of this paper, 'n' will be reserved for denoting the size of the input, when the unary representation is used for integers.

Lemma 7. *If $\mu \xrightarrow{\sigma} \overline{\mu}$, then σ can be rearranged into $\sigma = \sigma_1^1 \cdots \sigma_1^{l_1} \sigma_2^1 \cdots \sigma_2^{l_2} \cdots \sigma_k^1 \cdots \sigma_k^{l_k}$, for some sequences $\sigma_1^1, \cdots, \sigma_1^{l_1}, \sigma_2^1, \cdots, \sigma_2^{l_2}, \cdots, \sigma_k^1, \cdots, \sigma_k^{l_k}$ and integers $l_1, l_2, ..., l_k$, $1 \leq k \leq n$, such that*

1. $\forall 1 \leq i \leq k, 1 \leq j \leq l_i$ $\forall$ transition $t \in T$, $\#_{\sigma_i^j}(t) \leq 1$,
2. $\forall 1 \leq i \leq k, 1 \leq j, m \leq l_i, S(\sigma_i^j) = S(\sigma_i^m)$, and
3. $\forall 1 \leq i \leq k-1, 1 \leq j \leq l_{i+1}, 1 \leq m \leq l_i$ $S(\sigma_{i+1}^j) \subsetneq S(\sigma_i^m)$.

Proof: The result can be obtained by repeatedly applying Lemma 2. □
Notice that in the above lemma, $\sigma_i^1, \sigma_i^2, \cdots, \sigma_i^{l_i}$ $(1 \leq i \leq k)$ all comprise the same set of transitions, even though they may occur in different orders. However, as we will see later in the derivation of our results that the actual order by which transitions in $\sigma_i^j, 1 \leq j \leq l_i$, occur is not important. As a result, for ease of expression as well as for the sake of clarity, $\sigma_i^1, \sigma_i^2, \cdots, \sigma_i^{l_i}$ $(1 \leq i \leq k)$ will all be denoted as σ_i throughout the remainder of this paper.

Lemma 7 will play a vital role in the development of our polynomial time algorithm. The lemma allows us to rearrange an arbitrary firing sequence in a conflict-free Petri net into a 'canonical form' satisfying conditions (1) and (2) stated in the lemma. In words, (1) ensures that for each σ_i, no transition will appear more than once; hence, $|\sigma_i| \leq n$, for all i. (2) is a crucial property upon which our subsequent analysis will heavily rely. It simply says that $S(\sigma_1)S(\sigma_2)\cdots S(\sigma_k)$ forms a 'shrinking' sequence of sets in the sense that if a transition, say t, occurs in σ_i, for some i, then t is guaranteed to appear in $\sigma_j, \forall 1 \leq j \leq i$. Symmetrically, if t does not appear in σ_i, then it will never be found in $\sigma_j, \forall i \leq j \leq k$. Consider a place p and a segment σ_i. Suppose we let $\uparrow$ (respectively, $=$ and $\downarrow$) denote the case when $\Delta(\sigma_i)(p) > 0$ (respectively, $= 0$ and < 0) (i.e., upon the completion of σ_i the number of tokens in place p increases (respectively, remains the same and decreases). For convenience, we call such a symbol the *sign* of segment σ_i with respect to place p, throughout the rest of our discussion. Then the above lemma, in conjunction with the property of conflict-freedom, ensures that the pattern of signs (with respect to a given place) associated with $\sigma_1 \cdots \sigma_1 \sigma_2 \cdots \sigma_2 \cdots \sigma_k \cdots \sigma_k$ must exhibit the pattern of $\uparrow\uparrow \cdots \uparrow == \cdots =\uparrow\uparrow \cdots \uparrow == \cdots =$ or $\uparrow\uparrow \cdots \uparrow == \cdots =\downarrow\downarrow \cdots \downarrow == \cdots =$. (For example, no firing sequence in canonical form can have $\downarrow\downarrow \cdots \downarrow\uparrow\uparrow \cdots \uparrow$ as its pattern of signs.)

We are now in a position to show that if a formula is satisfiable, then there exists a short witness whose length is polynomial in n, where n is the size of the input. Before showing this, a couple of lemmas are in order.

Lemma 8. *Given a marking μ, a predicate $F(\sigma)$ and a $\overline{\sigma}$ with $\Delta(\overline{\sigma}) \geq \mathbf{0}$, let $\mu \xrightarrow{\sigma} \tilde{\mu}$ be the shortest (or one of the shortest) path satisfying*

1. $\mu \xrightarrow{\sigma} \tilde{\mu} \xrightarrow{\overline{\sigma}}$,
2. $F(\sigma)$=true, and
3. no token-free circuit is reachable from μ in the considered Petri net,

then $|\sigma| \leq 2(n+1)n^2$, where n is the size of the input (when the unary representation is used). The above bound holds even if an additional constraint $\mu \leq \tilde{\mu}$ is imposed.

Proof: Using Lemma 7, σ can be rearranged into $\overbrace{\sigma_1\cdots\sigma_1}^{l_1}\overbrace{\sigma_2\cdots\sigma_2}^{l_2}\cdots\overbrace{\sigma_k\cdots\sigma_k}^{l_k}$, for some integers $l_1, l_2, ..., l_k$ and sequences $\sigma_1, \sigma_2, ..., \sigma_k$ satisfying the conditions stated in Lemma 7. Suppose, to the contrary, that $|\sigma| > 2(n+1)n^2$. Then $\exists j$ such that $l_j > n+1$ (since $k \leq n$ and $|\sigma_i| \leq n$). Let h be the largest such j. In what follows, we show that the new (and shorter) sequence $\sigma' = \overbrace{\sigma_1\cdots\sigma_1}^{l_1}\overbrace{\sigma_2\cdots\sigma_2}^{l_2}\cdots\overbrace{\sigma_h\cdots\sigma_h}^{l_h-1}\cdots\overbrace{\sigma_k\cdots\sigma_k}^{l_k}$ is a legal firing sequence at μ and satisfies F as well. (Thus, a contradiction is obtained.)

In what follows, the proof will be for the case when the additional constraint $\mu \leq \tilde{\mu}$ is in effect. The case without the constraint is similar (in fact, simpler). We first show that σ' is executable at μ. Consider the following two cases:

I. $\forall p \in P, \Delta(\sigma_h)(p) \leq 0$. In this case, the removal of one σ_h does not affect the enabledness of the remaining sequence.

II. $\exists p \in P, \Delta(\sigma_h)(p) > 0$. Consider the following two subcases:

(a) $\forall j, h < j \leq k, \Delta(\sigma_j)(p) \geq 0$ (i.e., the number of tokens in p does not decrease in the subsequent segments). Then the new path resulting from deleting one σ_h will still be legal (with respect to place p).

(b) $\exists j, h < j \leq k, \Delta(\sigma_j)(p) < 0$ (i.e., the number of tokens in p decreases in at least one of the remaining segments). In this case, we have to make sure that upon the completion of $\overbrace{\sigma_1\cdots\sigma_1}^{l_1}\overbrace{\sigma_2\cdots\sigma_2}^{l_2}\cdots\overbrace{\sigma_h\cdots\sigma_h}^{l_h-1}$, place p has somewhat 'accumulated' a sufficient number of tokens so that the subsequent decrement can still be carried out. The sufficiency comes from the following facts:

- $\forall i, 1 \leq i \leq h, \Delta(\sigma_i)(p) > 0$. (As indicated before, the sign of token change must exhibit the pattern of $\uparrow\uparrow\cdots\uparrow==\cdots=\uparrow\uparrow\cdots\uparrow==\cdots=$ or $\uparrow\uparrow\cdots\uparrow==\cdots=\downarrow\downarrow\cdots\downarrow==\cdots=$. As a result, $\Delta(\sigma_h)(p) > 0$ and $\Delta(\sigma_j)(p) < 0$ (for some $j > h$) imply $\forall i, 1 \leq i \leq h, \Delta(\sigma_i)(p) > 0$.)
- $(l_1 + l_2 + \cdots + (l_h - 1)) - (l_{h+1} + \cdots + l_k) \geq (2(n+1)n - (l_{h+1} + \cdots + l_k)) - (l_{h+1} + \cdots + l_k) = 2(n+1)n - 2(l_{h+1} + \cdots + l_k) \geq 2(n+1)n - 2(n+1)(n-1) > 0$. (Recall that h is the largest such that $l_h > n+1$.)
- $\forall j, h < j \leq k, \Delta(\sigma_j)(p) \geq -1$. (Since $\mathcal{P}$ is conflict-free, no more than one transition can subtract from a place, unless each of such transitions is on a self-loop. Also in our model, $\varphi(p,t) \leq 1$, for every p and t.)

In view of the above discussion, we have $\Delta(\sigma')(p) \geq 0, \forall p$. Since no token-free circuit can be reached from μ, σ' (or one of its rearrangements) is executable at μ (Lemma 5). Furthermore, $\Delta(\overline{\sigma}) \geq 0$, in conjunction with the fact that no token-free circuit is reachable from μ, ensures that $\mu \stackrel{\sigma'}{\longmapsto} \mu' \stackrel{\overline{\sigma}}{\longmapsto}$, for some μ'.

We now show that $F(\sigma')$ remains true. First recall that $F(\sigma) = F_1 \wedge F_2 \wedge \cdots \wedge F_m$, where each term $F_i, 1 \leq i \leq m$, is of one of the following forms:

1. $\Delta(\sigma)(p) > 0$, for some place p.
2. $\#_\sigma(t) \geq (=, \leq, \text{or} >)\ d$, for some transition t and constant d.
3. $\#_\sigma(t_a) \geq \#_\sigma(t_b) + d$, for some transitions t_a, t_b and constant d.

As a result, we have the following:

1. From the above argument, it should be obvious that predicates of the form $\Delta(\sigma')(p) > 0$, where p is a place, remain true. (See I. and II.(a) and (b).)
2. Consider predicates of the form $\#_{\sigma'}(t) \geq d$. If $\#_{\sigma_h}(t) = 0$, then $\#_{\sigma'}(t) = \#_\sigma(t) \geq d$; otherwise, $\#_{\sigma'}(t) \geq l_h - 1 \geq (n+1) - 1 \geq d$. The '=' and $\leq$ cases are trivially true; the $>$ case is similar to that of the '$\geq$.
3. For the case $\#_{\sigma'}(t_a) \geq \#_{\sigma'}(t_b) + d$, consider the following three subcases:
   (a) $\#_{\sigma_h}(t_a) = 0$: trivial.
   (b) $\#_{\sigma_h}(t_a) = 1 = \#_{\sigma_h}(t_b)$: trivial.
   (c) $\#_{\sigma_h}(t_a) = 1$ and $\#_{\sigma_h}(t_b) = 0$.
   Due to the 'shrinking' property of $S(\sigma_1)S(\sigma_2)\cdots$, $\#_{\sigma_1\cdots\sigma_{h-1}}(t_a) \geq \#_{\sigma_1\cdots\sigma_{h-1}}(t_b)$ and $\#_{\sigma_{h+1}\cdots\sigma_k}(t_a) \geq \#_{\sigma_{h+1}\cdots\sigma_k}(t_b)(=0)$. We also have $\#\underbrace{\sigma_h\cdots\sigma_h}_{l_h-1}(t_a) \geq (n+1) - 1 \geq d$ and $\#\underbrace{\sigma_h\cdots\sigma_h}_{l_h-1}(t_b) = 0$. $\#_{\sigma'}(t_a) \geq \#_{\sigma'}(t_b) + d$ follows immediately from the above.

From the above discussion, we can construct a shorter path which also satisfies the conditions of the lemma – a contradiction. Hence, the lemma follows. □

Lemma 9. *Let $\overline{\mu} \overset{\sigma}{\longmapsto} \mu_1 \overset{\sigma_1}{\longmapsto} \mu_2 \overset{\sigma_2}{\longmapsto} \cdots \overset{\sigma_{m-1}}{\longmapsto} \mu_m$ be the shortest (or one of the shortest) path satisfying*

1. *$\forall 1 \leq i \leq m-1$, $\mu_i \leq \mu_{i+1}$,*
2. *$F(\sigma, \sigma_1, \ldots, \sigma_{m-1})$=true, and*
3. *no token-free circuit is reachable from μ.*

Then $|\sigma\sigma_1\sigma_2\cdots\sigma_{m-1}| \leq 2m(n+1)n^2$.

Proof: The proof is done by repeatedly applying Lemma 8. □

Combining Lemmas 6 – 9, we have the following theorem:

Theorem 10. *Given a conflict-free Petri net $\mathcal{P} = (N, \mu_0)$, if formula $\exists \mu_1 \leq \mu_2 \leq \cdots \leq \mu_m \exists \sigma_0, \sigma_1, \ldots, \sigma_{m-1}$ $(\mu_0 \overset{\sigma_0}{\longmapsto} \mu_1 \overset{\sigma_1}{\longmapsto} \mu_2 \overset{\sigma_2}{\longmapsto} \cdots \overset{\sigma_{m-1}}{\longmapsto} \mu_m) \wedge F(\sigma_0, \sigma_1, \ldots, \sigma_{m-1})$ is satisfiable in $\mathcal{P}$, then $\exists$ a witnessing path $\mu_0 \overset{\tau\sigma_0}{\longmapsto} \mu_1 \overset{\sigma_1}{\longmapsto} \mu_2 \overset{\sigma_2}{\longmapsto} \cdots \overset{\sigma_{m-1}}{\longmapsto} \mu_m$ such that τ is the sequence guaranteed by Lemma 1, and $|\ \tau\sigma_0\sigma_1\cdots\sigma_{m-1}\ | \leq p(n)$, for some fixed polynomial $p(n)$.*

So far we have shown that if a formula is satisfiable in a Petri net $\mathcal{P}$, then there exists a short 'witness' whose length is bounded by a polynomial. The next step of our strategy is to set up a linear programming instance in such a way that the instance has a solution iff the formula is satisfiable.

Suppose σ: $\mu_0 \stackrel{\tau}{\longmapsto} \mu_0' \stackrel{\sigma_0}{\longmapsto} \mu_1 \stackrel{\sigma_1}{\longmapsto} \mu_2 \stackrel{\sigma_2}{\longmapsto} \cdots \stackrel{\sigma_{m-1}}{\longmapsto} \mu_m$ is a path satisfying the formula given in Theorem 10. Without loss of generality, we let $\tau = t_1 t_2 \cdots t_r$ be the sequence guaranteed by Lemma 1. In our subsequent discussion, we will restrict our attention to Petri net (P, $\{t_1, t_2, ..., t_r\}$, φ', μ_0), where φ' is the restriction of φ to $\{t_1, t_2, ..., t_r\}$. For convenience, we also label the set of places as P=$\{p_1, p_2, ..., p_s\}$. Now we are ready to set up a set of linear inequalities to capture the essence of the above path. Since τ can be found in polynomial time (Lemma 1), segment $\mu_0 \stackrel{\tau}{\longmapsto} \mu_0'$ (more accurately, μ_0') will be computed in the beginning. As a result, only the suffix path starting at μ_0' needs to be expressed as a set of linear inequalities. The construction is done as follows:

1. $\mu_0' \stackrel{\sigma_0}{\longmapsto} \mu_1$

   We use variable $x_i^0, 1 \leq i \leq r$, to represent the number of occurrences of transitions t_i in path $\mu_0' \stackrel{\sigma_0}{\longmapsto} \mu_1$. For ease of expression, we let $x^0 = (x_1^0, x_2^0, ..., x_r^0)^T$ (i.e., the *transpose* of $(x_1^0, x_2^0, ..., x_r^0)$). Let A denote the incidence matrix of Petri net (P, $\{t_1, t_2, ..., t_r\}$, φ', μ_0). Then the inequalities corresponding to the given segment are

   (A1) $\mu_0' + A \cdot x^0 \geq 0$,

   (A2) $x^0 \geq 0$,

   (A1) is sufficient to imply the reachability of $\mu_1 (= \mu_0' + A \cdot x^0)$, for no token-free circuit can be reached from μ_0'. (A2) is trivial.

2. $\mu_j \stackrel{\sigma_j}{\longmapsto} \mu_{j+1}, 1 \leq j < m$.

   Let x_i^j be the number of occurrences of transition $t_i, 1 \leq i \leq r$, in σ_j. For ease of expression, we let $x^j = (x_1^j, x_2^j, ..., x_r^j)^T$. Then we include the following inequalities:

   (A3) $A \cdot x^j \geq 0$,

   (A4) $x^j \geq 0$,

   (A3) is sufficient to guarantee that μ_{j+1} is reachable. (A4) is again trivial.

Now that we have set up a system of linear inequalities to capture the computation of σ, it remains to do the same for the predicates. More precisely, we have the following:

(A5) Predicate $\Delta(\sigma_j)(p_i) > 0$.
For this type of predicate, include inequality $(A \cdot x^j)(p_i) \geq 1$.

(A6) Predicate $\#_{\sigma_j}(t_i) \geq (\leq, =, >) c$.
Add inequality $x_i^j \geq (\leq, =, >) c$.

(A7) Predicate $\#_{\sigma_j}(t_i) \geq \#_{\sigma_j}(t_k) + d$.
In this case, we include inequality $x_i^j \geq x_k^j + d$.

Given a Petri net $\mathcal{P}$ and a formula f, we define *ILP*($\mathcal{P}$, f) to be the set of linear inequalities listed above. It is not hard to see that *ILP*($\mathcal{P}$, f) can be constructed in polynomial time. Now we are in a position to present the following important theorem, which serves as the foundation upon which our polynomial time algorithm for the satisfiability problem relies. Based on the above discussion, the proof of the theorem should be straightforward.

Theorem 11. *Given a conflict-free Petri net $\mathcal{P} = (N, \mu_0)$, formula f:*

$$\exists \mu_1 \leq \mu_2 \leq \cdots \leq \mu_m \exists \sigma_0, \sigma_1, ..., \sigma_{m-1}$$
$$(\mu_0 \xrightarrow{\sigma_0} \mu_1 \xrightarrow{\sigma_1} \mu_2 \xrightarrow{\sigma_2} \cdots \xrightarrow{\sigma_{m-1}} \mu_m) \wedge F(\sigma_0, \sigma_1, ..., \sigma_{m-1})$$

is satisfiable in $\mathcal{P}$ iff $ILP(\mathcal{P}, f)$ has an integer solution.

Given the fact that integer linear programming is NP-complete, tractability of the satisfiability problem does not come free of charge, even with the help of Theorem 11. What makes a speed-up possible is through the application of a technique proposed by Esparza [5], taking advantage of certain nice properties offered by conflict-free Petri nets. The crux of the approach is that in a system of linear inequalities modeling the behaviors of a conflict-free Petri net, if a solution (over the reals) exists, then the ceiling of that solution is itself a solution, provided each of the inequalities satisfies certain constraints. As it turns out, all the inequalities introduced in $ILP(\mathcal{P}, f)$ meet such constraints. Based on the above idea, an integer-preserving transformation from integer linear programming to linear programming is possible, when reasoning about conflict-free Petri nets. Thus, the second hurdle (i.e., (N2)) stated earlier is overcome. As a consequence, a polynomial time algorithm follows immediately. To be more precise, we have:

Theorem 12. *Given a conflict-free Petri net $\mathcal{P} = (N, \mu_0)$ and a formula f, $ILP(\mathcal{P}, f)$ has an integer solution iff the following optimization problem has a solution (not necessary over the integers).*

$$Maximize \sum_{i=0}^{m-1} \sum_{j=1}^{r} x_j^i.$$

$$subject\ to \begin{cases} ILP(\mathcal{P}, f) \\ 0 \leq x_j^i \leq p(n), \forall 0 \leq i < m, 1 \leq j \leq r \end{cases}$$

where $x_j^i, 0 \leq i < m, 1 \leq j \leq r$ are those variables used in $ILP(\mathcal{P}, f)$, and $p(n)$ is the polynomial stated in Theorem 10.

Proof: Let $LP(\mathcal{P}, f)$ denote the above set of linear inequalities. Intuitively, $\sum_{i=0}^{m-1} \sum_{j=1}^{r} x_j^i$ amounts to the length of the path satisfying formula f in Petri net $\mathcal{P}$. According to Theorem 10, if $\mathcal{P} \models f$, then there exists a short witness whose length is bounded by $p(n)$. As a result, $LP(\mathcal{P}, f)$ has a solution (which maximizes the given function) if $ILP(\mathcal{P}, f)$ has an integer solution.

To prove the converse, it suffices to show that the optimal solution of $LP(\mathcal{P}, f)$ is an integer solution. Let $(x_1^0, x_2^0, ..., x_j^i, ...), 0 \leq i < m, 1 \leq j \leq r$, be the solution of $LP(\mathcal{P}, f)$. In what follows, we show that $(\lceil x_1^0 \rceil, \lceil x_2^0 \rceil, ..., \lceil x_j^i \rceil, ...)$ is a solution as well. To do so, recall that each inequality in $LP(\mathcal{P}, f)$ is of one of the following forms:

1. $x_j^i \geq (>, \leq, =) d$, for some variable x_j^i and integer constant d,
2. $x_j^i \geq x_k^i + d$, for some variables x_j^i, x_k^i and integer constant d,
3. $b + \sum_{i=1}^{r} a_{j,i} x_i^k \geq d$, where x_i^ks are variables, b and d (≥ 0) are integer constants, and $a_{j,i}$s, $1 \leq i \leq r$, are the components of the j-th row of the incidence matrix A. (This type of inequality comes from (A1) and (A3).)

Clearly, (1) and (2) remain after each variable is replaced by its ceiling. For case (3), first notice that due to the conflict-freedom property of $\mathcal{P}$, at most one component, say $a_{j,h}$, can be negative (and, if so, = -1). Hence, (3) can be rewritten as $b + \sum_{i=1..r, i \neq h} a_{j,i} x_i^k \geq (-a_{j,h}) x_h^k + d$, where $a_{j,h} = -1$. Clearly, $b + \sum_{i=1..r, i \neq h} a_{j,i} \lceil x_i^k \rceil \geq \lceil (b + \sum_{i=1..r, i \neq h} a_{j,i} x_i^k) \rceil \geq \lceil (-a_{j,h}) x_h^k + d \rceil = (-a_{j,h}) \lceil x_h^k \rceil + d$. Hence, $(\lceil x_1^k \rceil, \lceil x_2^k \rceil, ..., \lceil x_r^k \rceil)$ satisfies (3) as well.

In light of the above, the optimal solution of $LP(\mathcal{P}, f)$ must be an integer solution. This completes the proof of our theorem. □

Theorem 13. *The satisfiability problem for conflict-free Petri nets is P-complete.*

Proof. The upper bound follows immediately from Theorem 12 and the fact that linear programming is in P. Also, our path formulas are powerful enough to express unboundedness (see Section 5), the lower bound is obvious, given the fact that the boundedness problem is known to be P-complete for conflict-free Petri nets [15]. □

At this point, the reader should note that the above approach does not work for the reachability problem. Intuitively, for a marking μ to be reachable, it seems that equalities of the form $\mu_0 + A \cdot x = \mu$ have to be dealt with. Unfortunately, taking the ceiling of a non-integer solution of such an equation may not yield a valid solution. As a result, our analysis provides strong evidence as to why reachability is harder to analyze than boundedness for conflict-free Petri nets. Also, it is reasonably easy to show that the satisfiability problem for normal (sinkless) Petri nets is NP-complete. The proof closely parallels that of the boundedness problem for the two classes of Petri nets. (See [17] for details.)

5 Some applications

In this section, we demonstrate how a wide variety of problems (concerning conflict-free Petri nets) can be solved under a unified framework, using the main result derived in the previous section (i.e., Theorem 12). With the exception of the boundedness problem, all of our results are new. The reader is encouraged to consult [34] for related results concerning general Petri nets. (In [34], each of the following problems was shown to be solvable in EXPSPACE for general Petri nets.)

1. *Boundedness Problem.*
   Clearly, the existence of an unbounded path can be expressed as $\exists\mu_1 \leq \mu_2 \exists\sigma_0, \sigma_1((\mu_0 \xmapsto{\sigma_0} \mu_1 \xmapsto{\sigma_1} \mu_2) \wedge (\bigvee_{p\in P} \Delta(\sigma_1)(p) > 0))$.
2. *Fair Nontermination Problems.*
   Let $\mathcal{A}$ be a finite set of nonempty subsets of transitions. Given an infinite sequence of transitions $\sigma = t_1, t_2, ...$, let $\inf^T(\sigma)$ be the set of transitions occurring infinitely often in σ. In [16] the following 6 types of fairness were defined: σ is said to be
   - T1-fair iff $\exists$ A $\in \mathcal{A}$, $\exists i \geq 1, t_i \in$ A.
   - T1′-fair iff $\exists$ A $\in \mathcal{A}$, $\forall i \geq 1, t_i \in$ A.
   - T2-fair iff $\exists$ A $\in \mathcal{A}$, $\inf^T(\sigma) \cap$ A $\neq \emptyset$.
   - T2′-fair iff $\exists$ A $\in \mathcal{A}$, $\inf^T(\sigma) \subseteq$ A.
   - T3-fair iff $\exists$ A $\in \mathcal{A}$, $\inf^T(\sigma) =$ A.
   - T3′-fair iff $\exists$ A $\in \mathcal{A}$, A $\subseteq \inf^T(\sigma)$.

   The *fair nontermination problem* with respect to T1 (T1′, T2, T2′, T3, T3′, respectively) fairness is the problem of determining whether a given Petri net has an infinite type T1 (T1′, T2, T2′, T3, T3′, respectively) fair computation. In what follows, we show how the fair nontermination problem for each of the six notions of fairness is expressible by our path formulas. This will immediately yield the Ptime upper bound. Given a subset of transitions A, let v_A be a vector in N^r such that $v_A(i)$=1 (0) iff $t_i \in (\notin)\ A$. A Petri net is X-fair nonterminating iff
   - X=T1
     $\iff \exists\mu_1 \leq \mu_2\exists\sigma_0, \sigma_1((\mu_0 \xmapsto{\sigma_0} \mu_1 \xmapsto{\sigma_1} \mu_2) \wedge ((\bigvee_{A\in\mathcal{A}} \bigvee_{t_i\in A} 1 \leq \#_{\sigma_0}(t_i)) \wedge (\bigvee_{t\in T} \#_{\sigma_1}(t) > 0)))$
     $\iff \exists\mu_1 \leq \mu_2\exists\sigma_0, \sigma_1 \bigvee_{A\in\mathcal{A}} \bigvee_{t_i\in A} \bigvee_{t\in T}((1 \leq \#_{\sigma_0}(t_i)) \wedge (\#_{\sigma_1}(t) > 0))$, which is in disjunctive normal form.
   - X=T1′. For $A \in \mathcal{A}$, consider Petri net (P, A, φ_A, μ_0), where φ_A is the projection of φ on A. Then X=T1′
     $\iff$
     $\exists\mu_1 \leq \mu_2\exists\sigma_0, \sigma_1((\mu_0 \xmapsto{\sigma_0} \mu_1 \xmapsto{\sigma_1} \mu_2) \wedge (\bigvee_{t\in A} \#_{\sigma_1}(t) > 0))$.
   - X=T2
     $\iff \exists\mu_1 \leq \mu_2\exists\sigma_0, \sigma_1((\mu_0 \xmapsto{\sigma_0} \mu_1 \xmapsto{\sigma_1} \mu_2) \wedge ((\bigvee_{A\in\mathcal{A}} \bigvee_{t_i\in A} 1 \leq \#_{\sigma_1}(t_i)) \wedge (\bigvee_{t\in T} \#_{\sigma_1}(t) > 0)))$
     $\iff \exists\mu_1 \leq \mu_2\exists\sigma_0, \sigma_1 \bigvee_{A\in\mathcal{A}} \bigvee_{t_i\in A} \bigvee_{t\in T} ((1 \leq \#_{\sigma_1}(t_i)) \wedge (\#_{\sigma_1}(t) > 0))$.
   - X=T2′
     $\iff \exists\mu_1 \leq \mu_2\exists\sigma_0, \sigma_1((\mu_0 \xmapsto{\sigma_0} \mu_1 \xmapsto{\sigma_1} \mu_2) \wedge ((\bigvee_{A\in\mathcal{A}} \bigwedge_{t_i\notin A} \#_{\sigma_1}(t_i) \leq 0) \wedge (\bigvee_{t\in T} \#_{\sigma_1}(t) > 0)))$
     $\iff \exists\mu_1 \leq \mu_2\exists\sigma_0, \sigma_1 \bigvee_{A\in\mathcal{A}} \bigvee_{t\in T}((\bigwedge_{t_i\notin A} \#_{\sigma_1}(t_i) = 0) \wedge (\#_{\sigma_1}(t) > 0))$.

- X=T3
$\iff \exists \mu_1 \leq \mu_2 \exists \sigma_0, \sigma_1 ((\mu_0 \xmapsto{\sigma_0} \mu_1 \xmapsto{\sigma_1} \mu_2) \wedge ((\bigvee_{A \in \mathcal{A}} (\bigwedge_{t_i \notin A} \#_{\sigma_1}(t_i) \leq 0) \wedge (v_A \leq \#_{\sigma_1})) \wedge (\bigvee_{t \in T} \#_{\sigma_1}(t) > 0)))$
$\iff \exists \mu_1 \leq \mu_2 \exists \sigma_0, \sigma_1 \bigvee_{A \in \mathcal{A}} \bigvee_{t \in T} (((\bigwedge_{t_i \notin A} \#_{\sigma_1}(t_i) = 0) \wedge (\bigwedge_{r \in T} v_A(r) \leq \#_{\sigma_1}(r)) \wedge (\#_{\sigma_1}(t) > 0))$.
- X=T3′
$\iff \exists \mu_1 \leq \mu_2 \exists \sigma_0, \sigma_1 ((\mu_0 \xmapsto{\sigma_0} \mu_1 \xmapsto{\sigma_1} \mu_2) \wedge ((\bigvee_{A \in \mathcal{A}} v_A \leq \#_{\sigma_1}) \wedge (\bigvee_{t \in T} \#_{\sigma_1}(t) > 0)))$
$\iff \exists \mu_1 \leq \mu_2 \exists \sigma_0, \sigma_1 \bigvee_{A \in \mathcal{A}} \bigvee_{t \in T} ((\bigwedge_{r \in T} v_A(r) \leq \#_{\sigma_1}(r)) \wedge (\#_{\sigma_1}(t) > 0))$.

As a consequence, the nontermination problem for each of the above notions of fairness can be solved in P.

3. *(Strong) Promptness Detection Problem.*
The concept of ***(strong) promptness*** was introduced in [31] as a way to deal with systems communicating with the environment. Let T_I and T_E be two disjoint sets of transitions such that $T_I \cup T_E = T$. (T_I and T_E can be viewed as the sets of internal and external transitions, respectively.) A Petri net (P, T, φ, μ_0) is said to be
(a) *Strongly prompt* (with respect to (T_I, T_E)) iff $\exists k \in N, \forall \mu \in R(P, T, \varphi, \mu_0)$, $\forall w \in T_I^* : \mu \xrightarrow{w} \Longrightarrow |w| < k$, meaning that for every reachable marking μ, the longest sequence of internal transitions fireable in μ is of length $< k$, for some k.
(b) *Prompt* (with respect to (T_I, T_E)) iff $\forall \mu \in R(P, T, \varphi, \mu_0), \exists k \in N, \forall w \in T_I^* : \mu \xrightarrow{w} \Longrightarrow |w| < k$, meaning that for every reachable marking μ, there is no infinite sequence of internal transitions fireable in μ.
It is not hard to see that a Petri net is not prompt (strongly prompt) iff $\exists \mu_1 \leq \mu_2 \exists \sigma_0, \sigma_1 ((\mu_0 \xmapsto{\sigma_0} \mu_1 \xmapsto{\sigma_1} \mu_2) \wedge ((\bigwedge_{t \notin T_I} \#_{\sigma_1}(t) \leq 0) \wedge (\bigvee_{t' \in T} \#_{\sigma_1}(t') > 0)))$
The polynomial time upper bound for the (strong) promptness detection problem follows.

4. *Pairwise Synchronization Problem.*
In [28], the notion of ***y-distance*** was introduced. Given a Petri net $\mathcal{P}$ and a $y \in Z^r$, let D($\mathcal{P}$,y)=$sup_{\mu_0 \xmapsto{\sigma}} (|y \odot \#_\sigma|)$, where $\odot$ stands for the inner product. $\mathcal{P}$ is said to be ***y-synchronized*** if D($\mathcal{P}$,y) is finite. The ***y-synchronization problem*** is that of determining, given a Petri net $\mathcal{P}$ and a y, whether $\mathcal{P}$ is y-synchronized. Given two transitions t_i and t_j, the ***pairwise synchronization problem*** (with respect to t_i and t_j) is a special case of the y-synchronization problem with y defined as $y(i) = 1, y(j) = -1$, and $y(h) = 0, \forall h \neq i, j$. $\mathcal{P}$ is pairwise synchronized with respect to t_i and t_j iff for every path $\mu_0 \xmapsto{\sigma_0} \mu_1 \xmapsto{\sigma_1} \mu_2$, if $\mu_2 \geq \mu_1$, then $\#_{\sigma_1}(t_i) = \#_{\sigma_1}(t_j)$. Using our path formulas, $\mathcal{P}$ is not pairwise synchronized iff $\exists \mu_1 \leq \mu_2 \exists \sigma_0, \sigma_1 ((\mu_0 \xmapsto{\sigma_0} \mu_1 \xmapsto{\sigma_1} \mu_2) \wedge ((\#_{\sigma_1}(t_i) \geq \#_{\sigma_1}(t_j) + 1) \vee (\#_{\sigma_1}(t_j) \geq \#_{\sigma_1}(t_i) + 1)) \wedge (\bigvee_{t \in T} \#_{\sigma_1}(t) > 0))$
Hence, the pairwise synchronization problem can be solved in P.

5. *B-Fairness Detection Problem.*
Let $t_i, t_j \in T$, t_i, t_j are said to satisfy a B-fair relation (denoted by $t_i BF t_j$)

iff $\forall \mu \in R(\mathcal{P}, \mu_0)$ $\forall$ sequence σ such that $\mu \stackrel{\sigma}{\longmapsto}$, the following must hold: $\sigma(t_i) = 0 \Rightarrow \sigma(t_j) \leq k$ and $\sigma(t_j) = 0 \Rightarrow \sigma(t_i) \leq k$, for some constant k ([29]). It was shown in [29] that $t_i BFt_j$ iff $\forall \mu_1, \mu_2$ $((\mu_0 \stackrel{\sigma_0}{\longmapsto} \mu_1 \stackrel{\sigma_1}{\longmapsto} \mu_2) \wedge (\mu_2 \geq \mu_1)) \Rightarrow (\#_{\sigma_1}(t_i) = 0 \Leftrightarrow \#_{\sigma_1}(t_j) = 0)$. Hence, *not* $t_i BFt_j$ can be expressed as $\exists \mu_1 \leq \mu_2$ $((\mu_0 \stackrel{\sigma_0}{\longmapsto} \mu_1 \stackrel{\sigma_1}{\longmapsto} \mu_2) \wedge ((\#_{\sigma_1}(t_i) = 0 \wedge \#_{\sigma_1}(t_j) \geq 1) \vee (\#_{\sigma_1}(t_i) \geq 1 \wedge \#_{\sigma_1}(t_j) = 0)))$. As a result, B-fairness can be detected in P.

Acknowledgments: We thank the anonymous referees for their suggestions that improved the presentation of this paper.

References

1. Alimonti, P., Feuerstain, E. and Nanni, U. Linear time algorithms for liveness and boundedness in conflict-free Petri nets. *Proceedings of 1st Latin American Theoretical Informatics (LATIN'92)*, LNCS 583, pp. 1–14, 1992.
2. Baker, H. *Rabin's Proof of the Undecidability of the Reachability Set Inclusion Problem of Vector Addition Systems.* Memo 79, MIT Project MAC, Computer Structure Group, 1973.
3. Crespi-Reghizzi, S. and Mandrioli, D. A decidability theorem for a class of vector addition systems. *Information Processing Letters*, 3(3):78–80, 1975.
4. Desel, J. and Esparza, J. Reachability in reversible free choice systems. *Proceedings of the 8th Symposium on Theoretical Aspects of Computer Science (STACS'91)*, LNCS 480, pp. 384–397, 1991.
5. Esparza, J. A solution to the covering problem for 1-bounded conflict-free Petri nets. *Information Processing Letters*, 41(6):313–319, 1992.
6. Esparza, J. and Silva, M. A polynomial-time algorithm to decide liveness of bounded free choice nets. *Theoret. Comp. Sci.*, 102:185–205, 1992.
7. Garey, M. and Johnson, D. *Computers and Intractability: A Guide to the Theory of NP-Completeness.* W.H. Freeman & Company Publishers, San Francisco, California, 1979.
8. Grabowski, J. The decidability of persistence for vector addition systems. *Information Processing Letters*, 11(1):20–23, 1980.
9. Hack, M. The equality problem for vector addition systems is undecidable. *Theoret. Comp. Sci.*, 2:77–95, 1976.
10. Heiner, M. *Petri net based software validation prospects and limitations.* International Computer Science Institute, Berkeley, California. TR-92-022, March, 1992.
11. Hopcroft, J. and Pansiot, J. On the reachability problem for 5-dimensional vector addition systems. *Theoret. Comp. Sci.*, 8:135–159, 1979.
12. Howell, R. and Rosier, L. Completeness results for conflict-free vector replacement systems. *J. of Computer and System Sciences*, 37:349–366, 1988.
13. Howell, R. and Rosier, L. On questions of fairness and temporal logic for conflict-free Petri nets. In G. Rozenberg, editor, *Advances in Petri Nets 1988*, pages 200–226, Springer-Verlag, Berlin, 1988. LNCS 340.
14. Howell, R., Rosier, L., Huynh, D. and Yen, H. Some complexity bounds for problems concerning finite and 2-dimensional vector addition systems with states. *Theoret. Comp. Sci.*, 46:107–140, 1986.
15. Howell, R., Rosier, L. and Yen, H. An $O(n^{1.5})$ algorithm to decide boundedness for conflict-free vector replacement systems. *Information Processing Letters*, 25:27–33, 1987.

16. Howell, R., Rosier, L. and Yen, H. A taxonomy of fairness and temporal logic problems for Petri nets. *Theoret. Comp. Sci.*, 82:341–372, 1991.
17. Howell, R., Rosier, L. and Yen, H. Normal and sinkless Petri nets, in *'Proceedings of the 7th International Conference on the Fundamentals of Computation Theory,'* pp. 234-243, 1989. To appear in *J. of Computer and System Sciences*, 46, 1993.
18. Jones, N., Landweber, L. and Lien, Y. Complexity of some problems in Petri nets. *Theoret. Comp. Sci.*, 4:277–299, 1977.
19. Khachian, L. A polynomial algorithm for linear programming. *Doklady Akad. Nauk USSR*, 244, no. 5(1979), 1093-96. Translated in *Soviet Math. Doklady*, 20, 191-94.
20. Kosaraju, R. Decidability of reachability in vector addition systems. In *Proceedings of the 14th Annual ACM Symposium on Theory of Computing*, pp. 267–280, 1982.
21. Lipton, R. *The reachability problem requires exponential space.* Technical Report 62, Yale University, Dept. of CS., Jan. 1976.
22. Landweber, L. and Robertson, E. Properties of conflict-free and persistent Petri nets. *JACM*, 25(3):352–364, 1978.
23. Mayr, E. An algorithm for the general Petri net reachability problem. *SIAM J. Comput.*, 13(3):441–460, 1984. A preliminary version of this paper was presented at the *13th Annual Symposium on Theory of Computing*, 1981.
24. Murata, T. Petri nets: properties, analysis and applications. *Proc. of the IEEE*, 77(4): 541–580, 1989.
25. Peterson, J. *Petri Net Theory and the Modeling of Systems.* Prentice Hall, Englewood Cliffs, NJ, 1981.
26. Rackoff, C. The covering and boundedness problems for vector addition systems. *Theoret. Comp. Sci.*, 6:223–231, 1978.
27. Reisig, W. *Petri Nets: An Introduction.* Springer-Verlag, Heidelberg, 1985.
28. Suzuki, I. and Kasami, T. Three measures for synchronic dependence in Petri nets, *Acta Informatica*, 19:325–338, 1983.
29. Silva, M. and Murata, T. B-Fairness and structural B-Fairness in Petri net models of concurrent systems *J. of Computer and System Sciences*, 44:447–477, 1992.
30. Stockmeyer, L. The polynomial-time hierarchy. *Theoret. Comp. Sci.*, 3:1–22, 1977.
31. Valk, R. and Jantzen, M. The residue of vector sets with applications to decidability problems in Petri nets, *Acta Informatica*, 21:643–674, 1985.
32. Yamasaki, H. Normal Petri nets. *Theoret. Comp. Sci.*, 31:307–315, 1984.
33. Yen, H. A polynomial time algorithm to decide pairwise concurrency of transitions for 1-bounded conflict-free Petri nets. *Information Processing Letters*, 38:71–76, 1991.
34. Yen, H. A unified approach for deciding the existence of certain Petri net paths. *Information and Computation*, 96(1): 119–137, 1992.

A Colored Petri Net Model for a Naval Command and Control System

J. Berger
L. Lamontagne

Command and Control Division
Defence Research Establishment Valcartier - DREV
P.O. Box 8800, Quebec City, P.Q., Canada, G0A 1R0

Abstract. A colored Petri Net implementation of a high-level defence model is presented. The model is being used to represent the real-time aspects and behavior of critical system components and assess the performance of various weapon assignment decision policies. Simulation results for two candidate weapon allocation strategies are briefly discussed. A comparative analysis with an object-oriented simulation environment is then carried out in order to identify strengths and weaknesses of the approach related to system design.

1 Introduction

The use of Petri Net (PN) technology to tackle different aspects of command and control (C2) problems has gained popularity within the research community in recent years. Common decision-making applications lie in the specification and performance analysis of teams formed by individuals or entities sharing common resources and information [1]-[4]. In most cases, the objective is to assess and compare organizational structures, and various cooperation and coordination mechanisms governing these teams. However, few attempts have been made so far to investigate centralized decision-making strategies while considering generic component models of a command and control system.

This paper presents the modeling of a conceptual naval C2 system using the hierarchical Colored Petri Net (CPN) formalism [5]. It aims to demonstrate the feasibility of the approach in capturing attributes and real-time features of critical components and consequently assess the effectiveness of various weapon assignment policies. An attempt is made to determine how colored Petri Nets constitute a proper framework and a suitable simulation methodology to describe C2 systems.

Section 2 presents the Petri Net modeling of a conceptual naval ship combat system pictured as a high-level defence model. The basic elements are depicted and their main characteristics outlined. Simulation results are then summarized and briefly discussed in Section 3. Finally, Section 4 compares the CPN model with an object-oriented simulation environment previously developed at DREV in order to emphasize strengths and weaknesses of the approach related to system design.

2 Application Modeling

2.1 High-level Defence Model Overview

This subsection presents the fundamental components of a generic conceptual high-level defence model based on a system environment scheme. The environment accounts for the air threats while the system represents the force (ship unit). The model architecture considered, SARA (Situation Assessment and Resource Allocation), is pictured in Fig. 1.

The system environment scheme comprises six types of different processes or entities, namely, threat scenario, sensor, weapon, track manager, action manager, and the battle manager responsible for target evaluation and weapon assignment (TEWA). Message passing in SARA is depicted on Fig. 1 by self-explanatory underlined inscriptions, attached to directed edges binding two node components. The inscription delineates the nature of the interaction taking place between two elements.

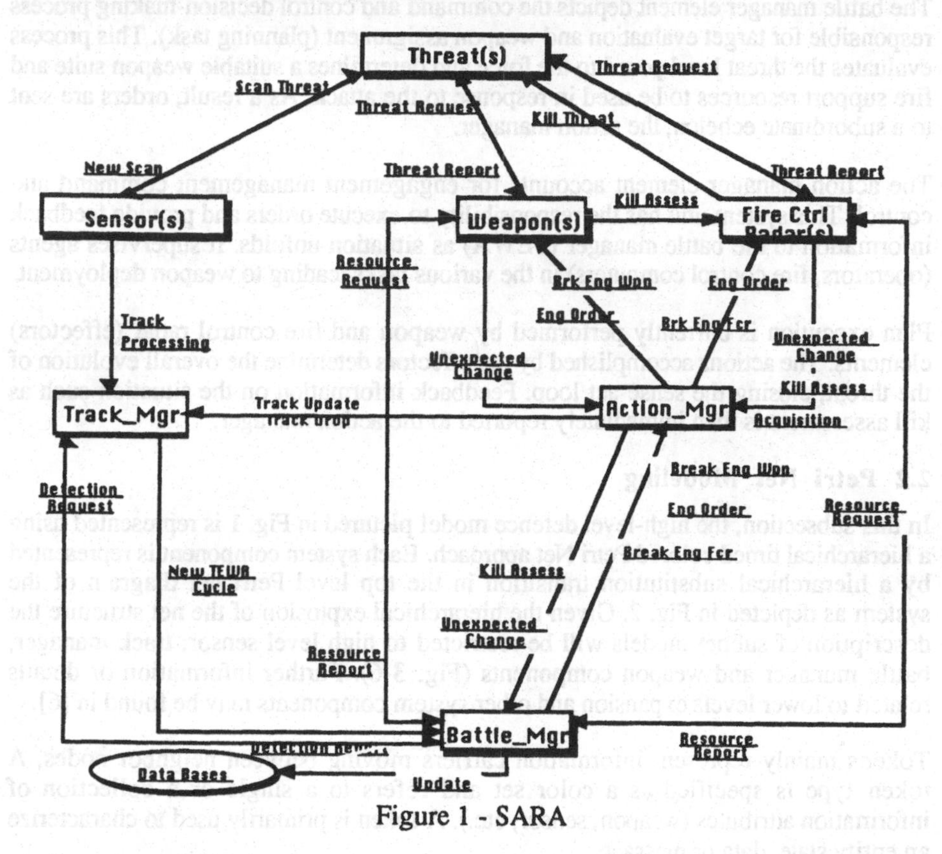

Figure 1 - SARA

The environment is pictured via the threat scenario element in which a collection of air threats namely anti-ship missiles emerging from hostile platforms, are directed toward a predetermined target (ship). The threat scenario characterizes the deterministic kinematical behavior of the air threats.

The system unit including the five remaining elements is made up of sensor(s), weapon(s), a track manager, a centralized TEWA decision-making process (battle manager), and an action manager. The following paragraphs of this section briefly explain how the system component operates and interacts with the environment, from detection to engagement management.

Targets are periodically detected by sensors responsible for data gathering and track generation (initiation, update, drop). Sensory information encoded as a set of tracks is then sent to the track manager as an input.

The track manager is responsible for carrying out track processing and data fusion, based on the multiple sensor outputs. The number of tracks obtained corresponds to the number of threats observed. As a track processing iteration is completed a new TEWA cycle is triggered.

The battle manager element depicts the command and control decision-making process responsible for target evaluation and weapon assignment (planning task). This process evaluates the threat level posed to the force and determines a suitable weapon suite and fire support resources to be used in response to the attack. As a result, orders are sent to a subordinate echelon, the action manager.

The action manager element accounts for engagement management command and control. This system unit has the responsibility to execute orders and provide feedback information to the battle manager (TEWA) as situation unfolds. It supervises agents (operators, fire control computers) in the various tasks leading to weapon deployment.

Plan execution is currently performed by weapon and fire control radar (effectors) elements. The actions accomplished by the effectors determine the overall evolution of the threat, closing the sense-act loop. Feedback information on the situation such as kill assessment is then immediately reported to the action manager.

2.2 Petri Net Modeling

In this subsection, the high-level defence model pictured in Fig. 1 is represented using a hierarchical timed colored Petri Net approach. Each system component is represented by a hierarchical substitution transition in the top level Petri net diagram of the system as depicted in Fig. 2. Given the hierarchical explosion of the net structure the description of subnet models will be restricted to high level sensor, track manager, battle manager and weapon components (Fig. 3-6). Further information or details related to lower levels expansion and other system components may be found in [6].

Tokens mainly represent information carriers moving between neighbor nodes. A token type is specified as a color set and refers to a single or a collection of information attributes (weapon, sensor, etc.). A token is primarily used to characterize an entity state, data or message.

Net places portrayed as token color set receivers designate a data store and message passing mechanism, and account for process state. Generally, places present two geometrical shapes: an oval, restricted to accepting entity state tokens and a circle, destined to acquire entity data (solid line) or message (dashed line) tokens. Dotted circular places constitute extra dummy nodes necessary to support simulation management and conflict resolution. Labels associated with places, refer to shared/fused (FG) places, or port nodes (B) to ensure interface (in/out) with other net diagrams.

Pictured as an information processor, a transition is generally used to represent a function or an action associated with a process entity, or a subnet abstraction. A delay

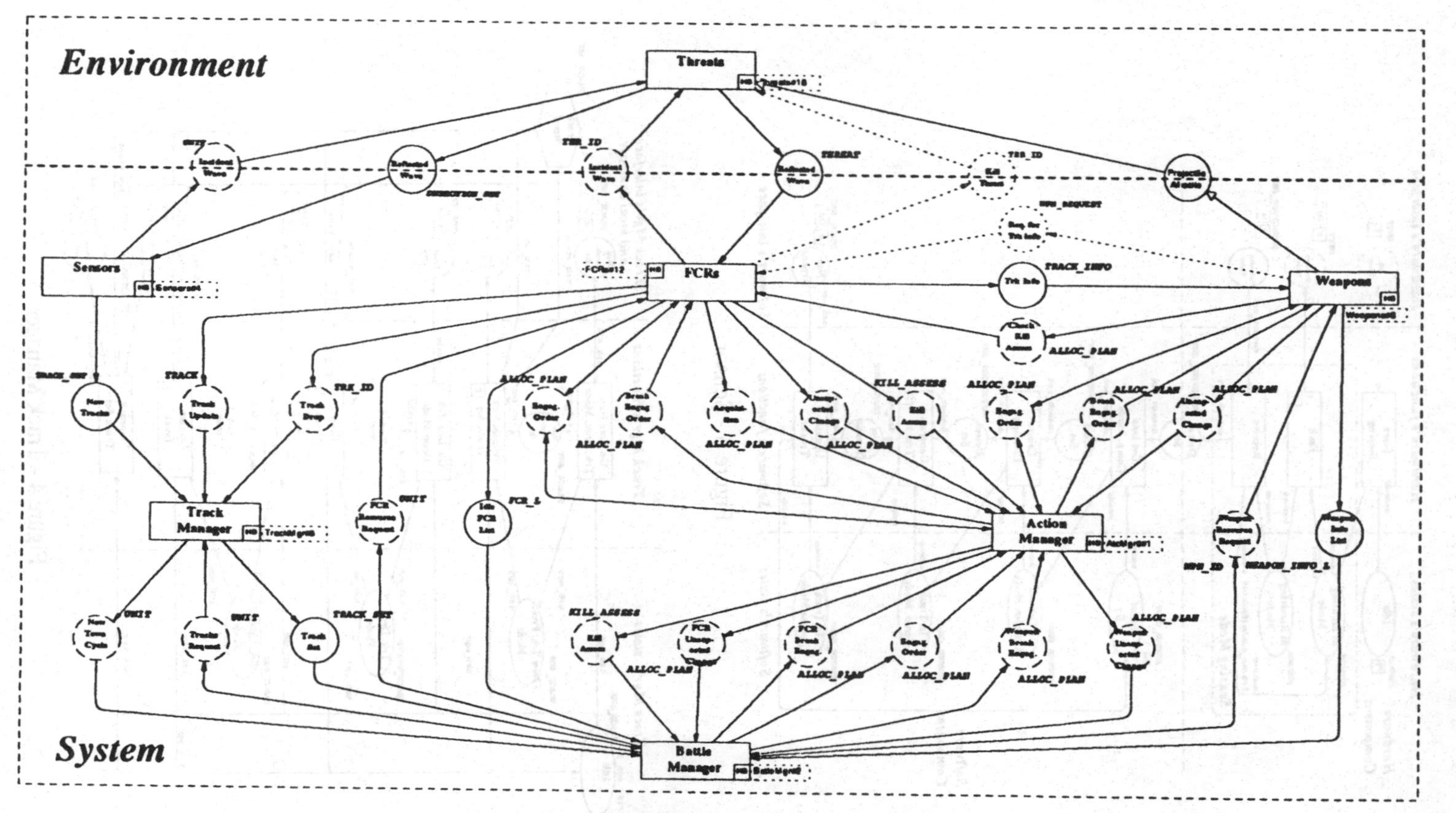

Figure 2 - Colored Petri Net Model (SARA)

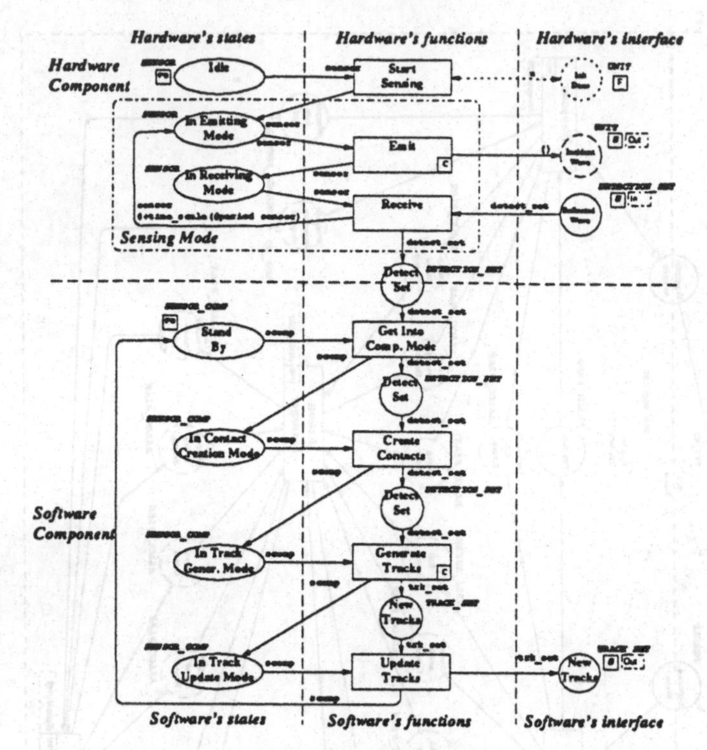

Figure 3 - Sensor

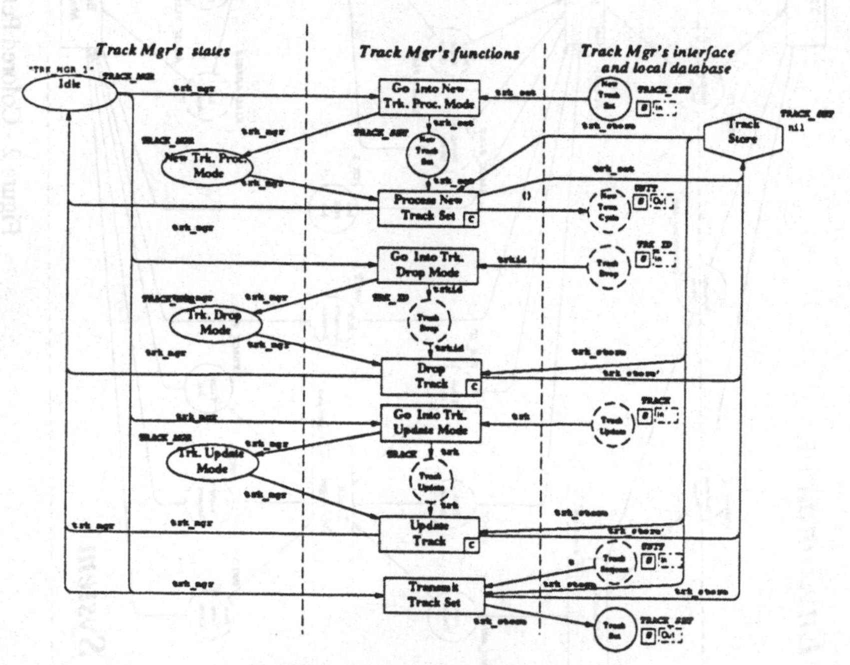

Figure 4 - Track Manager

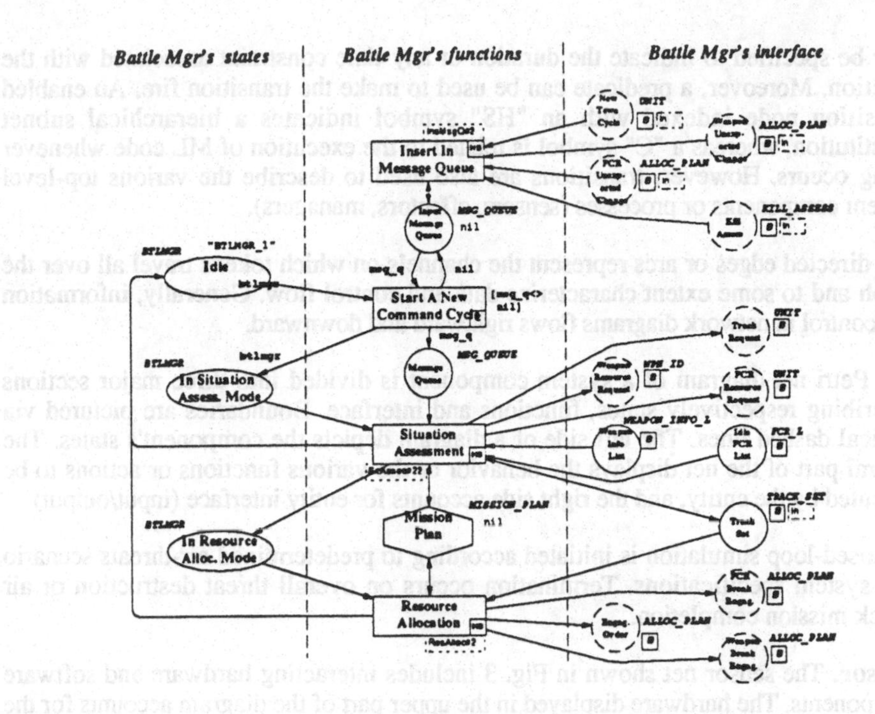

Figure 5 - Battle Manager

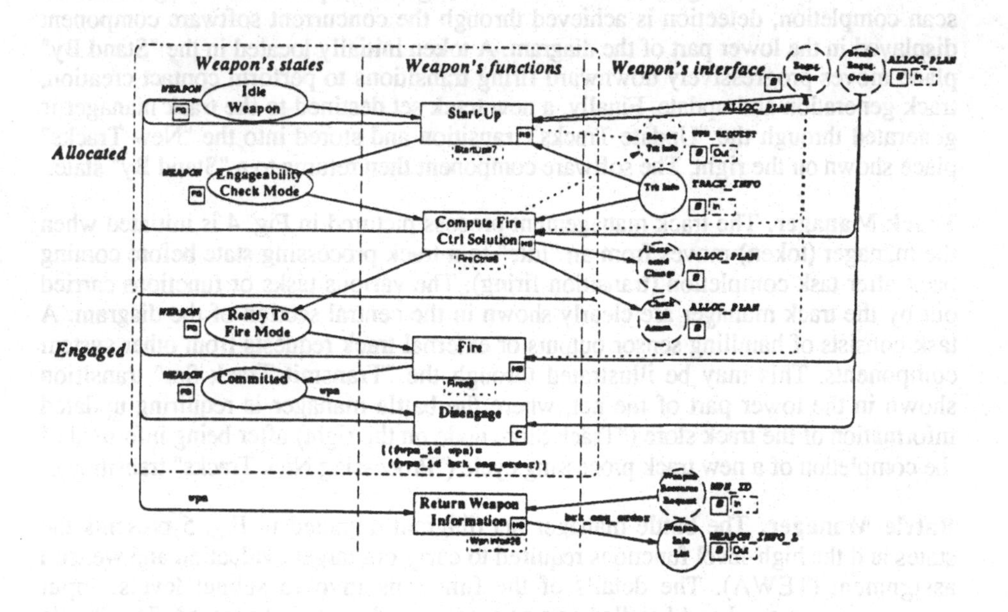

Figure 6 - Weapon

may be specified to indicate the duration or any time constraint associated with the function. Moreover, a predicate can be used to make the transition fire. An enabled transition node indexed with an "HS" symbol indicates a hierarchical subnet substitution, whereas a "C" symbol is related to the execution of ML code whenever firing occurs. However, transitions are also used to describe the various top-level system components or processes (sensors, effectors, managers).

The directed edges or arcs represent the channels on which tokens travel all over the graph and to some extent characterize data and control flow. Generally, information and control in network diagrams flows rightward and downward.

The Petri net diagram of a system component is divided into three major sections describing respectively states, functions and interface. Boundaries are pictured via vertical dashed lines. The left side of a diagram depicts the component's states. The central part of the net displays the behavior or the various functions or actions to be executed by the entity, and the right side accounts for entity interface (input/output).

A closed-loop simulation is initiated according to predetermined air-threats scenario and system specifications. Termination occurs on overall threat destruction or air attack mission completion.

Sensor. The sensor net shown in Fig. 3 includes interacting hardware and software components. The hardware displayed in the upper part of the diagram accounts for the scan process, where a "sensor" token initially marking the "idle" place on the left, moves periodically from an emitting to a receiving mode. The interactions with the environment (threats diagram) and the track manager are depicted on the right side. On scan completion, detection is achieved through the concurrent software component displayed in the lower part of the diagram. A token initially located in the "Stand By" place moves progressively downward firing transitions to perform contact creation, track generation and update. Finally, a new track set destined to the track manager is generated through the "Update Tracks" transition and stored into the "New Tracks" place shown on the right. The software component then returns to a "Stand By" state.

Track Manager. The track management process pictured in Fig. 4 is initiated when the manager (token) moves from an "Idle" to a track processing state before coming back after task completion (transition firing). The various tasks or functions carried out by the track manager are clearly shown in the central section of the diagram. A task consists of handling sensor outputs or external track requests from other system components. This may be illustrated through the "Transmit Track Set" transition shown in the lower part of the net, where the battle manager is requiring updated information of the track store ("Track Store node on the right) after being informed of the completion of a new track processing cycle ("Processing New Tracks" transition).

Battle Manager. The battle manager net diagram depicted in Fig. 5 presents the states and the high-level functions required to carry out target evaluation and weapon assignment (TEWA). The details of the functions involve subnet levels. Input messages are queued and handled as a new command cycle is triggered. The "battle manager" token initially located in the "idle" state then moves to the "In Situation Assess. Mode" state firing the "Start A New Command Cycle" transition. Thereafter, "Situation Assessment" is triggered and the "battle manager" token moves either downward to the "In Resource Alloc. Mode" state if replanning is necessary, or upward to an "idle" state to complete the decision cycle. When replanning is required, resource allocation is executed ("Resource Allocation" transition) modifying the current mission plan ("Mission Plan" place). As a result, engagement/disengagement orders are issued to the action manager and the battle manager becomes idle.

Weapon. The weapon net shown in Fig. 6 resembles in many respects the diagrams presented earlier. Resource states and functions are clearly described as "weapon" tokens (weapon instances) move independently downward from an "idle" to a "committed" state, subject to external interrupts.

The major obstacles encountered during system modeling were closely related to the design of proper mechanisms, involving the use of global variables and extra intermediate nodes to ensure conflict resolution in handling multiple enabled transitions, a critical task in achieving resource allocation. The overall three hundred node network which mainly focuses on the underlying high-level components of the system has been built using Design/CPN [7].

3 Simulation Results

Typical simulations have been conducted for various air-threat scenarios in order to determine single platform survivability. Based on a specific configuration for a single ship, the critical parameter representing input stimulus was characterized by threat intensities. Two weapon-target allocation strategies were considered: earliest intercept and random weapon assignment. The former consisted in pairing weapons to prioritized targets (earliest intercept) with maximum highest probability of kill. The latter randomly allocates weapons to targets according to constraints imposed by layer defence. Shoot-Look-Shoot doctrine prevailed for both policies.

Experiments have been conducted for simultaneous attack scenarios with respect to various air-threat intensities. Several series of simulation runs (30) have been performed for each scenario. Statistic estimators on ship survivability modeled as Bernoulli variables (survival/destruction) are briefly summarized in table I. For each intensity level of the threat a 95% confidence interval has been determined using a t-test. Results shown in table I proved to be compatible with simulation runs conducted using a more refined tool (object-oriented simulation environment) developed at DREV.

Threat Intensity (simultaneous attack)	Weapon Assignment Policy: Earliest Intercept	Random
4	0.967 ± 0.068	0.933 ± 0.093
5	0.833 ± 0.139	0.800 ± 0.149
6	0.310 ± 0.171	0.133 ± 0.127

TABLE I: Ship Survivability

Based on statistical analysis, Table I indicates similar performances for both strategies in low threat intensity. This is reflected by a reasonably high probability of ship survival for the investigated class of scenarios, diminishing the impact of reaction time as a critical factor. However, as threat size increases, the earliest intercept policy shows better performances. Hence, the Petri Net model can be used to determine the critical size of the threat in order to select the most suitable option among candidate decision policies. Tradeoffs between the quality of resource allocation and real-time measures of performance such as execution time, responsiveness and graceful adaptation may also be easily investigated.

4 Comparative Analysis

An evaluation of the CPN model put forward is made possible by examining an earlier implementation of the same conceptual high-level defence model based upon an alternate paradigm. The object-oriented discrete-event simulation environment recently developed at DREV [8], in order to investigate TEWA performances, represents the best candidate for comparison purposes. The simulator is implemented in Sim++ and currently runs on Sun platforms.

The Petri Net implementation turns out to offer the proper ingredients to represent satisfactorily problem domain modeling and considerably facilitates system design. It captures information/data flows and system concurrency through synchronous and asynchronous processing activities and properly encapsulates hierarchical behavior, a critical feature for a C2 system. The simulator constitutes an excellent framework for architecture analysis, minimizing the effort required to modify model structure and network connectivity. It is also fairly simple to reconfigure functional activities and change communication links between key elements. Most of these characteristics make the CPN approach definitely more attractive.

The colored Petri Net approach allows the designer to focus on the problem models at hand instead of the details of the implementation. However, preliminary experiment for simple system design with Design/CPN tends to suggest that even though a large variety of possible implementation is often feasible, the success of the method largely depends on which aspect of the system is intended to be modeled beforehand. Except for small systems, a single net accounting for all aspects is believed to be very unlikely. Consequently, if a complete and very detailed model is intended to be represented, the approach might be far more successful for simple system components such as a specific military subsystem.

The main advantages shown by the CPN approach over the object-oriented simulation environment can be summarized as follows: rapid prototyping, formal specification, system modularity (avoid spaghetti effect), explicit representation of concurrence (highly suitable for real-time systems), easily manageable for small size network, formal analysis capability, shorter development cycle, and maintenance. On the other hand, the weaknesses perceived by the authors are basically related to symbolic treatment limitations, granularity of the modeling, lack of flexibility concerning the implementation of sophisticated numerical algorithms, and restricted capabilities offered by the tool at hand for system design. A special emphasis on those aspects may lead to reject CPN over an object-oriented approach.

Generally, the CPN approach considered facilitates the identification and the analysis of the critical timing factors and impact on overall system performance. For instance, it allows the designer to identify bottlenecks as the problem size grows, determine reaction time (stimulus-response delay) for various critical processes, predict overall system responsiveness and test alternative concepts from threat detection to weapon release. These tasks can be accomplished much more easily due to the graphical representation of the system. The CPN approach also presents a capability to achieve formal analysis in order to prove system properties and carry out "what-if" simulation.

5 Summary and Conclusion

A high-level timed colored Petri Net model representing basic attributes and characteristics including real-time features for major components of a generic ship combat system has been successfully implemented. On this basis, simulations were conducted in order to assess two weapon assignment decision policies. An interesting observation is certainly the compatibility of the simulation results provided by the Petri Net model with those obtained using a detailed object-oriented simulation environment previously developed at DREV. Finally, a comparative analysis on system design was carried out between the Petri Net model and the simulation environment.

High-level timed colored Petri Net technology appear to be a suitable approach to formally specify concurrent system design to satisfy functional and time requirements. The underlying formalism supports rapid prototyping to test concepts and alternatives and allows the designer to focus hierarchically on problem domain modeling rather than implementation issues. Moreover, the technique presents a capability to perform "what-if" simulation and carry out formal analysis in order to prove system properties. However, the success of the colored Petri Net approach as a proper framework and a suitable simulation methodology to describe C2 systems largely depends on which aspect and the desired level of details the targeted system is intended to be modeled.

References

[1] Coovert, M. D., Salas, E. "Modeling Team Performance with Petri Nets", Symposium on Command and Control Research, pp. 288-296, 1990.

[2] Grevet, J.-L., Levis, A. "Coordination in Organizations with Decision Support Systems", Symposium on Command and Control Research, 1988.

[3] Boettcher, K. L., Levis, A. "Modelling and Analysis of Teams of Interacting Decisionmakers with Bounded Rationality", IEEE Transactions on Systems, Man, and Cybernetics, SMC-12, pp. 334-344, 1982.

[4] Tabak, D., Levis, A. "Petri Net Representation of Decision Models", IEEE Transactions on Systems, Man, and Cybernetics, SMC-15, 1982.

[5] Jensen, K. "Coloured Petri nets: Basic Concepts, Analysis Methods and Practical Use", EATCS Monographs on Theoretical Computer Science, Berlin; New York: Springer-Verlag, 1992.

[6] "On the Use of Petri Nets in Naval Command and Control Systems", Informission Ltée, Contract W7701-1-0665/01-XSK, DREV, August 1992.

[7] Jensen, K., et al. "Design/CPN: A Reference Manual", Meta Software Corporation, 125 Cambridge Park Drive, Cambridge, MA, 1991.

[8] Berger, J. "Target Evaluation and Weapon Assignment Demonstrator Design", DREV M-3138, 1992, UNCLASSIFIED.

Petri Net based Specifications of Services in an Intelligent Network - Experiences gained from a Test Case Application

Carla Capellmann and Heinz Dibold

Deutsche Bundespost Telekom, Research Center
P.O.B. 100003, D-6100 Darmstadt
E-MAIL: dibold@fz.telekom.de

Abstract. This paper summarises a project, which studied the feasibility of the Open Petri Net Method (OPM) - a specification method using Hierarchical Coloured Petri Nets as specification language - in the telecommunication field, especially for the specification of services in an Intelligent Network (IN). To this end an IN service test case was defined. The application of OPM is illustrated, as well as the resulting HCPN model of the IN test case service. The main project results are presented: they are promising, but in view of the special requirements valid in the telecommunication area, the usage of Petri nets for specification purposes still faces some problems.

1 Introduction

1.1 Background information

In the telecommunication area, the Intelligent Network (IN) is understood as a concept for the fast and economical provision of new services going beyond the limits of conventional telephony. Starting from the assumption that each IN service can be made up of single service-independent functional components, a logical and a physical architecture for service provision are defined. Details and a list of IN services known to date are e.g. given in [AMS89] and [Dibo90].

Within the framework of the IN architecture, there arises the problem of defining the individual service components and of describing their interaction. This description has to be made independently of the technology of the network components, for which reason each one of them is regarded as "black box" with an exactly defined behaviour. Due to the highly distributed and concurrent nature of telecommunications and the increased complexity of flexible services, a method as well as a fitting specification language is needed with emphasis on the following characteristics:

- to describe distributed systems with their inherent concurrency,
- to cope with complex systems by powerful structurisation means,
- to lead to declarative specifications, which are independent of implementation-technology, and
- to use formal, executable, analysable, but still comprehensible system models.

Comparable problems related to the implementation-independent description of the functional behaviour of telecommunication systems and services have already been treated in [Dibo88, Dibo92a and Dibo92b]. Due to the good experiences gained, we started a project to find out, whether the Open Petri Net Method elaborated at the Deutsche Bundespost Telekom Research Center would too be suitable for the description of services and service constituents in an IN. To this end, we applied OPM to the Universal Access Number Test Case (UANTC), a fictitious IN service with typical requirements, but limited in the amount of details to be modelled. The specification of UANTC gained and an interpretation of the project's results are the subject of this contribution.

1.1 Structure of this paper

This paper is structured as follows: Chapter 2 briefly introduces OPM and HCPNs. Chapter 3 presents selected parts of the UANTC specification, which is the outcome of a strict application of OPM to the IN service test case, with emphasis on the resulting HCPN system model. In this, the main characteristics of UANTC itself are also illustrated. Chapter 4 summarises our judgement about the feasibility of OPM and HCPNs for the specification of services in an IN.

2 OPM and HCPNs

The Open Petri Net Method (**OPM**) was developed at the Research Center of the Deutsche Bundespost Telekom to *support the first phases of the system (software) life cycle* [Dibo92a, Dibo92b]. For the *analysis* of the functional characteristics of the system under consideration, the OPM provides *guidelines,* which were primarily derived from the methods Structured Analysis [DeMa78], Essential Systems Analysis [MP84] and Strategies for Real-Time System Specification [HP87]. The analysis is based on *hierarchical functional decomposition* with special emphasise on recognising the *essential* system requirements, which are *independent of implementation technology*. The OPM supplements these guidelines by one analysis step, which aim is to ascertain the maximum permissible *concurrency* of system activities predetermined by the task setting.

In OPM the functional system behaviour, ascertained in the analysis phase, is described by means of an "event definition" part and an executable "system model". The *event definition* links the *concrete* interactions needed between the system and its environment in the real world to the *abstract* events of the system model. The *system model* is based on *hierarchical coloured Petri nets* (HCPNs) [Jens92] for which a tool for editing and simulation [Design/CPN] is commercially available on different computer platforms. In contrast to basic (low-level) Petri nets [Pete81], HCPNs enable the editing of a hierarchically structured, compact and comprehensive system model with easy usage of complex data manipulation functions and *full Turing power* (using the functional language SML [HMT87, Wiks87]). The main advantage of using Petri Nets as system model is a *one to one correspondence* between the elements of the analysis phase and the elements of the describing system model. This avoids the commonly recognised semantic gap when traversing from the analysis to the specification phase.

Other features of HCPNs, like physical consumption and production of tokens, locality, referential transparency, declarative description of data and data operations, freeness of side-effects, and the inherent concurrency, help to minimise the introduction of features in the system model, which are only due to the specification language and not to the system's task setting (avoidance of over specification).
In the first phases of the system life cycle, there exists no formal description of the needed system behaviour. Hence it is important, that the evolving system description can be "verified" at any stage against one's informal ideas about the system under question by means of *simulation* of the executable HCPN system model. Due to its conceptual simplicity, its graphical representation, and the compact and natural description of data and data operations, the HCPN system model and its simulation also forms an *unambiguous communication basis* between the system analyst(s) and the people of the application domain.

In the course of our project, OPM with its *seven main steps* was applied to UANTC, a test case based on a fictitious universal access number service. The essential parts of the resulting specification, thus also revealing the characteristics of UANTC itself, are presented in the next chapter.

3 Application of OPM to UANTC

3.1 Main steps of the method

OPM consists basically of *seven main steps*. These steps and their results with regard to UANTC will be shortly described in the following.

Step 1

Definition of task setting, system and relevant system environment.

System: Universal Access Number Test Case service, UANTC.

Aim of the Universal Access Number Test Case (UANTC), which was defined in the context of an EURESCOM[1] project, is to give a realistic example of the typical requirements of an IN service, without going unnecessarily into details.

Task setting

Main purpose of the UANTC service is the translation of an Universal Access Number (UAN), dialled by a service user[2], into a concrete terminating line number, depending on origin of call as well as time and date of call origination. The service is offered by the service provider[3] with the following service features:

- Feature "ONE" (one number): For the distribution of calls according to call origination, the network is divided into three distinct geographic zones: north, middle and south.
- Feature "TDR" (time distributed routing): For the distribution of calls according to time and day, the week is divided into two alternating time zones: business (e.g. Monday to Friday from 08:00 to 18:00) and leisure.
- The assignment of a terminating line in the above service features "ONE" and "TDR" is modifiable on-line by the service subscriber[4] during its subscription. To authorise the service subscriber for modification, he has to use an authorisation code which he has to get from the service provider in the course of subscription.

The UANTC is able to interact with other service features, which are defined for incoming calls to terminating lines, i.e. in our test case call queuing, which is defined as follows:

- Feature "QUE" (queuing of calls): Queuing of arriving calls is offered for a group of terminating lines, with one line's directory number dedicated to address the group. Arriving calls for the group are offered to available terminating lines of the group. If more than one line of the group is available, the served terminating line is chosen at random. If none of the terminating lines of the group is available, the call is queued. The queue length is fixed to ten. Calls are taken off the queue as soon as a line in the group becomes available.
- Modification of the parameters of the "QUE" feature are out of the scope of this service test case.

---

[1] European Institute for Research and Strategic Studies in Telecommunication Systems with headquarters in Heidelberg/Germany, founded in 1991 by 23 public network operators of 18 European countries

[2] *Service user:* Person (or entity) who actually uses a service

[3] *Service provider:* Person (or entity) who offers services for service subscription, usually on a commercial basis

[4] *Service subscriber:* Person (or entity) who subscribes to a service offered by the service provider, usally on a contractual basis.

Comment:

The UANTC actually comprises *two* different functionalities: a one dimensional translation of a given UAN into a destination number as function of call origin, and a two dimensional translation as function of time and origin of the call. The result of the translation is possibly further influenced by existing terminating line features.

Relevant system environment

While the service specification under question is basically meant for purposes of the service provider, the relevant environment for the system's functionality was found to consist of *service subscriber* and *service user*.
Additionally, the investigation of the task setting showed, that the UANTC functionality has to be able to interact with other service features. This means, that the system itself can not be purely treated as black box, but has to be composed on top of the functionality of feature *subsystems*, which in turn can be treated as black boxes, offering a defined functionality to their respective environments.
The corresponding subsystems were found to be the proper *UAN service* and the *terminating line feature*.

Step 2

Ascertainment of the relevant interactions between system and environment (system stimuli and responses) and fixing of their concrete appearance in the real world (event definition).

Comment:

The *event definition* links the items of the (logical) system model to the concrete interactions needed in the real world. It is essential for the correct functioning of the system, which is implemented according to the specified functionality, in its concrete real-world environment. Since we are not going to really implement the service as a system for real-world usage, the event-definition part is mainly skipped in this test case.

System stimuli (just one of five shown due to space limitations)

Name: **SubscrReq** from service subscriber

Purpose: Request for subscription to UANTC service.

Required parameters:
Identity of service subscriber; translation scheme, also indicating if one or two dimensional translation is wanted.

Concrete appearance:
Subscription request will be issued from the service subscriber to the service, i.e. to administrative people of the service provider responsible for the UANTC service, by means of *form sheets*, with a one or two dimensional translation table. By issuing the corresponding translation table, the service subscriber chooses between a one or a two dimensional translation for his UANTC service subscription.

System responses (just one of five shown due to space limitations)

Name: **SubscrConf** to service subscriber

Purpose: Confirmation of UANTC service subscription for the service subscriber.

Required parameters:
Identity of service subscriber; UAN assigned to service subscription; authorisation code for on-line access to service subscription.

Concrete appearance:
Preferably in written form, mailed to the service subscriber.

Step 3

Ascertainment of essential (required independently of the implementation) activities and essential memories of the system with a view to the required system responses and interfaces.

Comment:

The "essential" characteristics can be distinguished from the inessential ones by considering the system strictly independent of the implementation technology. For this reason a *perfect* system technology is assumed. This means that the active units can perform any number of actions fail-safe without measurable time consumption, whereas the passive system parts can store data of any number and any structure. The system environment is to be considered real, that means not idealised.

Instead of fixing the ascertained essential activities and essential memories by writing them down in just another list, it is more convenient to model them step by step into the evolving HCPN system model, annotated by comments were appropriate. Essential activities are modelled by transitions (thin bordered rectangular boxes), while essential memories are modelled by places (ellipses).
E.g. in the UANTC model, the following *essential system activity* was found and documented on page "profile #200" (Fig. 3) of the HCPN model (note, that only the "good" cases are modelled, to keep the model simple):

"subscribe to UAN service": all actions needed to process a service subscriber's request for subscription to the UANTC service. The described functionality of this activity consists partly of actions to be done by people belonging to the service provider's staff and partly of actions to be implemented by means of hardware and software.

The following *essential system memory* was found during system analysis and modelled on page "profile #200" (Fig. 3):

"avail.UANs": means for the system to remember which UANs are actually available for new service subscriptions.

Step 4

Ascertainment of essential activities for data management in essential system memories (initialisation and time-triggered activities).

The result of elaborating the needed (distributed) initial system state is reflected in the HCPN model by means of initial marking regions (underlined text close to a place, e.g. "99" for place "avail.Code" and "0123" for "avail.UAN" on page "profile #200" (Fig. 3). We restricted the available codes and numbers in this test case to one single item each, in order to enforce, that the given ones are assigned to the first subscription).
Since the test case should be kept simple, there are no essential activities for data management, i.e. which are not directly triggered by stimuli from the environment, in the HCPN model. One such candidate would be a time-triggered activity to monitor actual (system) time in order to change the marking of the place "time zone" on page "uanFeature #1000" from "Business" to "Leisure" and vice versa, according to the definition given by the service provider.

Step 5

Ascertainment of maximum concurrency of all essential activities given by the system's task setting.

Analysis and documentation of the maximum allowable concurrency of essential activities due to the task setting are a very important prerequisite to ensure economical and technical feasible system implementations. This point is widely neglected in other specification methods, partly due to the fact, that many of them use specification languages which are not able to express concurrency adequately.

The outcome of concurrency analysis for UANTC is reflected in the HCPN model, where all transitions are *by default* concurrent to each other, except if sequences or conflicts are *explicitly* modelled. E.g. it is modelled, that only one of the terminating line feature activities can occur, after the outcome of the UAN number translation has become available.

Step 6

Description of relationship between external or temporal events (system stimuli) and the essential activities (system responses) in the form of an operative system model on the basis of hierarchical coloured Petri nets, such that

- ***all essential activities are assigned to the event which triggers them directly***
- ***all essential memories are combined to object groups according to their causes in the system environment***
- ***the maximum possible concurrency of the essential activities is expressed.***

The result of this step is the *HCPN model for UANTC* (see Chapter 3.2).
Note: In practice, steps three to six are usually not strictly applied in sequence, but applied iteratively.

Step 7

Checking of the system description at any stage against one's informal ideas about the system behaviour under question by means of symbolic execution of the system model.

In the early phases of the system life cycle, there exists no formal system description against which the evolving specification can formally be checked. Hence it is very important to use executable specification models, to illustrate what kind of behaviour the system model worked out so far really describes. Especially when specifying distributed systems, the "simulation" of concurrent behaviour by arbitrary execution of concurrent activities helps a lot to avoid and/or detect specification errors, since human beings tend to think sequentially.
A "test run" of the elaborated UANTC specification was derived by symbolic execution of its HCPN model, which corresponded directly to the given message sequence charts (MSCs) of "good cases" as fixed for the test case (due to length restrictions of this paper it has to be skipped here).

3.2 HCPN model of UANTC service

This chapter presents selected parts of the HCPN, which forms the executable system model for the required UANTC service behaviour, as it was analysed according to the guidelines of OPM.
Since the given HCPN is highly self-explanatory and should be comprehensible for participants of Petri Net Conferences, we added mainly comments to ease the understanding of the underlying telecommunication procedures. For the exact syntax and semantics of HCPNs, we have again to refer to [Jens92] or other sources. The graphical conventions we used in our modelling are given in the Annex.

3.2.1 Basic model

The following diagrams represent the basic HCPN model of UANTC, i.e. describing the functionality according to the simplified definitions and *restrictions* of our test case.
Note, that the UANTC model is not only capable to describe the service functionality for interactions with *only one* user or subscriber at a time, but rather demonstrates the

system's capability for *concurrent* processing of different stimuli and their mutual influence, e.g. when accessing critical resources.

The UAN service and its environment

Page number 1 of the HCPN model, which constitutes the highest abstraction level, depicts the system and its environment, as worked out in the analysis phase (Fig. 1). System and environment interact via a (logical) interface.

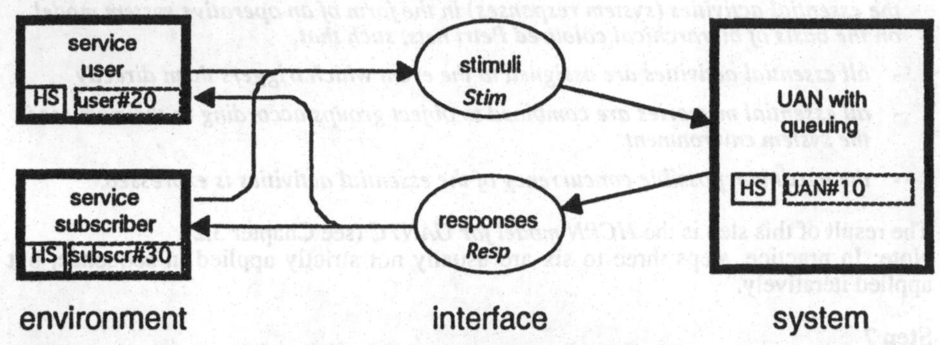

Fig. 1. UAN service and environment (page #1)

The description of the exact wanted functionality of the system as well as the expected functionality of the environment is hidden on this abstraction level by means of subnets "UAN" (on diagram page number 10), "user" (page 20) and "subscriber" (page 30), denoted in this diagram by the fat-bordered transitions "UAN with queuing", "service user" and "service subscriber".

Subnet "UAN"

The UAN service functionality is grouped into two subnets: "usage of service" and "profile management" (Fig. 2). This reflects the two different kinds of interactions with the UAN: The service can be invoked for actual usage or the profile of a service subscription can be managed, i.e. created, modified, cancelled.

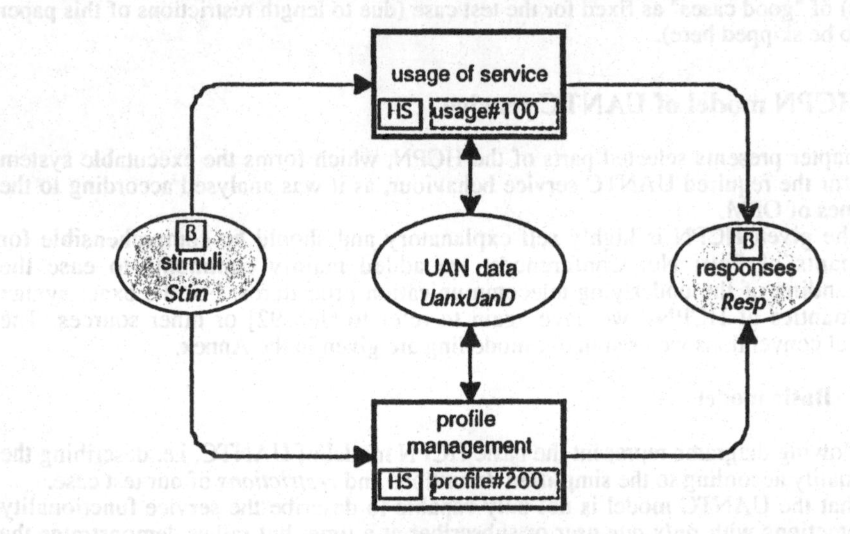

Fig. 2. Subnet UAN (page #10)

Both subnets have access to place "UAN data". This place models the memory of the system needed to hold the data that determine the customised services of the different subscribers. Functions of "profile management" are responsible for the management of UAN data, whereas functions of "usage of service" just interpret UAN data.
Note, that grey shaded ellipses with a "B" box ("stimuli", "UAN Data" and "responses") model subnet ports, which denote the corresponding places on upper level hierarchy pages. In connection with a subnet, arcs with arrows in both directions denote, that token *may* flow in both directions, but not necessarily the same token and not necessarily at the same time (in contrast to a "read arc" connected to a transition).

Subnet "profile"

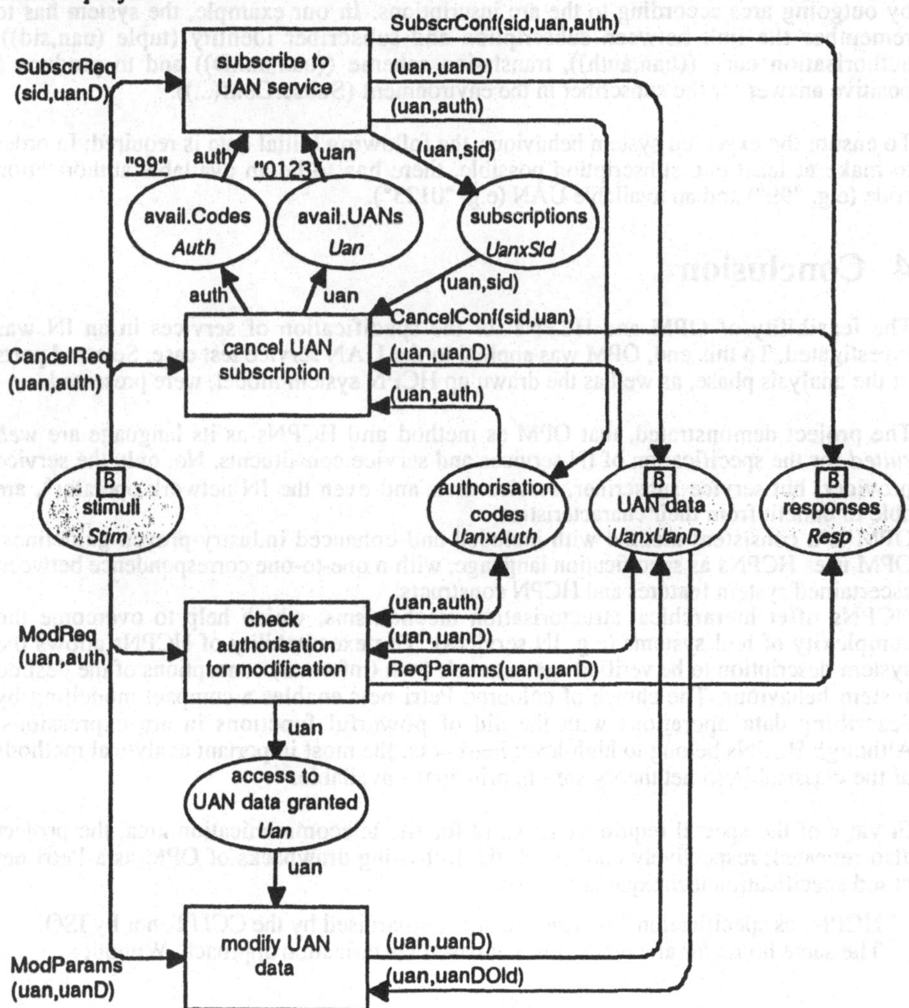

Fig. 3. Subnet profile (page #200)

Subnet "profile" (Fig. 3) specifies creation, modification and cancellation of UAN service subscriptions by means of the corresponding transitions. This is an example of a diagram page on the lowest abstraction level, where some part of the overall needed system functionality is formally described without any further reference to other subnets

(other diagrams will be presented at the conference but are skipped here due to space limitations).

In the case of a service subscription, "subscribe to UAN service" will be triggered by a stimulus "SubscrReq(...)"[5]. A subscription is possible, i.e. the transition "subscribe to UAN service" is enabled, if there is at least one UAN number and at least one authorisation code available. If this is the case, the transition may occur, i.e. the subscription request is processed:
In an *uninterruptable* action, the corresponding token on places connected by incoming arcs are consumed, in this case "SubscrReq(...)" is removed from stimuli, UAN "0123" and code "99" will no longer be available. Then token are produced on places connected by outgoing arcs according to the arc inscriptions. In our example, the system has to remember the link between subscription and subscriber identity (tuple (uan,sid)), authorisation code ((uan,auth)), translation scheme ((uan,uanD)) and to produce a positive answer for the subscriber in the environment (SubscrConf(...)).

To ensure the expected system behaviour, the following initial data is required: In order to make at least one subscription possible, there has to be an available authorisation code (e.g. "99") and an available UAN (e.g. "0123").

4 Conclusion

The feasibility of OPM and HCPNs for the specification of services in an IN was investigated. To this end, OPM was applied to the UAN service test case. Selected parts of the analysis phase, as well as the drawn up HCPN system model, were presented.

The project demonstrated, that OPM as method and HCPNs as its language are *well suited* for the specification of IN services and service constituents. Not only the service provider, but service subscriber, service user, and even the IN network operator[6], are able to benefit from their characteristics:
OPM is a consistent method with included and enhanced industry-proven guidelines. OPM uses HCPNs as specification language, with a one-to-one correspondence between ascertained system features and HCPN constructs.
HCPNs offer hierarchical structurisation mechanisms, which help to overcome the complexity of real systems (e.g. IN services). The executability of HCPNs allows the system description to be verified against one's own (informal) conceptions of the desired system behaviour. The choice of coloured Petri nets enables a compact modelling by describing data operations with the aid of powerful functions in arc expressions. Although HCPNs belong to high-level Petri nets, the most important analytical methods of the classical Petri net theory are - in principal - available.

In view of the special requirements valid for the telecommunication area, the project also revealed, respectively confirmed, the following drawbacks of OPM as a Petri net based specification technique in this field:

- HCPNs as specification language are not standardised by the CCITT, nor by ISO. The same holds for any other Petri net based specification approach. Without

---

5 Note, that this kind of token is not a function, but denotes a stimuli or response, composed of the signal name, followed by its parameters. This kind of modelling is due to Design/CPN's "union" construct and needed, whenever token with a different amount of parameters are to be allowed to share one and the same place. We decided to enclose the parameters in parentheses, since this kind of signal looks quite familiar to the common telecommunication engineer.

6 *Network operator:* Person (or entity) providing the IN network and its resources for the actual execution of services.

standardisation (which seems to be the main acceptance criteria applied by public network operators and hence telecommunication equipment manufacturers), there will be no chance for a *widespread* usage of Petri nets in the telecommunication area.

- The language SML, used in the Design/CPN tool for the description of data and data operations, is very unfamiliar in the telecommunication area, where ASN.1 can now be seen as a defacto standard.
- Triggered from developments in the Telecommunication Management Network (TMN) field, the specification and implementation of services in an IN too is likely to become object-oriented. Since the strength of OPM using HCPNs lies in a functional decomposition technique, this poses some problems. Though HCPNs are still feasible to describe the needed functionality of elementary objects, the usage of HCPNs arise nontrivial questions in describing the behaviour of objects, which are based on a composition of other objects, in a *pure* object-oriented way.

Acknowledgement

Thanks are due to our colleagues B. Hebing and E. Prinz, who supported the drawing up of the presented HCPN model, as well as to our EURESCOM partners, which supported the drawing up of the UAN test case.

References

[AMS89] Ambrosch, W.D.; Maher, A. and Sasscer, B.: The Intelligent Network. A Joint Study by Bell Atlantic, IBM and Siemens. Springer-Verlag, 1989.

[DeMa78] DeMarco, T.: Structured Analysis and System Specification. Yourdon Press, New York, 1978.

[Design/CPN] Design/CPN Manual. Meta Software Corporation. Cambridge, USA, 1991.

[Dibo88] Dibold, H.: A Method for the Support of Specifying the Requirements of Telecommunication Systems. Proceedings of the International Zurich Seminar, Zurich, 1988, pp. 115-122.

[Dibo90] Dibold, H.: Intelligente Netze - Einführung und Grundlagen. Der Fernmelde-Ingenieur, 44 (1990) 4.

[Dibo92a] Dibold, H.: Hierarchical Coloured Petri Nets for the Description of Services in an Intelligent Network. Proceedings of the International Zurich Seminar, Zurich, 1992, pp. 165-178.

[Dibo92b] Dibold, H.: Die Offene Petrinetz-Methode zur Analyse und Darstellung des funktionalen Verhaltens verteilter Systeme. Fachbericht des GI/ITG Arbeitskreises "Formale Beschreibungstechniken für verteilte Systeme", Springer-Verlag, Berlin, 1992, pp. 195-221.

[HMT87] Harper, R.; Milner, R.; Tofte, M.: The Semantics of Standard ML, Version 1. Technical Report ECS-LFCS-87-36, University of Edinburgh, LFCS, Department of Computer Science, University of Edinburgh, The King's Buildings, Edinburgh EH9 3JZ, August 1987.

[HP87] Hatley, D.J.; Pirbhai, I.A.: Strategies for Real-Time System Specification. Dorset House Publishing, New York, 1987.

[Jens92] Jensen, K.: Coloured Petri Nets: Basic Concepts, Analysis Methods and Practical Use, Volume 1. EATCS Monographs on Theoretical Computer Science, Springer-Verlag, Berlin, 1992.

[MP84] McMemamin, S.M.; Palmer, J.F.: Essential Systems Analysis. Yourdon Press, New York, 1984.

[Pete81] Peterson, J.L.: Petri Net Theory and the Modeling of Systems; Prentice Hall Inc., Engelwood Cliffs, N.J. 07632, 1981.

[Wiks87] Wikström, Å.: Functional Programming using Standard ML. Prentice-Hall International Series in Computer Science, 1987. ISBN 0-13-331968-7, ISBN 0-13-331661-0 Pbk.

On Net Modeling of Industrial Size Concurrent Systems

Ludmila Cherkasova, Vadim Kotov
Hewlett-Packard Laboratories
1501 Page Mill Road, Palo Alto, CA 94303, USA

Tomas Rokicki
Computer Science Department
Stanford University, Stanford, CA 94305, USA

Abstract. The paper discusses the modeling of OLTP (*On-Line Transaction Processing*) using colored Petri nets and the Design/CPN tool. We have performed industrial-sized simulation of tens of thousands of transactions running on databases with millions of records. OLTP applications were chosen because the OLTP workloads emphasize update-intensive database services with up to thousands of concurrent transactions per second. The goal of the experiment was to develop a methodology for the net modeling of OLTP-like applications with respect to the ability of different computer systems to meet the OLTP benchmarks requirements. The effort spent on the successful modeling of such large systems resulted in some methodological conclusions which required us to avoid straightforward approaches to the net modeling and to develop a combined technique of using nets together with embedded programming for simulation and analytical modeling.

1 Introduction

The Petri nets theory is approaching the stage where it can serve as a both formal and intuitive basis for developing modeling environments that efficiently support the rigorous specification, design, validation, and analysis of complex distributed systems. Modeling tools and environments based on net theory may prove to be especially useful at the initial stages of a design process when the most important key hypotheses and decisions about system structure and function should be made and checked by rapid prototyping.

During the last decade there has been increasing interest in building Petri net tools and net-based modeling systems. Most of these tools were initiated in research institutions and dealt with particular systems and problems of an "academic size" which served primarily as case studies for proving the credibility of these tools. The next natural step is to apply the net modeling methodology and techniques to realistic problems, to prove that net modeling is able to change the style of the design and analysis of large industrial systems.

An ideal modeling environment should include the following basic capabilities:
- means for adequately detailed and comprehensive (for different stages of the design) *specification* of the system,
- means for *formal analysis* of system properties, and
- *simulation* vehicles for performance evaluation.

We used Design/CPN™ [Jensen87, Pinci-Shapiro91] to model the *TPC Benchmark*™ *A* of OLTP workloads [Gray91]. Based on *Hierarchical Colored Petri Nets*, Design/CPN provides all of the mentioned capabilities. However, it is not a trivial task to combine these capabilities in one model for systems of such complexity if we want the model to execute reasonably quickly and supply trustworthy results.

The main difficulty in the practical implementation of Petri nets for modeling contemporary large systems is the necessity to take into consideration the huge number of concurrent non-deterministic activities and relations occuring in these systems. With straightforward net modeling, it is projected, for example, that intensive testing of firing conditions for a large number of transitions, exhaustive bindings of variables in high-level nets, and colored tokens in colored Petri nets rapidly lead to a time and space complexity explosion. Thus, the straightforward modeling of large systems by nets can result in non-executable models that are unacceptable for solving practical tasks.

Our main goal was to overcome this problem using a mixed strategy of combining nets with direct programming modeling. The net features were used when they provide a helpful insight into the concurrent structure of the modeled system. Efficient procedures of searching, matching, sorting, etc., have been coded that has reduced significantly the number of constructed and checked bindings.

Another important issue was the proper use of hierarchy and refinement that allowed us to construct the model in a compositional way. This made it possible to develop a technique of rapid prototyping for constructing a chain of related models with the ability to modify, update and refine them.

The process of the net model design was split into steps starting with a net specification model at the beginning that preserved basic structural features of the modeled system and finishing with a behavioral programming model oriented toward performance evaluation at the end. This strategy has allowed us to understand the real power and limits of net modeling. The gained experience and accumulated libraries of "standard" net modeling functions are now being successfully used for the modeling of distributed systems that are logically more complex than OLTP.

2 On-Line Transaction Processing

On-Line Transaction Processing characterizes a category of information systems with multiple interactive terminal sessions, intensive I/O and storage workload, and large volumes of data stored in databases.

TPC Benchmark A (TPC-A), that is standardized as a specification of OLTP workload, is stated in the form of a hypothetical *bank* that has multiple *branches* with multiple *tellers* at each branch. The bank has many customers, each of which has an *account*. The database represents the cash position of each entity (branch, teller, and account) in correspondent records and a history of recent transactions run by the bank. The transaction represents the work done when a customer makes a deposit or withdrawal against his account. The transaction is performed by a teller at some branch.

TPC-A performance is measured in terms of *transactions per second* (*tps*). An interesting feature of TPC-A Benchmark is a scaling rule that requires that the number of branches, tellers, and accounts be proportional to the reported transactions per second. The main requirements of TPC-A are the following:
- one branch, 10 tellers, and 100,000 account records per reported tps,
- response time under 2 seconds for 90% of the transactons, and
- a truncated exponential random distribution of transaction arrivals.

Concurrent systems under consideration should be able to run thousands of tps.

3 Colored Petri Nets and Design/CPN

In colored Petri nets (CPN) [Jensen87], *color* refers to the types of data associated with tokens and is comparable to data types in programming languages. The version of colored Petri nets used in Design/CPN incorporates variables (representing the binding of identifiers to specific colored tokens), arc inscriptions (expressions), and code associated with transitions.

In Design/CPN, colored Petri nets become a graphical programming language with rich specification and simulation possibilities. The programming language through which colored Petri nets specify desired operations in arc expressions and transition codes is ML.

A hierarchical CPN contains a number of hierarchically interrelated subnets called *pages*. In the net, they are represented by hierarchical substitution transitions. A hierarchical transition can be replaced by an associated page, giving a more detailed description of its internal activity or state.

To study large-scale systems of an arbitrary size, we should have scalable models: while the size of the modeled system increases substantially, the size of the model should remain constant as much as possible and modeling time should grow as little as possible. We will see later that colored Petri nets are quite scalable with relation to the model size, but they are not well scalable with relation to the modeling time.

Scalability of colored Petri nets can be provided by the use of colored tokens. They provide means to fold multiple transitions that represent identical or similar system components or activities into one transition whose different instances are enabled by tokens assigned different color values. In the similar way, arc expressions resulting in different values during an execution vary the weights of arcs, allowing a single piece of net to treat different but related (by their color set) tokens in different ways. By parameterizing the size of color sets, one can adjust the same CPN structure to model related systems with different numbers of components.

These few ideas are the foundation of CPN and are powerful enough to develop advanced methodology and techniques to support different stages of distributed system design: specification, rapid prototyping, validation and performance analysis.

4 Specification Model of OLTP

This model of OLTP was intended to specify a typical, basic schema of transaction processing implemented on different multiprocessor architectures, and to evaluate the performance of these implementations, selecting those which satisfy the TPC Benchmark A requirements.

The model incorporates two main parts:
- the functional schema of the OLTP transactions that is maximally independent of possible implementations and uses a common schema of the data access with typical I/O procedures, and
- the variable part that depends on the specific architectural features of a database and its implementation on a specific configuration of a computer system.

The central point of the functional part of the model is related to the locking mechanisms which prevent multiple transactions from simultaneously accessing the same database records. Each transaction, referring to a specific account, locks the access to this account record, then to the teller record associated with this transaction, and, finally, to the corresponding branch record. When the transaction commits (or completes), it unlocks these records.

A straightforward specification of the locking mechanism for accounts, tellers and branches is shown in Figure 1. Here the place *Account_DB* holds colored tokens from the color set *ACCOUNTS*. Each token represents an individual account record from the accounts database. Places *Teller_DB* and *Branch_DB* are organized in a similar way.

The transition *LockAccount* has *Account_DB* as one input place and *P1* as the second input place. Its input arcs are assigned arc expressions which are variables *Account_ID(t)* and *t* of the type (color) respectively *ACCOUNTS* and *TRANS*. The function *Account_ID(t)* selects an account identifier which matches with account record of the transaction *t*. While the transition *LockAccount* fires, it re-

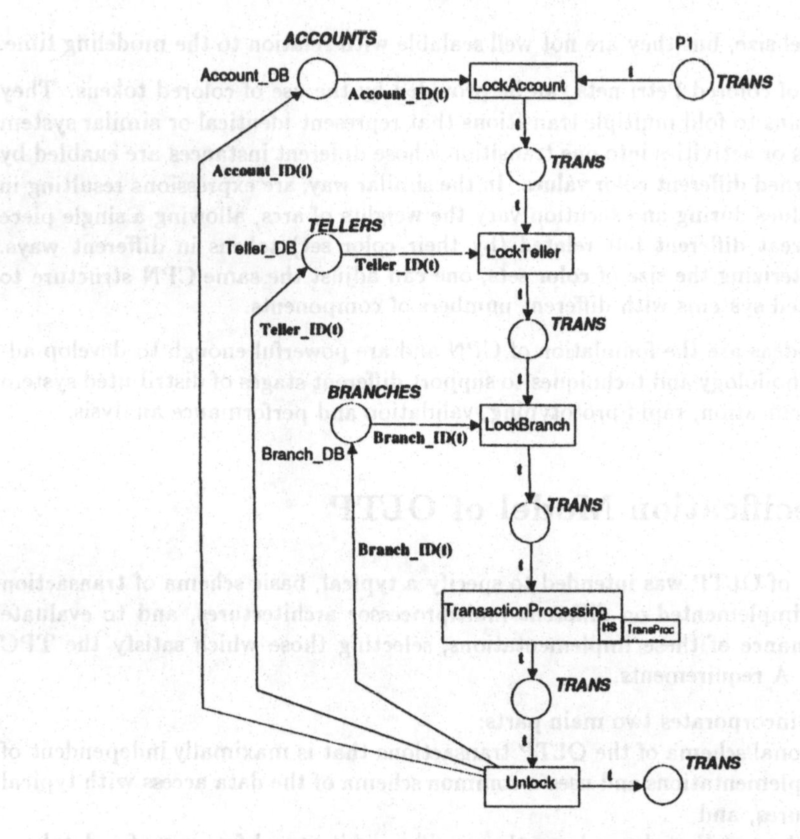

Figure 1: A colored Petri net representing locking mechanisms

moves the correspondent token from *Account_DB*, hence locking the access of all other oncoming transactions that are referring to this account identifier.

The transitions *LockTeller* and *LockBranch* are arranged in the similar way. The transaction tokens that have passed through the locking transitions invoke the compound transition *TransactionProcessing* which, in fact, represents all the lower levels of the hierarchical model. The tokens emerging from the compound transition represent completed transactions and enable the transition *Unlock* which returns tokens, that were removed from the places *Account_DB*, *Teller_DB*, and *Branch_DB*, back to these places.

5 Specification Model versus Simulation Model

This modeling schema, while natural for specification purposes, has a major drawback during simulation while modeling systems with a large number of components and higher rates of transactions (for example, hundreds of branches, thousands of terminals, millions of accounts and thousands of transactions running in the system). If the arriving rate for transactions increases, the number of tokens that queue in input places of transitions *LockAccount*, *LockTeller*, and *LockBranch* increases. This leads to a significant increase of the time and memory required for testing the transition enabling because, in general, every pair of two tokens chosen from the two input places has to be tested for the enabling. The simulation of the model simply fails under such large parameters.

The way the simulation engine works is by first finding all enabled transitions, then selecting one of these to fire, and then by firing that enabling. To find an enabled transition, it must find, for some transition, a binding of tokens to the variables and expressions on the input arcs such that the guard condition on the transition, if any, is true. The general way to do this is to consider, for each transition, every possible combination of tokens from all input places to that transition, and check that a legal binding is possible and that the guard condition (if any) is true.

For example, to enable the transition with two input places and a marking having n and m tokens in them, $n \times m$ combinations (bindings) should be constructed and checked. In addition, when firing a transition and adding or removing tokens from a place with k tokens in it, the representation of multisets used by Design/CPN requires an additional k comparisons. Whenever k, n, m are numbers exceeding several tens, the simulation stage might require unacceptably large modeling time even to simulate a few hundred transactions running through the system. In the case of OLTP applications, the size of a more or less representative experiment should be tens of thousands of transactions at a rate of hundreds of transactions per second. These problems are further exacerbated by the presence of timing delays in the model.

To overcome these problems and to make our simulation model efficient, we use several methods that transform and modify the specification model, preserving the net behavior in time. These methods use both programming and Petri net properties. The key idea behind these methods is that the firing rule of the transitions is straightforward and general; by using our specific knowledge of the nets we are building, we can modify nets to dramatically reduce the amount of brute-force search for legal bindings that must otherwise be performed.

As an example, consider the locking requirements as depicted in Figure 1. The simulation engine selects a particular token representing a transaction and then considers all possible lock tokens, looking for a match.

Instead of using individual tokens for each account record lock, we use a single token to represent the entire lock table. A fragment that models getting the record lock using hash tables is shown in Figure 2. For the transition that used to either add a token to or remove a token from *Account_DB*, we now have it remove the single lock table token from the place *AccountLockTable* and then replace it (add it back) with another lock table token representing the modified lock table (see Figure 2). Thus the transition *LockAccount* has the place *AccountLockTable* as its input and output place. The transitions that used to remove a token from the lock table has now its guard *notlocked*? extended to check that the lock is indeed available, since the presence of the lock table does not indicate in itself that a particular record lock is available. In addition, the code region of the transition is extended to indicate that a particular record is locked. The fragments related to the transitions *LockTeller* and *LockBranch* are transformed in a similar way. The transition *Unlock* that used to add a token are extended with a code region that modifies the lock table token to indicate that a lock has been freed. The net change to our model is the addition of

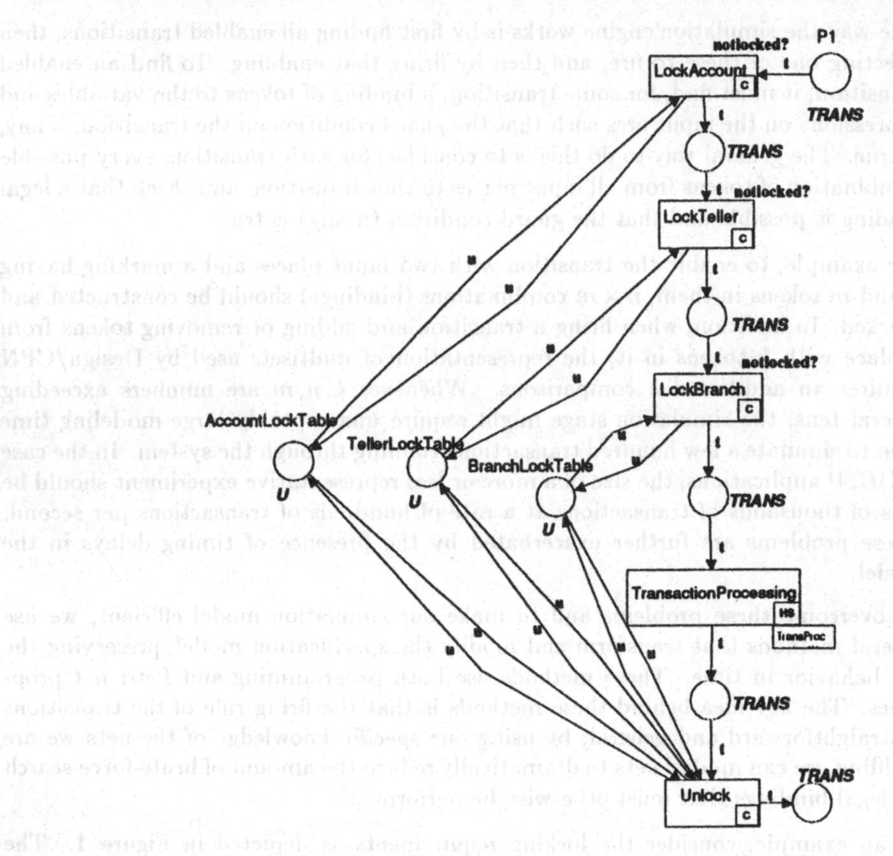

Figure 2: A CPN model of locking mechanisms using hash tables

a few arcs, the guard expressions, and the code regions. In addition, a small amount of code was written to implement the hash table. But the simulation time improved tremendously; before, if we had m database records and n waiting transactions, the simulator had to consider $m \times n$ possible bindings at each step with the old model; with the new model, only n bindings need be considered. Since m is typically in the millions and n is typically just a few, this yields a dramatic speedup. In addition, it is a general scheme that can be (and has been) used in many other portions of the model.

When we deal with timed CPN only the bindings that are available at the current simulation (model) time enable transitions. However, first, all the possible bindings, independently of their token time stamps, are constructed and checked with respect to matching their arc and guard constraints. Thus having a large number of tokens currently unavailable by time but which participate in the construction of bindings adds unnecessary complexity and slows down the simulation. We can reduce the number of bindings by considering only the tokens available for the current simulation time by using the following equivalent net transformation, shown in Figure 3. In the modified net shown in Figure 3b the transition *TimeFilter* lets pass the tokens from the place *TransReq* (transaction requests) to the place *CurrentTransReq* (current transaction requests) which are currently available (ready) in a model with respect to time. In this way, the number of bindings to enable the transition *ProcessRequest* is essentially reduced.

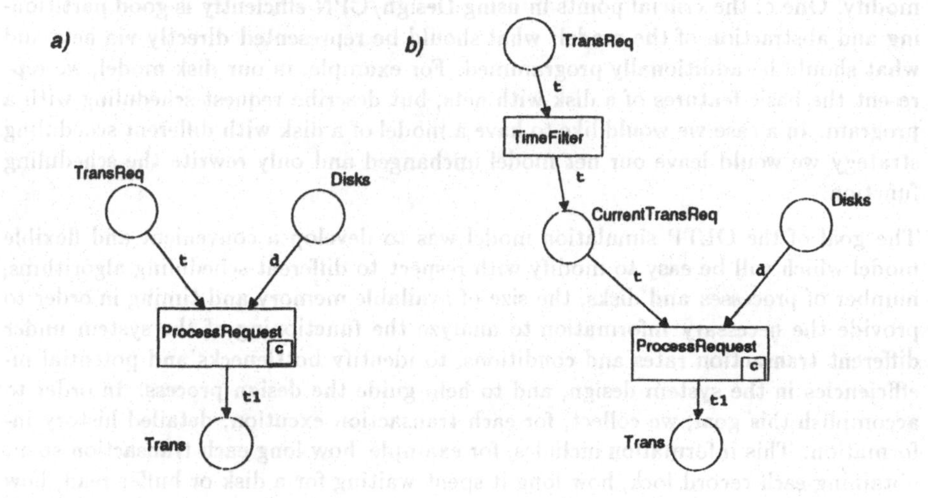

Figure 3: An equivalent net transformation reducing the number of possible bindings in timed CPN model

6 Simulation Model and Simulation Results

The core of our OLTP simulation model consists of 9 pages including:

1) a global declaration node (containing descriptions of types, colors, variables, reference variables, functions, predicates and functions used in the model),

2) an initialization page which allows us to run the model under different parameters,

3) the overall transaction diagram, specifying a logical schema of transaction processing,

4) a record locking and access schema describing the locking and access mecha-

nism for account, teller and branch record,

5) a disk interface page which prepares (augments) disk access requests to be processed, and then collects and distributes ready data,

6) a disk model based on HPUX^{TM} scheduling strategy,

7) a page cleaning model using "second chance" algorithm,

8) a group commit and timeout diagram, specifying a strategy for writing log information to disk and commiting the transaction, and

9) a page for collecting results and statistics.

The simulation model briefly presented here contains a fair amount of programming. The global declaration node specifying types, colors, variables, reference variables, functions and predicates used in the model has 1000 lines of code. However, the simulation model we built still is in many ways universal and flexible to use and modify. One of the crucial points in using Design/CPN efficiently is good partitioning and abstraction of the model: what should be represented directly via nets and what should be additionally programmed. For example, in our disk model, we represent the basic features of a disk with nets, but describe request scheduling with a program. In a case we would like to have a model of a disk with different scheduling strategy we would leave our net model unchanged and only rewrite the scheduling function.

The goal of the OLTP simulation model was to develop a convenient and flexible model which will be easy to modify with respect to different scheduling algorithms, number of processes and disks, the size of available memory and timing in order to provide the necessary information to analyze the functioning of the system under different transaction rates and conditions, to identify bottlenecks and potential inefficiencies in the system design, and to help guide the design process. In order to accomplish this goal, we collect, for each transaction excution, detailed history information. This information includes, for example, how long each transaction spent obtaining each record lock, how long it spent waiting for a disk or buffer read, how long it took to group commit, etc. We also collected history information on disk requests, including how long each request spent in the disk queue and how long it took for the physical disk to satisfy the request. We developed a few simple tools to analyze files containing these history traces, perform some simple statistical analysis, and illustrate the results graphically.

The simulation results are consistent with the results of the analytical modeling, and coincide with the performance results of the real system (HP 3000/977 running MPE/XL).

In this paper, we concentrated on a presentation of a uniprocessor model for OLTP applications, its simulation results and validation. However, this model might be used in a broader context. The hierarchical pages might be considered as submodels composing the whole model. In the ideal case, different subpages represent different parts of the designed system. Whenever we need to change or substitute some part with a new or different one, it would be convenient to substitute or redesign a correspondent hierarchical page instead of changing the whole model.

We follow this methodology for modeling OLTP applications on multiprocessor systems by extending our OLTP model for a uniprocessor.

7 Conclusion

We have presented our experiences in modeling an on-line transaction processing system using colored Petri nets and the Design/CPN tool. We have performed industrial-sized simulations of tens of thousands of transactions running on databases with millions of records. Our success in this modeling task shows that Petri nets and currently available Petri net tools provide a viable modeling environment for the performance analysis of large systems. However, the use of nets in the modeling was not straightforward: nets were combined with code (when it did not harm the model clarity and preserved the basic net modeling paradigm) to overcome the time consuming direct imitation of system activities . The design of some components of the net model evolved from a "net-like" specification model at the beginning that preserved basic structural features of the modeled system, to a behavioral programming model oriented towards performance evaluation at the end. This strategy has allowed us to understand the real power and limits of net modeling. Further study is intended to investigate the methods of augmenting the net modeling environment by formal analysis capabilities.

References

[Gray91] The Benchmark Handbook for Database and Transaction Processing Systems. Edited by Jim Gray, Morgan Kaufmann Publishers Inc., San Mateo, California, 1991.

[Jensen87] Jensen K. Coloured Petri Nets: A High Level Language for System Design and Analysis. In *Advances in Petri Nets 1990*, Springer Verlag, LNCS, vol.483, 1991, pp. 342-416.

[Pinci-Shapiro91] Pinci V., Shapiro R.M. An Integrated Software Development Methodology Based on Hierarchical Colored Petri Nets. In *High-Level Petri Nets*, Springer Verlag, 1991, pp. 649-666.

Analysis of the TMS320C40 Communication Channels Using Timed Petri Nets

David A. Hartley and David M. Harvey

Coherent and Electro-Optics Research Group
Liverpool John Moores University, Byrom Street, Liverpool, L3 3AF, U.K.

Abstract. The Texas Instruments' TMS320C40 Digital Signal Processor's communication ports and their associated DMA channels have been modelled using Timed Petri Nets. The Petri Net implementation is discussed, and the performance of the communication ports/DMA channels are evaluated. Analysis of the simulation results indicate that the single DMA bus provides insufficient bandwidth to drive the communication ports at maximum speed. During split mode operation the requirement to exchange the bus token reduces the communication bandwidth.

1 Introduction

A multiprocessor image processing system is currently being designed within the Coherent and Electro-Optics Research group at Liverpool John Moores University. This is based upon the Texas Instruments' TMS320C40 Digital Signal Processor which provides on-board hardware to support communication between co-operating processors. Although there is a software simulator for this device, it currently does not allow the simulation of multiple processor systems, therefore it was necessary to develop a model of the TMS320C40's communication hardware, so that an analysis of its communication performance could be undertaken.

The model is based upon petri nets which have previously been used to evaluate computer systems including the modelling of individual processor architectures [1], and at higher levels, system components such as processors, memories and interconnection networks [2]. The petri net implementation of the TMS320C40 represents the devices communication ports and the DMA coprocessor. Through simulation, the bandwidths of the communication ports have been determined under varying conditions.

2 TMS320C40 Communication Ports and DMA Coprocessor

The TMS320C40 [3] contains communication hardware which enables processors to be simply joined together to form a parallel processing system. Communication between processors is performed through six 12-bit bi-directional ports, each capable of transferring data at 20 MBytes/sec. Two processors can communicate with each other by directly connecting a communication port of one processor to a communication port of another processor. Bus ownership is performed using a *TOKEN*[1] passing protocol [4] whereby the processor possessing

---

NOTE 1: *TOKEN* indicates that a TMS320C40 has control of the communication bus. It should not be confused with petri net 'tokens'.

the *TOKEN* has ownership of the bus. If the receiving processor requires to transfer a word it requests the *TOKEN* from the transmitting processor. The processor with the *TOKEN* will transfer bus ownership via the *TOKEN* either immediately, or after it has completed transferring a word over the bus. *TOKEN* ownership is fair in that the processor with the *TOKEN* will release it after one word transfer even if it wants to transfer multiple words. Data is buffered from the communication ports using 8-word deep input and output FIFOs.

A 6-channel DMA coprocessor is incorporated within the TMS320C40 to off-load the burden of transferring data away from the CPU. A separate DMA address and data bus is contained within the TMS320C40 enabling concurrent CPU and DMA accesses. However, conflicts can occur in certain situations and these are resolved through programmable arbitration. The DMA channels provide the usual unidirectional memory to memory transfers. This configuration is known as Unified mode. The DMA channels can also be tied specifically to the communication ports to provide 12 DMA channels. This configuration is called Split mode, and allows data to be transferred to and from each communication channel. The 6 DMA channels arbitrate for use of the single DMA bus using either priority arbitration, where DMA channel 0 has the highest priority, or via a rotating priority.

Operations performed by the TMS320C40 can be defined in multiples of the processor's H1 clock. The H1 clock has a cycle time of 40ns for a 50MHz device and is representative of the instruction cycle time. The DMA coprocessor can perform a read access in one H1 cycle, irrespective of whether data is held in internal or external memory; a write access takes one H1 cycle for data stored in internal memory, and 2 cycles for data stored in external memory.

3 Petri Net Simulation Software

The petri net simulation software was written in the C programming language and the current version runs on SUN workstations and IBM compatible personal computers. The software implements timed transitions using 3-phase firing [5]. The firing times used for the TMS320C40 model are specified in multiples of the processor's H1 clock. The simulator provides selectable arbitration schemes for handling conflicts. Methods for resolving which transition to fire include randomly picking a transition, using a priority based scheme, or firing the transition with the shortest firing time [5]. Three types of arbitration can be specified for each conflict - random, fixed or rotating priority.

4 TMS320C40 Petri Net Model

The TMS320C40 petri net model consists of 6 identical sub-nets representing each communication port/DMA channel. Each sub-net can further be divided into 4 sections consisting of DMA channel arbitration, DMA channel, communication bus word transfer, and communication bus ownership (*TOKEN* passing).

4.1 DMA Channel Arbitration

Arbitration between the 6 DMA channels for control of the single DMA bus is modelled using 6 transitions, each having 2 input places and one output place. The first input place is shared by all 6 transitions and contains a single token representing 'DMA Bus Available'. The second input place indicates a 'DMA Bus Request' and a token is deposited in this place when a DMA channel wishes to use the DMA bus. Possible conflicts which occur when DMA channels compete for the single 'DMA Bus Available' token are resolved using either rotating or fixed priority. DMA arbitration is programmable within the TMS320C40 and was set to rotating priority during the simulations. The selected transition will deposit a token into the corresponding output place representing a 'DMA Bus Grant'. After the DMA channel has performed a read or write access, it will return a token back to the 'DMA Bus Available' place.

4.2 DMA Channel

The DMA channel model shown in figure 1, box A, can be divided into 2 sections. The top section represents the Primary channel and performs memory to communication port transfers. The bottom section represents the Auxiliary channel and performs communication port to memory transfers. In Unified mode only one section is active, whereas both sections are active in Split mode.

Consider the case where DMA channel 0 is in Unified mode and is performing memory to communication port transfers. Data stored in memory which needs to be transferred to the communication port is represented by tokens in place P1. The DMA channel is allowed to transfer data to the communication port output FIFO (P3) provided that the output FIFO is not full. The state of the output FIFO is held in place P4. Eight tokens in P4 indicate an empty FIFO whilst no tokens indicate a full FIFO. Tokens in P1 and P4 will cause transition T4 to fire resulting in a token being placed in the 'DMA Bus Request' place P338. If the channel is the only one desiring to use the DMA bus or it wins arbitration for the bus, then a token will be placed in place P339 ('DMA Bus Grant'). This enables transition T1, which after firing will have moved a token (word) from memory (P1) to the DMA channels internal register (P2). A token is also deposited in place P338 indicating that a second DMA transfer is requested. Once granted, transition T3 fires transferring a token to the output FIFO (P3). This completes a DMA word transfer. The firing times of transitions T1 and T3 represent the time required to perform a read and write access, and are specified in multiples of the processor's H1 clock cycle. Performing a Unified transfer from the communication port to memory follows a similar sequence of events. The state of the input FIFO is held in a place contained within the communication bus word transfer section (section 4.3).

In Split mode, both DMA sections may request the DMA bus. If the DMA channel wins arbitration of the DMA bus then a decision must be made whether to allow the Primary or Auxiliary channel to perform a transfer. Transitions T1 and T2 share an input place (P9) which causes a conflict if both transitions are enabled. This conflict is resolved using rotating arbitration.

4.3 Communication Bus Word Transfer

Figure 1, box B, shows the model to perform a word transfer over a communication bus. A token representing a data word is placed in the

Fig. 1. TMS320C40 Communication Channel and DMA Channel Petri Net.

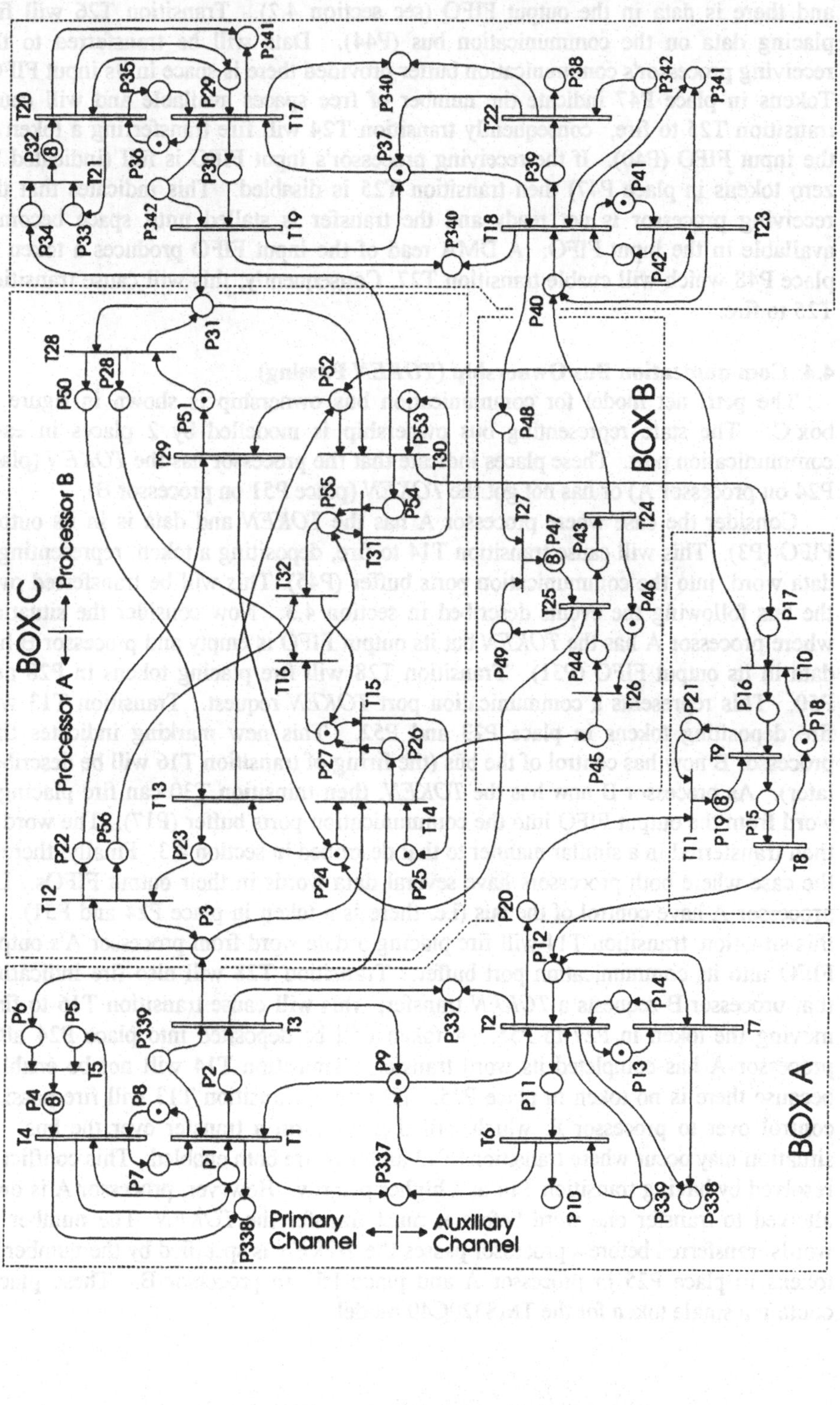

communication port's buffer register (P45) provided the processor has the *TOKEN*, and there is data in the output FIFO (see section 4.2). Transition T26 will fire placing data on the communication bus (P44). Data will be transferred to the receiving processor's communication buffer provided there is space in its input FIFO. Tokens in place P47 indicate the number of free spaces available and will cause transition T25 to fire, consequently transition T24 will fire transferring a token to the input FIFO (P40). If the receiving processor's input FIFO is full (indicated by zero tokens in place P47) then transition T25 is disabled. This indicates that the receiving processor is not ready and the transfer is stalled until space becomes available in the input FIFO. A DMA read of the input FIFO produces a token in place P48 which will enable transition T27. Consequently, this will cause transition T25 to fire.

4.4 Communication Bus Ownership (*TOKEN* Passing)

The petri net model for communication bus ownership is shown in figure 1, box C. The state representing bus ownership is modelled by 2 places in each communication port. These places indicate that the processor has the *TOKEN* (place P24 on processor A) or has not got the *TOKEN* (place P51 on processor B).

Consider the case where processor A has the *TOKEN* and data is in its output FIFO (P3). This will cause transition T14 to fire, depositing a token, representing a data word, into the communication ports buffer (P45). This will be transferred over the bus following the events described in section 4.3. Now consider the situation where processor A has the *TOKEN* but its output FIFO is empty and processor B has data in its output FIFO (P31). Transition T28 will fire placing tokens in P28 and P50. This represents a communication port *TOKEN* request. Transition T13 will fire depositing tokens in place P23 and P52. This new marking indicates that processor B now has control of the bus (the firing of transition T16 will be described later). As processor B now has the *TOKEN*, then transition T30 can fire placing a word from the output FIFO into the communication ports buffer (P17). The word is then transferred in a similar manner to that described in section 4.3. Finally, there is the case where both processors have several data words in their output FIFOs. Let processor A have control of the bus (i.e. there is a token in place P24 and P51). In this situation, transition T14 will fire placing a data word from processor A's output FIFO into its communication port buffer. Transition T28 will also fire indicating that processor B requests a *TOKEN* transfer; this will cause transition T16 to fire, moving the token in P27 to P55. A token will be deposited into place P24 after processor A has completed its word transfer. Transition T14 will not be enabled because there is no token in place P25. Therefore, transition T13 will fire passing control over to processor B, which will then perform a transfer over the bus. A situation may occur where transitions T13 and T14 are both enabled. This conflict is resolved by having transition T14 at a higher priority. However, processor A is only allowed to transfer one word before it must transfer the *TOKEN*. The number of words transferred before a processor passes the *TOKEN* is specified by the number of tokens in place P25 in processor A and place P53 in processor B. These places contain a single token for the TMS320C40 model.

5 Simulation of Unified Transfers

The performance of the DMA coprocessor/communication ports operating in unified mode were evaluated for variations in the number of channels, number of words transferred and whether data was stored in internal or external memory. The number of words transferred and channels used depends on the quantity of tokens placed in a DMA channel's memory place (for example, P1 for DMA coprocessor 0). Internal or external memory is modelled by varying the firing times of the DMA read and write transitions. For unified transfers it is assumed that one processor is broadcasting data to 'n' other processors, where ($1 \leq n \leq 6$).

5.1 Unified Transfer Simulation Results for Data Stored in Internal Memory

The simulation results for data stored in internal memory are shown in figure 2. This indicates that the transfer latency increases exponentially with increase in the number of words transferred. The transfer times when using 2 channels are only 2 cycles more than when using one channel. This can be explained by viewing the markings of the output FIFO places for channels 0 and 1. The first 60 simulation cycles for the 2-channel transfer is shown in figure 3. This indicates that each DMA channel transfers a word to the relevant output FIFO every 4 cycles. As the TMS320C40 takes 5 cycles to transfer a word over a communication bus, then the output FIFOs eventually become full and therefore, both communication channels are continuously transferring data.

Fig. 2. Unified mode transfer latency with data stored in internal memory. The latency for using 1-2 channels is virtually identical.

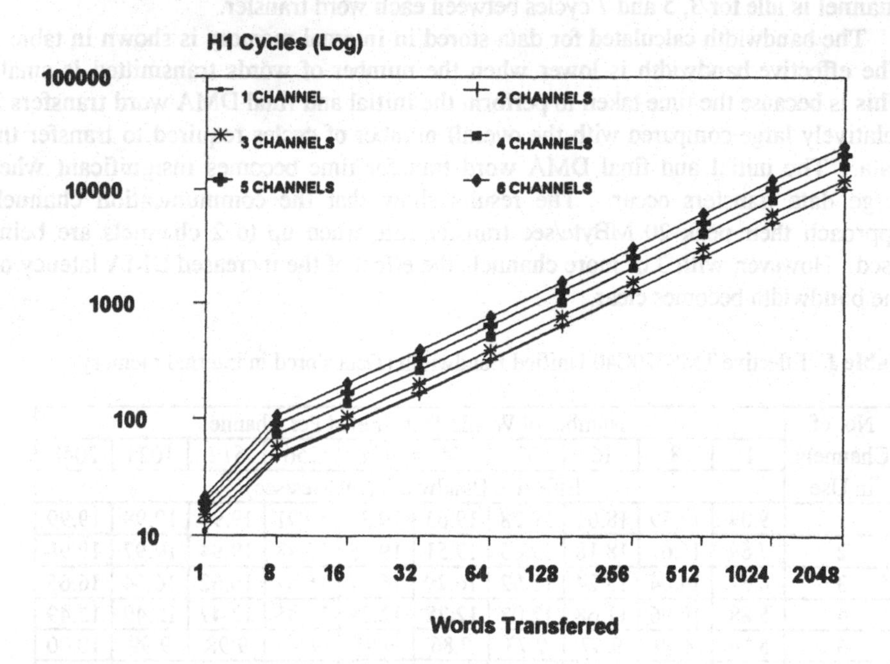

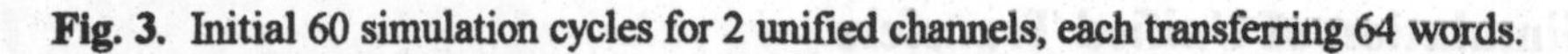

Fig. 3. Initial 60 simulation cycles for 2 unified channels, each transferring 64 words.

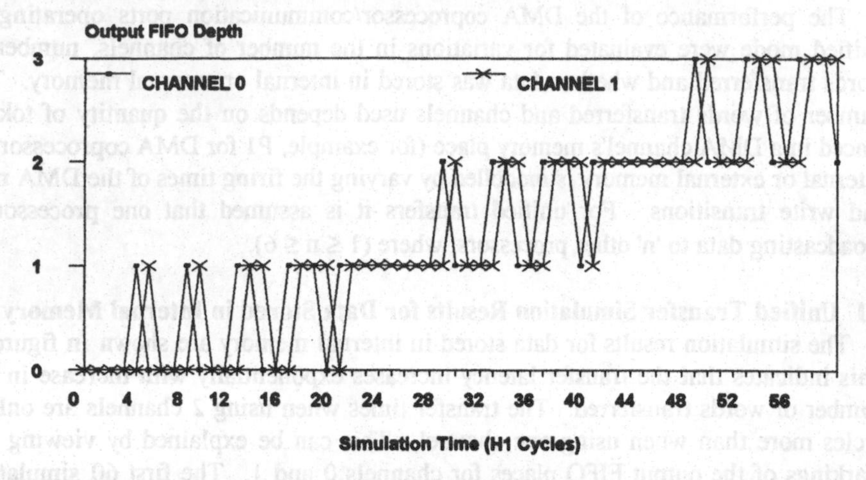

However, if there are more than 2 channels operating, the results show that communication latency rises with increasing number of channels. Analysing the markings of the output FIFOs with 3 channels operating revealed that each DMA channel writes to the relevant output FIFO every 6 cycles. This means that each communication bus is idle for one cycle between word transfers, because the DMA channels cannot supply data to the output FIFOs in sufficient time. For 4, 5 and 6 channels the time taken to transfer each word from memory to the output FIFO increases to 8, 10 and 12 cycles respectively. Therefore, each communication channel is idle for 3, 5 and 7 cycles between each word transfer.

The bandwidth calculated for data stored in internal memory is shown in table I. The effective bandwidth is lower when the number of words transmitted is small. This is because the time taken to perform the initial and final DMA word transfers is relatively large compared with the overall number of cycles required to transfer the data. The initial and final DMA word transfer time becomes insignificant when large data transfers occur. The results show that the communication channels approach their peak 20 MByte/sec transfer rate when up to 2 channels are being used. However, with 3 or more channels the effect of the increased DMA latency on the bandwidth becomes clear.

Table I. Effective TMS320C40 Unified Bandwidth (Data stored in internal memory).

No. of Channels in Use	Number of Words Transmitted Per Channel									
	1	8	16	32	64	128	256	512	1024	2048
	Effective Bandwidth (MBytes/sec)									
1	9.09	17.39	18.61	19.28	19.63	19.81	19.91	19.95	19.98	19.99
2	7.69	16.67	18.18	19.05	19.51	19.75	19.88	19.94	19.97	19.98
3	6.67	14.04	15.24	15.92	16.29	16.47	16.57	16.62	16.64	16.65
4	5.88	10.96	11.68	12.08	12.28	12.39	12.45	12.47	12.49	12.49
5	5.26	8.99	9.47	9.73	9.86	9.93	9.97	9.98	9.99	10.00
6	4.76	7.62	7.96	8.14	8.26	8.29	8.31	8.32	8.33	8.33

5.2 Unified Transfer Simulation Results for Data Stored in External Memory

The results obtained for data stored in external memory differed from their corresponding internal memory times by an additional H1 processor cycle. This additional cycle is because an external write takes 2 cycles, whereas an internal write takes only one cycle. Therefore, the latency and corresponding bandwidths are virtually identical to the internal memory results.

6 Simulation of Split Mode Transfers

Simulation of split mode transfers assume that every processor is using 'n' channels, where ($1 \leq n \leq 6$). Simulations were performed for data stored in internal and external memory.

6.1 Split Transfer Simulation Results for Data Stored in Internal Memory

The simulation results for data stored in internal memory are shown in figure 4. The results show that up to 4 channels can be in use without affecting the communication latency. Latency increases whenever there are 5 or 6 channels in use. When using 2, 3 or 4 channels, the latency only increases by 3, 6 and 7 cycles compared with using only a single channel. Analysing the output FIFO markings indicated that when using one, 2 or 3 channels, the output FIFOs eventually became full, and therefore, the communication channels were constantly busy. With 4 channels, the output FIFOs always contain at least one word but never exceed 2 words, thus the communication channels always remain busy. However, with 5 channels, the output FIFOs never contain more than one word and for the majority of the time they contain no words. The reason why the output FIFOs never fill up is because of the competition between the DMA channels for the single DMA bus. This competition is exasperated by the fact that individual DMA channels may have input and output channels competing for the bus. The worst case delay from an output FIFO DMA request to the data being written into the output FIFO is 20 cycles which occurs when the auxiliary DMA channel obtains access to the bus before the primary channel.

The minimum time between successive transfers from the output FIFO to the communication port buffer assuming continuous alternate word transfers is 16 cycles, therefore the communication port is not being used all the time which causes the increased latency. The percentage of transfer time that the communication bus is idle can be calculated from transition firing statistics. This produced figures of approximately 32% and 43% for 5 and 6 channels respectively. The effective communication port bandwidths calculated from the simulation results are shown in table II. As explained in section 5.1, the effective bandwidth is reduced when the number of words transmitted is low due to the initial and final DMA word transfers. The results for one to 4 channels clearly indicate the effect of bi-directional transfers on the communication bandwidth. This reduction in bandwidth is caused by the constant exchanging of the bus *TOKEN* after every word transfer. With 5 or 6 channels in use, the reduced bandwidth is due to the increased DMA latencies.

Fig. 4. Split mode transfer latency with data stored in internal memory. The latency using 1-4 channels is virtually identical.

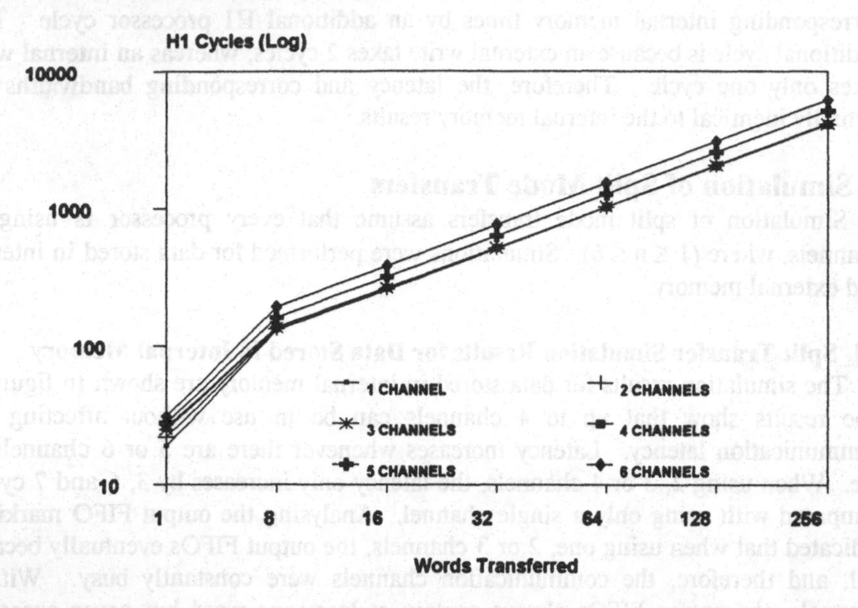

Table II. Effective TMS320C40 Split Mode Bandwidth (Data stored in internal memory).

Number of Channels in Use	Number of Words Transmitted Per Channel						
	1	8	16	32	64	128	256
	Effective Bandwidth (MBytes/sec)						
1	10.53	12.21	12.36	12.43	12.46	12.48	12.49
2	9.09	11.94	12.21	12.36	12.43	12.46	12.48
3	8.00	11.68	12.08	12.28	12.39	12.45	12.47
4	7.69	11.59	12.03	12.26	12.38	12.44	12.47
5	7.14	9.94	9.97	9.98	9.99	10.00	10.00
6	6.45	8.29	8.31	8.32	8.33	8.33	8.33

6.2 Split Transfer Simulation Results for Data Stored in External Memory

The simulation results for data in external memory show that the transfer times for one, 2 or 3 channels differs from the internal memory results by only one, 2 and 3 cycles respectively. Analysis indicates that the output FIFOs eventually fill up for these channels which ensures that the communication ports are continuously busy. However, for 4, 5 or 6 channels, the results show that the latency has increased over the internal memory results in section 6.1 by an average of 22%, 25% and 25%. Observing the output FIFOs indicates that they are empty for most of the time. This is caused by the competition for the DMA bus which is intensified because of the extra cycle required for each external write. The detrimental effect of this is that the communication bus is idle for approximately 15%, 46% and 55% of the time when using 4, 5 and 6 channels. The effective bandwidth calculated from the simulation results are shown in table III. For one to 3 channels, the bandwidth is limited by the exchanging of the bus *TOKEN* after each word transfer. For 4 to 6 channels, the bandwidth is reduced because of the increased DMA latencies.

Table III. Effective TMS320C40 Split Mode Bandwidth (Data stored in external memory).

Number of Channels in Use	Number of Words Transmitted Per Channel						
	1	8	16	32	64	128	256
	Effective Bandwidth (MBytes/sec)						
1	10.00	12.12	12.31	12.40	12.45	12.48	12.49
2	8.33	11.77	12.12	12.31	12.40	12.45	12.48
3	7.14	11.43	11.94	12.21	12.36	12.43	12.46
4	6.90	9.94	9.97	9.98	9.99	10.00	10.00
5	6.45	7.96	7.98	7.99	8.00	8.00	8.00
6	5.71	6.64	6.65	6.66	6.66	6.67	6.67

7 Conclusions

Petri nets provided a powerful method for simulating the communication aspects of the TMS320C40. This enabled the performance of the TMS320C40's communication ports to be analysed, which currently cannot be performed satisfactorily using the commercial TMS320C40 simulator. The conflict arbitration enhancements to the petri net model enabled the TMS320C40's fixed and rotating priority schemes to be modelled effectively.

In unified mode, the simulation results showed that peak communication channel bandwidth can be approached when up to 2 channels are used. However, when more channels are in operation the DMA coprocessor is unable to supply data to the output FIFOs in sufficient time to ensure that the communication ports remain busy. The split mode simulation results showed that communication channel bandwidth is limited by the exchanging of the bus *TOKEN*. This occurs when up to 4 channels are used with data stored in internal memory and when up to 3 channels are used when data is stored in external memory. If data is stored in internal memory and 5 or 6 channels are used then communication bandwidth is limited due to the increased latency of the DMA coprocessor. With data in external memory, the DMA coprocessor limits the communication bandwidth when 4 or more channels are used, therefore the results indicate that the single DMA bus is unable to provide sufficient bandwidth to the communication channels.

References

1. Peterson, J.L., "Petri Nets", Computing Surveys, Vol. 9, No. 3, pp. 223-252, Sept 1977.
2. Adiga, A.K. and Deshpande, S.R., "Evaluation of Effectiveness of Circuit Based and Packet Based Interconnection Networks via Petri Net Models", Proc. 1987 International Conference on Parallel Processing, pp. 533-541, 1987.
3. Texas Instruments Inc., TMS320C4x User's Guide, 1991.
4. Dally, W.J. and Song, P., "Design of a Self-Timed VLSI Multicomputer Communication Controller", Proc. International Conference on Computer Design, pp. 230-234, 1987.
5. Marsan, M.A., "Stochastic Petri Nets: An Elementary Introduction", Advances in Petri Nets, G. Rozenberg (ed), Springer-Verlag, pp. 1-29, 1989.

Protocol Optimization for a Packet-Switched Bus in Case of Burst Traffic by Means of GSPN

Guenter Klas
Siemens Corporate Research & Development
Otto-Hahn-Ring 6, D-8000 Munich 83, Email: klas@ztivax.uucp

Abstract

With multiprocessor systems for data base servers, high bandwidth *packet-switched* bus interconnection systems are currently used. We explore the design space relevant for optimal bus pipeline management strategies. Under various burst traffic assumptions on the bus, robust optimal parameters for the bus pipeline depth as well as for the pipeline slot allocation policy can be computed by means of a Generalized Stochastic Petri Net model.

KEYWORDS: Multiprocessor Systems, GSPN, Performance Modeling, Bus Pipeline

1 Introduction

Current architectures for systems like data base servers are often shared memory multiprocessor systems. They consist of multiple CPUs with local memory connected to a main memory by means of a *system bus*. In order to reduce the frequency of access to main memory, a second level cache memory (L2 cache) is situated between the CPU and the main memory. If the L2 cache is operated in *copy back mode*, a *cache coherency* protocol is needed to keep data consistent in the system, since multiple copies of the same data are available in several L2 caches. Then, on the system bus only transactions of the type Read or Copy Back originating from CPUs can be observed. In order to increase the bus bandwidth, packet-switching is employed to allow the interleaving of transactions. The bus and memory system behaves like a processing pipeline with a maximum degree of transaction interleaving also denoted *pipeline depth*. The allocation of pipeline slots to requesting agents is part of the *pipeline management policy (PMP)*. Various related performance models were proposed in the literature, more or less on a relatively abstract level with emphasis on different aspects like cache coherency (e.g., simple analytical model for throughput bounds in [1]), real time performance (e.g., queueing model in [7], arbitration priorities and message passing (e.g., relatively detailed GSPN model in [3]).

The aim of this study is to consider optimal pipeline depths and pipeline management policies of a multiprocessor system bus under various bus traffic loads. As a major result we show that burst traffic has a severe impact on the performance of a bus pipeline and, therefore, Poisson load assumptions are insufficient (In [2] the difference between Poisson arrivals and deterministic arrivals was pointed out). However, the optimum PMP proposed is optimal independend of the degree of burst traffic. The results are achieved by means of a Generalized Stochastic Petri Net model (GSPN) [6]. In Section 2, we outline the bus protocol. Section 3 is devoted to the modeling approach. In Section 4, the pipeline depth and the PMPs are defined. Section 5 introduces the characterization of the bus request arrival process. Section 6 defines the evaluation measures. Numerical results are presented in Section 7. Some conclusions follow in Section 8.

2 The Bus Protocol

Fig. 2.1a depicts a block diagram of a multiprocessor system and Fig. 2.1b outlines the components involved in the bus / memory subsystem. The bus is divided into three subbuses for data, addresses and control information. The packet-switching mechanism is outlined in Fig. 2.2.
a) Bus arbitration: a bus agent requests a pipeline *slot ID* from the set {1, ...,MAXSLOT} by a personal request line r. The slot ID is granted by the *arbiter* by a radial line g. Bus agent requests are granted in FIFO order.
b) Address transmission: after having got a slot ID, the bus agent transmits c_a address cycles on the A-bus as soon as this bus is idle.
c) Data transmission: in case of a copy back write the bus agent requests the D-bus after having been assigned a slot ID. In case of a data read response, the main memory requests the D-bus. This bus is granted as soon as available. A data transfer is restricted to a packet of c_d bus cycles.
d) Accessing and releasing a slot ID: A slot ID is *released* in the control circuit of the bus agent that used it after the write memory access in case of a copy back and after the end of the data transfer in case of a read memory access. The fact "slot ID released" is not immediately communicated to the arbiter. Instead, if the arbiter feels that most of his free slot IDs have gone, it is allowed to get as the next (highest priority) customer the A-bus to poll for released slot IDs. These c_p A-bus cycles are denoted polling cycles. The arbiter's strategy of when to start polling and for how long (namely c_p cycles) to give bus agents the chance to communicate released slot IDs is denoted *pipeline management policy*. Any slot ID that is *released* at the start of polling is afterwards known to the arbiter as *free*. In the sequel we also use the more accurate term *free, but not polled* instead of *released*.
e) Keeping data consistent: Especially for memory reads it is e.g. necessary after the address transmission to decide whose turn it is to furnish the data, either the memory (in normal case) or a bus agent's L2 cache (intervention). All bus agents *snoop* in their L2 cache in order to check the cache line

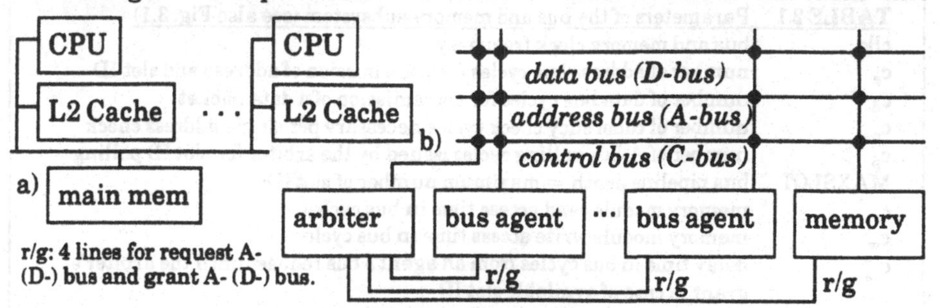

Fig. 2.1 a) Block diagram of shared memory multiprocessor, b) Bus / memory subsystem

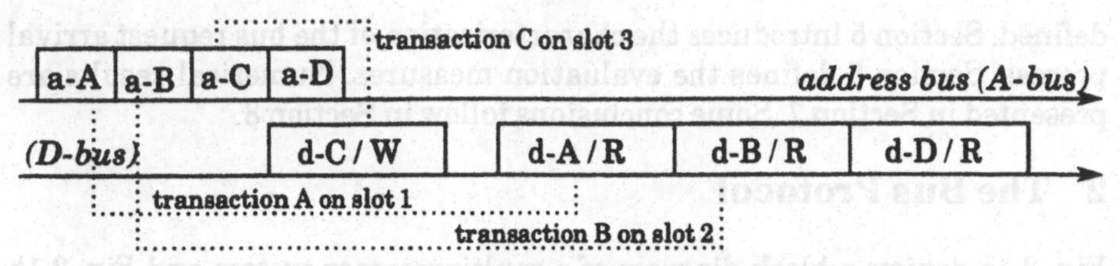

d: data, a: address, A,B,C,D: transactions, R: read response data, W: write data
pipeline depth*: maximum number of transactions concurrently pending on the bus = maximum degree of transaction interleaving = number of *pipeline slots

Fig. 2.2 The packet-switching mechanism and the bus pipeline

status. All address transmissions are checked in FIFO order. The result is broadcast during c_c *coherency cycles* on the C-bus.
***f) Memory access:* We assume a 4-way interleaved shared main memory. A read access may only begin after the coherency check occurred.**
***g) Interface buffer space:* Each bus interface chip, e.g., at the memory modules, has a buffer of MAXSLOT storage places, each for an entire data packet. The technical system parameters are stated in Table 2.1.**

3 The Modeling Approach

With the GSPN used [4], we have state dependent exponential service times and discrete probability distributions for random switches and generalized inhibitor arcs with arc weight k (default is 0). If the number of tokens in the place attached to the inhibitor arc exceeds the threshold k, the transition connected to the inhibitor arc is blocked. A place P of capacity n is written as P,n. The current number of tokens in place P is referred to as #P. In order to reduce the Petri net model complexity, we perform two approximations. (i) We replace the shared main memory as *approximation* by a load dependent flow equivalent server. The equivalent service rates $r_{M,r}(n)$ and $r_{M,w}(n)$ were computed separately for read and write requests. $r_{M,r}(n)$ is the service rate available to reads, if there is a total of n memory requests pending at the main memory. (ii) We ignore copy back write requests. This is justified if the probability that a modified cache line has to

TABLE 2.1	Parameters of the bus and memory subsystem (see also Fig. 3.1)
clk	bus and memory clock frequency
c_a	number of address bus cycles for transmission of address and slot ID
c_d	number of data bus cycles for transmission of a data packet
c_c	number of coherency check cycles necessary per single address check
c_p	number of A-bus polling cycles issued by the arbiter for slot ID polling
MAXSLOT	bus pipeline depth = maximum number of slot IDs
c_r	memory module read access time in bus cycles
c_w	memory module write access time in bus cycles
c_{rg}	delay time in bus cycles from an agent's bus request until the arbiter's grant in case of available slot ID
c_{ifb}	delay time for writing a read response data packet into the receiver

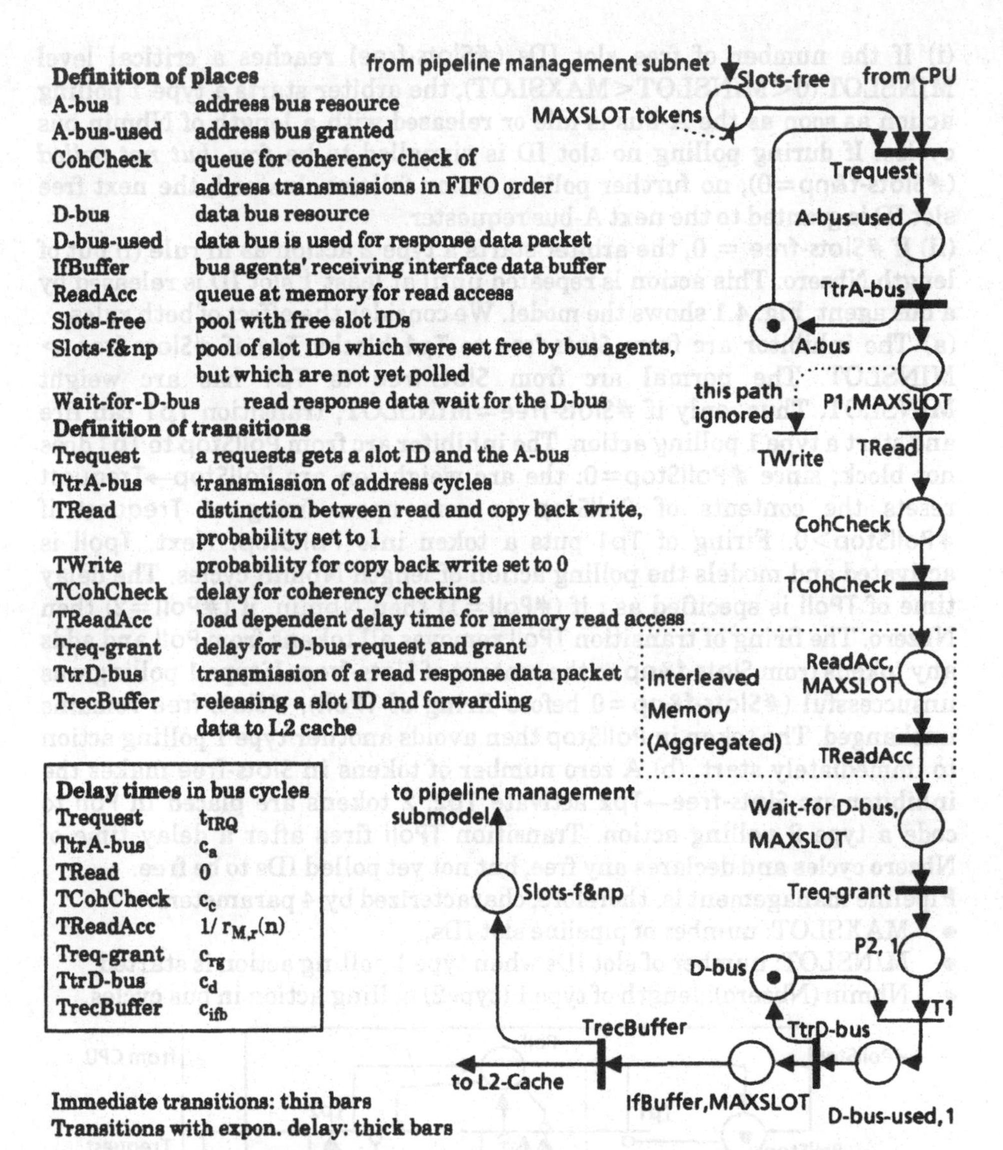

Definition of places
A-bus address bus resource
A-bus-used address bus granted
CohCheck queue for coherency check of address transmissions in FIFO order
D-bus data bus resource
D-bus-used data bus is used for response data packet
IfBuffer bus agents' receiving interface data buffer
ReadAcc queue at memory for read access
Slots-free pool with free slot IDs
Slots-f&np pool of slot IDs which were set free by bus agents, but which are not yet polled
Wait-for-D-bus read response data wait for the D-bus
Definition of transitions
Trequest a requests gets a slot ID and the A-bus
TtrA-bus transmission of address cycles
TRead distinction between read and copy back write, probability set to 1
TWrite probability for copy back write set to 0
TCohCheck delay for coherency checking
TReadAcc load dependent delay time for memory read access
Treq-grant delay for D-bus request and grant
TtrD-bus transmission of a read response data packet
TrecBuffer releasing a slot ID and forwarding data to L2 cache
Delay times in bus cycles
Trequest t_{IRQ}
TtrA-bus c_a
TRead 0
TCohCheck c_c
TReadAcc $1/r_{M,r}(n)$
Treq-grant c_{rg}
TtrD-bus c_d
TrecBuffer c_{ifb}
from pipeline management subnet
Slots-free
from CPU
MAXSLOT tokens
Trequest
A-bus-used
TtrA-bus
A-bus
this path ignored
P1,MAXSLOT
TWrite
TRead
CohCheck
TCohCheck
Interleaved Memory (Aggregated)
ReadAcc, MAXSLOT
TReadAcc
Wait-for D-bus, MAXSLOT
to pipeline management submodel
Slots-f&np
Treq-grant
P2, 1
D-bus
T1
TrecBuffer
TtrD-bus
to L2-Cache
IfBuffer,MAXSLOT
D-bus-used,1
Immediate transitions: thin bars
Transitions with expon. delay: thick bars

Fig. 3.1 Open model for pipeline processing of read requests on packet switched bus

be written back to main memory upon reading a new cache line into a bus agent's L2 cache is low. Fig. 3.1 depicts the GSPN model with the part for the pipeline management omitted.

4 Pipeline Management Policy (PMP)

After a read access, a slot ID is released by placing a token into place Slots-f&np. Fig. 3.1 does not show how these IDs are transferred to place Slots-free. It is up to the arbiter to start polling cycles on the address bus in order to ask bus agents for released IDs. Polling is defined by 2 rules.

(i) If the number of free slot IDs (#Slots-free) reaches a critical level MINSLOT (0<MINSLOT<MAXSLOT), the arbiter starts a type 1 polling action as soon as the A-bus is idle or released with a length of Nbmin bus cycles. If during polling no slot ID is signalled to be *free, but not polled* (#Slots-f&np=0), no further polling action follows. Instead, the next free slot ID is granted to the next A-bus requester.
(ii) If #Slots-free = 0, the arbiter starts a type 2 action as in rule (i) but of length Nbzero. This action is repeated until at least 1 slot ID is released by a bus agent. Fig. 4.1 shows the model. We consider the effect of both rules.
(a) The inhibitor arc from Slots-free to Tp1 blocks Tp1 if #Slots-free > MINSLOT. The normal arc from Slots-free to Tp1 has arc weight MINSLOT. Thus, only if #Slots-free=MINSLOT, transition Tp1 can fire and start a type 1 polling action. The inhibitor arc from PollStop to Tp1 does not block, since #PollStop=0: the arc weight on arc PollStop→Trequest resets the contents of PollStop to zero upon firing of Trequest if #PollStop>0. Firing of Tp1 puts a token into PollStop. Next, Tpoll is activated and models the polling action of length Nbmin cycles. The delay time of TPoll is specified as : if (#Poll=1) then Nbmin, if (#Poll=2) then Nbzero. The firing of transition TPoll removes all tokens from Poll and adds any tokens from Slots-f&np to the content of Slots-free. If type 1 polling was unsuccessful (#Slots-f&np=0 before firing of TPoll), #Slots-free remains unchanged. The token in PollStop then avoids another type 1 polling action to immediately start. (b) A zero number of tokens in Slots-free makes the inhibitor arc Slots-free→Tp2 activate Tp2. 2 tokens are placed in Poll to code a type 2 polling action. Transition TPoll fires after a delay time of Nbzero cycles and declares any free, but not yet polled IDs to be free.
Pipeline management is, therefore, characterized by 4 parameters:

- MAXSLOT: number of pipeline slot IDs,
- MINSLOT: number of slot IDs when type 1 polling action is started,
- Nbmin (Nbzero): length of type 1 (type2) polling action in bus cycles.

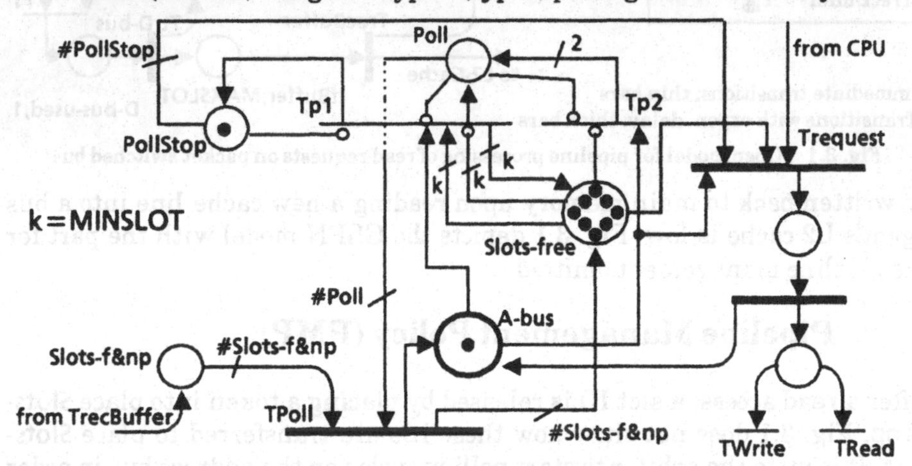

Fig. 4.1 Model of the pipeline management policy

A large number of slot IDs avoids the arbiter to be a system bottleneck. However, every bus agent needs buffer places as many as available slot IDs, a buffer place having the size of a data packet. In dependence on the threshold level MINSLOT, provisions for collecting released slot IDs are taken by the arbiter at a more or less early point of time. The longer a polling action lasts, the higher the chance that a release of a slot ID falls into the time period of the polling action. Type 1 polling should not too excessively use the A-bus in order not to hinder requests from A-bus access.

5 Model of the Request Arrival Process

Fig. 5.1 shows request streams of equal mean rate, but of increasing burstiness. In order to more accurately characterize a certain kind of burst traffic, we introduce the burst traffic model shown in Fig. 5.2. Each high traffic period of mean length t_{high} is followed by a low traffic period of mean length t_{low}. During high traffic, the mean interrequest time $t_{IRQ,h}$ is rather small whereas during low traffic it is large ($t_{IRQ,l}$). Traffic is characterized by the mean length of a time frame t_{frame} (Fig. 5.2) and by m, the mean number of arrival requests per bus cycle. Let r be the ratio $t_{IRQ,l} / t_{IRQ,h}$. We characterize the degree of burst traffic by the burst factor f_b as follows:

$f_b = z^2 - z$, with $z = (r-1)/(m.t_{IRQ,l}-1)$, $f_b \geq 0$. (5.1)

(i) $r=1$. The interrequest time is constant during the frame. Thus, no burst traffic appears and $f_b = 0$ in (5.1).

(ii) The length of the high traffic period t_{high} tends to the length of the whole frame t_{frame}, i.e., no burst traffic. Then, $r = m.t_{IRQ,l}$ or the mean number of requests during $t_{IRQ,l}$ equals $t_{IRQ,l} / t_{IRQ,h} \rightarrow f_b = 0$.

Given f_b and $u = m.t_{IRQ,l}$, the ratio r of the interrequest times can be computed from (5.1) as $r = (2+u)/2 + \mathrm{sqrt}(u^2/4 + f_b u^2)$. (5.2)

Furthermore, t_{high} can be formulated as $t_{high} = t_{frame} (u-1) / (r-1)$. (5.3)

Table 5.1 shows f_b, r, and the value of t_{high} for a medium sized request load. The burst model of Fig. 5.2 can be easily modeled by a 2-state Markov modulated Poisson process as shown in Fig. 5.3.

6 Measured Characteristics

The pipeline management policy is assessed by

a) throughput := throughput in MBytes / s on the D-bus,

b) %Slots-not-free := (MAXSLOT-mean of free slot IDs) / MAXSLOT,

c) %Slots-f&np := (mean of released slot IDs) / MAXSLOT,

d) ro-A (ro transmit address) := utilization of TtrA-bus,

e) ro-D (ro transmit data) := utilization of TtrD-bus for data transmission,

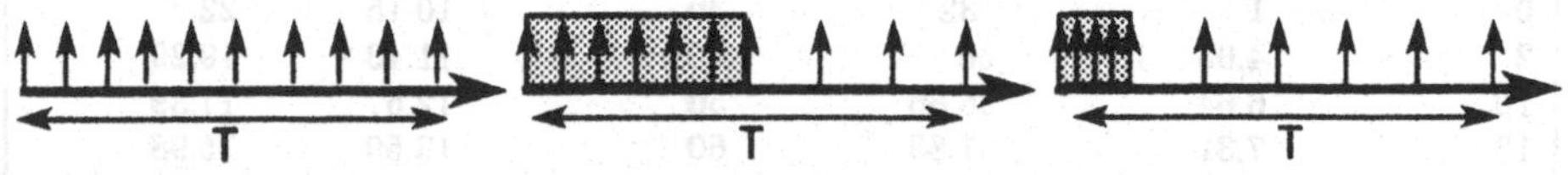

Fig. 5.1 Request streams with increasing burst behaviour from left to right

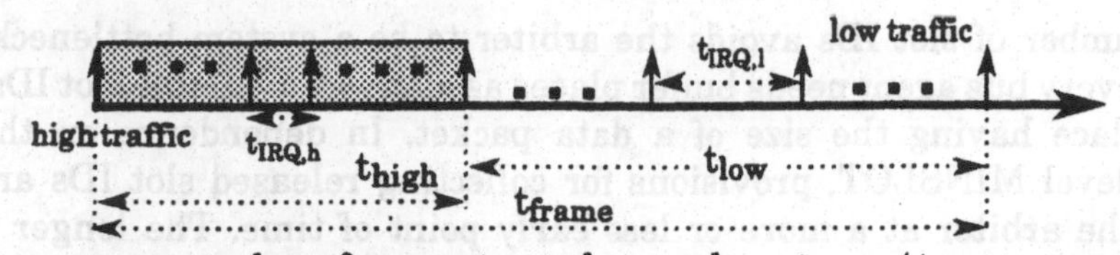

m: mean number of requests per bus cycle, r: $t_{IRQ,l}$ / $t_{IRQ,h}$, u: m.$t_{IRQ,l}$

Fig. 5.2 The burst traffic model with alternating high and low traffic periods

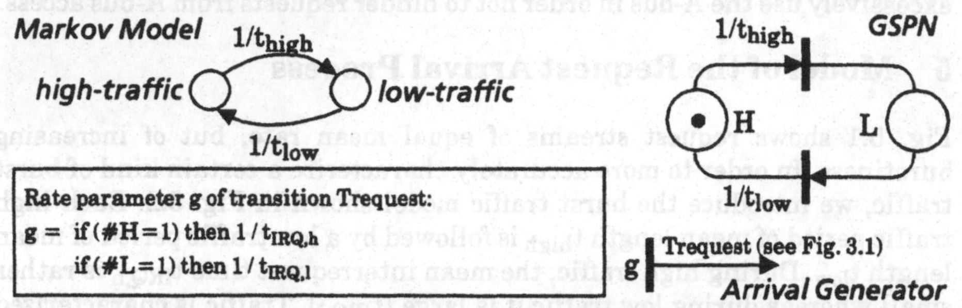

Fig. 5.3 Burst model as a Markov modulated Poisson process

f) ro-C (ro cohcheck) := utilization of TCohCheck for coherency checking,
g) ro-M (ro memory) := utilization of main memory := Re(TReadAcc) / $\max_n\{r_{M,r}(n)\}$ with Re(.) as effective transition firing rate,
h) ro-G (ro grant data) := utilization of Treq-grant,
i) Rm := Requests per memory module (memory is 4-way interleaved).

Thus, we determine the system bottleneck as the server with the highest utilization. Moreover, information is provided about the quality of the PMP, e.g., %Slots-f&np should be as small as possible.

7 Numerical Results

In Scenario I, a worst case request load prevails. Requests are modeled by a Poisson process. In Scenario II, a medium bus request load is assumed. However, the load varies in its degree of burst traffic. The chosen PMP is coded in the result figures as the triple *(MINSLOT, Nbmin, Nbzero)*. We fix base values for the system parameters as shown in Table 7.1. Scenario I: For a high load (t_{IRQ}=0.01), we compute from the GSPN models shown values for evaluation measures as a function of MAXSLOT (Fig. 7.1a). A larger pipeline depth MAXSLOT increases the possible data throughput.

Table 5.1 Burst traffic characterization for medium load

t_{frame} = 132 cycles, $t_{IRQ,l}$ = 16.66 cycles, m = 0.1515 requests per bus cycle.

f_b	r	t_{high}	f_b	r	t_{high}
0	1	132	30	10.15	22
2	4.05	66	40	11.43	19.29
10	6.64	35.66	50	12.57	17.39
13	7.31	31.88	60	13.59	15.98
20	8.62	26.40	65	14.07	15.39

At about MAXSLOT = 8, the bus / memory system starts tending towards saturation. From the various utilization values ro, it is possible to conclude that data transmission is the system bottleneck. For MAXSLOT = 16, only about 5 % of the released slot IDs are not yet polled as indicated by the curve %Slots-f&np. One may conclude that the pipeline management policy (2,1,4) performs quite well.

Next, we consider the policies (2,1,x), (2,4,x) and (4,1,x) where x stands for a variable value for Nbzero. We assume a Poisson worst case bus load. Fig. 7.1b shows that the number of polling cycles Nbzero, used in the case that no more free slot IDs are available, has a considerable influence on the bus performance. For all strategies holds that a single polling cycle (Nbzero = 1) is sufficient to collect released slot IDs. At the beginning this was not clear, since a longer polling period increases the chance to be able to collect released slot IDs. Moreover, even for the number of polling cycles used for type 1 polling (Nbmin), a single cycle seems to be sufficient too. This follows from the fact that policy (2,1,x) performs better than (2,4,x). All together, among the policies investigated policy (4,1,1) performs best.

From Fig. 7.1b one might conclude that the influence of the threshold value MINSLOT is not that high as a comparison of policies (2,1,x) and (4,1,x) indicates. This is in fact true as shown in [5], where we investigated the sensitivity of a policy (x,1,2) with respect to the threshold value MINSLOT.

Scenario II: For medium bus load, namely 0.15 requests per bus cycle, we consider the sensitivity with respect to the degree of burst traffic. We restrict the analysis to 3 representative values for the burst factor, namely f_b = 0, 13, and 65 (Fig. 7.2a). For Scenario I and II hold:

- Policy (4,1,1) can be considered as the optimal one.
- The number of polling cycles Nbmin should be chosen as 1.

The effect of burst traffic can be seen in the strong decrease of the bus throughput with increasing burst factor (the request arrival rate is constant). The reason might be that for burst traffic, the number of free, but not yet polled slot IDs (%Slots-f&np) is substantially higher than for Poisson traffic. The following situation may often occur: If a burst of requests arrives, there is soon a lack of slot IDs. Thus, the throughput is decreased. As soon as the first slot ID is released, it is polled and set free. There is a chance that the requests stemming from the high traffic period release their slot IDs during a low traffic period. As then no intensive

Table 7.1 Base values for the system under consideration

clk	40 MHz	data bus width	64 bit
c_a	2	$r_{M,r}(1)$	0.1052
c_d	4	$r_{M,r}(2)$	0.161
c_c	2	$r_{M,r}(3)$	0.2
c_p	either Nbmin, or Nbzero	$r_{M,r}(4)$	0.229
MAXSLOT	$\in \{1, ..., 16\}$	$r_{M,r}(5)$	0.251
c_r	$1 / r_{M,r}(n)$	$r_{M,r}(6)$	0.2753
c_{rg}	3	$r_{M,r}(7)$	0.296
c_{ifb}	2	$r_{M,r}(n \geq 8)$	0.313

polling is necessary, the sojourn time of slot IDs in state *free, but not polled* increases. This hypothesis is, in fact, supported if we consider the following aspects for various burst factors,

(i) the stationary probability distribution of the number of free slot IDs,

(ii) the conditional probability for the event (no slot ID available in a high traffic period of the burst model) denoted P{no slots free & high traffic}.

Fig. 7.2b depicts these evaluation measures for MAXSLOT=8. With increasing burst factor f_b, the probability, that during the high traffic period a lack of slot IDs occurs, increases too (see horizontal lines). Thus, the potential traffic from the high traffic period is partially lost and this decreases the throughput. On the other hand, increasing f_b (→longer low traffic period!) augments the chance that, on an average a larger number of slot IDs is available (e.g., the probability values for 5 slot IDs). However, the majority of slot IDs is available during the low traffic period which does not much contribute to the throughput. At the start of the short burst period, all the remaining slots are quickly occupied and then the rest of the burst traffic is lost. The GSPN model was numerically solved. For MAXSLOT=16 and Poisson input it comprises about 190.000 states.

8 Conclusions

The choice of a stochastic model is in this case justifiable since bus request arrivals and congestion at the main memory are well captured by a probabilistic approach and have a significant "randomization" effect. A burst traffic model was proposed to take into account non-Poisson bus request streams. The results indicate that in the case of hierarchical models, where submodels (like the bus/memory system) are evaluated based on estimated arrival streams, it is important to characterize the arrival streams by an appropriate technique including measurements.

Acknowledgement: Thanks are due to Christoph Wincheringer for contributing to the model of the bus and the pipeline management policy and to four anonymous referees for comments.

References

[1] M. Dubois, "Throughput Analysis of Cache-based Multiprocessors with Multiple Buses", IEEE Transactions on Computers, vol. 37, no. 1, Jan. 88.

[2] M. A. Holliday, M. K. Vernon, "Exact Performance Estimates for Multiprocessor Memory and Bus Interference", IEEE Trans. on Computers, vol. C-36, no. 1, Jan. 1987.

[3] G. Klas, C. Wincheringer, "A Generalized Stochastic Petri Net Model of Multibus II", CompEuro 92, The Hague, The Netherlands, 4-8 May, 1992, pp.406-411.

[4] G. Klas, R. Lepold, "TOMSPIN, a Tool for Modeling with Stochastic Petri Nets", CompEuro 92, The Hague, The Netherlands, 4-8 May, 1992, pp.618-623.

[5] G. Klas, "Protocol Optimization for a Packet-Switched Bus in Case of Burst Traffic by Means of GSPN", Extended version of this paper, 1992, available from the author.

[6] M. A. Marsan, G. Balbo and G. Conte, "A Class of Generalized Stochastic Petri Nets for the Performance Analysis of Multiprocessor Systems", ACM Trans. Comput. Syst., vol. 2, no. 2, May 1984, pp.93-122.

[7] M. H. Woodbury, "Performance Modeling and Measurement of Real-Time-Multiprocessors with Time-Shared Buses", Proc. of IEEE, CH2351-5/86.

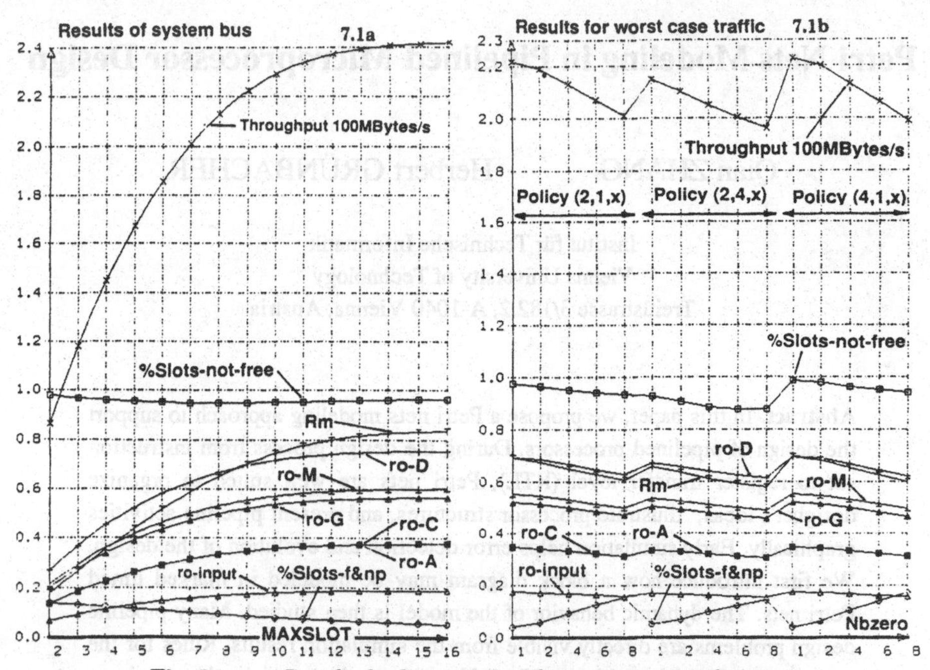

Fig. 7.1a **Results for Scenario I, polling strategy equals (2,1,4)**

Fig. 7.1b **Results for Scenario I: various pipeline management policies**

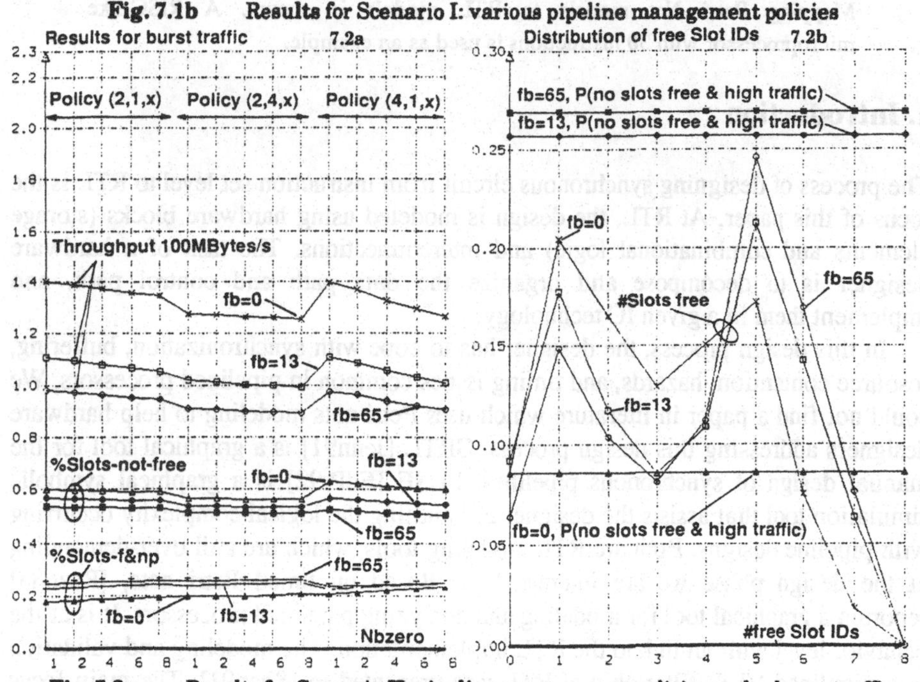

Fig. 7.2a **Results for Scenario II: pipeline management policies under burst traffic**

Fig. 7.2b **Results for policy (4,1,1) : Distribution of free slot IDs and P{no free slots in high traffic period} for MAXSLOT = 8**

Petri Nets Modeling in Pipelined Microprocessor Design

Qian ZHANG Herbert GRÜNBACHER

Institut für Technische Informatik
Vienna University of Technology
Treitlstrasse 3/182.2, A-1040 Vienna, Austria

Abstract. In this paper, we propose a Petri nets modeling approach to support the design of pipelined processors. During the design process from instruction set to register transfer level (RTL), Petri nets are well suited to organize designer's ideas, illustrate processor structures, and present pipeline activities graphically. Early simulation helps error detection and evolution of the design. We first introduce how a block diagram may be modeled in marked timed Petri nets. The dynamic behavior of the model is then studied. Many pipeline design problems are directly visible from the simulation results. Rules for the assignment of initial markings of Petri Nets for synchronous circuits are given. Mapping Petri Net models to RTL models is shown. A RISC-like microprocessor with 30 instructions is used as an example.

1. Introduction

The process of designing synchronous circuit from instruction set level to RTL is the focus of this paper. At RTL, the design is modeled using hardware blocks (storage elements and combinational logic) and interconnections. The task of a hardware designer is to decompose and organize the data path and control path, and implement them in a given IC technology.

In this design process, the designer has to cope with synchronization, buffering, resource contention, hazards, and timing issues common in pipelined processors. We could not find a paper in literature which uses Petri nets modeling to help hardware designers addressing this design process. GRTL [Jenn91] is a graphical tool for the manual design of synchronous pipelines. FUSE [EdPr92] is a graphical symbolic simulation tool that assists the designer in handling the logical complexity occurring with pipeline designs. Both are RTL modeling tools, which are still over-demanding at the design phase we are interested in. Based on timed Petri nets, [Razo88] reported a graphical tool for modeling and analyzing pipelined processors. It is at the behavior level without link to the RTL implementation. The modeling and validation of a pipelined VLSI filter chip at RTL was presented in [Shap91]. The main focus was on logical validation, not on assisting in the design process itself. [HoVe85] and [Razo88] used Petri Nets for the performance evaluation of pipelined machines. A comprehensive summary on Petri Nets and applications is in [Mura89]. General pipeline design principles are discussed in [HePa90] and [Kogg81].

In this paper, we consider the processor design above RTL as a stepwise refinement process. At the beginning, the designer concentrates on sequential system behavior, i.e., the execution of instructions without pipeline conflicts. As the result, the initial construction of the system is specified. Then the pipeline is checked whether hazards exist and how to add interlock circuitry. The structure of the processor will be refined and simulated iteratively till satisfaction.

Typical for this design process is that it is above the RTL but including timing considerations. At this design phase, it is natural that the designer wants to suppress the detailed coding of instructions, addressing modes, even semantics of instructions. The only concern is to examine how the information flows through the structure.

Petri nets modeling is rather useful during this design process. The model is specified at such an abstract level, that the simulation is independent of the implementation of the logic operations. The synchronization mechanism of Petri nets frees the designer from clocking considerations. If signal delay stages are missing, they are directly visible from the simulation result. When a hardware block is working on data coming from different instructions, it is possible that interlock circuitry is needed.

The paper is arranged as follows. The modeling of hardware block diagrams using timed marked Petri Nets is presented in section 2. Section 3 introduces the dynamic behavior of the model and simulation techniques. From the simulation result, many problems of pipeline designs are visible, as illustrated in section 3 and 4. Mapping from Petri Nets model to RTL model is discussed in section 5. Finally, conclusion and future research are given in section 6.

2. Timed Marked Petri Nets Modeling of Hardware

The definition and behavior of Petri nets and subclasses of Petri nets, methods to cope with time information within Petri Nets are explained in [Mura89], [HoVe85], [Shap91], [JeRo91]. Here we only address related interpretations and representations within our context.

2.1 Modeling of Hardware Modules

Petri nets are bipartite graphs of two kinds of nodes: places and transitions. Tokens are resident in places. If transitions are fired, they remove tokens from input places and produce tokens in output places. In our context, a hardware block is modeled as one or more transitions which represent transformations of the system. Signals are modeled as tokens. Interconnect lines are modeled as places plus arcs. The distribution of tokens on the places represents the state of the system. Transitions will change the token distribution, so that the system state is changed from one to another.

Modeling interconnections as places indicates that the functionality of plain wires is extended in the time domain from instantaneous transmission to queuing.

Modeling signals as primitives emphasizes the data flow aspect of the dynamic behavior of the hardware. In the rest of the paper, we do not distinguish between place or signal queue, token or signal.

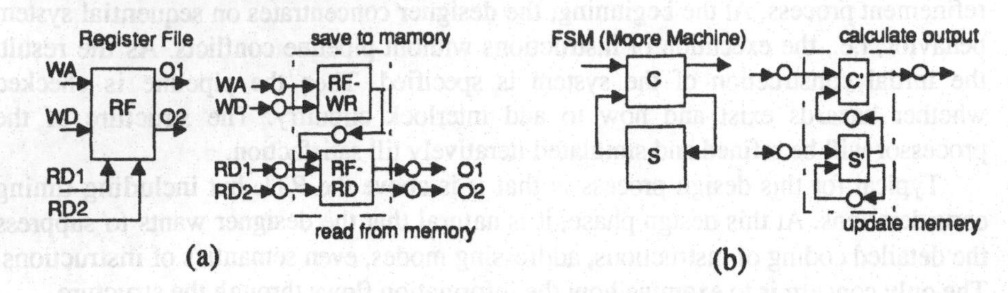

Figure 1. Modeling a register file and a finite state machine.

Two examples are given in figure 1. For the register file (RF), the model has two transitions. The write transition (RF-WR) accepts the write back address (WA), write back data (WD) and saves the data to memory. The read transition (RF-RD) accepts two read addresses (RD1, RD2), reads the data from memory and outputs the data. We do not further divide the read transition into two transitions, because we know that, in synchronous pipelined machines, two read addresses are issued at the same time. For finite state machines (Moore type), the Petri Net model also has two transitions. One calculates the output signal values based on current inputs and memory content. The other updates the memory content.

Please note that we use blocks instead of bars for transitions, to make the drawing easy to read for hardware designers.

Clearly the result is a marked Petri net. That is, every signal queue has only one input arc and one output arc.

2.2 Timing Representation

In our model, we have to consider clocking and timing in more detail. A link must be established in the time domain between firing of Petri Nets and clocking of synchronous circuits, keeping in mind that the Petri Nets model will ultimately be mapped to a RTL model.

In synchronous circuits, the clock signal is a sequence reference and also a time reference [Seit80]. As a sequence reference, the clock signal "ticks" define successive instants at which system state changes may occur. As a time reference, for accounting element and wiring delays in paths from the output to the following input of clocked elements.

In the design phase we are addressing, it is sufficient to concentrate on the sequencing function of the clock signal. The period between clock ticks is assumed to be long enough to account for the longest delay from the output to the following input of clocked elements. Therefore, in the RTL model, we measure time along an integer axis representing the ticks of the global clock.

In marked Petri nets, the enabling and firing of transitions is ruled by the appearance of tokens in its input places. The synchronization of the system is decentralized. Time may be assigned to transitions.

In our Petri nets model, we assign a processing time to every transition. This processing time can only be integers (including zero). It represents how much time (measured in processing time unit (*PTU*)) it takes for a transition to calculate the output once it is fired.

The processing time of transitions servers as a vehicle to link state changes of the Petri nets model to those of the RTL model. Because the state of the Petri nets model changes only at integer time coordinates, it makes it easy to map the Petri nets model to a RTL model. Further, we assign *PTU*s in such a way, that a transition of *PTU=n* implies that its RTL implementation takes *n* RTL machine state changes to calculate the output.

In one phase clocking, one *PTU* implies one clock cycle.

A construction rule of the model is that the *PTU* value of one transition should be independent of input signal values. If the processing time of a hardware block is dependent on input, we use multiple transitions to model it.

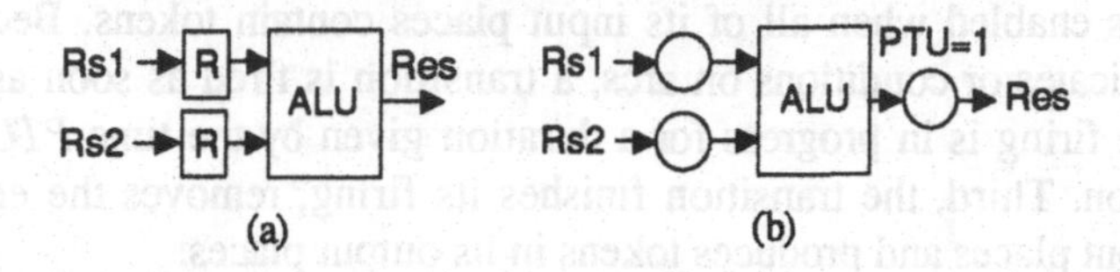

Figure 2. One possible implementation of the ALU (a) and its model (b). R in (a) stands for register.

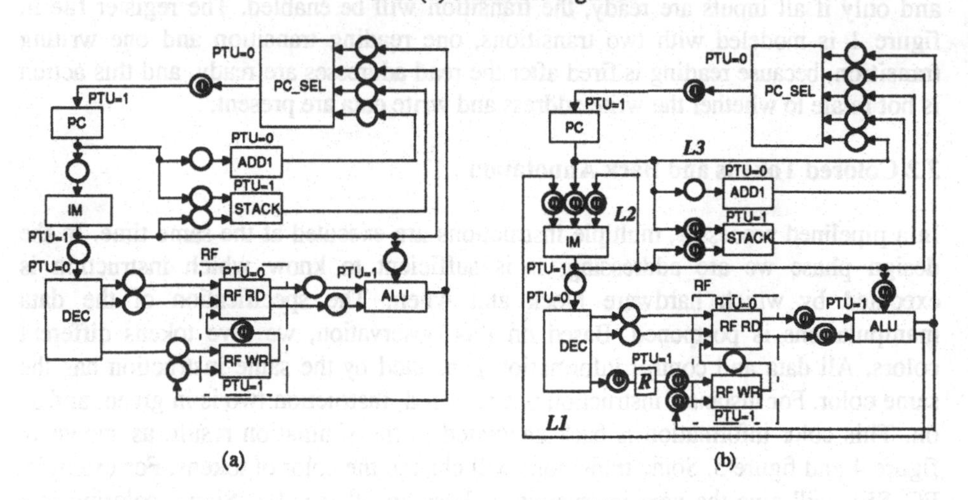

Figure 3. (a) Initial design model for the PIC processor. (b) Refined structure of (a). Interlock circuitry is added. In both (a) and (b), initial markings are shown.

The processing time of transitions is determined by the timing requirements in the IC design library and the designer's intention how the block will be implemented. In the Xilinx [Xili91] FPGA design library, for example, it takes one clock cycle (one phase clocking) to write into a register. Therefore, the *PTU* for a register is 1. If an ALU block is designed as combinational logic with latched input or output, then we assign *PTU*=1 to this ALU. If, however, the ALU is so complicated that it needs two levels of logic with register in between, it has *PTU*=2. An example is shown in figure 2. For combinational logic blocks we always have *PTU*=0.

A RISC-like microprocessor PIC [PIC90] is used as an example. The processor has been implemented using Xilinx FPGAs. The initial model for PIC is shown in figure 3(a).

3. Dynamic Behavior of the Model

3.1 Firing Rules

The firing of a transition is a multiphase action carried out in three steps. First, a transition is enabled when all of its input places contain tokens. Because we don't assign predicates or conditions on arcs, a transition is fired as soon as it is enabled. Second, the firing is in progress for a duration given by the time *PTU* associated to the transition. Third, the transition finishes its firing, removes the enabling tokens from its input places and produces tokens in its output places.

This firing rule enforces another construction rule of the Petri Nets model. That is, the inputs and outputs of one transition are logically related, in the sense that if and only if all inputs are ready, the transition will be enabled. The register file in figure 1 is modeled with two transitions, one reading transition and one writing transition, because reading is fired after the read addresses are ready, and this action is not relate to whether the write address and write data are present.

3.2 Colored Tokens and Back Annotation

In a pipelined processor, multiple instructions are executed at the same time. In the design phase we are addressing, it is sufficient to know which instruction is executed by which hardware block and when. The specification of the data manipulations is postponed. Based on this observation, we give tokens different colors. All data and control information generated by the same instruction has the same color. For instance, instruction one is in red, instruction two is in green, and so on. This color information is back annotated to the simulation result, as shown in figure 4 and figure 5. Some transitions will change the color of tokens. For example, PC_SEL will give the next instruction address an other color. Signal coloring is a representation technique for simulation, which depends on the designers' interpretation of the model.

When we assign an initial token to the signal queue in front of the program counter (PC) and simulate the design of figure 3(a), the trace in figure 4 is generated.

From inspection of figure 4, the following statements can be obtained.

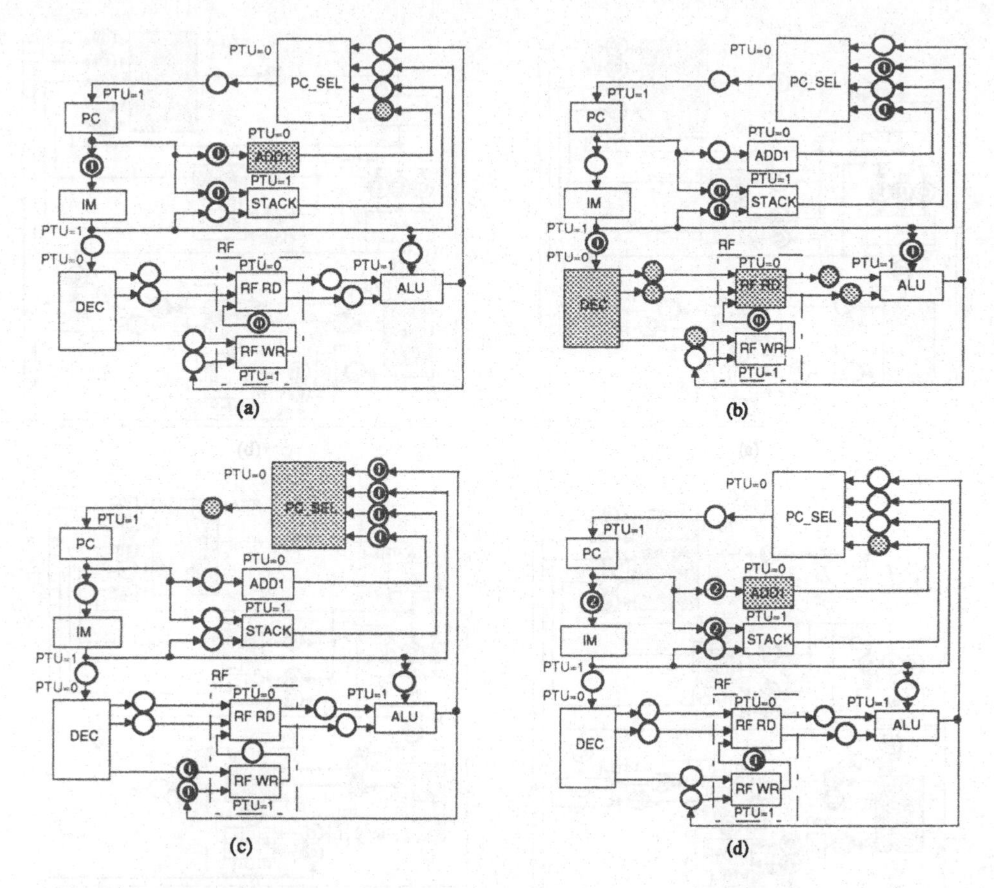

Figure 4. A set of simulation pictures. Initial markings are shown in figure 3(a). Picture (a), (b), (c), and (d) correspond to simulation time 1, 2, 3, and 4, respectively. Shaded boxes represent that the transitions are enabled immediately after the current simulation time. Shaded signal queues represent that tokens are moved to these places. Symbols ❶, ❷ represent all the data and control signals generated by instruction 1, 2, respectively.

- It takes 4 cycles to execute one instruction.
- The machine is not pipelined.
- There is no hazard because every transition is working on the data and/or control signals from the same instruction (same colored tokens).
- The bottleneck of the machine is in PC_SEL, because it needs signals which arrive in simulation time 1, 2, and 3.

Up to this point, there are many ways to refine the architecture. If the instruction set [PIC90] could be changed, one would not allow the option of PC being the destination of executions. Then there would be no connection between ALU and PC_SEL. If we re-organize the structure, we may add interlock circuitry to allow one block working on different instructions. This re-organization is shown in figure 3(b). The reason to do this comes from another simulation trace, discussed below.

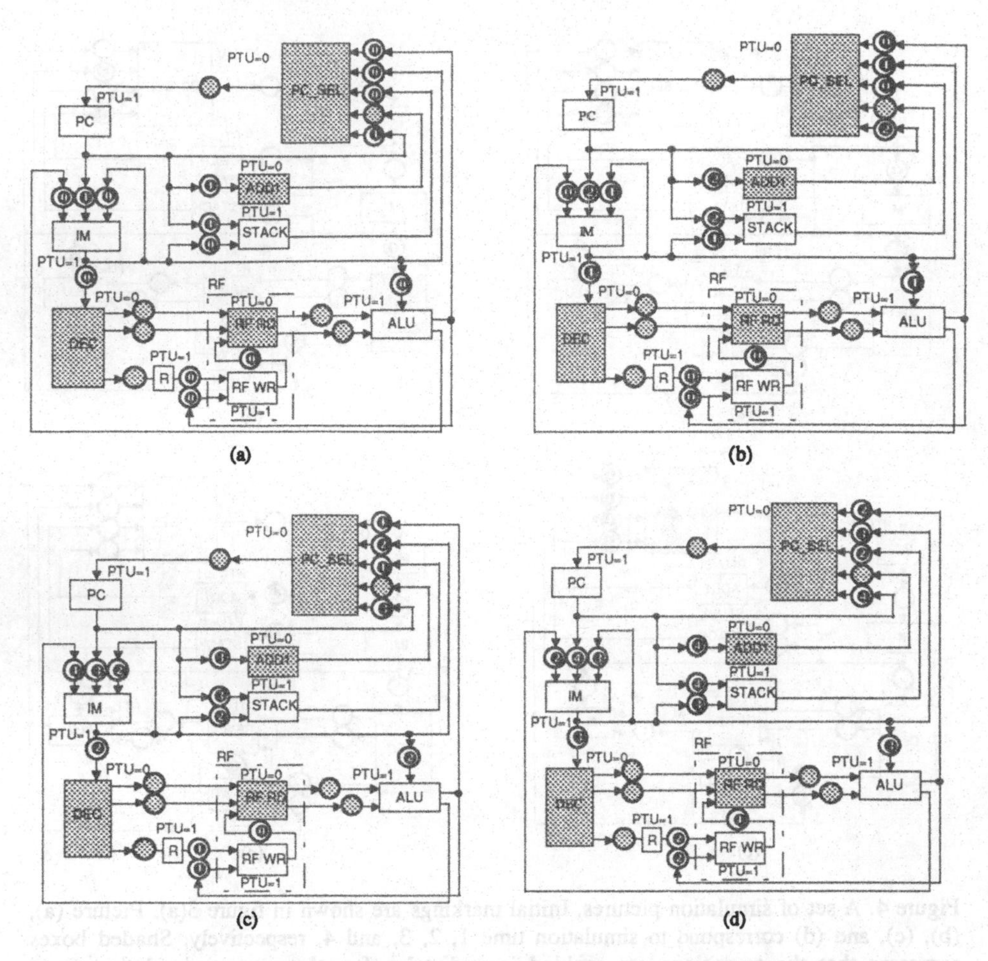

Figure 5. Another set of simulation pictures after the design is refined. Initial markings are shown in figure 3(b). Picture (a), (b), (c), and (d) corresponds to simulation time 1, 2, 3, and 4, respectively. Symbols ❶, ❷ represent all the data and control signals generated by instruction 1, 2, respectively.

If we assign initial markings to figure 3(a) following the rule, that input places of *PUT*=0 elements get no initial token, input places of *PTU*=1 blocks get one initial token, and input places of *PTU* > 1 blocks get more initial tokens with respect to their *PTU* values, we get a Petri net model analogous to figure 3(b). A simulation would give us another set of pictures (which is not shown here). From this set of pictures, we find that PC_SEL and IM work on differently colored tokens. This implies that there are hazards at these points. Therefore, interlocks L1, L2, and L3 have to be inserted (see figure 3(b)). L1 is for the test-and-skip instructions to transfer the test result. L2 is to test irregular instructions and to inform the IM to issue NOPs when necessary. L3 is for situations where the current instruction address should be kept. Also, one register is added in front of RF-WR address port for balancing the delay paths, because the simulation results illustrate that RF-WR write back data and write back address are of different color.

The simulation result of figure 3(b) is shown in figure 5.

4. Inspection of the Simulation Results

From the simulation shown in figure 5, we know:

- It still takes four clock cycles to execute one instruction.
- The machine is four stage pipelined. Four instructions are executed at the same time in the processor.
- Hazards rise at transitions where input signal queues have differently colored tokens. For example, at IM, PC_SEL, and RF, there are possible hazards.
- Interlock circuits are added to stall the pipeline until the hazard is resolved. For the instruction memory, we have to consider every possible combination of *three* sequential instructions. The same is true for the PC_SEL block. For RF, we have to consider whether the second instruction will use data calculated from the first one.

Possible ways to resolve hazards are discussed in [HePa90], [Kogg81], such as inserting NOPs as next instruction and feedthrough techniques to solve data dependencies.

We can further refine the structure (add feedthrough, for example) till the designer is satisfied all and possible pipeline hazards are solved.

5. Transformation to the RTL Model

It is relatively straightforward to transform the Petri nets model to a RTL level model. By adding delay elements or enable signals, and global clock, we can implement the functionality of signal queues.

Two kinds of timing information is given in the simulated Petri Nets model. First is the user-assigned *PTU* for every transition. Second are the number of signals residing in signal queues, the duration of the individual signal in a signal queue, and the empty period of the signal queue.

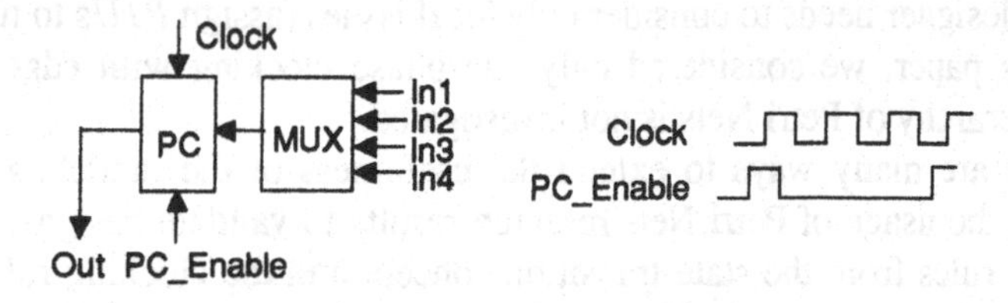

Figure 6. Implementing the PC and PC_SEL part in figure 3(a) and figure 4.

For the PC and PC_SEL part in figure 3(a) and figure 4, for example, the circuit in figure 6 implements the queuing behavior.

After the structure of the processor is defined, and all pipeline hazards are resolved, we go to the usual RTL design tools to specify details of the processor. Hardware description languages can be also adopted for further verification of the implementation. The Petri Nets simulation pictures are used as a guidance in the later design phases.

6. Conclusion

This paper described a promising application of Petri Nets modeling for high level synchronous pipelined microprocessor designs. By concentrating on key issues of hardware design and modeling the hardware structure using Petri Nets, it is possible to allow the designer to reason his design above RTL.

Textbooks and papers on pipelining at the computer architecture level hardly give an insight into real hardware blocks. On the other side, RTL (hardware) description of pipeline design make the pipeline problems like hazards and interlocks hardly visible. The approach introduced in this paper tries to fill the gap in between. We work on the hardware blocks and interconnection, but represent the activity of the circuits visible at the architecture level.

The key in the modeling approach is the level of abstraction. The distribution of tokens is interpreted as the state of the machine. Enabled transitions will change this distribution. The machine state is transferred from one to another by firing of transitions. State changes take time which is modeled as *PTU*s of blocks. The *PTU*s can only be integers (including zero). This machine state concept is completely different from conventional RTL state concept.

For our example, the initial markings shown in figure 3(a) correspond to non-pipelined execution. The initial markings shown in figure 3(b) correspond to pipelined execution, and the machine issues a new instruction every clock cycle. Other methods to assign initial markings are also possible, e.g., a pipelined machine issuing a new instruction every two clock cycle. The number of pipeline stages is determined by the structure of the machine and initial markings.

In the RTL model, clock is the global synchronization mechanism. This forces the designer to schedule hundreds of events through "when these latches are open, ..., when those latches are closed; ...; when phase A goes high and phase B goes low, ...". Petri Nets distributed synchronization mechanism changes the situation because designer needs to consider only local issues (assign *PTU*s to transitions).

It this paper, we considered only one phase clocking with edge triggered flip-flops. Hierarchy of Petri Nets is not investigated.

There are many ways to extend the usefulness of the modeling approach. For example, the usage of Petri Nets research results to validate designs; the abstraction of design rules from the state transition concept and the clocking rules in electronic circuits designs; the automatic transformation of Petri Net models to RTL models; performance evaluation at early design phase; and tools to support this modeling technique.

References

[EdPr92] W. Eder and H. Pristauz, "FUSE - A graphical simulation environment for pipelined designs". In Proceedings of the 1992 European Simulation Multiconference (ESM 92), SCS June 1-3, 1992, pages 445-450.

[HePa90] John L. Hennessy, David A. Patterson, Computer architecture: A quantitative approach, 1990, Morgen Kaufmann Publishers, Inc.

[HoVe85] M. Holliday and M. Vernon, "A generalized timed Petri Net model for performance analysis". In Proceedings of the International Workshop on Timed Petri Nets, Torino, Italy, July 1-3, 1985, pages 181-190.

[Jenn91] G. Jenning, "GRTL - A graphical platform for pipelined system design". In Proceedings of the 1991 European Conference on Design Automation (EDAC 1991), IEEE February 25-28, 1991, pages 424-428.

[JeRo91] K. Jesen, G. Rozenberg (Eds), High-level Petri Nets, theory and applications. Springer-Verlag 1991

[Kogg81] Petr M. Kogge, The architecture of pipelined computers. 1981, Hemisphere Publishing Corporation.

[Mura89] T. Murata, "Petri Nets: properties, analysis and applications". In Proceedings of the IEEE, Vol. 77, No. 4, April 1989, pages 541-580.

[PIC90] PIC 16C5x series EPROM-based 8-bit CMOS microcontrollers, Microchip Technology Inc.

[Razo88] R. R. Razouk, "The user of Petri Nets for modeling pipelined processors". In Proceedings of the 25th Design Automation Conference, June 1988, pages 548-553.

[Seit80] C. L. Seitz. System Timing. In C.Mead & L.Conway, Introduction to VLSI Systems, Chapter 7, 1980, Addison-Wesley.

[Shap91] R. M. Shapiro, "Validation of a VLSI chip using hierarchical colored Petri Nets". Microelectronics and Reliability, Special Issue on Petri Nets, Pergaman Press 1991.

[Smit85] Connie U. Smith, "Robust models for the performance evaluation of software/hardware designs". In Proceedings of the International Workshop on Timed Petri Nets, Torino, Italy, July 1-3, 1985, pages 172-180.

[View91] VIEWLogic workview references, Viewlogic Systems, Inc., 1991.

[Xiln91] The programmable gate array data book, Xilinx Inc., San Jose, 1991.

Springer-Verlag and the Environment

We at Springer-Verlag firmly believe that an international science publisher has a special obligation to the environment, and our corporate policies consistently reflect this conviction.

We also expect our business partners – paper mills, printers, packaging manufacturers, etc. – to commit themselves to using environmentally friendly materials and production processes.

The paper in this book is made from low- or no-chlorine pulp and is acid free, in conformance with international standards for paper permanency.

Lecture Notes in Computer Science

For information about Vols. 1–610 please contact your bookseller or Springer-Verlag

Vol. 611: M. P. Papazoglou, J. Zeleznikow (Eds.), The Next Generation of Information Systems: From Data to Knowledge. VIII, 310 pages. 1992. (Subseries LNAI).

Vol. 612: M. Tokoro, O. Nierstrasz, P. Wegner (Eds.), Object-Based Concurrent Computing. Proceedings, 1991. X, 265 pages. 1992.

Vol. 613: J. P. Myers, Jr., M. J. O'Donnell (Eds.), Constructivity in Computer Science. Proceedings, 1991. X, 247 pages. 1992.

Vol. 614: R. G. Herrtwich (Ed.), Network and Operating System Support for Digital Audio and Video. Proceedings, 1991. XII, 403 pages. 1992.

Vol. 615: O. Lehrmann Madsen (Ed.), ECOOP '92. European Conference on Object Oriented Programming. Proceedings. X, 426 pages. 1992.

Vol. 616: K. Jensen (Ed.), Application and Theory of Petri Nets 1992. Proceedings, 1992. VIII, 398 pages. 1992.

Vol. 617: V. Mařík, O. Štěpánková, R. Trappl (Eds.), Advanced Topics in Artificial Intelligence. Proceedings, 1992. IX, 484 pages. 1992. (Subseries LNAI).

Vol. 618: P. M. D. Gray, R. J. Lucas (Eds.), Advanced Database Systems. Proceedings, 1992. X, 260 pages. 1992.

Vol. 619: D. Pearce, H. Wansing (Eds.), Nonclassical Logics and Information Proceedings. Proceedings, 1990. VII, 171 pages. 1992. (Subseries LNAI).

Vol. 620: A. Nerode, M. Taitslin (Eds.), Logical Foundations of Computer Science – Tver '92. Proceedings. IX, 514 pages. 1992.

Vol. 621: O. Nurmi, E. Ukkonen (Eds.), Algorithm Theory – SWAT '92. Proceedings. VIII, 434 pages. 1992.

Vol. 622: F. Schmalhofer, G. Strube, Th. Wetter (Eds.), Contemporary Knowledge Engineering and Cognition. Proceedings, 1991. XII, 258 pages. 1992. (Subseries LNAI).

Vol. 623: W. Kuich (Ed.), Automata, Languages and Programming. Proceedings, 1992. XII, 721 pages. 1992.

Vol. 624: A. Voronkov (Ed.), Logic Programming and Automated Reasoning. Proceedings, 1992. XIV, 509 pages. 1992. (Subseries LNAI).

Vol. 625: W. Vogler, Modular Construction and Partial Order Semantics of Petri Nets. IX, 252 pages. 1992.

Vol. 626: E. Börger, G. Jäger, H. Kleine Büning, M. M . Richter (Eds.), Computer Science Logic. Proceedings, 1991. VIII, 428 pages. 1992.

Vol. 628: G. Vosselman, Relational Matching. IX, 190 pages. 1992.

Vol. 629: I. M. Havel, V. Koubek (Eds.), Mathematical Foundations of Computer Science 1992. Proceedings. IX, 521 pages. 1992.

Vol. 630: W. R. Cleaveland (Ed.), CONCUR '92. Proceedings. X, 580 pages. 1992.

Vol. 631: M. Bruynooghe, M. Wirsing (Eds.), Programming Language Implementation and Logic Programming. Proceedings, 1992. XI, 492 pages. 1992.

Vol. 632: H. Kirchner, G. Levi (Eds.), Algebraic and Logic Programming. Proceedings, 1992. IX, 457 pages. 1992.

Vol. 633: D. Pearce, G. Wagner (Eds.), Logics in AI. Proceedings. VIII, 410 pages. 1992. (Subseries LNAI).

Vol. 634: L. Bougé, M. Cosnard, Y. Robert, D. Trystram (Eds.), Parallel Processing: CONPAR 92 – VAPP V. Proceedings. XVII, 853 pages. 1992.

Vol. 635: J. C. Derniame (Ed.), Software Process Technology. Proceedings, 1992. VIII, 253 pages. 1992.

Vol. 636: G. Comyn, N. E. Fuchs, M. J. Ratcliffe (Eds.), Logic Programming in Action. Proceedings, 1992. X, 324 pages. 1992. (Subseries LNAI).

Vol. 637: Y. Bekkers, J. Cohen (Eds.), Memory Management. Proceedings, 1992. XI, 525 pages. 1992.

Vol. 639: A. U. Frank, I. Campari, U. Formentini (Eds.), Theories and Methods of Spatio-Temporal Reasoning in Geographic Space. Proceedings, 1992. XI, 431 pages. 1992.

Vol. 640: C. Sledge (Ed.), Software Engineering Education. Proceedings, 1992. X, 451 pages. 1992.

Vol. 641: U. Kastens, P. Pfahler (Eds.), Compiler Construction. Proceedings, 1992. VIII, 320 pages. 1992.

Vol. 642: K. P. Jantke (Ed.), Analogical and Inductive Inference. Proceedings, 1992. VIII, 319 pages. 1992. (Subseries LNAI).

Vol. 643: A. Habel, Hyperedge Replacement: Grammars and Languages. X, 214 pages. 1992.

Vol. 644: A. Apostolico, M. Crochemore, Z. Galil, U. Manber (Eds.), Combinatorial Pattern Matching. Proceedings, 1992. X, 287 pages. 1992.

Vol. 645: G. Pernul, A M. Tjoa (Eds.), Entity-Relationship Approach – ER '92. Proceedings, 1992. XI, 439 pages, 1992.

Vol. 646: J. Biskup, R. Hull (Eds.), Database Theory – ICDT '92. Proceedings, 1992. IX, 449 pages. 1992.

Vol. 647: A. Segall, S. Zaks (Eds.), Distributed Algorithms. X, 380 pages. 1992.

Vol. 648: Y. Deswarte, G. Eizenberg, J.-J. Quisquater (Eds.), Computer Security – ESORICS 92. Proceedings. XI, 451 pages. 1992.

Vol. 649: A. Pettorossi (Ed.), Meta-Programming in Logic. Proceedings, 1992. XII, 535 pages. 1992.